AF167544

High Pressure Chemistry, Biochemistry and Materials Science

NATO ASI Series

Advanced Science Institutes Series

A Series presenting the results of activities sponsored by the NATO Science Committee, which aims at the dissemination of advanced scientific and technological knowledge, with a view to strengthening links between scientific communities.

The Series is published by an international board of publishers in conjunction with the NATO Scientific Affairs Division

A Life Sciences **B Physics**	Plenum Publishing Corporation London and New York
C Mathematical and Physical Sciences **D Behavioural and Social Sciences** **E Applied Sciences**	Kluwer Academic Publishers Dordrecht, Boston and London
F Computer and Systems Sciences **G Ecological Sciences** **H Cell Biology** **I Global Environmental Change**	Springer-Verlag Berlin, Heidelberg, New York, London, Paris and Tokyo

NATO-PCO-DATA BASE

The electronic index to the NATO ASI Series provides full bibliographical references (with keywords and/or abstracts) to more than 30000 contributions from international scientists published in all sections of the NATO ASI Series.
Access to the NATO-PCO-DATA BASE is possible in two ways:

– via online FILE 128 (NATO-PCO-DATA BASE) hosted by ESRIN, Via Galileo Galilei, I-00044 Frascati, Italy.

– via CD-ROM "NATO-PCO-DATA BASE" with user-friendly retrieval software in English, French and German (© WTV GmbH and DATAWARE Technologies Inc. 1989).

The CD-ROM can be ordered through any member of the Board of Publishers or through NATO-PCO, Overijse, Belgium.

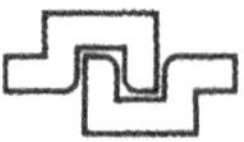

Series C: Mathematical and Physical Sciences - Vol. 401

High Pressure Chemistry, Biochemistry and Materials Science

edited by

R. Winter
Ruhr-University of Bochum,
Bochum, Germany

and

J. Jonas
University of Illinois,
School of Chemical Sciences,
Urbana, Illinois, U.S.A.

Springer Science+Business Media, B.V.

Proceedings of the NATO Advanced Study Institute on
High Pressure Chemistry, Biochemistry and Materials Science
Aquafredda di Maratea, Italy
September 20 – October 3, 1992

A C.I.P. Catalogue record for this book is available from the Library of Congress.

ISBN 978-0-7923-2290-0 ISBN 978-94-011-1699-2 (eBook)
DOI 10.1007/978-94-011-1699-2

Printed on acid-free paper

Table of Contents

PREFACE

This monograph, which is the outcome of the ASI on High Pressure Chemistry, Biochemistry, and Materials Science, illustrates new developments in the field of high pressure science. In fact, for chemists, biochemists, and materials scientists, pressure as an experimental variable represents a tool which provides unique information about systems of materials studied. It is interesting to note how the growth of the high pressure field is also reflected in the content of the recent ASI's dealing with this field. The ASI High Pressure Chemistry held in 1977 was followed by the ASI High Pressure Chemistry and Biochemistry held in 1986, and the coverage of the present ASI also includes applications to materials science.

In view of the teaching character of the ASI, it is natural that main contributions to this volume present overviews of the different subfields or applications of high pressure research. In contrast, contributed papers offer more specialized aspects of various high pressure studies. The various contributions to this volume make clear the impressive range of fundamental and applied problems that can be studied by high pressure techniques, and also point towards a major growth of high pressure science and technology in the near future. This ASI focused mainly on advances achieved in the six years since the previous ASI devoted to the high pressure field. The organization of this volume is as follows. The main lectures covering the three main areas of high pressure applications to chemistry, biochemistry, and materials science are followed by contributed papers. A summary of a panel discussion on the future of high pressure science and technology concludes this volume.

The editors gratefully acknowledge input of the organizing committee and express their thanks to all lecturers and contributors to this volume for their efforts in preparing their lectures and manuscripts. On behalf of all ASI participants, we express our gratitude for the generous financial support provided by the Scientific Affairs Division of the North Atlantic Treaty Organization.

Bochum, Germany — R. Winter
Urbana, Illinois, U.S.A — J. Jonas

January 1993

HIGH PRESSURE EQUATIONS OF STATE: THEORY AND APPLICATIONS

Marvin Ross
Physics Department
Lawrence Livermore National Laboratory
P.O.Box 808/L-299
Livermore, CA 94550, USA

ABSTRACT. This article reviews theoretical equations of state of solids and dense liquids and their applications to studies of melting, shock compression, matter at extreme conditions and planetary interiors.

TABLE OF CONTENTS

R. Winter and J. Jonas (eds.), High Pressure Chemistry, Biochemistry and Materials Science, 1–41.

1.0 Introduction

The present review is concerned with theoretical equations of state at high pressure. The equation of state (EOS) of a system describes the relationship between the thermodynamic variables. These properties can be directly related to the forces between atoms by the methods of quantum mechanics and statistical mechanics. The application of pressure offers a means by which the lattice constant or density may be varied resulting in changes in properties, including transitions to new structures or phases and modifications in electronic configurations. Theoretical techniques provide a capability for understanding changes on an atomistic level.

Recent advances in theory have come about as a result of rapid advances in computational capabilities with concurrent advances in high pressure experimental techniques. Accurate measurements are now carried out by static and dynamic methods to pressures of several megabars and thousands of degrees, and have stimulated new theoretical directions.

Basically the theories of matter at high pressures are similar to those at 1 atmosphere. But in practice their implementation typically requires certain approximations. As a result, theories which are adequate at ambient conditions may fail when tested under more severe conditions. In the present review we have restricted our coverage to those theories which have proved useful at high pressure. The topics chosen are necessarily limited by space constraints and reflect the interests of the author. However, additional references have been included for those readers seeking a broader perspective (Eliezer et al. 1986; Godwal et al. 1984; Young, 1991).

2.0 The Theory of Solids

The major contribution to the equation of state of a solid at high pressure comes from the ground state electronic energy, E_0, to which must be added a vibrational (E_{vib}) contribution arising from zero point and thermal motion. Contributions to the change in electronic energy from thermally excited electrons may be included by neglecting electron-phonon interactions and adding the term energy (ΔE_{el}). Thus,

$$E(V,T) = E_0(V) + E_{vib}(V,T) + \Delta E_{el}(V,T) \tag{1}$$

$$P(V,T) = P_0(V) + P_{vib}(V,T) + \Delta P_{el}(V,T). \tag{2}$$

2.1 Local Density Approximation

Current theoretical methods for determining E_0 use the local-density approximation (LDA) (Lundquist and March, 1983, Devreese and Van Camp, 1984). This method simplifies the many-body fermion problem into a mean-field approximation in which each electron moves in the self-consistent average field of all the electrons and the nuclei (V_{en}). The complicated many-body exchange and correlations included in the average field are conveniently approximated by a functional $\varepsilon_{xc}(\bar{r}, [\rho])$ of the local electron density. The total ground state energy is:

$$E_0 = T[\rho] + \int d\bar{r}\, V_{en}(\bar{r})\,\rho(\bar{r}) + \frac{e^2}{2}\int d\bar{r}' \int d\bar{r}' \frac{\rho(\bar{r})\,\rho(\bar{r}')}{|\bar{r} - \bar{r}'|} + \int d\bar{r}\,\rho(\bar{r})\,\varepsilon_{xc}(\bar{r}, [\rho]) + U_{ion\text{-}ion} \tag{3}$$

$$\rho(r) = \sum_j |\phi_j(r)|^2 \ .$$

One simple form of ε_{xc} which has had widespread use is the Wigner interpolation expressed as

$$\varepsilon_{xc} = -0.916/r_s - 0.878/(r_s + 7.8) \text{ Ry, where } r_s = (3/4\,\pi\,\rho)^{1/3} \tag{4}$$

and ρ is the electron density.

The first term is the local exchange and the second is the local correlation potential. The LDA approximation is believed valid in the limit of small gradients of the local electron charge density. The terms in Eq. 3 are respectively

$$E_o = T_e + U_{en} + U_{ee} + U_{xc} + U_{ion\text{-}ion} \tag{5}$$

where T_e is the total electron kinetic energy and the remaining terms are the ion-electron (U_{en}), electron-electron (U_{ee}), electron exchange and correlation (U_{xc}) and $U_{ion\text{-}ion}$ is the ion-ion Coulomb energy. The energy is minimized using a wavefunction gradient method ($\partial E/\partial\phi_j = 0$) which leads to one-electron Schrödinger equations in which the potential field is the same for each electron:

$$\left\{-\frac{\hbar^2}{2m}\nabla^2 + V_{en}(r) + e^2\int dr' \frac{\rho(r')}{|r-r'|} + \frac{\partial U_{xc}}{\partial\rho(r)}\right\}\phi_j(r) = \varepsilon_j\,\phi_j(r) \quad , \tag{6}$$

ϕ_j and ε_j are the single-particle wavefunction and eigenvalues. The methods for solving these equations are referred to as "Electron Band Theory." The pressure can be obtained from the virial theorem,

$$PV = \frac{2T}{3} + \frac{U}{3} , \tag{7}$$

where U the total potential energy is the sum of U terms in Eq. 5.

In general, the local density approximation works surprisingly well for solids and even for free atoms and small molecules. Using only a small fraction of the computer time required by Hartree-Fock electron-band calculations, LDA can accurately describe the lattice constant, cohesive energy and compressibility of the full range of metals, given only the atomic number (Skriver 1984). It has also been capable of providing accurate predictions of many high pressure phase transitions (McMahan 1986, Yin and Cohen 1982) and equations of state (Fig. 1a).

2.2 Lattice Thermal Energy -The Grüneisen Model

The total zero point and lattice thermal energy is given by

$$E_{vib} = \sum_{\alpha=1}^{3N}\left[\frac{1}{2}h\nu_\alpha + \frac{h\nu_\alpha}{e^{h\nu_\alpha/kT} - 1}\right] \tag{8}$$

The summation is over the 3N normal modes (ν_α) of the crystal, N being the total number of atoms. If the interatomic potential is known, then ν_α can be calculated from

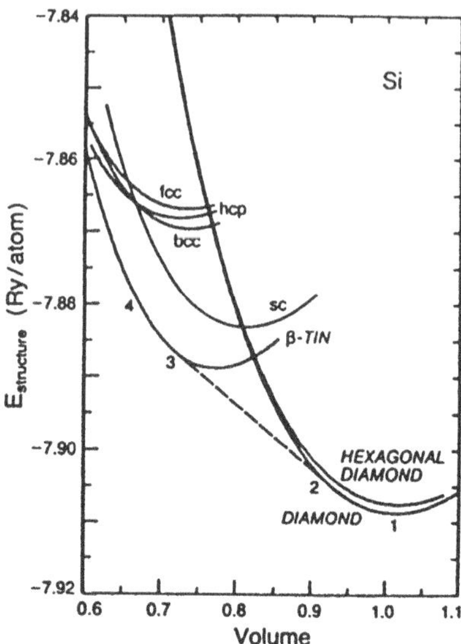

Fig. 1a. Total energy curves for various assumed crystal structures of Si as a function of volume normalized to the observed volume. The dashed line is the common tangent between the diamond and white tin phases. (Yin and Cohen 1982).

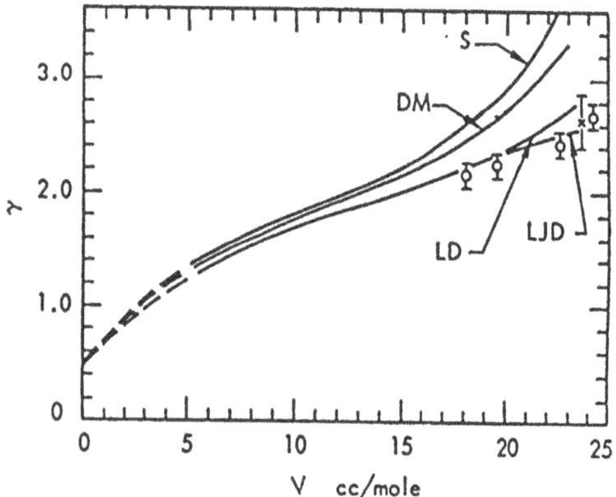

Fig 1b. Comparison of g as a function of voume at 65° K calculated by means of the Slater formula (S), Dugdale-McDonald (DM), lattice dynamics (LD), and the LJD cell model (LJD). The circles are the results of Monte Carlo calculations. The x is an experimental value.

lattice dynamics (Maradudin et al. 1971) leading to the Mie-Grüneisen equation of state (Girafalco 1973) (neglecting electronic thermal properties),

$$E = E_0 + E_{vib} \,, \tag{9}$$

$$P = -\frac{dE_0}{dV} + \frac{\gamma}{V} E_{vib} \tag{10}$$

The Grüneisen constant γ is obtained by assuming that γ_α for each mode

$$\gamma_\alpha \equiv -\frac{d\,\ell n\,\nu_\alpha}{d\,\ell n\,V} \tag{11}$$

can be replaced by its average over all the normal modes, $\gamma = \overline{\gamma_\alpha}$. In the classical limit, for a system of harmonically vibrating atoms $E_{vib} = 3\,NkT$, and the pressure becomes

$$P = Po + 3\,\gamma\,N\,k\,T/V \quad . \tag{12}$$

In the Debye model (Goodstein, 1975) the sum in Eq. 8 is replaced by an integral over a continuum distribution of frequencies.

$$\sum_{\alpha=1}^{3N} \rightarrow \frac{9N}{\nu_m^3} \int_0^{\nu_m} \nu^2 d\nu \tag{13}$$

This leads to a modified expression for E_{vib}; however, the remaining formalism (Eqs. 8–12) is unchanged. For either model γ must be determined as a function of volume. With the exception of simple metals and small molecules, this procedure is generally not feasible. Instead, approximations are introduced which bypass a knowledge of the microscopic physics and rely on macroscopic variables — the pressure-volume curve — to obtain useful approximations for γ. Several models for γ can be expressed in a single equation:

$$\gamma = -\frac{V}{2} \frac{\partial^2 \left(P_0\, V^{2m/3}\right)/\partial V^2}{\partial \left(P_0\, V^{2m/3}\right)/\partial V} + \frac{1}{3}(m-2) \tag{14}$$

where m = 0 is for the elastic continuum model of Slater; m = 1 is obtained from a harmonic expansion of the static lattice energy of Dugdale and McDonald (Rice et al. 1958); and m = 2 is based on the vibration of an atom in a spherically cell. Fig 1b compares values of γ for argon as a function of volume for several models. All calculations were made using an exponential-six potential (Holt and Ross, 1979).

A number of useful empirical relations have been offered which bypass these models. These assume γ/V = constant or $\gamma = CV + \frac{1}{2}$, where C is a constant fitted at normal density. $\gamma = \frac{1}{2}$ is the limiting case for lattice of positive ions in a background degenerate electron gas in which both T and V $\rightarrow 0$. In the perfect gas limit of T $\rightarrow\infty$, $\gamma = \frac{2}{3}$. A particularly large number of applications of the Mie-Grüneisen equations are to be found in Zharkov and Kalinin (1971).

2.3 Thermal Electron Properties

Contributions to E_{el} and P_{el} from thermally excited electrons can be obtained from finite-temperature self-consistent band calculations for electrons in a rigid lattice, using Fermi statistics to determine the energy level occupation numbers (McMahan and Ross 1977). Accordingly

$$\Delta E_{el}(V,T) = \sum_i \varepsilon_i(V)[n_i(T) - n_i(0)] \tag{15}$$

$$\Delta P_{el}(V,T) = -\sum_i \frac{d\varepsilon_i(V)}{dV}[n_i(T) - n_i(0)] \tag{16}$$

where $\varepsilon_i(r)$ and ni are the energy and occupation number of the electron energy level. $n_i(T)$ is typically approximated by the Fermi-Dirac distribution function. For kT « 0.1 E_F, the Fermi energy, the electron-thermal components E_{el} and P_{el} may be obtained from low-temperature expansions for the free electron gas.

$$\Delta E_{el}(V,T) = \frac{\pi^2}{6} N(E_F)(kT)^2 + \tag{17}$$

and

$$\Delta P_{el}(V,T) = \frac{\pi^2}{6} \frac{\partial N(E_F)}{\partial V}(kT)^2 + \tag{18}$$

where $N(E_F)$ is the density of electron states at the Fermi energy.

2.4 Applications

In this section two applications to high pressure research are described which illustrate the theories in sections 2.1 to 2.3. These are the s-d transition in cesium and the reduction of shock wave data to isotherms for calibration of pressure standards for the diamond-anvil cell.

2.4.1 s-d Electron Transfer in Cesium

Many of the interesting high-pressure structural phase transitions and irregularities in the equation of state can be understood in terms of systematic changes with compression in the electronic structure (McMahan 1986). In most metals, valence d electrons are relatively localized, and are closer in to the nucleus than s-electrons. Thus compression raises the energy of the s-state and transferring them into d-states causing changes in the compressibility and structural stabilities. Cesium with a single s-like conduction electron is probably the most striking example of this effect. A well known example is the isostructural electronic transition (fcc-fcc) Cs metal undergoes at room temperature and 42.2 kbar with a 9% volume change.

The s-d transition can influence the equation of state in ways that may be described by the Gruneisen model (Eq. 12). Po, the T=0 static lattice pressure calculated from LDA electron band theory, shown by the dashed line in Fig. 2, illustrates none of the features of illustrated the experimental results. The 0-K isotherm, is by itself not

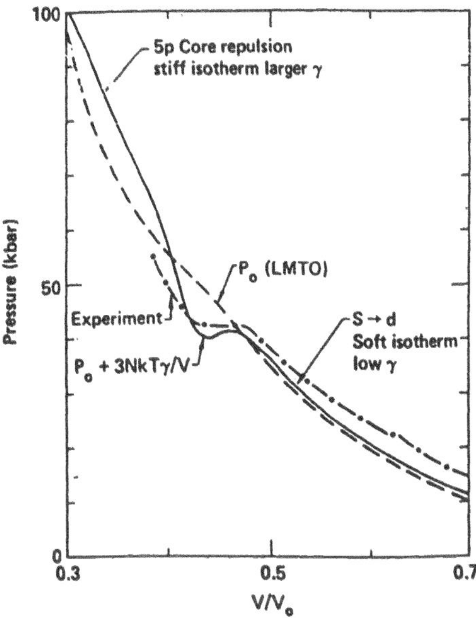

Fig. 2. Cs 298-K isotherm. Pressure as a function of relative volume, V/Vo. Po is the 0-K LMTO theoretical curve. The solid line is the 300-K isotherm calculated using Eq. 12 (Gloetzel and McMahan 1979).

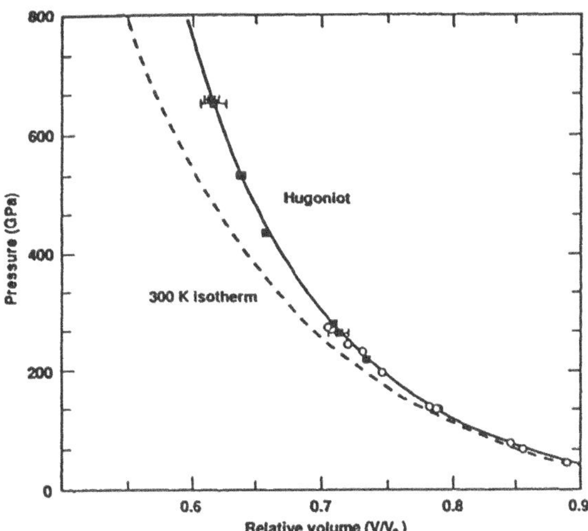

Fig. 3. Theoretical Hugoniot (solid line) for platinum and 300-K isotherm (dashed line), as compared against the Hugoniot data (squares and circles) (Holmes et al. 1989).

sufficient to drive the transition. The second term is the thermal lattice pressure and γ may be approximated by the Slater model (Eq. 14, m = 0). The calculated isotherms, using Eq. 12 are shown in Fig. 2 and predict a first order phase transition.

The essential feature to note is that γ is related to the derivatives of Po(V). In the case of Cs Glotzel and McMahan (1979) found that in the vicinity of the observed transition (V/Vo=0.43) a pocket of pure 5 d states in the Brillouin zone near the X symmetry point passes through the Fermi surface and causes a softening in Po. This softening is sufficient to cause a negative Grüneisen parameter which contributes a negative contribution to the thermal pressure leading to the van der Waals loop implying a first-order isostructural phase transition near 42 kbar as observed. This has been verified by the experimental work of Boehler and Ross (1984). At higher pressure, where the transition is largely over, γ becomes positive, and there is additional stiffening of the pressure.

2.4.2 Calibration of Diamond-Anvil Pressure Standards

Room temperature isotherms for the four metals Cu, Mo, Pd, and Ag reduced from shock data provided the original ruby calibration scale to 1 Mbar (Mao et al. 1978). By measuring the room temperature lattice parameters for these metals in a diamond-anvil cell by x-ray and recording the function of the shift in the ruby R_1 flouresence line, a comparison of the two data sets provided the pressure versus ruby R_1 line shift. Above 1 Mbar the ruby becomes increasingly unsatisfactory as a calibrant material, and an increasing reliance has been placed on the use of metals from which the lattice parameter can be obtained by x-ray methods, and the pressure obtained from room temperature isotherms obtained from shock data by the methods outlined in the previous sections.

Platinum has proven to be a useful pressure standard for multi-megabar research. Fig. 3 shows the shock compression curve (Hugoniot) for platinum (Holmes et al. 1989) obtained by combining a theoretical zero-temperature pressure-volume isotherm calculated by electron band theory with contributions from electron and ion thermal pressures as functions of volume. The isotherms obtained from shock reduced data are accurate to about 10% in pressure at a given density and can be used as pressure EOS standards for diamond-anvil cell static compression experiments.

3.0 Theory of Dense Liquids

During the past twenty years, the theory of liquids has made rapid advances and a number of accurate theories have been developed and applied to a wide range of problems dealing with molecules, liquid metals and dense plasmas. These developments have been periodically reported in a number of excellent books (Shimoji 1977, Hansen and McDonald 1986 and Hafner 1987). In the present review, we limit our discussion to those theories which have proven useful for high pressure applications.

At ambient conditions the molecular motion of an atom in a liquid is restricted to interatomic separations near the potential minimum. The application of high pressures shifts the mean intermolecular separation to shorter distances and it is the repulsive wall of the potential that then dominates the motion with the result that approximations valid at ambient conditions often become much less useful with increasing density. At sufficiently high temperatures all fluids dissociate and ionize to become plasmas. Although fluid theories are in principle also applicable to plasmas, the major problem in dense plasmas is in properly treating the long range coulomb forces acting between particles. In the case of partially ionized systems, the excited electronic states must be included in a thermodynamically self-consistent manner.

Developments in the theory of dense liquids at high pressure has been driven by a need to interpret experimental data. A particularly rich source has been shock wave experiments on liquids. The earliest application of Monte Carlo calculations to real systems was carried out by Wood and Parker (1957) on shock compressed liquid argon and later by Ross and Alder (1967). These calculations are relatively expensive and with the development of perturbation theory, the most widely used models have been based on those employing the hard sphere system.

3.1 Intermolecular Forces

On the most fundamental level the forces between molecules are manifestations of the electromagnetic interactions between electrons and nuclei of the entire system (Torrens 1972). For many systems the sum of these interactions may be expressed as effective intermolecular potentials. The connection between interatomic forces and EOS is provided by thermodynamics and statistical mechanics (McQuarrie 1976, Goodstein 1975).

Some typical rare gas pair potentials are shown in Fig. 4. One of the best known functional forms, the Lennard-Jones (LJ) potential, for a pair of atoms separated by a distance r can be written as

$$\phi(r)_{LJ} = \varepsilon\left[\left(\frac{r^*}{r}\right)^{12} - 2\left(\frac{r^*}{r}\right)^{6}\right] \tag{19}$$

where ε and r^* are the depth and position of the potential well. The attractive r^{-6} term represents the long-range dipole-dipole interaction. In particular the r^{-12} repulsion has been found to be too stiff and theory tells us that an exponential repulsion should be the proper form. A more useful function, particularly for examining high pressure data, has been the exponential-six (exp-6) potential,

$$\phi(r)_{exp\text{-}6} = \frac{\varepsilon}{\alpha - 6}\left\{6\exp\left[\alpha(1 - r/r^*)\right] - \alpha(r^*/r)^6\right\}. \tag{20}$$

The parameter α regulates the stiffness of the interatomic repulsion. For most systems α ~13-14. An exception is molecular hydrogen which has a rather soft repulsion and an α value of 11.1. In this case, at sufficiently high density the attractive r^{-6} may become unrealisticly large compared to the repulsion and a damping term needs to be introduced. Higher accuracy for the potential can be achieved by introducing additional parameters, higher multipole attractive interactions or three-body terms (Barker 1986).

The potential energy in an ionic crystal may be expressed as the sum of a Coulomb and a van der Waals type potential. For example, the potential between ions i and j with charge Z_i and Z_j may be written as

$$\phi(r_{i,j}) = \frac{Z_i Z_j e^2}{r_{i,j}} + \phi(r_{i,j})_{exp-6}. \tag{21}$$

Metals constitute a particular difficulty because of the inherently many-body nature introduced by itinerant electrons. For the case of simple polyvalent metals practical pair-potentials have been derived from theory (Shimoji 1977). Consider a system of ions placed in a neutralizing background electron gas. The electrons will tend to accumulate near the ions, screening the ion-ion interaction. A formal treatment of this problem leads to an effective pairwise interaction potential for a pair of ions

$$\phi(r) = \frac{Z^2 e^2}{r} + \frac{1}{2\pi^2}\int_0^\infty \frac{\sin qr}{qr}\,\phi_{BS}(q)\,q^2\,dq. \tag{22}$$

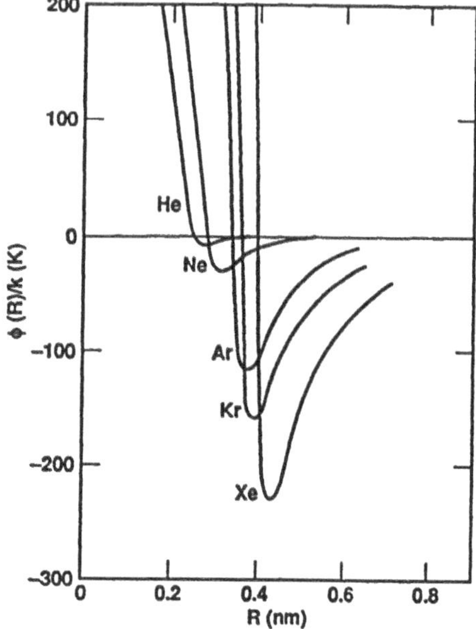

Fig. 4. Comparison of rare-gas pair potentials.

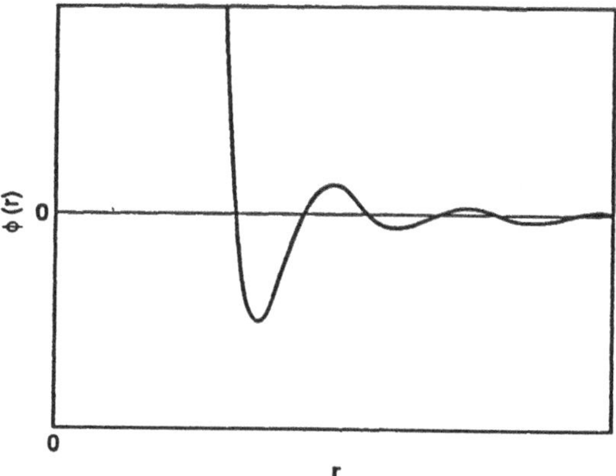

Fig. 5. A typical metallic pair potential.

The first term is the bare-Coulomb repulsion between ions, and the second is the screening correction due to the modulations in electron density near the ions. $\phi_{BS}(q)$ is a function of the electron-ion potential and dielectric constant and is in general density dependent. A typical metal ion-ion potential is shown in Fig. 5. The electrons piling up near the ion leads to a diminished charge elsewhere and to spatial oscillations in the potential.

Apart from the density dependence, the simple metal pair potentials are similar to those of rare gases. The potentials of both have minima roughly near the average nearest neighbor separation, and a steeply repulsive behavior at small interatomic spacings. The metal interatomic repulsion tends to be less steep. For covalently bonded systems such as carbon or quartz, three-body potentials are required (Tersoff 1989).

3.2 Statistical Mechanics of Fluids

The equation of state of a fluid can be obtained from the canonical partition function Z, and the Helmholtz free energy

$$A = -kT \ln Z \tag{23}$$

where

$$-\log Z = \log N! - \frac{3}{2} N \log \frac{2\pi mkT}{h^2} - \log Q \tag{24}$$

and

$$Q = \int \cdots \int \exp\left[-\sum_{i<j} \frac{\phi(\bar{r}_{ij})}{kT} \right] d\bar{r}_1 \ldots d\bar{r}_N$$

The potential energy is taken as the sum of pairwise additive potentials. An expression for the total energy is obtained by differentiating with respect to 1/kT,

$$E = \frac{3}{2} NkT + \frac{1}{Q} \int \cdots \int \sum_{i<j} \phi(\bar{r}_{ij}) \exp\left[\sum_{i<j} \frac{\phi(\bar{r}_{ij})}{kT} \right] d\bar{r}_1 \ldots d\bar{r}_N \tag{25}$$

The first term corresponds to the kinetic energy of an ideal gas and the second part arises from interactions between the particles. The energy and pressure equations for a spherically symmetric potential can be written in terms of the radial distribution function g(r):

$$E = \frac{3}{2} NkT + 2\pi\rho N \int \phi(r)\, g(r)\, r^2\, dr \tag{26}$$

and

$$P = \frac{NkT}{V} - \frac{2\pi N\rho}{3V} \int r^3 \frac{d\phi}{dr} g(r)\, dr \tag{27}$$

where

$$g(r) = \frac{V^2}{Q} \int \ldots \int \exp\left[- \sum_{i<j} \frac{\phi(r_{ij})}{kT} \right] d^3r_3 \ldots d^3r_N \qquad (28)$$

The radial distribution function measures the probability of finding a particle at a given distance from a fixed particle in the fluid. $\rho = N/V$ a typical form for g(r) is shown in Fig. 6.

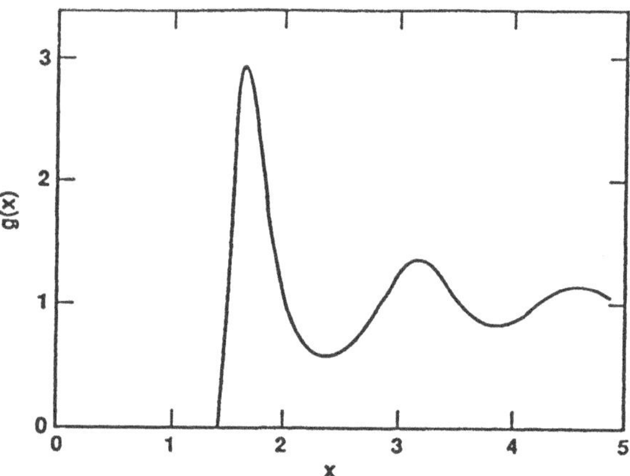

Fig. 6. A typical g(r) function for a liquid.

3.3 Computer Simulations

The most important advances in the theory of fluids have come from computer simulations carried out by Monte Carlo (MC) and molecular dynamics (MD) methods (Ciccotti and Hoover 1986, Allen and Tildesley 1987), and the development of approximate, but computationally fast and accurate models. In both methods, N molecules are confined to a box of volume V at temperature T. Computer size and speed limits N to several thousand particles. But these practical limitations are continually decreasing. To minimize surface effects periodic boundary conditions are employed. This consists of filling three dimensional space by replications of the original cell in which the molecules in each cell occupy the same relative positions. In MD properties are simulated using numerical methods to solve Newton's equations for N particles. MC employs probablistic or stochastic methods and is limited to equilibrium properties. MD is the more versatile and is applicable to the study of non-equilibrium systems (Hoover 1991).

Computer simulations should be viewed as experiments producing exact data for idealized model systems. For the case of strongly coupled degenerate plasmas with densities comparable to those found in high-energy astrophysics and laser-fusion-compression experiments, computer simulations are the only 'experimental' data. Simulations have been used to determine the pair distribution, thermodynamic and transport properties over a wide range of conditions. These results have been extremely valuable in the development and testing of computationally simpler approximate methods, such as perturbation theories and integral equations. The books by Hafner (1987) and Shimoji (1977) contain many useful applications to real liquids.

3.4 Hard Spheres

One of the most useful model systems has been that of hard spheres. The pressure data for the fluid may be expressed as

$$\frac{P}{kT\rho} = \left(1 + \eta + \eta^2 - \eta^3\right)/(1-\eta)^3 \quad , \tag{29}$$

Here η is the hard sphere packing fraction $\eta = (\rho/6)\pi d^3$. The Helmholtz free energy A_{hs} in excess of that for a perfect gas is then

$$\frac{A_{hs}}{NkT} = \frac{4\eta - 3\eta^2}{(1-\eta)^2} \quad , \tag{30}$$

Simple analytic expressions exist for exact thermodynamic results on other model systems such as the Lennard-Jones (Ree 1981) and inverse-power potentials (Young and Rogers 1984) and the one component plasma (Stringfellow et al. 1990).

3.5 One Component Plasma (OCP)

The one-component plasma (OCP) is an idealized classical system of ions, with immersed in a uniform neutralising electron background. The system is characterized by the Coulomb coupling parameter $\Gamma = (Ze)^2/a_s kT$ where $a_s = (3V/4\,\pi\,N)^{1/3}$ is the ion sphere radius. The OCP is a reasonable model for the description of super-dense, completely ionized matter typical of extreme astrophysical conditions (e.g. white dwarfs and neutron stars). Under these conditions the electrons are highly degenerate ($kT \ll E_F$) and play the role of a uniform background fluid. At sufficiently high density, the energy of such a system may be written as

$$E = 3/2NkT + E_{eg} + U_{ocp} \tag{31}$$

where U_{ocp} includes the Coulomb energy of the ion and electron interactions and E_{eg}, the electron gas term includes the kinetic, exchange energy and correlation energy (see Eq. 4).

$$E_{eg} = 2.21/r_s{}^2 - 0.916/r_s - 0.878/(r_s + 7.8) \text{ Ry.}$$

Extensive Monte Carlo simulations have yielded accurate values of the excess thermodynamic properties over the entire fluid range. Simple analytic fits (Stringfellow et al. 1990) to the excess internal energy and configurational Helmholtz free energy from the Monte Carlo (MC) data are

$$\frac{U_{ocp}}{NkT} = a\Gamma + b\Gamma^s + c \tag{32}$$

and

$$\frac{A_{ocp}}{NkT} = a\Gamma + \frac{b}{s}\Gamma^s + (3+c)\ln\Gamma - \left[a + \frac{b}{s} + 1.1516\right] , \tag{33}$$

where $a = -0.8992$, $b = 0.596$, $s = 0.3253$, and $c = -0.268$.

The OCP has been used as a reference system to model liquid metals, dense plasma and ion mixtures for systems, such as the liquid alkali metals, in which the interactions between ions is relatively soft (Ross et al. 1981).

3.6 Perturbation Theory

The idea of representing a fluid as hard spheres or charged ions moving in a background of interatomic interactions forms the basis of the van der Waals model and perturbation theory. The rationale being that if the free energy and pair distribution function of a reference system are well known, as for example from computer simulations, then the properties for a model system can be computed as corrections to those of the reference system. For example, dividing the pair potential into a hard sphere repulsion for r<d and an attraction, $u_1(r)$ leads to an expression for the free energy of the system

$$A = A_o - \frac{N\rho}{2}\int_o^\infty d^3r\, g(r)\, u_1(r) \quad . \tag{34}$$

where A_o is the free energy of the reference system and the second term is the correction term. The "closeness" of reference and model fluids largely determines the accuracy of the theory. For the case of a hard sphere reference there is no unique way to choose the value of a diameter. Weeks, Chandler and Anderson (WCA) (1971) chose the break at the potential minimum. The WCA scheme is useful at ambient conditions, but become unrealistic as the density is increased and the diameter approaches or exceeds the average interatomic spacing. At packing fractions near the freezing value for hard spheres the reference potential becomes too repulsive and predicts a metastable reference system in the fluid phase. To overcome this problem, Kang et al. (1985) have developed a modified version of WCA theory which equates the break point to the nearest neighbor distance (a) in an equivalent fcc lattice at high density when a<d, but otherwise retains the original WCA approximation. This modification leads to a good description of the thermodynamics of dense fluids. A second approach is based on variational theory. For an interacting fluid the excess Helmholtz free energy can be expressed in terms of the Gibbs-Bogolyubov inequality:

$$\frac{A}{NkT} \leq \frac{A_o(z)}{NkT} + \frac{\rho}{2kT}\int_o^\infty \left[\phi(r) - \phi_o(r)\right] g_o(r,z)\, 4\pi r^2 \; dr \; . \tag{35}$$

This equation states that the free energy of the interacting system is bounded above by the free energy of a reference system (A_0) plus the difference in energy between the actual system and the reference, averaged over all configurations of the reference. z is the reference coupling parameter. As an example for the hard sphere system, the diameter is chosen as the coupling parameter which is varied to minimize the Helmholtz free energy. For most materials the hard-sphere potential is unrealistically stiff in comparison with a realistic intermolecular potentials and the pressures obtained are typically too high. This deficiency has been remedied by the use of soft-sphere reference systems (Ross 1979, Young and Rogers 1984). The excess Helmholtz free energy in soft-sphere variational theory is written as:

$$A = A_{hs}\,(\eta) + A\,(\eta)\,NkT + A_{int} + A_{qm} \tag{36}$$

where the term

$$A\,(\eta) = -\left(\eta^4/2 + \eta^2 + \eta/2\right) \tag{37}$$

is introduced to compensate for the softness of the repulsive part of the potential and makes the reference free energy approximate that of an inverse-12 system. The intermolecular interaction energy of the system is,

$$A_{int} = \rho N/2 \int_d^\infty \phi(r)\, g_{hs}(r,\eta)\, d^3r. \tag{38}$$

and $g_{hs}(r, \eta)$ is the hard sphere radial distribution function in the Percus-Yevick approximation. The term A_{qm} incorporates the first order Wigner-Kirkwood expansion and is given by

$$A_{qm} = \frac{\beta h^2 \rho N}{96\pi^2 m} \int_d^\infty \nabla^2 \phi(r) g_{hs}(r,\eta)\, d^3r \tag{39}$$

A useful analytic for $g_{hs}(r,\eta)$ has been presented by Smith and Henderson (1970). The hard-sphere diameter d is determined by minimizing the free energy and energy and pressure are determined by computing the temperature and volume derivatives of the total Helmholtz free energy.

4.0 Melting at High Pressure

An important application of liquid theory, made possible by the development of accurate perturbation theories, has been the calculation of melting curves requiring a minimum of computational effort. The melting curve is determined by the condition that the Gibbs free energy, pressure and temperature are equal in the solid and liquid phases. That is $G(P_s,T) = G(P_l, T)$ for $P_s = P_l$ and $T_s = T_l$.

Perturbation theory provides a useful model for calculating accurate liquid properties. In order to obtain accurate solid properties it is necessary to implement harmonic lattice dynamics calculations with small, but important, anharmonic corrections. The solid free energy is given by,

$$A(V,T) = U(V,0) + kT \sum_{i=1}^{N} \left\{ \frac{h\nu_i}{2KT} + \ln\left[1 - \exp\left(-\frac{h\nu_i}{kT}\right)\right]\right\} + \Delta A(V,T) \tag{40}$$

where ν_i are the normal mode frequencies calculated using Lattice dynamics. ΔA, the anharmonic correction is the difference between calculations using the full and harmonic potentials. It can be treated as a perturbation energy and approximated by a spherical cell mean field model in which each atom is allowed to move in the potential field of its stationary neighbors. The anharmonic correction to the Helmholtz free energy of the solid may be written as

$$\Delta A = \frac{N}{2} \int_{cell} [V(R) - V_{ho}(R)] \exp[-\beta V_{ho}(R)]\, R^2\, dR \int_{cell} \exp[-\beta V_{ho}(R)]\, R^2\, dR \tag{41}$$

$$+ \frac{N}{2} \int_{cell} [V(R) - V_{ho}(R)] \exp\left[\frac{m\omega R^2}{\hbar}\right] R^2\, dR \int_{cell} \exp\left[-\frac{m\omega R^2}{\hbar}\right] R^2\, dR$$

$V(R)$ and $V(R)_{ho}$ are the spherically averaged full and harmonic potentials, where R is the displacement from the center of the cell. The exact expressions for V(R) are given by Young and Ross (1981). Theoretical calculations of the He, H_2, Ne and Ar melting curves have been carried out with these equations. The intermolecular potential, of the exp-6 form, was obtained by fitting to the experimental static isotherm. The calculated results for argon (Zha et al. 1986) are shown in Fig. 7 and agreement between theory and experiment along the melting curve is generally very good, limited only by the scatter in the experimental data.

For many systems accurate calculations of the free energy are difficult to make, and semiempirical models are useful to predicting melting curves. Two of these are the Lindemann law for melting and the rule of constant packing fraction for freezing. Both arise from the view that on a microscopic level the spatial configuration of atoms at melting is invariant at all temperatures. The Lindemann law assumes that a solid melts when the amplitude of the atomic vibrations exceed a certain fraction (about 10%) of the lattice spacing. This leads to the relationship

$$\frac{\langle R^2 \rangle}{a^2} = \frac{3kT_m (\hbar/k)^2}{M\Theta_E^2 \, a^2} , \tag{42}$$

where T_m is the melting temperature, a is the nearest neighbor spacing, Θ_E is the Einstein temperature, and M is the atomic mass. The liquid analog of the Lindemann law states that the liquid packing fraction:

$$\eta = \frac{\pi N d^3}{6V_l} ,$$

is a constant along the freezing curve d is the hard sphere diameter determined from the liquid perturbation theory. The calculated values of $\langle R^2 \rangle / a^2$ and η are seen in Table I to be nearly constant and these empirical relations are valid for Ar.

The availability of high-pressure isotherms and melting data for He (Vos et al. 1990), Ne (Vos et al. 1991), and Ar (Zha et al. 1986) provide an opportunity for testing the theory of corresponding states. The melting curves of these elements have now been calculated with exp-6 potentials with a varying between 13.0 and 13.2. The consistency of the stiffness parameter indicates that the properties of the three elements "correspond." This can be checked by scaling the He and Ar data by ratios to bring them onto the Ne temperature and pressure scales. For example, the Ar melting pressures and temperatures may be scaled to those of Ne by the expressions

$$P(Ar \rightarrow Ne) = P(Ar) \frac{\varepsilon/r_m^3 (Ne)}{\varepsilon/r_m^3 (Ar)} \tag{42a}$$

and

$$T(Ar \rightarrow Ne) = T(Ar) \frac{\varepsilon/k (Ne)}{\varepsilon/k (Ar)} \tag{42b}$$

The scaled Ar and He melting curves are compared with the Ne data in Fig. 8. The Ne and Ar data coincide, showing that these two elements obey a corresponding states relationship, at least over the range 0-60 kbar. The He melting data falls only slightly below the Ne and Ar data.

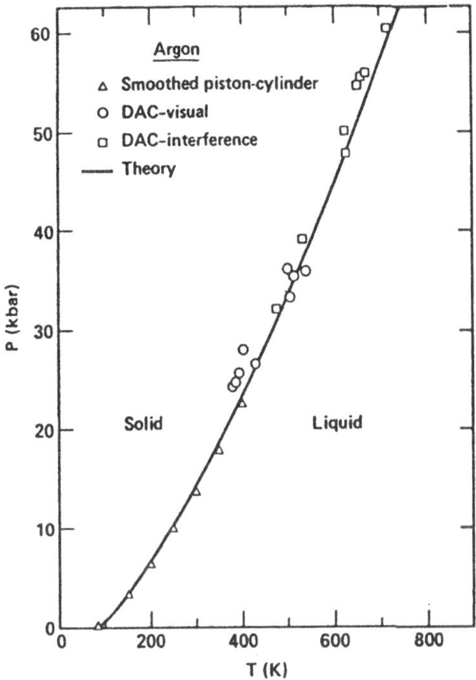

Fig. 7. Experimental and theoretical Ar melting curves.

T (K)	P (kbar)	V_s (cm^3/mol)	$\Delta V/V_s$	$\Delta S/R$	η	$\sqrt{\langle R^2\rangle/a^2}$
83.81	0.094	24.86	0.136	1.718	0.455	0.096
100	0.81	24.25	0.113	1.561	0.459	0.097
200	6.75	21.25	0.071	1.241	0.465	0.101
300	14.31	19.41	0.060	1.164	0.464	0.102
400	23.36	18.07	0.054	1.120	0.464	0.103
600	44.56	16.22	0.047	1.078	0.463	0.103
800	70.03	14.92	0.044	1.054	0.463	0.103
1000	98.65	13.95	0.041	0.965	0.463	0.103
1500	183.0	12.24	0.038	0.947	0.462	0.103
2000	283.1	11.11	0.034	0.932	0.461	0.105
2500	401.5	10.23	0.032	0.920	0.461	0.105
3000	531.6	9.56	0.031	0.915	0.460	0.105

Table I. Theoretical melting curve for Ar and comparison with simple models.

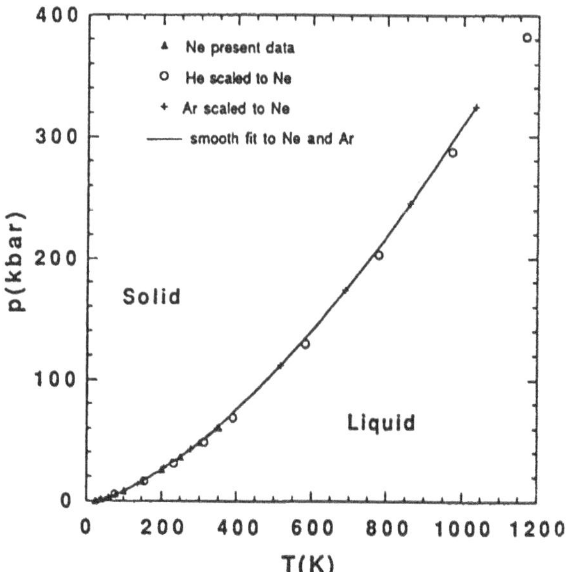

Fig. 8. Neon melting data compared with scaled helium and argon data.

The melting behavior of the alkali halides stands in sharp contrast to the situation with relatively simple monatomic fluids. In these materials the melt is believed to undergo a continuous pressure-induced structural rearrangement from that of a relatively open structure at low pressure to one with more densely packed ordering at high pressure. At low pressure, the liquid structure of an alkali halide consists of alternating shells of positive and negative ions in which the short-range repulsions are balanced by the long-range Coulomb interaction and each atom has about four to five nearest neighbors. However, at high pressure the repulsive forces dominate with the result that the atomic packing more closely resembles that of an inert-gas fluid with twelve nearest neighbors of mixed charge (Adams, 1976; Ross and Rogers, 1985). This change in the liquid structure leads to a decrease in the volume change on melting and is responsible for the large curvature observed in the P - T plane (Pistorius, 1965). In many of the alkali halides, particularly the chlorides and the iodides. The melting temperature of the NaCl structure appears to be approaching a maximum suggesting that the liquid density is nearly that of the solid. This behavior indicates that the liquid undergoes a continuous transition from a an open to a more closely packed structure.

The Lindemann law and the rule of constant liquid packing fraction assume that the atomic arrangement of the two coexisting phases scale with compression. That is, the relative arrangement of the atoms taken along the melting line would always be identical if properly scaled. These theories are valid for describing the melting of monatomic systems, but must fail for alkali halides over the range in which the fluid is undergoing changes in coordination number.

5.0 Matter at Extreme Conditions

Much of the experimental work on dense liquids and plasmas can be organized according to the phase diagram shown in Fig. 9. Shock Hugoniot data for dense liquids have been obtained to pressures and calculated temperatures as high as several Mbar and several tens of thousands of degrees Kelvin (Ross 1985). These results have provided important information about intermolecular forces and atomic and electronic processes at extreme conditions. Gases have been compressed using explosive shock tubes to achieve non-ideal plasma states in Ar, Xe and Cs (Fortov 1982). Hot expanded metals (Hensel and Uctmann, 1989) are intermediate between liquids and non-ideal plasmas. This range is, at present, very difficult for theoretical and experimental study. Experiments require the combination of high temperatures and pressures while studying materials in a reactive state. The principal impediment to theoretical calculations on expanded metals stems from a lack of understanding of the localized itinerant electronic transition. Higher pressures than those attainable in dynamic-impact experiments using conventional drivers, such as chemical explosives and two-stage guns, can only be obtained using nuclear explosives and large lasers. Conventional drivers are currently limited to pressures between 1Mbar for liquids and about 10 Mbar for dense metals. Pressures up to 570 Mbar have been obtained by the use of nuclear explosives (Avrorin et al. 1987).

5.1 Shock Compressed Liquids

In a shock wave experiment, a supersonic wave is passed through the material under investigation (Rice et al. 1958, Al'tshuler 1965, Zeldovich and Raizer 1966). A measurement of the shock wave and mass velocity are sufficient to determine the P, V, E of the compressed state. Figure 10 illustrates schematically a one-dimensional, steady, planar shock wave moving at a velocity U_S generated by accelerating a piston from rest to some final constant velocity U_p referred to as the mass or particle velocity. The width of the discontinuity is generally a few molecular mean-free-path lengths. The thermodynamic pressure, volume and energy (P, V, E) behind the shock front are related to the initial properties (P_0, V_0, E_0) in front of the shock wave by the dynamic variables U_S and U_p through the Rankine-Hugoniot conservation equations:

Conservation of mass,

$$\frac{V}{V_0} = 1 - \frac{U_p}{U_s}, \tag{43a}$$

conservation of momentum,

$$P - P_0 = U_S\, U_p/V_0, \tag{43b}$$

and conservation of energy,

$$E - E_0 = \frac{1}{2}(P + P_0)(V_0 - V). \tag{43c}$$

Thus a measurement of U_S and U_p, with a knowledge of initial conditions, is sufficient to determine the P, V, E of the compressed state. The resulting P–V curve is known as a Hugoniot and represents the locus of points that can be reached from an initial condition. It can be calculated theoretically by using a known EOS and satisfying Eq. 43c. Shock compression is typically accompanied by a large rise in temperature which must be obtained from a separately from theory or a measurement of the emitted radiation. A large collection of shock data has been assembled by Marsh (1980).

Shock wave experiments on liquids provide theorists with an unusual set of conditions and phenomena. The most unique features are the high temperature

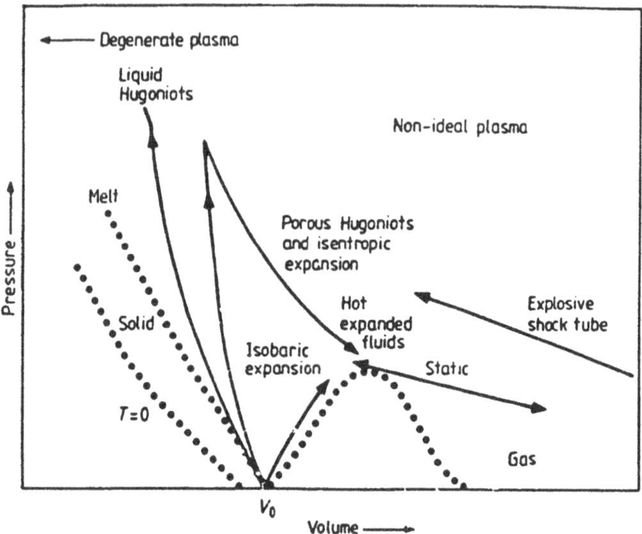

Fig. 9. Schematic of the phase diagram (dotted curves) with several of the experimental trajectories of particular interest to this review.

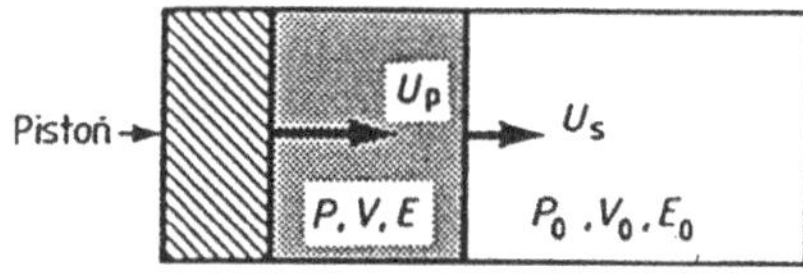

Fig. 10. Schematic of a shock wave traveling at a velocity U_S, driven by a piston moving with velocity U_p. In an experiment the piston is typically a metal plate accelerated to a constant velocity by means of a high explosive or a laser.

generated. The large kinetic energy allows atoms to interact at small interatomic separations. For simple liquids, such as argon or xenon, final pressures of the order of 100 GPa (1 Mbar) may achieved at temperatures up to 30,000 K and at two to three times normal liquid densities. These conditions are sufficient to thermally excite electrons into the conduction band and create a metallic-like or dense plasma state. In this section, we provide several examples of experiment and theory.

5.1.1 Hydrogen and Helium

Shock wave experiments have been made for deuterium to 750 kbar and about four times liquid density, (Nellis et al. 1983 and van Thiel et al. 1974). The data shown in Fig. 11, has been fitted using soft-sphere perturbation theory to obtain effective intermolecular potentials (Ross et al. 1983). In these experiments, temperatures up to 7000 K are achieved thereby probing the repulsive potential to intermolecular separations to about 1.5 A. This is approximately equal to the nearest neighbor separation of molecules in the vicinity of the metallic transition, estimated to be near 3-4 Mbar. The shock data provides less information about the potential at low energies, near the attractive well, and this information is best obtained from static experiments. Isotherms of solid normal hydrogen have been determined from single-crystal x-ray diffraction measurements to maximum pressures of 37 GPa (Hemley et al. 1990). Liquid helium has been shock compressed to 550 kbar and 22,000 K (Nellis et al. 1984). All the available high pressure equaton of state and melting data can be fitted with an exp-6 potential (Ross and Young, 1986). Figure 12 compares isotherms for solid helium predicted by several model pair potentials to diamond-anvil data (Mao et al. 1988).

For both H_2 and He, the effective potential determined from shock and static data are much softer than the two-body potential providing evidence for significant many-body effects. As an example, for helium the pair-potential from shock data is a in good agreement with the static data (Mao et al. 1988) while the two-body potentials obtained from theory (Ceperely and Partridge, 1986) and low density gas phase data (Aziz et al. 1979) are in poor agreement.

5.1.2 Xenon

Under shock loading conditions electrons are thermally excited into the conduction band. A theoretical model has been developed which combines soft-sphere variational theory (Ross 1968, Ross and McMahan 1980) for the fluid with an electron gas model, Fermi-statistics are used to calculate thermal electron excitation. The terms in the excess free energy model equations are written as:

$$A = A_{int} + A_{eg} \tag{44}$$

$$A_{int} = A_{hs}(\eta) + A(\eta)NkT + \frac{N^2}{2V}\int_d^{\infty} \phi(r) g_{hs}(r,\eta)\, d^3r\,,$$

$$A_{eg} = N\mu - NkT \int \ln(1 + \exp[(\mu - \varepsilon)/kT])\, \rho(\varepsilon)\, d\varepsilon$$

A_{int} is the excess free energy from fluid variational theory. The exp-6 potential is used for the interatomic interactions and is assumed to be unaffected by the electronic excitation. This approximation can be justified by the fact that at very high pressure the

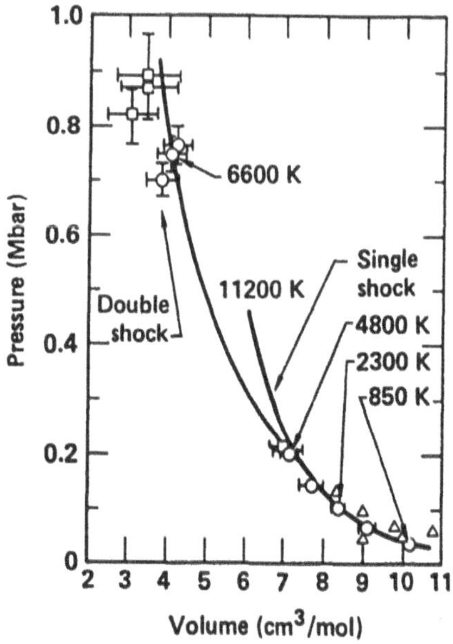

Fig. 11. Single and double-shock deuterium Hugoniots. Some calculated temperatures are shown.

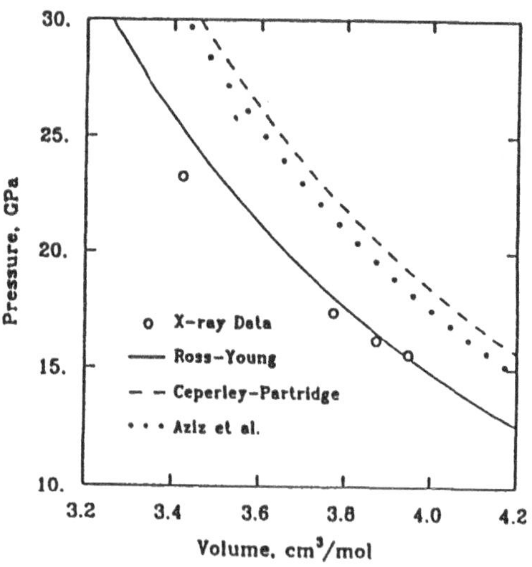

Fig. 12. Pressure-volume equation of state for solid hcp helium at 300 K (from Mao et al. 1988).

equation of state of xenon and its neighboring elements metallic iodine, cesium and CsI are similar. A_{eg} is the free energy of an electron gas and μ is the chemical potential and $\rho(\varepsilon)$ is the density of states. The model assumes spherical valence and conduction bands separated by a gap in the density of states over a range of energies. Typically the energy band gap, decreases with increasing density. For most cases the band gap is obtained from a separate set of electron band theory calculations. The closure of the band gap, observed in solid xenon near 1.4 Mbar leads to metallization (Reichlin et al. 1989 and Goettel, 1989).

As a result of the high temperatures attained, a shock experiment with liquid xenon, up to 30,000 K electrons are thermally excited into the conduction band. Under these conditions a semiconductor model is applicable for calculating the number of valence electrons N_e thermally excited:

$$N_e(T,V) = 2(g_v g_c)^{1/2}\left[\frac{2\pi kT}{h^2}\right]^{3/2} \frac{V}{N}\left(m_v^* m_c^*\right)^{3/4} \exp\left[-\frac{\Delta E(V)}{2kT}\right], \quad (45)$$

where $\Delta E(V)$ is the volume-dependent band gap. The total energy and pressure of the fluid are:

$$E(V,T) = E_{int}(V,T) + \Delta E(V) N_e(V,T) + E_{eg}(V,T), \quad (46a)$$

and

$$P(V,T) = P_{int}(V,T) - \frac{\partial \Delta E}{\partial V} N_e(V,T) + P_{eg}(V,T), \quad (46b)$$

E_{int} and P_{int} are the energy and pressure of the fluid calculated using variational theory. E_{eg} and P_{eg} are the thermal energy and pressure of the free electrons and holes. The effective masses, m_V*, m_C* are approximated as m_e, the free-electron mass. The band degeneracies are g_V and g_C. An important feature is that for the case p and d-states, the large degeneracy of factor ($g_V = 3$, $g_C = 5$) in the prefactor lends a big boost to the degree of excitation.

Figure 13 shows experimental shock wave data (Hugoniots) for xenon compared with theoretical calculations. Curves A and B. were calculated using two slightly different volume-dependent band gaps obtained from band theories. Calculations that do not include electronic excitation (the pure insulator case) are shown by the uppermost curve (C). By absorbing some of the shock energy, the excited electrons act as thermal sinks, keeping the temperature down and lowering the pressure. The volume dependence of the energy gap in the second term of the pressure expression, also lays an important role. As a result of them 5p-5d band gap with decreasing volume, excited electrons undergo a decrease in their partial pressures, lowering the pressure further. The thermal excitation of valence p-electrons to d levels has also been observed in shock-compressed liquid Ar. Electron-band calculation predict that the 3p valence band and 3d conduction band will cross and argon metallizes at a volume of 4.5 cm^3/mol and a pressure of about 5 Mbar (McMahan 1986).

5.1.3 Molecular Nitrogen

The pressure-induced metallization of diatomic molecules has been of considerable interest ever since Drickamer and coworkers first observed the metallization

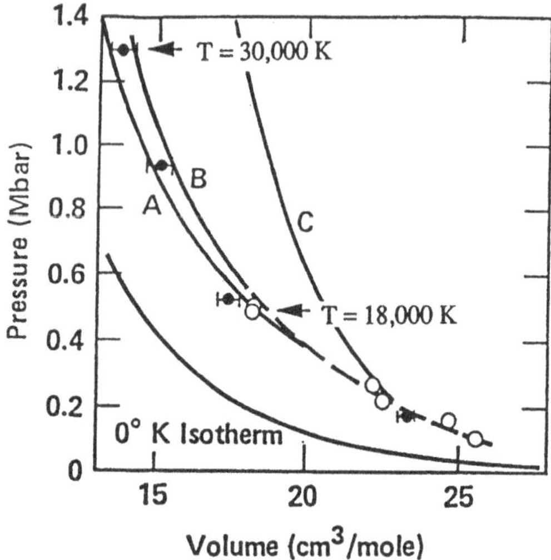

Fig. 13 Shock pressure against volume for liquid Xe. The initial volume (V_0) is 44.4 cm^3 $mole^{-1}$. Theoretical curves A and B include electronic excitation. Curve C is the insulator case (Nellis et al. 1982).

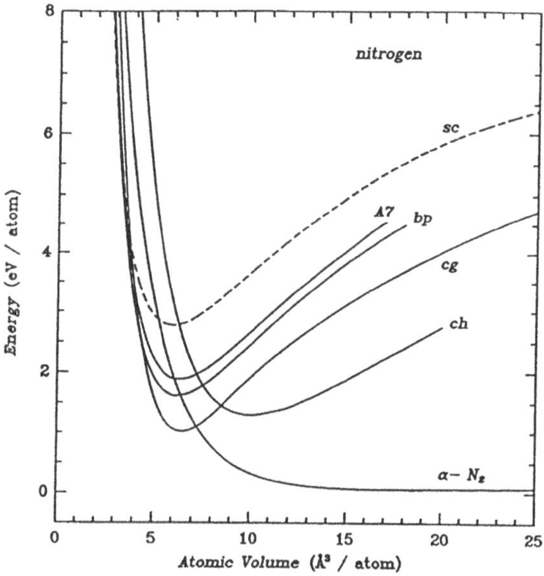

Fig. 14. Polymeric phases of nitrogen. Calculated total energy per atom for the simple cubic (sc, dashed curve) phase threefold-coordinated arsenic (A7), black-phosphorus (bp), and cubic gauche (cg) phases of nitrogen, and for a twofold-coordinated chain-like (ch) phase of nitrogen, as a function of atomic volume. The zero of energy corresponds to the minimum energy of the diatomic a - $N_2(Pa\bar{3})$ phase. (Mailhiot et al. 1992)

of iodine (see Lynch and Drickamer 1966). They noted the onset of metallization by a closure of the optical gap and a sharp rise in the electrical conductivity near 180 kbar. Subsequently, it has been found that iodine remains diatomic at that pressure and undergoes a continuous structural change while in a molecular metallic state, finally becoming a monatomic metal near 210 kbar (Takemura et al. 1982). In the case of molecular hydrogen the evidence to date is incomplete, but what does exist strongly indicates that it too passes through an intermediate molecular metallic state as it converts from a molecule to a monatomic metal.

Molecular nitrogen is an interesting case that has been studied extensively both theoretically and experimentally and with both static diamond-anvil and shock wave techniques. Room temperature diamond-anvil measurements have been made for solid nitrogen up to 1.8 Mbar with no indication from the Raman spectra that a change from diatomic bonding has occurred. However total energy calculations (Fig. 14), by Mailhiot et al. (1992), using the LDA method, predict that above 500 kbar the stable form is a cubic gauche distortion of simple cubic (cg) which is a single bonded, threefold-coordinated polymeric structure. This suggests that the molecular form found experimentally above this pressure is actually metastable. The two lowest energy structures lying above the diatomic are cg and a two-fold coordinate chain-like structure (ch). While the chain has a lower energy above a volume of 8.5 A it never crosses the diatomic curve preventing the possibility that it may exist as a stable high pressure structure. A general feature of the curves is a diatomic lattice ground state whose total energy is increasing with density and higher energy states with energies that are decreasing relative to the ground state, leading to a crossing at some high pressure.

Liquid molecular nitrogen has been shock compressed to 0.8 Mbar and a temperature of about 14,000 K (Fig. 15). Theoretical calculations using a simple model neglecting dissociation show good agreement with experimental Hugoniots (15a) and shock temperatures (15b) up to 0.32 Mbar and 7000 K. Above this pressure nitrogen undergoes a transformation to a new state in which the chemical bonds are broken or rearranged. Shock experiments do not provide such information which must be gleaned from theory. Calculations made with models which do not include the dissociation energy (curve A) or a dissociation energy of the gas phase diatomic Do = 9.76 eV (curve B), predict Hugoniot curves which are much to stiff. A model, discussed below, which explains the data (curves C and D) includes an effective dissociation energy D_{eff} that decreases with volume. Curves C and D.are calculated using two slightly different functions of D_{eff}.

In the original analysis of the shock data, done prior to the total energy calculations of Mailhiot et al., the introduction of a volume dependent dissociation energy was justified by the following considerations. In the gas phase 9.76 eV are required to break the chemical bond and remove the atoms to infinite separation. However, in the dense fluid, dissociation of the dimer allows a recombination of atoms into units with larger cooordination numbers. This clustering returns some of the energy expended in bond breaking and decreases the total energy change required for dissociation. This is a cohesion energy of the cluster and will have a density dependence.

The model used to calculate the shock-compression of liquid nitrogen considers a reacting mixture of atoms and molecules (Ross 1987)]. Let x is the fraction of dissociated molecules. A, the Helmholtz free energy per two atoms, is

$$A = (1-x)\, A^0_{N2} + x\, A^0_{2N} + Amix + Aint + (1-x)\, Deff, \qquad (47)$$

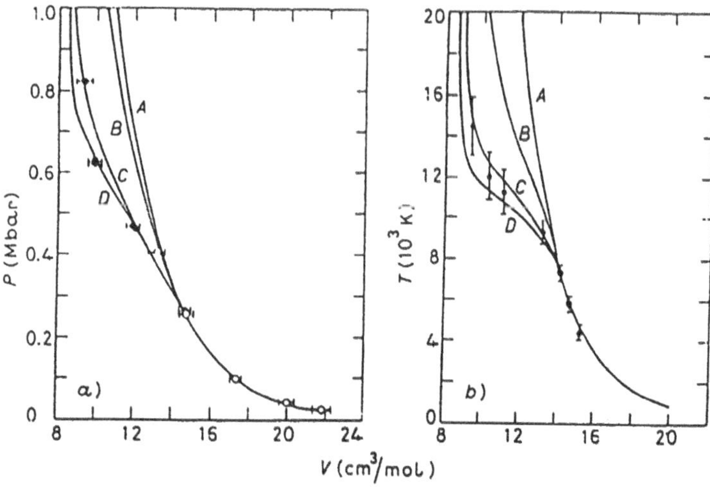

Fig. 15. Liquid-nitrogen shock Hugoniots (a)) and shock temperatures (b)), V_0 = 34.7 cm3/mol. Experimental data (Nellis et al. 1991) Theoretical curve A does not include dissociation. B includes dissociation to atoms using a constant value of the dissociation energy (9.76 eV). C and D calculated using a volume dependent dissociation energy (D_{eff}) which includes a binding energy of the monatomic phase.

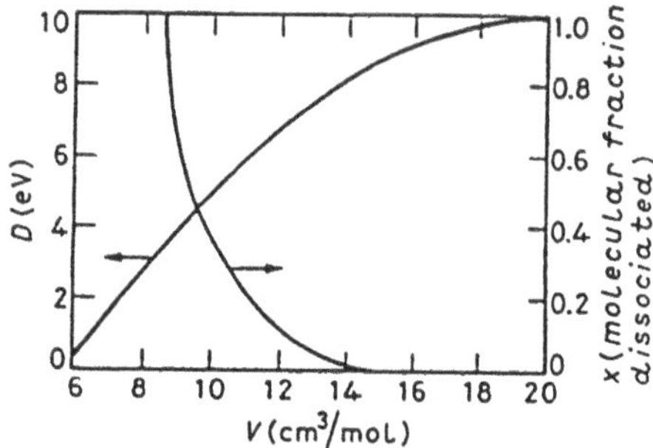

Fig. 16. Effective dissociation energies (D_{eff}) and fractions of dissociated molecules (x) calculated along the nitrogen Hugoniot.

where A^0_{N2} and A^0_{2N} are respectively, the free energies of the isolated molecule and of two atoms. A_{mix} is the free energy of mixing. A_{int} is the interaction energy of the fluid calculated using soft-sphere fluid variation theory (Eq. 36). The N_2 -N_2 potential was determined from the shock data below 0.25 Mbar. Minimization of A determines x. By making several simplifying approximations regarding the N-N and N-N_2 interatomic forces the following expression is obtained for the pressure:

$$P = (1+x)\,RT/V + P_{int} - xdD_{eff}/dV \tag{48}$$

D_{eff} enters into the pressure as a consequence of the volume dependence of the effective dissociation energy. In this regard it is similar to the volume dependence of the energy-band gap in xenon (Eq. 46). As in the case of xenon it leads to an an increase in the degree of dissociation and a considerable softening of the Hugoniot. Calculated values of the dissociation energy and the fraction of molecules dissociated are plotted in Fig. 16. The conclusions drawn from analyzing the shock data are consistent with the work by Mailhiot et al., although the exact correspondence between their solid structures and the high temperature species present in a dense shock experiment remain unclear. They suggest that a possible candidate would be some linear or polymeric form since at atomic volumes larger than 8.5 Å/atom the chain-like structure has the lowest energy above the diatomic. This is in the volume range probed by the high shock experiments and the high temperatures attained are likely to produce smaller molecular units.

To gain a deeper understanding, molecular orbital calculations were made for pairs of nitrogen dimers (Winter 1992). They show that the difference in energy between the ground (N_2-N_2) and lowest singly excited state (N_2-N_2^*) decreases with decreasing intermolecular separation. The separation at which crossing occurs is equivalent to a solid volume of about 8 Å/atom in agreement with the work of Mailhiot et al.

5.2 Shock-compressed metals: aluminum

For the case of metals, long-range Coulomb interaction and partially filled conduction bands lead to interatomic forces that are inherently many-body, unlike the simple two-body interactions between small molecules. As a result, practical calculations for the thermodynamic properties of metals present a difficult problem. For simple nearly free electron metals, such as aluminum the pseudopotential electron theory provides a formalism which leads to effective two-body forces (Moriarty et al. 1984). Aluminum which has been extensively studied up to nearly 10^4 Mbar provides a good example of the available theories applied to shock compressed metals (Young et al. 1985).

For a system of nearly free electrons and screened ions, the Helmholtz free energy for a simple liquid metal may be described by the same soft-sphere variational theory used for Xenon (Eq. 44) in which the free electron gas term is written as

$$A_{eg} = N\mu - NkT \int \ln(1 + \exp\,[(\mu - \varepsilon)/kT])\,\rho(\varepsilon)\,d\varepsilon + A_{corr}\,(T,\mu\,Z) + A_{ex}\,(T,\mu\,Z).$$

A_{eg}, includes electron correlation (corr) and exchange energies (ex). The use of Fermi-Dirac statistics permits application of the theory to finite temperatures. The theory can be generalized to include thermal excitation of core electrons as well as the excitation of conduction electrons. An effective interatomic potential can be determined by fitting parameters in the electron-ion pseudopotential to experimental and theoretical isotherms.

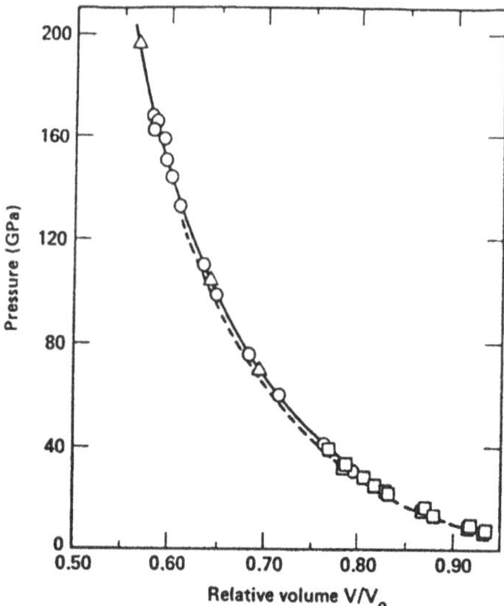

Fig. 17a. The Hugoniot curve for aluminum. The solid line was calculated with a fluid model and the dashed line with a solid model.

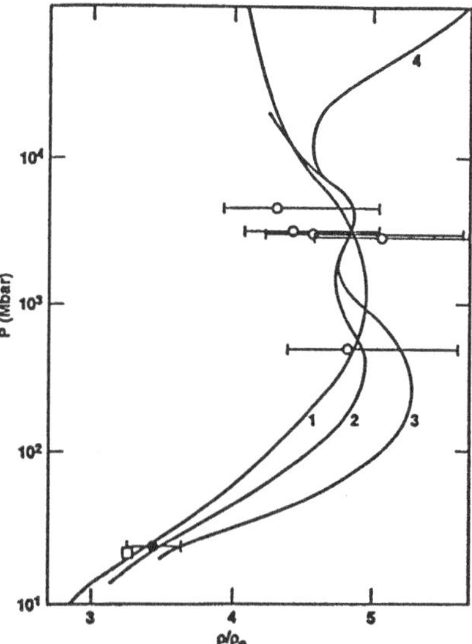

Fig.17b. Experimental data (of pressure versus compression ρ/ρ_0) for shock compressed Al compared to a Thomas-Fermi calculation (1) and two LDA-type models (2) and (3). Note the oscillations in the latter. Thermal radiation effects (4) appear above 10^4 Mbar.

A plot of shock pressure versus compression (Fig. 17a) shows that both the calculated liquid and solid Hugoniots agree with experiment. Although melting is predicted to occur at 1.2 Mbar and 4700 K, the P-V curves show no change in slope. This illustrates the difficulty for detecting melting along the Hugoniot. The method of sound speed measurements, developed by Brown and McQueen (1980) represents a sensitive indicator of the loss of shear strength in going from solid to liquid. Using this method, aluminum melting has been detected at 1.3 Mbar in agreement with theoretical predictions (Moriarty et al. 1984).

Underground nuclear explosions have been used to generate pressures of up to 10^3 Mbar in shock experiments with metals, an attempt to discover possible shell effects (Avrorin et al. 1987). Figure 17b shows some experimental data for aluminum. They are compared to theoretical calculations made with a temperature-dependent Thomas-Fermi and local density model which includes electron structure. Although the experiments are not sufficiently accurate to distinguish between theories, the predicted effect of shell structure on the calculated Hugoniot is seen clearly.

5.3 Dense Plasmas - The Saha-Boltzmann Ionization Equation

At very high temperatures gas molecules dissociate and atoms are found in multiply ionized states (Ebeling et al. 1977, Zeldovich and Raizer 1966). Under conditions of local thermodynamics equilibrium, and negligible interatomic interactions the ratio of the densities of the different states of ionization is given by the Saha equation. N_r, the number density, of atoms in the rth stage of ionization is related to that of the (r + 1)th stage, N_{r+1}, by

$$\frac{N_{r+1}}{N_r} N_e = \frac{u_{r+1}}{u_r} \frac{2(2\pi\, mkT)^{3/2}}{h^3} \exp\left(\frac{-I_r}{kT}\right),$$

where N_e is the free electron density, u_r is the internal partition function of the rth stage, I_r, is the energy required to remove an electron from the ground state of the r-times ionized atom m is the electron mass, and h is Planck's constant. The Saha equation can be derived by minimizing the Gibbs free energy of a reacting mixture. The pressure, for negligible interactions, is that of a perfect gas.

Ionization of atoms begins at values of kT much lower than the ionization potential. The reason here is that the statistical weight of the free electron state is very large. With exception of the alkali metals, which have very low ionization potential, the first ionization potentials of the majority of atoms and molecules vary between 7 and 15 eV ($I/k \sim 80{,}000$–$170{,}000$ K). Ionization usually begins at thermal energies of the order of a few tenths of the ionization potential. The degree of ionization increases with temperature. When the temperature is of the order of several tens of thousands of degrees practically all of the atoms are singly ionized and the gas contains multiply ionized atoms.

The effects of small nonideal contributions from interparticle interactions is to lower the ionization potential and therefore enhance ionization. The corrections are collectively responsible for the phenomenon of "pressure ionization." For the conditions under which the Saha equation has been derived, nonideal effects only produce a slight enhancement of the degree of ionization. The Saha equation is widely used for astrophysical applications where the properties of stellar and interstellar matter are determined by the chemical composition and the degree of ionization and state of excitation.

Dense plasmas properties are generally calculated using the 'chemical' and 'physical' models. The chemical model is the simpler and more widely used. It is a philosophical descendent of the Saha and Debye models, described in a series of papers by Graboske et al (1969, 1971). Gryaznov et al (1980) have extended the model to higher densities.

5.3.l Chemical Model. The chemical model a logical extension of fluid perturbation theory applied to dense plasmas. It is intuitively appealing in that it assumes the partition function to be separable into distinct terms which permit the Helmholtz free energy (A) to be written as a sum of contributions:

$$A = A_{trans} + A_{coul} + A_{hs} + A_{internal} \tag{49}$$

A_{trans} is the translational free energy of ions and electrons. A_{coul} is the contribution from the Coulomb interaction of all non-bound particles. It includes electron exchange and correlation. A_{hs} is a hard-sphere free energy and serves to simulate ion-ion repulsions. $A_{internal}$ is the free energy of bound electrons in the ion. The electron energy levels are determined by solving the Schrödinger equation in a sphere, outside of which the potential is infinite and the wavefunction is zero. The sphere may be equal to the average atomic volume or determined by the hard-sphere diameter. Electrons with positive energy are considered free and contribute to translational and Coulomb terms. As a result, discontinuities in the thermodynamic properties arise when energy levels undergo changes of sign and electrons are transferred from internal bound to translational free states. Recently Hummer and Mihalas (1988) have proposed a method which avoids these discontinuities. The free energy is determined by minimizing with respect to the composition of chemical species and hard-sphere diameter. The value of the hard-sphere diameter as the atom sphere is used as the outer boundary for solving the Schrödinger equation of the bound electrons. In the chemical model all the degrees of freedom are coupled and thermodynamically self-consistent calculations have been carried out. The major limitation stems from the artificial nature of the separation between bound and free electrons.

5.3.2 Physical Model. These difficulties are avoided by the more rigorous 'physical' approach, which represents a consistent quantum statistical-mechanical treatment of the many-body problem that does not make distinctions between bound and free electrons (Rogers, 1991). In practice, the theory is much more complicated than the chemical model, making the construction of a suitable computer program extremely laborious. The starting point of the theory is an activity expansion of the grand partition function. The leading terms are the Saha ionization equilibrium equation and Debye-Hückel correction. Higher-order terms resemble many-body virial and cluster coefficients for Coulomb potentials. The theory allows for (i) formation of electron-nucleus composites, (ii) coupling of heavy ions and (iii) moderate (gas-like) coupling between electrons. To carry out numerical calculations, it is necessary to obtain multi-electron bound and scattering states. This is accomplished through the introduction of effective electron-ion pseudo-potentials. The parameters in the pseudo-potential are determined from experimental spectroscopic data, and plasma screening effects are then added to the long-range tail.

6.0 High Pressure Studies of Planetary Matter

For the purpose of discussion, the planets may be categorized into two groups: The terrestrial planets, Earth, Mars, Mercury and Venus which are made up of the rocky and solid-like material (Fe, metal oxides, silicates, etc.) and the Jovian planets, Jupiter, Saturn, Uranus and Neptune, consisting of an inner rock core estimated to be of several earth masses surrounded by a large volume of dense fluid. Jupiter and Saturn are made mainly of a solar mixture of H_2 and He (7 parts H_2 to 1 part He by number). Uranus and Neptune which are smaller and are at lower temperatures have retained considerable amounts of Ice in addition to their primordial H_2 and He. "Ice," is a term used to denote a planetary mixture of H_20, NH_3 and CH_4. Pluto is believed to be mainly made of Ice. A useful survey on the subject of planetary interiors is the book by Hubbard (1984).

6.1 Phase Diagram of Iron

A knowledge of the iron equation of state and phase diagram is of great importance for constraining the temperature and density at the center of the Earth. Seismological studies indicate that the core of the Earth is iron mixed with elements of lower atomic weight. Also, the presence of these elements implies that the melting temperature of pure iron at the pressure of the liquid outer core-solid inner core boundary (3.3 Mbar) is an upper bound to the temperature at this boundary inside the Earth. Until recently, discussion about the phase diagram of iron under Earth's interior conditions was highly speculative, but the advent of the megabar diamond anvil cell and of high-precision shock wave experiments has begin to deliver experimental data in this range.

Brown and McQueen [1980, 1986] measured the sound velocity of iron along the shock Hugoniot and found a continuous rise of velocity with pressure in the ε phase up to 2.00 Mbar, where the estimated temperature was 4250 ± 150 K. At this pressure and temperature the sound speed showed a sharp discontinuity, indicating a transition presumed to be at the γ–ε phase boundary. At 2.43 Mbar and an estimated temperature of 5250 ± 350 K the sound speed showed another discontinuity, presumed to be a transition to the liquid, because the measured sound speed corresponded to the expected bulk sound speed of the compressed liquid.

Boehler (1986) reported diamond anvil measurements of the ε–γ (hcp-fcc) and melting curves to 40 GPa. Figure 18 shows the phase diagram of iron, combining the results of Boehler with the sound velocity measurements of Brown and McQueen. Boehler predicted that the γ phase ends in a triple point near 75 GPa, and he speculated that the shock anomaly at 2.0 Mbar must correspond to a new high pressure solid phase. Theoretical calculations (Ross et al., 1989), predict that this anomaly may be caused by a new high temperature bcc phase, referred to here as α'. This model predicts the cp-bcc (ε–α') phase boundary as shown in Fig. 18. The total Helmholtz free energy of the solid is written as the sum of a volume-dependent mean field energy, a pair-potential static lattice energy, an electronic thermal free energy, and a lattice vibrational free energy:

$$E_o(V) = \frac{1}{2}E_b\left[(V_o/V)^{1/3}\right]$$

and

$$A(V,T) = E_o(V) + \frac{1}{2}\sum_{ij}\phi(r_{ij}) + A_{el}(V,T) + A_{vib}(V,T) \tag{50}$$

For simplicity, all of the cohesion is included in E_o and that the potential $\varphi(r)$ is purely repulsive. The electronic thermal free energy is obtained from band structure calculations of the electron density of states [Boness et al., 1986; Bukowinski, 1977].

$$\phi(r) = \varepsilon\,\frac{\exp(-r/r_m)^k}{(r/r_m)}$$

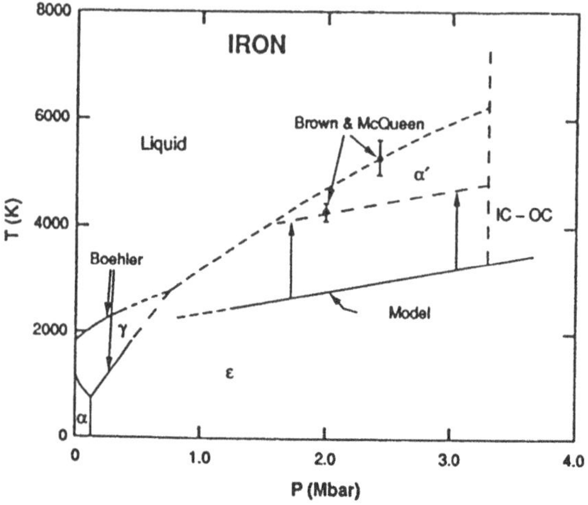

Fig.18. Phase diagram of iron up to Earth core conditions. The experimental phase boundaries of Boehler (1986) are shown to 400 kbar and are extrapolated to a e-g-liquid triple point. The melting curve is then smoothly drawn through the Brown and McQueen (1980, 1986) shock anomaly at 2.43 Mbar. The theoretical cp-bcc phase boundary from the model is shown, and the arrows indicate a uniform shift necessary to have this boundary intersect the 2.00 Mbar shock anomaly.

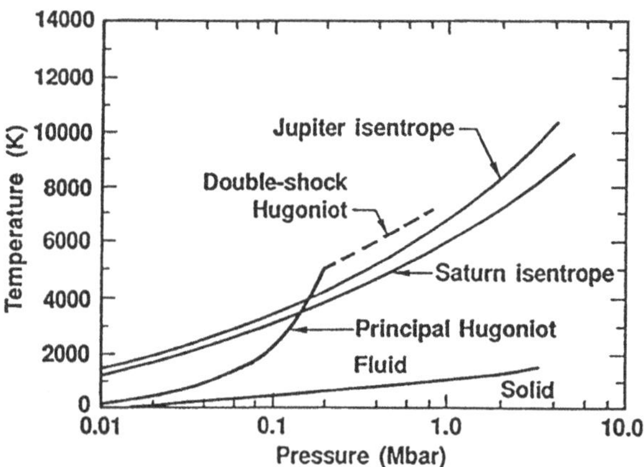

Fig. 19. Experimental Hugoniots compared with Saturn and Jupiter isentropes. The metal is believed to exist above 3 Mbars.

In the expression for Eo, V_c and E_b are the volume and energy of the minimum point respectively. The potential is a simple repulsive exponential form with coulombic behavior at small r. These expressions have been chosen to agree qualitatively with typical functions calculated from pseudopotential theory. The model parameters are determined by fitting the experimental shock Hugoniot, isotherm, and melting curve of iron.

The model predicts bcc to be stable at high temperatures and that the cp-bcc phase boundary has a positive dT/dP slope. The theoretical shock Hugoniot crosses the cp-bcc phase boundary at 1.30 Mbar and 2400 K. Although this differs from the 2.0 Mbar experimental shock anomaly, given the approximate nature of the model, the agreement is good. If the calculated cp-bcc phase boundary slope is correct and the boundary is shifted to intersect the shock anomaly at 2.0 Mbar, then the first appearance of the bcc α' phase is predicted to occur at 1.50 Mbar and 4000 K on the melting curve.

The conclusion drawn is that a high-pressure bcc phase is possible and that the 2.0 Mbar shock anomaly may be the transition to this phase.

6.2 Hydrogen and Helium at Megabar Pressure: An Application to Modeling the Giant Planets

Hydrogen and helium are the simplest elements and are also the major constituents of the two largest planets, Jupiter and Saturn. It is believed that the Jupiter and Saturn have an outer mantle of fluid molecular hydrogen and helium (7% He by atom fraction), a middle layer of helium and metallic hydrogen at a pressure above 3 Mbar surrounding a solid central inner core made of silicate and metal oxides. Pressures and temperatures in the hydrogen-helium mixture reach a maximum of about 45 Mbar, and 20,000 K at the fluid-solid core boundary for Jupiter, and 10 Mbar and 10,000 K for Saturn. The interest in the EOS of the planetary mixture stems from a need to explain why Jupiter is cooling at a rate consistent with planetary evolution models, while Saturn appears to be cooling more slowly than predicted. It has been suggested by Smoluchowski (1967) and Salpeter (1973) that a phase separation in the fluid mixture in Saturn and subsequent release of gravitational energy by the helium rich droplets may provide the additional source of energy. Since the original conjectures, there have been a number of attempts to calculate the phase separation of the mixture.

An essential feature of attempts to model the Giant planets is the need for accurate theories of the fluid mixture and interatomic potentials to calculate the equations of state at high temperature and pressure. The great interest in these two elements provides a useful framework for illustrating many of the current theories of dense plasmas. At low pressures hydrogen and helium are molecular insulators. Solid hydrogen is believed to become a molecular metal near 1.5 Mbar and it is predicted to become a monatomic metal between 3 or 4 Mbar. Solid helium is predicted to become a metal above 35 Mbar. Shock wave experiments have been made for hydrogen and helium, and effective intermolecular potentials obtained from the data (Section 5.1.1). The usefulness of shock data to planetary modeling is illustrated in Fig. 19, which plots the temperatures attained by shock experiments along with temperatures calculated along the giant planet isentropes.

6.2.1 Mixtures of H_2 and He

The EOS in the homogeneous, fully mixed region, is usually well approximated by the composition average of the pure components. In general, there will be small nonideal contributions to the free energy that determine the immiscibility or unmixing. If these are positive, then the system can lower its free energy by separating into pure components. Predictive calculations require very accurate equation of state models and, in particular, require a discouragingly accurate knowledge of the interatomic forces between unlike species. Soft-sphere perturbation theory has been successfully applied to fluid mixtures in terms of a one-fluid van der Waals model. In this approximation, an effective intermolecular interaction for a hypothetical pure fluid is constructed in terms of a composition averaged set of potential parameters. This model considerably simplifies calculations. For the exp-6, the effective parameters are

$$\alpha = \sum_{ij} x_i x_j \alpha_{ij} \varepsilon_{ij} \frac{\left(r^*_{ij}\right)^3}{\varepsilon \left(r^*\right)^3} .$$

$$\left(r^*\right)^3 = \sum_{ij} x_i x_j \left(r^*_{ij}\right)^3 .$$

and

$$\varepsilon = \sum_{ij} x_i x_j \varepsilon_{ij} \frac{\left(r^*_{ij}\right)^3}{\left(r^*\right)^3} .$$

In these equations x_i denotes the mole fraction of i th component and r^*_{ij} and ε_{ij} are the potential parameters of the interaction between a molecule of species i and j. Thus the effective one-component parameters r* and ε are then composition dependent. For a particular P and T, the coexisting phases of a mixture can be determined by finding the pair of compositions x_1 and x_2 at which the Gibbs free energy of mixing

$$G_m\ (P,T,x) = G(P,T,x) - xG(P,T,1) - (1-x)G(P,T,0) \tag{51}$$

has a common tangent. Van den Bergh and Schouten (1988) calculated the miscibility of the insulating mixture. They used pair potentials fitted to shock wave data for the pure component and H_2-He potential parameters adjusted to fit low pressure miscibility experimental data. For pressures up to 1 Mbar and temperatures to 2500 K, the components would be completely miscible.

6.2.2 Mixtures of Pressure-ionized H-He

Attempts to calculate the phase separation of the H-He phase generally start with the assumption that both hydrogen and helium are pressure-ionized so that the electronic structure is approximately that of free or weakly screened electrons in the presence of positive ions. Stevenson (1975) used a hard-sphere mixture model for the ions with an ion-ion pseudopotential to account for electron screening, and predicted that a mixture containing 7% by number would phase separate at a temperature of about 7000 K at 8 Mbar. Calculations made for the fully or nearly fully ionized mixture either through the bare Coulomb interaction or a screened interaction (linear response theory) have all arrived at predictions similar to those of Stevenson.

6.2.3 Two Component Plasma (TCP)

A useful starting point for illustrating the modeling of pressure-ionized hydrogen and helium mixtures are the Monte Carlo simulations of Stringfellow et al. (1990) for the one-component plasma (OCP). The usual independent variables in these simulations are r_s, and $\Gamma_e = (e^2/ r_s k_B T)$. The excess Helmholtz free energy for the OCP, including E_{eg} is the total energy of the electron gas, is written as;

$$A_{OCP} = E_{eg} + k_B T\, f_{ocp}\,, \tag{52}$$

where f_{ocp} is the interaction part of the ocp free energy given in Eq. 33. Ichimaru et al. (1988) have shown that accurate free energies for a binary ionic mixture can be obtained from Monte Carlo simulations of the OCP by using the "linear-mixing rule" (LM) for the interaction energy of the mixture,

$$f_{LM}(x) = x f_{ocp}\left(Z_2^{(5/3)}\Gamma_e\right) + (1-x)\, f_{ocp}\left(Z_1^{(5/3)}\Gamma_e\right) \tag{53}$$

where Z_1 and Z_2 are the two ionic charges and x is the number fraction of type 2 ions. The excess free energy for a two-component plasma (TCP) is now,

$$A_{TCP} = k_B T[\, x\ln x + (1-x)\ln(1-x)] + E_{eg} + k_B T\, f_{LM}\,, \tag{54}$$

where the first term includes the entropy of mixing. The calculated phase separation temperatures at 10.5 Mbar, are plotted as the dotted curve in Fig. 20 (labeled OCP-LM). In order to check the validity of the LM rule, the same calculations were carried out, but using the published results from the Monte Carlo simulations of Hubbard and DeWitt (1985). In these simulations, the ions interact via a screened Coulomb interaction; the Lindhard function is used to approximate the dielectric response of the electron gas to the presence of the bare ions. The phase separation temperatures obtained using the Hubbard and DeWitt free energies are shown as the dashed curve in Fig. 20 (labeled Lindhard). A comparison of this curve to the OCP-LM curve indicates that the LM rule works quite well and also that the use of Lindhard screening makes very little difference in terms of the phase separation temperatures.

6.2.4 Ion-Sphere Model

In the ion-sphere model of the two-component ionic solid, the total energy is written as

$$E_{IS} = E_{eg} + E_{mad}\,, \tag{55}$$

where E_{mad} is the Madelung energy that describes the total energy of a two component system of ions in a neutralizing electron gas background.

$$E_{mad} = (\alpha_M / r_s)\left[x Z_2^{(5/3)} + (1-x)\, Z_1^{(5/3)}\right] \tag{56}$$

and α_M is the Madelung constant. Eq. 56 is analogous to the LM rule given in Eq. 53, but applied to the 0 K solid. If r_s is in units of Bohr radii then, for the ion-sphere model, $\alpha_M = 1.8$ Ry, This value is very nearly equal to the true Madelung constants for both bcc and fcc lattices. The expression for the excess Helmholtz free energy in the ion-sphere model is

$$A_{IS} = k_B T[\, x\ln x + (1-x)\ln(1-x)] + E_{eg} + E_{mad}\,, \tag{57}$$

and the Gibbs free energy of mixing becomes,

$$\Delta G_{IS}(x) = \Delta H_{IS}(x,T=0)+k_BT[x\ln x+(1-x)\ln(1-x)] \; . \qquad (58)$$

The phase separation temperatures for the ion-sphere model at 10.5 Mbar are plotted as the dot-dashed line in Fig. 20 (labeled Ion-sphere). The difference between this curve and the curve labeled OCP_{LM} at x = 0.07 is very close to 1000 K. The corresponding difference in the critical temperatures is approximately 2000 K.

6.2.5 First Principles LDA

Recently first principles LDA calculations (see Section 2.1) have been made of the electronic structure, equation of state and enthalpy of mixing for H-He alloys of a number of different compositions in bcc and fcc crystals (Klepeis et al. 1991). These are fully quantum mechanical self-consistent calculation and makes no assumption regarding the state of pressure ionization. The major approximation in the LDA method is that the exchange and correlation energy is given by a free electron functional in terms of the local electron density . The accuracy of LDA methods, which are well established, offer the possibility of accurately describing both constituents.

The primary result of the calculations are given in Fig. 21 by the solid line which shows the enthalpy of mixing per atom at T = 0 K and 10.5 Mbar,

$$\Delta H\,(x) = H(x) - xH(x = 1) - (1 - x)H(x = 0). \qquad (59)$$

Here H(x) = E(x) + PV(x) is the enthalpy per atom and x is the number fraction of helium atoms, E(x) is the total energy per atom, P is the pressure, which is held constant, and V(x) is the volume per atom. The solid curve is a polynomial fit to all of the calculated values (both bcc and fcc). For comparison values of ΔH_{IS} at 0 K for the Ion-Sphere model are also shown. They are significantly smaller than those obtained from the LDA.

The calculated ΔH is a very slow varying function of pressure. ΔH(x) is virtually unchanged between 5 and 10 Mbar. ΔH (x = 1/2) is nearly constant between 5 and 20 Mbar, and that above 20 Mbar it decreases very slowly with increasing pressure. He is observed to metallize near 35 Mbar, but nothing striking occurs to ΔH and its value remains large.

In Section 6.2.4, it was shown that the ion-sphere approximation, which neglects the thermal energy of the ions, is in good agreement with the more accurate Monte Carlo based predictions. Thus, a simple model for calculating the Gibbs free energy of mixing for the LDA case based on these results is assumed,

$$\Delta G_{LDA}(x) =\Delta H_{LDA}(x,T=0)+k_BT[x\ln x+(1-x)\ln(1-x)] \qquad (61)$$

The values for ΔH_{LDA} are based on the a to the solid curve in Fig. 21.

Through a series of common tangent constructions, the phase separation temperature can be determined for all values of x. The results of this calculation, are shown as the solid curve in Fig. 21. The critical phase separation temperature is found to be T = 40,000 K ± 10,000 K at a critical concentration of x_c = 0.43. The phase separation temperature at x = 0.07 (relevant to Jupiter and Saturn) is T=15,000 K ± 3000 K.

The large value obtained for the mixing enthalpy, which is responsible for the high critical temperature, stems from the fact that the helium derived electrons are more

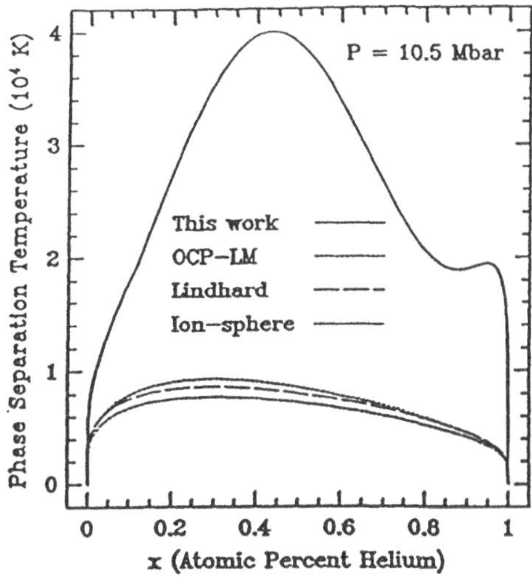

Fig. 20. The solid curve is the calculated phase separation temperatures at 10.5 Mbar (based on the results of total energy calculations discussed in Section 6.2.5. The remaining curves are obtained from three different plasma models (see text).

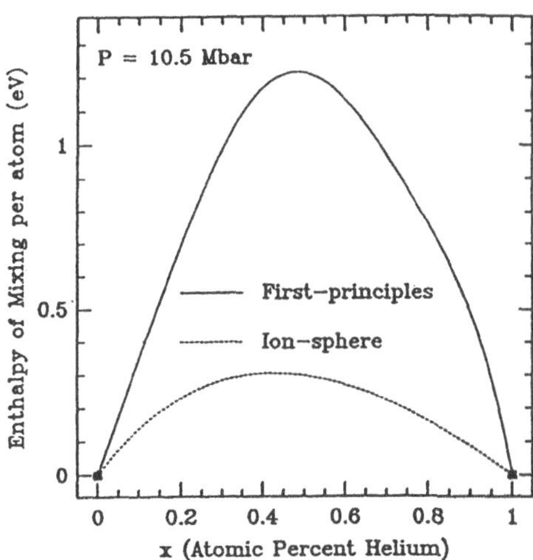

Fig. 21. The enthalpy of mixing per atom at T = 0 K (Eq. 59) obtained from total energy calculations at 10.5 Mbar, temperature of 0 K.

tightly bound to the nucleus and therefore have a lower energy relative to the more itinerant hydrogen derived electrons. The effect manifests itself in the electron-band structure as two distinct bands. One which is entirely below the Fermi level, composed primarily of helium 1s orbitals, and another which intersects the Fermi level composed primarily of hydrogen 1s orbitals. This energy separation inhibits the formation of the alloy. In a pressure-ionized mixture of free electron and ions there would be no distinction between helium and hydrogen electrons. In view of these large deviations from free electron behavior, it is likely that any perturbation theory treatment starting with free-electron states will break down for the helium-derived electronic states.

The relatively large, positive values of ΔH imply that a high temperature must be reached before entropy dominates and the fully mixed phase becomes energetically favored over the coexistence of a helium-rich phase and a helium-poor phase. In order to directly address the question of phase separation, a simple model is constructed which extends the total energy calculations to finite temperatures. The uncertainty in the model can be estimated by a comparison with accurate results based on computer simulations.

It has been argued that the model for mixing neglects the size difference of the atoms. This matter can be partially accounted for by using a mixing entropy function in terms of volume fractions rather than atomic fractions. This will lead to a smaller mixing entropy and to larger miscibility temperatures. The difficulty in applying such a model comes in unambiguously defining atomic sizes from electron density plots. Several estimates indicate any changes will be minor.

In terms of planetary physics, the result confirms that the fluid interior of Saturn has phase separated since the maximum temperature in the fluid is estimated to be only 10,000 K. The estimated temperatures in the fluid interior of Jupiter range from 10,000 K near the surface to 20,000 K at the central core, (Grossman 1980) Thus the calculations predict that partial phase separation has probably occurred in Jupiter. In view of this prediction, the fact that the currently successful evolutionary models of Jupiter do not need to invoke phase separation may indicate either a failure of these models or some inherent limitation of the present results. Alternatively, it has been suggested that phase separation may have occurred too late in the evolution of Jupiter to provide a significant internal energy source up to the present time. In either case, new evolutionary calculations are needed to resolve this problem.

Acknowledgment

This work was performed under the auspices of the U.S. Department of Energy by the Lawrence Livermore National Laboratory under contract No. W-74045-ENG-48.

References

Adams, D.J. (1976) J. Chem. Soc. Faraday Trans. II 72, 1372.

Al'tshuler, L.V. (1965) Sov. Phys. – Usp 8, 52-91.

Allen, M.P. and Tildesley, D. J. (1987) Computer Simulation of Liquids, Oxford: Clarendon Press.

Avrorin, E.N., Vodolaga, B.K., Voloshin, N.P., Kovalenko, G.V., Kuropatenko, V.F., Simonenko, V.A. and Chernovolyuk, B.T. (1987) Sov. Phys. JETP66, 347.

Aziz, R.A., Nain, V.P.S., Carley, J. S., Taylor, W. L. and McConville (1979) J. Chem. Phys. 70, 4330.

Barker, J. A. (1986) Phys. Rev. Lett. 57, 230-233.

Boehler, R. (1986) Geophys. Res. Lett. 13, 1153.

Boehler, R. and Ross, M. (1984) Phys. Rev. B29,3673.

Boness, D.A., Brown, J.M. and McMahan, A.K. (1986) Phys. Earth Planet. Inter 42, 227.
Brown, J.J. and McQueen, R.G. (1980) Geophys. Res. Lett. 7, 533.
Brown, J.J. and McQueen, R.G. (1986) J. Geophys. Res.,91,7485.
Bukowinski, M.S.T. (1977) Phys. Earth Planet Inter. 14, 333.
Ceperley, D.M. and Partridge,H. (1986) J.Chem. Phys. 84, 820.
Ciccotti, G. and Hoover, W.G. (1986) Molecular-Dynamics Simulation of Statistical-Mechanical Systems, Amsterdam, North Holland.
Devreese, J.T. and Van Camp, P. (1984) Electronic Structure, Dynamics, and Quantum Structural Properties of Condensed Matter, New York: Plenum.
Ebeling, W., Kraeft, W.D., Kremp, D. (1977) Theory of Bound States and Ionization Equilibrium in Plasmas and Solids, Berlin: Akademie Verlag.
Eliezer, S., Ghatak, A., Hora, H. (1986) An Introduction to Equations of State: Theory and Applications, Cambridge: Cambridge University Press.
Fortov, V.E. (1982) Sov. Phys. Usp. 25, 781.
Girafalco, L.A. (1973) Statistical Physics of Materials, New York: Wiley.
Gloetzel, D. and McMahan, A.K. (1979) Phys. Rev. B20, 3210.
Godwal, B.K., Sikka, S.K. and Chidambaram, R. (1983) Physics Reports 102, 121-197.
Goettel, K.A., Eggert, J.H., Silvera, I.F. and Moss, W.C. (1989) Phys. Rev. Lett. 62, 665.
Goodstein, D.L. (1975) States of Matter, New Jersey, Prentice Hall.
Graboske, H.C., Harwood, D.J. and DeWitt, H.E. (1971) Phys. Rev. A3 1419.
Grossman, A.S., Pollack, J.B., Reynolds, R.T., Summers, A.L. and Graboske, H.C. (1980), Icarus 42,358.
Gryaznov, V.K., Zhernokletov, M.V., Iosilevskii, I.L. and Fortov, V.E. (1980) Sov. Phys.-JETP 51 288.
Hafner, J. (1987) From Hamiltonians to Phase Diagrams, Berlin: Springer-Verlag.
Hansen, J.P. and McDonald, I.R. (1986) Theory of Simple Liquids, London, Academic.
Hensel, F. and Uctmann, H. (1989) Ann. Rev. Phys. Chem. 40.
Hemley, R.J., Mao, H.K., Finger, L.W., Jephcoat, A.P., Hazen, R.M. and Zha, C.S. (1990) Phys. Rev. B 42,6458.
Holmes, N. C., Moriarty, J. A., Gathers, G. R. and Nellis, W. J. (1989) J. Appl. Phys. 66, 2962.
Holt, A.C. and Ross, M. (1970) Phys. Rev B1, 2700.
Hoover, W.G. (1991) Computational Statistical Mechanics, New York: Elsevsier.
Hubbard, W.B. (1984) Planetary Interiors, Van Nostrand Rienhold Co.
Hubbard, W.B. and DeWitt, H.E. (1985) Astrophys. J. 290, 339.
Hummer, D.G. and Mihalas, D. (1988) Astrophys. J. 331, 794.
Ichimaru, S., Iyetomi, H. and Ogata, S.(1988), Astrophys. J. Lett. 334, L17.
Japhcoat, A.P., Mao, H.K., Finger, L.W., Cox, D.E., Hemley, R.J., and Zha, C.S. (1987), Phy. Rev. Lett. 49, 2670.
Kang, G.S., Lee, C.S., Ree,T., Ree, F.H. (1985) J. Chem. Phys. 82, 414
Klepeis, J., Schafer, K.J., Barbee, T.W. III, and Ross, M (1991) Science 254, 986.
Lynch, R.W. and Drickamer, H.G. (1966) J.Chem.Phys. 45,1020.
Lundquist and March (1983), Theory of the Inhomogeneous Electron Gas, New York: Plenum.
Mailhiot, C. et al. (1992), Phys. Rev. B, in press.
Mao, H.K., Bell, P.M., Shaner, J.W. and Steinberg, D.J. (1978) J. Appl. Phys. 49, 3276.

Mao, H.K., Hemley, R.J., Wu, Y., Jephcoat, A.P., Finger, L.W., Zha, C.S. and Bassett, W.A. (1988) Phys. Rev. Lett. 60, 2649.

Maradudin, A.A., Montroll, E.W., Weiss, G.H., Ipatova, I.P. (1971) Theory of Lattice Dynamics in the Harmonic Approximation, New York: Academic Press.

Marsh, S.P. (1980) LASL Shock Data Berkeley: University of California Press.

McMahan, A.K. (1986a) Physica 139 & 140B, 31.

McMahan, A.K. (1986b) Phys. Rev. B. 33, 5344.

McMahan, A.K. and Ross, M. (1977) Phys. Rev. B. 15, 718.

McQuarrie, D.A. (1976) Statistical Mechanics, New York: Harper & Row.

Moriarty, J.A., Young, D.A. and Ross, M. (1984) Phys. Rev. B30, 578.

Nellis, W.J., van Thiel, M. and Mitchell, A.C. (1982) Phys. Rev. Lett. 48, 816.

Nellis, W.J., Mitchell, A.C. and van Thiel, M. (1983) J. Chem. Phys. 79, 1480.

Nellis, W.J., Holmes, N.C., Mitchell, A.C., Trainor, R.J., Governo, G.K., Ross, M and Young, D.A. (1984) Phys. Rev. Lett., 53, 1248.

Nellis, W.J., Radousky, H.B., Hamilton, D.C., Mitchell, A.C., Holmes, N.C. Christianson, K.B. and van Thiel, M. (1991) J. Chem. Phys. 94, 2244..

Pistorius, C. (1965) J. Phys. Chem. Solids 26, 1543

Ree, F.H. (1981), J. Chem. Phys. 73, 5401.

Reichlin, R., Brister, K. E., McMahan, A. K., Ross, M., Martin, S., Vohra, Y. K. Ruoff. A. L., (1988) Phys. Rev. Lett. 62, 669.

Rice, M.H., McQueen, R. G. and Walsh, J.M. (1958) Solid State Phys. 6, 1.

Rogers,F.H. (1991) High Pressure Equations of State:Theory and Applications Amsterdam,Holland. Ed. Eliezer,S. and Ricci,R.A. p.77.

Ross, M. (1968) Phys. Rev. 171,777.

Ross, M. (1979) J. Chem. Phys.71, 1567.

Ross, M. (1987) J.Chem. Phys. 86, 7110

Ross, M. (1985) Rep. Prog. Phys. 48, 1.

Ross, M. and Alder, B.J. (1967) J.Chem. Phys. 46, 4203.

Ross, M., DeWitt, H. E. and Hubbard, W. B. (1981) Phys. Rev. B. 24, 1016-1020.

Ross, M. and McMahan, A.K.,(1980) Phys. Rev. B21,1658.

Ross, M., Ree, F.H. and Young, D.A. (1983) J.Chem.Phys.,79,1487.

Ross, M. and Rogers, F.J. (1985) Phys. Rev. B31, 1463.

Ross, M. and Young, D.A. (1981) 79, 1487.

Ross, M. and Young, D.A.(1986) Phys. Lett. A. 118,463.

Ross, M., Young, D.A. and Grover, R.(1990) J. Geophys. Res. 95, 21713.

Salpeter, E.E.,(1973) Astrophys. J. Lett. 181, L83.

Shimoji, M. (1977) Liquid Metals, London: Academic Press.

Skriver, H.L. (1984) The LMTO Method, Berlin: Springer-Verlag.

Smith, W.R. and Henderson, D.(1970) Mol. Phys. 19, 411

Smoluchowski,R. (1967) Nature 215, 691.

Stevenson, D.J., (1975) Phys. Rev. B12, 3999.

Stringfellow, G.S., DeWitt, H.E. and Slattery, W.L. (1990) Phys. Rev. A 41, 1105.

Takemura, K., Minomura, S., Shimomura, O., Fujii,Y. and Axe, J.D. (1982) Phys Rev. B 26, 998.

Tersoff, J. (1989) Phys. Rev. B 39, 5566.
Torrens, I.M. (1972) Interatomic Potentials, New York: Academic.
van den Bergh, L.C. and Schouten, J. A., (1988) J. Chem. Phys. 89,2336.
van Thiel, M., Hord, L.B., Gust, W.H., Mitchell, A.C., D'addario, M., Boutwell, K., Wilbarger, E., and Barrett, B. (1974) Earth Planet. Inter. 9, 57.
Vos,W.L., van Hinsberg, M.G.E. and Schouten, J.A. (1990) Phys. Rev. B41,6106.
Vos, W.L., Schouten, J.A., Young, D.A. and Ross, M. (1991) J. Chem. Phys. 94, 3835-3838.
Weeks, J.D., Chandler, D. and Anderson, H.C (1971) J.Chem.Phys. 54,5422
Winter, N.W. (1992) Unpublished results.
Wood and Parker (1957) J.Chem. Phys. 27, 720.
Yin, M.T. and Cohen, M.L. (1982) Phys. Rev. B 26, 5668.
Young, D.A. (1991) Phase Diagram of the Elements: Berkeley: University of California Press.
Young, D.A. and Rogers, F.J. (1984) J. Chem. Phys. 81, 2789.
Young, D.A. and Ross, M. (1981) J.Chem. Phys. 74, 6950.
Young, D.A., Wolford, J.K., Rogers, F.J. and Hokan, K.S. (1985) Phys. Lett. 108A, 157.
Zharkov and Kalinin (1971), Equations of State for Solids at High Pressure and Temperatures (New York: Consultants Bureau).
Zeldovich, Ya.B. and Raizer, Yu.P (1966) Physics of Shock Waves and High Temperature Hydrodynamic Phenomena, Vol. 1 and 2, New York, Academic.
Zha, C-S, Boehler, R., Young, D.A. and Ross, M. (1986) J. Chem. Phys. 85,1034.

SOME RECENT APPLICATIONS OF PRESSURE TUNING SPECTROSCOPY

H. G. DRICKAMER, Z. A. DREGER and J. M. LANG
School of Chemical Sciences, Dept. of
Physics and Materials Research Laboratory
University of Illinois (Urbana)
Urbana, IL 61801

ABSTRACT In this paper we discuss recent applications of luminescence spectroscopy to resolve problems concerning phosphors in crystals or dissolved in polymers.

On the first two examples we treat intramolecular excitations for molecules dissolved in polymers and explain mechanisms of energy dissipation due to changes of molecular conformation.

The second set of examples involve ZnS phosphors with various dopants. We establish the characteristics of the donor and acceptor states via photoluminescence, and the trap characteristics by means of thermoluminescence. We treat very briefly thermoluminescence in crystalline coronene.

Introduction

Over the last 15 years or so there have been extensive measurements of x-ray scattering, optical absorption spectra, both electronic and molecular, Raman scattering etc in DAC. Until very recently there have been very few luminescence studies and these have been highly specialized and not primarily of interest to chemists.

In this paper we discuss examples of both photoluminescence and thermoluminescence work of interest in both chemistry and materials science. The first section will be devoted to the former topic and the second section to the latter.

Photoluminescence

The photoluminescence studies presented here involve organic molecules or metallorganic complexes in the crystalline state or dissolved in polymeric media. A brief statement of the characteristics of such excitations and emissions will be useful in the discussions of individual studies. Fig 1 shows a schematic potential energy diagram. The horizontal axis is a configuration coordinate - a coordinate along which some molecular

R. Winter and J. Jonas (eds.), High Pressure Chemistry, Biochemistry and Materials Science, 43–65.

vibration, rotation or translation, or some process of interest for the particular absorption or emission takes place. The vertical axis is energy. The ground state (S_0) is an harmonic potential well as is the excited state (S_1). Excitations and emissions are vertical on such a diagram in concordance with the Franck-Condon principle. An excited electron can dispose of vibrational energy very rapidly and emit from the lowest vibrational level of the excited state. This process typically occurs in 1-10 x 10^{-9} seconds and is called fluorescence. It can also return to the ground state thermally either directly (the weak coupling case) or via intersystem crossing (strong coupling). Englman and Jortner [1] have analyzed these cases in detail. In the weak coupling case the rate of thermal dissipation depends on the energy difference between S_1 and S_0, increasing with decreasing energy difference. For strong coupling the height of the energy barrier between S_1 and T_1 is controlling as shown in Fig 1. In general the intersystem crossing is to a state of different multiplicity and the $T_1 \rightarrow S_0$ radiative process is formally forbidden. For organic molecules this emission (phosphorescence) appears only at low temperature. Where heavy atoms are present (e.g. Br^-, I^-, Ru(II), Re(I) etc.) the spin-orbit coupling may permit significant emission from T_1 to S_0 at room temperature with a lifetime which may vary from milliseconds to hundreds of nanoseconds.

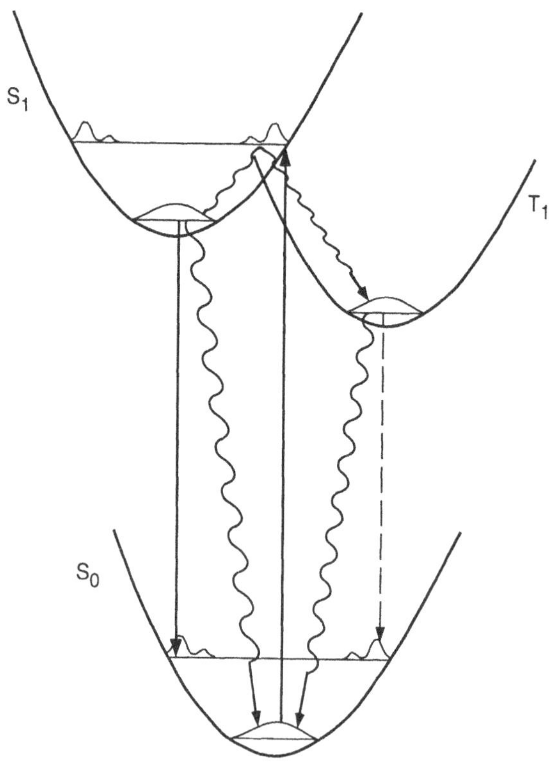

Figure 1 Schematic configuration coordinate diagram

Ruthenium (II) Polypyridyl Complexes

Ruthenium II polypryridyls, especially $Ru(bpy)_3^{2+}$ (tris) have been widely studied as photocatalysts, photosensitizers, chemi and electroluminescent systems and for energy transfer processes. The related ion $Ru(bpy)_2(py)_2^{+2}$(bis) has been less generally studied but the comparison of the two ions is of interest [2]. Their structure is shown in Fig 2. The only difference is one missing bond between the pyridines.

The luminescence efficiency of the bis complex is 5-50 times less than the tris depending on the environment. We show here how pressure can help understand this. In Fig 3

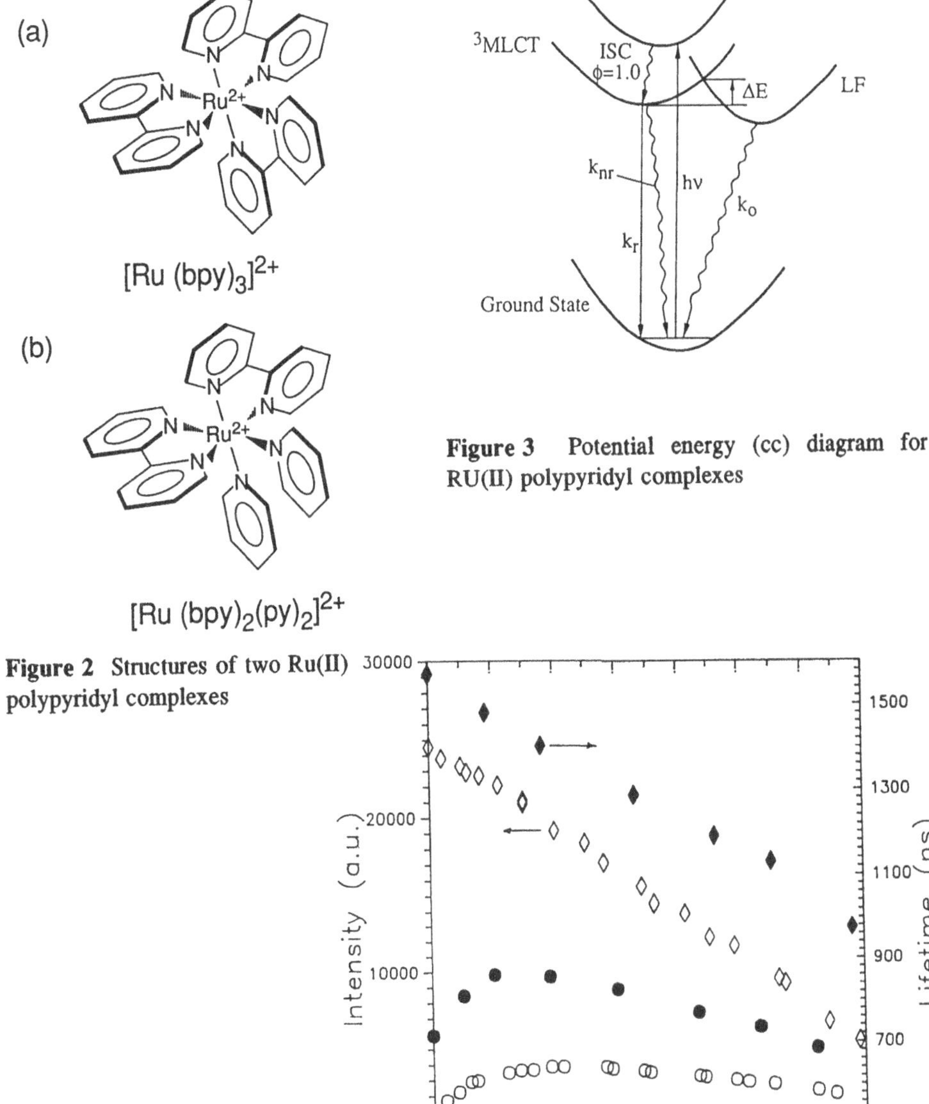

Figure 2 Structures of two Ru(II) polypyridyl complexes

Figure 3 Potential energy (cc) diagram for RU(II) polypyridyl complexes

Figure 4 Intensity and lifetime vs pressure: Ru(II) polypyridyl complexes in PAA

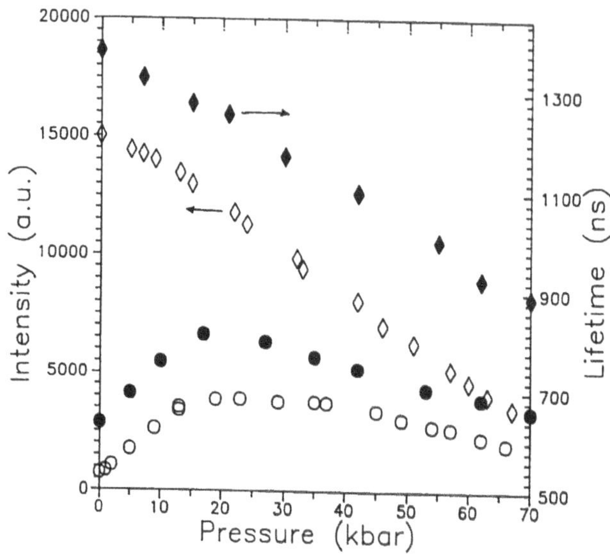

Figure 5 Intensity and lifetime vs pressure: Ru(II) polypyridyl complexes in PVP

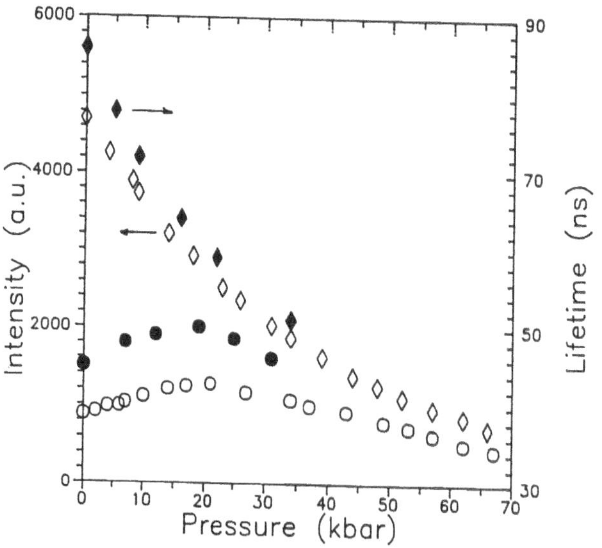

Figure 6 Intensity and lifetime vs pressure: crystalline Ru(II) polypyridyl complexes

we show an energy diagram applicable to either ion. The excitation is a metal to ligand charge transfer (MLCT) to a singlet state which decays very rapidly with unit efficiency to an excited triplet-^{3}MLCT. This triplet can emit radiation, it can decay thermally, or it can undergointersystem crossing to an excited state on the Ru(II) and decay thermally to the ground state. Most papers, based on temperature data at ambient pressure, have attempted to account for the difference in efficiency in terms of differences of ΔE for intersystem crossing. However k_{nr} is also very temperature dependent, and the results differ greatly and are often contradictory. We have measured the energies of the luminescent peaks, the change of efficiency with pressure and the lifetimes in two polymeric media (PVP-polyvinyl pyridine) and (PAA = polyacrylic acid) as well as in the crystalline solid [2]. The energies are given in Table I and the efficiencies and lifetimes in Fig 4-6.

We discuss first the tris complex in the three media. In all cases there is a monotonic shift to lower energy and a decrease in efficiency and lifetime. Since the changes in intensity and lifetime are essentially proportional one can eliminate changes in k_r as the cause. A second possibility is an increase in intersystem crossing. However the effect of pressure on the ligand field splitting should increase the energy of the excited LF state and therefore ΔE. This effect would not be large since the covalent Ru-N bonds are not very compressible. Moreover, the shift to lower energy of the emission implies a decrease in energy of the ^{3}MLCT state and therefore a further increase in ΔE and thus a higher intensity of emission. The most acceptable solution is a decrease in efficiency via the energy gap law as the energy of the emission decreases with pressure. As can be seen, the intensity decrease in proportional to the shift in all three media.

The behavior of the bis complex is more interesting. There is an increase in efficiency and in lifetime with pressure in the low pressure region, followed by a decrease above 15-25 kbar depending on the medium. We consider first the region up to the maximum. The emission peak show only a very small shift in the polymer and a distinct red shift in the crystal. In no case could these shifts increase the yield significantly via the energy gap law. One might expect a modest increase in ΔE in the polymers because of increase in the LF excited state energy, but this could not be sufficient to account for the observed increases of a factor of 3-5 in efficiency. In the crystal the increase in ΔE would be augmented by the decrease in energy of the ^{3}MLCT state, but the crystal actually shows the smallest increase in efficiency and lifetime.

The most likely solution is that the lower efficiency for the bis complex is due to the dissipation of excitation via the vibrational motion of the pyridines which are less restricted than the bipyridyl molecules. This motion should be inhibited as the medium in compressed and becomes more rigid. The excitation is probably initially localized but could easily be transferred in the relatively long lifetimes observed. At higher pressure the decrease in efficiency observed for the bis complex is due to the increase in k_{nr} by action of the energy gap law. At high pressure the difference in efficiency between the bis and tris tends toward a constant value much smaller than the initial value and probably indicates a difference in ΔE between the two complexes.

It is of interest to note that the lifetimes are longest in PAA where the coupling between the ion and the environment is weakest, and very much shorter in the crystal where the coupling is strongest.

These results demonstrate how high pressure studies can establish a mechanism for a process where a large body of ambient pressure results had left it ambiguous.

Table I

ν_o and $\Delta\nu$ vs pressure

Tris bipyridyl (Ru(II))

Medium	ν_o	P (kbar) $\Delta\nu$ 15	30	45	60
PAA	16260	-60	-135	-210	-290
PVP	15870	-100	-250	-390	-540
Crystal	16190	-290	-600	-820	-980

Bis bipyridyl - pyr_2 (Ru(II))

Medium	ν_o	P (kbar) $\Delta\nu$ 15	30	45	60
PAA	16030	+50	-20	-80	-130
PVP	15570	+40	-30	-120	-250
Crystal	15800	-480	-780	-1000	-1150

ν_o and $\Delta\nu$ given in cm^{-1}

Flexible Dye Molecules

There exist a large class of molecules which have a single ground state conformation and two or more conformations in excited electronic states. E.g., they may be planar in the ground state and may have a planar excited state, but, within the excited state lifetime, they can assume a twisted conformation. Rettig [3] has recently reviewed much of the literature on these molecules with extensive references both to original papers and other reviews. In general, the excitation involves a transfer of charge from one part of the molecule to another ($D \rightarrow A$) with a large increase in dipole movement. The twisting may further increase the dipole moment.

One can think of three possible situations regarding the luminescence. They may emit only from the planar excited state, only from the twisted state or from both. These molecular are widely used as probes of local conditions in fluid and polymeric media and have potential for organic electronic devices.

We have initiated a study of the effect of pressure on the luminescent properties of these molecules in polymer matrices. The three molecules studied so far are shown in Fig 7. They are: Julolidine-malononitrile (JDMN), Polyester yellow (PY) and p-N,

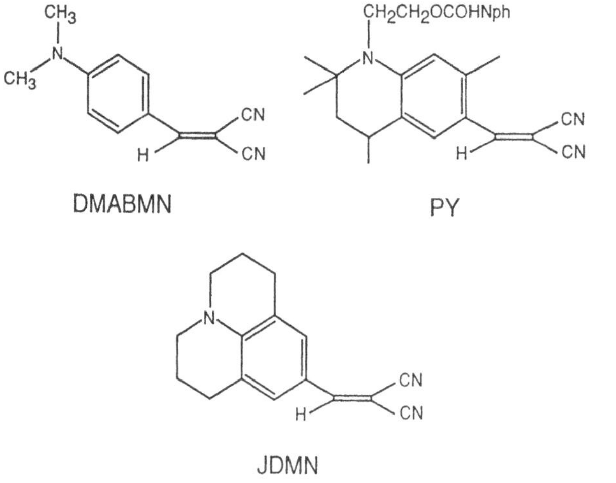

N´-dimethylaminomalononitrile (DMABN). These have been extensively used by Loutfy and coworkers [4-6] to probe polymerization and structural properties of polymers. In the course of their work they have established accurate quantum yields for these molecules in one or two polymers. We could then establish quantum yields in other polymers by comparison of emission intensity in an atmospheric holder of fixed geometry. Further, from the change of intensity with pressure we could measure the effect of pressure on the quantum yield. This is essential for our analysis as discussed below. In addition, the absorption and emission characteristics were consistent with the requirements of our high pressure luminescence apparatus. The sources of the molecules, purifications procedures etc. are discussed in our original papers [7-10].

DMABMN PY JDMN

Figure 7 Structure of three flexible dye molecules

These molecules have in common an essentially planar ground state and both a planar and twisted excited state. They emit only from the planar excited state. For all three molecules there is a malononitrile group which can twist in the excited state. In addition, DMABN can exhibit a rotation of the $(CH_3)_2N$ group, while PY can exhibit rotations either at the C-N-phenyl group or with smaller probability in the polymer matrix, at the C-N bond to the ring nitrogen.

We studied the fluorescence, (and absorption) characteristics in four polymeric media: polymethylmethacrylate (PMMA), polyvinylacetate (PVAc), polyvinylchloride (PVCl) and polystyrene (PS). For PMMA and PS in some cases more than one molecular weight was used. For some matrices there was a modest molecular weight effect, but we shall not emphasize this feature in the present overview.

The luminescent efficiencies at one atmosphere for the three probes were in the order JDMN > PY > DMABMN roughly in the ratio 10/3.5/1, with some variation with the medium. The quantum efficiencies at one atmosphere were not high in any medium but were a factor 3-5 lower in PS than for the other polymers.

We discuss first JDMN where only one twist is possible. The situation can be described in terms of the schematic configuration coordinate (cc) diagram shown in Fig 8. The horizontal axis is the angle of twist while the vertical axis is energy. The solid potential wells represent

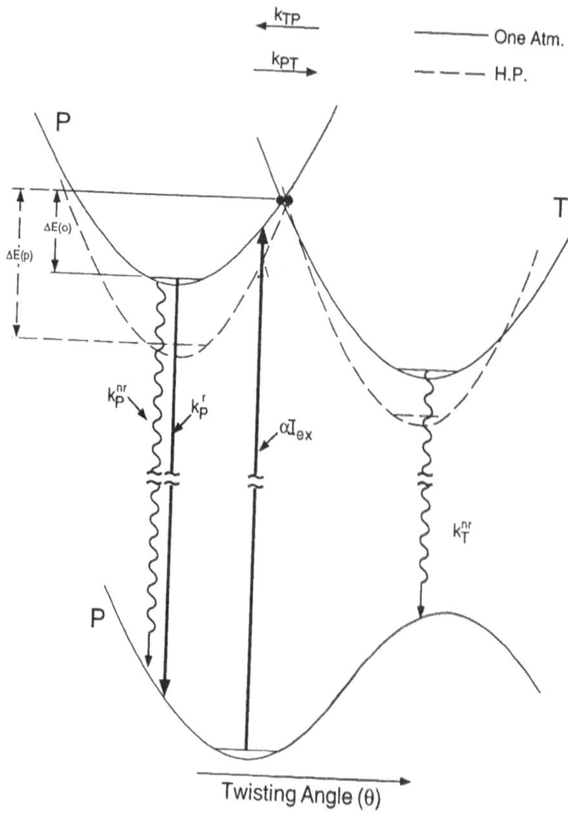

Figure 8 Schematic cc diagram along the angle of twist normalized plot of emission intensity vs pressure

the situation at one atmosphere. The absorption is given by αI_{ex}. k_p^r and k_p^{nr} are the rates of radiative and nonradiative energy loss from the planar state (P). k_{PT} and k_{TP} are the rates of crossing from planar to twisted state and vice-versa. k_T^{nr} represents the rate of nonradiative loss of energy from the twisted state (T). The analysis is given in ref. [7] and [8] and only the main results outlined here.

The quantum efficiency is a function of k^r assured independent of pressure, k_p^{nr} and a coefficient k_{nr}' given by

$$k_{nr}' = \frac{k_{PT}}{1 + \frac{k_{TP}}{k_T^{nr}}} \quad (1)$$

k_{nr}' can be extracted from the experimental data from the relationship:

$$\frac{k_{nr}'(p)}{k_{nr}'(0)} = \frac{r^{-1} - \Phi(0)\,\Phi_{max}^{-1}}{1 - \Phi(0)} \quad (2)$$

where $r = \frac{\Phi(p)}{\Phi(0)}$ and Φ_{max} is the maximum quantum efficiency obtained. When Φ_{max} approaches one this reduces to

$$\frac{k_{nr}'(p)}{k_{nr}'(0)} = \frac{r^{-1} - \Phi(0)}{1 - \Phi(0)} \quad (3)$$

In Fig. 9 we exhibit the relative change of intensity of emission with pressure for JDMN in four polymers. Except for PS, the modest drop at high pressure can be associated with an increase in the nonradiative rate k_p^{nr} [8]. For all three molecules the behavior is PS, especially at the higher pressures is anomalous and has been only partially explained. In this review we are concerned in all cases only with the increase in luminescence efficiency at pressures below that at which the efficiency maximizes - in general below 20-25 kbars.

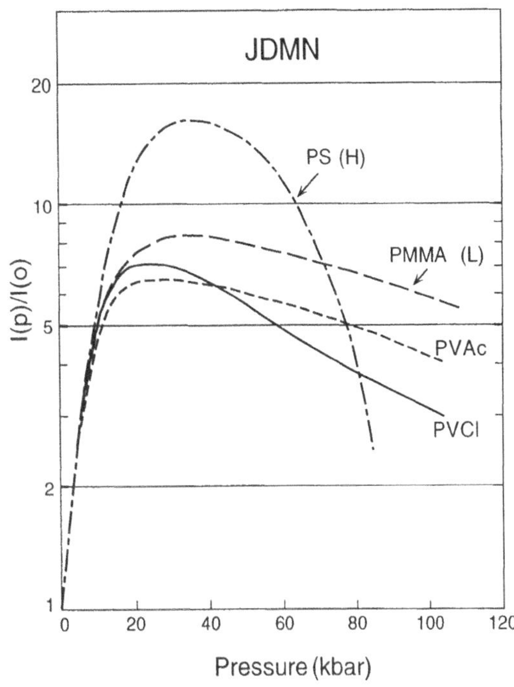

Figure 9 Normalized plot of emission intensity vs pressure - JDMN in various polymers

For JDMN in PMMA, PVCl and PVAc the quantum yield at the maximum is 0.93 to 0.98 so one can use eq. (3) to extract $\frac{k'_{nr}(p)}{k'_{nr}(0)}$. For PS the maximum quantum efficiency was 0.7. While there is some pressure effect on nonradiative rates it has been found that the use of various values of Φ_{max} just above the experimental maximum resulted in only small differences in the value of $\frac{k'_{nr}(p)}{k'_{nr}(0)}$ at any pressure, which did not affect the conclusions.

In Fig. 10 we plot $\ln\left[\frac{k'_{nr}(p)}{k'_{nr}(0)}\right]$ vs. pressure. We analyze the linear plots using an Eyring type argument. k'_{nr} as defined in eq. 1 is a combination of three rate constants. Nevertheless we write:

$$k_{nr}' = \frac{k_B T}{\hbar} \exp\left[- \frac{\Delta G^{\ddagger\prime}}{RT} \right] \tag{4}$$

Then

$$\frac{\partial \ln k_{nr}'}{\partial P} = - \frac{"\Delta V"}{RT} \tag{5}$$

Where k_{nr}' is dominated by a single rate it is an activation volume $\Delta V^{\ddagger}$. Otherwise it is some combination of volumes associated with the twist. For JDMN we extract "ΔV" values from 4.5 - 6.0 cc/mol (average ~5.5 cc/mol). They seem to correlate with the initial emission energy, but it is not yet possible to say much about this variation.

It must be kept in mind that pressure couples directly to the totally symmetric coordinate (volume) so the "ΔV" values are not represented directly along the angle of twist coordinate of Fig. 8.

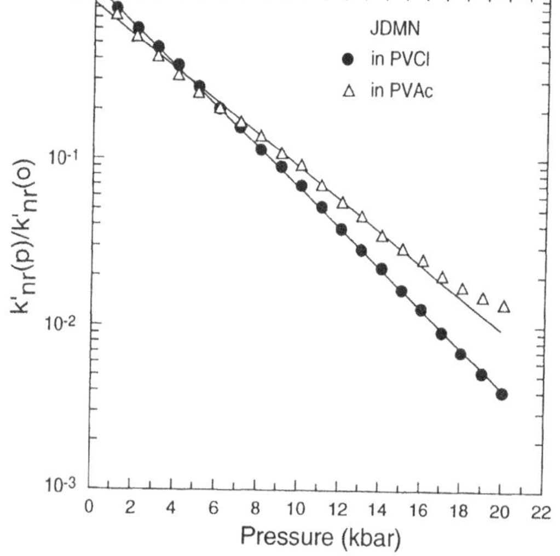

Figure 10 ln $[k_{nr}'(p) / k_{nr}'(0)]$ vs pressure for JDMN in PVAc and PVCl

In order to extract more information about the rate controlling step we plot (Fig. 11) $\ln\left[\frac{k_{nr}'(p)}{k_{nr}'(0)}\right]$ vs the emission peak energy. Except for PS(H) at very low pressure quite good linear plots were obtained. The lines in Fig. 11 represent

$$k_{nr}'(p) = k_{nr}'(0) \exp - \left(\frac{\delta E}{k_B T}\right) \tag{6}$$

where $\delta E = k_B T$ is the value of energy which provides a change in $k_{nr}'(p)$ of e^{-1}. The fact that this coincides well with the data indicates that the decrease in k_{nr}' with pressure is due only to the shift of the luminescence peak and the consequent increase of the energy barrier. From eq. 1

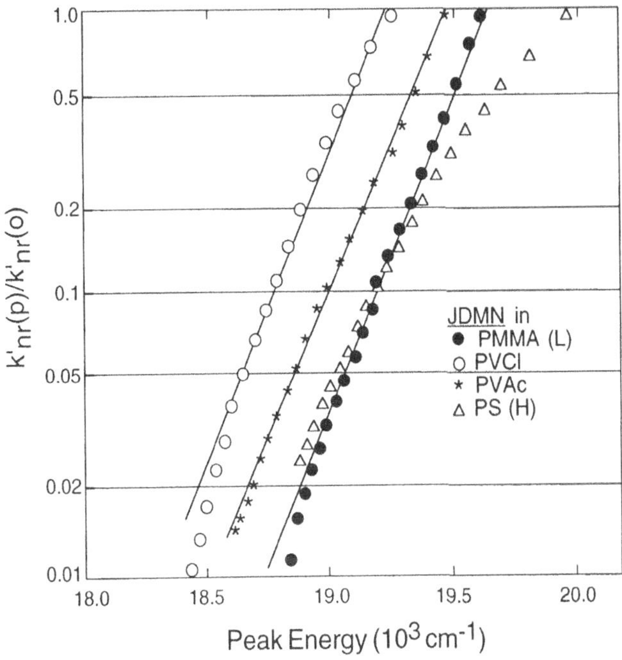

Figure 11 ln $[k_{nr}'(p) / k_{nr}'(0)]$ vs emission peak energy - JDMN

this implies that the second term in the denominator is negligible and for these materials (except PS(H)) in this pressure region, i.e. $k_{nr}' \cong k_{PT}$. Thus in this case "ΔV" is a true activation volume $\Delta V^{\ddagger}$.

This result implies that the energy of intersection of the P and T wells is essentially pressure independent except for PS(H). One then requires that the potential wells "stiffen" as shown in Fig 8 which implies that a smaller change in angle occurs with a given vibration at high pressure. The wells, of course, must be anharmonic if the force constant changes with compression. One may further infer that for PS(H) the intersection point of the P and T wells decreases in energy with compression in the very low pressure region.

For PY and DMABMN there are two possible twists as mentioned above. One must now consider two different configuration coordinates each with an atmospheric pressure cc diagram like Fig 8. Fig 12 and 13 present normalized plots of relative intensity vs pressure. The increase is greatest for DMABMN which had the lowest quantum yield at one atmosphere. For PY Φ at the maximum is over 0.9 for all materials except PS(H) where it is ~ 0.75, so in this last case it is necessary to use eq. 2 to establish $\frac{k_{nr}'(p)}{k_{nr}'(0)}$. For DMABMN, Φ at the maximum ranges from 0.33 for PS(L), 0.38 for PS(H), 0.48 for PVCl, 0.66 for PMMA to 0.68 for PVAc, so it was necessary to use eq. 2 for all media.

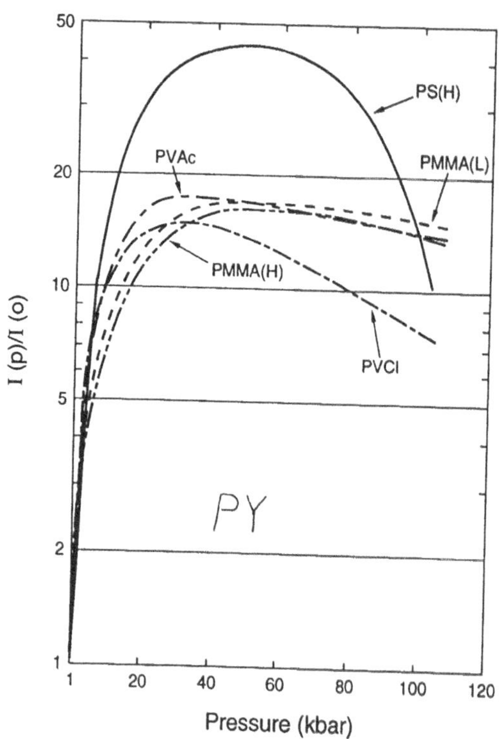

Figure 12 Normalized plot of emission intensity vs pressure - PY in various polymers

Fig 14 and 15 show typical plots of $\ln\left[\frac{k_{nr}'(p)}{k_{nr}'(0)}\right]$ vs pressure. The data can clearly be resolved into two straight lines. Both compounds in the other media gave equally unequivocal results. It is evident that we are dealing with two competing processes, one dominant at low pressure and one at higher pressures. For short hand purposes we refer to the process dominant at low pressure as (A) and the process where the rate changes more slowly with pressure as (B).

It is evident that one can apply eq. (4) and (5) to each process separately using the same arguments as applied for JDMN. For process (A) we obtained for PY values of "ΔV_1" in the range 16-23 cc/mol with an average of ~ 21 cc/mol. For process (A) in DMABMN the range was 12-20 cc/mol with an average of ~ 15.5 cc/mol. For process (B) in PY the values of "ΔV_2" ranged from 2.6-4.1 cc/mol with an average of 3.3 cc/mol, while for process (B) in DMABN the corresponding range was 2.6 to 3.5 cc/mol with an average value of ~ 3.0 cc/mol.

The next step is to see if one of these processes can be identified with the malononitrile twist observed for JDMN. In Fig 16 and 17 we plot $\ln\left[\frac{k_{nr}'(p)}{k_{nr}'(0)}\right]$ vs emission peak energy. The lines represent eq. (6) as for JDMN. It is clear that process (B) can be identified with the malononitrile twist. By the arguments used above, "ΔV_2" is then a true activation volume.

The situation with respect to process (A) is more complex.

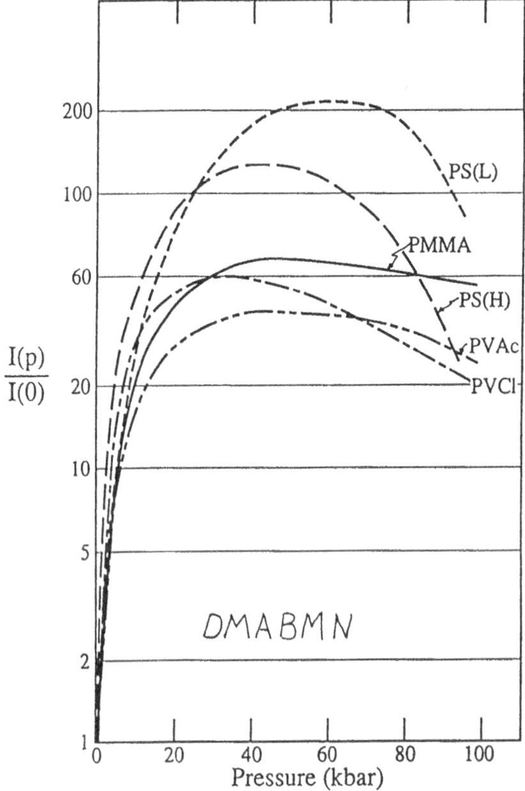

Figure 13 Normalized plot of emission intensity vs pressure - DMABMN in various polymers

It is not a present possible to resolve the various contributions to "ΔV". It is hoped that some of the experiments outlined below will aid in this process and in our overall understanding of flexible molecules.

(1) A useful step would be to study pressure effects in polymeric media with molecules which emit only from the twisted state as well as those which emit from both conformations.

(2) The measurement of Φ as a function of temperature at various pressures could give further information about barrier heights.

(3) It is possible that lifetime measurements as a function of pressure could be useful.

(4) To resolve the anomalies observed for (PS) it would be useful to use other selected polymers. PS is the most polarizable and least polar of the media, and has the bulkiest side groups. Perhaps polyvinylnaphthalene as a medium would reveal useful information.

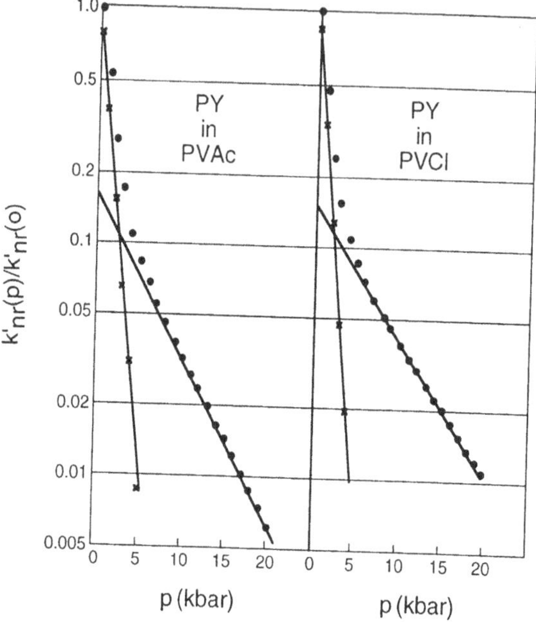

Figure 14 ln $[k_{nr}'(p) / k_{nr}'(0)]$ vs pressure - PY in PVCl and PVAc

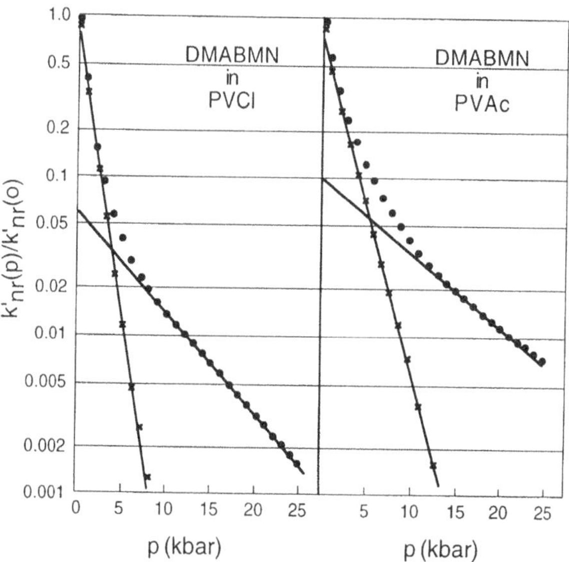

Figure 15 ln $[k_{nr}'(p) / k_{nr}'(0)]$ vs pressure - DMABMN in PVCl and PVAc

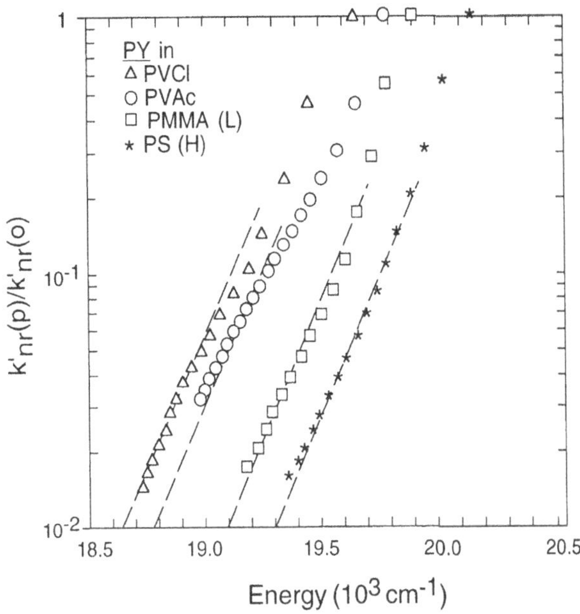

Figure 16 ln $[k_{nr}'(p) / k_{nr}'(0)]$ vs emission peak energy - PY

Thermoluminescence

Thermoluminescence is a technique for measuring trap depths in phosphors. An understanding of trap depths is of both scientific and technological importance. For many phosphors the simple description in terms of excitation followed by either emission or thermal decay is not accurate. An electron can be excited to the conduction band or to a state of high mobility, then trapped at some sort of defect or lattice anomaly. It may further be thermally reexcited to the conduction band, and ultimately reach a recombination center (donor) from which state it emits light while transferring to an acceptor which may or may not be the state from which it originated. Most phosphors used in TV screens, monitors etc make use of this process. There has been in the past only one previously published attempt at high pressure thermoluminescence [11] and that was not very successful.

A brief description of the technique involved is in order. In our case a Merrill-Bassett triangular DAC in an aluminum heating jacket contains the sample, a ruby chip and a pressure medium. It is placed in an insulated cooling box and connected to a sapphire window for excitation via the 325nm (3.82eV) line of a He-Cd laser. The emission is taken out the back of the DAC with a light pipe to a photomultiplier. The box is filled with liquid nitrogen and cooled for 20 minutes. There is no emission since all electrons are in traps too deep to be emptied at 77K. The nitrogen is removed very rapidly and the cell is heated at a constant rate to room temperature. The heating rate is controlled by a thermocouple at the edge of the diamond. This couple activates a control device. As the temperature increases electrons are released from the traps and light is emitted. When the traps are emptied the light dies off. The

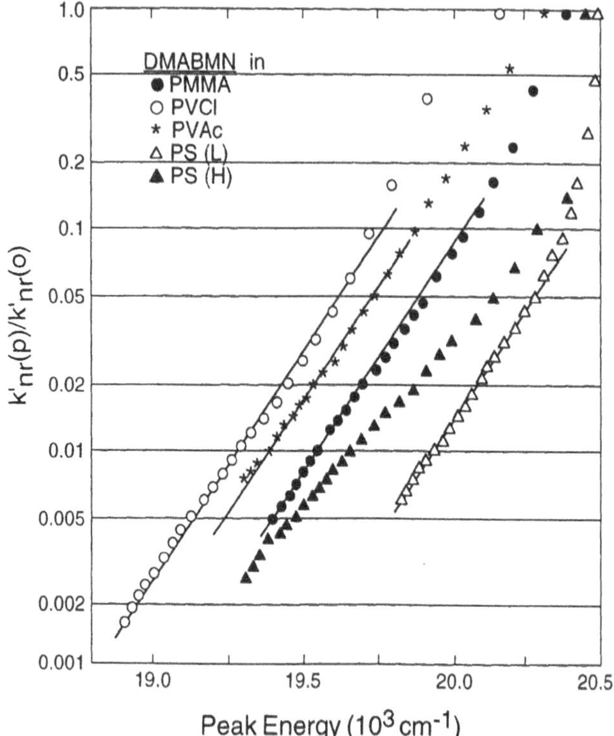

Figure 17 ln $[k_{nr}'(p) / k_{nr}'(0)]$ vs emission peak energy - DMABMN

resulting plot of emission intensity vs. temperature is called a glow curve. The pressure is then increased at room temperature and the process is repeated.

There are a number of ways of extracting the trap depth from glow curves. By comparison with computer generated glow curves it has been established that the most generally accurate approach is to use the temperature maxima (T_{max}) from glow curves generated by two different constant heating rates. When a glow curve clearly contains a large distribution of traps, it is possible to heat to a temperature T_1, recool and heat to a temperature T_2 etc. One can then get a reasonable approximation to the trap depths by using the initial slope of each glow curve. A more detailed discussion of the techniques can be found in the papers to which we refer below, and in references contained therein.

We first discuss two types of ZnS phosphors with different dopants. Both types are of scientific and technological importance.

The first example involves ZnS doped with Cl^- and Cu^+ [12]. There are three samples with Cu^+ concentrations (mol %) of 0.03, 0.05 and 0.08. In each case the Cl^-/Cu^+ ratio is ~8. The excess Cl^- induces Zn vacancies. At ambient pressure the excitation is from the top of the

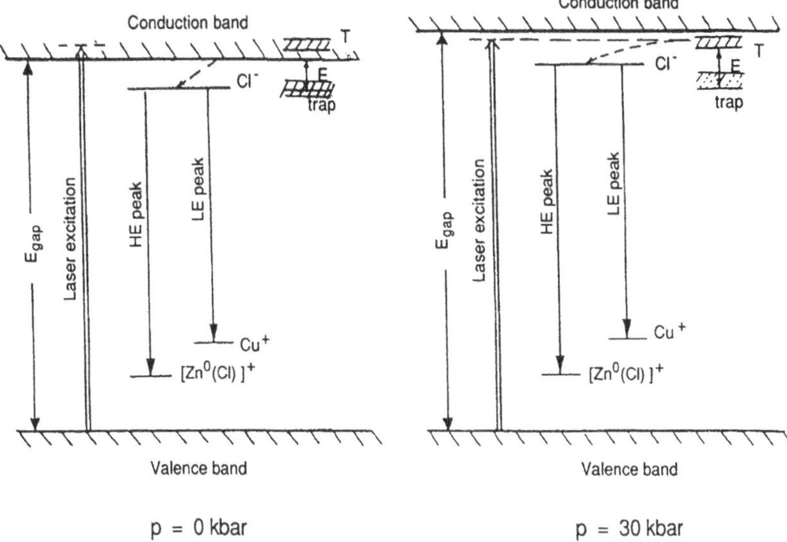

Figure 18 Schematic band structure of doped ZnS at 0 and 30 kbar

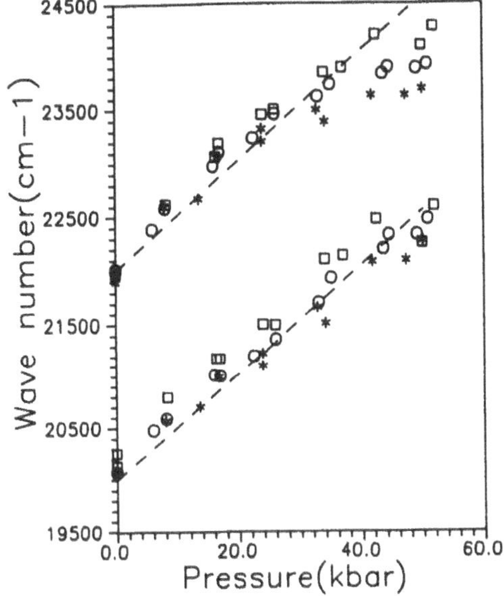

Figure 19 Peak shift vs pressure - blue and green emission in ZnS:Cu:Cl

valence band to the bottom of the conduction band. There are two peaks in the emission spectrum, one at ~22,000 cm^{-1} (2.72eV) (blue) corresponding to emission $Cl^- \rightarrow Zn(0)$ and one at ~ 20,000 cm^{-1} (2.47eV) (green) due to $Cl^- \rightarrow Cu^+$ emission. Because of the higher efficiency of the latter process, with increasing Cu^+ content the green peak becomes more dominant. The left hand side of Fig. 18 shows the situation at ambient pressure. In Fig. 19 we exhibit the shifts of the two emission peaks with pressure. The dashed line indicates the change of energy gap with pressure. Clearly the Cl^- donor is pinned to the bottom of the conduction band while the Cu^+ is pinned to the top of the valence band. From the shift of the blue peak it is evident that the Zn vacancy is closely associated but not completely pinned to the top of the valence band.

The trap depth as a function of pressure is shown in Fig. 20. Initially the trap depth increases with increasing energy gap which indicates that the trap energy is independent of the conduction band energy. At a certain pressure, which is lowest for the highest dopant concentration the trap depth ceases to follow the gap and levels off.

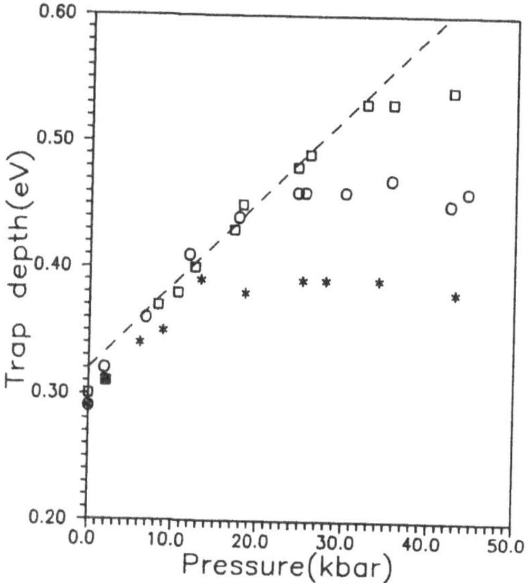

Figure 20 Trap depth vs pressure - ZnS:Cu:Cl

This behavior can be explained in terms of the concept of "deep levels" and the diagram on the right of Fig. 18. The concept of a "deep level" was developed to describe trapping centers in Si and especially GaAs. These levels are independent of the valence or conduction band and were first thought to be in the middle of the gap - hence their name. It is now known that they can be resonant in energy with one of the bands but still in no way coupled to the band because of symmetry and related considerations. For a more detailed discussion of these levels see Ref. [12] and references therein.

The gap is initially 3.66 eV while the exciting light is the 3.82 eV line of a He-Cd laser. The gap increases in energy by 6.5 meV/kbar. Above ~20-25 kbar the laser can no longer excite an electron to the conduction band so that excitation is to the band of "deep levels" exposed by the movement of the conduction band. The greater the concentration of dopant, and especially of Cl^-, the broader the band of deep levels and the lower is the pressure at which they become available. As stated above, "deep levels" have been studied extensively in GaAs but this is the first conclusive evidence for them in II-VI compounds, specifically in ZnS. There are technological as well as scientific implications.

A second phosphor, also technologically important involves ZnS with a relatively large amount (~2.4%) of Mn^{+2} and smaller amounts (0.34%) of Cl^- and (0.08%) Cu^+ [13]. The

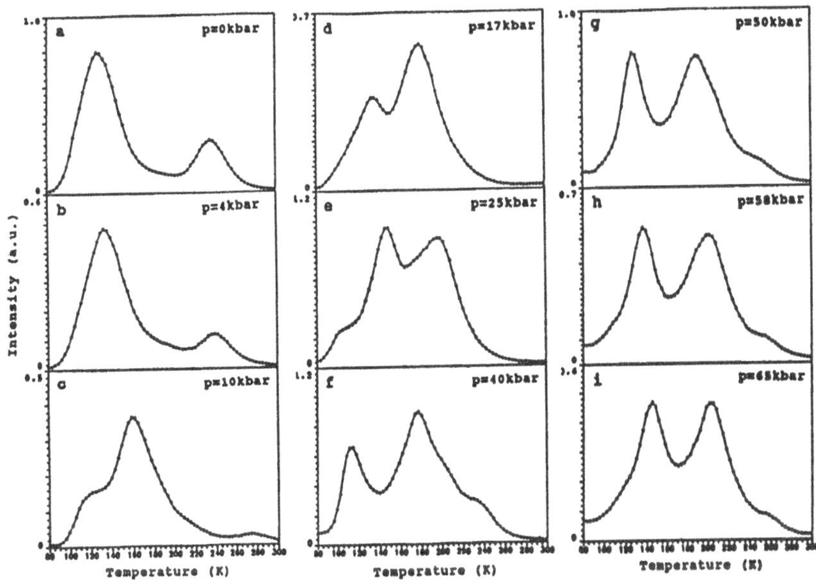

Figure 21 Thermoluminescence spectra at nine pressures - ZnS:Mn:Cu:Cl

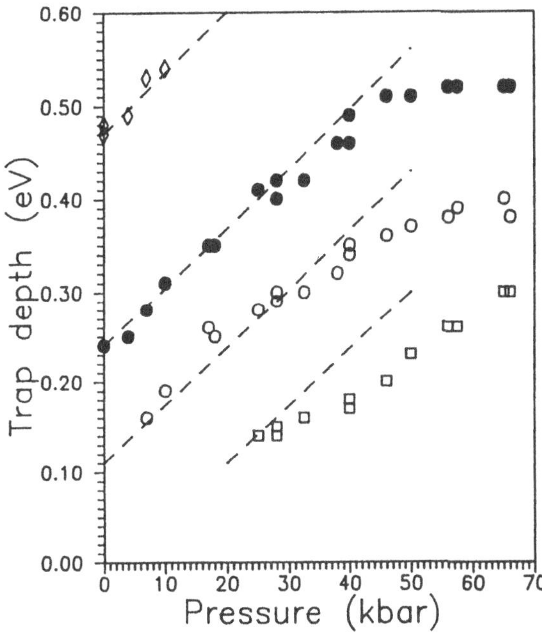

Figure 22 Trap depths vs pressure - ZnS:Mn:Cu:Cl

emission is from an excited state of Mn^{+2} to the ground state and lies in the yellow region of the spectrum (17,000 cm^{-})(2.10eV). This d $\rightarrow$ d emission shifts to lower energy in a manner consistent with ligand field theory.

Fig 21 shows the thermoluminescence spectra at nine pressures. At ambient pressure there is a large peak at low temperature and a less intense one at considerably higher temperature. With increasing pressure the peaks shift to higher temperature and fade as they can no longer be emptied at 300K. New shallow traps move in, first near 10 kbar and then near 25 kbar. Fig 22 shows the trap depths as a function of pressure. The two intermediate peaks behave much like those in ZnS:Cu:Cl where the traps are associated with Cl^{-}. A sample was prepared in our laboratory containing significant amounts of Mn^{+2} and very little Cl^{-}. The highest, and apparently the lowest depth traps appeared, but the intermediate traps were very weak. Apparently the deepest trap, and possibly the shallowest are associated with Mn^{+2} dopants, while the intermediate depth traps are associated with the Cl^{-}.

CORONENE

Thermoluminescence studies also have applicability to organic crystals. An example in the study of the coronene crystal [14]. In Fig 23 we show the molecular structure of this highly symmetric molecule and the broad thermoluminescence peak. Clearly there is a distribution of traps which covers a wide range of energies. One can separate these into some sort of ordering by the partial heating method described above.

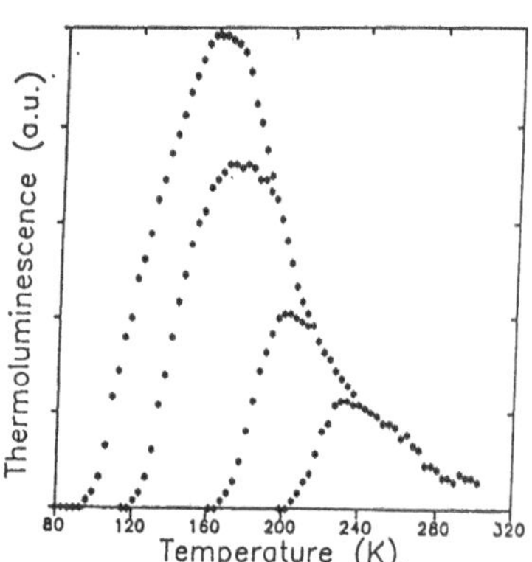

Figure 23 Structure of coronene and thermoluminescence spectrum

An important consideration is the nature of the excitation and the trap. The lowest energy trap depth decreases in energy with increasing pressure. If the excitation is to a conducting or highly mobile level, the basic interaction should be between an electron and a molecular dipole. In this case the interaction should be proportional to R^{-4} where R is the electron - dipole distance. One can approximate the factor $\left[\frac{R(o)}{R(p)}\right]^4$ by $\left[\frac{V(o)}{V(p)}\right]^{4/3}$. In Fig 24 we show a plot of trap depth vs. $\left[\frac{V(o)}{V(p)}\right]^{4/3}$ which indicates that the electron-dipole interaction indeed describes the results. This study demonstrates that high pressure thermoluminescence can be used to establish the nature of electronic events in organic molecular crystals.

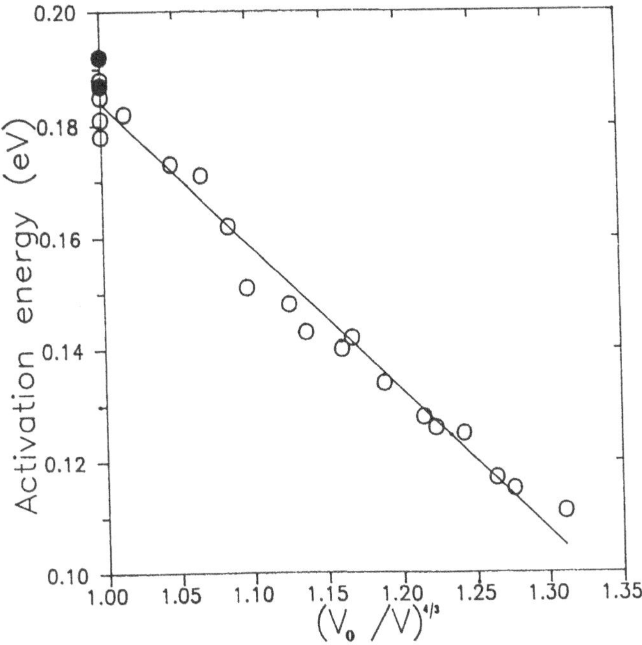

Figure 24 Trap depth vs $[V_0 / V]^{4/3}$ - crystalline coronene

Summary

We have treated a variety of studies where luminescence is demonstrated to be a useful probe of phenomena in rigid phases. The problems have been treated only in outline, and some are only in their early stages. It is hoped that the results are sufficiently clear to illustrate the power and versatility of high pressure luminescence in chemistry and materials science.

Acknowledgement

It is a pleasure to acknowledge the support of the Materials Science Division of the Department of Energy under Contract DEFG02-91ER45439.

References

[1] Englman, R and Jortner, J. (1970)
"The Energy Gap Law for Radiationless Transitions in Large Molecules
Molecular Physics 16 145-164.

[2] Lang, J. M., Dreger, Z. A., and Drickamer, H. G. (1993)
"High Pressure Studies of Photoluminescence Properties of Ruthenium (II) Polypyridyl Complexes"
J. Phys. Chem. (submitted) (and references therein).

[3] Rettig, W. and Baumann, W. (1992)
in "Photochemistry and Photophysics VI, J. Rabek ed.
CRC Press pp 79-134 (and references therein).

[4] Loutfy, R. O. and Law, K. Y. (1980)
"Electrochemistry of Intramolecular Charge Transfer
p, N, N-dialkylaminobenzylidinemalononitriles"
J. Phys. Chem. 84 2803-08.

[5] Law, K. Y. and Loutfy, R. O. (1981)
"Spectroscopy of Dyes in Polymer Matrices"
Macromolecules 14 587-91

[6] Loutfy, R. O. (1986)
"Fluorescence Probes for Polymer Free Volume" in Photophysical and Photochemical Tools in Polymer Science"
M. A. Winnik (ed) D. Reidel pale. pp 429-48 (and references therein).

[7] Dreger, Z. A., Lang, J. M. and Drickamer, H. G. (1992)
"High Pressure Effect on the Twisted Intramolecular Charge Transfer Fluorescence and Absorption of p - N, N - dimethylaminobenzilidenemalononitrile (DMABMN)
Chemical Physics 166 1-16 (and references therein).

[8] Dreger, Z. A., Lang, J. M. and Drickamer, H. G.
"High Pressure Study of Flexible Dye Molecules in Solid Polymeric Media I Jolulidineamalononitrile"
Chem. Phys. (submitted).

[9] Dreger, Z. A., Lang, J. M. and Drickamer, H. G.
"High Pressure Study of Flexible Dye Molecules in Solid Polymeric Media II Polyester Yellow LG-LSW"
Chem. Phys. (submitted).

[10] Dreger, Z. A., Lang, L. M. and Drickamer, H. G.
"High Pressure Study of Flexible Dye molecules in Solid Polymeric Media III DMABMN reanalyzed"
Chem. Phys. (submitted).

[11] Hook, J. W. (III) and Drickamer, H. G. (1978)
"High Pressure Studies of Thermoluminescence of Doped ZnS Phosphors"
J. App. Phys. 49 2503-08.

[12] Lang, J. M., Dreger, Z. A. and Drickamer, H. G. (1992)
"High Pressure Studies of Photoluminescence and Thermoluminescence of ZnS:Cu:Cl Phosphors using Laser Selection Excitation"
J. Phys. Chem. 96 85-90.

[13] Lang, J. M., Dreger, Z. A. and Drickamer, H. G. (1992)
"High Pressure Thermoluminescence and Photoluminescence Study of ZnS:Mn:Cu:Cl Phosphor"
J. App. Phys. 71 1914-18.

[14] Dreger, Z. A., Lang, J. M. and Drickamer, H. G. (1991)
"Thermoluminescence of Charge Traps Under High Pressure in Crystalline Coronene"
J. Lumin. 50 279-86.

HIGH PRESSURE CHEMISTRY IN THE SOLID STATE; PRESSURE INDUCED MOLECULAR REARRANGEMENTS IN RIGID MEDIA

H. G. DRICKAMER
School of Chemical Sciences, Dept. of
Physics and Materials Research Laboratory
University of Illinois (Urbana)
Urbana, IL 61801

ABSTRACT

In this paper we present the results of a series of studies of molecular rearrangements in rigid media-crystalline and polymeric. These have been selected from a considerably more extensive set of data because they are representative of various regularities which have been observed.

These regularities are presented and summarized with the hope of introducing some systematics and possibly some degree of predictability into solid state chemistry.

Introduction

In rigid media the most probable chemical reactions induced by pressure are polymerization and change of molecular conformation. In this paper we present a series of studies of high pressure molecular conformation studies. The examples are selected to illustrate a series of principles or regularities which, it is hoped, can provide a basis for the systematics of pressure induced reactions in solids.

The diagnostic tool used is the electronic spectra. Changes in energy, intensity or number of electronic absorption peaks, are, for the systems studied here, a powerful, usually a decisive, indication of the molecular conformation. Ideally it would be desirable to supplement these measurements with x-ray diffraction. However, it is not at present possible to perform useful x-ray studies at high pressure on molecules of the complexity necessary to illustrate the relevant principles.

The studies are grouped according to which principle, or principles they illustrate. There is, however one theme which applies in a degree for all the systems. In a gas a molecular rearrangement is a stochastic process; i.e. the distribution of molecules between conformation A and B can be expressed on terms of the free energy difference between the conformations and the temperature. In a liquid the same considerations apply but ΔG will, in general, be different as the two conformations interact differently with their environment. Consider, however, a strongly bound solid. No molecule can change conformation unless every molecule in the crystal does so.

R. Winter and J. Jonas (eds.), High Pressure Chemistry, Biochemistry and Materials Science, 67–77.

In the language of physics, the change of conformation involves a first order phase transition. In weakly bound molecular crystals in polymers or glasses etc, the extent of cooperativity may vary from nearest neighbor to macroscopic (first order) or be intermediate, depending on the strength of interaction with the medium and the complexity of the rearrangement.

T vs P and Inter-vs Intramolecular Effects

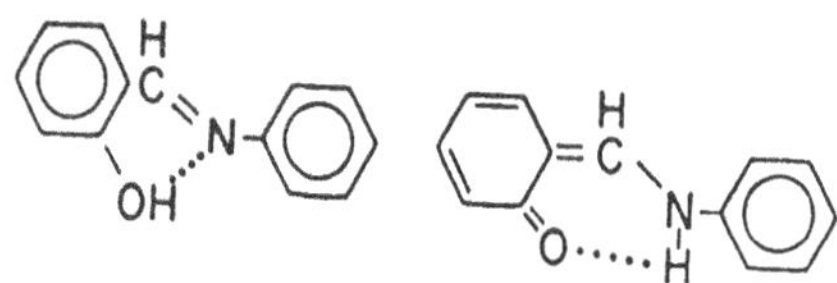

Figure 1 Structure of enol and Cis-keto forms of Salicylidine anils.

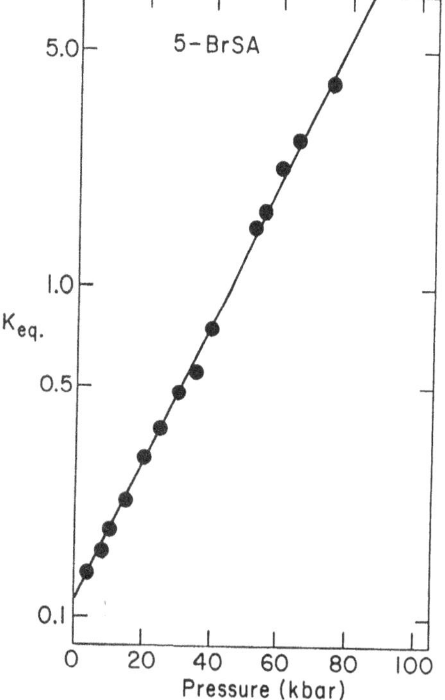

Figure 2 Equilibrium constant for transformation from enol to cis-keto form vs pressure for a salicylidineanil.

The first example discussed illustrates two principles. The salicylidine anils (Fig. 1) in the ground state at one atmosphere and 25°C have the enol structure as shown. They absorb light in the near UV range of the spectrum. When they are heated at ambient pressure, a new absorption peak appears in the visible. By 150 - 180°C the visible peak shows 6-12% of the intensity of the UV peak. At higher temperatures other effects intervene. A combination of NMR and IR indicates that the new peak is due to the formation of the cis-keto structure; ie the hydrogen transfers to the nitrogen. Because of the color change, these compounds are labelled "thermochromic". The process is reversible.

With increasing pressure at 25°C a new peak grows in the visible spectrum at the same location i.e., these compounds are also piezochromic. However by 60 kbar the new peak is dominant and by 100 kbar the conversion is essentially complete [1]. High pressure IR measurements demonstrate that this is indeed the cis-keto structure. Again the process is reversible. As shown in Fig. 2 the conversion is continuous with pressure as for a stochastic reaction in a fluid medium. The ratio of peak areas can be taken as a measure of the equilibrium constant.

The first point to be noted is that increasing T and P have the same effect. This is contrary to a common notion that temperature and pressure are conjugate variables. Increasing temperature T increases the probability of passing over a barrier of a given height. Pressure can perturb the relative energy of two potential wells either to increase or decrease the barrier height, and thus either to oppose or augment the effect of increasing temperature.

A second principle is also illustrated. From the pressure derivative of ln K one can extract the volume change in the reaction, in this case 2cc/mol (4.5-5 $Å^{3}$/molecule). This is very large compared with the change in molecular geometry involved in the transfer of a hydrogen. In general the ΔV of a reaction in a condensed phase reflects differences in packing and intermolecular forces, more than differences in molecular geometry. This argument applies to liquids as well as rigid phases. A further point can be made concerning kinetics in solution. It is common to extract an "activation volume" $\Delta V^{\pm}$ from the pressure coeffieient of the reaction velocity constant. This $\Delta V^{\pm}$ is more likely to reflect differences in intermolecular interactions, i.e. polarizability, polarity, solvation etc. between the activated complex and reactants than the details of the molecular geometries of the two states.

Effect of the Medium

In this section we discuss two examples of the effect of the medium on the extent of cooperativity of the rearrangement.

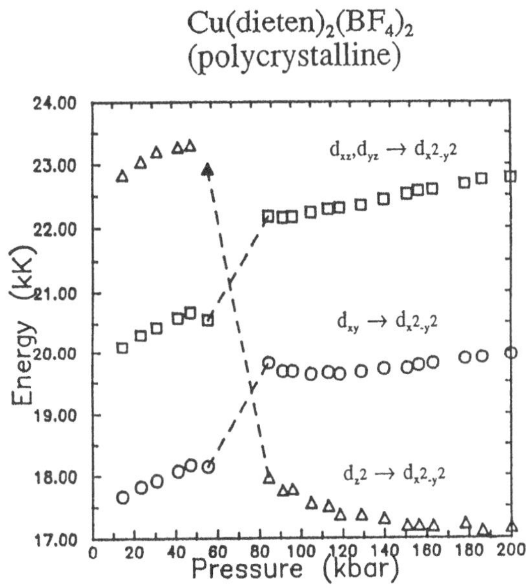

Figure 3 Peak shift vs pressure-crystalline Cu(dieten)$_2$BF$_4$ - note transformation at 65-80 kbar.

In the first of these we consider Cu(II) complexed with ethylene diamine and its derivatives [2]. For ethylene diamine there are four nitrogens ~2Å from the Cu(II) with the counter ions 2.3 - 2.6 Å above and below. While the symmetry is strictly D_{4h} the Cu(II) is exposed to an essentially octahedral environment. The spectrum then contains two relatively intense excitations ($d_{xy} \rightarrow d_{x2-y2}$ and $d_{x2y2} \rightarrow d_{x2-y2}$) and one ($d_{z2} \rightarrow d_{x2-y2}$) which is much less intense due to weaker overlap. In this essentially octahedral symmetry the weak peak is at the lowest energy.

If Cu(II) complexes with diethylethylene diamine, because of its larger bulk the counterions now lie 3.3 - 4 Å from the Cu(II), and off center. The symmetry is now essentially square planar and the weak $d_{z2} \rightarrow d_{x2-y2}$ peak lies on the high energy side.

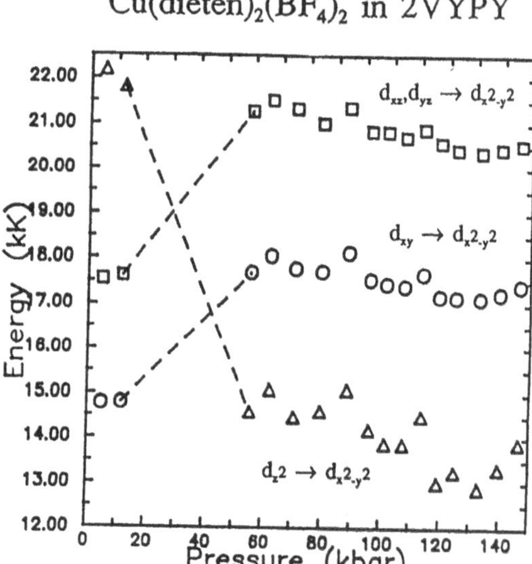

Figure 4 Peak shift vs pressure for $Cu(dieten)_2BF_4$ dissolved in polyvinyl pyridine.

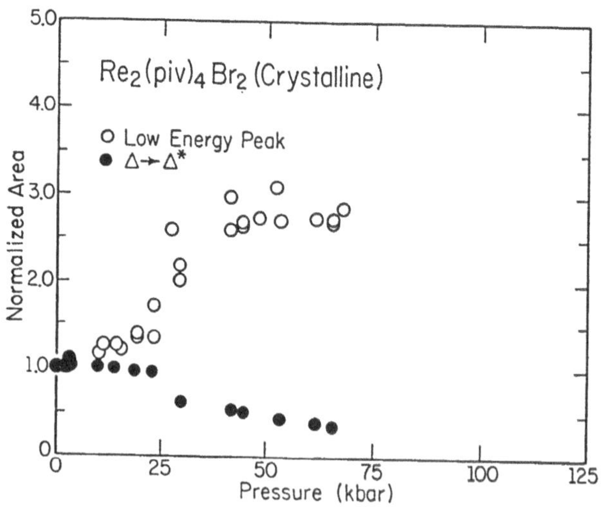

Figure 5 Conversion of $Re_2(piv)_4Br_2$ vs pressure (crystal).

We have studied diethylethylenediamine complexes using a variety of counterions [2]. Here we present results for the BF_4^- complex which is representative.

Fig 3 shows the peak shifts as a function of pressure. The major feature is that between 65-80 kbar the weak peak shifts from the highest to the lowest energy position indicating that compression has introduced essentially octahedral symmetry. In a perfect single crystal the transition could well be discontinuous (first order) so the extent of cooperativity is large.

In Fig 4 we compare the behavior of the same complex dissolved in polyvinyl pyridine. The same rearrangement occurs, but now it initiates below 20 kbar and extends over a range of 40-50 kbar. In the less densely packed medium the extent of cooperativity is significantly less.

A second example of the effect of the medium involves complexes of the form $Re_2(piv)_4X_2$ where the pivalates ($(CH_3)_3CO_2$) bridge across the Re - Re bond and (X = Cl or Br) is attached axially to each rhenium [3]. An important feature of these metal-metal bonds is $\delta \rightarrow \delta^*$ excitation which is sensitive to the environment. The pivalates are large and bulky. The chloride and bromide form orthorhombic crystals with nearly identical lattice parameters so that the molecules with bromide

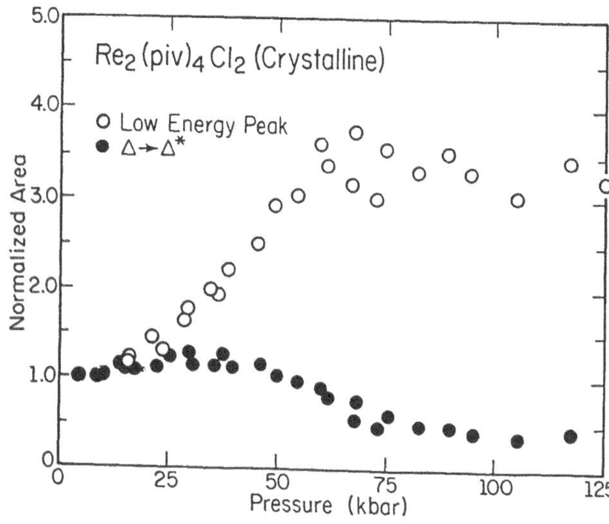

Figure 6 Conversion of $Re_2(piv)_4Cl_2$ vs pressure (crystal).

attached are more crowded in the crystal than the chlorides.

With increasing pressure a new peak grows in the spectrum on the low energy side of the δ - δ^* excitation. It must have a higher transition moment as it grows much more rapidly than the δ - δ^* peak fades. From the standpoint of this paper, the main features are illustrated in Fig 5 and 6. In the bromide the transition is essentially discontinuous (first order) while in the chloride it stretches over 25-30 kbar which indicates a lower extent of cooperativity in the less crowded crystal. Incidentally both complexes exhibit the same transition when dissolved in PMMA, but in this medium it runs continuously over the entire pressure range like a reaction in a fluid.

Although it is aside from the main thrust of this paper, the nature of the change of conformation is of interest. It has been shown [4] that when a given ligand assumes a monodentate rather than a bidentate conformation the δ - δ^* excitation appears at a lower energy with significantly higher intensity. The high pressure IR spectra of these pivilates exhibit free C = 0 and C - 0 stretches not present at one atmosphere just as one would anticipate for a monodentate conformation. It is perhaps surprising that so drastic a change of conformation is reversible, although with considerable hysteresis.

Cation		Dihedral Angle	Symmetry
(Regular Tetrahedron)	—	90°	T_d(A)
Cs_2^+	—	67.9°	D_{2d}(B)
$[C_6H_5-CH_2-N^{\oplus}H(CH_3)_3]_2$	(tmba)	66.6°	D_{2d}(B)
$[C_6H_5-CH_2-CH_2-NH(CH_3)H^{\oplus}]_2$	(nmpH)	0°	D_{4h}(D)

Figure 7 Counterions used with $CuCl_4^{-2}$

Range of Conformation

Frequently a given complex ion can assume a range of conformations depending on the nature (e.g. the electron withdrawing power) of the counterion. Since Cu(II) is a d^9 ion and especially subject to Jahn-Teller distortion, it is particularly useful in this regard. The ion Cu Cl_4^{-2} can assume arrangements from nearly tetrahedral (dihedral angle -

90°) to square planar [5] (dha = 0°). The former conformation minimizes the $Cl^- - Cl^-$ repulsion while the latter maximizes the Cu - Cl bonding. There exist several dozen complexes with dihedral angles ranging from 0 - 70° with differing electronic spectra, so the spectrum is a quite accurate diagnostic of the conformation.

Here we discuss the three counterions shown in Fig 7. The first two have minimal electron withdrawing power so the dihedral angle is large. The third is strongly hydrogen bonding so it is square planar [5]. Several other counterions of intermediate character have also been studied. As can be seen from Fig 8, for the distorted tetrahedron one may expect to see two peaks at relatively low energy, while the square planar complex exhibits three peaks centered at relatively high energy. The effect of pressure on the spectrum of the Cs^+ complex is shown in Fig 8 along with a schematic picture of the energy levels and transitions associated with different symmetries. In Fig 9 we plot the peak locations as a function of pressure for the two distorted tetrahedral complexes. We note that a nearly discontinuous distortion towards square planar occurs between 60-80 kbar. The degree of transformation is a little smaller in the Cs^+ than for the organic counterion. By comparison with a series of one atmosphere spectra, the dihedral angle for the Cs^+ is estimated at 35-40° and for the $(tmba)^+$ about 30-35°.

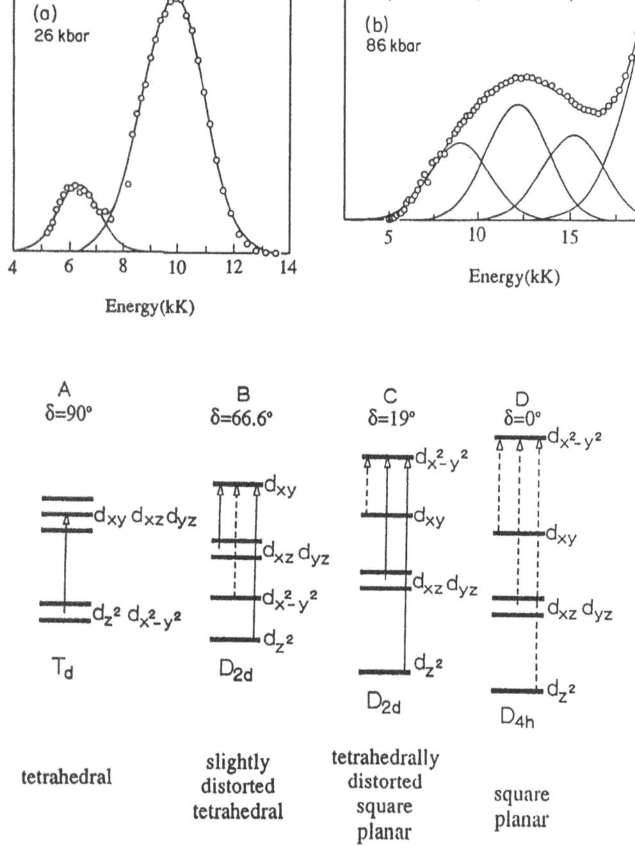

Figure 8 Energy states, and transitions for various configurations of $CuCl_4^{-2}$ and spectra for Cs_2CuCl_4 at 26 and 86 kbar.

Finally, we consider the behavior of the complex with the strongly hydrogen bounding $(nmpH)^+$ counterion. The hydrogen bonding removes electron density from the Cl^- and permits

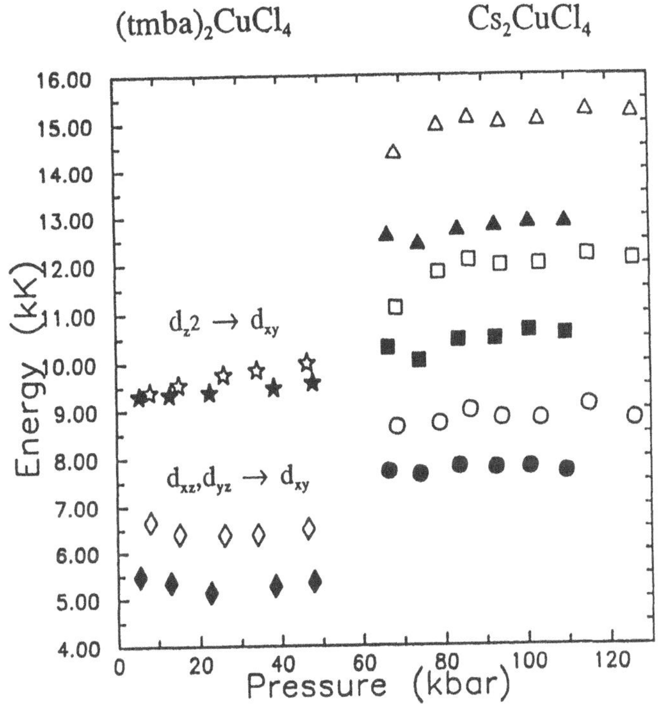

Figure 9 Peak energies vs Pressure - Cs_2CuCl_4 and $(tmba)_2CuCl_4$. Closed symbols - Cs_2CuCl_4 Open symbols - $(tmba)_2CuCl_4$

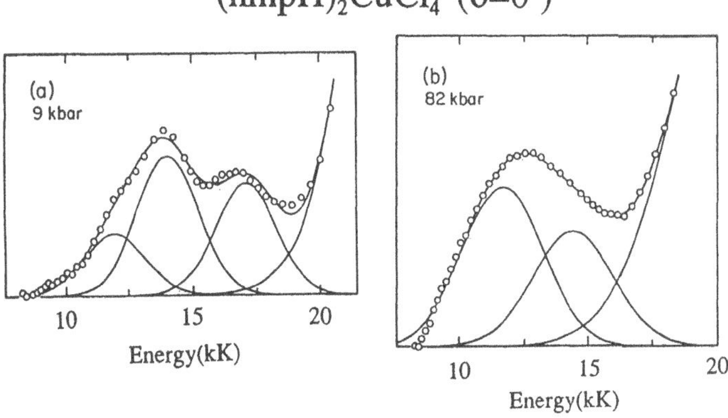

Figure 10 Spectra of $(nmpH)_2CuCl_4$ at 9 and 82 kbar.

the stabilization of the square planar (dihedral angle = 0°) configuration with three peaks at relatively high energy (Fig 10). At high pressure the center of the spectrum has shifted strongly to lower energy. It is fit with two peaks, although there may well be three present. By comparison with atmospheric pressure spectra the dihedral angle is in the neighborhood of 30°. This rearrangement is also essentially discontinuous in the range 70-80 kbar.

The principle exhibited by this study is that where there is a wide range of conformations at low pressure, the range will be significantly smaller at high pressure. This is because with compression, the effects of the relatively weak attractive forces, polarizability, hydrogen bonding etc. are relatively diminished and the configuration is determined primarily by packing and repulsive forces.

Continuous vs State to State Conformation Changes

The final group systems we present undergo a configuration change of a different sort. In this case Cu(II) is coordinated to five nitrogens furnished by diethylene diamine and dipyridylamine. Depending on the anion, the complexes vary in symmetry around the Cu(II) from nearly trigonal bypyramidal (TBP) to nearly square pyramidal (SQP). These are defined structurally by a N-Cu-N dihedral angle which can vary from 120° (TBP) to 165° (SQP). At an angle near 145° the nature of the ground state changes. In TBP symmetry one expects an intense peak at lower energy and two weak peaks at higher energy. In strictly SPQ symmetry one observes an intense allowed peak at a higher energy with two weak (forbidden) peaks at lower energy. From spectra at ambient pressure for a series of complexes with SQP ground state, but differing degrees of distortion, the amount of distortion correlates with the intensity of the intermediate energy peak. In this study we discuss three complexes NO_3^- (dha = 137°), ClO_4^- (dha = 152°) and Cl^- (dha = 159°)[6]. The first has a distorted TBP ground state while the latter two are distorted SQP. It is of importance to note that Muetterties [7] postulated that for systems with a distorted ground state of a given symmetry there should be continuum of allowed configurations, which defines a reaction pathway with increase or decrease of distortion.

The Cl^- complex is most nearly regular SQP with an initial dha = 159°. As can be seen in Fig 11 (closed squares, upper pressure scale) the intensity of the intermediate peak decreases with pressure starting almost immediately. This indicates a movement towards true SQP symmetry. ClO_4^- also has an SQP ground state, but more distorted (dha = 152°). As shown by the open squares and lower pressure scale [Fig 11], at ~55-60 kbar it is at the same point as the Cl^- at 0-10 kbar and follows along some 50 kbar as behind it.

The NO_3^- complex initially has a distorted TBP symmetry. In the region 50-70 kbar it undergoes a transformation, presumably to a distorted SQP ground state (Fig12). At 80 kbar the spectrum corresponds to that of Cl^- at ~20 kbar, and the intensities follow along, but some 60 kbar out of phase. Evidently where there is a change of ground state there are two discrete configurations. The characteristics of the transformation between them depends on the degree of change of conformation, the height of the energy barrier between conformations, and the rigidity of the medium. In the case of a change of degree of distortion of a given ground state, a continuous change in conformation defining a reaction pathway in the sense of Muetterties is possible.

Summary

A recapitulation of the regularities we have noted is as follows:

(a) Molecular rearrangements in liquids are stochastic processes involving only nearest neighbor interaction. In a tightly bound solid it may be necessary for all molecules to transform together, i.e. the cooperativity is macroscopic and one has a first order phase transition. In molecular crystals, glasses, polymeric media and intercalates the extent of cooperativity may be at either extreme or at some intermediate value. In the language of physics a higher order phase transition may be involved.

(b) The extent of cooperativity depends on the nature of the rearrangement and on the

$[Cu(dien)(bipyam)]Cl_2 \cdot 2H_2O$ (filled symbols)

$[Cu(dien)(bipyam)](ClO_4)_2 \cdot H_2O$(open symbols)

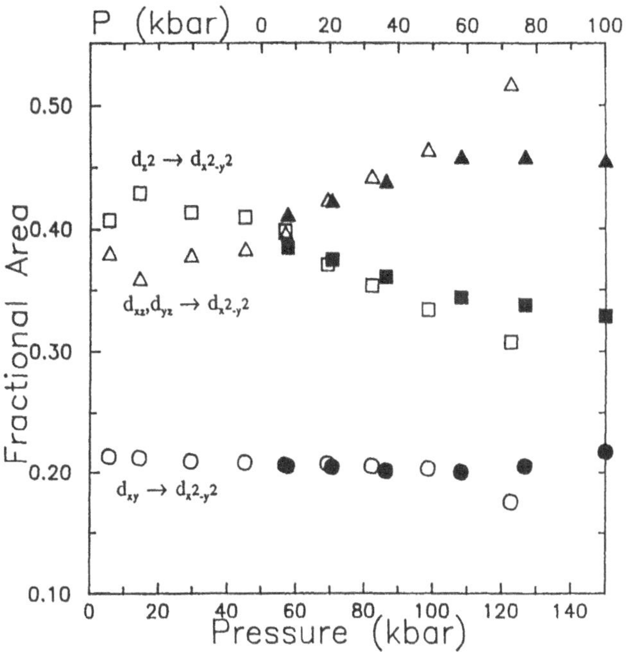

Figure 11 **Pressure dependence of the fractional areas associated with excitations for [Cu(dien)(bipyam)] $(ClO_4)_2 \cdot H_20$ (open symbols lower scale) and [Cu(dien)(bipyam)] $Cl_2 \cdot 2H_20$ (filled symbols - upper scale) vs pressure.**

medium. A rearrangement which is discontinuous in the crystal may involve a lesser extent of cooperativity in a polymeric matrix or glass.

(c) The probability that a given rearrangement is induced by pressure depends more on differences in intermolecular (interionic) interactions and packing than on differences in bond lengths between configurations.

(d) Temperature and pressure are not, in general, conjugate variables. Pressure may act either to inhibit or accelerate the effect of temperature.

(e) Frequently a given complex, e. g. $CuCl_4^{-2}$ exists in a wide range of geometries at ambient pressure depending on the extent of hydrogen bonding, van der Waals forces etc. provided by the surroundings. At high pressure the range of geometries is much smaller as packing and repulsion dominate over the weaker attractive forces at higher compression.

(f) In general there appear to be two modes by which pressure induced isomerizations and consequently, piezochromism, can occur. There can be a transformation from one distinct conformation to another due to the relative stabilization of the potential well associated with conformation B with respect to that of A. This process may be either an equilibrium process if the energy barrier is small, or involve a first or higher order phase transition depending on the extent of cooperativity demanded by the type of transformation and the environment. Such

[Cu(dien)(bipyam)]$(NO_3)_2$ (open symbols)

[Cu(dien)(bipyam)]Cl_2•$2H_2O$ (filled symbols)

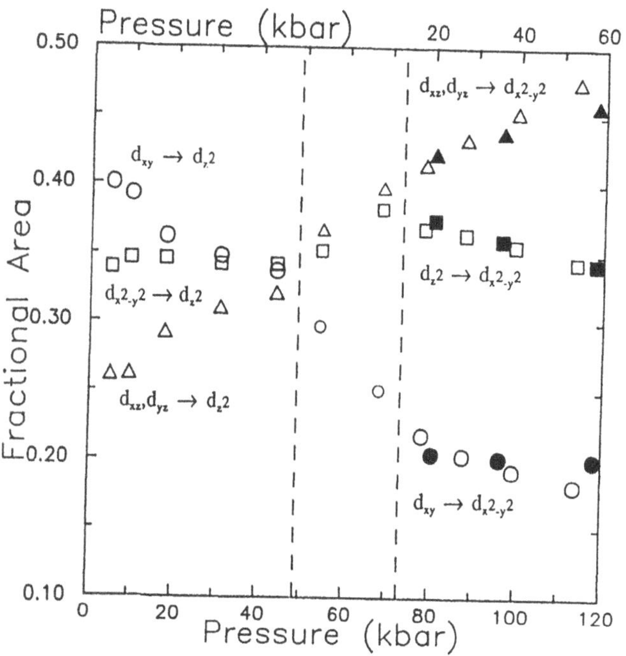

transformations are characterized by a change in the electronic character of the ground state. If there is no distinct change in the electronic character of the ground state, a series of compounds may follow each other along a reaction pathway as they undergo a similar piezochromic transformation over different pressure ranges.

There is, of course, no guarantee that this set of rules is either universal or complete. It is hoped that it may help to introduce some systematics into solid state chemistry, and possibly also some degree of predictability.

Figure 12 **Pressure dependence of the fractional areas associated with excitations for [Cu(dien)(bipyam)] $(NO_3)_2$ (open symbols - lower scale) and [Cu(dien)(bipyam)] $Cl_2 \cdot H_20$ (filled symbols, upper scale) vs pressure.**

Acknowledgement

I should like to acknowledge the central role of my students, E. N. Hockert, K. L. Bray, R. T. Rojinski, and T. L. Carroll to this work, as well as the essential cooperation with my colleague Professor John Shapley on the rhenium complexes.

It is a pleasure to acknowledge the continuing support of the Materials Science Division of the Dept. of Energy under Contract DEFG02-91ER45439.

References

[1] Hockert, E. N. and Drickamer, H. G. (1977) "High Pressure Optical Studies of Crystalline Anils and Related Compounds" J. Chem. Phys. 67 5178-88.

[2] Bray, K. L., Schmitt, E. A., Hendrickson, D. N. and Drickamer, H. G. (1989) "Piczochromism: Pressure Induced Rearrangements of Thermochromic and Related Cu(II) Complexes Containing Asymmetrically Substituted Ethylenediamine" J. Am. Chem. Soc. 111 2849-52.

[3] Roginski, R. T., Carroll, T. L., Moroz, A., Whittlesey, B. R., Shapley, J. R. and Drickamer, H. G. (1988) "High Pressure Spectroscopic Studies of Dihalotetrapivalato Dirhenium (III) Metal-Metal Quadruply Bonded Complexes" Inorg. Chem. 23 3701-6.

[4] Hopkins, M. D., Gray, H. G., and Miskowski, V. M. (1987) δ - δ^* Revisited: What Energies and Intensities Mean, Polyhedron 6 705-710.

[5] Bray, K. L. and Drickamer, H. G. (1990) "Piezochromism: Pressure Induced Isomerizations in the Tetrachlorocuprate Ion in the Solid State with Various Cations" J. Phys. Chem. 94 2154-59.

[6] Bray, K. L. and Drickamer, H. G. (1989) "Piezochromism in Cu(II) Cation Distortion Isomers: Pressure Induced Changes in the Solid State Geometry of Diethylenetriamine (2,2′ dipyridylamine) Cu(II) Complexes" J. Phys Chem. 93 7604-09.

[7] Muetterties, E. L. and Guggenberger, L. J., (1974) Idealized Polytropal Forms. Description of Real Molecules Referenced to Idealized Polygons or Polyhedra in Geometrical Reaction Path Form, J. Am. Chem. Soc. 96 1748.

RECENT DEVELOPMENTS IN DIAMOND-ANVIL CELLS

D.J. DUNSTAN
Department of Physics
University of Surrey
Guildford,
Surrey, GU2 5XH
England

ABSTRACT. Diamond-anvil cells have been simplified and miniaturised to the point where they can be added with little difficulty to a wide range of experiments. This allows high pressure, or the control of interatomic spacing, to be more generally applied in solid state physics than hitherto. The principles of the diamond-anvil cell are presented together with details of its operation in practice. Particular attention is paid to the choice and design of gasket, and to methods of pressure calibration.

1. Introduction

The diamond-anvil high-pressure cell (DAC) has revolutionised high-pressure experimentation in the last few years. In place of apparatus weighing tonnes, or even hundreds of tons, and able to generate a few GPa (tens of kbar), we now have apparatus which can be held in the palm of the hand and yet can generate hundreds of GPa (well into the megabar region) in complete safety (Jayaraman, 1986). This advance is not without its disadvantages, of which the major ones are the small size of the high-pressure volume, the sample space, and the associated difficulties in gaining access to the sample and the pressure gauge.

A very wide range of experiments have nevertheless been performed in diamond-anvil cells. Many of these have been done by high-pressure specialist laboratories, and demonstrate the possibilities (Jayaraman, 1986). More and more, however, the DAC is being used in laboratories with other specialities, where high pressure is seen as just one more perturbation to apply to the sample, like temperature or magnetic field. For these laboratories, the DAC should ideally be a commercial instrument, with an operating manual, which can be used with existing apparatus in the same way as a cryostat or a magnet. This paper is therefore a review of recent developments in DAC technique, of its capability in the hands of specialists certainly, but with an emphasis on the advances in practice and in theory which make it a reliable tool for the routine generation of high pressure in the hands of non-specialists.

We review briefly the principles of operation of the DAC, and some of the designs of DAC which have been developed for different applications. The design and construction of diamond-anvil cells has been discussed in detail recently (Dunstan and Spain, 1989), the operation as well (Spain and Dunstan, 1989) and we avoid unnecessary duplication of the material in those papers.

R. Winter and J. Jonas (eds.), High Pressure Chemistry, Biochemistry and Materials Science, 79–99.

The gasket is a crucial component for high-pressure operation, and is almost infinitely variable, and so we devote considerable attention to it (Section 2.2). Choice of pressure calibrant is an important subject and is discussed in Section 3. We conclude in Section 4 with a very brief discussion of experiments that can be carried out in the diamond-anvil cell.

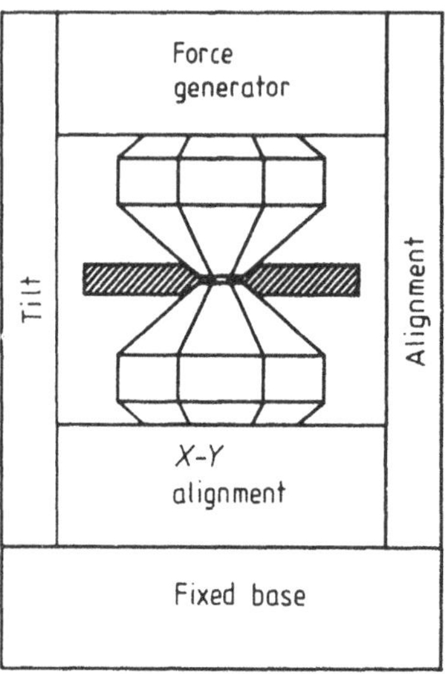

Figure 1. A schematic diagram of a generalised diamond-anvil cell, showing the gasket and diamonds at the centre, surrounded by the functional units of the cell. *Reproduced from Dunstan and Spain (1989).*

2. Pressure Generation in the DAC

High pressure is achieved by the application of a large force to a small area:

$$P = \frac{F}{A} \qquad [1]$$

In the DAC, the force may be of the order of a tonne or more, and the area can be a tenth of a square millimetre or less. For these numbers, Eqn.1 then gives a pressure of a megabar (100GPa). The diamonds transfer the force from a large area (the table of the diamond) where the pressure is much lower, within the realm of ordinary engineering stresses, to the small high-pressure area, the culet. The design of the DAC is therefore concerned with the generation and control of the force and its transfer to the diamonds, while the transfer of the pressure to the sample is the rôle of the gasket. We will consider these functions separately, starting with the diamonds and then the gasket, and concluding with a discussion of the cell once we have established what we require of it.

2.1 THE DIAMOND ANVILS

Diamond is a hard material, in fact the hardest known. Hardness is equivalent to the pressure the material will sustain without either plastic deformation or brittle failure, and the high hardness of diamond is due, firstly, to its very high elastic moduli (around 5Mbar) and the high elastic deformation it will withstand before yielding plastically, over 30%, according to Ruoff et al (1991b). Being a hard material means that diamond is strong in compression, but will fail unpredictably under tensile stress. The design of the anvils so that they will be subject principally to compressive stresses is not the concern of the user. Given the optimum anvil design, Adams et al (1992) have shown that the design of the backing can affect the tensile stresses considerably but we will consider this below. The principal concern of the user is the culet diameter; also the number of sides, any bevels, coatings etc which are to be specified when the diamond is ordered.

Hard materials fail when there is sufficient elastic energy present to create a crack. The elastic energy derives from the product of the stress and the volume under stress. Quite generally, the smaller the volume, the higher the stress before failure; this is because the crack initiation energy is independent of the size of the object. Thus, in grinding a hard material in a pestle and mortar, it will be found that it is quite easy to reduce it to a certain particle size, at which the applied stress times the particle volume is no longer sufficient to initiate cracks. Beyond this point, the stress, or force on the pestle, needs to be increased in proportion to the desired decrease in particle size.

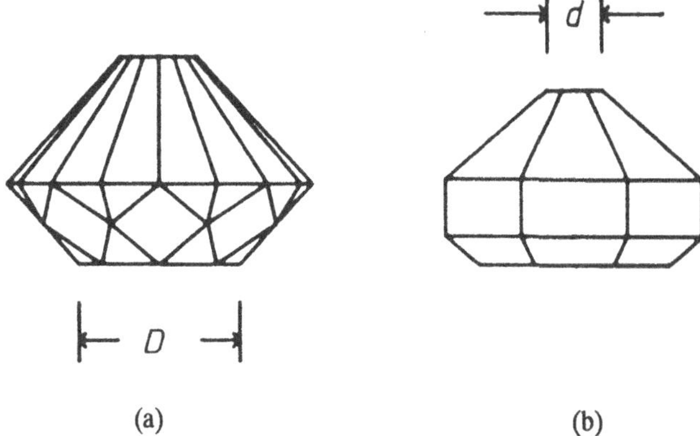

Figure 2. Diamond anvils: (a) the brilliant cut and (b) the Drukker Dubbledee eight-sided design. The table diameter D and the culet diameter d are marked.

Applying this principle to diamond anvils, the stress is highest at the culet, and decays back down the diamond in proportion to the increasing area. If the rated pressure is made inversely proportional to the culet area (see Figure 2), the total elastic energy in the diamond is approximately independent of rated pressure. Also, the volume of the most highly stressed part of the diamond, near the culet where the diamond is subject to a stress of the order of the peak pressure, decreases as the cube of the culet radius; if the rated pressure increases as the inverse square, then the elastic energy in this crucially important volume actually decreases in diamonds rated for higher pressures. The empirical rule given by Dunstan and Spain (1989) is

$$P_{max}\,(\mathrm{GPa}) = \frac{12.5}{d^2\,(\mathrm{mm})} \qquad [2]$$

The consequence of this rule is that the force required in a diamond-anvil cell is independent of the intended pressure range, and is about 1 tonne. The use of beveled culets, together with very careful design and use of the cell, allows the use of forces up to about three or four tonnes, but for a given pressure this allows the culet to be only twice as large (giving eight times the sample volume—see Section 2.2.1). Bell et al (1984) show data with a force up to seven tonnes.

When ordering diamonds, then, the first decision is the culet diameter, and this depends only on the intended pressure range. Then comes the overall size of the stone. One-third carat stones are usual for relatively low pressures, as this provides a table large enough that the one tonne force does not overstress a seating of ordinary metals (beryllium-copper, hardened steel, etc.). For megabar work, smaller diamonds are often preferred to reduce both the risk and the cost of breakage [although the group of Ruoff use ⅓-carat stones (Baublitz et al, 1981)]. Together with the higher forces (up to three or four tons) used in some megabar cells, the smaller stones require a much harder seating for the diamond. For example, three tons on a 2mm table generates 100kbar pressure on the seating and this can only be supported by a hard material such as tungsten carbide or sapphire.

Diamonds are classed as Type I and Type II according to the nitrogen content and distribution. Impurities can also give absorption bands and luminescence bands. In the smaller sizes, artificial stones are available as well as natural. The anvil manufacturer's advice should be sought on these matters, although there is also room for experimentation. For example, it has generally been believed that Type I diamonds, strengthened by nitrogen impurities, were necessary for multi-megabar work, but Ruoff et al (1991b) recently reported 4.16Mbar with Type II diamonds.

The standard commercial anvil is eight-sided and has no bevels (Figure 2.b). Specifying sixteen sides, or bevelling the culet, add marginally to the cost but protect the stone to some extent against accidental damage (chipping). On the other hand, these modifications make the cell rather harder to align accurately as the view through the two stones is more confusing.

Another modification to the standard diamond that should be considered is metallisation. The anvil manufacturers can lay down coatings adapted to various purposes. For example, diamonds sold as heat sinks to the semiconductor industry have coatings that provide for a secure brazed or soldered joint to the semiconductor device. The author uses such coatings on the table of the diamond to solder the diamonds into the cells (Dunstan, 1991). This is very important for cryogenic use: glues do not survive cryogenic cycling very well, and other methods of mounting the diamonds have their own disadvantages (Dunstan and Spain, 1989). Patterned metallisation coatings on the flanks and culet of the diamond can be used to provide electrical contacts to samples (Grzybowski and Ruoff, 1984; Hemmes et al, 1989).

It should be noted that many of these modifications need not be ordered from the anvil manufacturers but can be made by the user. Boehler, for example, does high-temperature high-pressure experiments in DACs, and these experiments damage the diamond culet in each run. Rather than returning the stone each time to the manufacturers for re-polishing, he regrinds them himself, and puts on bevels etc (Boehler, 1992). Similarly, Hemmes et al (1989) use metal coatings evaporated in their own laboratory for electrical feedthroughs. However, both the metal and the evaporation conditions have to be carefully chosen to ensure adequate adhesion—for example, Hemmes et al use Pd evaporated onto diamond held at 300°C.

2.2 THE GASKET

In many ways, the gasket is the most interesting part of a diamond anvil cell. It is prepared afresh for each experiment, and so the experimenter can modify it according to the needs of the moment. Also, it is the least understood part of the cell, and perhaps has the greatest potential for future development.

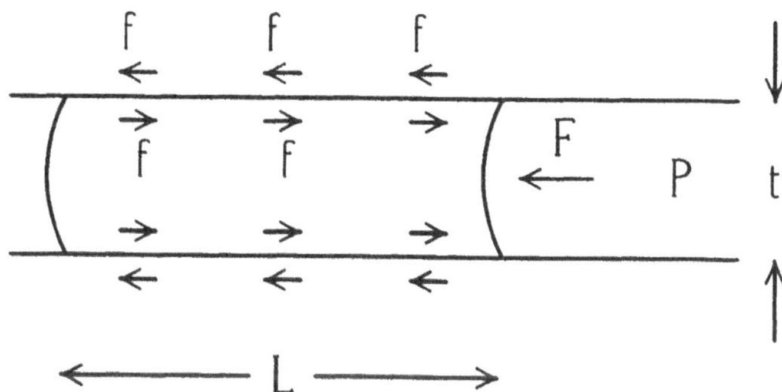

Figure 3. A soft medium is shown being pushed through a narrow gap. The forces per unit width and the pressure are indicated; f is the shear force per unit length and $F=fL$ is the total force required. We see that the force required to cause plastic shear increases with the length but is independent of the thickness. Consequently, the pressure P required to generate this force increases as the reciprocal of the thickness, $P=Ft^{-1}$, and so the pressure is proportional to the aspect ratio L/t of the gasket.

2.2.1 *Theory.* It is surprising that a metal gasket with a yield strength of perhaps 10-20kbar is capable of supporting pressures in the sample hole of ten, a hundred or even more times this value. The answer to this apparent paradox was found by Schroeder and Webster (1949), who were interested in the very high forces required to press coins—much higher than expected from the area of the coin and the yield strength of the metal. They showed, what is perhaps intuitively obvious: that it is very hard to squeeze even a soft material out of a narrow gap. Under high pressure, the friction between the gasket and the anvil is sufficient that the gasket is stationary at the anvil surface, and is squeezed out by plastic shear deformation. The pressure difference across an element required to reach the shear yield strength is inversely proportional to the width of the gap, and so the hydrostatic pressure rises along the gap linearly (see Fig.3):

$$\frac{\partial P}{\partial z} = \frac{2}{\sqrt{3}} \frac{k}{t} \qquad [3]$$

where k is the stress for yielding under uniaxial load (Dunstan, 1989). Analysis shows, of course, that the higher the pressure required the thinner the gasket must be. Since the culet radius decreases as

$$r \propto \frac{1}{\sqrt{P}} \qquad [4]$$

from Eqn.2, the result is that

$$t \propto P^{-\frac{3}{2}} \qquad [5]$$

2.2.2. *Thick and Thin Gaskets.* An important practical consequence of the analysis is the distinction between thick and thin gaskets (see Dunstan, 1989, for a full analysis). If the relationship between the hydrostatic pressure in the sample space and the gasket thickness is such that Equation 3 simply gives a monotonically increasing pressure in the gasket from the edge of the culet to the sample space, the gasket is referred to as *thick*. In this case, pressure can only be increased by making the gasket thinner, by advancing the diamonds, so that metal flows outwards and the sample hole increases in diameter (see Figure 4.a). This is a very unsatisfactory situation, as now the distance over which the pressure gradient of Eqn.3 is integrated for the sample pressure is reduced. This requires the gasket to be even thinner, so the hole must expand further still. Instability can occur, the gasket bursts and the diamonds may be broken. Worse, any asymmetry will result in the gasket hole expanding asymmetrically, encouraging instability.

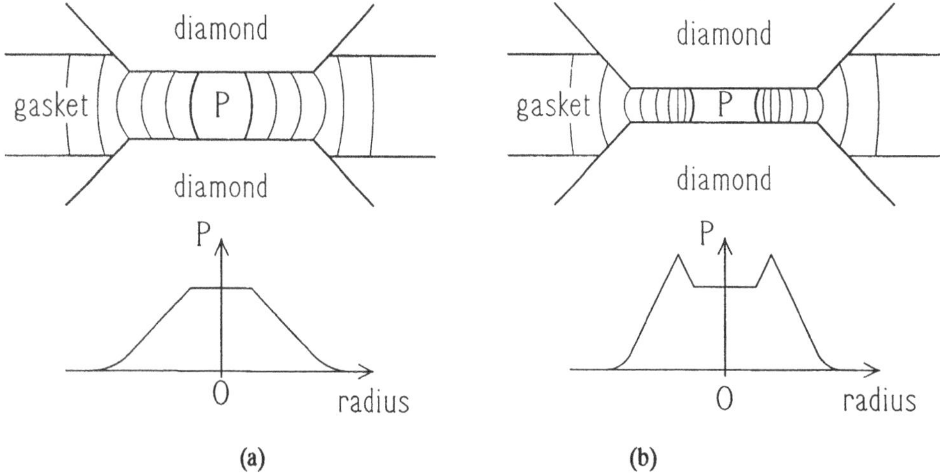

Figure 4. The gasket plastic flow patterns are shown schematically for (a) the thick gasket and (b) the thin gasket, for a hydrostatic pressure P_0 in the sample space. Below are shown the hydrostatic pressure profiles across the cell diameter.

If, however, the gasket is thinner to start with, the pressure increases faster from the edge, according to Eqn.3, and so reaches a peak before *decreasing* again to the sample hole. Now, when the diamond is advanced to increase the pressure, metal flow outwards from the peak pressure down the gradient of Eqn.3—it is extruded outwards as before—but metal within the peak pressure contour is extruded inwards, *reducing* the size of the sample hole (Figure 4.b). This is a completely stable situation, and the cell should always be run in this régime, the *thin* gasket.

2.2.3. *Gasket Construction.* There are very many ways to construct DAC gaskets. Given the small dimensions and high accuracy required, the author's preference is for methods that separate the delicate steps so that only one difficult operation at a time is required. The first option is pre-indentation. The diamonds are pressed into thicker stock metal than the required gasket thickness, before the sample hole is drilled. This enables a free choice of gasket thickness, rather than being restricted to stock metal thicknesses. It work-hardens the active part of the gasket, and provides massive support, increasing the pressure range for a given thickness (Dunstan, 1989). It allows *automatic* selection of the correct gasket thickness, by the application of the same load as will be used in the high-pressure experiment. Finally, and most important, it gives automatic positioning of the gasket on the diamond when the cell is loaded. Non-preindented gaskets have none of these advantages and require that both the sample and the gasket are correctly positioned on the diamond simultaneously. With pre-indented gaskets, the only operations that require precision and careful handling are the drilling of the sample hole in the centre of the indentation, and the placing of the sample on the centre of the diamond culet. Each of these operations is completely separate.

Many cells are provided with gasket guide pins; the gasket is drilled to slide over the pins, providing automatic location. It should not be forgotten, however, that ordinary engineering tolerances are more than 10μm, so that the accuracy of location of the gasket by the guide pins is no better than this. With a non-preindented gasket, the sample hole is drilled with reference to guide-pin holes, and the centering of the sample hole on the diamond is subject to an accumulation of errors (gasket to pin location, sample hole to guide-pin hole, and even the location of the diamond relative to the guide pins).

The only disadvantage of pre-indentation is the risk of diamond breakage. The load is as great as in a high-pressure experiment, and the flow of the metal is entirely outwards which puts the maximum tensile forces on the diamond. The risk can be reduced by applying the load progressively, returning to no load in between. In principle, this should make no difference as the stress field while the metal is undergoing plastic deformation is unique and not dependent on the past; in practice it does seem to reduce diamond breakage. Some workers drill a hole in the centre of the indent; each time the hole collapses during progressive indentation it is redrilled. This does reduce the load required for pre-indentation and the tensile stresses on the diamonds.

The conventional gasket is made from flat sheet metal; popular materials are stainless steel, high-carbon steel (clock spring), beryllium copper and Inconel. Stainless steel has been used at low pressures (it is standard in the author's laboratory) and also at multi-megabar pressures (Bell et al, 1984). Be:Cu and Inconel have the disadvantage that heat treatment is required after pre-indentation to obtain the desired properties. Carbon steel also requires careful heat treatment but a very wide range of properties can be obtained in a controlled manner (by hardening and then drawing the temper, as in traditional tool-making). Stainless steel has the advantage of simplicity and repeatability, since no heat treatments are given. It is, however, important to buy the right grade; we find that cold-rolled 316 material is satisfactory but not as good as an unidentified material obtained from our workshops some years ago and now unfortunately used up! In principle, from Eqn.3, the stronger materials, carbon steel, Be:Cu and Inconel will give greater gasket thicknesses for a given pressure, but data given in the literature shows no significant differences. This is possibly due to an increase in strength of the stainless steels under severe cold-working at high pressure.

Many variations on the standard gasket are possible. Instead of a simple parallel sample hole, a hole with a lip has been used to provide masking for optical transmission experiments (Kobayashi, 1985) and also to support the sample in laser-heated high-temperature high-pressure experiments (Boehler and Chopelas, 1991). Machining a lip at such small dimensions is not easy, and the author prefers to achieve the same geometry by laminating the gasket from two thicknesses of stock sheet (Leong et al, 1992). The two sheets are glued together for pre-indentation and a pilot hole is drilled for the sample hole. The laminations are then separated and the sample hole is enlarged in one of them, leaving the other to provide the lip. Both the gasket guide pins and the indentation ensure accurate alignment when the two laminations are re-assembled during the loading of the cell.

Many authors have sought ways to make the gasket electrically insulating. The Bridgman anvil cell uses an insulating gasket as standard, usually a mixture of a powder such as MgO and a weak matrix such as epoxy resin (Bridgman, 1952). It is easy to run wires through such a gasket, for electrical measurements on the sample in the high-pressure volume. The author is not aware of any successful work with this kind of gasket in the DAC. Nor have strong plastics such as Kevlar made satisfactory gaskets; such materials appear to embrittle at high pressure. Many authors have, instead, found ways to insulate the surface of a metal gasket so that wires can be trapped between the diamond and the insulator giving electrical access to the sample. We have found that pressing in fine alumina powder, followed by spinning on low-viscosity epoxy resin, provides both insulation and sealing (Leong et al, 1992). Photoresist, CVD alumina, and many other materials can also be used, but few authors have achieved both insulation and sealing (see Spain and Dunstan, 1989 and references therein).

Until recently, it was thought that electrical measurements in the DAC were demonstrable but nearly impossible (Spain and Dunstan, 1989). This is probably still true in the megabar range with an inert gas pressure medium. But under 100kbar, it is no longer true. None of the process steps used in the author's laboratory (Leong et al, 1992) are any harder to do than the processes required for an conventional optical experiment, and for measurements up to about 50kbar, using argon as the pressure medium, we have a success rate of about 50%. This is sufficient to make electrical work in the DAC almost routine (up to 50kbar) and opens up a much higher range of pressure to the many experiments currently done in hydraulic equipment up to say 10kbar. Most device structures, such as semiconductor Hall samples, can easily be accomodated in the DAC; they can be prepared by photolithography to dimensions down to microns without affecting the physics of the device. The major exception is the semiconductor laser diode, which requires lengths in the range of 200μm to 1mm in order to operate.

2.3 THE CELL AND THE DRIVE

The cell is required to support the diamonds, to provide for alignment, and to transmit the drive force to the diamonds while retaining alignment. Provision must also be made for the experiment which is to be carried out on the samples under pressure. The drive force may be generated internally in the cell or it may be generated elsewhere and transmitted to the cell. Since provision must also be made for pressure measurement, and this is usually done optically, a cell which allows for optical experiments may be termed a general-purpose cell (Section 2.3.1.). Also, electrical measurements may be made in general-purpose cells, as the special requirements for making electrical feedthroughs into the sample space are entirely to do with the gasket and diamonds. On the other hand, some experiments require specialised, purpose-built cells and these are discussed separately in the following sections.

Compared with the diamonds and the gasket, the requirements for the cell are fairly simple and purely mechanical, except as discussed below. To fulfil these mechanical requirements for different users does not require a tremendous diversity of cell designs; instead, there should be one optimum design which may be capable of taking minor modifications for different applications. Most laboratories do indeed use their basic cell design for a wide diversity of experiments. The prospective user does not therefore have to worry too much about which design to adopt, provided only that the proposed cell will fit in his equipment and that previous users have found no obvious vices. We have reviewed design and construction elsewhere (Dunstan and Spain, 1989), and so make only a few comments here.

2.3.1. *General-Purpose Cells.* A marked aspect of the evolution in DAC design has been in the direction of miniaturisation. Early cells typically had dimensions of tens of centimetres and weights up to a kilogram. For example, the cell used by the group of Ruoff and co-workers (Baublitz et al, 1981) is 5cm×5cm×9cm. Such large cells require purpose-built equipment such as cryostats to operate them in. The micro-cell reported by Eremets and Timofeev (1990) in contrast has an outside diameter of 10mm and a weight of under 9gm and has been taken up to 110GPa (Eremets and Shirokov, 1991). Essentially, since the diamond-gasket assembly is only a few mm across, there is no need for the cell to be much larger. The author's own miniature cell (Dunstan and Scherrer, 1988) is a fairly conventional design, incorporating diamond alignment mechanism controlled by screws. However, the screws can be small because the force for the diamonds is not transmitted through them, and so the cell was easily be kept down to 19mm diameter and a weight of 50gm to fit in standard cryostats and magnets (Figure 5). The special design feature of the cell of Eremets et al which enables it to be little larger than the absolute minimum is that the use of bolts and screws is avoided, together with the large wall thicknesses required for bolts to screw into. The tilt alignment, which in most cells requires the use of bolts, is provided by the counter-rotating wedge principle proposed by Dunstan and Spain (1989).

Cells may be divided into three classes. *Clamp cells* have no internal provision for force generation; instead the force is applied in some kind of press, away from the experiment, and clamped. Each pressure change requires the cell to be returned to the press. An example is the miniature cell of Eremets *et al.* (1990, 1991). *Internally-driven cells* have screws, levers, hydraulic or gas rams inside the cell body. *Externally-driven cells* are a kind of cross between these two types. They have a cell body that could be loaded in a press and clamped, but in addition they have a drive which connects them to the force generator. This method enables pressure changes to be made in an experiment without disturbing the cell, while keeping the screws, levers, hydraulic or gas rams or whatever well away from the experiment (e.g. outside the cryostat). An example is the author's own design of a miniature cryogenic cell (Dunstan and Scherrer, 1988; Wilkinson and Dunstan, 1990); the complete system including drive is shown in Fig.5.

Externally-driven cells are ideal for loading with inert gases (argon or helium) under cryogenic conditions since the very feature which enables pressure changes at low temperature enables closure of the cell while immersed in liquid argon or liquid helium (Spain and Dunstan, 1989). Clamp cells, in contrast, require special equipment for rare gas loading; several authors have used high-pressure loading rigs where the cell is loaded with gas at several kbar.

General-purpose cells may be modified extensively for particular experiments. Components are built for stiffness (lack of deformation under load) rather than for ultimate strength (load at failure). It is not generally appreciated how much metal may often be removed from a component without affecting its stiffness significantly (while severely reducing its ultimate strength). Holes may be drilled almost anywhere in a DAC without compromising its performance at all, to the point where it may resemble Swiss cheese. For light scattering at 90°, for example, a small

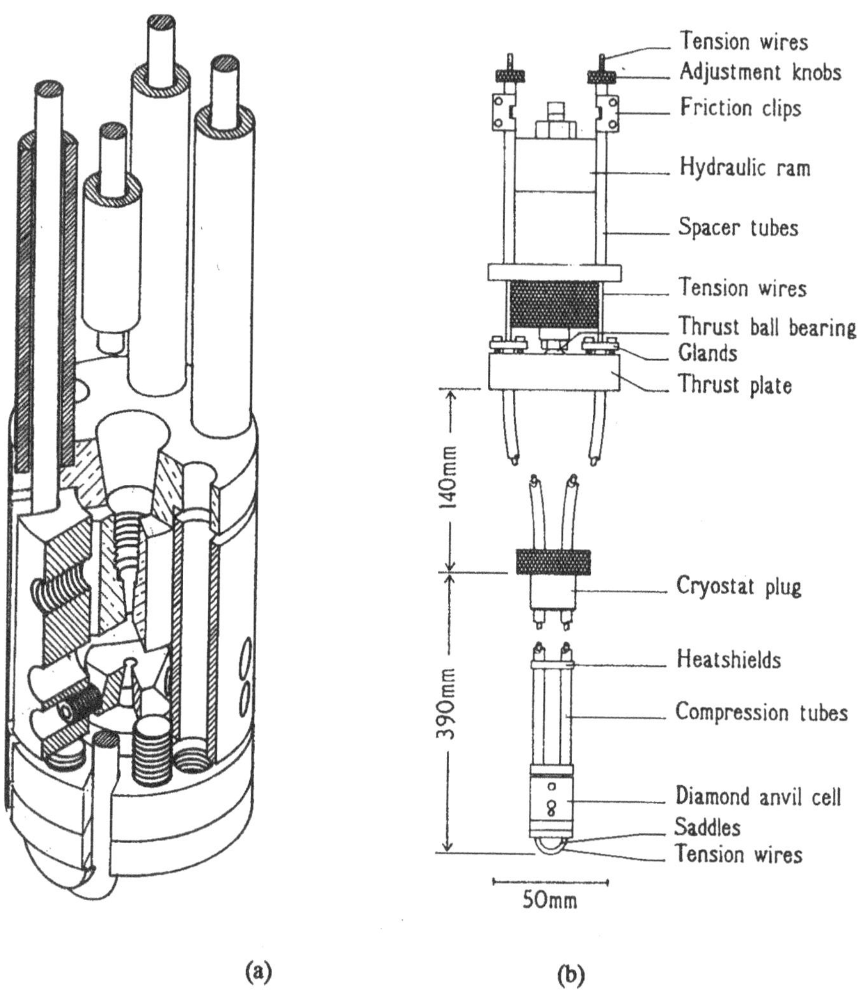

Figure 5. The author's diamond anvil cell (a) and external drive mechanism (b), built for insertion in a standard helium flow cryostat with a 2cm bore, and allowing for pressure changes *in situ*. Optical access is axial, using the end window of the cryostat. *Figure 5.b is reproduced from Dunstan and Scherrer (1988).*

parallel hole to input a laser beam at 45° can be drilled through the cell body and the diamond backing, and a second, tapered hole at 90° to the input hole can be added as long as it avoids any screws or other small components. The only difficulty is to calculate refraction so that the optical axes of the two holes intersect at the sample position.

Adams et al (1992) have used finite element analysis to calculate the stresses in the diamond and in its backing plate for a wide range of geometries. They obtained a startling result, that the crucially important tensile stresses in the diamond are *reduced* if much of the backing plate is cut away so as to give a much larger optical aperture than is conventionally used, up to *f*/1. They established that friction between the diamond and the backing was sufficient to prevent slippage as the load is increased, so that the diamond and the backing deform as a single unit, and this is the key to understanding their result. Fig.6 is an attempt to interpret the results of Adams et al. With a solid backing, the diamond acts like a beam and is bent (Fig.6.a), creating tensile stresses in the region of the table. A cutaway backing and sufficient friction to avoid slippage allows a stress pattern as in Fig.6.b to develop; in principle this need have only compressive stresses in the diamond. However, the actual stress pattern must be very dependent on the details of the diamond-backing interface; for example, the author solders the diamonds to the backing and this introduces a thin layer of solder with a low shear strength at the interface. This would allow some degree of slippage as the load is applied, and whether there is still an advantage in cutting away the backing is then less clear. This interpretation leads to the speculation that perhaps the diamond would best be supported by the lower bevel, rather than by the table itself.

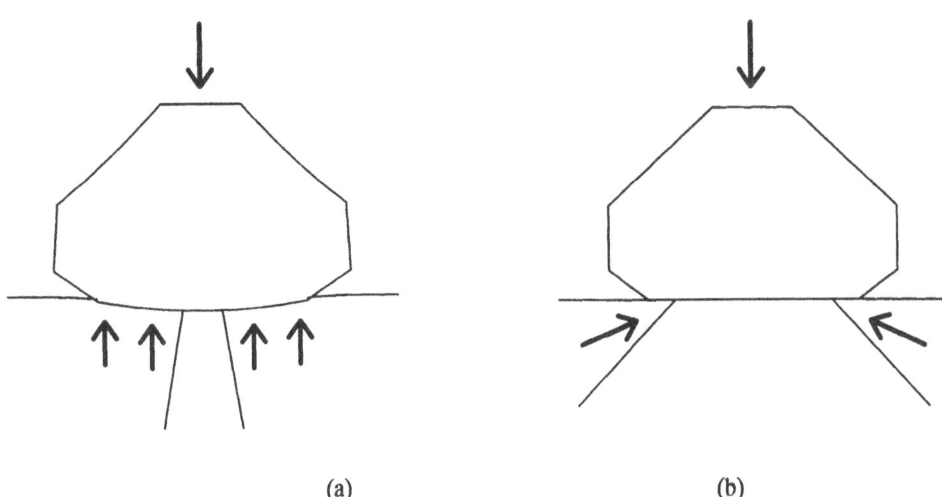

Figure 6. In (a), a section through a diamond and a conventional backing plate, the deformation that occurs with the nearly solid backing is shown schematically. The bending of the diamond is exaggerated to make clear the presence of tensile stress in the table. In (b), the stresses with a severely cutaway backing are shown; the three convergent forces can leave much reduced tensile stresses compared with (a).

2.3.2. *Specialised Cells for High Temperatures.* While the highest temperatures, several thousand degrees, can be generated at high pressure by laser irradiation of the sample (see Section 4.1), for many problems it is sufficient to heat the cell up to much more moderate temperatures. Moreover, resistive heating of the whole diamond-gasket-sample assembly removes uncertainty about the temperature uniformity and thermodynamic equilibrium of a laser-heated sample. Barnet et al (1973), for example, heated their entire cell up to 700°C at 300kbar. Adams (1992) takes two approaches. For temperatures up to about 800°C, a resistive heater may be placed about the anvils in a standard cell in the same way as Barnet et al (1973), and Sung (1976), who achieved

300kbar at 1000°C with water-cooling for the body of the cell. This requires some consideration of materials; for example, a maraging stainless steel is preferable to beryllium-copper or tool steel, and tungsten carbide may be used to support the diamonds.

For higher temperatures, up to the likely onset of diamond graphitisation around 1600°C, Adams (1992) uses a graphite heater in a ceramic ring around the diamonds, in a tungsten piston-cylinder insert which goes into a standard cell body for force generation. Still higher temperatures are obtained by laser heating of the sample in a conventional cell (see Section 4).

2.3.3. *Specialised Cells for Very High Pressures.* The principal feature of cells used above a megabar (100GPa) is that they all have a much longer piston-cylinder assembly than standard cells (see Bell et al, 1984; Baublitz et al, 1981). This helps to decouple any asymmetric forces arising in the drive from the diamonds; it also reduces the amount the piston can tilt in the cylinder. It does not improve the lateral control of the piston; to do this other mechanical principles are required such as diaphragms (Dunstan and Spain, 1989).

The problem is that mechanical tolerances and clearances can be reduced below about 10μm only with difficulty. But 10μm is large on the scale of the culet diameter for multi-megabar pressures (typically in the region of 50μm). Guidance based on flexure instead of sliding, as in the Huxley micropositioner, is capable of giving sub-micron accuracy with no accurately machined components (Dunstan and Spain, 1989). To our knowledge, no such cell has been built, but the author believes that ultra-high precision cells should use this principle instead of the sliding piston-cylinder assembly.

2.3.4. *Specialised Cells for X-Ray Diffraction.* X-ray diffraction poses some particular problems for DAC design and construction. In contrast to most optical experiments, it is desirable to detect out-going radiation over a wide angle, in principle up to 180°. The sample is very small for an X-ray experiment, so that diffracted intensities are low; for this reason many workers now use synchroton radiation for work which, in a larger sample, could be done using a laboratory X-ray source. However, there are metals which are transparent to X-rays and this can make design easier; beryllium is an example and has reasonable mechanical properties.

Two approaches have been taken. One is to make a fairly standard cell, but using beryllium for the diamond seatings and to support the beryllium in such a way as to leave a large aperture for the X-rays. The cell produced by Adams (1992b) is a standard two-plate cell (Merrill-Bassett) but with large-aperture beryllium seatings. The other approach is to take advantage of the ability to cut large holes in a component without affecting its stiffness much. Adams et al (1992) calculate that a slot can be cut in place of the hole under the diamond without significantly compromising the support given to the diamond. The slot can allow full 180° access.

2.4 PRESSURE MEDIA

The pressure medium surrounding the sample is there to maintain hydrostatic conditions. It would ideally be a fluid, but under high pressure, and particularly at cryogenic temperatures, fluids have a tendency to solidify. The non-hydrostatic component of the strain in the sample is then essentially determined by the shear strength of the solidified pressure medium. This opens up a complete new field similar to metallurgy. The shear strength of the solidified medium is likely to depend on the pressure and on the temperature, but also very much on impurities, on the thermal history and on the plastic deformations which it has undergone. Burnett et al (1990) have reviewed and tested experimentally the popular pressure media and their conclusion is, not surprisingly, that solid helium is the best option. However, the author's laboratory uses exclusively argon, with the intention of changing to helium when we have any evidence of a need to; the issue is not completely clearcut.

The traditional pressure medium is a eutectic alcohol mixture, 4:1 methanol and ethanol. This freezes to quite a strong solid at high pressure and is quite unsuitable as a cryogenic medium. In the author's laboratory, it is also disliked because it is actually harder to load than the inert gases. This may seem surprising, but it is because it is loaded by hand, with the sample and gasket exposed on one diamond. A drop of liquid is carefully placed on the sample hole and the cell has to be assembled and clamped before the liquid evaporates. The drop can easily run out, or wash the sample away, or an air bubble can easily remain trapped. Some workers cool the cell with ice before doing alcohol loading. In contrast, inert gas loading is entirely automatic, done in a loading jig, and requires no delicate manipulations by the operator.

Argon is a very convenient medium because its boiling point is rather above that of liquid nitrogen. A cell which provides for pressure changes at low temperature need only be loaded with the samples, assembled but without any force on the diamonds, and inserted in a can filled with argon gas. This is condensed with liquid nitrogen, and then the load is applied to the cell (Spain and Dunstan, 1989). Helium is loaded similarly, but requires a proper bath cryostat, rather than the uninsulated can which is all that argon loading requires.

Burnett et al compare the axial strains observed in ruby and in a GaAs/AlGaAs quantum well by the broadening of photoluminescence lines under pressure. They conclude that xenon is much worse than argon, methanol-ethanol and helium. Helium gives no perceptible broadening—that is, no perceptible non-hydrostatic behaviour. They make all their pressure changes at room temperature; pressure changes at low temperature result in increased broadening in all cases except with helium-loading.

As remarked above, we have not yet required to change to helium loading, even in studies of quantum well samples similar to those in which Burnett et al (1990) observe significant broadening with argon. This is all the more surprising since we routinely make our pressure changes at cryogenic temperature. Shear strains in the medium may however depend critically on minor details of the experimental procedure. We flush the loading can with argon and pump it out before filling, to maintain the purity of the argon. Then we carry out experiments only while increasing pressure in the cell. This is important since we use thin gaskets, as defined in Section 2.2.2 (see Figure 4). The sample hole in a thin gasket does not change its shape very much as pressure is raised, as both the diamonds and the circumference of the hole are all moving inwards. This will minimise shear strains. A thick gasket has an expanding circumference. While decreasing pressure the cell must leak—how it does so is not clear, but there is no other way for it to lose pressure—and it is likely that it does so asymmetrically. In any case, loss of argon from the corner of the sample hole must set up large shear strains. Burnett et al remark that solid helium contracts by a factor of four between 50 bar and 20 kbar at 4K; while the pressure medium is contracting so fast the gasket will be thin as the sample hole contracts. This should be taken into account when designing the gasket.

Shear strains in the pressure medium are also related to the reliablility of pressure gauges (see Section 3). A medium that can support a shear strain is also capable of supporting a hydrostatic pressure gradient. In summary, I recommend using argon as the pressure medium, provided that data is only recorded while increasing pressure, and that the sample is also the pressure gauge (see Section 3), and preferably that the experiment is capable of detecting shear strains of a magnitude that would vitiate the results. If these conditions are not met, the added complexity of helium loading becomes well worth while.

3. Pressure Calibration

The calibration of the hydrostatic pressure and the detection of non-hydrostatic stresses in the sample are perhaps the most controversial subject in DAC work. For some years, the shift of the R1 and R2 photoluminescence lines of ruby (Al_2O_3:Cr) has been considered to be the primary pressure gauge for the DAC—indeed, one of the easiest ways to get an experimental paper rejected by the referees is to use another gauge instead of ruby. The pressure-induced shift in the ruby luminescence has been carefully calibrated by Barnet et al (1973) and is 0.365 nm/GPa (linear for work below say 20GPa; at higher pressures a non-linear term should be added—see Bell et al, 1984). The advantages of the ruby scale are that ruby is reproducible, that PL in the DAC is an easy experiment, and that only a very small amount of ruby need be used. It is indeed common to scatter several tiny pieces of ruby around the sample space, to detect non-hydrostaticity (Bell et al, 1984). For these reasons, ruby probably is the ideal pressure gauge, in the pressure range of about 10-200GPa. Outside this range, other gauges are preferable.

3.1 CALIBRATION AT LOW PRESSURES

In the low pressure range, below 10GPa, the ruby shift is simply too small for accurate work. It requires a very good spectrometer, and experienced users, to measure the wavelength of a PL line to the accuracy of 0.01nm required just to get the pressure to better than 1kbar (0.1GPa). Yet this is only 1% accuracy at 100kbar. Barnet et al (1973), taking extreme precautions with a good spectrometer, still obtained only a reproducibility of 0.3kbar. Worse still, the ruby and the sample are spatially separated, and so non-hydrostaticity can appear as a pressure error. At cryogenic temperatures, where the pressure medium is solid, it is possible for strains in the pressure medium to give different hydrostatic pressures in the sample and the ruby simply because of their different elastic moduli and geometries. In the author's view, it is far preferable to use a more sensitive gauge, and, moreover, one that can be integrated with the sample under study.

An example is the use of semiconductor gauges when studying semiconductors under pressure. Direct semiconductor band-gaps are much more sensitive to pressure than ruby (values typically of 100 nm/GPa), and a suitable semiconductor is often integrated with the sample by modern epitaxial growth techniques. Some workers dislike this approach because the values of pressure coefficient for a given semiconductor appear to be sample dependent—at least, the values given in the literature are widely dispersed, outside the claimed experimental error. Take, for example, GaAs. Values in the literature for its direct gap pressure coefficient range from 9.6 meV/kbar to 12.6 meV/kbar (see Venkateswaran, 1986, and references therein); some values are plotted in Fig.7.a against time so that the effect of improvements in techniques may be judged, compared with sample-to-sample scatter.

The lattice constant of GaAs (at 0GPa) is given in the data books to six significant figures, as 5.65325(2)Å at 300K. The band-gap is given to four significant figures, 1.424(1)eV at 300K, and to six figures, 1.51914eV at 0K. These values are properties of the bulk perfect crystal, and are affected in only very small, known ways by normal levels of impurities and defects. Now, the pressure coefficient of the band-gap is due firstly to the elastic constants which relate the lattice constant to the pressure, and secondly to Schrödinger's equation which in principle yields the band-gap from the positions and potentials of all the atoms in the crystal. It is completely implausible that the band-gap and lattice constant at one pressure (0GPa) should be virtually sample-independent and yet that at any other pressure either should vary as much as shown in Figure 7.a. And most of the ruby data points in Figure 7.a. obtained using ruby fluorescence as

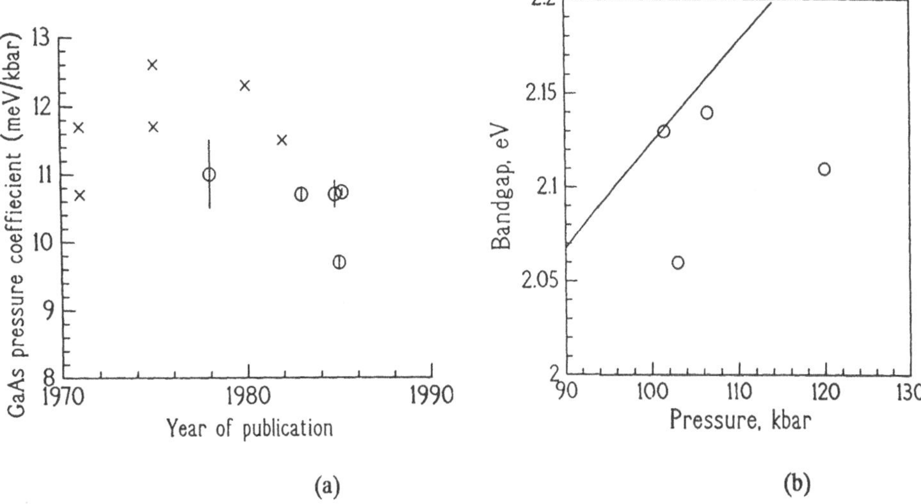

Figure 7. In (a), some values from the literature for the pressure coefficient of the band-gap of bulk GaAs are plotted against the year of publication. The data point indicated by circles correspond to values obtained in the diamond-anvil cell using ruby calibration and the eror bars on these data are those of the authors; the crosses were obtained by other techniques. In (b), the band-gap energy of InP at the phase transition is plotted against the phase transition pressure. The solid curve is the accepted dependence of the band-gap on pressure.

the pressure gauge agree on a value of 10.7meV/kbar within error, giving no evidence for sample to sample scatter. From this data, the problem with ruby calibration is the size of the experimental error, which makes small differences in pressure coefficient in different structures undetectable (Venkateswaran et al, 1986), and the evidence for sample-to-sample scatter in GaAs is weak as well as theoretically unlikely.

In Figure 7.b we have plotted the available data on the phase transition of InP under pressure. The solid line shows the dependence of the band-gap on pressure; the data points show both the band-gap and the pressure at which the phase transition was observed. Now, an unvarying dependence of the band-gap on lattice constant with sample-to-sample variation of the phase transition pressure should leave all the data points on the curve. A constant phase transition pressure and sample-to-sample variation in the pressure dependence of the band-gap would leave the data points in a vertical straight line. The most plausible interpretation of this data is that the phase transition pressure can vary by some ten kilobar, being reduced by axial strains in the pressure medium, that the band-gap energy is the most reliable pressure gauge, and that the ruby can easily disagree with the sample by again some ten kilobar. This data is a strong argument for using helium as the pressure medium for phase transition experiments (see Section 2.4), but again is not evidence that semiconductors show a sample-to-sample variation in their band-gap pressure dependence.

The use of a semiconductor gauge is illustrated in Figure 8. This is taken from a study in which we wished to investigate the pressure coefficients of quantum wells as a function of

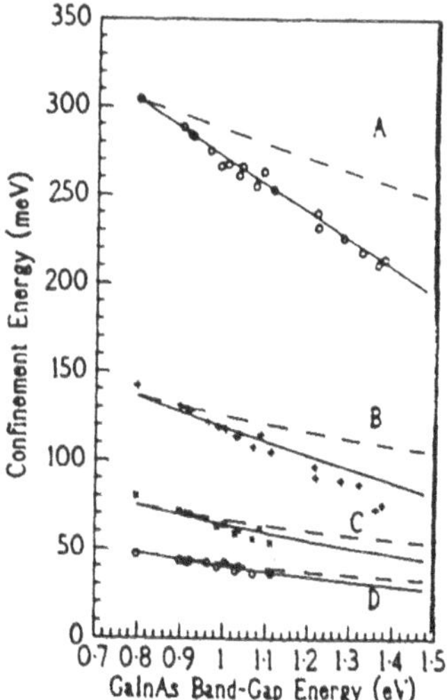

Figure 8. The confinement energies, the difference between the quantum well band-gap energy and the bulk InGaAs band-gap energy, is plotted against the latter. The curves are various theoretical fits. *Reproduced from Lambkin et al (1988).*

quantum well width (Lambkin et al, 1988). In the limit of wide wells, the coefficient is that of the bulk well material because the confinement energy is small and any changes with pressure can be neglected. In the limit of very narrow wells, the coefficient must approach that of the bulk barrier material since the limit of a narrow well is a non-existent well. In between these limits, there is much interesting physics. However, the range of pressure coefficients expected is less than the error bars in Figure 7.a for a bulk semiconductor.

By integrating the different quantum wells under study into a single sample, *differences* in the pressure coefficients may be measured very accurately (Venkateswaran, 1986). The photoluminescence experiment samples a large area of the sample (about 50μm square). Within this, the various quantum wells are separated by only a few hundred Angstroms. They all therefore experience the same strain environment—if one is subject to a uniaxial stress, or a hydrostatic pressure gradient, then so are the others. Differences in their behaviour under pressure are therefore real, and not attributable to pressure inhomogeneities. Plotting their emission energies, not against pressure, but against each other, shows up these differences very sensitively. This may be seen in Fig.8, where the energy differences between InGaAs quantum wells and bulk InGaAs is plotted against the energy of the InGaAs; pressure is a dummy variable. Over a pressure range in which the bulk InGaAs shifts 300meV, to 1.1eV, the quantum well D has shifted only about 10meV less (the difference has dropped from 50 to 40meV)—that is, its pressure coefficient is only 3% less. Yet this difference can be measured from Fig.8 accurately.

3.2 CALIBRATION AT HIGH PRESSURES

Ruoff and coworkers have offered two major criticisms of the ruby scale at multi-megabar pressures (above 250GPa). As the ruby PL moves to higher energy (shorter wavelength), the band-gap of diamond moves to lower energy, and the two cross at about 3eV at a pressure of 250GPa. Above this pressure, the diamond is opaque to the ruby emission and ruby is therefore useless as a pressure gauge. Ruoff et al (1991a) give the formula

$$E_L(eV) \leq 3.95 - 0{\cdot}0051\,P(GPa) \qquad [6]$$

as the high-energy, short-wavelength limit for optics in the DAC. They suggest that authors who have reported pressures above 250GPa using ruby have mistaken another emission for ruby PL.

Secondly, Ruoff and Vanderborgh (1991) point out that ruby becomes chemically reactive at high pressure. In particular, they suggest that in experiments to find metallic hydrogen at high pressure, the apparent metallisation observed is in fact due to a reaction between the hydrogen and the ruby, resulting in the deposition of free aluminium metal in the cell.

As a satisfactory calibrant at multi-megabar pressures, Ruoff and coworkers use measurement of the lattice parameter of a metal using X-ray diffraction. A major difficulty with this method is the need to collimate the X-ray beam to a diameter of a few microns, to obtain the required spatial resolution. To obtain a reasonable intensity despite such collimation, Ruoff and coworkers use a synchroton source. Using a gasket made of molybdenum, without a sample hole, and the Birch I equation of state to determine the pressure. Ruoff (1992) was able to demonstrate the probable generation of over 5Mbar (560 GPa) at the centre of the gasket.

3.3 CALIBRATION—THE MORAL

While it is agreeable to incorporate a reliable and accurate pressure gauge in an experiment, it is certainly not generally a scientific necessity. Notice that in the semiconductor experiment discussed in Section 3.1 the pressure does not need to be known, for the science is done by relating one band-gap shift to another. In interpretation of band-edge crossovers, Wolford et al (1986) and Venkatswaran al (1986) were concerned only with the energies of the optical transitions, and the pressure merely provides a convenient x-axis for their graphs.

In Ruoff's experiments, what matters, and what he measures by X-ray diffraction, is the atomic separation or lattice constant of his gasket. Obtaining a pressure from that is only to satisfy curiosity. Similarly, in structural studies of crystalline phase transitions by X-ray diffraction or by Raman under pressure, the theoretician does not want to know the pressure—again, he wants to know the lattice constant as this is what appears in his theories. On the other hand, in geophysics, the pressure is the true variable of interest. Rock at some depth in the earth has properties which depend on the weight of the overlying strata—that is, the pressure.

The moral is therefore that the need for pressure calibration should be thought through carefully for each experiment, and that if it is not necessary then undue effort should not be wasted on it.

4. Experiments in the Diamond-Anvil Cell

4.1 HIGH-PRESSURE EXPERIMENTS.

Experiments that can be carried out in the DAC were reviewed by Jayaraman (1986) and Spain and Dunstan (1989). Optical experiments have always been the easiest, because the diamonds are transparent. Photoluminescence (PL) is the simplest, and is done routinely for ruby pressure calibration. It is worth noting that of the various spectroscopies that observe the density of states of a sample or its absorption rather than its emission, absorption is facilitated by the masking provided by laminated gaskets, as described in Section 2.2.3, and then it is as simple as PL.

Photoluminescence excitation spectroscopy is hampered by the need to use a tunable laser for excitation; a tungsten lamp cannot be focused down to obtain sufficient intensity incident on the small sample in the DAC. Tunable lasers have narrow wavelength ranges, around 200meV, and since the pressure in the DAC can move features of interest in the PLE spectrum of a semiconductor sample by an order of magnitude more than this the experiment with a laser becomes tedious as dyes have to be frequently changed. The group of Chandrasekar at Missouri have recently demonstrated that photoreflectance is possible (Rockwell et al, 1991); a tungsten lamp can be used for this since in contrast to PLE a bright specular reflection from the sample is used. The chief difficulty is in detecting and aligning the optics for the specular reflection from the sample rather than any of the other reflections from the diamond surfaces etc.

Measurements of importance to geophysics (high temperature at high pressure) can be done purely optically. Laser heating of a sample in a DAC can be used to give temperatures above 2000K at pressures up to a megabar for periods of hours. First used by Ming and Bassett (1974), laser heating has been improved recently in a number of ways (Boehler and Chopelas, 1991; Boehler, 1992). A wider choice of lasers shows the advantage of a wavelength that is heavily absorbed by the sample, for example the CO_2 laser for many minerals and oxides. It is preferable to record the emission spectrum of the heated sample rather than using a single-wavelength pyrometer. And a gasket geometry that holds the sample away from the diamond (a lipped gasket, similar to an optical absorption gasket) is an advantage as the argon pressure medium can then insulate the sample from the diamond.

4.2 UNIAXIAL STRAIN

Precisely those features of the diamond anvil cell which enable the diamonds to survive the stresses of high-pressure generation also make it well adapted to uniaxial strain generation. Conventional uniaxial strain cells usually have hardened-steel opposed anvils with no provision for alignment apart from the accuracy of fabrication, and indeed no provision for monitoring the alignment. Inaccuracy in the alignment leads to non-uniform stress in the sample and encourages sample fracture. In such an apparatus, Eliseev et al (1984, 1986) studied the performance of semiconductor lasers under axial strain, and found that both tensile strain and compressive strain are beneficial—reducing values of threshold current. They also found that the lasers were damaged or disintegrated at axial strains above about 2.5kbar. We have used a completely unmodified diamond anvil cell as a uniaxial strain generator also for measurements on semiconductor lasers, and we have been able to take lasers repeatedly to 8kbar without any damage (until the catastrophic failure at maximum load). An example is shown in Figure 9 (Jones, 1992). The laser was mounted between the diamonds so that the strain was in the growth direction; to provide electrical contacts to the top and bottom faces a pair of thin steel leaves (25μm thick) were inserted between the laser and the diamonds. The laser was a quantum well device, which has already some of the beneficial advantages of compressive biaxial strain in the active region. The minimum in performance found by Eliseev et al (1986) at zero strain therefore occurs here at about 7 kbar, and the improvement observed here above 7kbar corresponds to the improvement found by Eliseev for tensile strain.

A diamond-anvil cell used in this way provides above all accurate alignment of the anvils, in the usual way using the interference fringes between the culets. The hardness of the diamond also reduces deformation at the edges of the sample and so reduces strain non-uniformity in the sample (the bulk modulus of diamond is about 5Mbar, and of steel, only 2Mbar), although we are not taking advantage of this ourselves because of the steel contacting leaves. Diamond was

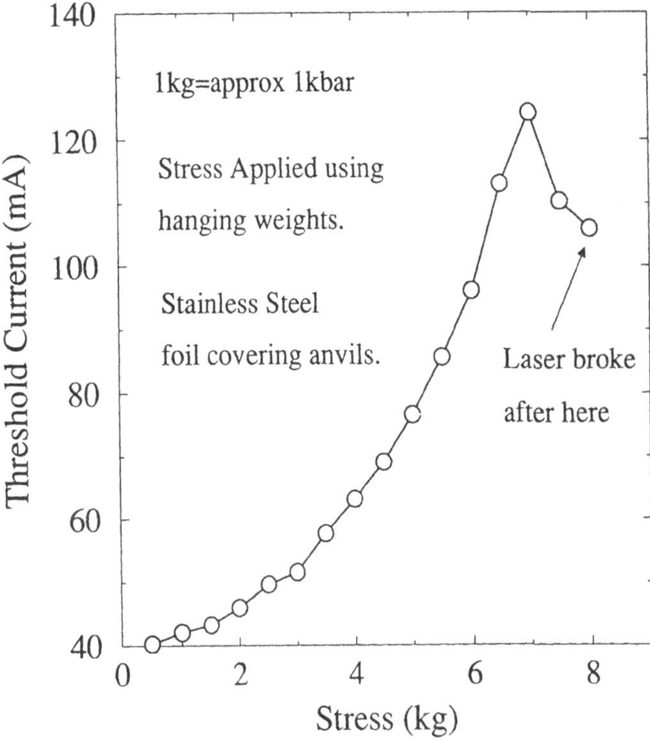

Figure 9. The variation of the threshold current of a quantum well InP-based semiconductor laser under uniaxial strain in a diamond-anvil cell. High threshold currents are undesirable, and the peak at 7kbar corresponds to the peak observed in bulk lasers at zero strain by Eliseev et al (1986). *From Jones (1992).*

also used for uniaxial anvils by Patel et al (1973) but without a DAC for controlled alignment. On the other hand, a DAC is built for tonne forces rather than a few kilograms, and piston-cylinder friction becomes important in calibrating the stress. Nevertheless, this result—the relatively high pressure reached without damage to the laser—shows that diamond-anvil cell technology could usefully be introduced into the design of uniaxial stress generators.

Acknowledgements

I am grateful to many other workers in the field for exchanging details of techniques, in particular to Dr D.M. Adams, Prof. J.M. Besson and Dr M. Leroux. As coworkers, I wish to thank Dr A.D. Prins, Dr J.D. Lambkin, Dr V.A. Wilkinson, D. Leong, J. Sly and H. Feyrit. I am grateful to G. Jones for providing the uniaxial strain data of Fig.9 prior to publication. I would particularly like to acknowledge the generosity and enthusiasm with which the late Prof. I.L. Spain introduced me to diamond-anvil cell technology and experimentation.

References

Adams, D.M. (1992a) *High-temperature high-pressure diamond anvil cells,* in the Collected Abstracts of the HPG-92 IUCr Workshop on X-Ray and Neutron Diffraction at High Pressures (Washington DC, 1992).

Adams, D.M. (1992b). See the brochures of Diacell Ltd.

Adams, D.M., Christie, A.G. and Newman, A.J. (1992) *Computer-assisted design optimisation of diamond-anvil cell performance,* Measurement Sci. & Tech. (to be published).

Barnet, J.D., Block, S. and Piermarini, G.J. (1973) *An optical fluorescence system for quantitative pressure measurement in the diamond-anvil cell,* Rev. Sci. Instrum. **44**, 1-9.

Baublitz, M.A., Arnold, V. and Ruoff, A.L. (1981) *Energy dispersive x-ray diffraction from high pressure polycrystalline specimens using synchroton radiation,* Rev. Sci. Instrum. **52**, 1616-1624.

Bell, P.M., Mao, H.K. and Goettel, K. (1984) *Ultrahigh pressure: Beyond 2 megabar and the ruby fluorescence scale,* Science **226**, 542-544.

Boehler, R. (1992) *Advances in high temperature research in a diamond cell,* Proceedings of the XIII AIRAPT Conference on High Pressure Science and Technology (Bangalore, 1991) 591-600).

Boehler, R. and Chopelas, A. (1991) *A new approach to laser heating on high pressure mineral physics,* Geophys. Rev. Lett. **18**, 1147-1150.

Burnett, J.H., Cheong, H.M. and Paul, W. (1990) *The inert gases Ar, Xe and He as cryogenic pressure media,* Rev. Sci. Instrum. **61**, 3904-3905.

Bridgman, P. (1952) *The Physics of High Pressure* (Bell, London).

Dunstan, D.J. (1989) *Theory of the gasket in diamond anvil high-pressure cells,* Rev. Sci. Instrum. **60**, 3789-3795.

Dunstan, D.J. and Scherrer, W. (1988) *Miniature diamond-anvil high-pressure cell,* Rev. Sci. Instrum. **59**, 627-630.

Dunstan, D.J. and Spain, I.L. (1989) *The technology of diamond anvil high-pressure cells: I. Principles, design and construction,* J. Phys. E: Sci. Instrum. **22**, 913-923.

Dunstan, D.J. and Wilkinson, V.A. (1990) *Miniature cryogenic diamond anvil cell,* High Pressure Research **5**, 794-796.

Eliseev, P.G., Sverdlov, B.N. and Shokhudzhaev, N. (1984) *Reduction of the threshold current of InGaAsP/InP heterolasers by unidirectional compression,* Sov. J. Quantum Electron. **14**, 1120-1121.

Eliseev, P.G., Sverdlov, B.N., Ismailov, I. and Shokhudzhaev, N. (1986) *Influence of anisotropic deformation on the radiative characteristics of GaInAsP/InP injection lasers. I. Lasing threshold, polarisation, and watt-ampere characteristic,* Sov. J. Quantum Electron. **16**, 1046.

Eremets, M.I., Krasnovskij, O.A., Struzhkin, V.V., Timofeev, Ju.A. and Shirokov, A.M. (1990) *Method of low-temperature optical measurements with diamond anvil cells,* High Pressure Research **5**, 880-882.

Eremets, M.I. and Shirokov, A.M. (1991) *Miniature diamond anvil cells,* Proceedings of the XIII AIRAPT Conference on High Pressure Science and Technology (Bangalore, 1991) 763-765.

Eremets, M.I. and Timofeev, Ju.A. (1990) *Diamond anvil microcell,* Proceedings of the IV International Conference on High Pressure in Semiconductor Physics (Chalkidiki, Greece) ed. D.S. Kyriakos and O.E. Valassiades (Aristotle University, Thessaloniki, Greece), p.156.

Grzybowski, T.A. and Ruoff, A.L. (1984) *Band-overlap metallisation of BaTe*, Phys. Rev. Lett. **53**, 489-492.

Hemmes, H., Driessen, A., Kos, J., Mul, F.A., Griessen, R., Caro, J. and Radelaar, S., (1989) *Synthesis of metal hydrides and in situ resistance measurements in a high-pressure diamond anvil cell*, Rev. Sci. Instrum. **60**, 474-480.

Jayaraman, A. (1986) *Ultrahigh pressures*, Rev. Sci. Instrum. **57**, 1013-1031.

Jones, G. (1992) private communication.

Kobayashi, T. (1985) *Diamond-anvil high-pressure cell for optical spectroscopy at low temperature*, Rev. Sci. Instrum. **56**, 255-259.

Lambkin, J.D., Dunstan, D.J. and O'Reilly, E.P. (1988) *The pressure dependence of the band offsets in a GaInAs/InP multiple quantum well structure*, J. Crystal Growth **93**, 323-328.

Leong, D., Feyrit, H., Prins, A.D., Wilkinson, V.A., Homewood, K.P. and Dunstan, D.J. (1992) *Laminated gaskets for absorption and electrical measurements in the diamond anvil cell*, Rev. Sci. Instrum. (in press).

Ming, L.C. and Bassett, W.A. (1974) *Laser heating in the diamond anvil press up to 2000°C sustained and 3000°C pulsed at pressures up to 260 kilobars*, Rev. Sci. Instrum. **45**, 1115-1118.

Patel, N.B., Ripper, J.E. and Brosson, P. (1973) *Behaviour of threshold current and polarisation of stimulated emission of GaAs injection lasers under uniaxial stress*, IEEE J. Quant. Electr. **9**, 338-341.

Rockwell, B., Chandrasekhar, H.R., Chandrasekhar, M., Ramdas, A.K., Kobayashi, M. and Gunshor, R.L. (1991) *Pressure tuning of strains in semiconductor heterostructures: (ZnSe epilayer)/(GaAs epilayer)*, Phys. Rev. B**44**, 11307-11314.

Ruoff, A.L. (1992) *X-ray diffraction in the 500GPa range: Prospects for attaining 1TPa*, Proceedings of the XIII AIRAPT Conference on High Pressure Science and Technology (Bangalore, 1991) 769-778.

Ruoff, A.L., Luo, H. and Vohra, Y.K. (1991a) *The closing diamond anvil optical window in multimegabar research*, J. Appl. Phys. **69**, 6413-6416.

Ruoff, A.L., Luo, H., Vanderborgh, C.A. and Vohra, Y.K. (1991b) *Generating near-earth-core pressures with type IIa diamonds*, Appl. Phys. Lett. **59**, 2681-2682.

Ruoff, A.L. and Vanderborgh, C.A. (1991) *Hydrogen reduction of ruby at high pressure: Implications for claims of metallic hydrogen*, Phys. Rev. Lett. **66**, 754-757.

Schroeder, W. and Webster, D.A. (1949) *Press-forging thin sections: Effects of friction, area and thickness on pressures required*, J. Appl. Mech. **16**, 289-294.

Spain, I.L. and Dunstan, D.J. (1989) *The technology of diamond anvil high-pressure cells: II. Operation and use*, J. Phys. E: Sci. Instrum. **22**, 923-933.

Sung, C.M. (1976) *New modification of the diamond anvil press: A versatile apparatus for research at high pressure and high temperature*, Rev. Sci. Instrum. **47**, 1343-1346.

Venkateswaran, U., Chandrasekhar, H., Chandrasekhar, M., Vojak, B.A., Chambers, F.A. and Meese, J.M., (1986) *High pressure studies of GaAs-$Ga_{1-x}Al_xAs$ quantum wells of widths 26 to 150Å*, Phys. Rev. B**33**, 8416-8423.

Wolford, D.J., Keuch, T.F., Bradley, J.A., Gell, M.A., Nimmo, D. and Jaros, M. (1986) *Pressure dependence of GaAs/$Al_xGa_{1-x}As$ quantum-well structures: The determination of valence-band offsets*, J. Vac. Sci. Tech. B**4**, 1043-1050.

APPLICATIONS OF DIAMOND-ANVIL CELLS TO MATERIALS SCIENCE

D.J. DUNSTAN
Department of Physics
University of Surrey
Guildford,
Surrey, GU2 5XH
England

ABSTRACT. Diamond-anvil cells allow the study of condensed matter as the interatomic spacing is changed over a wide range. This causes both quantitative and qualititative changes in material properties. Semiconductors undergo phase transitions at relatively low pressures, and they provide many examples of the use of high pressure in the range up to 10GPa. The pressure range up to 100GPa is widely used and examples are discussed from geophysics and fundamental condensed matter physics. The multi-megabar region (over 100GPa) is still difficult and some of the problems in working at these pressures are reviewed.

1. Introduction

Compared with other methods of generating high hydrostatic pressures, the diamond-anvil cell (DAC) offers only a tiny experimental volume, capable of taking a microgram or less of sample. In return, it provides both a dramatically higher pressure range, and the ability to carry out a very wide range of experiments in standard apparatus. An illustration of the flexibility in use is the world record static pressure of over five megabar, in a diamond-anvil cell, measured by X-ray diffraction at a synchrotron (Ruoff, 1991a). Any other high-pressure equipment capable of being transported to a synchrotron facility would be limited to tens of kilobar. Not every diamond-anvil cell can be taken up to five megabar routinely, of course, but as another illustration of the flexibility and convenience of the DAC, I have travelled by air with, in my pocket, a loaded and pressurised diamond-anvil cell for an optical experiment in a foreign laboratory (Dunstan et al, 1988). This would be illegal as well as unwise with any other form of high-pressure generator.

To what use in materials science can this tremendous pressure range be put? Briefly, the value of hydrostatic pressure as a perturbation is that it varies the lattice constant or interatomic separation of a material. Whether one is calculating the electronic band structure of a semiconductor, or predicting what crystal structure a chemical compound should adopt, the lattice constant will appear in the theory. It is relatively difficult to check whether constants are correctly included in a theory; however, if the constants are made to vary, the theoretical predictions for the dependence of any other observable on the constant may be verified. Nature provides four ways to vary the lattice constant of a compound. Replacing one of the elements, moving up or

R. Winter and J. Jonas (eds.), High Pressure Chemistry, Biochemistry and Materials Science, 101–119.

down the Periodic Table, gives discrete changes together with a chemical change that may be due only to the change in atomic size but may also be due to the extra filled orbitals or the change in nature of the valence electrons. Alloying, partially replacing an element by one higher or lower in the Table, makes the change in lattice constant continuous instead of discrete, but still has the chemical change of uncertain effect and also adds any effects of alloy disorder. Changing the temperature of a sample changes the lattice constant through thermal expansion but also changes the phonon density which can be a more profound effect. In semiconductors, for example, roughly 80% of the change in band-gap with temperature is due to the electron-phonon interactions, and only 20% is due to the thermal expansion. Only hydrostatic pressure allows the lattice constant to be changed continuously, without changing anything else, in a single sample.

Any experiment that is worth doing at a fixed lattice constant is worth doing as a function of lattice constant. There are of course some constraints. Not every experiment can be done under pressure, notably surface science of the kind normally done under high vacuum. But if an experiment can be done under high pressure, it is safe to say that something interesting will come out of it, something that will clarify the understanding of the ambient pressure science. There is, consequently, far more work on materials under high pressure than can possibly be adequately reviewed in one short paper, and so I have chosen to pick, somewhat arbitrarily, three areas of work only. This is not to suggest that others are of lesser importance.

A large body of work on semiconductors exists, most of which has been carried out under 100kbar—this pressure range is readily accessible to non-specialists and includes many interesting effects. Some examples are discussed in Section 2. The pressure range up to 1Mbar has been used for scientific study of a very wide range of materials and properties and we review some of the work on the simplest atoms and molecules briefly in Section 3. Finally, the multi-megabar pressure range has been applied to a few problems but the major challenge is the pressure experiment itself. High pressure techniques are outside the scope of this article, and so I make only brief reference to the work in this pressure range in Section 4.

2. High pressure studies of semiconductors

Semiconductors are the materials underlying the information revolution, as coal and iron underlay the industrial revolution. For applications, it is usually their properties at ambient pressure and temperature which are required, but to understand them, experiments at low temperature or high pressure can be very helpful. We begin, in Section 2.1, with a discussion of some high-pressure experiments which are done for this purpose, and then go on to work of more purely high-pressure interest.

Much of the middle of the Periodic Table provides semiconducting elements and compounds with a very simple crystal structure. Diamond, silicon, germanium and one form of tin all take up the diamond structure. Most of the III-V and II-VI compounds and alloys, such as GaAs and InGaAsP, ZnS and ZnSSe, have the same crystal structure except that alternate atoms are cations and anions; this is the zinc-blende structure. (It should be remarked that many of the II-VI compounds can have the (111) planes stacked differently so as to create the wurtzite hexagonal crystal structure; this is not very different from the zinc-blende structure.) All of these compounds and alloys have very similar properties, varying only quantitatively with average atomic weight and with ionicity. In particular, the electronic band structure is very similar in all these elements, compounds and alloys, and in fact it turns out to depend largely on lattice constant, secondarily on ionicity, and very little on any other aspect. The strong dependence on lattice constant makes high pressure a very useful tool to study these materials.

2.1. BAND CROSSOVERS IN SEMICONDUCTORS

2.1.1 *Semiconductor Band Structure.* The band structure of a tetrahedral semiconductor has a maximum in the valence band at the Brillouin zone centre, the Γ point, but has three minima in the conduction band at the Γ point and at or near the X and L points (Fig.1.a). Which of these conduction-band minima lies lowest in energy determines whether the semiconductor is direct-gap and interacts strongly with light (useful in optoelectronics), or indirect and useful only for microelectronics. In the II-VI compounds and most of the III-V's, the Γ minimum lies lowest and the semiconductors are direct. In silicon the six X minima are lowest, while in germanium the four L minima are lowest, and these two materials are indirect semiconductors.

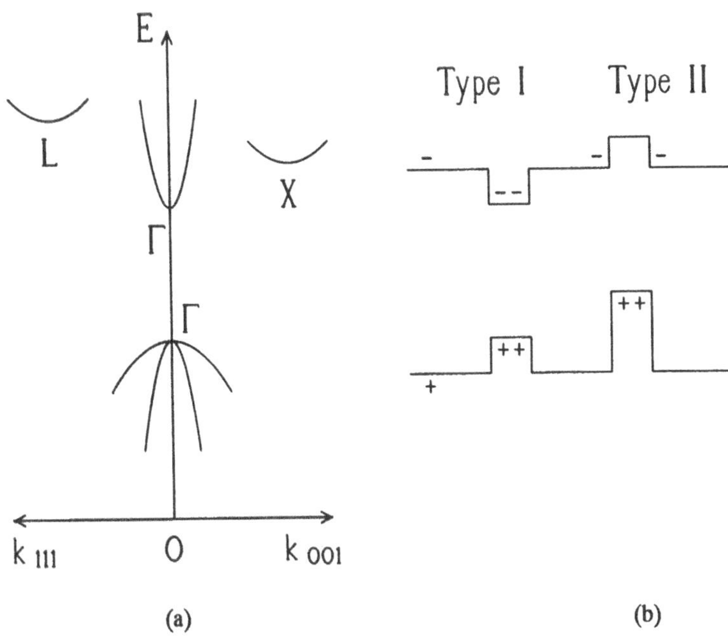

Figure 1. The band structure of a typical semiconductor is shown schematically. In (a), the allowed electron energies are plotted against momentum, *k*, in two directions, in the neighbourhood of the conduction and valence band extrema. The diagram shows the conduction band Γ minimum lower than the X and L minima so that this is an indirect semiconductor. In (b), the energies of the conduction and valence band edges are plotted against position for two of the structures discussed in the text, a Type I and a Type II quantum well. Some conduction-band electrons (-) and valence-band holes (+) are marked.

Under pressure, the valence band maximum moves little (relative to an absolute energy scale) and for any experiments in which only relative movements between the conduction and valence bands are important, the valence band maximum may be taken as a stationary energy reference point. The three conduction-band minima move at characteristic rates with pressure:~10meV/kbar for the Γ minimum, 5mev/kbar for the L minima and -2meV/kbar for the X minima (Paul and Warschauer, 1963). These values vary only by about a factor of two in different compounds. Evidently, under pressure, the semiconductor can be changed dramatically. A direct-gap compound can have its band-gap changed by about 1eV under the modest pressure of 100kbar. Even more dramatically, it may be driven indirect if the Γ minimum is driven up above the X or L minima. Similarly, germanium under pressure becomes like silicon when the L minima rise under pressure above the X minima. Only under pressure can the effects of the nature of the atoms on the electronic band structure be distinguished from the effects of the lattice constant.

Two very important applications of high pressure have been studied. The positions of the indirect minima in direct-gap semiconductors can be unambiguously determined to high accuracy. And the relative positions of the band edges in different materials, crucial to the properties of heterostructures, can likewise be determined unambiguously and accurately.

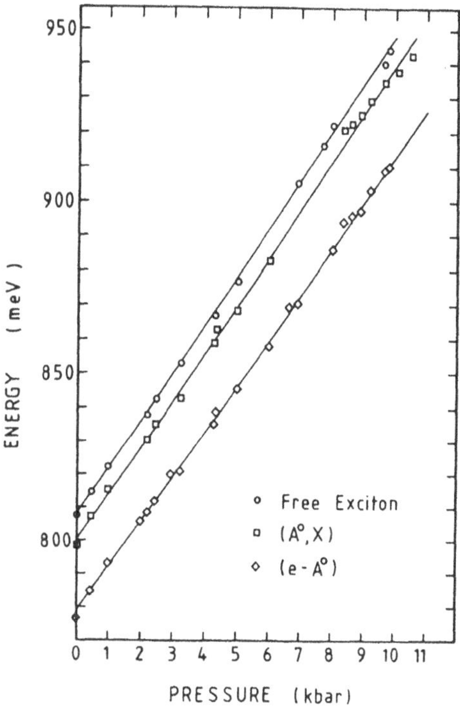

Figure 2. A plot of the photoluminescence peak energies in bulk GaSb against pressure. No luminescence could be observed above 10.7 kbar. *Reproduced from Warburton et al. (1991).*

2.1.2. *Indirect Minima in Bulk Semiconductors.* Prior to high-pressure determinations, the positions of the indirect minima in direct gap materials had to be calculated theoretically, guessed from reflectance spectroscopy in which the identification of the features was uncertain, or deduced indirectly from the performance of devices or structures in which these minima play a rôle. An example is the Gunn diode, made from GaAs, in which oscillations occur due to the transfer of electrons from the Γ minimum to the higher minima under a high electric field. The reliability of these methods is illustrated by the long-standing controversy whether the X or L minima in GaAs were the lower.

High pressure can determine the positions of the higher minima accurately if they can be observed once the Γ minimum rises above them. This may require the use of heterostructures. An example is the pressure dependence of the band-to-band photoluminescence of GaSb, shown in Fig.2. This Γ-related luminescence peak energy rises at 13.8meV/kbar but is quenched completely at about 10kbar due to crossover with the L minima. From this data, little can be said about the exact position of the L minima at ambient, and nothing about the X. A quantum well of InGaSb with GaSb barriers gives much more revealing results (Figure 3). The quantum confinement of the carriers prevents the luminescence quenching above crossover; now we observe luminescence due to electrons in the indirect minima recombining with holes in the valence band. At first the pressure coefficient is 3.9mev/kbar, an L-related value, and at higher pressure there is a second crossover and a pressure coefficient of -1.4meV/kbar, an X-related value. These parts of the curve may be extrapolated back to 0kbar, and this extrapolation gives the 0kbar positions of the indirect minima to an accuracy of a few meV. Thus the high-pressure experiment is capable of both identifying the minima and measuring their energies with spectroscopic accuracy.

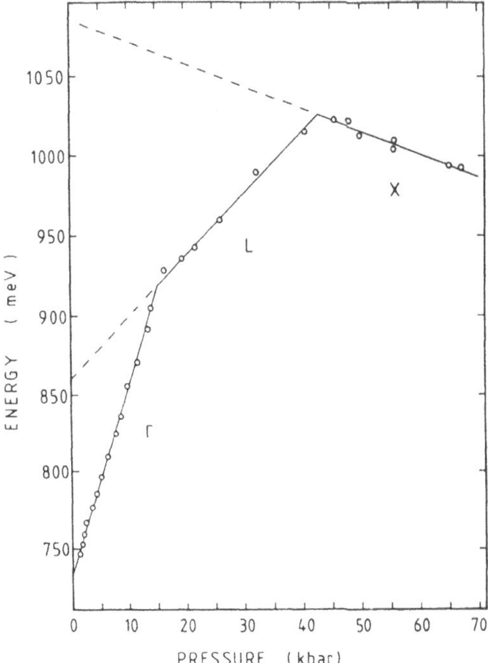

Figure 3. A plot of the InGaSb quantum-well emission energy against pressure. The transition changes from Γ to L to X character with increasing pressure. The dashed lines show how the data can be extrapolated to measure the ambient pressure indirect energy gaps. *Reproduced from Warburton et al (1991).*

In this case, of course, it is the L and X energies in InGaSb that has been measured, after correction for quantum confinement, not the values for GaSb. The method is quite general, however. For GaSb values, either several alloy compositions of InGaSb could be used and the results extrapolated to zero In content. Alternatively, quantum wells of GaSb in a larger gap barrier such as GaAsSb could be used and the GaSb X and L energies measured directly. It should however be noted that the method fails to obtain the position of a level which, as pressure is raised, never becomes the lowest—we shall see below that the L minimum in GaAs is an example.

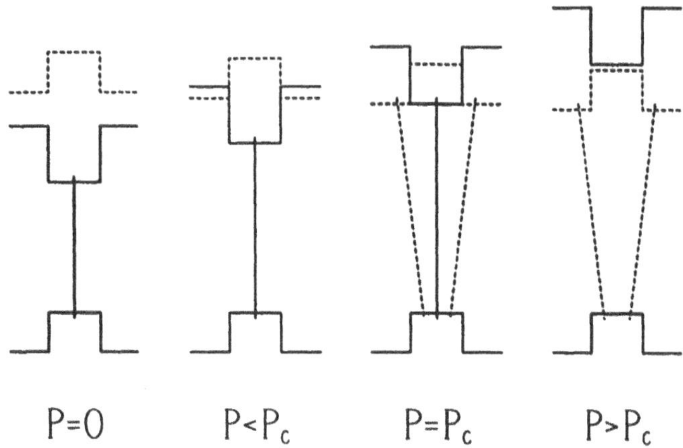

Figure 4. The principle of the high-pressure determination of band-offsets in (AlGa) is shown. At ambient pressure, the GaAs quantum well is Type I and the PL transition is direct in momentum space and in real space (vertical solid line). The X minimum, however, is Type II. The entire Γ conduction band edge rises with pressure, until at crossover ($P=P_c$) the Γ minimum in the well crosses the X minima in the barrier. Now the PL is indirect (slanting broken lines) and its pressure coefficient becomes negative. Extrapolation back to 0 kbar gives the indirect gap of the barrier material *less* the valence-band offset.

2.1.3. *Band Line-ups in Heterostructures.* Modern semiconductor epitaxial growth enables the chemical composition of a crystal to be changed from one atomic monolayer to the next. This allows the fabrication of structures with very thin layers in which quantum effects can be observed, quantum wells and superlattices (Fig.1.b and Fig.4). One key question that needs to be answered for each combination of semiconductors is the way in which their band structures line up against each other at such a junction. The two band-gaps are known, and of course their difference is known, but the proportion of this difference appearing in the step in the valence band and in the step in the conduction band is not known, and cannot be predicted theoretically with any certainty. Pressure experiments provide the most direct and reliable measurements of these band-offsets. In 1986, two groups reported determinations of the GaAs/AlGaAs band offset ratio using high pressure; Wolford et al found 32% ± 2% of the band-gap difference in the valence band and Chandrasekhar et al reported a value of 30% ± 4%. Previous determinations using other methods had ranged from 15% to 50%, the contrast with the unambiguity of the high pressure results and the good agreement between the two high-pressure values is striking.

The principle of the high pressure determination of band offsets is shown in Figure 4. The key point is that while the structure is Type I (see Fig.1.b) for the conduction band Γ minimum, providing quantum wells which trap both electrons and holes, it is Type II for the X minima so that the X minima in the AlGaAs barrier are below the X minima in the well. When the entire Γ band edge, throughout the structure, is lifted using pressure, the well minimum crosses the barrier X leaving the latter the lowest electron level in the structure. The photoluminescence peak changes character at this point to X-related and so moves down with further increases in pressure at the characteristic rate of about -1 meV/kbar. Extrapolation back to 0 kbar gives an energy which is just the indirect band-gap of the barrier material (which is known) less the valence band offset. Minor corrections have to be made for quantum confinement energy, but the theory required for this is well known and reliable.

For these experiments, superlattices have to be used. The $X_{barrier}$—Γ_{well} radiative transition is indirect both in real space and in momentum space, and so is forbidden and very weak. Using a superlattice both breaks the translational symmetry in real space and so weakens the selection rule in k-space, and helps to push the electron wavefunctions in the barrier into spatial overlap with the hole wavefunctions in the well.

This method gave such good results that it may seem surprising that no further determinations of band offset ratios between pairs of semiconductors have been made by the same method. This is because no other pair has yet been found in which the barrier X minimum occurs below the well X minimum. The InGaSb/GaSb system referred to above is an example of the more common situation—as seen in Figure 3, the well Γ minimum undergoes crossover against the well L minima and then the well X minima before reaching the barrier X or L minima. This gives therefore no information on the band offsets, except to put a minimum value on the valence band offset so that the barrier indirect minima are indeed above the well indirect minima. Other examples of this situation are GaSbAs/GaAs (Prins et al, 1991, 1992) and InGaAs/GaAs (Wilkinson et al, 1991).

Reliable values for other combinations of semiconductors in the III-V heterostructure system may nevertheless be obtained for alloy compositions in the vicinity of GaAs. Taking a GaAs/AlGaAs superlattice structure, with its crossover between the well Γ and the barrier X minima, small amounts of the other Group III or Group V elements can be alloyed into the well without changing the situation. In doing so, the photoluminescence at 0 kbar will be shifted by the change in the band-gap of the well. But the energy of the barrier X minima is unaffected by the composition of the well, and so the energy of the X-related emission above crossover will be shifted by the change in the valence band offset (Fig.5). A graph, for various amounts of the alloying element, of the energy of the X-related emission against the energy of the 0 kbar emission gives a straight line whose gradient is the band offset ratio of the alloy against GaAs. Minor corrections have to be made, as before, for quantum confinement energies etc. We have reported results using this method for the system InGaAs/GaAs, using InGaAs/AlGaAs superlattices, and obtained a ratio of Q_v=40% (Wilkinson et al, 1991). The principal difficulty of this method lies lies in the sample growth. Growth of GaAs/AlGaAs is very well developed and conditions for very high quality structures have been found; this is necessary to obtain clear, bright, indirect X-related emission above crossover. Adding a second alloy to the structure creates difficulties of growth optimisation which are by no means trivial, and which have been studied far less than optimisation of the GaAs/AlGaAs binary-ternary alloy system.

The model-solid theory of van de Walle and Martin (1987) is frequently used to predict band offset ratios that have not yet been measured unambiguously. Their theory has two parts. First, the absolute energy positions of the valence band edges of the binary compounds are calculated. These values can be adjusted as necessary to agree with experiment where data is available. Secondly, absolute energies for alloy valence band edges are obtained by treating the alloy as a superposition of two binaries, the binary with the smaller lattice constant being dilated and the

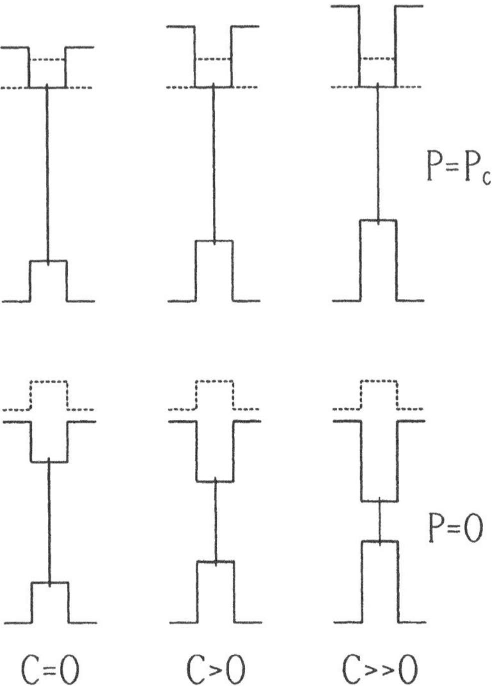

Figure 5. The principle of the method for finding the band offsets between GaAs and GaAs-based alloys where there is not a Γ-well—X-barrier crossover. Various concentrations of the alloying element are introduced into a GaAs quantum well in AlGaAs barriers. The PL energy at 0 kbar is reduced by the shifts of both the conduction and valence band edges; this is shown in the band-structures in the bottom row for none of the alloying element (*C*=0) and for two other concentrations. The band-structure at crossover is seen in the upper row, and the PL energy at crossover is reduced as *C* increases by the shift in the valence band only. The three crossover pressures in the top row are different; in this diagram we have neglected the small shifts of the X minima with pressure and with alloying.

larger one being compressed. The binary valence-band edges are moved according to the dilation or compression, and then the alloy value is found by linear interpolation (weighted averaging). The valence band offset for the heterojunction is the difference between the absolute values for the two materials, and can be expressed as a proportion of the band-gap difference since the bulk band-gaps are known. This theory predicts transitivity—the valence band offset between A and B, and between B and C add up to the offset between A and C. Transitivity can be verified by the high-pressure experiments above. By adding indium to the GaAs wells, and by adding phosporus, the InGaAs/GaAs and GaAsP band offsets can be found. By adding indium and phosphorus simultaneously, the InGaAsP/GaAs offset is obtained. If transitivity is obeyed, this value will be deducable from the InGaAs and GaAsP values; then extrapolation to pure InP would be justified to obtain an InP/GaAs offset for comparison with the binary value given by the first part of the van de Walle theory.

2.2. BULK MODULI OF SEMICONDUCTORS

Any theoretical analysis of the results of high-pressure measurements requires a knowledge of the bulk modulus of the material, to relate pressure to volume change. High-pressure measurements provide several ways to measure the bulk moduli of materials; three methods using pressures in the 100kbar range have been applied to semiconductors. The most reliable traditional way to obtain the elastic moduli of solids is to study the propagation of acoustic waves, but this method has two drawbacks. It uses small deformations only, so that while it can give very accurate values for the elastic constants or moduli at zero pressure, second-order elastic constants, differential moduli, or the change in the elastic properties with pressure are less well established. Secondly, acoustic methods require a bulk sample. Many modern semiconductor compound and alloy materials are available only as very thin epitaxial layers grown epitaxially on a substrate of a single crystal with a similar lattice constant.

The simplest way to measure the bulk modulus of a sample in a DAC is to observe its contraction under pressure. This requires a bulk sample, rather than an epitaxial layer, but is unambiguous and straightforward. An example is the study make by Tanaka and Maeda (1986) of selenium, in which they observed the contraction of the sample under pressure with a microscope. They improved the accuracy of the experiment by digitising the image and carrying out some image processing. The stiffening of the crystal with pressure was very clearly observed, so that they obtained the bulk modulus B and the differential bulk modulus B' to an accuracy of nearly 1%.

This approach is not applicable to epitaxial semiconductor layers, for the in-plane contraction of the sample is dominated by the substrate. Prins and Dunstan tried to measure the bulk moduli of epitaxial layers of InGaAsP and InGaAs grown lattice-matched on InP by comparing samples constrained by their substrates against detached layers. These layers were lattice-matched and therefore unstrained at ambient pressure, and the strain introduced by any difference between the contraction of substrate and the layer would have a hydrostatic component. The shift of the band-edge luminescence with pressure would then be different for the detached and substrate-constrained sample. Figure 6 shows that no difference outside experimental error could be observed (Prins and Dunstan, 1988, 1989). This result was in accordance with Keyes' scaling law, an empirical rule which states than in a system of chemically similar compounds such as the III-V compounds the bulk modulus depends only on the lattice constant. It was, however, unexpected that Keyes' rule was so accurately obeyed.

The sensitivity of this kind of experiment could be increased considerably if the strain in the layer on its substrate could be measured, instead of the difference between the mounted and the detached samples. Rockwell et al (1991) did this in a ZnSe epitaxial layer grown on a GaAs substrate by measuring the heavy-hole—light-hole splitting, which is proportional to axial strain. ZnSe has a slightly larger lattice constant than GaAs, so that pseudomorphic epitaxial growth puts the epitaxial layer under biaxial compression. The lower bulk moduli of the ZnSe is expected to relieve this compressive strain under pressure, as the ZnSe compresses more than the GaAs. Zero strain was indeed observed, at 36.2kbar, and a tensile strain was induced in the ZnSe at higher pressures. Rockwell et al (1991) were however unable to use this data to compare accurately the bulk moduli of ZnSe and GaAs as the splitting of the heavy and light hole states did not behave as predicted theoretically—effectively, it behaved as if the ZnSe softened under pressure, going into tensile strain much faster than predicted. Instead, they took the established values for the bulk moduli, and they introduced a pressure-dependent term into the expression for the splitting of the holes to account for their results. Nevertheless, their results demonstrate that with a well-behaved and properly understood strain gauge in the epitaxial layer, differences in both the bulk modulus

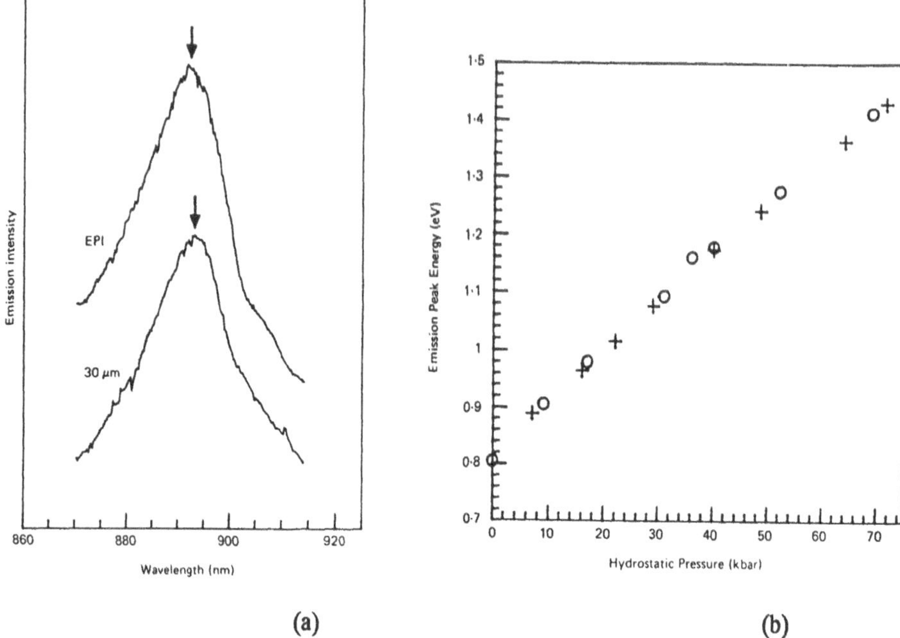

(a) (b)

Figure 6. The spectra in (a) are from samples of InGaAsP epitaxial layers grown on InP. The spectrum marked "30μm" is from a layer still on its substrate, thinned to 30μm thickness for loading in the DAC. The spectrum marked "EPI" is from a layer which has been removed from its substrate. In (b), the peak positions are plotted against pressure. No significant difference can be observed between the sample on its substrate (o) and off (+). *Reproduced from Prins and Dunstan (1988).*

and the differential bulk modulus between the layer and the substrate could be found very accurately by high-pressure measurements.

2.3 PHASE TRANSITIONS IN BULK SEMICONDUCTORS

Other crystal structures for these elements and compounds are possible and have completely different properties—for example, tin can be a metal. In the case of tin, the phase transition between the two structures can be observed simply by varying the temperature; the phase transition temperature being -13°C. Similarly, diamond, which is metastable at low temperatures, will graphitise at temperatures above about 1600°C (or lower in the presence of catalysts).

In most cases, other crystalline forms of these elements and compounds can be created only using high pressure. The four-fold co-ordinated diamond structure is very open—if the atoms are treated as equal touching spheres then only 34% of the volume is occupied, and the packing fraction rises to a maximum of 75% if the cations (anions) are much larger than the anions (cations) so that they touch. Under high pressure, it is energetically favourable for the diamond structure to collapse to a denser phase. Phase transition pressures have been measured for many of the diamond and zinc-blende semiconductors (Fig.7). From inspection, the phase transition pressure decreases with increasing lattice constant, and also increases with ionicity. However, there is considerable scatter around these trends. There is a considerable body of work on the phases obtained above the phase transition pressures shown in Figure 1; we return to this subject in Section 3.

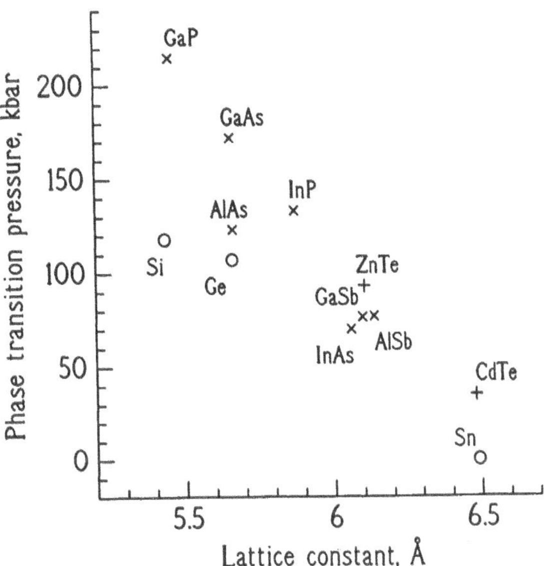

Figure 7. The phase transition pressures of some tetrahedral semiconductors are plotted against the lattice constant. Among the Group IV elements (o), diamond is omitted as the phase transition has not been observed up to over 5Mbar. The III-V data (×) and the II-VI data (+) show the small effect of increasing ionicity.

2.4 PHASE TRANSITIONS IN SEMICONDUCTOR HETEROSTRUCTURES.

A fascinating field of study was opened up by Weinstein and co-workers at SUNY in the study of the phase transition pressures of component layers of epitaxial heterostructures. Weinstein et al (1986) studied AlAs/GaAs multi-layer structures and superlattices; for a recent review of their work see Weinstein et al (1991).

2.4.1 Observation of Superpressing. The SUNY group found that above the phase transition pressure of AlAs, P_t^{AlAs} = 124 kbar, the behaviour of the AlAs layers depends on their thickness. For thick AlAs layers, as the thickness is decreased the pressure required to achieve the transition increases, with

$$\Delta P_t \sim \frac{2000}{t} \qquad [1]$$

where P_t is in kbar and t in Å. This phenomenon is called overpressing, or superpressing. Once P_t reaches the GaAs value, the entire heterostructure undergoes the phase transition together, at

P_t^{GaAs} = 172 kbar. It is noteworthy that, according to Weinstein et al, the GaAs layers show no underpressing; independent of their thickness they transit at 172kbar.

These results may be understood in a similar way to the stability of pseudomorphic epitaxial strained layers, in terms of a *critical thickness*, a concept introduced in semiconductor epitaxial growth to explain when a strained layer would relax by plastic deformation (see Hu, 1991, for a critical assessment of theories of critical thickness). The elastic energy in a strained layer is proportional to its thickness, and in a thick layer the elastic energy ensures that the layer relaxes towards an unstrained state. But the interface between the relaxed layer and the substrate has a higher energy than the strained layer interface, and this increase in interface energy is nearly independent of layer thickness. So the layer cannot relax until, at the critical thickness, there is enough stored elastic energy to create the higher energy interface. Weinstein explain their data in a similar way. The disordered interface between transited AlAs and T_d GaAs has a higher energy than the low-pressure ordered interface, and this energy is provided by the product of the superpressure ΔP_t and the layer volume, accounting for the t^{-1} in Eqn.1. If, on the other hand, the two layers transit together, the interface energy remains unchanged.

This model predicts that the GaAs layers should transit below P_t^{GaAs}—displaying *underpressing*—and short-period superlattices should transit as a whole, at a pressure intermediate between the values of P_t for the two constituents. In the AlAs/GaAs system, Weinstein et al (1991) do not observe such behaviour, and they account for this by proposing that the interface always becomes disordered at the phase transition.

The system studied by Weinstein et al is lattice-matched: the difference in lattice constant between AlAs and GaAs is negligible. It is interesting to consider what effect built-in strain should have on the observed phase transition pressure. Most semiconductor epitaxial combinations are not lattice-matched, and in a superlattice the component of larger lattice constant will be under biaxial in-plane compression. The other component may be relaxed, if it has the same lattice constant as the substrate, but commonly it is grown under tension balancing the compression in the compressed layers (this is known as a freestanding superlattice). Now, generally, the larger the lattice constant, the lower the phase transition pressure, and so one would expect generally to be able to superpress the compressed layer as it can be supported above its P_t by the stable untransited small-lattice-constant component. However, biaxial compression, for a Poisson's ratio $\nu=⅓$, may be resolved into two-thirds hydrostatic compression and one-third shear strain, and since the layer is therefore under hydrostatic pressure as-grown, at 0kbar external pressure, the external pressure required for the phase transition should be lowered.

In a study of CdTe/ZnTe superlattices we found superpressing of the compressed component, CdTe, together with underpressing of the tensile component, ZnTe (Prins et al, 1990; Dunstan et al, 1991). The superlattices transited at pressures rather below the predicted values for the equivalent homogeneous alloys $Cd_xZn_{1-x}Te$, giving 25-30kbar of superpressing in the CdTe and 30-45 kbar of underpressing in the ZnTe layers. We interpreted this in terms of a very high energy for the interface if one component alone transited.

2.4.2. Discussion of Superpressing. It is difficult to reconcile these very different results—a complete absence of underpressing in one case (GaAs), and more than the theoretically predicted underpressing in the other (ZnTe). The crystal structure is the same and the chemical difference (III-V or II-VI) would not be expected to change the results. There is no question but that with sufficiently thin layers (say one monolayer) the superlattices would behave as the equivalent alloys. But the structures studied by Weinstein et al include GaAs layers as thin as 25Å (with 60Å AlAs layers) which is only ten monolayers. Our samples contained six and eleven monolayers of ZnTe. There were no significant differences in sample preparation. Both groups removed the superlattices from their substrates. Indeed, we found no difference in phase transition pressure on or off the substrate (Gil et al, 1989).

One major difference between the two systems is the growth quality. Epitaxy in the AlAs/GaAs system is very well developed, and one can expect the samples to be almost perfect crystallographically, limited only by the GaAs substrates. Interfaces should be flat to one monolayer, and although Weinstein et al (1991) does not give it, photoluminescence characterisation would give linewidths below a couple of meV. In contrast, II-VI epitaxy is difficult. There are no high-quality lattice-matched substrates, and so the superlattices were grown on a relaxed buffer layer on GaAs substrates. The plastic relaxation in the buffer layer reduces the crystallographic quality seriously (see Dunstan, 1991). The interface quality in the II-VI materials is less well controlled, and the photoluminescence linewidths are around 20meV. So the III-V superlattices should be much nearer perfect. However, how this might affect the issue of underpressing is not clear. If the AlAs/GaAs displayed underpressing, and the CdTe/ZnTe did not, one might relate it to sample quality, since it could be argued that a high defect density could facilitate the formation of more defects at a transited/untransited interface. We have, however, the opposite problem.

We believe that the phenomenon is directly related to strain—and of most importance is the strain after the phase transition of one component. The volume reduction at the transition is about 18%, which corresponds to a reduction in the unit cell length of about 6%. This is the same as the difference in lattice constants between CdTe and ZnTe. In the CdTe/ZnTe samples, when the CdTe transits, it therefore goes into lattice-match. So the coherence of the II-VI interface need not be broken. But now we have a high energy interface, and all this energy is available to encourage the phase transition in the ZnTe—that is, to induce underpressing.

In contrast, in the AlAs/GaAs samples, when the AlAs transits, it departs from lattice-match by 6%. It is quite unlikely that the coherence of the interface can be maintained against 6% tensile strain (for example, growth at such large strains is very difficult). But if coherency is lost, a high density of dislocations and other defects will be created, and probably epitaxial orientation will be lost during plastic deformation of the AlAs by 6%. This is likely to create a situation where the subsequent transition of the GaAs (and a further 6% of plastic deformation, in the GaAs layers) will not restore a low energy interface. Consequently, the energy of the interface after the AlAs transition is not available to cause underpressing in the GaAs.

2.4.3. *Uses of Superpressing.* The phenomenon of superpressing is not merely interesting in its own right, but has some practical uses. It is well known in systems in which the phase transition pressure is *negative* (Navrotsky and Phillips, 1975), and enables materials which do not exist in bulk form to be studied. Thus, heteroepitaxial thin layers of MnSe can be grown in ZnSe, with the MnSe in the zinc-blende structure rather than its normal 0kbar rocksalt structure (Kolodziejski et al, 1986). Similarly, heteroepitaxial α-tin can be stabilised on suitable substrates (Farrow et al, 1981; Yuen, 1991). More generally, our results on the strained superlattices show that almost any semiconductor can be superpressed since the compressive strain resulting from heteroepitaxy on a substrate with smaller lattice constant and higher P_t does not prevent superpressing. This could be very valuable in the study of materials in which some interesting effect is predicted to occur, but only at a pressure above the bulk P_t. An example is the pressure-induced crossover of the direct and indirect bandgaps in InP—the study of this crossover has been very fruitful in GaAs (see Section 2.1) and should be equally useful in InP.

An almost trivial but very useful application of superpressing is the greater accuracy that can be obtained in the measurement of bandgap pressure coefficients when they can be measured over a wider range. For example, in our measurements on CdTe in CdTe-ZnTe superlattices, superpressing nearly doubled the range over which we could measure the CdTe band edge, from 35kbar to 64kbar (Gil et al, 1989).

3. Materials up to 1 Megabar

A very large body of work exists on the crystalline phases and equations of state (pressure-volume relationships) of simple materials in the pressure range up to 100GPa. This is both for comparison with theory, and also for the very practical reason that most of the Universe is composed of the four materials, hydrogen, helium, iron and ice, in this pressure range or above Ross, 1990). Understanding planets and stars therefore requires experimental data on these four materials at high pressure (and, preferably, high temperature). Even an understanding of the earth's mantle requires knowledge of the crystalline forms of silicates at pressures up to a few hundred kilobar and temperatures up to 2000K (see, e.g., Christensen, 1988).

3.1 HIGH PRESSURE PHASES OF TETRAHEDRAL SEMICONDUCTORS

The semiconductors discussed in Section 2 provide an example of the richness of the behaviour to be found at high pressure, even in relatively simple systems. While the first phase transition in these materials can be recognised optically as they usually become metallic, the study of the high-pressure phases requires X-ray diffraction. This gives both the crystal structure and the lattice constant, enabling both the phase thransitions themselves and the equations of state of the various phases to be obtained. X-ray diffraction from the small volumes of sample in the diamond-anvil cell is by no means trivial, yet methods such as angle dispersive powder diffraction using synchrotron radiation (Nelmes et al, 1991) allow very detailed information to be obtained.

Silicon goes to the β-tin structure at 11GPa, then at 13-16GPa it goes through another transition to simple hexagonal. At 40GPa this in turn transforms to hexagonal close-packed, and in a study up to a megabar Duclos et al (1987) found yet another transition to face-centred cubic, at 78GPa. Most high-pressure phases exist at high pressure only, but some can be recovered at ambient pressure—that is, the phase transition displays sufficient hysteresis that the high pressure phase can be metastable at 0kbar. Hu et al (1986) found that on decompression of the β-tin phase of silicon yet another phase is found at ambient pressure, with a body-centred cubic structure. Diamond is, of course, another example of a metastable crystal structure; whether natural or artificial it is made under pressure, but is stable almost indefinitely at STP. But diamonds are not forever!

The full phase diagram for silicon remains to be obtained, since the transitions depend on temperature as well as on pressure. This may be seen from the results of Yu et al (1978), who found several high-pressure phases of InSb, in agreement with previous authors. At around 20 kbar InSb, like silicon, transforms from the tetrahedral ambient pressure phase to a tetragonal (β-tin) phase, then at 60kbar there is a second transition to an orthorhombic phase. But with a temperature range only up to 100°C, Yu et al were able to identify a low-temperature orthorhombic phase and distinguish a hexagonal orthorhombic phase occuring at higher temperature.

Caution has to be applied to attempts to predict the phase transition pressures theoretically, because of hysteresis. Besson et al (1990) show that the equilibrium thermodynamic phase line from the ambient pressure form of GaAs to the first high pressure form is at only 11±2 GPa, while the phase transition is observed at 17.5 GPa (see also Weinstein et al, 1991). On the other hand, theoretical predictions of the equations of state appear to be quite successful. The theoretical isothermal equations of state of high-pressure phases of silicon calculated using a linear muffin-tin orbital electron-band theory by Neethiulagarajan and Balasubramanian (1991) agree well with the experimental curves, while Cheong and Chang (1991) obtain results consistent with experiment for both the equations of state and the phase transition pressures for tin using the self-consistent *ab initio* pseudopotential method.

3.2 HIGH PRESSURE PHASES OF SIMPLE ATOMS AND MOLECULES

Turning to atoms and molecules still simpler in principle that the tetrahedral semiconductors, we find the same general rule, that the larger the atoms, the lower the pressure range in which results can be obtained. Generally, all materials are expected to become metallic at sufficiently high pressure as molecules approach each other sufficiently closely and disassociate, and electronic bands broaden and overlap. This has been studied in detail in iodine (I_2 at ambient pressure). Iodine becomes a metal at 18GPa, but Takemura et al (1982) showed that at this pressure there is no structural phase transition—that is, the metallisation is due to band overlap. At pressures above 21GPa, the iodine monomerises, so that at these higher pressures the crystal structure retains no trace of I_2, only iodine atoms.

Lighter atoms are harder to study as the pressure ranges are higher. For example, Siringo et al (1990) calculate that the metallisation occuring at 18 GPa in I_2 is expected at 33GPa in Br_2 and 67 GPa in Cl_2. Consequently, in a given pressure range less of the phase diagram of the lighter elements can be studied and fewer phase transitions are observed. Interest centres on the equations of state, and on metallisation. Grimsditch et al (1986) measured the equation of state of argon using Brillouin scattering up to 350 kbar. This pressure range was sufficient to show deviations from simple theory involving only pair potentials and they fitted their data using three-body potentials. Hydrogen and deuterium being much lighter, such deviations were not observed by Shimizu et al (1981) who measured the equations of state of these elements to 200 kbar at room temperature. Mao et al (1988) grew a single crystal of helium in a DAC under pressure and obtained phase transitions and the equation of state.

Turning to iron and the other materials relevant to the denser planets such as Earth, Boehler has shown that laser heating can be combined with high pressure to obtain the melting point of iron and compounds in the Fe-O-S system up to 1.2 MBar. This is quite a remarkable experiment, since temperatures up to 3000K are required. Similar temperatures but more modest pressures are used to study silicates, of relevance to the earth's mantle. Particularly important is the change in Si-O co-ordination from four to six occuring at about 240 kbar and responsible for the large increase in seismic velocity at 670 km depth. The measurements of Boehler and Chopelas (1991) in the DAC constrain the temperature at that depth to 1900 ± 100 K. Results of this sort fully justify the description of the DAC as "a window to the centre of the earth."

A fascinating variation on the simple atoms such as argon is provided by the fullerenes, and particularly the simplest, C_{60}. This is a spherical molecule, and interactions between the molecules is through van der Waals forces only. The molecule is thus similar to an inert gas atom, except several times the diameter. A high-pressure study has been reported by Duclos et al (1991) in which they observed the effects of pure hydrostatic pressure up to 20GPa, and also the effects of shear stress. They obtained the equation of state and found that the low-pressure phase is stable over the range of the experiment.

4. Experiments in the DAC above 1 Megabar

Much above 100GPa, the experimental difficulties are at present dominant and little in the way of scientific results have been reported. Although Ruoff (1991) has reported 560 GPa static pressure in a cell, measured by X-ray diffraction, this was a demonstration that such pressures can be attained quantitatively—that is, under control and unambiguously. Most scientific results are negative—hydrogen does not metallise up to two megabar; diamond does not metallise up to five megabar. Vohra and Ruoff (1990) have shown that the metals Pt, Fe, Re and W are stable in their low-pressure crystallographic forms to pressures between 2 and 3 Mbar. Much, too, is being learnt about the elastic behaviour, the band structure and the refractive index of diamond itself at these pressures (see Vohra, 1991), as a spin-off from solving the experimental difficulties of multimegabar generation.

Apart from the obvious points that the sample volume is much smaller at multimegabar pressures and the diamond more likely to be broken, what are the difficulties? Firstly, as pointed out by Ruoff (1991) the ruby pressure scale cannot be applied much above 2 Mbar and synchrotron X-ray diffraction is used instead on a relatively large volume of metal. This makes it harder to do any other experiment than X-ray diffraction, especially at temperatures other than 300K. No true sample hole nor hydrostatic medium has yet been used at these pressures. Thus Hemley and Mao (1990) stabilised their sample hole by having it 90% full of ruby with only 10% of sample; it is likely that the ruby then acted to some extent as gasket itself. Ruoff (1991) used no sample hole nor hydrostatic medium: the sample was part of the gasket. Further development is required before the relatively well-understood submegabar gasket and sample hole can be extended to much higher pressures. Ruoff (1991b) notes also that much unknown chemistry may occur at these pressures, and that one must always consider the possibility that the sample may react with the diamonds, the ruby or the X-ray marker, or the gasket. Nevertheless, attainable static pressures may yet reach 1TPa (Ruoff, 1991a), and unquestionably scientific results will follow.

References

Besson, J.M., Weill, G., Mansot, J.L. and Gonzalez, J. (1990) *Phase diagram of GaAs,* High Pressure Research **4**, 312-314.

Boehler, R. (1991) *Advances in high temperature research in diamond cells,* in Recent Trends in High Pressure Research: Proceedings of the XIII AIRAPT Conference, 591-600.

Boehler, R. and Chopelas, A. (1991) *A new approach to laser heating on high pressure mineral physics,* Geophys. Rev. Lett. **18**, 1147-1150.

Cheong, B.H. and Chang, K.J. (1991) *First principles study of the structural properties of Sn under pressure,* Phys. Rev. **B44**, 4103-4108.

Christenen, U. (1988) *Is subducted lithosphere trapped at the 670-km discontinuity?* Nature **336**, 462-463.

Duclos, S.J., Brister, K., Haddon, R.C., Kortan, A.R. and Thiel, F.A. (1991) *Effects of pressure and stress on C_{60} fullerene to 20GPa,* Nature **351**, 380-382.

Duclos, S.J., Vohra, Y.K. and Ruoff, A.L. (1987) *Hcp-to-fcc transition in silicon at 78GPa and studies to 100GPa,* Phys. Rev. Lett. **58**, 775-778.

Dunstan, D.J. (1991) *Relaxed buffer layers,* Semicon. Sci. Technol. **6**, A76-A79.

Dunstan, D.J., Gil, B. and Homewood, K.P. (1988) *Hydrostatic and uniaxial pressure coefficients of CdTe,* Phys. Rev. **B38**, 7862-7865.

Dunstan, D.J., Prins, A.D., Gil, B. and Faurie, J.P. (1991) *Phase transitions in CdTe/ZnTe strained-layer superlattices,* Phys. Rev. **B44**, 4017-4020.

Farrow, R.F.C., Robertson, D.S., Williams, G.M., Cullis, A.G., Jones, G.R., Young, I.M. and Dennis, P.N.J. (1981) *The growth of metastable heteroepitaxial films of α-Sn by metal beam epitaxy,* J. Crystal Growth **54**, 507-518.

Gil, B., Dunstan, D.J., Calatayud, J., Mathieu, H. and Faurie, J.P. (1989) *Electronic structure of cadmium-telluride—zinc-telluride strained-layer superlattices under pressure,* Phys. Rev. **B40**, 5522-5528.

Grimsditch, M., Loubeyre, P. and Polian, A. (1986) *Brillouin scattering and three-body forces in argon at high pressure,* Phys. Rev. **B33**, 7192-7200.

Hemley, R.J. and Mao, H.K. (1990) *Structural transitions in hydrogen and deuterium at ultrahigh pressures,* High Pressure Research **3**, 156-158.

Hu, J.Z., Merkle, L.D., Menoni, C.S. and Spain, I.L. (1986) *Crystal data for high-pressure phases of silicon,* Phys. Rev. **B34**, 4679-4684.

Hu, S.M. (1991) *Misfit dislocations and critical thickness of epitaxy,* J. Appl. Phys. **69**, 7901-7903.

Keyes, R.W. (1962) *Elastic properties of diamond-type semiconductors,* J. Appl. Phys. **33**, 3371-3372.

Kolodziejski, L.A., Gunshor, R.L., Otsuka, N., Gu, B.P., Hefetz, Y. and Nurmikko. A.V. (1986) *Two-dimensional metastable magnetic semiconductor structures,* Appl. Phys. Lett. **48**, 1482-1484.

Mao, H.K., Hemley, R.J., Wu, Y., Jephcoat, A.P., Finger, L.W., Zha, C.S. and Bassett, W.A. (1988) *High-pressure phase diagram and equation of state of solid helium from single-crystal X-ray diffraction,* Phys. Rev. Lett. **60**, 2649-2652.

Navrotsky, A. and Phillips, J.C. (1975) *Ionicity and phase transitions at negative pressure,* Phys. Rev. **B11**, 1583-1586.

Neethiulagarajan, A. and Balasubramanian, S. (1991) *Theoretical isothermal equation of state of the high-pressure phases of silicon,* Phys. Rev. **B43**, 13525-13527.

Nelmes, R.J., Hatton, P.D., McMahon, M.I., Piltz, R.O. and Crain, J. (1991) *Angle dispersive powder diffraction techniques for crystal structure refinement at high pressure,* in Recent Trends in High Pressure Research: Proceedings of the XIII AIRAPT Conference, 753-753.

Paul, W. and Warschauer, D.M. (1963) *Solids at High Pressure* (McGraw-Hill, New York) Ch.8, p.226.

Prins, A.D. and Dunstan, D.J. (1988) *A determination of the relative bulk moduli of GaInAsP and InP,* Philos. Mag. Lett. **B58**, 37-44.

Prins, A.D. and Dunstan, D.J. (1989) *Bulk moduli of GaInAsP and GaInAs by photoluminescence up to 100kbar,* Semicon. Sci. Technol. **4**, 239-340.

Prins, A.D., Gil, B., Dunstan, D.J. and Faurie, J.P. (1990) *CdTe/ZnTe strained layer superlattices under high pressure,* High Pressure Research **3**, 63-65.

Prins, A.D., Lambkin, J.D., O'Reilly, E.P., Adams, A.R., Dunstan, D.J., Pritchard, R., Truscott, W.S. and Singer, K.E. (1992) *Band offsets of strained GaAsSb/GaAs quantum wells from high pressure photoluminescence,* Physical Review B (to be published).

Prins, A.D., Lambkin, J.D., O'Reilly, E.P., Adams, A.R., Pritchard, R., Truscott, W.S. and Singer, K.E. (1990) *Band offsets in GaAsSb/GaAs strained-layer structures from high-pressure photoluminescence,* Proceedings of the 20th International Conference on the Physics of Semiconductors (Thessaloniki, Greece, 1990), ed. E.M. Anastassakis and J.D. Joannopoulos (World Scientific, Singapore) pp. 933-936.

Ross, M. (1990) *High pressure studies of planetary matter,* High Pressure Research **5**, 683-685.

Rockwell, B., Chandrasekhar, H.R., Chandrasekhar, M., Ramdas, A.K., Kobayashi, M. and Gunshor, R.L. (1991) *Pressure tuning of strains in semiconductor heterostructures: (ZnSe epilayer)/(GaAs epilayer),* Phys. Rev. **B44**, 11307-11314.

Ruoff, A.L. (1991a) *X-ray diffraction in the 500GPa range: Prospects for attaining 1TPa,* in Recent Trends in High Pressure Research: Proceedings of the XIII AIRAPT Conference, 769-778.

Ruoff, A.L. (1991b) *Megabar chemistry,* in "Recent Trends in High Pressure Research: Proceedings of the XIII AIRAPT Conference," 576-578.

Shimizu, H., Brody, E.M., Mao, H.K. and Bell, P.M. (1981) *Brillouin measurements of solid n-H_2 and n-D_2 to 200kbar at room temperature,* Phys. Rev. Lett. **47**, 128-131.

Siringo, F., Piccitto, G. and Pucci, R. (1990) *Metal-insulator transition of solid halogens under pressure,* High Pressure Research **3**, 162-164.

Takemura, K., Minomura, S., Shimomura, O., Fujii, Y. and Axe, J.D. (1982) *Structural aspects of solid iodine associated with metallisation and molecular dissociation under high pressure,* Phys. Rev. B**26**, 998-1004.

Tanaka, K. and Maeda, J. (1986) *Pressure-volume measurements by using diamond-anvil cells and an imaging system,* Rev. Sci. Instrum. **57**, 500-501.

Warburton, R.J., Nicholas, R.J., Mason, N.J., Walker, P.J., Prins, A.D. and Dunstan, D.J. (1991) *High-pressure investigation of GaSb and $Ga_{1-x}In_xSb$ quantum wells,* Phys. Rev. **B43**, 4994-5000.

Weinstein, B.A., Cui, L.J., Venkateswaran, U.D. and Chambers, F.A. (1991) *Enhanced Stability of heterostructures under pressure,* in NATO ASI Series B *Frontiers of High Pressure Research,* ed. H.D. Hochheimer and R.D. Etters (Plenum Press, New York) pp. 257-267.

Venkateswaran, U., Chandrasekhar, H., Chandrasekhar, M., Vojak, B.A., Chambers, F.A. and Meese, J.M., (1986) *High pressure studies of GaAs-$Ga_{1-x}Al_xAs$ quantum wells of widths 26 to 150Å,* Phys. Rev. **B33**, 8416-8423.

Vohra, Y.K. (1991) *Spectroscopic studies on diamond anvil under extreme static stress,* in Recent Trends in High Pressure Research: Proceedings of the XIII AIRAPT Conference, 349-358.

Vohra, Y.K. and Ruoff, A.L. (1990) *Phase transitions and equations of state at multimegabar pressures,* High Pressure Research **4**, 296-299.

van de Walle, C.G. and Martin, R.M. (1987) *Theoretical study of band offsets at semiconductor interfaces,* Phys. Rev. **B35**, 8154-8165.

Wilkinson, V.A., Prins, A.D., Dunstan, D.J., Howard, L.K. and Emeny, M.T. (1991) *Investigation of the band structure of the strained systems InGaAs/GaAs and InGaAs/AlGaAs by high pressure photoluminescence,* J. Electronic Materials **20**, 509-516.

Warburton, R.J., Nicholas, R.J., Mason, N.J., Walker, P.J., Prins, A.D., and Dunstan, D.J. (1991) *High-pressure investigation of GaSb and $Ga_{1-x}In_xSb$ quantum wells,* Phys. Rev. **B43**, 4994-5000.

Wolford, D.J., Keuch, T.F., Bradley, J.A., Gell, M.A., Nimmo, D. and Jaros, M. (1986) *Pressure dependence of GaAs/$Al_xGa_{1-x}As$ quantum-well structures: The determination of valence-band offsets,* J. Vac. Sci. Tech. **B4**, 1043-1050.

Yu, S.C., Spain, I.L. and Skelton, E.F. (1978) *Polymorphism and the crystal structure of InSb at elevated temperature and pressure,* J. Appl. Phys. **49**, 4741-4745.

Yuen, W.T., Liu, W.K., Stradling, R.A. and Joyce, B.A. (1991) *The growth and electrical properties of α-Sn thin films grown on InSb(110) and (111) substrates by molecular beam epitaxy,* J. Crystal Growth **111**, 943-947.

Ultra high pressure experiments on high-T_c superconductors

Rinke J. WIJNGAARDEN, E.N. VAN EENIGE, J.J. SCHOLTZ, D. TRISTAN JOVER AND R. GRIESSEN
Department of Physics and Astronomy, Free University,
De Boelelaan 1081, 1081 HV Amsterdam, The Netherlands

ABSTRACT. In this review we treat various selected properties of superconductors under ultra high pressure. Emphasis is given to the superconducting parameters T_c and H_{c2} in high temperature superconductors. The pressure dependence of T_c is explained using a phenomenological model.

1. Introduction

Superconductivity was discovered by Kamerlingh Onnes [1] in 1911; two years later he received for all of his low temperature work the Nobel prize. In his experiment (see fig.1) the resistance of mercury decreased to a very low value at the critical temperature $T_c = 4.20$ K. Much later (in 1962) Kim et al. [2] showed that a current in a superconducting ring of niobium decays with a half life of > 10^{380} year: the resistance must be very small indeed.

Another important aspect of superconductors is the diamagnetism (expulsion of magnetic field), discovered by Meissner and Ochsenfeld [3]. In type II superconductors, however, if the external field is large enough, it will penetrate the sample trough small normal areas, called flux-lines, which are created by the magnetic field. These normal areas have a diameter which is closely equal to the coherence length ξ. If the whole sample is completely covered by flux-lines, superconductivity can no longer exist: the magnetic field has reached the upper critical field H_{c2}.

A microscopic theory for superconductivity was completed in 1957 by Bardeen, Cooper and Schrieffer, now commonly called the BCS theory [4]. In this theory the electron-phonon interaction causes a net electron-electron attraction V. Although more elaborate expressions are sometimes useful, we give here a very simple formula for T_c:

R. Winter and J. Jonas (eds.), High Pressure Chemistry, Biochemistry and Materials Science, 121–146.

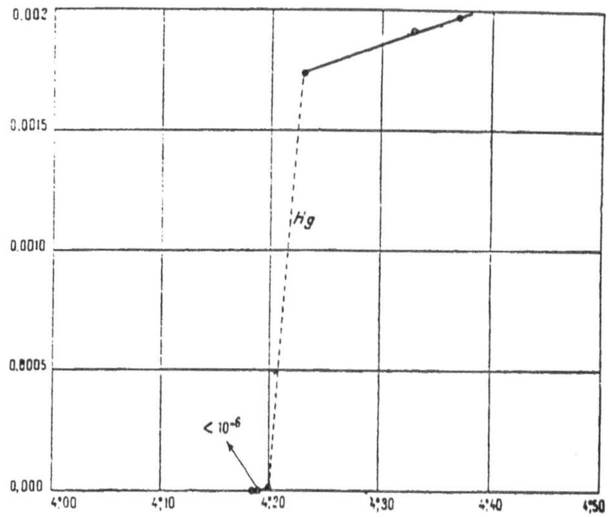

Figure 1. Resistive transition in mercury, measured in 1911 by G. Holst and H. Kamerlingh Onnes [1]. Vertically R(T)/R(T = 300K) and horizontally T in Kelvin are plotted.

$$k_B T_c \approx \hbar\omega \, e^{-\frac{1}{NV}} \tag{1}$$

where ω is the average phonon frequency and N is the density of states at the Fermi-level.

High pressure experiments have played an important role in superconductivity research; the first experiment was done in 1925 by Sizoo and Kamerlingh Onnes [5] up to 300 bar on tin and indium. The first organic superconductors were discovered using high pressure. After the discovery of high temperature superconductors of 30 K in La-Ba-Cu-O by Bednorz and Müller [6] in 1986 (who also received their Nobel prize two years later) high pressure experiments by Chu et al. [7] led to the discovery of $YBa_2Cu_3O_7$ with the spectacular critical temperature of 90 K.

In this lecture we will very selectively discuss the effect of high pressure on some of the "old" superconductors (section 2). Then we will focus on our main topic which is the new class of cuprate perovskite high T_c superconductors (section 3). After some general remarks, we discuss the structural properties, normal state properties and superconducting properties, all as a function of pressure. We end with a brief discussion of uniaxial experiments and some concluding remarks.

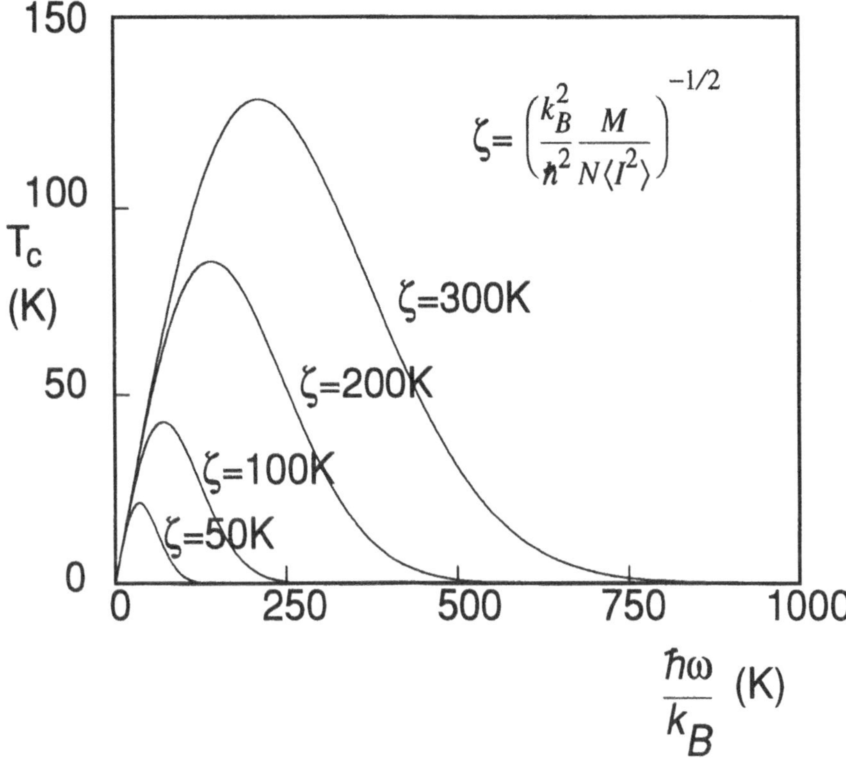

Figure 2. Critical temperature versus phonon frequency according to the simple BCS formula eqn (2), for various values of $\zeta = (k_B/\hbar)\,(M/N\langle I^2\rangle)^{-1/2}$ as indicated. As a function of pressure generally ω will increase and ζ will be constant.

2. "Old" superconductors under pressure

In general any material should become superconducting if sufficient pressure is applied. The overlap of electronic wave functions is increased by pressure, leading to metallization, while magnetization is suppressed because unpaired spins take more volume than paired ones. Because all non-magnetic metals are believed to become superconducting, all materials are expected to become superconductors at a high enough pressure. Nevertheless pressure is not in general beneficial for the superconducting phase. The electron-phonon coupling constant V of eqn. (1) can be written as $V = \langle I^2\rangle/M\omega^2$, where $\langle I^2\rangle$ is an average electron-phonon coupling matrix element and M is the ion mass. Substitution in eqn. (1) yields

$$k_B T_c = \hbar\omega\, e^{-\frac{M\omega^2}{N\langle I^2\rangle}} \qquad (2)$$

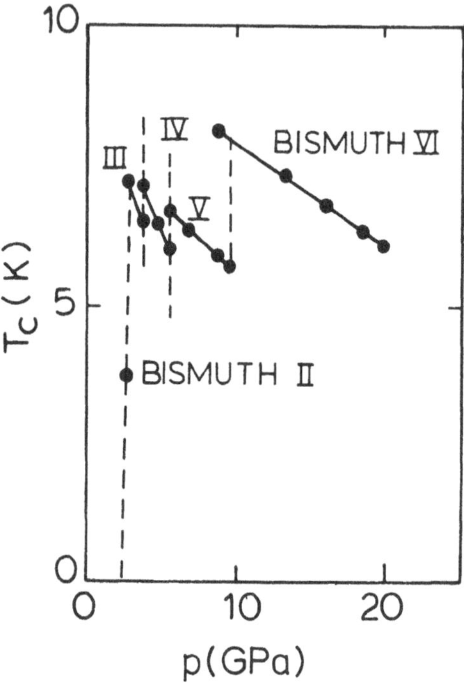

Figure 3. Critical temperature versus pressure for Bi (after ref. 8); the discontinuities correspond to structural phase transitions.

In the absence of phase transitions the potential will become stiffer as a function of pressure, leading to an increase of ω. In fig. 2 $T_c(\omega)$ is plotted. Clearly the application of pressure will finally lead to a vanishingly small value for T_c.

In fig. 3 the $T_c(p)$ for bismuth is shown [8]. This element is non-superconducting at ambient pressure (except as a thin film), but pressure induces superconductivity. The plot shows very clearly that T_c decreases with pressure until it increases stepwise at a structural phase transition, which is accompanied by a decrease in (some) phonon frequencies, (in general because the coordination number and hence the average bond length increases).

Even more complicated behavior can be observed for $T_c(p)$ since superconductivity is a rather subtle phenomenon. Some examples are mentioned in our review [9], where also a table of the superconducting T_c of the elements as a function of pressure may be found.

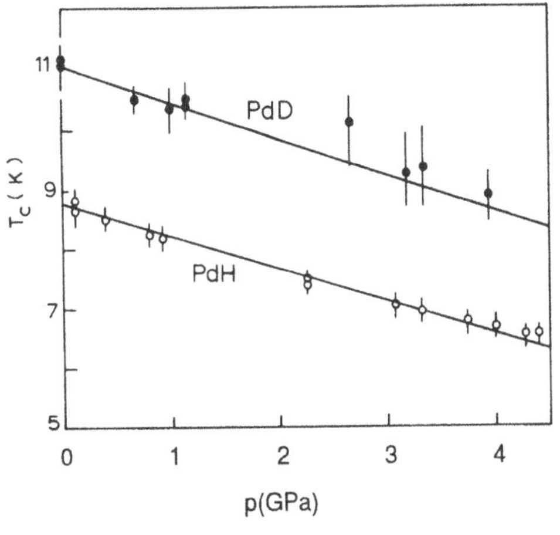

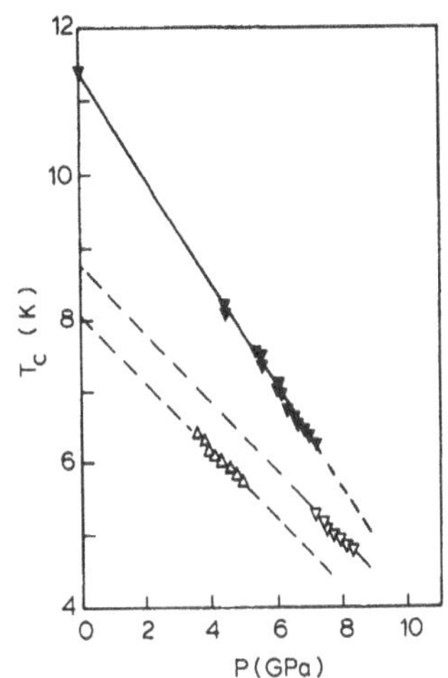

Figure 4. Pressure dependence of T_c of PdH (O), PdD (●), $Pd_{0.93}Ag_{0.07}H$ (Δ and ∇) and $Pd_{0.93}Ag_{0.07}D$ (▼). The latter sample was subjected to hydrostatic pressure, the former three to uniaxial pressure (after refs. 17).

2.1 Metallic hydrogen and some metal hydrides

From fig. 2 it is immediately clear that the highest T_c's may be obtained if $M/N\langle I^2\rangle$ is small. As suggested by Ashcroft [10] a prime candidate is solid atomic hydrogen, which has the lowest ion mass M and which also has a very strong electron phonon coupling $\langle I^2\rangle$, because the Coulomb interaction is not screened by core electrons [11]. Calculations by Ashcroft and many other authors yield very high values for T_c; a recent calculation gives ~ 230 K [12]. Regrettably at ambient pressure hydrogen solidifies in an insulating molecular solid. Although it is very difficult to calculate the transition pressure to an atomic metallic solid, values seem to converge around 500 GPa. This pressure has been reached in the laboratory [13], but not in hydrogen. Only in some recent dynamic high pressure experiments (by shock waves) some evidence has been obtained [14] of a closing band gap. However, this is still in the *molecular* phase. The only indication of the solid slowly transforming towards an *atomic* solid are experiments where the vibron (intramolecular excitation) frequency starts to decrease with increasing pressure above ≈ 100 GPa [15]. Even if hydrogen were pressurized at say 500 GPa it would be extremely difficult to check for superconductivity.

Ideas have been proposed to lower the metallization pressure of hydrogen by impurity doping [16]. Other candidates could be possibly the metal-hydrides. In these materials the hydrogen in the lattice is atomic hydrogen. However, the advantageous properties of the hydrogen are to some extend "diluted" by the mass and phonon frequencies of the metal ions. To date the highest T_c in metal hydrides is not higher than in other superconductors. In our group Hemmes et al. [17] have succeeded in synthesizing stoichiometric PdH by using a diamond anvil cell: at a pressure of ~ 4 GPa and room temperature the chemical potential of H_2 is larger than the chemical potential of H in PdH. Stoichiometric PdH is thus formed. At low temperature the diffusion constant is so small that even at ambient pressure PdH is stable, enabling the investigation of the superconducting T_c as a function of pressure, see fig. 4a. One striking feature is the higher T_c for the heavier isotope. Both this phenomenon and the value for $\partial T_c/\partial p$ can be explained using a slightly modified eqn. 2 (i.e. the Allen and Dynes modification of the MacMillan formula) and by taking both the zero point motion of the hydrogen isotope and the anharmonicity of the hydrogen-metal-ion potential into account [18]. Based on calculations of essentially $M\omega^2/\langle I^2 \rangle$ by Papaconstantopoulos et al. [19] we estimate that a $T_c \cong 30$ K might be possible for $Pd_{0.7}Ag_{0.3}H$, which probably can be synthesized around 50 GPa. To check this idea at more modest pressures $Pd_{0.93}Ag_{0.07}H$ was measured [17], see fig. 4b; clearly a higher T_c is obtained than for PdH. The difference in slope between the two isotopes is due to a different sample configuration, the $Pd_{0.93}Ag_{0.07}H$ sample was fixed to the diamonds (as were the PdH and PdD samples), and hence uniaxially compressed, while the $Pd_{0.93}Ag_{0.07}D$ sample came loose and was compressed hydrostatically. The difference in slope can be explained from this difference only.

3. High temperature superconductors

The typical crystal structure of high-T_c superconductors is exemplified by fig. 5. The CuO_2-layers are responsible for the superconducting behavior and are always present; the CuO-chains may be absent, or be replaced by other layers (e.g. TlO or BiO). As we will see later the structure governs the anisotropic compressibility. The number of charge carriers in the CuO_2-layers can be changed by doping with other elements or by changing the total amount of oxygen in the crystal. Also it may be changed by pressure or electrostatic fields. We define δ as the number of charge carriers in these planes per plane per planar copper atom. As a function of δ the compound may change from an anti-ferromagnet to a superconductor [20] (see fig. 6).

Immediately after the discovery of these materials [6], much technological interest was aroused due to the much larger potential for application if a cheap coolant such as liquid nitrogen could replace the expensive liquid helium. However, ironically the flux-lines in these materials can move relatively easy causing resistive loss. Recently at very low temperatures even macroscopic quantum tunneling of flux-lines in $YBa_2Cu_3O_7$ was discovered by members of

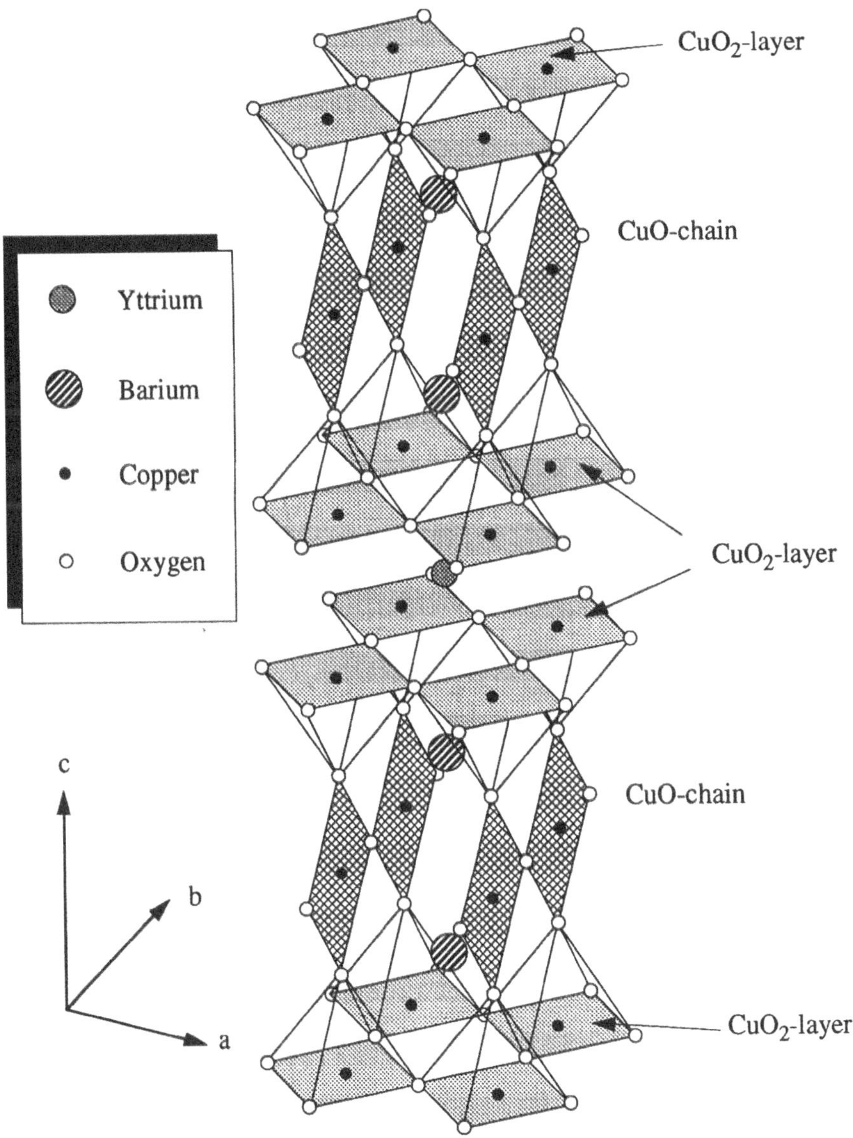

Figure 5. Crystal structure of $YBa_2Cu_3O_7$. The CuO_2 layers are responsible for the superconducting behavior.

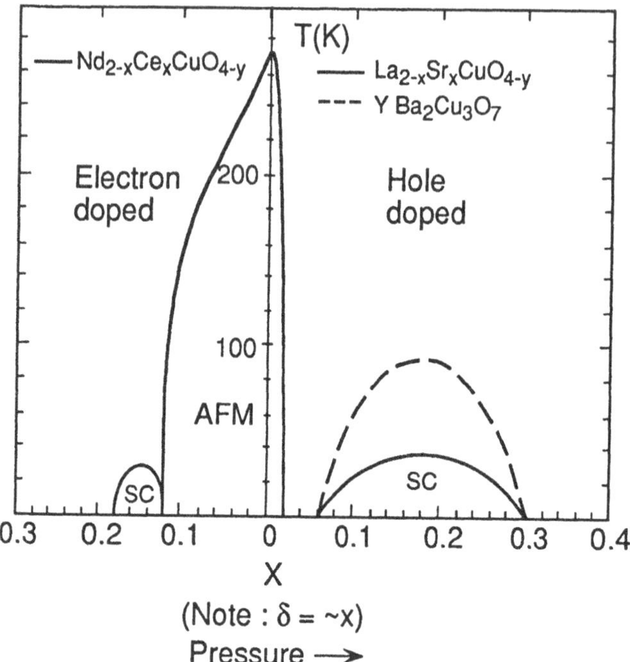

Figure 6. Plot of temperature versus doping. For the anti-ferromagnetic (AFM) phase the Néel temperature is plotted, for the superconducting phase (SC) T_c. (after ref. 20).

our group [21]. Of course, much technological effort is generated towards improving the pinning ("sticking") of the flux-lines.

3.1 Structural properties

Some recent reviews on structural properties were done by Fietz [22] and Schilling and Klotz [23], additional tables may be found in our review [24]. Here we briefly mention some general trends. These ceramic materials have a relatively high bulkmodulus B ranging from a mere 63 GPa in $Bi_2Sr_2Ca_1Cu_2O_8$ to 180 GPa in La_2CuO_4. The bulkmodulus is smaller if the number of CuO_2 layers per unit cell or the number of layers between two non-adjacent CuO_2 layers is larger. The compressibility along the c-lattice vector is largest, along the chains smallest. In particular in oxygen deficient samples the oxygen may form superstructures, which can be dependent upon previous treatment of the sample. For example Sieburger and Schilling [25] found that the $T_c(p)$ of $Tl_2Ba_2CuO_{6+y}$ is different if the pressure is changed at different temperatures. Oxygen ordering may also explain the different $T_c(p)$ curves measured on different samples of $YBa_2Cu_3O_{7-\delta}$, (see e.g. fig. 16 of [24]). For stoichiometric samples this phenomenon seems to be much less pronounced. In $YBa_2Cu_4O_8$, which can only be made stoichiometrically, very nice agreement exists between different determinations.

3.2 Normal state properties

Well above T_c, the resistivity varies linearly with temperature; however, Sundqvist and Andersson [26] have shown that correction for thermal expansion removes this linearity which hence seems to be fortuitous. With increasing pressure the resistivity decreases in all high-T_c superconductors (both "hole-" and "electron-doped") at a rate $\partial \ln\rho/\partial p$ = -0.1... -0.2 GPa^{-1}. This is much larger than expected from volume compression alone; for example $\partial \ln\rho/\partial p$ for copper is -0.02 GPa^{-1}. The decrease along the c-lattice vector is larger than along the planar lattice vectors. Hence pressure reduces the anisotropy of the resistivity tensor. For the Hall resistivity $\partial \ln R_H/\partial p$ = -0.15 GPa^{-1} is found [27] for $YBa_2Cu_4O_8$, while -0.08 is found [28, 29] for $YBa_2Cu_3O_7$. The pressure dependence of both resistivity and Hall resistivity are indicative of an increasing charge carrier concentration with pressure. Also the same follows from the pressure dependence of the Néel temperature: $\partial T_N/\partial p \cong + 200$ K/GPa for electron doped and $\partial T_N/\partial p \cong -5$ K/GPa for hole doped superconductors, both consistent with an increase of charge carrier concentration (equivalent with increased doping) with pressure. Gruneisen parameters γ derived from thermal expansion measurements and from Raman scattering [30] are γ = 1.5... 2.0, very normal values. For a more extensive discussion the reader is again referred to refs. [23] and [24].

3.3 Pressure dependence of T_c

Particularly in the early days of high-T_c superconductors, our experimental research was stimulated by theoretical models which predicted a large pressure dependence of T_c; a number of models is discussed in ref. [24]. For brevity we mention only one illustrative example: the Resonating Valence Bond (RVB) theory of Anderson [31]. In this theory a simple T_c formula is

$$k_B T_c \cong \delta \frac{t_\perp^2}{t_{//}} \tag{3}$$

where $t_\perp$ and $t_{//}$ are the transfer integrals for charge carrier transfer from one CuO_2 layer to an adjacent one ($t_\perp$) and within a single CuO_2 layer ($t_{//}$). If we suppose that the pressure dependence of the conductivity σ is governed by the pressure dependence of the charge carrier den-

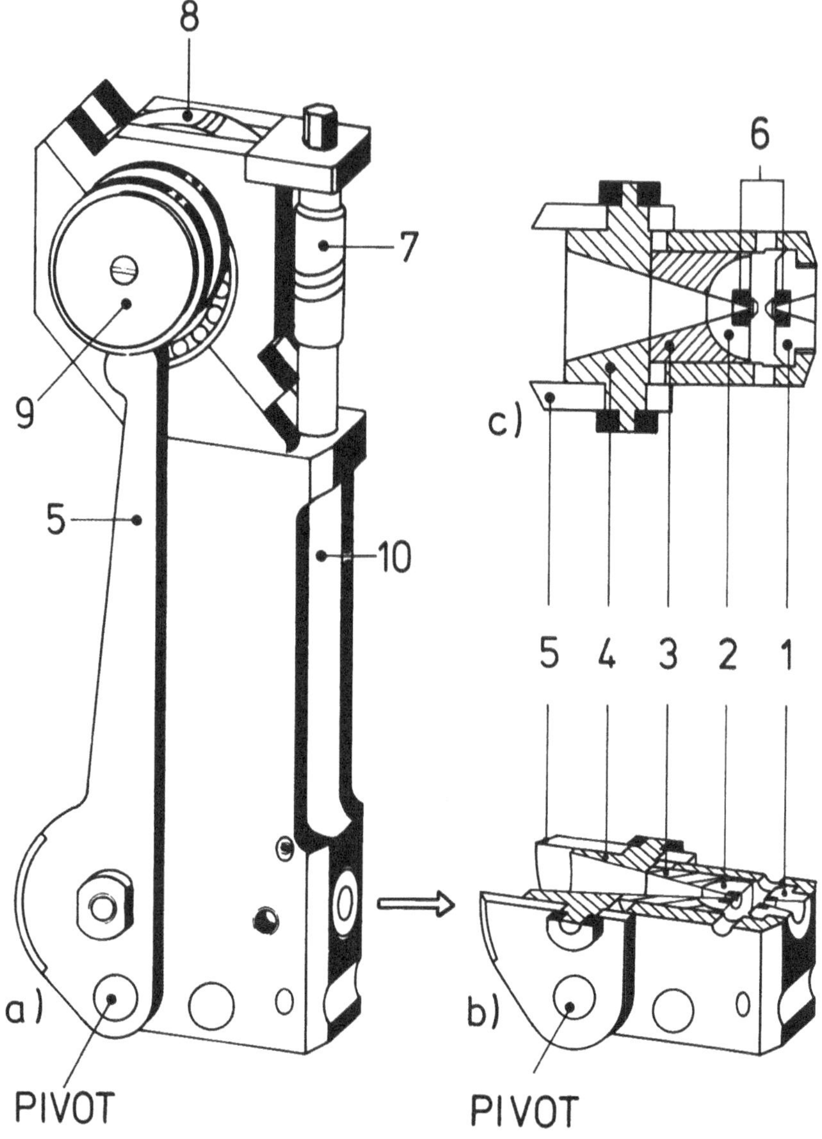

Figure 7. Schematic drawing of Diamond Anvil Cell I. Legend: 1-translation table for diamond alignment, 2-hemisphere for diamond alignment, 3-piston, 4 pushing block, 5-lever arms, 6-tungsten carbide diamond seats, 7&8-worm gear, 9-eccentric cam, 10-heat exchanger chamber.

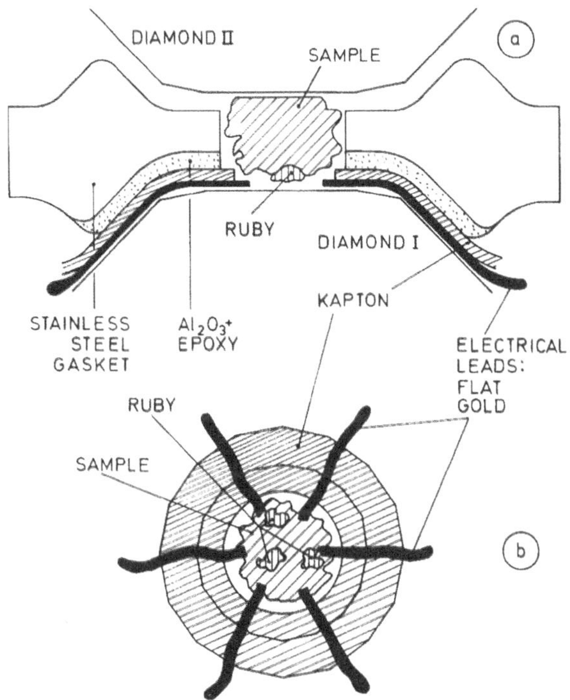

Figure 8. Cross-section (a) and top view (b) of the pressure chamber between the two diamonds. Typically the gasket hole diameter is 300 μm, wires are 25 μm.

sity δ and the transfer integrals $t_\perp$ then we have

$$\sigma(p) \propto t(p)\,\delta(p) \tag{4}$$

or

$$\frac{\partial \ln \sigma}{\partial p} = \frac{\partial \ln t}{\partial p} + \frac{\partial \ln \delta}{\partial p} \tag{5}$$

Using the determination of $\sigma_{//}(p)$ and $\sigma_\perp(p)$ by Konczykowski et al.[32] and our determination (see below) of δ(p) we find from eqn. (5) for $YBa_2Cu_3O_7$ $\partial \ln T_c/\partial p = 0.4\ GPa^{-1}$, whereas experimentally $\partial \ln T_c/\partial p < 0.01\ GPa^{-1}$. This may imply one or more of the following three possibilities: i) the experimental values for the anisotropic conductivity do not give the intrinsic material properties, e.g. due to sample inhomogeneities, ii) eqn. (4) does not hold, iii) the simple RVB-T_c formula eqn. (3) does not hold. Possibility i) seems to be the more unlikely explanation.

3.3.1 Experimental results on $T_c(p)$

Our experimental results were obtained using two different diamond anvil cells [33, 34] of which one can be operated in the bore of a 12 Tesla superconducting magnet. The other cell is displayed in fig. 7. The critical temperature and upper critical field are determined by a four point resistive method, using a low frequency lock-in technique to eliminate thermal voltages. A typical sample configuration is shown in fig. 8. Six wires in stead of four are used to have some redundancy. Pressure is measured by the ruby fluorescence method [35] using several small ruby chips dispersed on the sample; at low temperatures the ruby shift is corrected for temperature [36]. Temperature is measured by a Platinum resistor, corrected for magnetic fields where appropriate or by a Germanium resistor below 30 K in the absence of magnetic fields.

A number of measurements of our group are summarized in fig. 9. A broad sprectrum of qualitatively different behavior is observed: very large and very small values for $\partial T_c/\partial p$ occur, both increasing and decreasing with pressure. All this different behavior can be explained using a very simple phenomenological model. Let us assume for the moment that the charge carrier concentration increases with increasing pressure. (See section 3.2 and section 3.3.2, where we demonstrate that this is indeed the case). In fig. 10 we show the dependence of T_c on the charge carrier concentration at zero pressure as determined from chemical doping experiments and reviewed by Shäfer and Penney [37]. Clearly $T_c(\delta)$ has roughly the shape of an (inverted) parabola. If a compound has a charge carrier concentration δ at zero pressure which is left of the maximum, and pressure increases δ, then T_c comes closer to the maximum (this is the case for e.g. $CaLaBaCu_3O_7$). In more extreme cases, pressure will take T_c trough the maximum and down again (as observed for e.g. $YBa_2Cu_4O_8$). If at zero pressure δ is already large, then T_c may be seen to decrease immediately (as observed for e.g. $Bi_2Ca_2Sr_1Cu_2O_8$). To investigate this idea in more detail, we have carried out measurements of the upper critical field H_{c2} as a function of pressure. From H_{c2} we were able to deduce the pressure dependence of the charge carrier concentration. This is the subject of the next section.

3.3.2 Experimental determination of $H_{c2}(p)$ and the calculation of $\delta(p)$

Because the zero-temperature value of H_{c2} is around 100 Tesla, this value could not be mea-

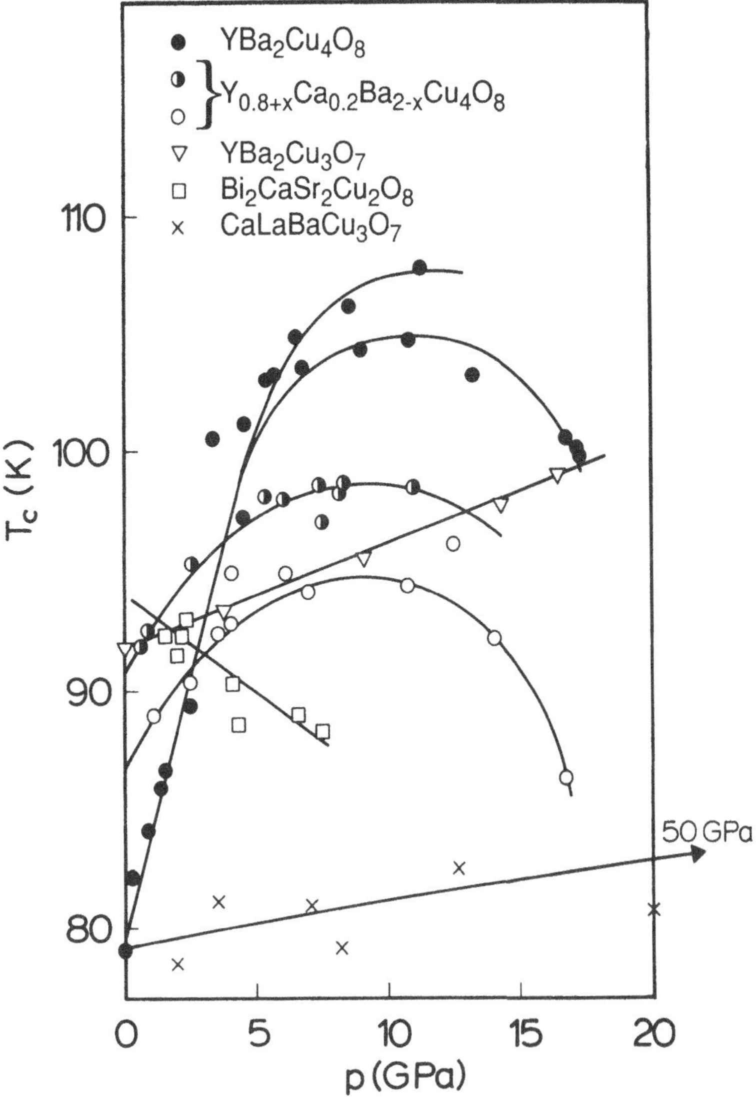

Figure 9. Superconducting critical temperature T_c versus pressure for a number of compounds investigated in our group.

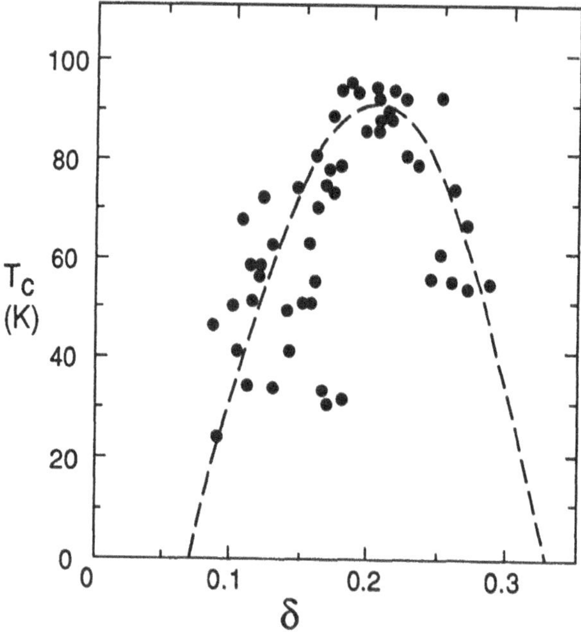

Figure 10. T_c versus charge carrier concentration for high-T_c superconductors with two CuO_2 layers. Datapoints from ref [37], the curve is $T_c = 90\,[1 - 60(\delta - 0.2)^2]$

sured directly. However with our 12 Tesla magnet the upper critical fields H_{c2} of $CaLaBaCu_3O_7$ and $YBa_2Cu_4O_8$ were determined close to T_c. Using the theory of Werthamer Helfand and Hohenberg [38] with spin-orbit coupling parameter $\lambda_{so} = 2$, the high temperature results could be extrapolated to the T = 0 values, which are shown in fig. 11. In the calculation of $\delta(p)$ these T = 0 values are used. It must be emphasized, that the values obtained for $\delta(p)$ are very insensitive to the method of extrapolation used for H_{c2} since in our calculation only $\partial \ln H_{c2}/\partial p$ enters, i.e. only relative changes of H_{c2} are of importance. In a previous NATO conference [39] we have discussed in detail how $\delta(p)$ may be found from $H_{c2}(p)$ and $T_c(p)$. Here we mention only the leading idea which is based on the following series of relations: the number of charge carriers determines the Fermi-wave vector, which is proportional to the Fermi velocity, which together with T_c determines the coherence length ξ, which determines H_{c2}. Hence H_{c2} and T_c determine δ. The full expression derived in ref. [39] is:

$$\frac{\partial \ln \delta}{\partial p} = -\frac{1}{3B} - \frac{1}{3}\frac{\partial \ln c}{\partial p} - \frac{1}{3}\frac{\partial \ln H_{c2}}{\partial p} + \frac{2}{3}\frac{\partial \ln T_c}{\partial p} \tag{6}$$

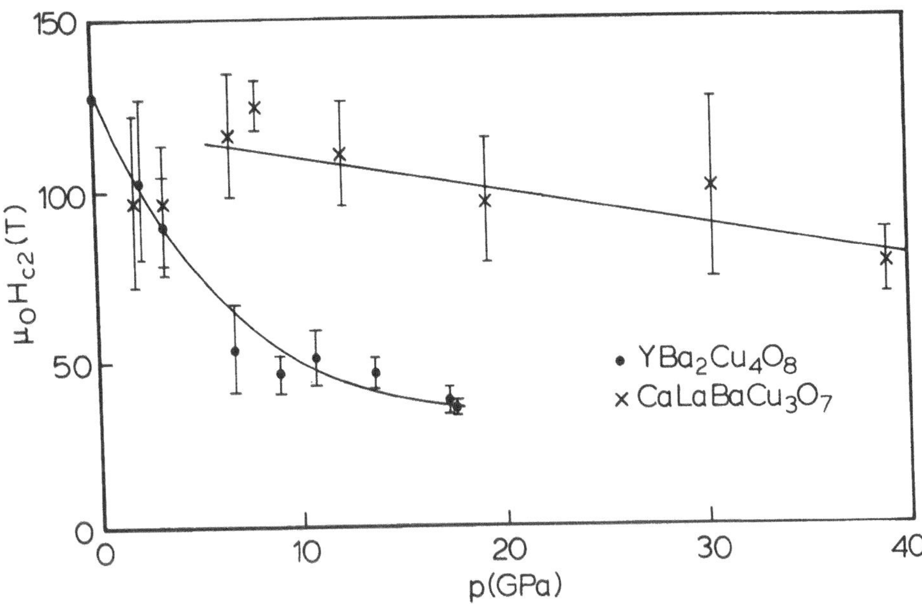

Figure 11. Pressure dependence of the upper critical field μ_0H_{c2} at T = 0 for $YBa_2Cu_4O_8$ and $CaLaBaCu_3O_7$. (calculated by extrapolation from experiment, see text).

Since δ at zero pressure can be read from fig. 10, eqn. (6) can be integrated to give $\delta(p)$. For both $YBa_2Cu_4O_8$ and $CaLaBaCu_3O_7$, $\delta(p)$ is roughly linear and $\partial\delta/\partial p = 0.009$ GPa^{-1} for $YBa_2Cu_4O_8$ and $\partial\delta/\partial p = 0.0005 GPa^{-1}$ for $CaLaBaCu_3O_7$. This difference explains the difference in slope of $\partial T_c/\partial p$ between these two compounds. Because of the high cost associated with the operation of the Diamond Anvil Cell in the 12 Tesla magnet we have to date not measured H_{c2} of other compounds. Nevertheless we believe that all behaviors shown in fig. 9 can be understood from (i) a pressure induced increase in δ and (ii) T_c following the inverted parabola in fig. 10.

The idea that pressure increases the charge carrier density is substantiated by several other publications [40 - 42]. We will apply this idea in the next sections. Let us finally remark that the difference in $\partial\delta/\partial p$ between $YBa_2Cu_4O_8$ and $CaLaBaCu_3O_7$ is caused by the difference in structure: in the tetragonal $CaLaBaCu_3O_7$ the charge transfer from the chains to the CuO_2 layers is very difficult as borne out by electronic structure calculations [43].

3.3.3. Relation between $\partial T_c/\partial x$ and $\partial T_c/\partial p$ in YBa_2Cu_3Ox

We now apply the findings of the previous section to explain $\partial T_c/\partial p$ for $YBa_2Cu_3O_x$. Murata

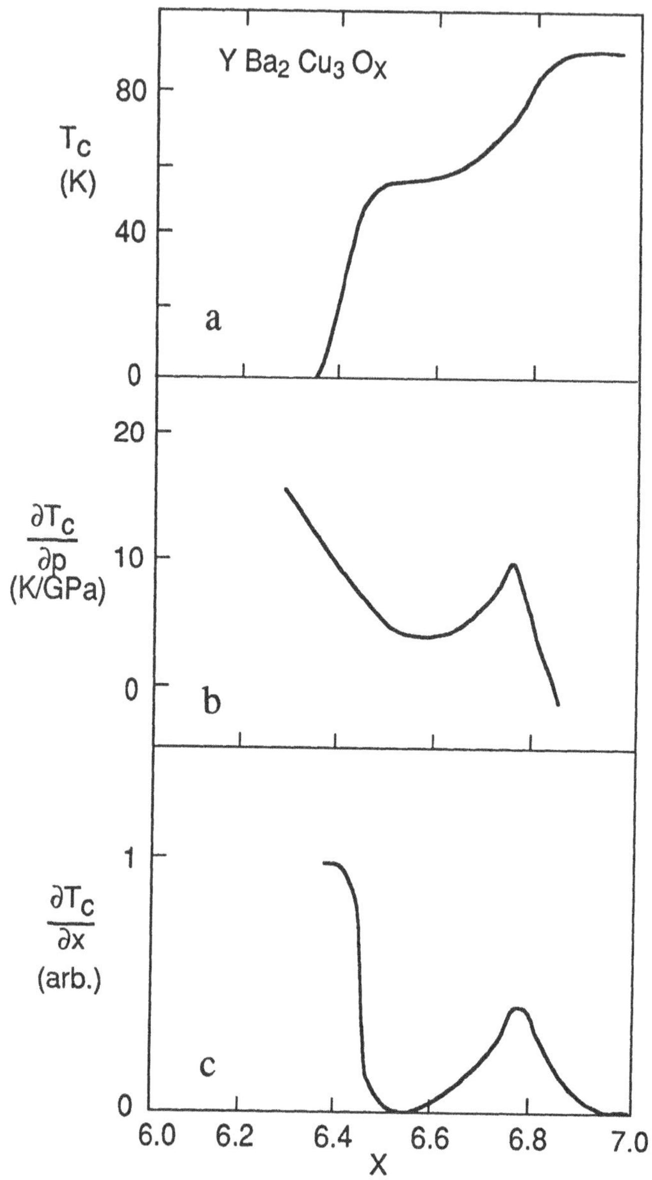

Figure 12. T_c (a) and $\partial T_c/\partial p$ (b) for $YBa_2Cu_3O_7$ after ref. [44]. In (c) $\partial T_c/\partial x$ is plotted. Note the similarity with $\partial T_c/\partial p$

et al. [44] have measured $T_c(x)$ and $\partial T_c/\partial p$ for different x in $YBa_2Cu_3O_x$, their main findings are summarized in fig. 12ab. If pressure increases the number of charge carriers, the effect of pressurizing a sample is equivalent to increasing x at zero pressure. Hence we expect $\partial T_c/\partial p$ to be large if $T_c(x)$ is steep. More precisely we expect $\partial T_c/\partial p$ to be proportional to $\partial T_c/\partial x$. From Murata et. al.'s data we constructed numerically $\partial T_c/\partial x$ as shown in fig. 12c. Clearly a striking similarity is observed with fig. 12b. We conclude that $\partial T_c/\partial p$ is indeed roughly proportional to $\partial T_c/\partial x$.

3.3.4. Relating between $\partial \ln T_c/\partial p$ and T_c

If we combine the parabolic dependence $T_c(\delta)$ and the linear dependence $\delta(p)$ then we find

$$T_c = T_0 \{1 - \beta [(\delta_1 - \delta_0) + \alpha p]^2\} \tag{7}$$

Here T_0 is the maximum T_c observed as a function of pressure (or doping) in a particular compound. β is the width of the parabola of fig. 10, which we assume for simplicity to be independent of compound and equal to 60; δ_1 is the charge carrier density at zero pressure and δ_0 is the charge carrier density corresponding to the maximum in $T_c(\delta)$. In other words: $T_c(\delta_0) = T_0$. Note that $\alpha \equiv \partial\delta/\partial p$. From eqn. (7) we find by simple calculus

$$\frac{\partial \ln T_c}{\partial p} = \sqrt{\frac{4\alpha^2 \beta T_0 (T_0 - T_c)}{T_c^2}} \; \mathrm{sgn}\,(\delta_0 - \delta) \tag{8}$$

In figure 13 we compare this simple relation with a wealth of experimental data [24] by very many groups. Using α as a fitting parameter (except for $YBa_2Cu_4O_8$, where $\alpha \equiv \partial\delta/\partial p = 0.009$ GPa^{-1} from section 3.3.2 was used) we are able to explain reasonably well the relation between $\partial \ln T_c/\partial p$ and T_c. In fig. 13 c the same eqn. (8) is used to explain the relation between $\partial \ln T_c/\partial p$ and T_c in several high pressure experiments by Mori [45]. Note that one set of data points in this plot refers to determinations on a single compound in one experiment at different pressures, while the data in fig. 13 ab refer to $\partial \ln T_c/\partial p$ at $p = 0$. Clearly our very simple relation (8) is able to explain most of the systematics in the $\partial \ln T_c/\partial p$ vs T_c plots.

Some prudence in applying the above is, however, well in its place. Of course, one would expect the curve $T_c(\delta)$ of fig. 10 to be itself modified as a function of pressure. Above we have

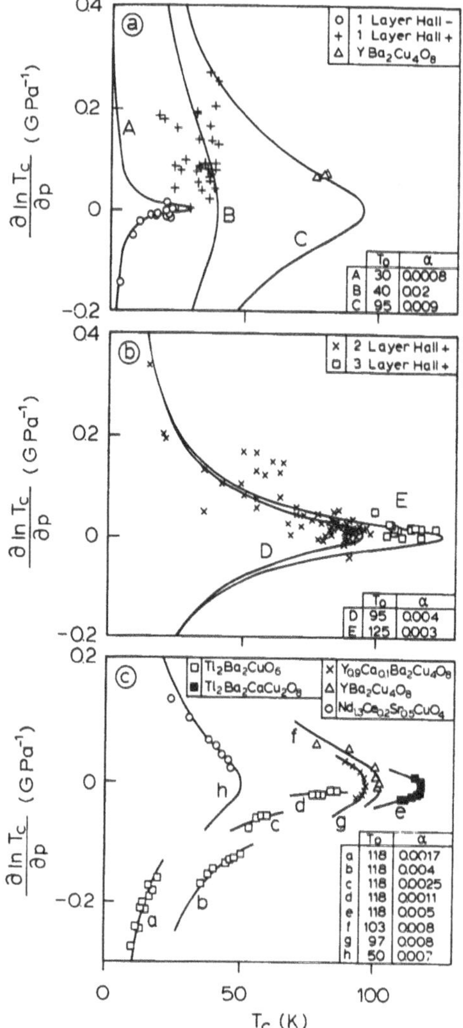

Figure 13. Datapoints: (a) and (b) summary of published data (as reviewed mainly in [24]) on $\partial T_c/\partial p$ at $p = 0$. The number of CuO_2 layers and the sign of the Hall resistivity is indicated (-: electron-like, +: hole-like) (c) $\partial T_c/\partial p$ at $p = 0$ and $p \neq 0$ after Mori [45].

Curves are fits to the datapoints using eqn. (8) with $\beta = 60$ and α (in GPa^{-1}) and T_0 (in K) as indicated.

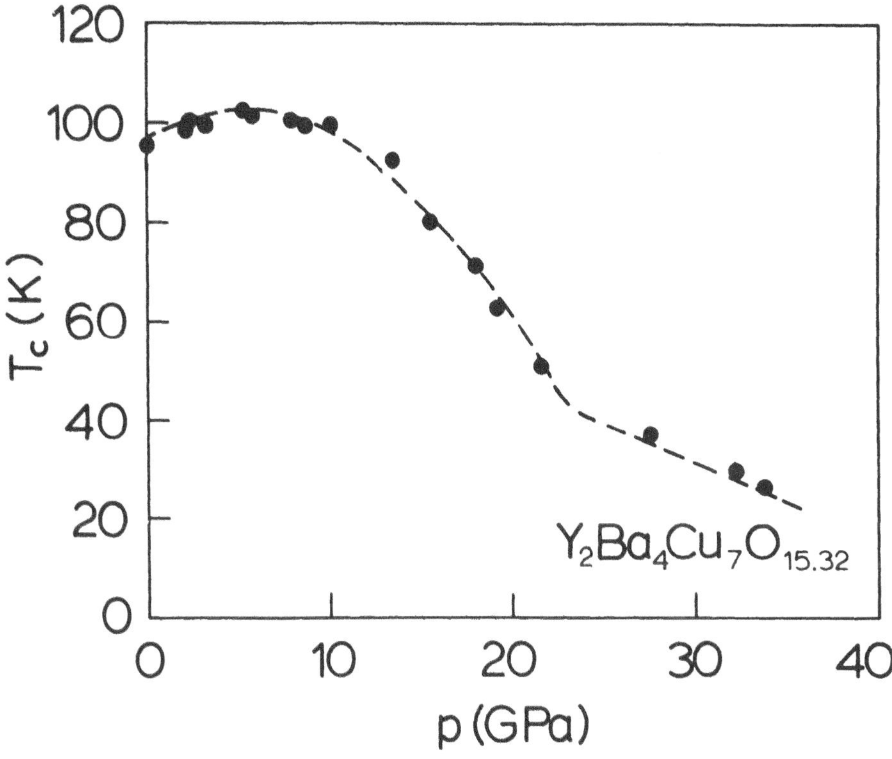

Figure 14: Pressure dependence of T_c for $Y_2Ba_4Cu_7O_{15.32}$. Note the kink at ~ 23 GPa. The dashed line is calculated in the dirty limit for a proximity-coupled multilayer of $YBa_2Cu_3O_7$ and $YBa_2Cu_4O_8$ (see text).

assumed this effect to be negligible. However, the high pressure experiments by Tanahashi [46] on $La_{2-x}Sr_xCuO_4$ clearly show that - at least for this compound - both T_0 and δ_0 are pressure dependent. The reader is referred to our discussion in refs. [39] and [47]. In fact it may well be that $\partial T_0/\partial p$ may convey more important information on the effect of e.g. transfer integrals on T_c than $\partial T_c/\partial p$ does. Further work in this direction seems well justified.

3.3.5. Comparison of different determinations of $\partial\delta/\partial p$

It is illustrative to compare a few determinations of the pressure dependence of the charge carrier density $\partial\delta/\partial p$ (see table 1). All values for $\partial\delta/\partial p$ were calculated by us. Determinations were made from H_{c2} and T_c (see section 3.3.2) and from $\partial \ln T_c/\partial p$ vs T_c (see section 3.3.4).

TABLE 1.
Pressure derivatives of the charge carrier density $\partial\delta/\partial p$ at $p = 0$. The references refer to the original data. Where a large number of papers contributed to a single determination we refer to ref. 24.

Compound	method	ref.	$\partial\delta/\partial p$ (holes/GPa)
$YBa_2Cu_3O_7$	Hall-resistivity	28, 29	0.017
	$\partial lnT_c/\partial p$ vs T_c	24	0.004
$YBa_2Cu_4O_8$	Hall resistivity	27	0.03
	H_{c2}	48	0.009
	$\partial lnT_c/\partial p$ vs T_c	24	0.009
$CaLaBaCu_3O_7$	H_{c2}	49	0.0005
	$\partial lnT_c/\partial p$ vs T_c	24	≤ 0.0005
$Nd_{2-x}Ce_xCuO_4$	$\partial lnT_c/\partial p$ vs T_c	24	0.0008
$La_{2-x}(Ba/Sr)_x CuO_4$	H_{c2}	46	0.016
	$\partial lnT_c/\partial p$ vs T_c	24	0.02

For the interpretation of Hall measurements we used $\partial ln\delta/\partial p = \partial lnR_H/\partial p - 1/B$ [39]; hence the pressure dependence of all charge carriers is taken into account and not only of those in the CuO_2 layers which are relevant for superconductivity.

In general a reasonable agreement exists between various determinations (except for the Hall measurements), although we certainly would not claim an error smaller than ~ 30%. The ratio between the highest and lowest value for $\partial\delta/\partial p$ is ~ 40, a significant difference, which is probably due to a difference in structure, as discussed above. Knowing $\partial\delta/\partial p$ we are now able to use pressure to adjust the charge carrier concentration in a compound to a desired value.

3.3.6. $T_c(p)$ for $Y_2Ba_4Cu_7O_{15}$

$Y_2Ba_4Cu_7O_{15}$ is a natural multilayer of $YBa_2Cu_3O_7$ and $YBa_2Cu_4O_8$. In fig. 14 the $T_c(p)$ for $Y_2Ba_4Cu_7O_{15.32}$ is shown, from a very recent experiment by Van Eenige et al. [50]. Clearly this behavior cannot be explained by T_c following a simple parabola. As discussed in [50] this behavior is due to the fact that inequivalent layers are present in $Y_2Ba_4Cu_7O_{15}$: each layer follows its own parabola and (because of the relative small anisotropy in the yttrium cuprates) the T_c of the sample is determined by the coupling of the T_c's of the individual layers trough the proximity effect. Using fits to the experimental curves for $T_c(p)$ of $YBa_2Cu_3O_7$ and $YBa_2Cu_4O_8$ (shifted in pressure to take into account that our $Y_2Ba_4Cu_7O_{15.32}$ sample had excess oxygen) and using the formula for the proximity effect in a multilayer, the drawn curve in fig. 14 was calculated. A very nice agreement to the data points is obtained. This clearly shows that $Y_2Ba_4Cu_7O_{15}$ is indeed a multilayer and that the structural unit which determines T_c is not a CuO_2 bilayer, because there is only one kind of CuO_2 bilayer in this compound. If this were determining T_c, we would not expect a kink. Of course the occurrence of a phase transition cannot be ruled out completely, but no evidence for such a transition in this compound exists to date. A very striking feature is the enormous decrease in T_c with pressure, probably unrivalled in any other compound.

3.4 Uniaxial pressure dependence of T_c

As mentioned in section 3.3 the high-T_c superconductors are highly anisotropic and e.g. the RVB formula eqn. (3) predicts a very different pressure dependence for T_c for compression in the ab-plane as opposed to compression along the c-lattice vector. Therefore uniaxial pressure experiments are very important. Several types of experiments have been performed: i) pressing directly uniaxially [51], ii) bending the substrate [52], iii) hydrostatically pressing a substrate with different compressibility from the sample [53], iv) thermal expansion measurements [54, 55, 56]. For ii) and iii) the sample must be prepared in several orientations with respect to the substrate. For iv) the Ehrenfest relation $\partial T_c/\partial p_i = T_c \, \Delta \alpha_i/\Delta C_p$ is used to calculate the uniaxial compressibility $\partial T_c/\partial p_i$ from the thermal expansion coefficients $\Delta \alpha_i$;Δ–C_p is the jump in specific heat at T_c. In table 2 a summary is given of the results obtained so far. Apparently for $YBa_2Cu_3O_7$ the $\partial T_c/\partial P_i$ for the a and b lattice vectors are large and opposite in sign; this implies that the small hydrostatic $\partial T_c/\partial p$ value is due to a cancellation of terms and hence may be sensitive to small changes in the lattice.Clearly the situation is not completely clear yet and more experimental work needs to be done.

TABLE 2.
Uniaxial pressure derivatives $\partial T_c/\partial_{pi}$ at p = 0.

<table>
<tr><th>Compound</th><th>method</th><th>ref.</th><th>$\partial T_c/\partial p_a$ (K/GPa)</th><th>$\partial T_c/\partial p_b$ (K/GPa)</th><th>$\partial T_c/\partial p_c$ (K/GPa)</th></tr>
<tr><td>$YBa_2Cu_3O_7$</td><td>uniaxial press</td><td>51</td><td>-</td><td>-</td><td>+0.8</td></tr>
<tr><td></td><td>thermal expansion</td><td>55</td><td>-3.4</td><td>+3.7</td><td>~ 0</td></tr>
<tr><td></td><td>thermal expansion</td><td>54</td><td>-1.9</td><td>+2.2</td><td>~ 0</td></tr>
<tr><td></td><td>film on substrate bending</td><td>52</td><td colspan="2">0.5</td><td>-</td></tr>
<tr><td>$YBa_2Cu_3O_{6.5}$</td><td>thermal expansion</td><td>55</td><td>-1.5</td><td>+3.5</td><td>+2.1</td></tr>
<tr><td>$GdBa_2Cu_3O_7$</td><td>film on substrate in hydrostatic medium</td><td>53</td><td>-3.1</td><td>+0.38</td><td>+3.5</td></tr>
<tr><td>$La_{1.88}Sr_{0.12}CuO_4$</td><td>thermal expansion</td><td>56</td><td colspan="2">+6.2</td><td>-6.7</td></tr>
</table>

4. Conclusion

High pressure experiments play an important role in the physics of superconductivity. As a very recent example we mention the negative pressure dependence of T_c in the buckminster fullerenes, which led to the replacement of the small alkalides by the larger ones, indeed increasing T_c. In the field of high-T_c superconductors at least a phenomenological understanding is emerging. Building upon this we expect to be better able to determine the important physical parameters and their optimal values, thereby again contributing to the advancement of the physics of superconductivity by the use of high pressure experiments.

Acknowledgements
We thank K. Heeck for his expert help with developing very sensitive resistive measurement techniques. This work is part of the research program of the Stichting voor Fundamenteel Onderzoek der Materie (FOM) which is financially supported by NWO.

References:

1. H. Kamerlingh Onnes, Leiden Comm. 120b, 122b, 124c (1911)
2. Y.B. Kim, C.F. Hempstead and A.R. Strnad, Phys. Rev. Lett. 9, 306 (1962)
3. W. Meissner and R. Ochsenfeld, Naturwissenschaften 21, 787 (1933)
4. J. Bardeen, L.N. Cooper and J.R. Schrieffer, Phys. Rev. 108, 1175 (1957)
5. G.J. Sizoo and H. Kamerling Onnes, Leiden Comm. 180b (1925)
6. J.G. Bednorz and K.A. Müller, Z. Phys. B 64, 189 (1986); see also Rev. Mod. Phys. 60, 585 (1988)
7. C.W. Chu, P.H. Hor, R.L. Meng, L. Gao, Z.J. Huang, Y.Q. Wang, Phys. Rev. Lett. 58, 405 (1987)
8. M.A. Il'ina and E.S. Itskevich, JETP Lett. 11, 218 (1970) and H. Kawamura and K. Tachikowa, J. Phys. Soc. Jap. 57, 3647 (1988)
9. R.J. Wijngaarden and R. Griessen, “Superconducting Materials under Pressure” in J. Evetts (ed.) “Concise Encyclopedia of Magnetic and superconducting materials”, Pergamon Press, Oxford, 1992, p. 583
10. N.W. Ashcroft, Phys. Rev. Lett. 21, 1748 (1968)
11. A.A. Abrikosov “Fundamentals of the theory of metals” North Holland, Amsterdam (1988) p. 372
12. See e.g. T.W. Barbee III, A. Garcia and M.L. Cohen, Nature 340, 369 (1989)

13. See e.g. W.C. Moss, J.O. Hallquist, R. Reichlin, K.A. Goettel and S. Martin, Appl. Phys. Lett. 48, 1258 (1986)

14. W.J. Nellis, A.C. Mitchell, P.C. McCandless, D.J. Erskine and S.T. Weir, Phys. Rev. Lett. 68, 2937 (1992)

15. R.J. Hemley and H.K. Mao, Phys. Rev. Lett. 61, 857 (1988)

16. A.E. Carlsson and N.W. Ashcroft, Phys. Rev. Lett. 50, 1305 (1983)

17. H. Hemmes, A. Driessen, R. Griessen and M. Gupta, Phys. Rev. B39, 4110 (1989) and H. Hemmes, A. Driessen, J. Rector and R. Griessen, J. Phys.: Cond. Matter 1, 8123 (1989)

18. R. Griessen and D.G. de Groot, Helv. Phys. Acta 55, 699 (1982)

19. D.A. Papaconstantopoulos, E.N. Economou, B.M. Klein and L.L. Boyer, Phys. Rev. B 20, 177 (1979)

20. M.B. Maple, N.Y. Ayoub, J. Beille, T. Bjornholm, Y. Palichaouch, E.A. Early, S. Ghamaty, B.W. Lee, J.T. Markert, J.J. Neumeier, G. Nieva, L.M. Paulius, I.K. Schuller, C.L. Seaman and P.K. Tsai in Proc. Int. Conf. on Transp. Prop. of Superconductors, Rio de Janeiro 1990, World Scientific, Singapore 1990

21. R. Griessen, J.G. Lensink and H.G. Schnack, Physica C 185-189, 337 (1991)

22. W.H. Fietz, H.A. Ludwig, B.P. Wagner, K. Grube, R. Benischke and H. Wühl, "'Elastic properties of high temperature superconductors derived from high pressure experiments" in H.D. Hochheimer and R.D. Etters (eds.), "Frontiers of High Pressure research", NATO ASI series B 286, Plenum Press, New York (1991) p. 433

23. J.S. Schilling and S. Klotz, "The influence of high Pressure on the superconducting and normal state properties of high temperature superconductors" in D. Ginsberg (ed.) "Physical Properties of high temperature superconductors", World Scientific, Singapore (1992) vol. 3

24. R.J. Wijngaarden and R. Griessen, "High Pressure Studies" in "Studies of High Temperature Superconductors" A.V. Narlikar (ed.), Nova Science Publ., New York (1989) Vol. 2 p. 29

25. R. Sieburger and J.S. Schilling, Physica C 173, 403 (1991)

26. B. Sundqvist and B.M. Andersson, Solid State Comm. 76, 1019 (1990)

27. C. Murayama, Y. Iye, T. Enomoto, A. Fukushima, N. Mori, Y. Yamada and T. Matsumoto, Physica C 185-189, 1293 (1991)

28. T. Hiraoka, Jap. J. Appl. Phys. 28, L 1135 (1989)

29. I.D. Parker and R.H. Friend, J. Phys. C: Solid State Phys. 21, L 345 (1988)

30. K. Syassen, M. Hanfland, K. Strössner, M. Holtz, W. Kress, M. Cardona, U. Schröder, J. Prade, A.D. Kulkarni and F.W. de Wette, Physica C 153-155, 264 (1988)

31. P.W. Anderson, Science 235, 1196 (1987)

32. M. Konczykowski, R. Rullier-Albenque and G. Collin, Physica C 153-155, 1365 (1988)

33. H. Hemmes, A. Driessen, J. Kos, F.A. Mul, R. Griessen, J. Caro and S. Radelaar, Rev. Sci. Instr. 60, 474 (1989)

34. J.J. Scholtz, A. Driessen, R. v.d. Berg, H. van Groen, H. Verhoog, J.J. de Kleuver, R.J. Wijngaarden and R. Griessen, High Pressure Research 5, 874 (1990)

35. H.K. Mao, J.Xu and P.M. Bell, J. Geophys. Res. 91, 4673 (1986)

36. I.F. Silvera and R.J. Wijngaarden, Rev. Sci. Instrum. 56, 121 (1985)

37. M.W. Shäfer and T. Penney, Eur. J. Solid State Inorg. Chem. 27, 191 (1990)

38. N.R. Werthamer, E. Helfand and P.C. Hohenberg, Phys. Rev. 147, 295 (1966)

39. R.J. Wijngaarden, J.J. Scholtz, E.N. van Eenige and R. Griessen, "High T_c superconductors under very high pressure" in H.D. Hochheimer and R.D. Etters (eds.) "Frontiers of high pressure research" NATO ASI Series B 286, Plenum Press, New York (1991) p. 399

40. P. Allenspach, J. Mesot, U. Staub, A. Furrer, H. Blank, H. Mutka, C. Vettier, E. Kaldis, J. Karpinski and S. Rusiecki, Eur. J. Solid State Inorg. Chem. 28, 627 (1991)

41. Y. Yamada, T. Matsumoto, Y. Kaieda and N. Mori, Jap. J. Appl. Phys. 29, L250 (1990)

42. J.D. Jorgensen, S. Pei, P. Lightfoot, D.G. Hinks, B.W. Veal, B. Dabrowski, A.P. Paulikas and R. Kleb, Physica C 171, 93 (1990)

43. M. Gupta and R.P. Gupta, Physica C 171, 465 (1990)

44. K. Murata, Y. Honda, H. Oyonagi, Y. Nishihara, H. Ihara, N. Terada, R. Sugise, M. Hirabayashi, M. Tokumoto and Y. Kimura, MRS Int'l. Mtg on Adv. Mats 6, 681 (1989)

45. N. Mori, H. Takahashi and C. Murayama, Supercond. Sci. Technol. 4, S 439 (1991)

46. N. Tanahashi, Y. Iye, T. Tamegai, C. Murayama, N. Mori, S. Yomo, N. Okazaki and K. Kitazawa, Jap. J. Appl. Phys. 28, L762 (1989)

47. R.J. Wijngaarden, E.N. van Eenige, J.J. Scholtz and R. Griessen, Physica C185-787 (1991)

48. J.J. Scholtz, E.N. van Eenige, R.J. Wijngaarden and R. Griessen, Phys. Rev. B45, 3077 (1992)

49. J.J. Scholtz, E.N. van Eenige, R.J. Wijngaarden, R. Griessen and D.M. de Leeuw, Physica C 198, 231 (1992)

50. E.N. van Eenige, R. Griessen, K. Heeck, H.G. Schnack, R.J. Wijngaarden, J.-Y. Genoud, T. Graf, A. Junod and J. Muller, Europhys. Lett. 20,41 (1992)

51. M.F. Crommie, A.Y. Liu, A. Zettl, M.L. Cohen, P. Parilla, M.F. Hundley, W.N. Creager, S. Hoen and M.S. Sherwin, Phys. Rev. B39, 4231 (1989)

52. G.L. Belenky, S.M. Green, A. Roytburd, C.J. Lobb, S.J. Hagen, R.L. Greene, M.G. Forrester and J. Talvacchio, Phys. Rev. B44, 10117 (1991)

53. S.L. Bud'ko, O. Nakamura, J. Guimpel, M.B. Maple and I.K. Schuller, Physica C 185-189, 1947 (1991)

54. C. Meingast, O. Kraut, T. Wolf, H. Wühl, A. Erb and G. Müller-Vogt, Phys. Rev. Lett. 67, 1634 (1991)

55. H. Wühl, R., Benischke, M. Braun, B. Frank, O. Kraut, R. Ahrens, G. Bräuchle, H. Claus, A. Erb, W.H. Fietz, C. Meingast, G. Müller-Vogt and T. Wolf, Physica C 185-189, 755 (1991)

56. W. Schnelle, O. Hoffels, E. Braun, H. Broicher and D. Wohlleben, In: R. Kossowsky, B. Raveau and S. Patapis (eds.) "Physics and Material Science of High-Tc superconductors II", NATO-ASI series, Kluwer Academic Publ., Dordrecht, 1992

FLUID ALKALI METALS AT HIGH TEMPERATURES AND PRESSURES

F. HENSEL
Institute of Physical Chemistry
and Materials Science Center
Philipps-University of Marburg
D-3550 Marburg, FRG

ABSTRACT. The demands of new technologies and the continuing interest in the metal-insulator transition have resulted in an increased demand for reliable measurements of the properties of fluid alkali metals up to very high temperatures and pressures. These fluids are typical examples of materials whose electronic structures strongly depend on the thermodynamic state of the system. The most striking manifestation of this state dependence is certainly the metal-nonmetal transition which occurs when the fluid is expanded by heating to its liquid-vapour critical point. The existence of this transition noticeably influences the thermodynamic features of the vapour-liquid phase transition of metals.
The paper discusses recent experimental results including equation of state data, electrical, optical and magnetic properties, and neutron scattering measurements of $S(Q,\omega)$ with special emphasis on the change in these properties in the nonmetal to metal transition region.

Introduction

The study of the properties of alkali metals has attracted considerable experimental as well as theoretical attention during the last two decades. Part of this interest is motivated by the large number of current and potential applications of fluid alkali metals. They may be used in metal vapour topping cycles in nuclear electric power plants in outer space and also to increase the efficiency of fossil, fission and fusion power plants on Earth. They may act as a heat transfer medium in solar power systems, and they may be involved in liquid metal magnetohydrodynamic systems coupled to solar power. In heat pipes, heat may be transferred by passing alkali metal vapour from hot to cool ends of a pipe. Alkali metals, because of their high latent heats of evaporization and heat transfer-coefficients, are prominent candidates for working fluids in such applications. For the selection of a particular liquid metal and for the design and efficient operation of the technical processes, the precise knowledge of the physical and thermodynamic properties of those liquids is essential. The necessity for safety analysis and risk assessment requires to model the evolution of hypothetical accidents which can be postulated to occur. Consequently, one needs to know the properties, in particular equation of state data, up to conditions far above the proposed temperatures and pressures of normal operation, preferably up to and beyond the liquid-vapour critical point temperatures.
Experimental research on such properties is complicated, because fluid alkali metals are difficult to experiment with, because, as Table 1 shows, a combination of high temperature and pressure is required to bring the sample anywhere near its critical point.

R. Winter and J. Jonas (eds.), High Pressure Chemistry, Biochemistry and Materials Science, 147–166.

Table 1: Critical temperatures, pressures and densities of alkali metals

Metal	T_c/K	p_c/MPa	ρ_c/g cm^{-3}	Reference
Cesium	1924	9.25	0.38	Jüngst et al. (1985) [1]
Rubidium	2017	12.45	0.29	Jüngst et al. (1985) [1]
Potassium	2178	15.0	0.17	Hensel et al. (1991) [2]
Sodium	2485	24.8	0.30	Binder (1984) [3]
Lithium	3500	38.0	0.11	Ohse (1985) [4]

The critical temperatures T_c and pressures p_c are low enough to be studied under static conditions only for Cs, Rb, K and eventually Na. The estimated location of the critical point of lithium is in fact outside the range of conventional measurements under static, equilibrium conditions. Only transient experiments such as shock waves, exploding wires, and laser heating are reaching temperatures and pressures high enough to explore the critical region of lithium. But these methods are less accurate than static experiments. This means that there is much interest in methods of estimating the critical conditions of metals. The different estimation techniques are all based on the relation between the critical data and other thermophysical properties. The most common basis is the principle of corresponding states which is well established for molecular fluids, but its validity for fluid metals is open to serious questions. The theoretical assumption underlying the principle of corresponding states, namely, that the nature of the interaction between the particles of the fluid is independent of the thermodynamic state of the fluid, is violated for metals.

The fundamental difference in dealing with the liquid-vapour phase transition of metallic fluids is that the molecular and electronic structure of the vapour phase is completely different from that of the liquid. For example, far below the liquid-vapour critical point (e.g. near the triple point), when the density difference between the coexisting phases is large, liquid alkali metals conduct electricity but their coexisting vapours do not. Observations of this sort signal that the electronic wave function has changed from that of almost completely free electrons to that of electrons whose distribution is dominated by the atomic and molecular details. In such a situation, near the triple point, the liquid-vapour phase transition coincides precisely with a metal-nonmetal transition. Both the density and the electronic structure change on passing from one phase to the other. The liquid phase is a metallic conductor and the interparticle interactions are dominated by Coulomb potentials [5]. In contrast, the coexisting vapour phase is nonconducting. Its paramagnetic susceptibility follows the Curie law [6] which indicates that the overwhelmingly dominant species present in the vapour phase under this condition is the highly polarizable atomic monomer interacting through weak van der Waals forces.

However, both types of interactions change with the thermodynamic state. Nearer to the critical point, the liquid density is much less and the vapour density much greater. For example, approaching the critical region of cesium along the coexistence curve, the measured DC-electrical conductivity σ of the liquid drops sharply [7] whereas the measured magnetic susceptibility of the vapour shows a strong diamagnetic deviation [6] which is direct evidence of a high amount of diamagnetic molecular associations like Cs_2 and, probably, larger clusters in the nonmetallic dense vapour phase. This shows that the diatomic structure of cesium vapour persists to relatively high densities. The interesting

question to emerge, therefore, is up to what density of the liquid the pairing mechanism present in the vapour can survive the passage to the metallic state.
A particularly great deal of effort has been devoted to the study of the metal-nonmetal transition of liquid alkali metals. With a single electron per atom they closely resemble the expanded crystals with half full bands considered by Mott [8] in his original discussion of the metal-nonmetal transition due to correlation. Most of the effort has focussed on cesium and rubidium because of their relatively low critical temperatures. The general subject of low density alkali metals has repeatedly been reviewed in the literature [7], [9], [10], [11], [12]. Therefore, in the following we select for attention a few new results with an emphasis on recent work on the dynamic structure, electrical conductivity and equation of state data of low density alkali metals.

Metal-Nonmetal Transition

The problem of the interrelation of the metal-nonmetal transition and the liquid-vapour phase transition in fluid alkali metals has received considerable theoretical attention in the past. The pioneering study of Landau and Zeldovitch [13] suggested the possibility of separate first-order electronic and liquid-vapour transitions in fluid metals. Subsequent theoretical attempts to model the statistical mechanics of the metal-nonmetal transition in fluids reach similar conclusions but are still insufficient to provide a clear-cut answer to this question from theory [14], [15], [16], [17], [18], [8].
From the experimental [2] standpoint it is now unquestionable that there is no sharp (first-order) electronic transition except across the liquid-vapour phase change for fluid cesium, i.e. the liquid vapour phase separation tends to separate the nonmetallic and metallic fluids and it seems that the variation in density is the most influential factor governing the variation of the electronic structure in course of the metal-nonmetal transition. This is most vividly demonstrated by data such as those displayed in Figures 1 and 2 which show a selection of the most reliable existing experimental results for the electrical conductivity (Figure 1) and the density (Figure 2) of the coexisting liquid and vapour phases of cesium up to the critical point. Far below the critical point, when the density of the liquid phase is large, the electrical conductivity of liquid cesium exhibits characteristics typical for materials with metallic electron concentrations.
Perhaps the most obvious of these is the fact that the behaviour of the DC electrical conductivity in this range can be explained within the framework of the Ziman theory for the nearly free electron metal [19], [20] (dashed line in Figure 1). The application of this theory is justified when the average distance between scattering of the electrons by the disordered liquid structure, the electron mean free path, is substantially larger than the average near neighbour distance of the atoms. The dashed curve in Figure 1 has been calculated by employing the measured values of the static structure factor [21] together with the model pseudopotential form factor given by the Ashcroft "empty core" potential [22]. The calculation took into account exchange and correlation effects described by Sham [23]. Comparison between calculated and experimental conductivity values shows that the nearly free electron theory provides a good account of the variation of the conductivity for temperatures lower than 1100°C, i.e. at densities of the coexisting liquid higher than

1.3 g/cm^3. The failure of the nearly-free-electron theory at densities lower than 1.3 g/cm^3 reflects most probably the increased importance of electron correlations in this region.

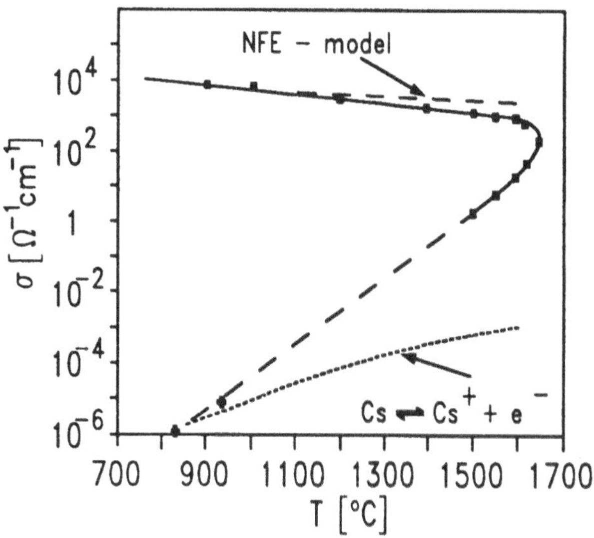

Figure 1: The experimental electrical conductivity data of cesium along the liquid-vapour coexistence curve in comparison with the values calculated for the nearly-free-electron model (dashed line) and for the thermal equilibrium ionization fraction of monoatomic vapour (dotted line).

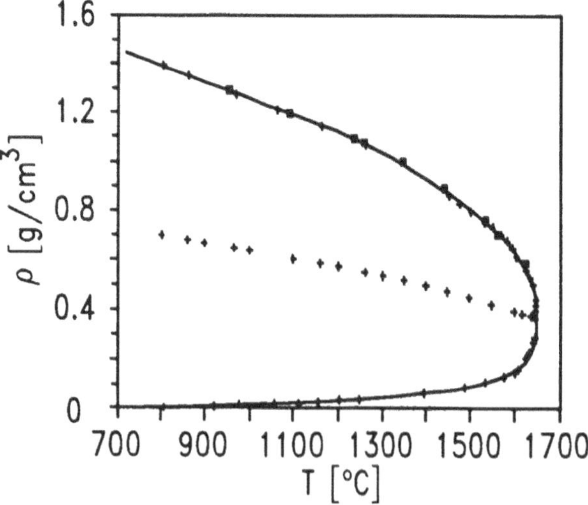

Figure 2: The coexistence curve of fluid cesium together with the curve of mean density (diameter).

Magnetic susceptibility [6], NMR [24] and optical reflectivity [25] studies of low-density liquid cesium have yielded evidence of strong electron-electron-correlations in the form of susceptibility enhancements and antiferromagnetic spin fluctuations. Figure 3 shows for example, the temperature dependence of the total magnetic mass susceptibility, expressed as χ_g, susceptibility per gram for cesium in the liquid and vapour phase along the coexistence curve.

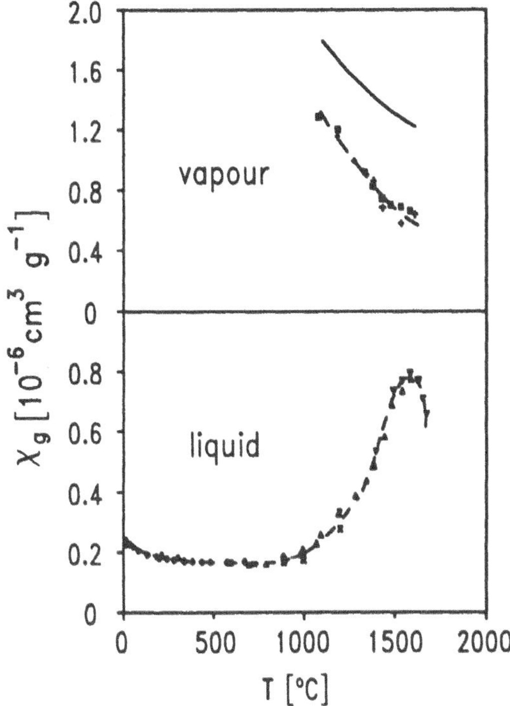

Figure 3: Total mass susceptibility of the coexisting liquid and vapour phases of cesium versus temperature (the critical temperature is $T_c = 1651$°C).

χ_g of liquid cesium exhibits a low temperature dependence at low temperatures followed by a striking increase at temperatures higher than 1100°C or densities lower than 1.3 g/cm^3. A maximum is reached at a temperature near, but clearly below the critical temperature $T_c = 1651$°C. This behaviour of χ_g reveals that significant changes in the electronic structure of cesium developes at low density. To obtain the electronic paramagnetic contribution from the total susceptibility χ_g, a correction for diamagnetism must be made, using the ionic core diamagnetism, known from studies of ionic solids and aqueous solutions [26] and values of the electronic diamagnetic susceptibility estimated from theory [27]. The dominant feature of the density-dependence of the paramagnetic susceptibility is as for χ_g a relatively sharp rise as the density falls below about 1.3 g/cm^3, and a decrease in the lowest density range, i.e. at the highest temperatures covered by the measurements shown in Figure 3.

Enhancement of the electronic susceptibility by electron correlation effects was described by Brinkman and Rice [28]. This description explicitly introduces the role of the atoms. The essential idea of this theory is that due to the large mutual repulsion energy of electrons occupying the same atom the majority of atoms is instantaneously singly occupied leaving on a small fraction doubly occupied. Since metallic properties require instantaneous double occupation of sites as the electrons move through the system, the effective number of electrons is reduced to twice the number of double occupied sites, or alternatively, the effective mass m^* in the highly correlated metal is enhanced according to $m^*/m_e = 1/2 \cdot f$. The combined analysis of the magnetic susceptibility [6] and Knight shift data [29] of expanded cesium reveals a correlation enhancement of the effective mass $m^* = 5 \cdot m_e$ at a density of 0.8 g/cm^3. This is the density where the maximum in χ_g occurs in Figure 3, the corresponding temperature of the coexisting liquid is near, but clearly below, the critical temperature T_c. Mott [8] has argued that the Ziman conductivity should be reduced by a factor $(m^*/m_e)^{-2/3}$ in the range where the effective mass is enhanced by correlations . This is qualitatively in agreement with the data presented in Figure 1.

A more direct study of the onset of correlation enhancement of the effective mass m* in expanded liquid cesium for densities smaller than 1.3 g/cm^3 can be made by measuring the optical reflectivity and extracting m^* from the plasma frequency $\omega_p^2 = 4\pi n_e e^2/m^*$. Reflectivity spectra [25] $R(\omega)$ versus $\hbar\omega$ for liquid cesium are shown in Figure 4.

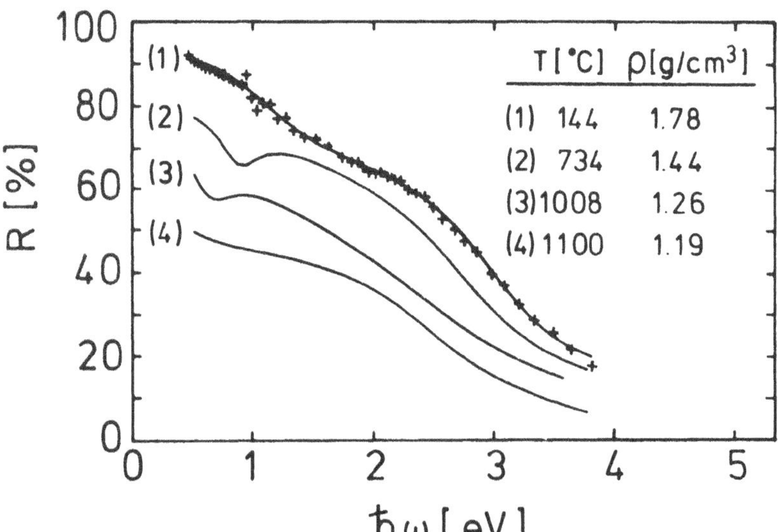

Figure 4: Examples of reflectivity spectra $R(\hbar\omega)$ of liquid cesium behind a sapphire window at different constant temperatures or densities.

These data encompass the range from the metal near the melting point (T = 144°C, ρ = 1.78 g/cm^3) to the slightly expanded metal (T = 1100°C, ρ = 1.19 g/cm^3). The onset of corrosion of the sapphire windows by liquid cesium limited the measurements to the temperature range smaller than 1100°C.

The optical reflectivity of liquid cesium at high density is very similar to that of a free electron metal. A relatively high reflectivity is observed at lower frequencies, i.e. in the infrared and visible range, as well as a decrease to low reflectivity values in the near ultraviolett. As it is expected for fluid materials the spectra are rather smooth and structureless as a function of $\hbar\omega$, but show pronounced changes with temperature.

Since the photon energy range is limited, the usual Kramers-Kronig inversion can not be used to obtain $\epsilon_1(\omega)$ and $\epsilon_2(\omega)$ from the reflectivity spectra. The alternative is a classical oscillator fit method [30]. This method is based on the use of a simple analytical form for the dielectric function, which consists of a sum of contributions from damped oscillators and the contribution of free electrons within the nearly free electron Drude model. A typical example for the quality of the fit is illustrated for one spectrum in Figure 4.

In Figure 5 examples of $\epsilon_1(\omega)$ and $\epsilon_2(\omega)$ derived by the oscillator-fit method of the R(ω)-spectra are displayed. In terms of the above mentioned interest in the evolving electronic structure, the ϵ_1-curves are of special interest because near the plasma resonance, $\hbar\omega_p$, ϵ_2 is very small so that $\hbar\omega_p$ is determined quite accurately by the condition $\epsilon_1(\hbar\omega_p)=0$ (see the inset in Figure 5). These $\hbar\omega_p$-values are plotted in Figure 6 in the form $(\hbar\omega_p)^2$ versus density. It is seen from a glance at Figure 6 that for densities ρ smaller than 1.3 g/cm^3 the optical mass determined from the equation $\omega_p{}^2=4\pi n_e e^2/m^*$ begins to increase. As noted above, the apparent onset of increased optical mass enhancement coincides with changes in density dependence of a number of other properties (electrical conductivity, magnetic susceptibility and Knight shift). It seems therefore reasonable to identify the density value of about 1.3 g/cm^3 as the onset of significant modification of electronic properties by electron correlation effects. The density of this onset is about $^2/_3$ that of the solid or the liquid at the melting point.

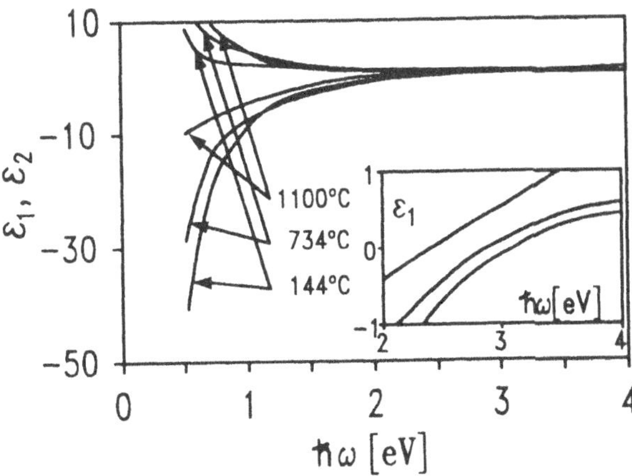

Figure 5: Examples for $\epsilon_1(\omega)$ and $\epsilon_2(\omega)$ of liquid cesium at different constant temperatures or densities. The inset displays the energy range around $\epsilon_1(\hbar\omega_p)=0$.

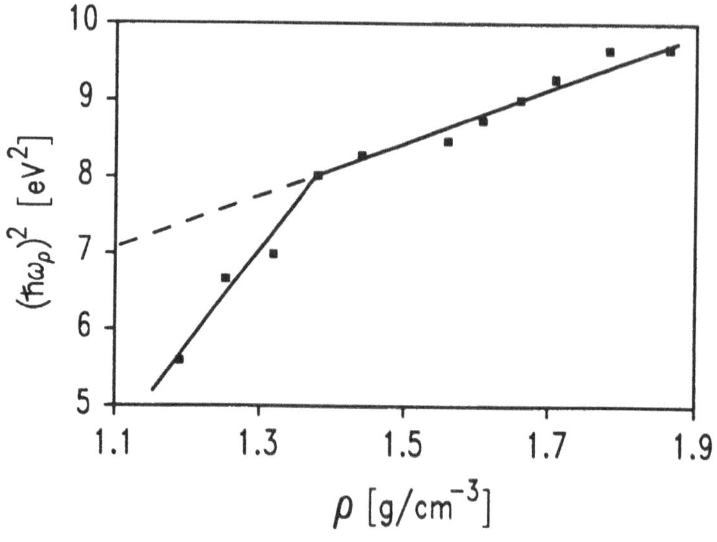

Figure 6: Density dependence of the plasma frequency of liquid expanded cesium.

Like the masses obtained from the magnetic susceptibility, the optical masses at high densities near the melting point of liquid cesium are reasonably close to that of the solid and agree well with the free electron values, i.e. $m^* = 1.15 \cdot m_e$. This observation shows that the loss of long range order at the melting transition is insufficient to cause a major change in the electronic structure. This is because, like most liquids, alkali metals close to the melting point posses a large degree of correlation in the atomic positions [21]. Many electronic properties like electrical conductivity, magnetic susceptibility, optical reflectivity etc. then resemble those of the corresponding solids. For liquid cesium this remains basically true for densities larger than 1.3 g/cm^3. The generally accepted description regards the structure of liquid cesium in that density range as built up of single screened ions, each diffusively uncoupled from each other. This is in accord with the essentially monoatomic arrangement typifying the bcc crystalline state of cesium.

However, as noted above this correspondence between liquid and solid applies only at high densities. When the liquid evaporates, the low density vapour cannot be regarded as monoatomic. There is clear experimental evidence that dimerization is common in the alkali metal (M) vapour phase. M_2-molecules are formed in which the M-atom is closer to its molecular partner than to any other atom in the vapour, and the paired electrons are confined to their home molecule. Direct experimental information about stable species such as Cs_2 present in cesium vapour is available from measurements of the optical absorption [31], [32]. Figure 7 shows a selection of such data for the saturated vapour phase of cesium in the temperature range between 547°C and 947°C, i.e. at densities between $5.7 \cdot 10^{-4}$ g/cm^3 and 0.013 g/cm^3. The dominant species contributing to the spectra are the neutral atom and the neutral dimer.

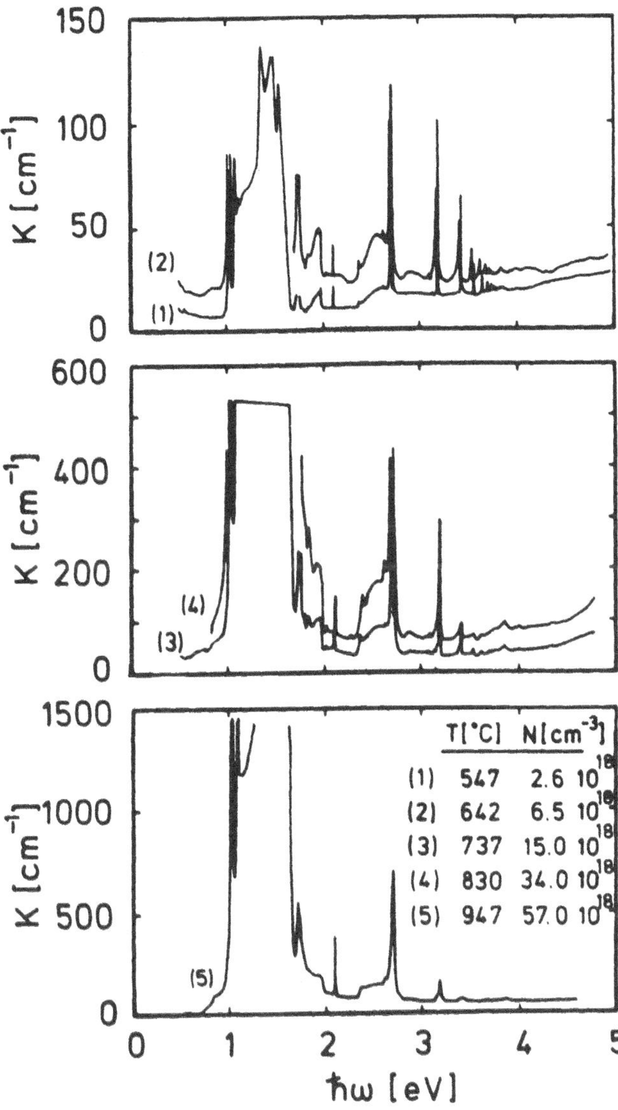

Figure 7: Absorption spectra of cesium vapour between 547°C and 947°C i.e. at densities between $5.7 \cdot 10^{-4}$ g/cm^3 and 0.013 g/cm^3.

The atom contributes a series of fundamental electron absorption lines due to the transition $6^2S_{1/2} \rightarrow n^2P_{1/2,3/2}$. In addition to this dipole allowed s-p transitions, the quadrupol allowed $6^2S_{1/2} \rightarrow 5^2D_{3/2,5/2}$ lines are also detectable in the absorption spectrum. The dominating feature of the spectrum is a number of Cs_2-molecular electronic bands with maxima at about 2.6 eV, 1.9 eV and between 1.0 and 1.65 eV. These experimental data clearly show that a major effect of heating the coexisting vapour phase to higher temperature, i.e. toward the critical region , is an increase in the relative concentration of the neutral dimer. This reflects a competition between increasing density which tends to increase the dimer concentration, and increasing temperature which tends to decrease their concentration by thermal dissociation.

The inference of a high concentration of dimers in the saturated cesium vapour close to the critical point is also consistent with the behaviour of the magnetic susceptibility (Figure 3) which shows a strong diamagnetic deviation from the Curie law on approaching the critical region along the coexistence curve. An interpretation of the susceptibility data in terms of a crude model in which the total vapour susceptibilities are split into a mixture of paramagnetic monoatomic Cs and diamagnetic Cs_2 molecules suggests more than 25% dimer concentration at a temperature of about 1600°C and a corresponding density of 0.15 g/cm^3 [6] which is only about 40% of the critical density. Consequently one can expect, that the dimer concentration will increase further on approaching the critical point. However it must be pointed out that this crude model of the vapour susceptibility does not consider the possible occurence of higher diamagnetic molecular associations or the formation of the Cs_2^+-molecule. The formation of such charged clusters like Cs_2^+ or Cs_n^+ must however play an important role in the high density cesium vapour.

This is seen from a glance at Figure 1 where the measured electrical conductivities of the vapour are shown in comparison with those calculated (dotted line) for the thermal equilibrium ionization fraction of monoatomic cesium employing the vacuum ionization potential of 3.89 eV (Saha equation). The agreement between experimental and calculated values is sufficient only at very low densities where the overwhelmingly dominant species present in the vapour is the atomic monomer [6]. At higher densities the conductivity can no longer be described on the basis of the assumption that the vapour is monoatomic. This is not surprising because with higher vapour densities, species like Cs_2 and higher clusters will occur. These clusters have ionization potentials much lower than the corresponding atoms so that much higher degrees of ionization should be expected than those simply calculated by the Saha equation. This, on the other hand, will result in more stable ionized clusters Cs_n^+ which certainly must have a strong effect on all properties in the critical region.

Be that as it may, the optical and the magnetic data suggest that the diatomic structure of cesium persists up to relatively high densities. The question to emerge, therefore, is up to what density and temperature the pairing mechanism present in the vapour can survive the condensation to the liquid state. This question is based on the view that the mechanism for the nonmetal- to metal transition which occurs in diatomic molecules with increasing density, may begin with metallization of the diatomic system by an overlap of the valence and conduction bands which at higher density is followed by a gradual dissociation into a monoatomic state [33]. The existing experimental data do not provide information related to this question. As noted above, liquid cesium at high densities shows characteristics typical of a nearly free electron metal. As the density is reduced below about 1.3 g/cm^3, qualitative changes develop. The measured electrical conductivity is clearly lower than the nearly-free electron value. The failure of the nearly free electron theory reflects the increased importance of electron correlation which is signaled by changes in a number of properties in this region.

It must be emphasized, however, that the preceeding discussion of the qualitative changes in the optical reflectivity and magnetic susceptibility in liquid cesium with decreasing density has been largely based on concepts familiar from the study of solid metals at low temperatures. The influence of the high temperatures near the critical point has not been taken into account.

A complementary theoretical approach developed by Alexeev and Jakubov [34] takes as the starting point the description of the low-density, high-temperature vapour as a weakly ionized plasma. Hernandez [35], [36] has adopted this point of view and calculated numerical results for the concentrations of the various polyatomic species as the density of the vapour is increased. This approach provides a consistent description of the magnetic properties up to the critical point and predicts, especially, an important role for the diamagnetic dimers in the critical region.

The theory of the dense plasma can, at least conceptually, be extended through the critical region into the low density liquid and predicts, for example, substantial spin pairing at liquid densities [37]. This would be consistent with the decrease in the susceptibility when the density falls below 0.8 g/cm^3. However, it must be emphasized that Warren et al. [29] were able to show that the electronic spin susceptibility in this density range follows the Curie law expected for one electron per atom obeying classical Boltzmann statistics. The latter is consistent with the view that cesium is highly correlated also for densities below 0.8 g/cm^3 and it is not necessary to invoke a substantial fraction of paired spins to explain the behaviour of the susceptibility in this range of the subcritical liquid.

Static and Dynamic Structure

It is evident from the foregoing that the structure of liquid alkali metals at high densities can be regarded as monoatomic whereas a high amount of diamagnetic molecular associations like Cs_2 and probably larger clusters are present in the vapour. The challenge to the understanding of the structure arises approaching the region near the liquid-vapour critical point where remnants of both limits (diatomic and monoatomic) must be present. The diamagnetic associates present in the dense vapour might well survive the passage to the metallic liquid. Thus, while liquid alkali metals might be regarded as monoatomic near the melting point, the viewpoint might be rather different for the lower density liquid regime where dimers or higher clusters can be the appropriate subunits.

This problem has been a primary motivation for neutron scattering measurements of the static structure factors of fluid Cs und Rb up to the critical region [21], [38].

Typical results are presented in Figure 8 for Cs in form of the Fourier-Transforms of the static structure factors S(Q), i.e. the pair correlation functions g(R), at various temperatures and densities ranging from the melting point to the critical region. With decreasing density or increasing temperature a number of changes in g(R) are noteworthy.

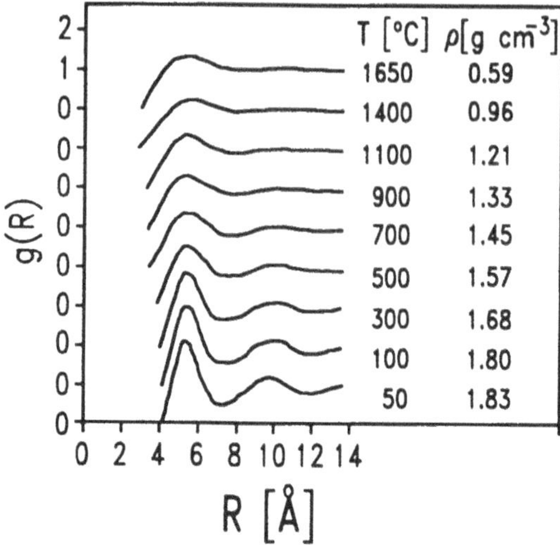

Figure 8: The pair distribution function g(R) of expanded liquid cesium at conditions near the liquid-vapour coexistence line.

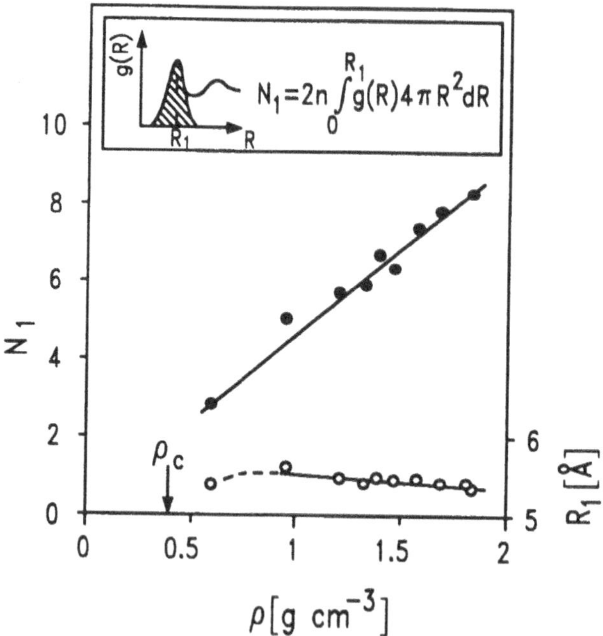

Figure 9: Average number N_1 (•) of nearest neighbours for liquid Cs and the average distance R_1 (O) of nearest neighbours for liquid Cs as a function of the density. (The inset shows the method employed to determine N_1).

First, the intensity of the main peak of g(R) is strongly reduced and broadened while the peak position R_1 shifts only slightly towards higher R.
Now the pair correlation function g(R) is related to the radial distribution function $N(R)=4\pi R^2 ng(R)$ which determines the number of neighboring atoms N(R)dR in a spherical shell of radius R and thickness dR centered on a particular atom of interest. The average coordination number N_1 is determined by the area under the first peak in g(R), whereas the average nearest neighbour-distance R_1 is given by the position of this peak. Analysis of data such as those of Figure 8 shows that for Cs, N_1 tends to decrease as the density is decreased by thermal expansion (Figure 9). It is particularly noteworthy that the position of the first peak in g(R), namely R_1, remains virtually constant while N_1 decreases by about a factor of three. The volume of the liquid increases with increasing temperature by about the same factor over this range. This indicates that clustering occurs as the density decreases, keeping many atoms highly coordinated. The average coordination number N_1 then decreases because of an increase in the number of atoms on the surface of clusters. The average nearest neighbour-distance R_1 tends to increase at the low density- or high temperature region, respectively . This is in accord with the suggestion that liquid cesium in this density region consists of two atomic species corresponding to long and short distances of atomic separation.
Some interesting insight into the question whether the pairing mechanism, present in the dense vapour phase of alkali metals, can survive the condensation to the liquid state can be obtained by examining recent coherent inelastic neutron scattering spectra for liquid rubidium [39], [40] which extend near to the critical region. Typical results for the coherent dynamic structure factor $S(Q,\omega)$ at constant monumentum transfer $Q=1\text{Å}^{-1}$ are presented in Figure 10 for rubidium as a function of energy transfer $\hbar\omega$ at various temperatures along the liquid-vapour coexistence curve. Figure 10 includes also data near the melting point at 50°C from the early work of Copley and Rowe [41], which clearly show that the metal rubidium, unlike dense Lennard-Jones (inert gas) systems, exhibits distinct collective excitations. These excitations extend over a relatively large Q-region. The excitations are still visible for the higher temperatures 800°C, 1100°C and 1400°C, but the widths of the peaks are so large that they are no longer resolvable and only broad shoulders are indicated.
The presence of such modes for liquid metals, and their absence in inert gas liquids, has been attributed to the characteristic features of the effective metallic interaction potential. In fact, molecular dynamics calculations [42], [43], [44], using typical metallic interaction potentials, are in excellent agreement with the experimental results for liquid Rb near the melting point [42] and up to temperatures of at least 1400°C [43]. Figure 11 shows the dispersion relation of the collective excitation of expanded rubidium at temperatures of about 70°C and 1400°C obtained by the molecular dynamics simulation of Hoshino et al. [43] which employes an effective pair potential derived by the pseudopotential theory. They calculated the static structure factor in their modified hypernetted chain approximation and the dynamic structure factor in the viscoleastic approximation.

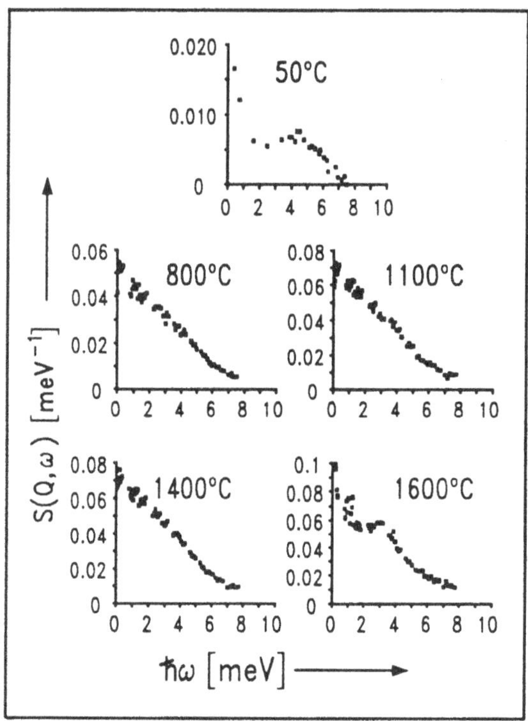

Figure 10: The dynamic structure factor S(Q,ω) of expanded liquid rubidium at constant momentum transfer $Q = 1Å^{-1}$ and at selected temperatures near the liquid-vapour coexistence line [39],[41].

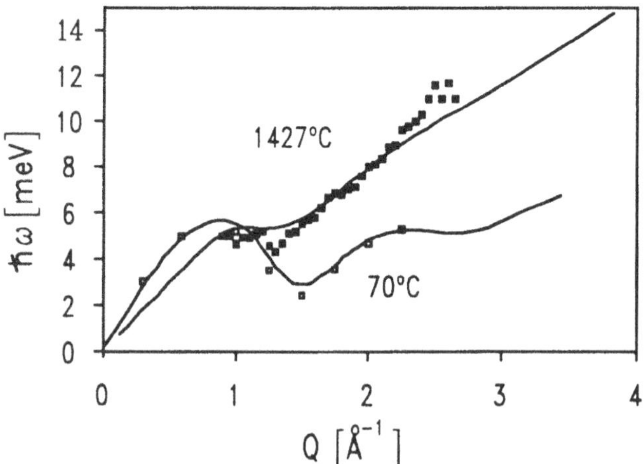

Figure 11: Dispersion Relation of the current correlation function. Theoretical values [43] (lines) are compared with experimental data [40], [41].

The dispersion obtained from the experimental $S(Q,\omega)$-data in the same temperature range are compared with the theoretical data in Figure 11. The experimental dispersion relation was determinded from the maximum of the longitudinal current density correlation function $J_l(Q,\omega) = \omega^2 \cdot S(Q,\omega)/Q^2$. Examples of the behaviour of $J_l(Q,\omega)$ for two Q-values and two temperatures are shown in Figure 12. It is obvious that $J_l(Q,\omega)$ is more useful to determine the dispersion quantitatively because the detailed structure at smaller $\hbar\omega$ present in $S(Q,\omega)$ is lost in this respresentation. As noted above, the dominant effect of thermal expansion on the structure of fluid alkali metals is a reduction of the average coordination number rather than an increased near neighbour distance. The latter grows only very little. The consequences of these structure changes are reflected in Figure 11. The minima of the dispersion curves shift slightly to lower Q values as the maxima of S(Q) and the free particle dispersion becomes more pronounced at higher temperature (lower density) as one would expect. More interesting is the excellent agreement between the calculated and the experimental dispersion relation at the high temperature of 1400°C and a density of about three times the critical density which indicates that the mean forces given by the metallic binding and the screening by the nearly free electrons is still controlling the dynamics of liquid rubidium at this conditions. This is no longer the case, if the temperature is increased still further (see e.g. in Figure 10, T=1600°C, the corresponding density is about twice the critical density), a clear change in the shape of $S(Q,\omega)$ is observed. $S(Q,\omega)$ at $Q = 1Å^{-1}$ exhibits a well defined excitation peak around $\hbar\omega = 3.2$ meV.

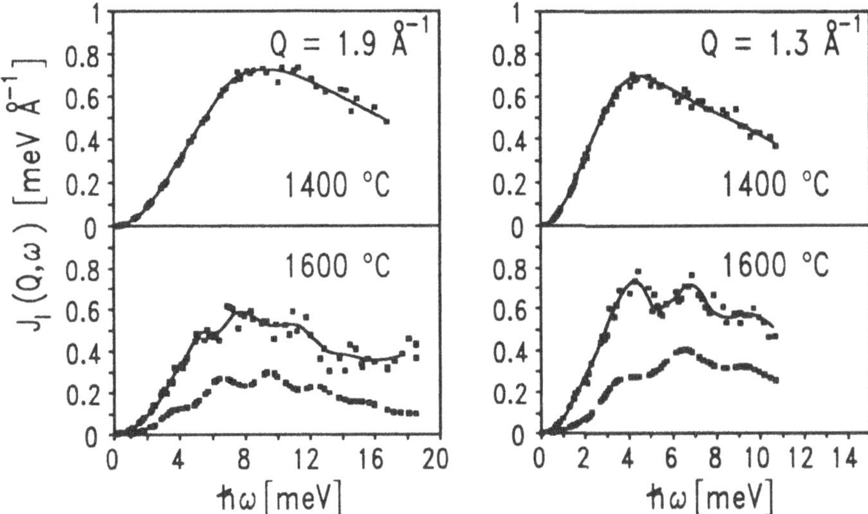

Figure 12: Comparison of the experimental and calculated longitudinal current correlation function $J_l(Q,\omega) = \omega^2 \cdot S(Q,\omega)/Q^2$ for two different Q-values and temperatures.

One way of interpreting this observation is to propose that the peak is due to an optic-type mode, i.e. optic vibration, in which two species tend to move in opposite directions. The intensities of such excitation peaks must depend strongly on the momentum transfer Q.

This can be seen at a glance by considering the model for scattering from a particle executing harmonic oscillations about a centre for $k_BT > \hbar\omega_0$:

$$S(Q,\omega) = \sum_{n=-\infty}^{\infty} I_n(y)\, e^{-y}\, \delta(\omega - n\omega_0)\ ,\ y = \frac{\hbar^2 Q^2 k_B T}{M\,(\hbar\omega_0)^2} \tag{1}$$

Herein $I_n(y)$ are modified Bessel-Functions of the first kind and n-th order, $\hbar\omega_0$ is the energy difference between adjacent energy-levels of the oscillating particle and the δ-function under the sum ensures conservation of energy during the interaction between the neutron and the molecule. k_B is the Boltzmann-constant and M the mass of the particle.
The Q-dependence of the intensities of the excitation peaks of this model is shown for $\hbar\omega$=3.2 meV at a temperature of 1600°C in Figure 13. It is evident that the intensities have a maximum near 1Å^{-1} in agreement with the experimental observations [39]. For Q-values larger than 1Å^{-1}, the intensity becomes so small that the peaks are no longer resolvable, and a knowledge of the relative intensity of the peaks is required in order to analyse the data.

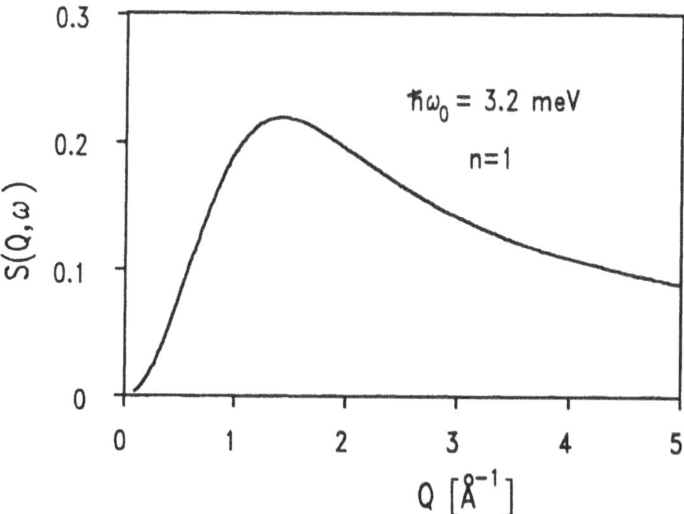

Figure 13: The Q-dependence of the intensities of the excitation peaks for a particle executing harmonic oscillations about a centre for $\hbar\omega_0$=3.2 meV as taken from the experimental result, T=1600°C

In order to demonstrate that the experimentally observed variation of the observed scattering laws with monumentum transfer Q is qualitatively consistent with the predictions of the harmonic oscillator model we compare in Figure 12 the experimental

and calculated longitudinal current correlation function $J_l(Q,\omega)=\omega^2\cdot S(Q,\omega)/Q^2$ for two different Q-values. The shapes of the two curves are surprisingly similar.
This observation is consistent with the view that paired units with intramolecular dynamic effects are present in liquid expanded rubidium under these conditions.

Asymmetry in the Liquid-Vapour Coexistence Cureve

The most significant experiments relevant to the effect of the assymetry of the liquid-vapour interparticle interaction of fluid alkali metals are those on the coexistence curves of cesium and rubidium [1].
Despite of the extreme conditions at the critical points of these metals, the liquid-vapour coexistence curves have been measured to a resolution $\tau=|T-T_c|/T_c\approx 5\cdot 10^{-4}$, which is close enough to demonstrate the important differences between metallic and insulating molecular fluids. Figure 2 shows a plot of the coexisting liquid (ρ_L) and vapour (ρ_V) densities together with the curve of the mean densities $\rho_d=(\rho_L+\rho_V)/2$ versus temperature for the metal cesium.
The diagram demonstrates an interesting consequence of the strong thermodynamic-state-dependence of the effective particle interaction in metallic fluids as the critical region is traversed. Fluid metals violate the hundred year old emperical law of rectilinear diameter [46] over a surprisingly large temperature range. By contrast, the deviations from this law are extremely (mostly immeasurably) small for the coexistence curves of essentially all nonmetallic one-component fluids [47].
The law states that the locus of the tie-line mid-points $\rho_d=(\rho_L+\rho_V)/2$ is a linear function of T. Since both ρ_L and ρ_V approach the limiting density ρ_c at the liquid-vapour critical point, the law can be written

$$\rho_d-\rho_c=D_1\tau \tag{2}$$

with D_1 constant. Modern theory of liquid-vapour critical phenomena based on certain solvable lattice models [48], thermodynamic arguments [49], and renormalization group studies [50] permits calculation of the diameter anomaly including effects of large scale density fluctuations. The theory predicts that the temperature derivative of the diameter $d\rho_d/dT$, diverges at least as fast as the constant-volume specific heat c_V. That is, as the reduced temperature τ goes to zero, the diameter varies as

$$\rho_d-\rho_c=D_0\tau^{(1-\alpha)}+D_1\tau+\ldots \tag{3}$$

where $\alpha=0.11$ is the same exponent that describes the behaviour of the constant-volume specific heat c_V. Since $(1-\alpha)=0.89$ is not very different from unity, the true singularity is difficult to separate from the analytic temperature term. The coefficient D_1 does not even have to be much larger than D_0 for the analytic term to dominate the entire range accessible to experimentation. The latter causes the difficulty in observing the $(1-\alpha)$-

singularity for most metallic fluids. And it was only with high-precision experiments on Ne, N_2, C_2H_4 and C_2H_6 [47] that is was possible to reveal that singularities with the predicted exponent $(1-\alpha)$ do indeed exist for nonconducting [51] molecular fluids, i.e. that $D_0 \neq 0$ for these fluids as well. Analysis of these data [51] has led to the suggestion that many-body interactions lead to the anomalous $\tau^{(1-\alpha)}$-term in these fluids. In particular, it is believed that the symmetry-breaking present in these fluids due to many-body dispersion forces may be understood in terms of a thermodynamic state dependent effective pair interaction. Consequently, there seems to be a natural connection between this explanation and the observation of very large amplitudes of the diameter anomalies in fluid metals where the occurence of the metal-nonmetal transition implies a strong variation of the interparticle interaction.

It is evident from a glance at Figure 14 that the diameter anomaly for fluid cesium is characterized by an exponent $(1-\alpha)$. The anomaly is so strong that a nearly pure power-law behaviour is seen over several decades in the reduced temperature τ. It is certainly tempting to speculate [5] that the strong dependence on density of the effective interparticle potentials in the range of the gradual metal-insulator transition is responsible for the large amplitude D_0 of the $\tau^{(1-\alpha)}$-term in alkali metals.

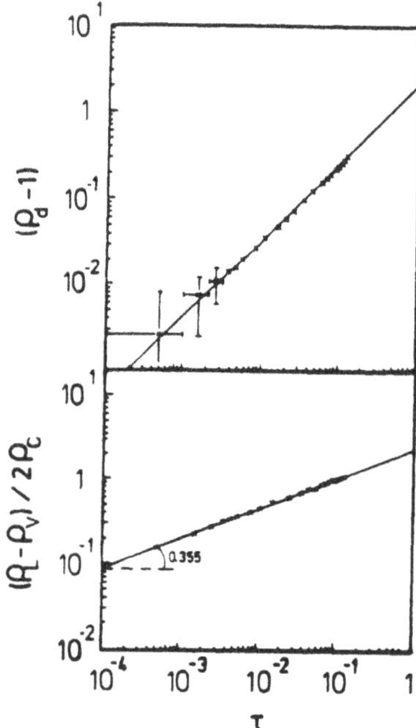

Figure 14: Power-law analysis of the order parameter and the diameter of cesium.

The presence of the strong liquid-vapour interparticle assymetry in metals does not, however, imply that the shapes of their coexistence curves cannot be described with the same exponents β as molecular fluids in equation (4)

$$\rho_L - \rho_V = B\,\tau^{\beta} \qquad (4)$$

which describes the divergence of the order parameter approaching the critical point. Any speculation that the critical points of metals could fall into a different universality class than the insulating fluids is clearly not true as is demonstrated by Figure 14 for cesium. The parameter β lies between 0.35-0.36, a value only slightly higher than that found for the three-dimensional Ising-model. Since these values are determined by single power-law fits for the range of temperatures accessible experimentally for cesium it would be modified by allowance for corrections because the data do not extend to the true critical region. The presence of strong state-dependent interactions in metals becomes visible mainly in the behaviour of the diameter. The main difference between the coexistence curves of molecular and metallic fluids is the magnitude of the coefficient D_0 in equation (3).

Ackowledgement

Financial support by the Deutsche Forschungsgemeinschaft and the Fonds der Chemischen Industrie is gratefully acknowledged.

References

[1] Jüngst S., Knuth B. and Hensel F. (1985), Phys. Rev. Lett. **55**, 2160
[2] Hensel F., Stolz M., Hohl G., Winter R., and Götzlaff W. (1991), Journal de Physique **IV**, Colloque C5, supplement au Journal de Physique I, Vol. 1, 191
[3] Binder H. (1984) Doctoral Thesis, University of Karlsruhe, FRG
[4] Ohse W. (1985) "Handbook of Thermodynamic and Transport Properties of Alkali Metals", Blackwell Scientific Publications, Akademic Press, Oxford
[5] Goldstein R., and Ashcroft N. W. (1985), Phys. Rev. Lett. **55**, 2164
[6] Freyland W. (1979), Phys. Rev. **B20**, 5140
[7] Freyland W. and Hensel F. (1985), "The Metallic and the Nonmetallic State of Matter", Editors: Edwards P. P. and Rao C. N., Taylor and Francis (London), p 93
[8] Mott N. F. (1974), "Metal-Insulator Transition", Taylor and Francis, London
[9] Yonezawa F. and Ogawa T. (1982), Prog. Theo. Suppl. **72**,1
[10] Hensel F. (1990), J. Phys.: Condens. Matter **2**, SA33-SA45
[11] Hensel F. and Uchtmann H. (1989), Ann. Rev. Phys. Chem. **40**, 61
[12] Stratt R. M. (1990), Ann. Rev. Phys. Chem. **41**, 175
[13] Landau L. and Zeldovitch G. (1943), Acta Phys. Chem.USSR **18**, 1940
[14] Mott N. F. (1978), Phil. Mag. **37**, 377
[15] Nara S., Ogawa T. and Matsubara T. (1977), Prog. Theo. Phys. **57**, 1474
[16] Ebeling W. and Sändig R. (1973), Ann. Phys. (Leipzig) **28**, 289
[17] Ebeling W., Kraeft W.D. and Kremp D. (1976), Theory of Bound States in Plasmas and Solids. Berlin: Akademie-Verlag

[18] Krumhansl J.A. (1965), in: "Physics of Solids at High Pressures", Tomizuka C.T. and Emrick R.M. (eds.), New York, Academic Press
[19] Faber T. E. (1972), "An Introduction to the Theory of Liquid Metals", Cambridge University Press
[20] Ziman J. M. (1961), Phil. Mag. **6**, 1013
[21] Winter R., Bodensteiner T., Gläser W. and Hensel F. (1987), Ber. Bunsenges Phys.Chem. **91**, 1327
[22] Ashcroft N. W. (1966), Phys. Lett. **23**, 48
[23] Sham J. C. (1965), Proc. R. Soc. **A283**, 33
[24] El-Hanany W., Brennert G. F. and Warren W. W. (1983), Phys. Rev. Lett. **50**, 540
[25] Knuth B., Hensel F., (1990) High Pressure Res. **5**, 552
[26] Meyers W.R., (1952), Rev. Mod. Phys. **24**, 15
[27] Komozowo H. and Matsudowa N., (1960) Prog. Theor. Phys. **23**, 433
[28] Brinkmann W. F. and Rice T. M. (1970), Phys. Rev. **B2**, 4302
[29] Warren W. W., Brennert G. F. and El-Hanany U. (1989), Phys. Rev. **B39**, 4038
[30] Verleur H. W. (1968), Opt. Soc. Am. **58**,1356
[31] Knuth B. (1990), Doctoral Thesis. Univ. Marburg
[32] Müller B. (1993), Doctoral Thesis. Univ. Marburg
[33] Ashcroft N. W. (1988), Zeitschrift für Physik. Chem.N.F. **156**, 41
[34] Alexeev V.A. and Jakubov I.T. (1983), Phys. Report **96**, 1
[35] Hernandez J. P. (1986), Phys. Rev. **A34**, A316
[36] Hernandez J. P. (1986), Phys. Rev. Lett. **57**, 3183
[37] Redmer R. and Röpke G. (1989), Contr. Plasma Phys. **29**, 343
[38] Franz G., Freyland W., Gläser W., Hensel F. and Schneider E. (1980), Proc. 4th Int. Conf. Liquid and Amorphous Metals, J.Phys. Colloq **C8 41**, 194
[39] Pilgrim W.-C. (1992), Doctoral Thesis Univ. Marburg
[40] Pilgrim C., Winter R., Hensel F., Morkel C. and Gläser W. (1991), Ber. Bunsenges. Phys. Chem. **95**, 1133
[41] Copley J. R. D. and Rowe J. M. (1974), Phys.Rev.Lett. **32**, 49
Copley J. R. D. and Rowe J. M. (1974), Phys.Rev. **A9**, 1656
[42] Rahman A. (1974), Phys. Rev. Lett. **32**, 52
Rahman A. (1974), Phys. Rev. **A9**, 1667
[43] Hoshino K.,Ugawa H. and Watanabe M. (1992), J. Phys. Soc. Japan **61**, 2182
[44] Kahl G., Kambayashi S. and Nowotny G. (1992), Proc. of the VIII Int. Conf. on Liquid and Amorphous Metals, Vienna
[45] Lovesey S. W., (1987) "Theory of neutron Scattering from Condensed Matter" Vol. 1, Clarendon Press, Oxford
[46] Cailletet L. and Mathias E. C. (1886), R. Acad. Sci. **102**, 1202
[47] Goldstein R. E., Parola A., Ashcroft N. A., Pestak M.W., Chen M.H.W., de Bruyn J.R. and Balzarin D.A., (1987), Phys. Rev. Lett, **58**, 41
[48] Rowlinson J. S. (1970), Adv.Chem.Phys. **41**, 1
[49] Mermin N. D. (1971), Phys. Rev. Lett. **26**, 957
[50] Nicoll J. F. (1981), Phys. Rev. **A24**, 2203
[51] Goldstein R. E. and Parola A. (1988), J. Chem. Phys. **88**, 7059

NEUTRON AND X-RAY SCATTERING OF FLUIDS AT HIGH PRESSURE AND HIGH TEMPERATURE

R. WINTER
Ruhr-University of Bochum
Institute of Physical Chemistry
Universitätsstraße 150
D-4630 Bochum 1
Germany

ABSTRACT. The purpose of this paper is to describe the high-pressure high-temperature neutron- and X-ray-scattering technique, the experimental set-ups and data evaluation procedures, which can be used for the investigation of the structure and dynamics of liquids up to temperatures of about 2000 K and to pressures of several hundred to thousand bars. For example, these techniques allow to determine the structural and dynamical properties of liquid metals, such as liquid cesium and mercury, and liquid semiconductors, such as liquid selenium and sulphur, from the melting point up to their liquid-vapour critical point.

1. INTRODUCTION

Neutron- and X-ray diffraction is a well-established probe for investigations of the microscopic structure of fluids. In the past, most of these studies have been performed at moderate conditions of pressure and temperature or at either high pressure or elevated temperature. The combined application of high temperature and pressure has received less attention. However, there are several reasons why pressure studies at elevated temperatures are of interest. At high temperatures and elevated pressures, both excitation phenomena and chemical reactions can give rise to new features in the structural, electronic and dynamical properties of fluids. For instance, a great deal of effort has been focused during the past years on the study of liquid metals up to their liquid-vapour critical point [1,2]. This effort has been motivated by the strong theoretical interest in the transition from a degenerated plasma (fluid metal) to a slightly ionized nonideal plasma (nonmetal), that occurs if a fluid metal such as cesium, rubidium or mercury, is expanded by heating it up to its liquid-vapour critical point. The occurrence of the metal to nonmetal transition implies that the effective interatomic cohesion must change dramatically during the expansion of the fluid system. Such a change may strongly effect the structural and dynamical features of the fluid. However, fluid metals are difficult to experiment with, because the high cohesive energy of metals places their critical points at very high temperatures and relatively high pressures. In order to investigate the structure

R. Winter and J. Jonas (eds.), High Pressure Chemistry, Biochemistry and Materials Science, 167–199.

and dynamics of these fluids up to their critical point (critical data e.g. of cesium: T_c = 1924 K, P_c = 92.5 bar, d_c = 0.38 gcm^{-3}, of Hg: T_c = 1751 K, P_c = 1673 bar, d_c = 5,80 gcm^{-3}), many technical difficulties have to be overcome, such as the containment of the highly corrosive metals at high pressures and temperatures.
It is the purpose of this paper to describe the high-temperature high-pressure neutron- and X-ray-scattering technique, the experimental set-ups and the data reduction procedures which allow to investigate the temperature and pressure dependence of the static and dynamic structure factor of liquid metals and semiconductors. As most of the investigations performed on these systems so far used the neutron scattering technique, we will mainly focus on this method. Whereas in the past theoretical descriptions of liquid metals and semiconductors could be tested mainly at the high densities near their triple point, these measurements open up the possibility of testing liquid state theory over a wide range of temperatures and densities, even up to conditions where the metallic properties of liquid metals must give way to an insulating state or where semiconducting liquids undergo chemical reactions or a transition to the metallic state.

2. BASIC SCATTERING THEORY OF SIMPLE LIQUIDS

The scattering of radiation by condensed matter may be related to the distribution of atomic positions if the wavelength λ of the radiation is of the order of magnitude of the interatomic spacings r. The distribution of scattered intensity contains information on the distribution of atoms, and the measured spectrum of energy transfer contains information on single and collective particle motion. A continuing objective in the theory of fluids has been to find and test the relationship between intermolecular forces and the microscopic structure and dynamics of the fluids [3,4].
Different kinds of radiation may be used for scattering experiments, such as electromagnetic (e.g. X-rays), electrons and neutrons. Differences arise due to the differences in the scattering properties of single atoms for each kind of radiation. X-rays can be considered as being scattered by the electron density seen from each nucleus. In the case of electrons, scattering takes place by interactions with electric charges, so that both electrons and nuclei play a part. However, because due to the strong interaction between the charged electrons and the sample, scattering takes place in the surface layers of the liquid only, so that the usefulness of this technique is often limited. In contrast, neutrons are scattered by the nuclei only and penetrate matter rather easily. They have the simplest scattering properties since they are scattered by a "point"-nucleus which is the atoms centre of mass.
Scattering experiments generally fall into two classes. First there are experiments in which only the angular distribution of the scattering intensity are determined, and secondly those in which both the angular distribution for the intensity and the energy transfer is measured. In the former experiment the intensity is related to the so-called static structure factor S(Q), and in the latter it is related to the dynamic structure factor $S(Q,\omega)$, where $\hbar Q$ and $\hbar\omega$ are the momentum and energy transferred to the particle in the scattering process.
The scheme of a scattering experiment is illustrated in Figure 1. If the incident and scattered wave vectors and energies are denoted by $(\mathbf{k}_0,\mathbf{k})$ and (E_0,E) respectively, then $\hbar\mathbf{Q} = \mathbf{k}_0-\mathbf{k}$ and $\Delta E = \hbar\omega = E_0-E$. The direction of a wave vector is the direction of

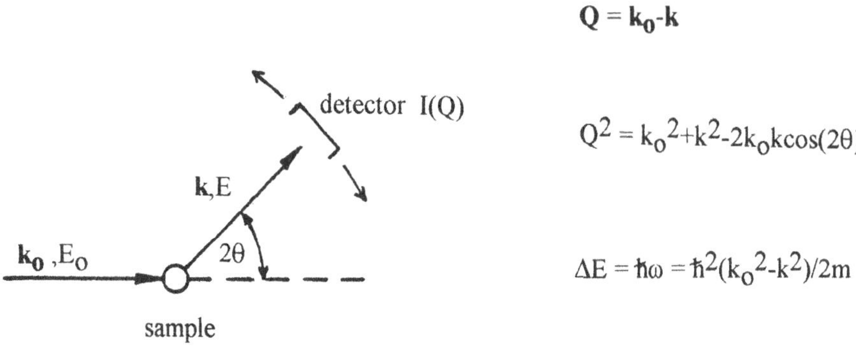

Fig. 1 Scheme of a scattering experiment

propagation of the plane wave and its magnitude is $2\pi/\lambda$. The ability of energy analysis conventionally is a unique feature of the neutron probe. The static and dynamic structure factors are the most useful for studying liquids [3,4] and will be discussed in more detail in the following.

The most important quantitity for dealing with the static structural problem of liquid order is the pair correlation function g(r). In simple atomic liquids, like liquid metals, only positional configurations are of interest. If the bulk liquid number density is n, then g(r) is defined such that if one sits on a particle at the origin $r = 0$, then the probability of finding a second particle at distance between r and r+dr is given by $4\pi r^2 g(r)dr$ and the density of atoms is found by multiplying this by n. This pair function is accessible via diffraction experiments. Let I(Q) be the intensity of radiation of wavelength λ, incident on the liquid sample and scattered through an angle 2θ, then in an elastic scattering experiment ($\Delta E = 0$, $|\mathbf{k_0}| = |\mathbf{k}|$) the magnitude of **Q** is given by

$$Q = (4\pi/\lambda)\sin\theta \tag{1}$$

The coherent scattering intensity I(Q) is proportional to the differential cross-section $d\sigma_{coh}/d\Omega$ for coherent scattering, which is defined as the flux scattered with a solid angle $d\Omega$ per unit incident flux, and which is related to the liquid structure factor by

$$I(Q) \propto d\sigma_{coh}/d\Omega = Nf^2 S(Q) \tag{2}$$

where N is the number of atoms in the sample and f is the scattering length per particle (atomic scattering factor $f_x(Q)$ for X-rays, scattering length b for neutrons). In turn, S(Q) is related to the pair correlation function g(r) by

$$S(Q) = 1 + n\int [g(r)-1]\exp(-i\mathbf{Q}\mathbf{r})d\mathbf{r} = 1 + n\int [g(r)-1][\sin(Qr)/(Qr)]4\pi r^2 dr \qquad (3)$$

Figure 2 shows measured data of S(Q) on liquid Cs near the melting point. The long wavelength limit of S(Q), i.e. S(0), can be related to fluctuations $<\Delta N^2>$ in the number of particles and thereby to the isothermal compressibilty χ_T:

$$S(0) = nk_B T\chi_T \qquad (4)$$

For liquid argon near its triple point, $S(0) \approx 0.06$, whereas for liquid metals $S(0) \approx 0.02$. This is in contrast to a gas, where $S(0) \approx 1$ except near the liquid-vapour critical point, where S(Q) diverges. The large r behaviour of g(r) as the critical point ($T = T_c$, $d = d_c$) is approached, is given by [5]

$$g(r)-1 \approx \text{constant} / r^{1+\eta} \qquad (5)$$

which leads to

$$S(Q) \approx \text{constant}/Q^{2-\eta} \qquad (6)$$

at $Q \to 0$ for 3 dimensional systems. Experimental evidence points to the fact that η is very small ($\eta \approx 0.03$) [5].

As seen in Figure 2a, S(Q) has pronounced oscillations at large Q. These primarily come from the hardness of the core of the interparticle interaction potential $\phi(r)$. In Figure 2b, its Fourier transform, g(r), is shown. There is a region inside the core diameter σ, where $g(r) = 0$, then g(r) rises sharply to its first peak followed by an oscillatory behaviour corresponding to successive shells of atoms centered on the one chosen as the origin. The steep rise of the first peak of g(r) mainly leads to the oscillations at large Q in S(Q).

In the most general theory of liquids, higher order distribution functions also appear, for example functions involving the positions of three or four particles. Information about e.g. the triplet correlation function $g_3(\mathbf{r}_1,\mathbf{r}_2,\mathbf{r}_3)$, the probability of finding a particle at $\mathbf{r}_3$, when there is another one at $\mathbf{r}_2$ and $\mathbf{r}_3$, respectively, can be gained from the pressure derivative of g(r):

$$k_B T[\partial g(r)/\partial p]_T = \int [g_3(\mathbf{r}_1,\mathbf{r}_2,\mathbf{r}_3)-g(r)g(r_{23})-g(r)g(r_{31})+g(r)]d\mathbf{r}_3 \qquad (7)$$

It can be seen from equation 7 that the experimental study of the pressure derivate of S(Q) and thus g(r) gives information on the three-body correlations, integrated over all values of the coordinates of the third particle. The appearance of the various pair functions in the square bracket implies that only the behaviour of the triplet function over the configurations, in which the third particle is close to the other two particles, is relevant.

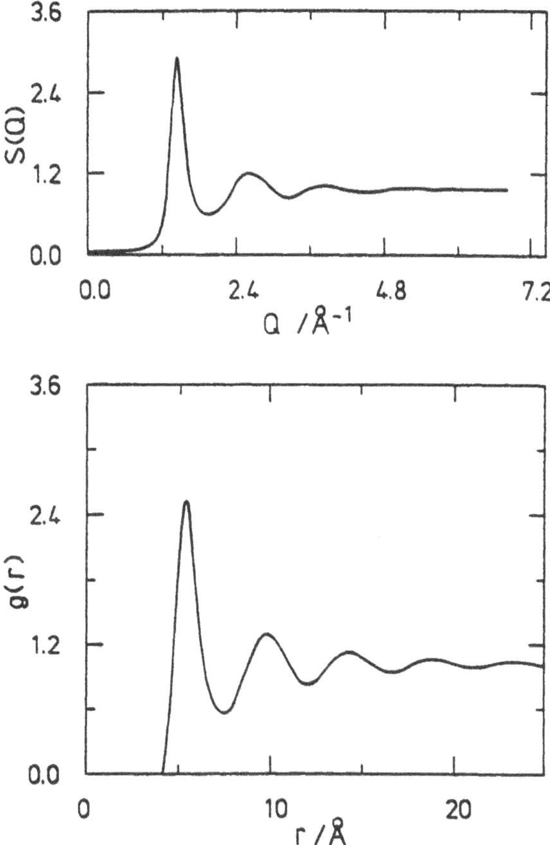

Fig. 2 a) Static structure factor S(Q) and b) pair correlation function g(r) of liquid Cs near its melting point (T = 303 K).

In order to include the motion of the particles, the pair correlation function was generalized by van Hove [6] and is expressed in the so-called van Hove space-time correlation function

$$G(\mathbf{r},t) = (1/N)\langle\sum_{i,j=1}^{N}\delta(\mathbf{r} + \mathbf{r}_i(0) - \mathbf{r}_j(t))\rangle \tag{8}$$

$G(\mathbf{r},t)$ describes the probability of finding an atom (i or j) at position $\mathbf{r}$ and time t, if there was an (other or same) atom at the origin at t = 0. Individual and collective motion of atoms in liquids can be described by this function. All the physical properties of the atomic many-body problem involved in the scattering process are hidden in the so-called scattering law $S(Q,\omega)$. As shown by van Hove, the relation between $S(Q,\omega)$ and $G(\mathbf{r},t)$ is

$$S(Q,\omega) = 1/(2\pi) \iint \exp(-i(\mathbf{Qr}-\omega t)G(\mathbf{r},t)d^3\mathbf{r}dt \tag{9}$$

Unlike X-ray scattering, the scattering lengths b_i for neutrons vary from isotope to isotope and depend on the relative orientation of neutron and nuclear spin. The averaging of the b_i over all these states may be expressed in terms of coherent and incoherent scattering lengths [4]

$$b^2_{coh} \equiv \langle b\rangle^2 \tag{10}$$

$$b^2_{inc} \equiv \langle b^2\rangle - \langle b\rangle^2 \tag{11}$$

by using the notation $\langle b_i^2\rangle = \langle b^2\rangle$ and $\langle b_ib_j\rangle - \langle b_i\rangle\langle b_j\rangle = \langle b\rangle^2$. The first part, which arises from the average scattering length $\langle b\rangle$ is coherent scattering, while the root-mean-square deviation of the scattering lengths ($\langle b^2\rangle - \langle b\rangle^2$) yields the incoherent scattering. It is only the average potential which can lead to interference and hence coherent scattering. Since the deviations from the average potential are randomly distributed, no interference can occur. If a sample consists of just one isotope of an element and this has zero nuclear spin, then $\langle b^2\rangle = \langle b\rangle^2$ and there is no incoherent scattering. By varying the isotopic composition of the sample, or by using polarized neutrons, it is possible to measure separately the coherent and incoherent differential cross-sections.
What is directly measured in a neutron scattering experiment is the doubly differential cross-section, which can be written with the notation above as

$$d^2\sigma/(d\Omega d\omega) = (k/k_0)\,[b^2_{coh}S(\mathbf{Q},\omega) + b^2_{inc}S_{inc}(\mathbf{Q},\omega)] \tag{12}$$

which is the flux scattered into solid angle $d\Omega$ about Ω and energy range $d\omega$ relative to the incident flux. It should be noted that at constant scattering angle 2θ the variables Q and ω are coupled, and the ranges of Q and ω are generally restricted by experimental factors.

$S_{inc}(Q,\omega)$ is the space-time Fourier transform of the so-called self correlation function $G_s(\mathbf{r},t)$,

$$G_s(\mathbf{r},t) = (1/N)\langle\sum_{i=1}^{N}\delta(\mathbf{r} + \mathbf{r}_i(0) - \mathbf{r}_i(t))\rangle \tag{13}$$

the index s (s: self) denoting that in this case only self motion of a single scattering atom is observed. In general, $G(\mathbf{r},t)$ may be described as a sum

$$G(\mathbf{r},t) = G_s(\mathbf{r},t) + G_d(\mathbf{r},t) \tag{14}$$

where $G_d(\mathbf{r},t)$ (d: distinct) describes the dynamical correlation between pairs of particles. This formalism can be easily generalized for the case of samples containing different atomic spezies.
$G_d(\mathbf{r},0)$ gives the well-known static pair correlation function g(r):

$$G_d(\mathbf{r},0) = ng(r) \tag{15}$$

Different types of single particle motions of atoms and molecules are possible which can be studied by incoherent inelastic neutron scattering, such as translational diffusion, rotational or vibrational motion [3,4,7]. E.g. for long times $t \to \infty$, corresponding to small ω, and $Q \to 0$, the simple Brownian diffusion approximation is valid

$$G_s(\mathbf{r},t) = (4\pi Dt)^{-3/2}\exp(-\mathbf{r}^2/4Dt) \tag{16}$$

and the diffusion coefficient D may be calculated from the incoherent (so-called quasi-elastic) scattering function $S_{inc}(Q,\omega)$:

$$S_{inc}(Q,\omega) = (1/\pi)DQ^2/[(DQ)^2+\omega^2] \tag{17}$$

Equation (17) represents a Lorentzian curve centred at $\omega = 0$ having a width at half height equal to $2DQ^2$.
For very short times ($t \to 0$), however, the force action between particles may be neglected and a description of free particle motion as always valid in an ideal gas, is approached, leading to

$$S_{inc}(Q,\omega) = [(2\pi)^{1/2}Qv_0]^{-1}\exp(-\omega^2/(2Q^2v_0^2)) \tag{18}$$

where $v_0 = (k_BT/m)^{1/2}$. Equation (18) provides a reasonable fit to data on simple liquids at wavelengths significantly shorter than the spacing between particles, typically for $Q\sigma >> 1$ (σ hard core diameter).
The scattering law for a simple isotropic harmonic oscillator is given by

$$S_{inc}(Q,m\omega_0) \approx (1/m!)(Q^2\langle u^2\rangle)^m\exp(-Q^2\langle u^2\rangle) \tag{19}$$

where $\langle u^2\rangle$ is the mean square displacement of the oscillator ($m = 0, 1, 2, \ldots$) [7].

Coherent neutron scattering yields information on the relative arrangement and relative motion of atoms in a liquid. At reduced wavenumbers, $S(Q,\omega)$ has a sharp peak centered at zero frequency and two more or less well-defined side peaks or high frequency shoulders, one on each side of the central peak. In the hydrodynamic limit ($Q \to 0$) the side peaks (Brillouin-lines) observed at long wavelength correspond to propagating sound waves, i.e. pressure fluctuations (see Figure 3). At shorter wavelengths, the sound waves become damped, and the high-frequency structure disappears when $Q\sigma \geq 2$, becoming only a single, Lorentzian like central peak. The central line in the hydrodynamic limit (Rayleigh line) represents a steadily decaying thermal mode (entropy fluctuations). By light scattering experiments, the Rayleigh-Brillouin spectrum at long wavelength ($\lambda \approx$ 5000 Å) can be measured. By inelastic neutron scattering, the question can be studied, whether the propagating density fluctuations characteristic of the hydrodynamic regime can still be supported at wavelengths comparable with the spacing between particles. In fact, rather well-defined collective excitations of the hydrodynamic type have been observed in neutron scattering experiments for several simple liquids near their melting point, such as in rubidium [8], cesium [9] and lead [10,11] for $Q \leq 1$ Å^{-1}, or e.g. in neon [12] for $Q \leq 0.15$ Å^{-1}.

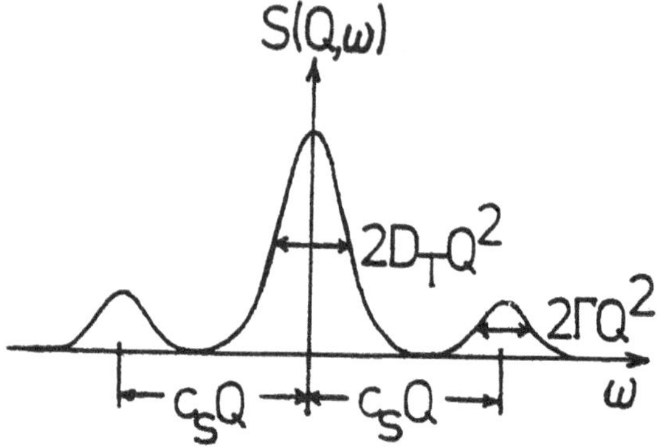

Fig. 3 Dynamic structure factor $S(Q,\omega)$ in the hydrodynamic limit $Q \to 0$ (D_T thermal diffusivity, Γ sound-attenuation coefficient, c_s adiabatic sound velocity) [4].

The first two even moments of $S(Q,\omega)$, which are defined generally as

$$<\omega^n> = \int_{-\infty}^{\infty} \omega^n S(Q,\omega) d\omega \tag{20}$$

are known exactly and are therefore useful for testing the reliability of the experimental data. They are given by

$$<\omega^0> = S(Q) \tag{21}$$

$$<\omega^2> = v_0^2 Q^2 \tag{22}$$

Another important correlation function in the theory of liquids is the so-called current correlation function. Its longitudinal part, which can be easily obtained from the scattering law $S(Q,\omega)$, is very useful for the description of the dynamics of a liquid:

$$J_l(Q,\omega) = (\omega^2/Q^2) S(Q,\omega) \tag{23}$$

Neutrons are unique in spanning a large and useful range of (Q,ω)-space, so that the above mentioned dynamical quantities can be measured. Energy transfers range from 10^{-6} eV, measured by spin-echo methods using reactor cold neutrons, even up to the eV energy transfers of a pulsed source resonance detector instrument.

3. NEUTRON SCATTERING

As X-ray sources and the X-ray diffraction technique is widely in use, we mainly focus on the method of neutron scattering. In the following table, some peculiarities of the neutron and X-ray scattering methods are presented.

TABLE 1 Peculiarities of the neutron and X-ray scattering technique

neutrons

- interact with the nuclei (centre of mass) of each atom
- neutrons penetrate most materials easily
- the wavelength range available (0.1 - 30 Å) is appropriate for probing both atomic scale structures and larger spacial correlations
- the scattering amplitudes for neutrons are independent of scattering angle and of neutron wavelength
  momentum transfers up to 100 $Å^{-1}$ are possible, so that good real space resolution can be obtained
- the scattering amplitude for neutrons is isotope dependent and may differ considerably for different isotopes thus permitting to examine the distribution functions involving particular atoms in chemical compounds
- the energy of thermal neutrons is similar to the energy of excitations and energy transfer can be determined over a wide range of reciprocal space
- the magnetic moment of the neutron provides a unique probe for the study of magnetic structures and excitations
- neutron sources are few and relatively weak, which often requires large sample volumes and limits the pressure range

X-rays

- see the electron distribution
- are easily absorbed by matter
- their scattering amplitude varies with scattering angle
- no separation of individual elements in chemical compounds (conventionally)
- dynamic distribution functions cannot be studied (conventionally)
- X-rays can observe special effects, e.g. bonding electron distributions
- X-ray sorces are many, and are relatively strong
- synchrotron radiation sources provide still higher spectral intensities, a tunable wavelength range, a well defined polarization, and by using anomalous dispersion, elements with similar number of electrons can be distinguised

Two types of neutron sources are in routine operation: nuclear reactors and pulsed neutron sources. In the reactors, the neutrons are produced in the core by fission and their energies are moderated. In thermal equilibrium with the moderator (e.g. D_2O), the neutrons have a Maxwellian distribution of energies which is peaked at about 25 meV for a 300 K moderator. Different moderators can operate at higher or lower temperatures in order to shift the peak in the distribution to shorter or longer wavelengths. They produce a steady neutron flux, dependent on the reactor power and on the shape and material of the moderator. High flux reactors, like that of the Institut Laue-Langevin at Grenoble, have a power level around 50 MW. Monochromators, like single crystals, are used to select one wavelength out of the Maxwellian energy distribution, and so pick out a chosen Q-value in diffraction, or a chosen (Q,ω)-value in inelastic scattering. In pulsed sources [7], such as ISIS at Rutherford Laboratory (U.K.), a spallation process of bombarding a heavy metal with protons takes place. At ISIS a pulsed beam of 750 MeV protons impinge on an uranium target producing many neutrons per proton. The neutrons so produced are partially moderated to yield sharp neutron pulses with a higher flux of high energy (epithermal) neutrons than that obtained from reactors. The broad energy span produced by a pulsed source allows a wide Q-range (up to 100 $Å^{-1}$) in diffraction to be measured. Large Q-values are important to obtain good resolution in real space since that is inversely proportional to Q_{max}. A Q_{max}-value of 50 $Å^{-1}$ would give real-space resolution of $2\pi/Q_{max} = 0.1$ Å.

On a reactor the incident wavevector is defined by a monochromator crystal and the scattered neutrons are counted either in a single detector which moves through a range of 2θ or in a multidetector which itself has a range of 2θ. On a pulsed source, the neutron pulses contain a spread of wavelengths and the scattered wave vector is determined by the neutron time-of-flight to a fixed detector at 2θ. The linear relationship between the wavelength of a neutron and its time of flight t may be obtained from de Broglie's hypothesis relating momentum p to wavelength λ: $p = m_n v = m_n(l/t) = h/\lambda$, where l is the total flight path of the neutron. On inelastic spectrometers, both k and E are determined and hence the complete sample scattering function S(Q,ω) is determined over a range in Q and ω and with a resolution determined by the details of the spectrometer design. E is determined from the relative angular position of the detector with respect to the incident beam. On reactor instruments, k_0 must be defined either by a crystal monochromator or choppers, whereas on a pulsed source instrument the time-of-flight can be used.

3.1 THE HIGH-TEMPERATURE HIGH-PRESSURE NEUTRON SCATTERING TECHNIQUE

As the neutron is a neutral particle and its total cross section (scattering plus absorption) is rather low for most materials, neutron scattering experiments can easily be performed in an environment where the sample is contained in a pressure cell, furnace etc., through which the incident and scattered neutrons can pass. In the design of a high-pressure and high-temperature autoclave for neutron-scattering investigations, several specific points have to be considered. The parasitic scattering of the sample environment materials (cell, furnace, heat shields, pressure vessel) exposed to the neutron beam must be kept small. However, a high mechanical strength for the pressure vessel is required and the materials found to be suitable for sample cells at these high temperatures are often limited, e.g. with regard to the high reactive nature of liquid alkali metals. Materials found to be suitable for sample containers are metals like vanadium, niobium, molybdenum, tungsten and tungsten-rhenium or insulators like quartz and sapphire. The choice depends on the chemical reactivity of the material and the maximum temperature desired. These cells cannot be constructed from these materials in such a way that they can withstand a high internal pressure at higher temperatures. This is demonstrated in Figure 4, where the

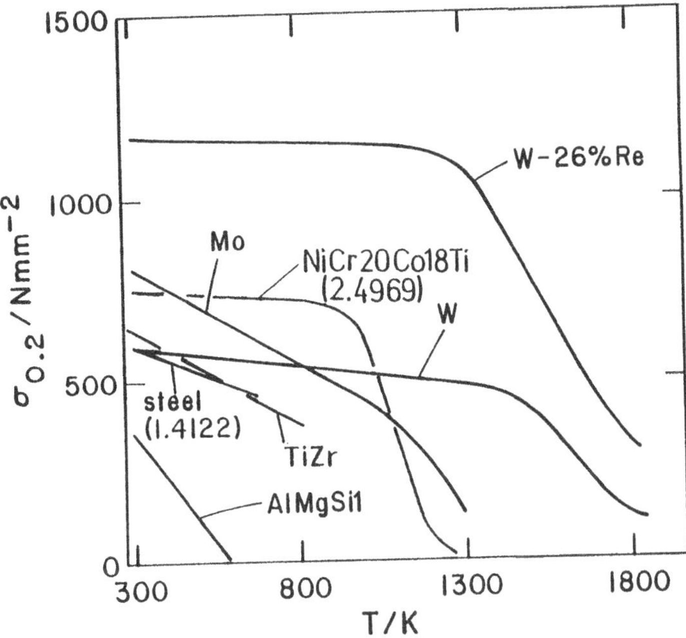

Fig. 4 Temperature dependence of the tensile strength $\sigma_{0.2}$/Nmm^{-2} (0.2 % permanent elongation load) of some high-pressure vessel materials.

tensile strength $\sigma_{0.2}$ of some metals usually used for pressure vessels and sample cells is shown as a function of temperature. The strong decrease of $\sigma_{0.2}$ for all materials at temperatures T > 1300 K makes the use of a special experimental arrangement necessary, in which the sample cell, together with the necessary furnaces and heat shields, is placed in a larger pressure vessel, which is filled e.g. with argon under the same pressure as the fluid inside the cell. This avoids any mechanical stress on the walls of the sample cell.

High-pressure vessels for neutron scattering experiments are often made from alumina, special steels, aluminium alloys or "zero-alloys", e.g. an alloy consisting of 34 % titanium and 66 % zirconium. In comparison to the former materials, the zero-alloys scatter neutrons only incoherently, as the mean scattering length for the constituent atoms is zero, thus yielding a homogeneous background scattering. In Table 2 the maximum working pressure and the attenuation of the incident neutron beam for some pressure vessel materials of a given geometry (outer diameter: 90 mm, inner diameter: 80 mm) are given. For a low maximum pressure and a weak scattering sample such as liquid cesium, which requires a high incident neutron flux, an aluminium alloy is appropriate. The parasitic Bragg-scattering peaks of this material for Q > 2.7 Å^{-1} do not disturb the structural information of the sample's neutron scattering pattern at high temperatures.

TABLE 2. Maximum working pressure and neutron beam attenuation for several high-pressure vessel materials (calculated for a cylinder of 90 mm outer diameter and 80 mm inner diameter).

Material	Al-alloy	Steel (1.4122)	TiZr-alloy	Al_2O_3
P_{max}/bar (300K)	200	450	510	160
n-beam attenuation/% (at 25 meV)	10	67	36	30

In order to achieve high temperatures, a resistance heating method normally is applied. Most of the suitable materials that can be used as heater elements have an appreciable neutron-scattering and absorption cross-section (see Table 3), so that a careful balance must be reached in the design of the heater element and heat shields between their attenuation of neutrons and their finite lifetime due to evaporation at the highest temperatures. One has also to keep in mind that most of the metals soften well below their melting point and some recrystallize at high temperatures. Care must also be taken so that adjacent materials (e.g. for isolation) do not chemically react with the metals at high temperatures. As can be seen from the vapour pressure data in Table 3, for the highest temperatures tungsten foils would be appropriate. The high neutron-beam attenuation of the material can be overcome by the use of foils of great thinness. The heat-shield assembly can be constructed from similar materials. Vanadium foils are suitable for lower temperatures and have the advantage of giving a homogeneous scattering background, refractory metal foils have to be used for higher temperatures.

TABLE 3. Physical properties of materials that can be used as high-temperature resistance heaters ($\sigma_{sc} = 4\pi b^2$ scattering cross-section, σ_a absorption cross-section; 1 barn = 10^{-24} cm^2).

Material	T_{melt}/K	Vapour pressure at 2300 K/mbar	σ_{sc}/barn	σ_a/barn (25 meV)
Vanadium	2170	$1.0 \cdot 10^{-4}$ (1770 K)	5.20	5.08
Niobium	2770	$2.5 \cdot 10^{-6}$	6.25	1.15
Molybdenum	2880	$2.0 \cdot 10^{-5}$	6.35	2.55
Tantalum	3270	$2.2 \cdot\cdot 10^{-8}$	6.02	20.60
Rhenium	3450	$2.3 \cdot 10^{-8}$	11.50	90.70
Tungsten	3680	$1.1 \cdot 10^{-9}$	4.86	18.40

Based on the consideratons above, a high-temperature and high-pressure autoclave has been constructed for neutron-scattering experiments on liquid alkali metals, which can be used up to a temperature of 2000 K and a pressure of 200 bar (see Figure 5). The sample container (2), which is mounted in the middle of the high-pressure vessel (1), consists of a thin-walled molybdenum cylinder of 0.25 mm thickness, 15 mm width and 80 mm height. It is connected via a molybdenum capillary with a liquid-metal reservoir container (3). The high-pressure vessel is made of an aluminium alloy of high tensile strength. It has a wall thickness of 5 mm in the neutron window beam height, which corresponds to a maximum working pressure of about 200 bar. We chose an aluminium alloy (AlMgSil) of high tensile strength and with only small fractions of other elements. This has a considerably increased mechanical strength, whereas the scattering and absorption cross-section is only slightly larger than that of pure aluminium. The pressure on the liquid metal is applied by compressed high-purity argon, which is pressure-balanced between the inside and outside of the sample cell by a small opening in the liquid reservoir container. The autoclave is heated internally by means of two concentric tungsten tubes (4) of 0.05 mm thickness. They are clamped or spot-welded at their lower end to a molybdenum ring and clamped at the top end to the main heater feed-throughs (7). Boron nitride or alumina spacers (9) are used to separate the heaters, to give an accurate alignment of the heater assembly and to reduce heat loss upwards. In the height of the neutron beam, heat conduction and radiation between the cell and the wall of the autoclave are reduced by several heat shields, made from thin Mo-foils of 0.025 mm thickness. Ceramic materials (zirconia powder or pyrophyllit) below and above the neutron window keep the heat convection low. A second furnace (6) wound on the molybdenum capillary controls the temperature profile along the sample cell. The temperature along the measuring compartment is controlled by two W-3%Re/W-25%Re thermocouples, which are attached at the top and bottom ends of the cylindrical sample cell. An effective water cooling system (8) above and below the neutron window keeps the temperature of the outer wall low. The high-pressure connection (12) and the electrical feed-throughs for thermocouples and heaters (10,7) are led through the top and bottom flanges of the autoclave, which are made from stainless steel. A temperature and water flowmeter control system with a current limiter protects the autoclave from overheating.

A similar experimental set-up can be used for measuring the structure factor of expanded molecular liquids, like liquid selenium or liquid sulphur. Due to the molecular structure of these fluids, these scattering experiments have to be performed up to high scattering angles. For this reason, the high pressure neutron window often is constructed from a TiZr-alloy, which is a purely incoherent neutron scattering material. Sample cells from quartz and sapphire have been employed, the latter ones for the higher temperatures. Single crystal sapphire cells can be used, which can be oriented in the neutron beam in such a way so that no sapphire Bragg peaks disturb the scattering pattern of the liquid sample [13].

Pulsed neutrons obtained from an accelerator-based pulsed neutron source often offer the additional advantage that diffraction experiments can be done at a single, fixed scattering angle by the time-of flight-technique. Thus, if windows are needed (e.g. in a pressure vessel), they need to cover only a limited angular range. More importantly, in the fixed-angle scattering geometry, shielding and collimation can be optimized in order to access the largest possible sample volume while avoiding scattering from the surrounding sample

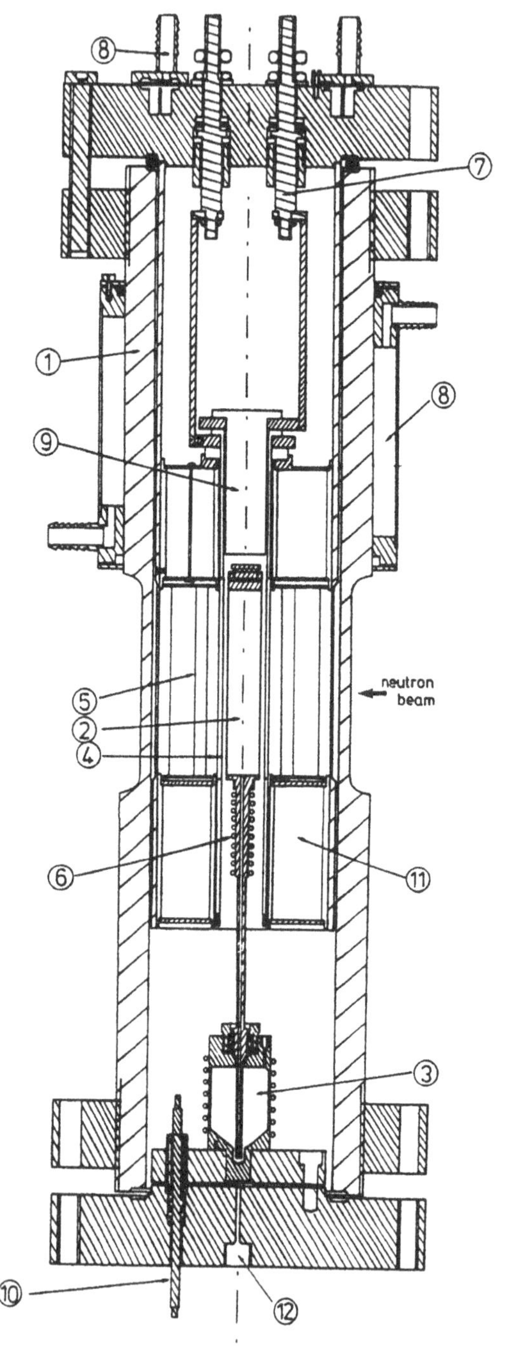

Fig. 5 High-temperature high-pressure autoclave for neutron scattering experiments on expanded liquid alkali metals.

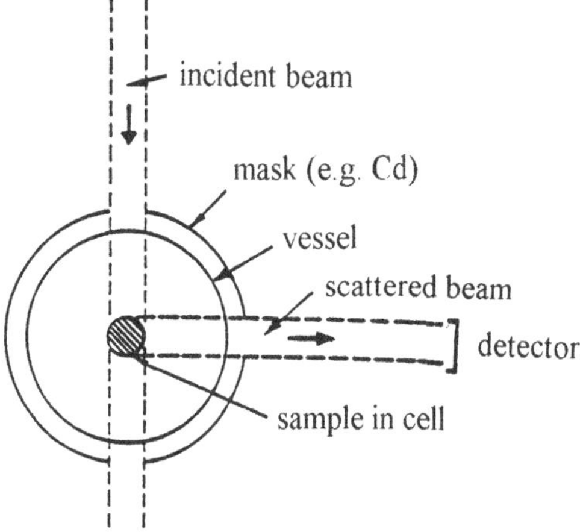

Fig. 6 Environment geometry for time-of-flight diffraction experiments at high pressures with collimation of both the incident and scattered beam at a fixed angle of $2\theta = 90^{\circ}$.

vessel. The geometry shown in Figure 6 is most useful, as the scattering angle has been chosen to be at $2\theta = 90^{o}$, so that the collimation defines a localized sample volume. For larger or smaller angles it becomes increasingly more difficult to eliminate scattering from the region surrounding the sample.

3.2 EVALUATION OF NEUTRON DIFFRACTION DATA

Our aim is to determine experimentally the static structure factor S(Q). What we actually measure is a neutron intensity I(Q), however, either as a function of angle for a given incident wavelength at a steady state (reactor) source, or at a given angle as a function of time-of-flight. To obtain $d\sigma_{coh}/d\Omega$, and hence S(Q) we must make allowance to absorption, incoherent and multiple scattering and then normalise the corrected intensity to give an absolute cross section. A further complication arises due to the background scattering from the furnace, pressure vessel etc. In order to deal with these corrections, a series of additional measurements, such as an empty cell, empty container and vanadium calibration measurement, has to be performed under the most closely similar experimental conditions as that of the sample run [14].

In a first step the scattering intensity has to be corrected for scattering from the background (i.e. scattering from the container materials) and for absorption in the sample and container. The corrected intensity I_s^{corr} for the samples s is given by

$$I_s^{corr} = (1/A_{S,SC})[I_{SCA}-I_S^B-(A_{C,SC}/A_{C,C})(I_{CA}-I_C^B)] \tag{24}$$

where I_{SCA} denotes the experimentally measured total scattering intensity of the sample S, surrounded by the sample cell C and autoclave A (including furnace and heat shields); I_{CA} is the corresponding intensity from the empty container. $I_S{}^B$ and $I_C{}^B$ give the corresponding background intensity for the sample or empty cell, respectively. The first subscript of the attenuation coefficients $A_{i,j}$ indicates the scattering being attenuated while the second subscript shows what is causing the attenuation. Thus, e.g. $A_{C,SC}$ is the attenuation of container scattering by container and sample. All absorption factors are angular and wavelength dependent. Often, computer routines such as that of Poncet [15] are used in calculating the $A_{i,j}$, which are based on the method of Paalman and Pings [16]. Another approach would be to calculate the attenuation coefficients as a function of wavelength using the measured attenuation of unscattered neutrons determined using the transmission monitor.

The observed scattering intensity is the sum of all contributions I_n of n-fold scattered neutrons. However, only the single-scattering contribution I_1 is directly related to the structure factor. In order to obtain I_1, the multiple-scattering fraction has to be substracted from the total scattering intensity. This contribution of the multiple-scattering to the total scattering intensity can roughly be estimated by the use of the quasi-isotropic approximation of Blech and Averbach [17]. These values are tabulated for different sample geometries [17-19].

The size of the sample volume is determined by two factors: the desirability of using a lot of scattering material in the beam in order to achieve good statistics and to shorten the experimental running time, and on the other side the necessity of having little material present so that absorption and multiple scattering effects are small. By the method described above, only the multiple scattering of the sample is considered, without multiple scattering processes occurring in the container and cross-scattering contributions from sample and containment materials. For simple experimental set-ups, these additional minor contributions can be numerically calculated by Monte Carlo techniques. If the absorption cross section is sufficiently large, multiple scattering corrections can be neglected. When the sample absorption cross section is small, multiple scattering can be reduced by adding an absorption material. Where this method is not applicable, as it generally is in hot fluids, the thickness of the sample should be chosen so that no more than about 10 % of the neutrons incident on the sample are scattered.

For complicated scattering geometries, the multiple scattering contribution and the self shielding and absorption factors are calculated by Monte Carlo methods, taking into account also such factors as beam size, beam intensity profile, detector solid angle etc. [20].

We now still have to apply the inelasticity or Placzek corrections. They consider that the static approximation does not hold, i.e. energy transfer occurs between the neutron and the sample, and that the detector effeciency $\varepsilon(k)$ depends on the neutron energy or wave number k of the scattered neutron. Furthermore, it has to be taken into account that the static structure factor S(Q) is defined by the integration of the dynamic structure factor $S(Q,\omega)$ according to

$$S(Q) = \int_{-\infty}^{\infty} S(Q,\omega)d\omega \qquad (25)$$

This integration is performed by the spectrometer at constant scattering angle 2θ, however, and not at constant Q-value, and is limited on the energy loss side to the value of the incident neutron energy. The most widely used method to correct for these effects is the one proposed by Placzek [21] and its extensions [20,22-26] to arbitrary counter efficiencies. Placzek showed that for nuclei much heavier than the neutron the correction adopts a form which is essentially independent of the detailed dynamics, and is related only to the nuclear mass, the sample temperature, the incident energy, and the geometry and efficiency of the neutron detection process.

The scattering intensities corrected so far are put on an absolute scale by comparison with the intensity from a specimen of known scattering cross-section. Vanadium is commonly used as a standard scatterer because it scatters almost totally incoherently, giving an isotropic scattering pattern over the whole Q-region [14].

A check of the whole correction procedure is possible, based on different limiting requirements of S(Q) that have to be fulfilled, such as $S(\infty) = 1$ and $S(0) = nk_BT\chi_T$, χ_T being the isothermal compressibility, which can be obtained from PVT-measurements, or the following sum rule

$$\int_0^\infty [S(Q)-1]Q^2dQ = 2\pi^2N \tag{26}$$

3.3 EXPERIMENTAL RESULTS AND DISCUSSION

As an example, Figure 7a shows a selection of measured structure factors S(Q) of expanded liquid cesium for different temperatures and densities near the liquid-vapour coexistence curve, covering the whole liquid range from the melting point up to nearly the critical point [26,27]. For $Q \leq 0.28$ Å^{-1}, the structure factors are extrapolated to $S(0) = nk_BT\chi_T$, with the number density n and isothermal compressibility χ_T taken from the most accurate PVT-data [1,2]. The neutron-diffraction experiments have been performed using the two axis diffractometer D4B at the high-flux reactor of the Institut Laue-Langevin in Grenoble. Figure 7b displays the corresponding Fourier-transform of the S(Q)-data, the pair correlation function g(r). With decreasing density or increasing temperature the following changes in S(Q) or g(r) of the liquid metal are apparent: The intensities of the first peak of S(Q) and g(r) are strongly reduced and broadened, whereas the peak positions Q_1 and r_1 shift only slightly towards lower Q- or higher r-values, respectively. Compared with a simple nonmetallic liquid such as argon, the oscillations of S(Q) for liquid cesium are drastically damped with increasing temperature. Only a broad first maximum is seen in S(Q) of cesium near the critical point, whereas in the case of liquid argon three well defined maxima and minima are observable near the critical point. This difference in behaviour can be attributed to differences in the repulsive part of the effective interaction potential $\phi(r)$ for liquid alkali metals and argon (see Figure 8). The repulsive part is much softer in metals than in argon. A continuous smearing out of the liquid structure is observed during the expansion of the liquid metal. The average nearest-neighbour distance r_1 is given by the position of the first peak in g(r), whereas the average

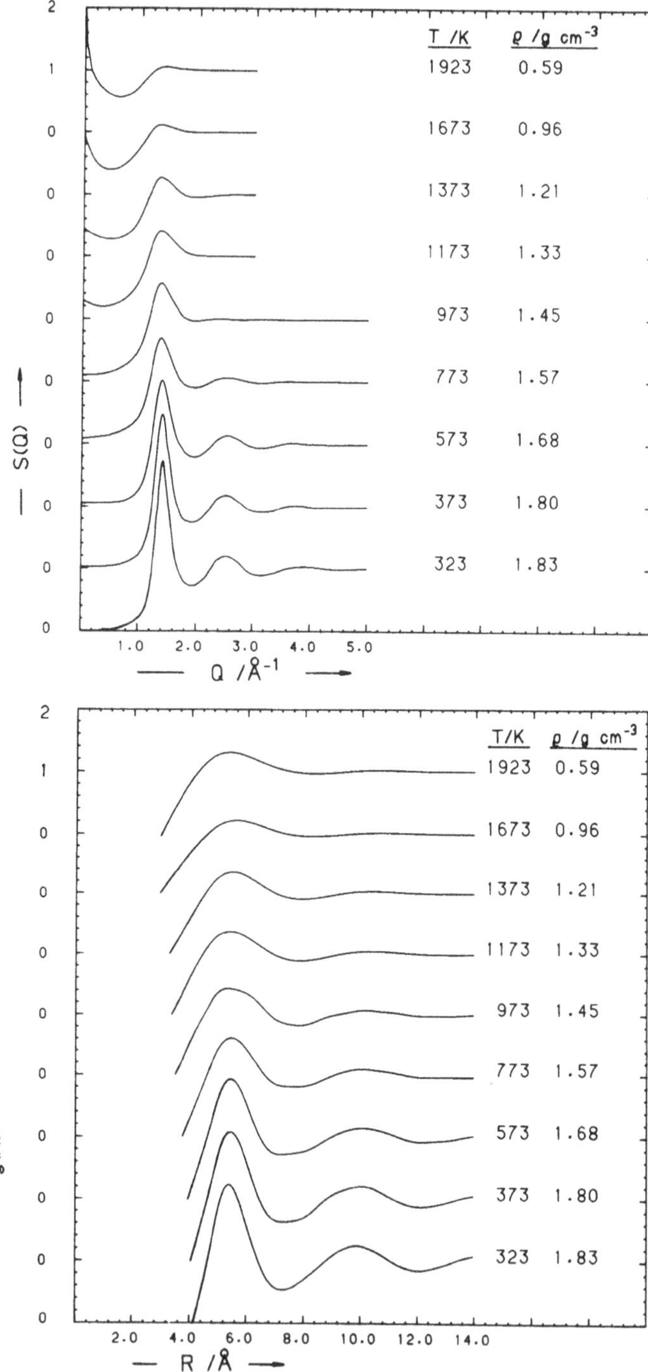

Fig. 7 a) Static structure factor S(Q) and b) pair correlation function g(r) of expanded liquid Cs at conditions near the liquid-vapour coexistence curve.

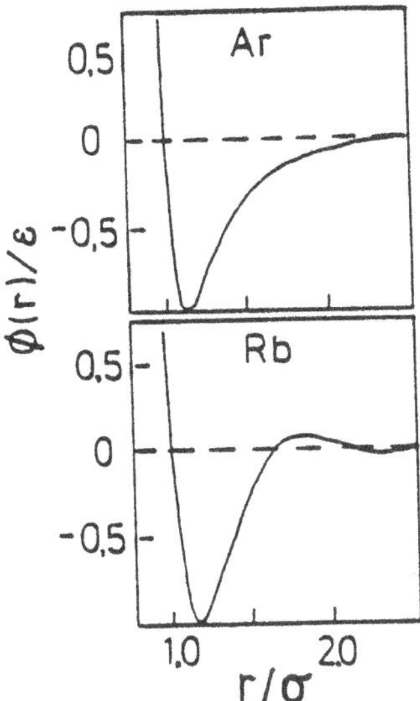

Fig. 8 Comparison of the effective pair potential of liquid Ar and liquid Rb near their melting point (ε depth of potential well).

coordination number N_1 can be obtained from the area under the first peak. The absolute value of N_1 depends sensitively on the method employed to define and integrate the first neighbour peak. Figure 9 shows that for liquid cesium N_1 tends to decrease linearly with decreasing density from 9 close to the melting point to roughly 3 in the critical point region, whereas the mean position r_1 of the nearest neighbours remains essentially constant. These data clearly demonstrate that the density reduction by expanding liquid cesium develops mainly from a decrease in the average number of nearest neighbours rather than from an expansion to larger interatomic distances. A similar behaviour has been observed for expanded liquid rubidium [28]. Recent Reverse Monte Carlo calculations have demonstrated that at the higher temperatures, a tendency towards atomic clustering takes place [29].

Figure 10 displays the height of the first maximum of S(Q), $S(Q_1)$, of a few liquid metals as a function of temperature, both quantities are reduced to their values at the melting point [27,28,30]. Obviously, no reduced correlation between the alkali metals and the polyvalent metals is observed, indicating that reduced correlations are unlikely to hold for the liquid metals as a group. However, the alkali metals can be reduced in corresponding regions of the phase diagram, which indicates that the thermodynamic state dependence of the effective interaction potential in metals may be similar within groupings of metals like the group of the alkali metals.

As is well known, electrical transport measurements are directly correlated with the structure of the liquid metal, i.e. the structure factor S(Q). In the nearly-free electron (NFE) approach, the electrical conductivity σ is described by the Faber-Ziman formula [31]

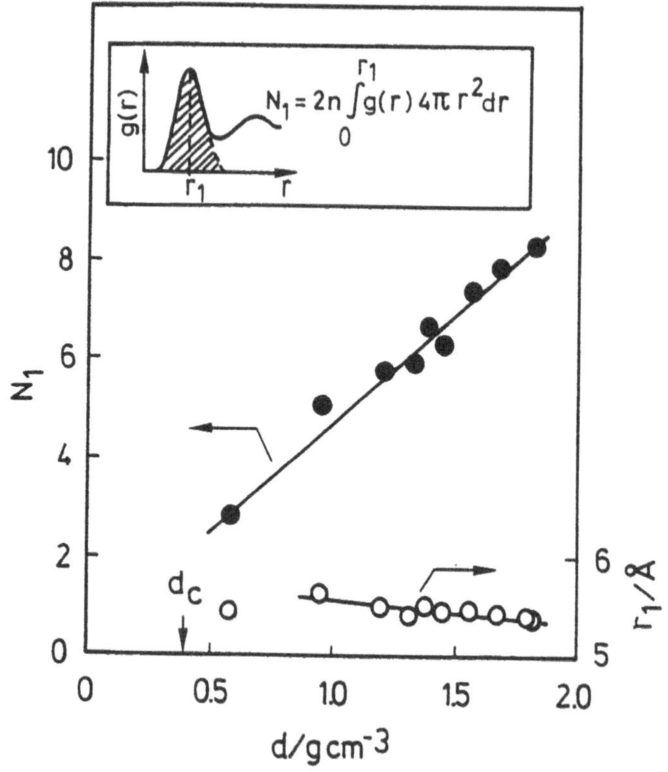

Fig. 9 Mean number N_1 and mean distance r_1 of nearest neighbours of expanded liquid Cs (inset: method employed here for calculating N_1).

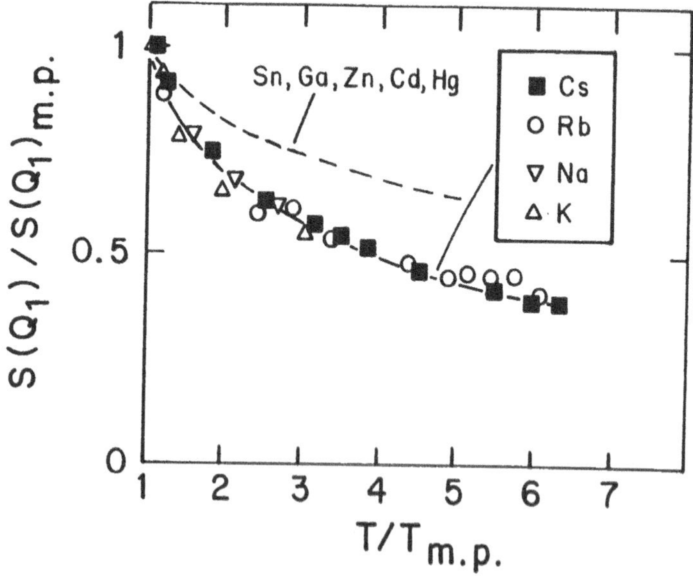

Fig. 10 The height $S(Q_1)$ of the first peak of $S(Q)$ for a few monovalent and divalent liquid metals as a function of temperature (both values are reduced to their values at the melting point).

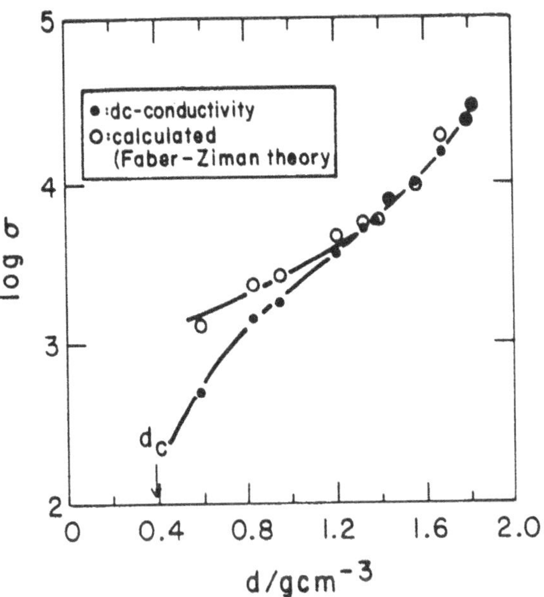

Fig. 11 Measured and calculated (Faber Ziman NFE-model) electrical conductivity $\sigma/\Omega^{-1}cm^{-1}$ of liquid cesium as a function of density.

$$\sigma^{-1} = 3\pi m^2 V_A/(4\hbar^3 e^2 k_F^6)\int_0^{2k_F} S(Q)\,|V(Q)/\varepsilon(Q)|^2 Q^3 dQ \qquad (27)$$

Where $V(Q)/\varepsilon(Q)$ denotes the screened ion pseudopotential, k_F the wavenumber of the electrons at the Fermi surface and V_A is the atomic volume. For calculating V(Q) we used the Ashcroft empty-core potential, combined with a density dependent dielectric function $\varepsilon(Q)$ which takes into account exchange and correlation effects [27]. Figure 11 shows the results of these calculations compared with the experimental values of the dc-electrical conductivity for expanded liquid cesium. The agreement is satisfactory for higher densities. The applied formalism starts to fail already at a density of about three times the critical density d_c. The region of densities, where these changes occur is the same as that in which magnetic data [32,33] indicate the presence of electron correlation effects, which are the precursors of the metal-nonmetal transition, which finally takes place near the liquid-vapour critical point. As can be revealed from the measurements of the magnetic susceptibility and from quantum statistical calculations [34,35], the formation of localized spezies, such as Cs atoms and dimers, must become effective near the critical point.

When the alkali metals are expanded at conditions near the coexistence curve, density and temperature are changed simultaneously. In order to understand the influence of each of these thermodynamic variables on the structural properties upon expansion in more detail, it is necessary to study separately the isobaric temperature and isothermic pressure dependence of S(Q). The effect upon S(Q) of increasing the temperature or pressure at

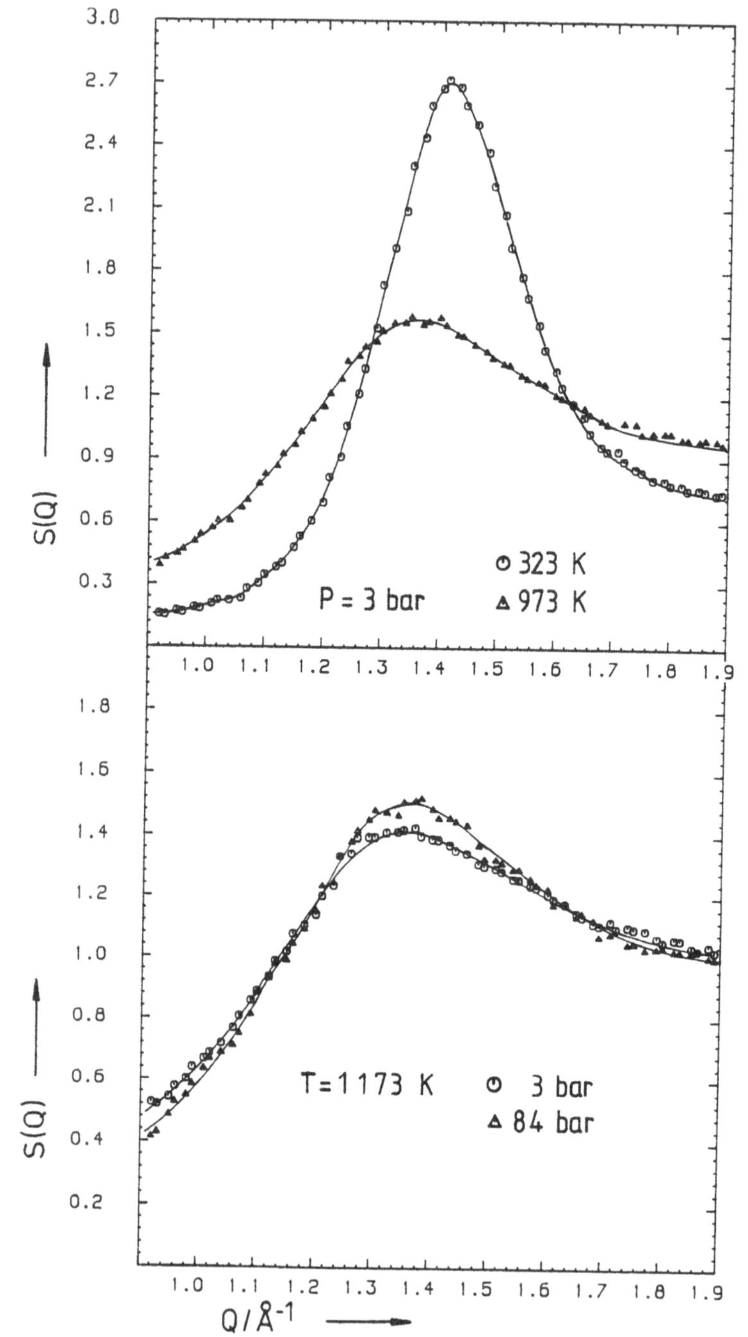

Fig. 12 Temperature and pressure effect on the region of the main peak of the static structure factor S(Q) of liquid Cs.

constant pressure and temperature, respectively, is illustrated for one example in Figure 12. It can be seen that the temperature increase broadens the main maximum of S(Q) and shifts its main peak position Q_1 to a smaller value, whereas the compression results in a slight shift of Q_1 to a higher value, while its peak height rises by a small amount. This example clearly shows that the temperature and pressure effects on $S(Q_1)$ and Q_1 and therefore also on the distance of nearest neighbours, r_1, have opposite signs and compensate each other to some extent during the expansion of the metal.
The isothermal pressure derivative of the structure factor S(Q) is of particular interest, because it is related to three-body correlations (see equation 7) and plays an important role in extracting quantitative information on the effective pair potential from experimental diffraction data [36].

In recent years, considerable amount of work has also been devoted to the theoretical calculation of the structural properties of expanded liquid alkali metals by using computer simulation, perturbation and integral equation techniques [37-47]. Generally, the calculations were performed on the basis of an effective pair potential $\phi(r)$ calculated by the pseudopotential theory with various types of pseudopotentials and dielectric functions. In agreement with experimental results it has been found that the characteristic features of the density dependence of the structure along the liquid-vapour coexistence curve are common features of the expanded liquid alkali metals. Differences can be observed in comparison to expanded divalent metals, like Hg, however [30]. The calculations also showed that the minimum of $\phi(r)$ becomes deeper with decreasing density due to a decrease of the screening of the ion cores, whereas the position of the minimum does not change very much. On approaching the critical point, often discrepancies with the experimental data have been observed at small Q-values, which can be explained by using the NFE-theory in describing the interatomic forces in the low density region.

In order to obtain also information about dynamical properties of the expanded liquid metals, such as the self-diffusion coefficient and collective excitations, the dynamic structure factor $S(Q,\omega)$ of liquid Rb has been measured up to 1900 K [50]. As known for long, and in contrast to results on dense Lennard-Jones systems, liquid alkali metals near their melting point exhibit distinct longitudinal collective sound mode excitations in a large region of the Q-ω-space [8,9,48], even up to Q-values higher than 1 $Å^{-1}$. As far as liquid Rb and liquid Cs at their melting point are concerned, these phonon-like collective modes have been observed as side peaks or shoulders in a wide range of wave numbers Q far outside the hydrodynamic region. In rare gases, sound damping is larger and acoustic waves are already overdamped at wavelengths that are several times larger than the mean interatomic distance [48,49]. The presence of such modes for liquid metals, and their absence in inert gas liquids, can be attributed to the special shape of the effective interaction potential of the fluid metal [51-54].
As an example, Figure 13 exhibits plots of $S(Q,\omega)$ at different temperatures for Q = 1 $Å^{-1}$. Broad shoulders around the central peak of $S(Q,\omega)$ are still indicated at Q-values around 1 $Å^{-1}$ for the higher temperatures of 1073 K up to 1673 K, i.e. the sound mode excitations seem still to be visible at these temperatures, though they are highly damped. In order to explore the possible existence of a dispersion relation of the collective excitations, the positions in energy of the maxima of the inelastic side peaks versus momentum transfer have to be known. In order to locate these values in spectra within the overdamping

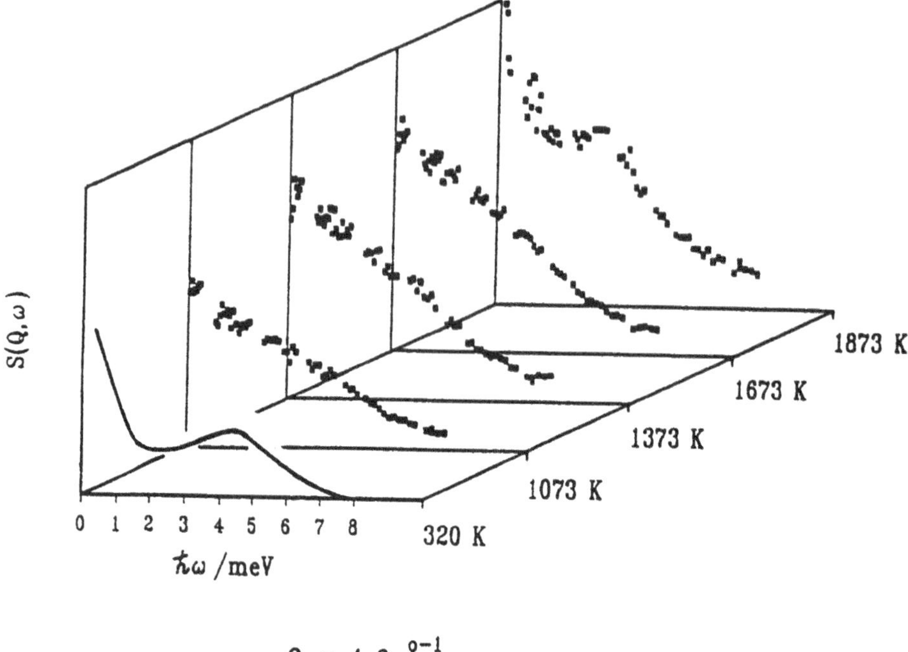

Fig. 13 Plot of the dynamic structure factor $S(Q,\omega)$ of expanded liquid Rb at 350 K [8] , 1073 K, 1373 K, 1673 K and 1873 K for $Q = 1$ Å^{-1}.

region, the longitudinal current correlation function $J_l(Q,\omega)$ is computed (see equation 23). This function always shows side peaks located at frequencies $\omega_m(Q)$. Although these frequencies cannot be identified with the true physical frequency of the excitation, they provide a first approximation valid to explore the spatial dispersion of the excitation, however. In Figure 14, the corresponding dispersion relation $\hbar\omega_m(Q)$ is shown for the lower four temperatures. For comparison, the corresponding data of Copley and Rowe for Rb at 320 K [8] are included. As expected, the minima of $\hbar\omega_m(Q)$ become shallower and their positions shift - similar to S(Q) - towards smaller Q-values with increasing temperature. At high Q-values, where $S(Q) \approx 1$, kinetic effects become important, and the curves tend to their corresponding ideal gas limiting value $\hbar\omega_m(\text{ideal gas}) = \hbar Q(2k_BT/m)^{1/2}$ [48].

The experimental results can be compared with recent molecular dynamics calculations of $S(Q,\omega)$ by Kahl et al. [53,54]. They used a typical metallic interatomic interaction potential based on the Ashcroft empty-core pseudopotential and used the Ichimaru-Utsumi parametrisation for the local field corrections to the dielectric function. As an example, Figure 15 shows the comparison between experimental and theoretical data for T = 1373 K. Good agreement has been found between theory and experiment up to 1700 K. Similar theoretical results have been obtained by Hoshino et al. [55]. These observations indicate that at these measured conditions of temperature and density, i.e. up to about three times the critical density, the collective ion-ion dynamics is still controlled by the "electron sea" of the expanded liquid metal.

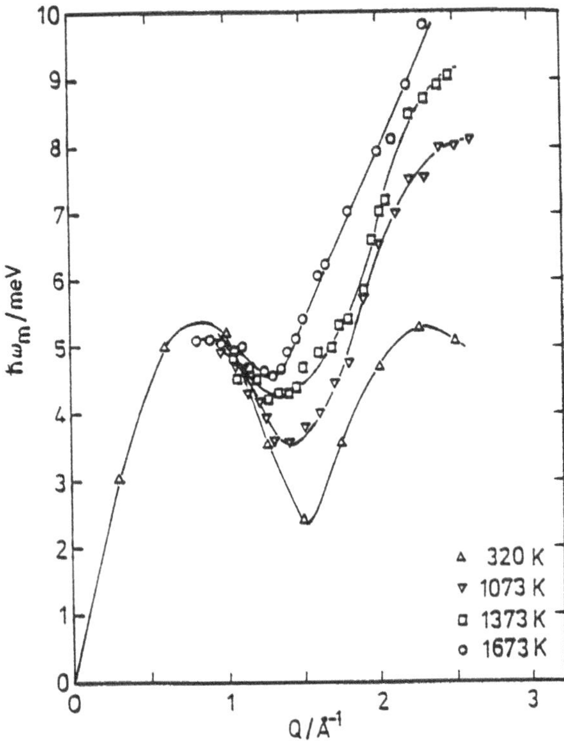

Fig. 14 The dispersion relation $\hbar_m\omega(Q)$ - as obtained from the maximum of the current correlation function $J_l(Q,\omega)$ - of liquid Rb at several temperatures (320 K [8] (Δ), 1073 K (∇), 1373 K (□) and 1673 K (o)) [50].

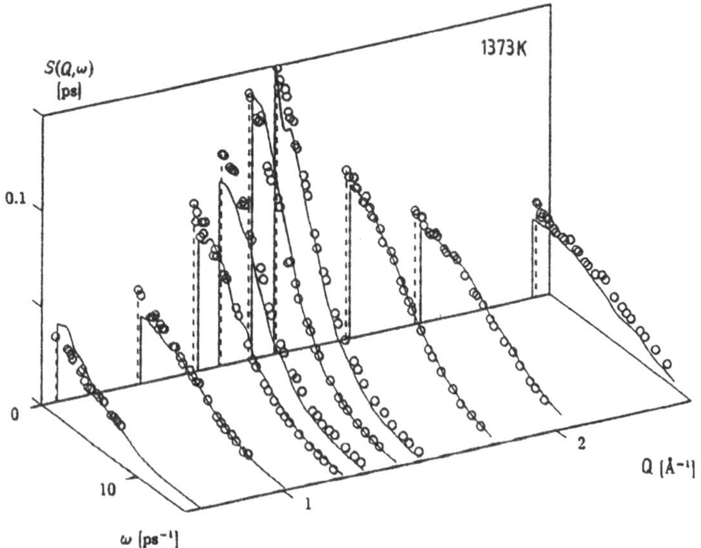

Fig. 15 Dynamic structure factor $S(Q,\omega)$ of liquid Rb for T = 1373 K. The full lines are the theoretical results [53,54], the symbols are experimental data.

At T = 1873 K and d ≈ $2d_c$, however, a change in the shape of S(Q,ω) of liquid Rb is observed, indicating a change in nature of the interatomic forces of the expanded metal. As can be clearly seen from Figure 13, a broad peak in S(Q,ω) has developed near 3 meV for momentum transfers around $1Å^{-1}$ out of the structureless scattering function observed for the temperatures 1073 K ≤ T ≤ 1673 K. The shape of S(Q,ω) and its Q-dependence might be explained by an additional contribution of simple oscillators with an energy of about 3 meV to the scattering law. The appearance of this peak thus probably indicates the transition from collective to more localized mode behaviour. Hence, also in dynamical properties, the transition towards localization at densities around twice the critical density seems to be visible.

Following recent theoretical results [56], the coherent central half width $\omega_{1/2}(Q)$ of S(Q,ω) in the vicinity of the structure factor maximum $S(Q_1)$ can be described in good approximation by

$$\omega_{1/2}(Q) = D_E Q^2 d(Q)/S(Q) \tag{28}$$

with D_E the Enskog self-diffusion coefficient (d(Q) contains spherical Bessel functions $j_n(Q\sigma)$; σ hard-core diameter). This equation is valid for about $Q_1 L \leq 1$ (L mean free path), i.e. up to a maximum temperature of about 1400 K. In Figure 16, D_E as obtained from the equation above is shown in comparison to the self-diffusion coefficient of the simple Enskog hard-sphere model

$$D_{HS} = (3/8)(n\sigma^2 g(\sigma))[k_B T/\pi m]^{1/2} \tag{29}$$

which considers uncorrelated binary collisions only [31]. The temperature dependent hard-core diameter σ(T) has been obtained by fitting the hard-sphere model in the Percus-Yevick approximation [48] to the experimentally obtained first maximum of S(Q). As can be clearly seen, up to 1200 K the experimentally obtained self-diffusion coefficient D_E of the expanded fluid metal is similar to that of a hard-sphere fluid with an effective hard-core diameter σ(T) as derived from $S(Q_1)$ [50]. Around 1400 K, small deviations are observed, which could be due to the limited validity of the method to extract D_E, however.

4. X-RAY DIFFRACTION

Recently, also X-ray diffraction measurements have successfully been carried out on expanded liquid semiconductors and metals at temperatures up to 2000 K and pressures from ambient up to about 2000 bar.

4.1 EXPERIMENTAL TECHNIQUE

As an example, the wonderful X-ray diffraction measurements of Tamura et al. [30,57] on expanded liquid Hg are presented. It is well-known, that also Hg is transformed from a metallic to a nonmetallic, semiconducting state, when it is heated up to its liquid-vapour

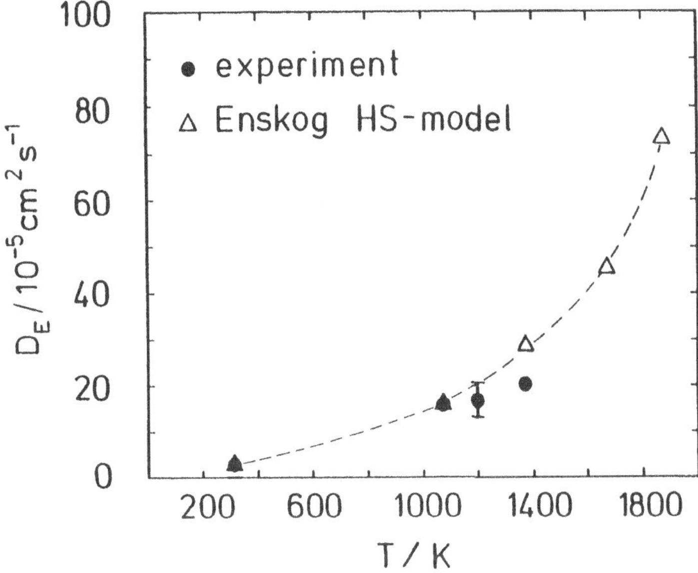

Fig. 16 Enskog self-diffusion coefficient D_E (●) of expanded liquid Rb and comparison with the results for a hard-sphere fluid (Δ).

critical point (T_c = 1751 K, P_c = 1651 bar, d_c = 5,80 gcm^{-3}). A number of measurements of physical properties, such as the electrical conductivity, thermopower, Hall coefficient, optical reflectivity and NMR have shown that, as the density is reduced, the metal-nonmetal transition starts to take place at a density of about 9 gcm^{-3} [1,2].
Tamura et al. made use of the energy-dispersive method, where white X-rays are used as the primary beam and the scattered photons are detected and energy-analyzed by a Ge solid state detector which was rotated around the vertical axis of the goniometer, and was connected with a multichannel pulse height analyzer. The energy-dispersive measurements were performed at several fixed scattering angles 2θ in order to cover a sufficiently wide range of the scattering wave number $Q = (4\pi \sin\theta / hc)E$ (E photon energy, h Planck's constant, c velocity of light). The diffracted X-ray photons were collected at 2θ of 7, 10, 15, 19, 26 and 39°, by operating a tungsten X-ray tube at 50 kV and 25 mA. The spectrum in the energy region 17-42 keV was used for the analysis. The angle settings correspond to regions of Q between 1 $Å^{-1}$ and 14 $Å^{-1}$.

The experimental conditions of high temperatures and high pressures were achieved with an internally heated high pressure vessel with an inner diameter of 58 mm and two flanges for the electrical feedthroughs and the gas inlet (see Figure 17). The primary and scattered X-ray beams were passing through beryllium windows of 5 mm thickness and 4 mm diameter. The vessel was pressurized with He gas, which has a low absorption constant for X-rays in the energy range of the experiment. Water cooling jackets around the outside of the vessel kept the temperature of its wall low. Hot liquid Hg must be contained in a cell made of special material being transparent for X-rays and resistant to chemical corrosion. A single crystal sapphire cell was developed for this purpose, with a sample thickness in the X-ray beam of 15 μm, which is connected with a liquid mercury

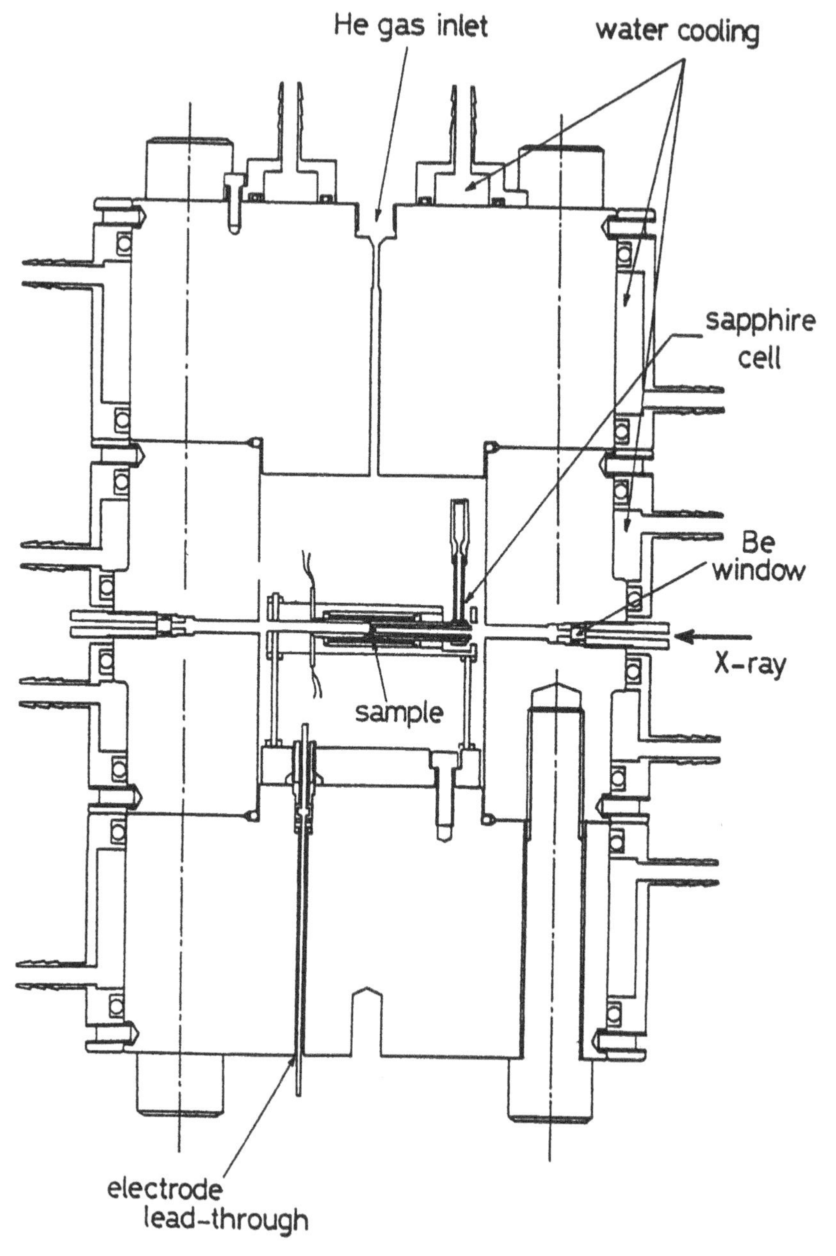

Fig. 17 Side view of the high-pressure high-temperature vessel of Tamura et al. for energy dispersive X-ray diffraction studies [30,57].

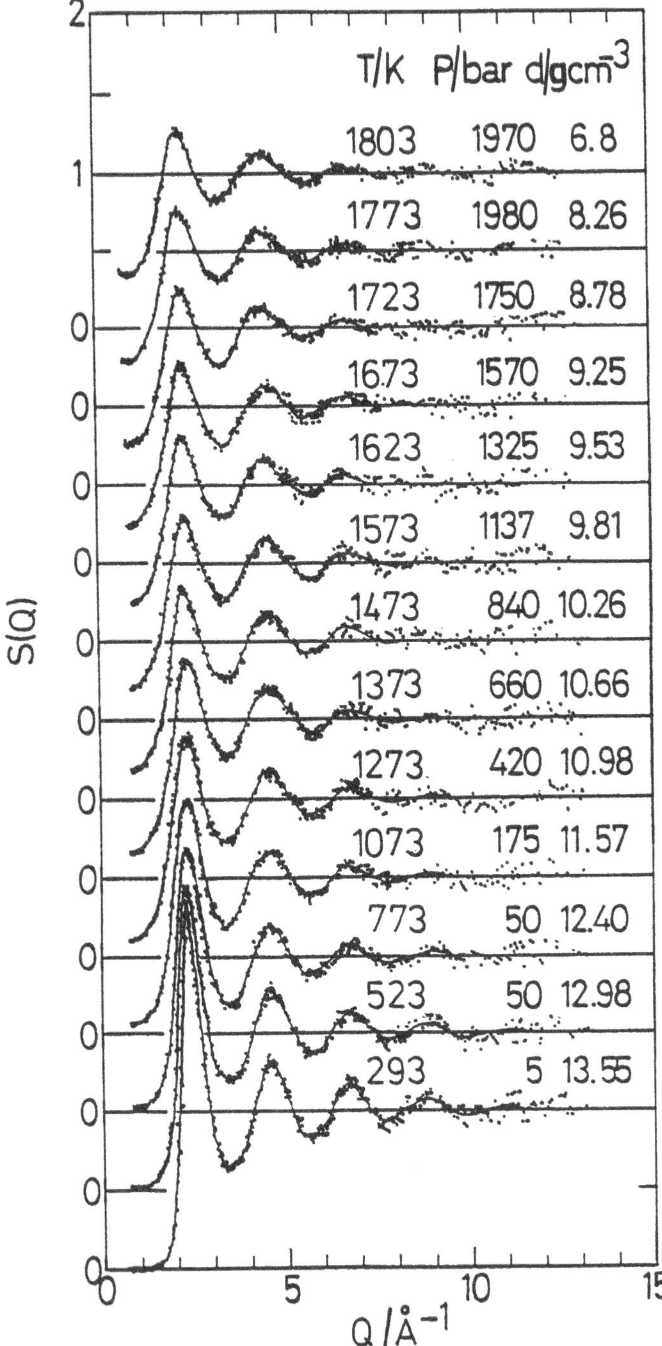

Fig. 18 Static structure factor S(Q) of expanded liquid Hg along the saturated vapour pressure curve [adopted from 30,57].

reservoir. Due to the high X-ray absorption coefficient of mercury in this energy range, the thickness of the sample cell has to be very small. The problem of Bragg reflections of the sample container was overcome by orienting the single crystal sapphire tube in such a way that no diffraction spots of the sapphire material appeared. The cell was heated by a resistance heater made of electrically isolated Mo wire, which is wound on a Mo tube around the cell. The temperature of the sample was measured by two Pt-30%Rh/Pt-6%Rh thermocouples which were located in holes of the Mo tube. The free space in the pressure vessel was filled with alumina powder to obtain good thermal insulation and also to prevent convection of the compressed He gas. The X-ray photon accumulating time at each scattering angle was about 30 minutes.

In order to obtain the structure factor S(Q) of expanded liquid Hg from the experimental scattering intensity, several data corrections had to be made, such as those taking into account the escape effect originated from the detector, the energy spectrum of the primary beam, the absorption by the sapphire cell, the compressed He gas, Be windows and liquid Hg itself, and finally Compton scattering from liquid Hg and the sapphire cell. For a detailed description of these correction procedures see [30,57].

4.2 RESULTS AND DISCUSSION

Figure 18 shows S(Q) data for expanded fluid Hg in the temperature and pressure range up to 1803 K and 1970 bar along the saturated vapour-pressure curve at densities ranging from 13.55 to 6.8 gcm^{-3}. As can be clearly seen, the oscillations of the static structure factor S(Q) of expanded mercury are damped and the peak widths become broader with increasing temperature and pressure. This tendency is especially observed for the first peak of S(Q). Figure 19 shows the mean coordination number N_1 together with the mean nearest neighbour distance r_1 as obtained from g(r) as a function of density. It is clearly seen, that N_1 decreases substantially and almost linearly with decreasing density in the metallic region, that is, down to about 10.5 gcm^{-3}. On the contrary, r_1 remains almost constant. When the metal-to-nonmetal transition is approached, however, slight deviations from the linear plots of N_1 become evident and r_1 sligtly increases [30,57].

5. SUMMARY

In this paper the general principles of neutron and X-ray scattering experiments on expanded liquids have been discussed and illustrated for expanded liquid metals. With current neutron- and X-ray scattering techniques, experiments can be performed up to temperatures of about 2000 K at pressures up to about 2000 bar. Undoubtedly, the future will bring further developments, coupled with an increased neutron flux and resolution at the neutron sources and the high intensity of photons at synchrotron sources, which allow an extension of the experiments to much higher pressures.

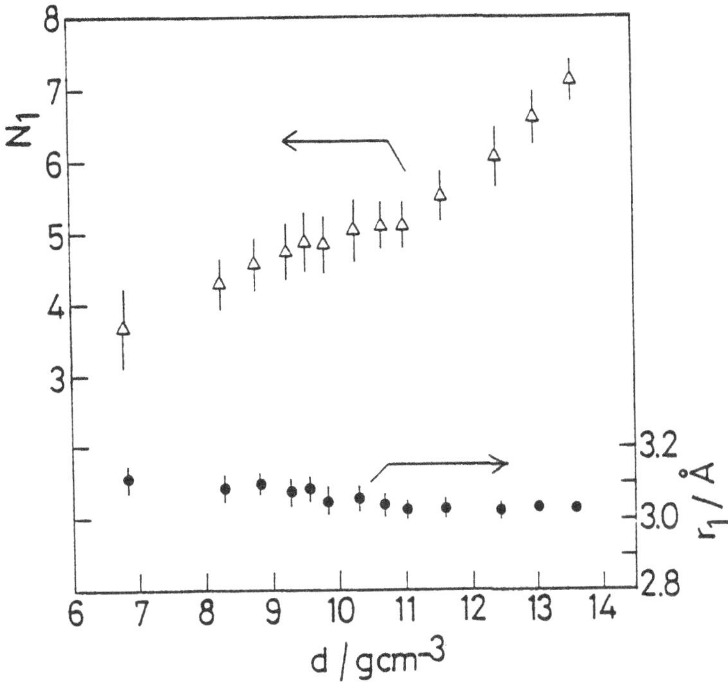

Fig. 19 Mean coordination number N_1 and mean first neighbour distance r_1 of expanded liquid Hg as a function of density (the values of N_1 have been obtained from g(r) using the method of symmetric main maximum) [30,57].

6. REFERENCES

[1] F. Hensel, J. Non-Cryst. Solids **117/118**, 441 (1990)

[2] F. Hensel and H. Uchtmann, Ann. Rev. Phys. Chem. **40**, 61 (1989)

[3] P.A. Egelstaff (1992) An Introduction to the Liquid State, Oxford University Press, Oxford

[4] J.P. Hansen and I.R. McDonald (1986) Theory of Simple Liquids, Academic Press, New York

[5] H.E. Stanley (1971) Introduction to Phase Transitions and Critical Phenomena, Clarendon Press, Oxford

[6] L. Van Hove, Phys. Rev. **95**, 249 (1954)

[7] R.J. Newport, B.D. Rainford and R. Cywinski (1988) Neutron Scattering at a Pulsed Source, Adam Hilger, Bristol

[8] J.R.D. Copley and M. Rowe, Phys. Rev. Lett. **32**, 49 (1974)

[9] T. Bodensteiner, Chr. Morkel and W. Gläser, Phys. Rev. A **45**, 5709 (1992)

[10] O. Söderström, J.R.D. Copley, J.B. Suck and B. Dorner, J. Phys. F. **10**, L151 (1980)
[11] K.-E. Larsson, Phys. Chem. Liq. **12**, 273 (1983)
[12] H. Bell, H. Moeller-Wenghoffer, A. Kollmar, R. Stockmeyer, R. Springer and H. Stiller, Phys. Rev. A **11**, 316 (1975)
[13] M. Edeling and W. Freyland, Ber. Bunsenges. Phys. Chem. **85**, 1049 (1981)
[14] R. Winter and T. Bodensteiner, High Pressure Res. **1**, 23 (1988)
[15] P.F.J. Poncet (1977) ILL-Report 77P0139S
[16] H.H. Paalman and C.J. Pings, J. Appl. Phys. **33**, 2635 (1962)
[17] I.A. Blech and B.L. Averbach, Phys. Rev. **137**, A1113 (1965)
[18] V.F. Sears, Adv. in Physics **24,** 1 (1975)
[19] G.H. Vineyard, Phys. Rev. **96**, 93 (1954)
[20] M.A. Howe, R.L. McGreevy and W.S. Howells, J. Phys.: Condens. Matter **1**, 3433 (1989)
[21] G. Placzek, Phys. Rev. **86A**, 377 (1952)
[22] J.L. Yarnell, M.J. Katz, R.G. Wenzel and S.H. Koenig, Phys. Rev. **A7**, 2130 (1973)
[23] C.G. Windsor (1981) Pulsed Neutron Scattering, Taylor & Francis, London
[24] W.S. Howells (1986) Rutherford Appleton Lab. Report RAL-86-042
[25] A.K. Soper and P.A. Egelstaff, Nucl. Inst. Meth. **178**, 415 (1980)
[26] R. Winter, T. Bodensteiner, W. Gläser and F. Hensel, Ber. Bunsenges. Phys. Chem. **91**, 1327 (1987)
[27] R. Winter and F. Hensel, Phys. Chem. Liq. **20**, 1 (1989)
[28] G. Franz, W. Freyland, W. Gläser, F. Hensel and E. Schneider, J. de Physique Coll. **41**, C8-194 (1980)
[29] V.M. Nield, M.A. Howe and R.L. Mc Greevy, J. Phys.: Condens. Matter **3**, 7519 (1991)
[30] K. Tamura and S. Hosokawa, J. de Physique IV Colloque C5, 39 (1991)
[31] M. Shimoji (1977) Liquid Metals, Academic Press, London
[32] W. Freyland, Phys. Rev. B **20**, 5104 (1979)
[33] W. El Hanany, G.F. Brennert and W.W. Warren, Phys. Rev. Lett. **50**, 540 (1983)
[34] R. Redmer and G. Röpke, Contrib. Plasma Phys. **29**, 343 (1989)
[35] R. Redmer, H. Reinholz, G. Röpke, R. Winter, F. Noll and F. Hensel, J. Phys. C: Condens. Matter **4**, 1659 (1992)
[36] R. Winter, F. Hensel, T. Bodensteiner and W. Gläser, J. Phys. Chem. **92**, 7171 (1988)
[37] R.D. Mountain, J. Phys. F **8**, 1637 (1978)
[38] M. Tanaka, J. Phys. F **10**, 2581 (1980)
[39] R. Evans and T.J. Sluckin, J. Phys. C **14**, 3137 (1981)
[40] A.M. Bratkovsky, V.G. Vaks, S.P. Kravchuk and A.V. Trefilov, J. Phys. F **12**, 1293 (1982)
[41] I.L. McLaughlin and W.H. Young, J. Phys. F **14**, 1 (1984)
[42] G. Kahl and J. Hafner, Phys. Rev. A **29**, 3310 (1984).
[43] K.N. Khanna and J.L. Bretonnet, Phys. Chem. Liq. **16**, 55 (1986)
[44] S. Naito and I. Yokoyama, Z. Phys. Chem. N. F. **156**, 513 (1988)
[45] J.A. Ascough and N.H. March, Phys. Chem. Liq. **21**, 251 (1990)

[46] K. Hoshino, N. Matsuda and M. Watabe, J. Phys. Soc. Jpn. **59**, 2027 (1990)
[47] K. Hoshino, N. Matsuda, H. Mori and M. Watabe, J. Non-Cryst. Solids **117/118**, 44 (1990)
[48] J.P. Boon and S. Yip (1980) Molecular Hydrodynamics, McGraw Hill, New York
[49] J.-B. Suck (1992) in: Recent Developments in the Physics of Fluids, Adam Hilger, Bristo
[50] R. Winter, C. Pilgrim, F. Hensel, C. Morkel and W. Gläser (1992), in: Proceedings of the International Conference on Liquid and Amorphous Metals (LAM8), Vienna, J. Non-Cryst. Solids, in press
[51] A. Rahman, Phys. Rev. A **9**, 1667 (1974)
[52] L. Sjögren, J. Phys. C: Solid State Phys. **12**, 425 (1979)
[53] S. Kambayashi and G. Kahl, Europhys. Lett. **18**, 421 (1992)
[54] G. Kahl, S. Kambayashi and G. Nowotny (1992) in: Proceedings of the Conference on "Liquid and Amorphous Metals" (LAM8), Vienna
[55] K. Hoshino, H. Ugawa and M. Watabe, J. Phys. Soc. Jpn. **61**, 2182 (1992)
[56] E.G.D. Cohen, P. Westerhuijs and I.M. de Schepper, Phys. Rev. Lett. **59**, 2872 (1987)
[57] K. Tamura and S. Hosokawa (1992) in: Proceedings of the Conference on "Liquid and Amorphous Metals" (LAM8), Vienna

SELECTED THERMODYNAMIC ASPECTS OF THE INFLUENCE OF PRESSURE ON POLYMER SYSTEMS

Erik Nies
Department of Polymer Technology
Eindhoven University
P.O.Box 513
5600 MB Eindhoven
The Netherlands

ABSTRACT The influence of pressure on a selection of physical/thermodynamic properties is explored. For single constituents the following properties are examined: a) the equation of state of single constituents and b) first order transitions, i.e. solid-liquid and solid-solid transitions in crystallisable polymers. Subsequently, for mixtures the influence of pressure on miscibility behaviour is elucidated for polymer solutions and blends. Furthermore the implications for synthesis and processing are referred to.
In the further work the bearing of pressure on the glass transition temperature and glassy state is shown. Especially the importance of formation history and physical ageing is clarified. The analysis of these data is facilitated, employing reduced variables. Finally, this method is also applied to address the frequency dependence of thermodynamic properties in general.

1. Introduction

The tractability of polymers is greatly related to the ease by which polymers can be formed to commercial products by a variety of processing techniques. In most processing conditions pressure, in combination with temperature, plays a prominent role. For instance, each individual small material element in a polymer melt which is subjected to e.g. an injection moulding process, perceives during the short process cycle a complicated formation history. Important variables defining this formation history are e.g. the temperature and pressure as well as the rates of cooling/heating and pressurizing sensed by the material element. For instance, during injection moulding cooling/heating rates of 100 $K.s^{-1}$ and pressurizing rates of $30MPa.s^{-1}$ are not exceptional. Furthermore, if the final product is used in the glassy state the formation history also influences the ultimate properties of the products. For instance the dimension stability of precision formed components are particularly sensitive to the formation history, i.e the precise course by which this glassy state was accomplished.
In the current contribution the influence of pressure on equilibrium as well as time dependent

R. Winter and J. Jonas (eds.), High Pressure Chemistry, Biochemistry and Materials Science, 201–224.

thermodynamic properties is addressed. The first sections deal with the equation of state of polymers and is concentrated on the control exerted by pressure on solid-liquid and solid-solid transitions occurring in crystallisable polymers. Before the equation of state behaviour of the glass region will be considered the influence of pressure on the thermodynamic properties of mixtures is examined. In mixtures also the occurrence of liquid-liquid phase transitions must be anticipated. From the examples it is evident that even in the fully disordered liquid state pressure can have a very intriguing and complicating role. The bearing of these considerations on the polymerisation in solution and bulk synthesis as well as the processing of polymer mixtures or blends is briefly illustrated.
The remaining sections deal with the situations in which the time dependence of thermodynamic properties become apparent. The discussion of this time dependence is greatly facilitated by the utilisation of the time-temperature/pressure superposition principle which affords a valuable separation of the influence of the different variables on viscoelastic properties in general and the time dependent thermodynamic properties in particular. The first examples comprise the influence of pressure and cooling rate on the glass transition temperature, the path dependence of the equation of state in the glassy state and the succeeding volume relaxation behaviour. Finally the high frequency analogue of the bulk compressibility and its course with pressure and temperature is illustrated.

2. Equilibrium Properties

2.1. Equation of state

It is well known that the energy U, the enthalpy H, the free energy A and the free enthalpy G are important basic thermodynamic functions of state. For example the latter two govern experiments conducted at constant volume V and temperature T and constant pressure P and temperature T respectively. From these state functions general conditions for equilibrium can be deduced [1,2], e.g.

$$\begin{aligned} &\text{at constant T and p} : dG_{T,P} \leq 0 \\ &\text{at constant T and V} : dA_{T,V} \leq 0 \end{aligned} \tag{1}$$

where the inequalities denote the direction of natural processes and the equalities are valid at equilibrium.

A one component system, the simplest conceivable example, can be in a one-phase, two-phase or three-phase condition. Examples of two-phase equilibria are the common transitions solid-liquid, liquid vapour, solid-vapour but also transitions amongst different solid crystalline structures. Coexisting phases in equilibrium are characterised by the following equilibrium conditions

$$
\begin{array}{lll}
\text{mechanical equilibrium:} & p' = p'' = \ldots = p & \\
\text{thermal equilibrium:} & T' = T'' = \ldots = T & \quad (2) \\
 & g' = g'' = \ldots & \\
\text{or} & \mu' = \mu'' = \ldots &
\end{array}
$$

where the superscripts denote different phases; g, expressed in intensive units, is the free enthalpy density which for a one component system is equal to the chemical potential μ.

The relation between volume, temperature and pressure is given by the equation of state. For a one phase condition this is also the only equation constraining the thermodynamic properties of the system. Typical examples of the equation of state of amorphous and crystalline polymers are shown in figure 1.

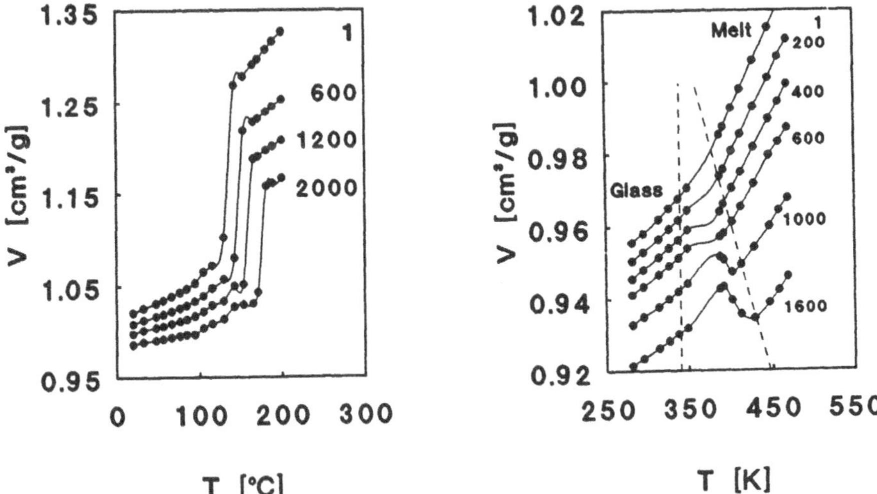

Figure 1. Equation of state data for a) linear polyethylene [3] and b) polystyrene [4]. Specific volume is drawn vs. temperature at indicated pressures [bar]. The lines are shown for convenience.

It should be remembered that only polymers with sufficient stereoregularity are able to crystallise. At sufficiently high temperatures the polymers are in the liquid state and the values of the expansion coefficient $\alpha = (1/V)(\partial V/\partial T)_p$ and compressibility factor $\beta_T = (-1/V)(\partial V/\partial p)_T$ are typical of ordinary liquids. At sufficiently low temperatures the crystallisable polymer is in the solid crystalline state. At the melting temperature the isobars show a discontinuity in the volume as well as in e.g. entropy and enthalpy. Under normal circumstances the polymer is not fully crystalline but part of the material remains in the liquid state which becomes glassy at even lower temperatures. In figure 1a. this can not be observed since the glass transition temperature of PE is at a much lower temperature. The glass transition is typical of amorphous polymers which are unable to crystallise (see figure 1b) and can be observed in the pVT diagram as a change in slope in the isobars or as a discontinuity in the expansion coefficient and compressibility factor. For polystryrene (PS) the pressure and temperature dependence of the compressibility factor is shown in figure 2 [5]. In the glassy state values for β_T and α are typical of solids. Several polymers show other transitions in the glassy state [6,7]. However, often other techniques have to be used to demonstrate their existence.

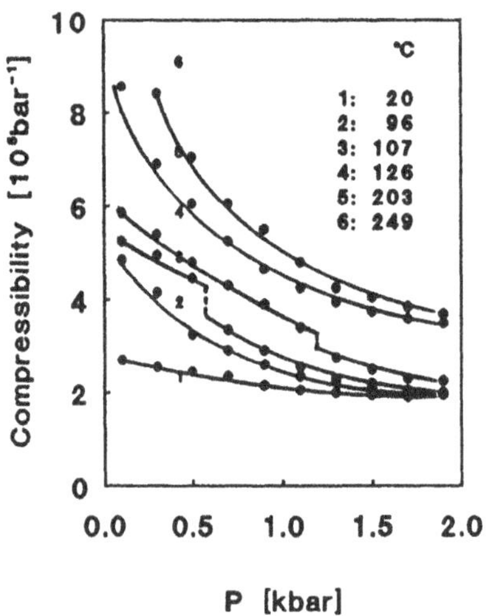

Figure 2. Isothermal compressibility of polystyrene vs pressure at indicated temperatures [5]. Lines are shown for convenience and the transitions are drawn too sharp.

2.2. Crystallisation and Melting Behaviour

Transitions like the solid-liquid transition are known as first order transitions because the first order derivatives of the basic thermodynamic functions of state show a discontinuity at the transition. At the glass transition, the second order derivatives of the basic functions of state, e.g. α and β_T, apparently possess a discontinuity. This can point to a second order phase transition. However, presently it is not settled whether the glass transition is (related to) a second order transition or is solely a consequence of kinetic and non-equilibrium effects. In the following the glass transition will be discussed in terms of kinetic and non-equilibrium effects.

From data as shown in figure 1a it can be be observed that the melting transition temperature shifts with pressure. The influence of pressure on the melting behaviour of small molecules and macromolecular materials is principally identical and can be calculated from the Clausius-Clapeyron equation [2]

$$dT_m/dp = T_m \, \Delta v/\Delta h \tag{3}$$

with Δv and Δh the changes in volume and enthalpy accompanying the melting transition.

In all studies on polymers where sufficient experimental details are available, eq. (3) is obeyed. Concentrating on the melting transition, the p,T phase diagram of linear polyethylene (PE), is shown schematically in figure 3 [8,9].

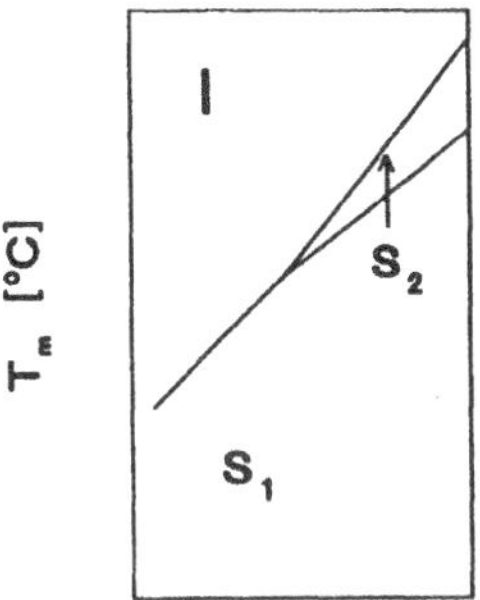

Figure 3. Schematic pressure-temperature phase diagram of polyethylene [8,9]. Solid lines indicate the solid-liquid and solid-solid transition curves. s_1 and s_2 denote orthorombic and hexagonal crystal structures respectively.

At sufficiently low temperatures PE is a crystalline solid possessing an orthorhombic unit cell structure denoted by s_1. An increase in temperature eventually results in the melting of PE which is accompanied by a change in all thermodynamic properties. At the transition temperature the solid and liquid phases are in equilibrium and in accordance with eq (3), application of pressure changes this transition temperature. In the case of PE an increase in melting temperature is observed, which is the most common situation since volume and enthalpy are in general larger in the liquid state than in the solid state. A few exceptions, including e.g. H_2O, show a negative slope for the melting curve because in those cases $\Delta v < 0$ [10]. At sufficiently high pressures and in a suitable temperature range a solid-solid phase transition from the orthorhombic to a second crystalline structure with a hexagonal unit cell is observed (solid phase s_2) [11]. In this new hexagonal phase the chains possess an enhanced mobility as observed with e.g. NMR [12]. Naturally, also this hexagonal phase melts upon heating. In the following the importance of this high-entropy, hexagonal phase for the crystallisation behaviour will be discussed. At a unique temperature and pressure (the triple point) three phases, i.e. liquid-orthorhombic and hexagonal phases, are in equilibrium.
So far only information concerning the crystalline unit cell has been presented. An important and typical observation for flexible polymer molecules is the tendency (certainly at moderate pressure) to crystallise by chain folding, giving rise to lamellar crystals [13]. The lamellar thickness is small compared to the length of the fully extended macromolecules. Although the

fully extended chain crystals are thermodynamically more stable, due to kinetic effects the chain folded crystals are preferentially formed. Evidence for the higher stability of the extended chain crystals is drawn from the thickening process occurring during the growth process itself or during subsequent heating (annealing) of the crystals.

At sufficiently elevated pressures another mode of crystallisation occurs and lamellae are formed, several microns thick, in which the molecules may reach full extension or chain fold only a few times [13,14]. At these elevated pressures and sufficiently high temperatures the crystals are in the hexagonal phase. Upon lowering the pressure and temperature these crystals transform to the orthorhombic phase.

It was thought that the equilibrium *melting phase diagram* shown in figure 3 was also determinative for the kinetically controlled crystallisation behaviour [12]. It has been shown only recently that also for pressures below the triple point, for which the orthorhombic phase is the thermodynamically stable structure, the *crystallisation* occurs via the (metastable) hexagonal phase [15]. Subsequently, these crystals transform to the thermodynamically more stable orthorhombic structure. Certainly, for pressures down to 1800 bar the initial crystallisation always occurs in the hexagonal phase. The pertinence of these recent findings on the general understanding of polymer crystallisation and melting phenomena is a matter of current investigations.

A far more complex phase diagram, shown in figure 4, is observed for polytetrafluoroethylene (PTFE) [9,16]. For this polymer four different polymorphs are known. The crystal phase denoted by s_3 is an orthorhombic structure and exists only at high pressure. Phase s_1 is a high temperature crystal, phase s_2 a low temperature, triclinic crystal and finally phase s_4 a trigonal crystal structure.

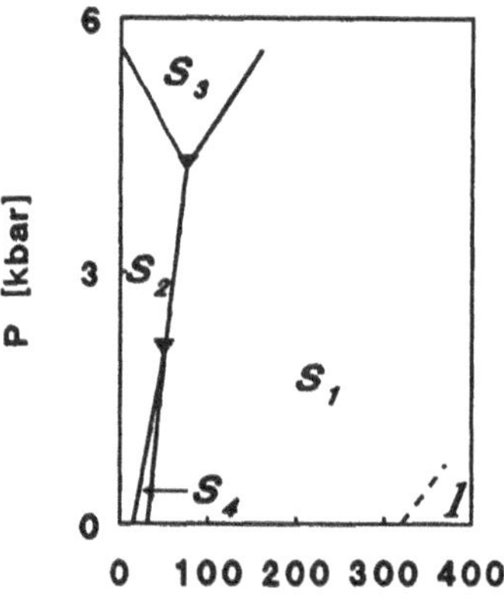

Figure 4. Schematic pressure-temperature phase diagram of poly(tetrafluoroethylene) [9,16]. Solid lines: solid-solid equilibria, dashed line: solid-liquid transition. ▾: triple points. The triple point $s_1/s_2/s_3$ is rather uncertain. Also the s/l transition is only drawn approximately.

Very recently, another interesting observation was made for isotactic poly(4-methyl-pentene-1) [17]. The polymer, crystalline under ambient conditions, transforms in a disordered phase upon increasing pressure in two different temperature regimes situated around 20 °C and 200 °C respectively. The high temperature pressure induced melting has been observed previously in non polymeric materials, e.g. in ice and silica, and is related to the negative change in volume accompanying the melting process. The low temperature phenomenon is found to be reversible. Furthermore, upon heating this amorphous solid a new crystalline structure is formed which upon cooling reverted to the amorphous solid. According to the authors the disordering found upon cooling at high pressures has not been observed previously. To complete the understanding of these phenomena more experiments are devised.

The complexity induced by pressure in the solid state region of the phase diagram is considerable and is related to the existence of different polymorphs the stability of which is determined by geometrical considerations typical of solid crystalline structures. However, this is not the only possibility by which pressure can complicate the phase diagram of polymer systems.

2.3. MIXTURES: MISCIBILITY BEHAVIOUR

If one is also interested in mixtures, additional transitions can be observed in the liquid state in which pressure can play an important role. Eqs. (1) are also applicable to mixtures and can be fitted to discuss the equilibrium amongst different phases for multi-component systems [1,2]. These conditions, which are very similar to those in one component systems, are

$$
\begin{aligned}
&p' = p'' = p''' = \ldots\ldots = p \\
&T' = T'' = T''' = \ldots\ldots = T \\
&\mu_i' = \mu_i'' = \mu_i''' = \ldots\ldots = \mu_i'^{\cdots'} \quad \text{for all components i}
\end{aligned}
\tag{4}
$$

One particular example of heterogeneous equilibrium is the coexistence of liquid phases differing in composition. To illustrate the principles, a binary mixture will be considered [18]. The thermodynamic functions can be expressed as intensive properties depending on the composition but not on the extension of the system. Correspondingly, the composition of the mixture is expressed in the corresponding intensive composition variable which is denoted by $x_2 = (1-x_1)$. Proper composition variables are, e.g. mole fraction, mass fraction, volume fraction etc. Suppose that a binary liquid mixture of composition x_{21}, with an intensive excess free enthalpy $\Delta g^{(1)}(x_{21})$, separates in two liquid phases with composition x_2' and x_2'' (see figure 5a). The free enthalpy of such a two-phase system is equal to $\Delta g^{(2)}(x_{21})$. Evidently, for the separation to proceed, the free enthalpy of the system must increase by an amount $\Delta g^{(2)}(x_{21})-\Delta g^{(1)}(x_{21})$, indicated by the arrow. According to the thermodynamic stability criteria Eqs. (1) such a process cannot happen spontaneously. This is further exemplified by the values of the chemical potentials which cannot show the identities required by the equilibrium conditions. Hence, a stable one-phase liquid mixture is characterised by a positive curvature of the $\Delta g(x_2)$ function or

$$\left[\frac{\partial^2 \Delta g}{\partial x_2^2}\right]_{P,T} > 0 \quad \textit{for any } x_2 \tag{5}$$

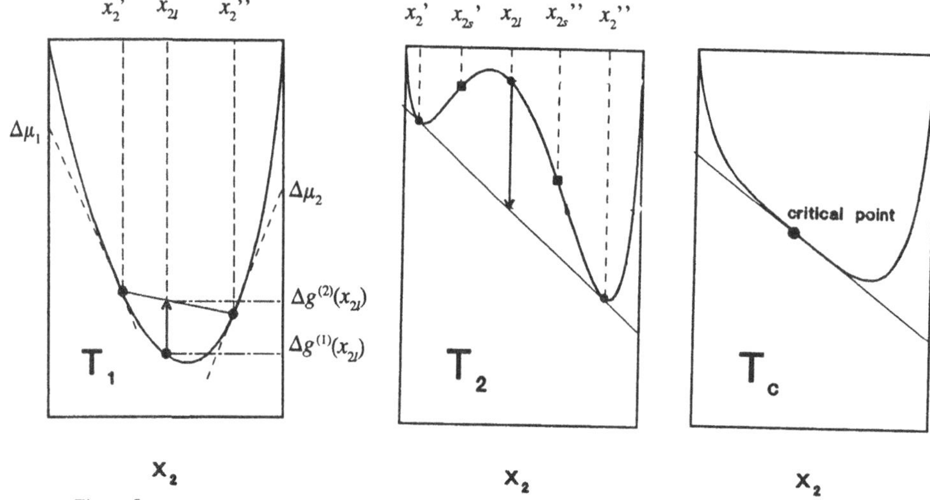

Figure 5. The excess free enthalpy Δg vs compositoon x_2 at three typical temperatures. a) the temperature T_1 in the homogeneous region, b) the temperature T_2 in the heterogeneous region and c) T_c the critical temperature. A discussion is given in the text.

If, however, the equilibrium situation is that of a liquid split up into two phases differing in composition, the equilibrium conditions (4) define a double tangent, touching the $\Delta g(x2)$ curve at two different x_2 values, corresponding to the concentrations of the two coexisting phases. To fulfil this condition it is obvious that the $\Delta g(x_2)$ curve must have a plait as in figur 5b. The concave part of the plait is bounded by two inflection points at compositions x_{2s}' and x_{2s}''.

Now, a system with overall composition x_2 can diminish its free enthalpy, by an amount indicated by the arrow, if it separates into two liquid phases with compositions x_2' and x_2'' (called coexisting or conjugate phases). Within the two compositions a one-phase system will separate spontaneously because any concentration fluctuation, however small, will diminish Δg and hence trigger further separation until a minimum value of Δg is reached. Such separations are called spinodal decomposition, the two points of inflexion are known as spinodal points. The spinodal condition is

$$\left(\frac{\partial^2 \Delta g}{\partial x_2^2}\right)_{P,T} = 0 \tag{6}$$

The condition for instability in composition in a liquid mixture is

$$\left(\frac{\partial^2 \Delta g}{\partial x_2^2}\right)_{P,T} < 0 \tag{7}$$

Therefore the composition range [x_{2s}', x_{2s}''] is called the unstable region. The ranges [0, x_2'[and]x_2'', 1] are stable regions. The remaining ranges [x_2', x_{2s}'[and]x_{2s}'', x_2''] are metastable because they are stable with respect to small concentration fluctuations. Metastable systems require some form of nucleation to reach their stable two-phase state, whereas unstable systems can do so without nucleation.
By appropriate changes in the thermodynamic variables the plait shown in figure 5b will diminish and eventually will disappear. Consequently, the coexisting and spinodal compositions gradually approach each other and eventually coincide (figure 5c). This is called the consolute or critical state and it can be shown that in addition to the spinodal conditions an extra conditions applies

$$\left(\frac{\partial^3 \Delta g}{\partial x_2^3}\right)_{P,T} = 0 \tag{8}$$

Consolute state, spinodal and coexisting concentrations are conveniently presented in a temperature-composition phase diagram (figure 6). The curve defined by the coexisting phases is called a binodal, the curve of spinodal points is the spinodal. Spinodal states can rarely be directly observed. They can be generated by extrapolation of light scattering data. The spinodal condition corresponds to infinitely large fluctuations of concentration and therefore to very large intensities of scattered light [18,19].
Depending on the molecular interactions the system may show different regions of limited miscibility at high and low temperatures. Schematic drawing of some frequently occuring phase boundaries are shown in figure 7. Miscibility gaps induced by increasing temperature are called lower critical solution temperature (LCST) gaps, a demixing with decreasing temperature is denote upper critical solution temperature (UCST) gap.

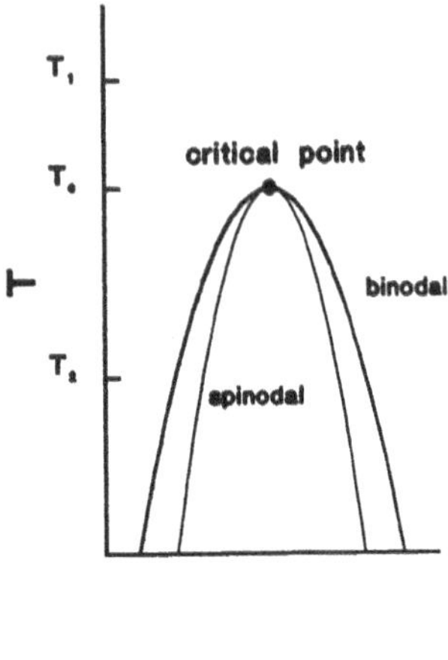

Figure 6. Temperature-composition phase diagram showing consolute state, liquid-liquid binodal and spinodal conditions.

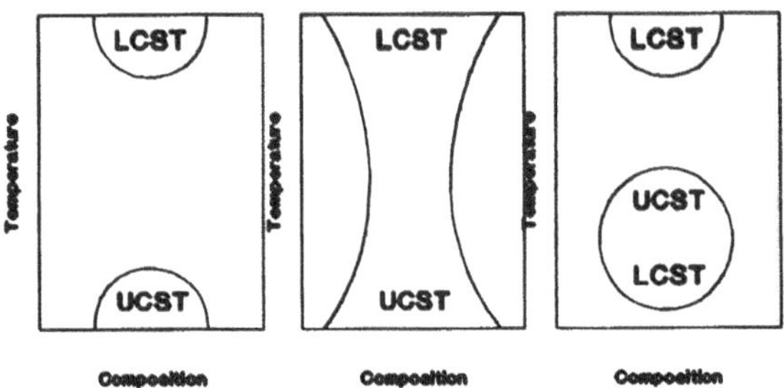

Figure 7. Schematic drawing of frequently occuring combinations of UCST and LCST miscibility gaps.

2.4. Influence of Pressure on Phase Stability

Now it is instructive to expose the influence of the liquid's compressibility on its stability with respect to composition fluctuations [20]. The stability condition (5) can be formulated as

$$\left(\frac{\partial^2 g}{\partial x_2^2}\right)_{T,p} = \left(\frac{\partial}{\partial x_2}\left(\frac{\partial g}{\partial x_2}\right)_{T,p}\right)_{T,p} = \left(\frac{\partial}{\partial x_2}\left(\frac{\partial g}{\partial x_2}\right)_{T,p}\right)_{T,V} + \left(\frac{\partial}{\partial V}\left(\frac{\partial g}{\partial x_2}\right)_{T,p}\right)_{T,x_2} \cdot \left(\frac{\partial V}{\partial x_2}\right)_{T,p} > 0 \qquad (9)$$

If one realizes that $(\partial g/\partial x_2)_{T,p} = (\partial a/\partial x_2)_{T,V}$ and one changes the order of differentiation in the last term, Eq. (9) becomes

$$\left(\frac{\partial^2 g}{\partial x_2^2}\right)_{T,p} = \left(\frac{\partial^2 a}{\partial x_2^2}\right)_{T,V} - \left(\frac{\partial p}{\partial x_2}\right)_{T,V} \cdot \left(\frac{\partial V}{\partial x_2}\right)_{T,p} > 0 \qquad (10)$$

Finally $(\partial V/\partial x_2)_{T,p}$ can be related to the compressibility, leading to the desired result

$$\left(\frac{\partial^2 g}{\partial x_2^2}\right)_{T,p} = \left(\frac{\partial^2 a}{\partial x_2^2}\right)_{T,V} - V\beta_T\left(\left(\frac{\partial p}{\partial x_2}\right)_{T,V}\right)^2 > 0 \qquad (11)$$

The first term at the rhs of eq (11) presents the contribution of concentration fluctuations occurring at constant volume, whereas the second term expresses the influence of the system compressibility. Since the separate factors in this term are all positive the compressibility contribution in all cases destabilizes the system. Given a sufficiently high compressibility β_T the system may even become unstable and show liquid-liquid phase separation. This occurs in the neighbourhood of the vapour-liquid transition of the mixture [21]. The disconnected LCST miscibility gaps presented in Figures 7a and 7b are due to this destabilizing effect of the compressibility. LCST phase behaviour can also occur due to the presence of strong interactions, in this case a closed miscibility gap is obtained. A second LCST gap due to compressibility effects may appear above the closed loop.

The shift of UCST or LCST critical coordinates with pressure is related to the volume and enthalpy of the mixture according to [22]

$$(\partial T/\partial p)_c = (\partial^2 \Delta v/\partial x_2^2)_{p,T} \,/\, (\partial^2 \Delta s/\partial x_2^2)_{p,T} = T\,(\partial^2 \Delta v/\partial x_2^2)_{p,T} \,/\, (\partial^2 \Delta h/\partial x_2^2)_{p,T} \qquad (12)$$

with Δv, Δs and Δh the intensive volume, entropy and enthalpy of the mixture.

Eq. (12) is very similar to the Clausius-Clapeyron relation describing two-phase equilibrium in a one-component system. According to Eq. (12) the shift of the critical temperature with pressure is related to the composition curvature of the excess volume and excess enthalpy at

$$(\partial x/\partial p)_c = \frac{(\partial^3\Delta h/\partial x_2^{\,3})_{p,T}\,(\partial^2\Delta v/\partial x_2^{\,2})_{p,T} - (\partial^2\Delta h/\partial x_2^{\,2})_{p,T}\,(\partial^3\Delta v/\partial x_2^{\,3})_{p,T}}{(\partial^2\Delta h/\partial x_2^{\,2})_{p,T}\,(\partial^4\Delta g/\partial x_2^{\,4})_{p,T}} \tag{13}$$

This shift is one of the very rare phenomena which is related to the fourth composition derivative of the free energy. Further progress can be made if one realises that the sign of $(\partial^2\Delta h/\partial x_2^{\,2})_{p,T}$ at a lower and an upper critical solution temperature is fixed [22]

$$\begin{aligned} &\text{at UCST conditions: } (\partial^2\Delta h/\partial x_2^{\,2})_c < 0 \\ &\text{at LCST conditions: } (\partial^2\Delta h/\partial x_2^{\,2})_c > 0 \end{aligned} \tag{14}$$

From these last two conditions, it can be concluded that UCST and LCST demixing are enthalpic and entropic driven phenomena respectively. Combining Eqs (12) and (14) it can be concluded that the shift of the critical temperature with pressure $(\partial T/\partial p)_c$ is determined by the sign of $(\partial^2\Delta v/\partial x_2^{\,2})_c$.

2.5. Polymer Solutions and Blends

All the above considerations are applicable to binary systems. However, polymer solutions and mixtures are not strictly binary systems (solvent + polymer or polymer + polymer) [23]. For example a synthetic homopolymer contains many components differing in chain length. Due to this polydispersity the critical state is not necessarily situated at the extreme of the miscibility gap [24]. Furthermore if e.g. non-linear polymers (short and long chain branching) or copolymers are considered matter are even more complicated. For the present purpose we will ignore these difficulties since most experimental results refer to model systems for which quasi-monodisperse polymer samples were used.

In all polymer solutions studied, LCST miscibility gaps shift to higher temperatures with increasing pressure, i.e. miscibility increases by applying pressure. A typical example, the system PS in methylacetate is considered in figure 8a [25]. The influence of pressure on UCST behaviour is shown in figure 8b [25]. In comparison with LCST demixing, the absolute change with pressure is generally much smaller.

The influence of pressure on the UCST demixing is not always that simple. For instance the miscibility can increase or decrease with pressure or can even revert with pressure. In addition, also a change in the molar mass of the polymer sample can influence dramatically the observed pressure dependence of the critical conditions. In figure 9 the pressure dependence of the critical temperature is depicted for the system polystyrene (PS)/ cyclohexane (CH) [26]. For a polymer sample with sufficiently high molar mass pressure initially improves miscibility, i.e. the critical temperature is lowered. However, at approximately 100 bar a minimum is reached and further pressurization results in an increase of the critical temperature.

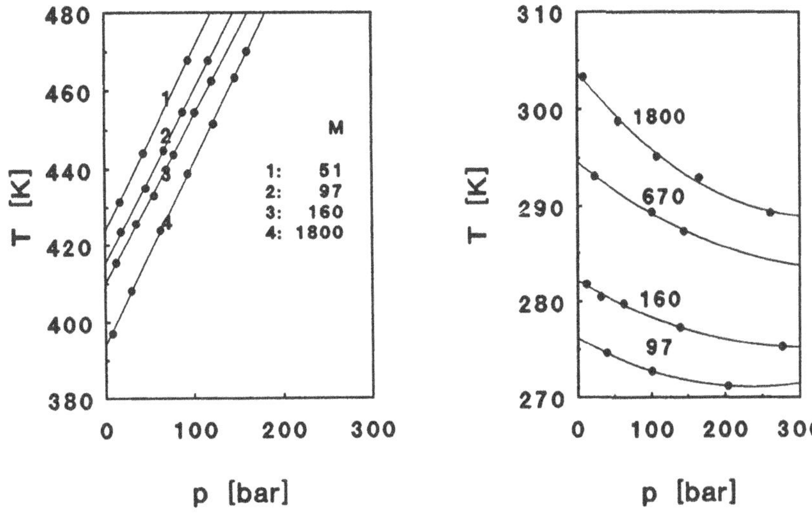

Figure 8. The influence of pressure on a) LCST and b) UCST critical temperatures for the system polystyrene/methylacetate at the indicated molar masses [kg/mol] [25]. Lines are drawn for convenience.

In figure 10 the influence of molar mass is illustrated [27]. For the pressure range considered in this study and in agreement with the data in figure (9), at high molar masses miscibility improves with pressures. However, if the molar mass is decreased the influence of pressure becomes less pronounced and eventually miscibility decreases by application of pressure. The influence of pressure has also been investigated for other systems. In many cases the observed behaviour is equally intriguing as that of PS/CH [25,26,28,29].

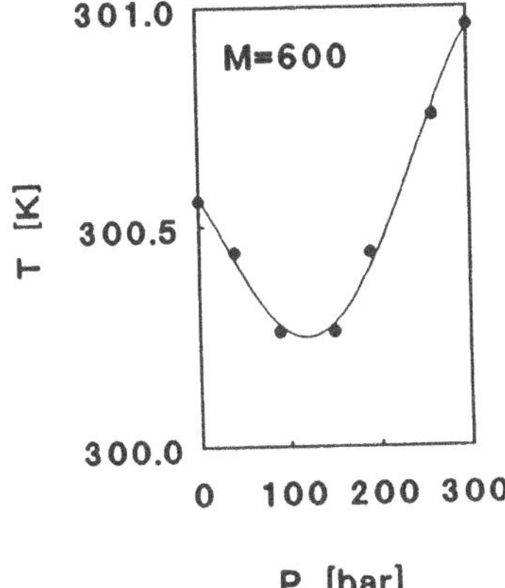

Figure 9. Influence of pressure on the UCST critical temperature for solutions of polystyrene (M_w = 600 kg/mol) in cyclohexane [26].

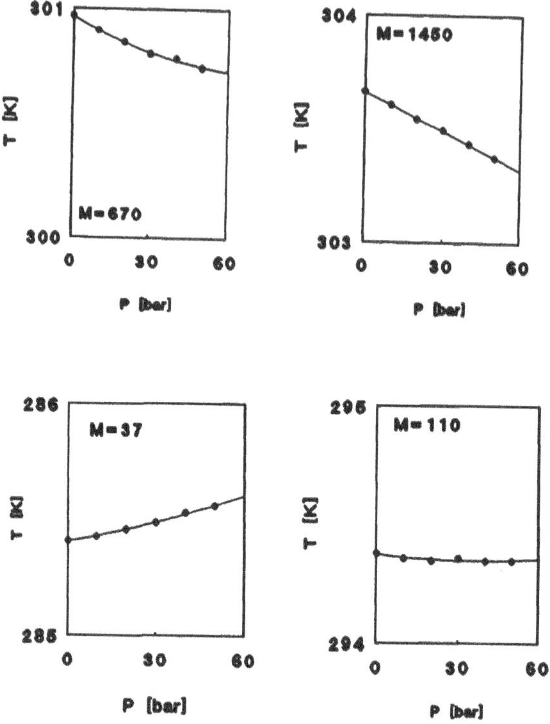

Figure 10. Influence of pressure on the UCST critical temperatures for different indicated molar masses of polystyrene [kg/m-ol] [27].

Also mixtures of polymers or polymer blends are prone to liquid-liquid phase separation. For these systems the observation of e.g. critical points is extremely hard and therefor eqs (12,13) are difficult to check. Nevertheless, the influence of pressure has been studied to some extent [30,31]. The complex influence of pressure is clearly illustrated in figure 11 for mixtures of polyethylacrylate and poly(vinylidenefluoride) [31].
Depending on the composition range miscibility is either enhanced or diminished and the shape of the miscibility gap changes significantly. These effects can be of significance to the practical problem of polymer blending. Quite sizeable pressures may build up during extrusion or other processing conditions. If a blend is a two-phase system at ambient pressure, the pressure built up during extrusion may be sufficient to homogenize the mixture and a possibly intended two-phase extrusion cannot be realized (situation A). At a different blend composition (B), the reverse effect may occur and jeopardize a one-phase extrusion. The consequences for the final product properties may be serious.

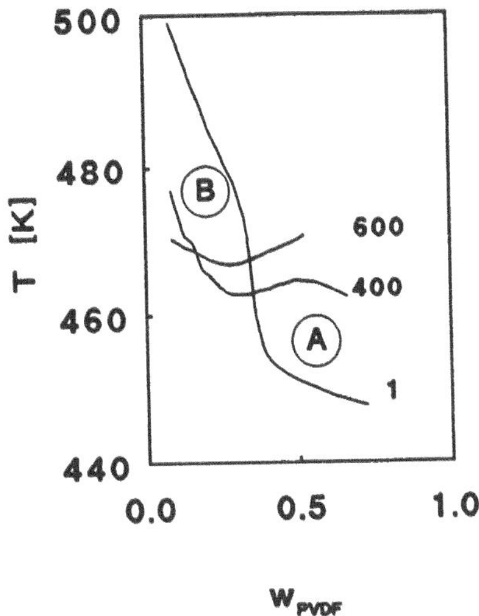

Figure 11. The temperature-composition phase diagram of polyethylacrylate and poly(vinylidenefluoride) for three indicated pressures [bar] [31]. The composition is expressed in mass fraction polyvinylidenefluoride w_{PVDF}. At atmospheric pressure A and B represent a heterogeneous and homogeneous system respectively.

So far the complications arising from the presence of vapour phases has not received much attention. From experiment it is known that a pure polymer is not very keen to evaporate and from theoretical considerations one can conclude that liquid-vapour equilibria probabely cannot exist in polymers [32]. The situation is different in solutions since the solvent surely has a liquid-vapour critical point. Some indications of the importance of liquid-vapour equilibria was already mentioned in the discussion of the destabilizing effect of the compressibility on liquid phase behaviour. It was shown that in the neighbourhood of the liquid-vapour transition of the mixture unstabilities with regard to composition fluctuations may induce LCST demixing. Such interferences can be important in the synthesis of polymers. A well known example is the free radical synthesis at high pressure of low density polyethylene (LDPE) in bulk ethylene at temperatures and pressures far above of the liquid-vapour critical point of ethylene. The production of LDPE takes place in autoclaves at pressures between 1400 and 3500 bar and temperatures up to 600 K. In the reaction approximately 10-35 % monomer is converted to polymer which is separated from the mixture at 150-300 bar. Thus, phase separations occur both under synthesis and recovery conditions and in this case the knowledge of the thermodynamics of the system becomes of practical importance. A detailed discussion can be found in references [18,33].

3. Glass Transition and Glassy State

3.1. Formation History and Physical Ageing

The study of equilibrium conditions has the enormous advantage that physical properties have unique values determined by a limited set of microscopic variables. An example considered previously: at a given pressure and temperature the volume of a single component liquid is uniquely determined by its equation of state. In the glass transition region and the glassy state this beautiful simplicity no longer holds [34,35]. To illustrate this consider the situation depicted in figure 12.

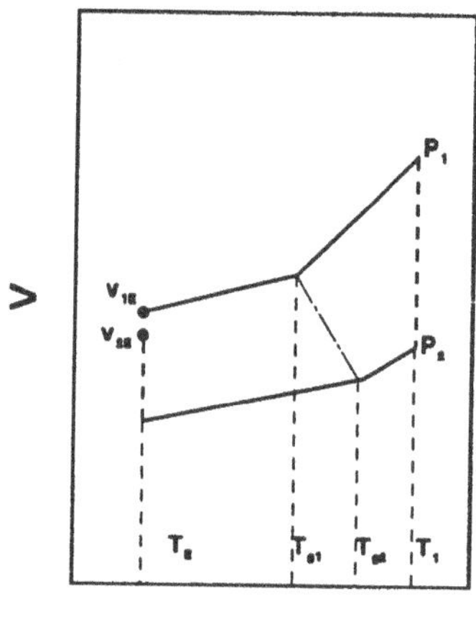

Figure 12. Schematic volume-temperature diagram for a glass forming polymer. Two different glass formation routes are shown and discussed in the text.

A polymer melt isobarically cooled at a fixed rate from temperature T_1 at P_1 to a temperature $T_e<T_g$. The polymer glass will be characterised by a volume V_{1e}. The polymer glass at T_1 and P_1 can also be prepared along an different route: first pressurise the melt at T_1 from P_1 to P_2, cool isobarically (at the same fixed rate) to T_e and finally reduce the pressure to P_1. In this case the volume of the polymer glass will be V_{2e}, i.e. different from V_{1e}. From this example it is clear that i) the formation history, i.e. the exact path followed to prepare a particular glassy sample, must be specified and thus ii) the glassy state is not an equilibrium state. In figure 12 it can be observed that the value of the glass transition temperature depends on the formation history and pressure in particular. Experimental details for PS are shown in figure 13a [4]. Furthermore, as depicted in fig. 13b also the cooling rate has an influence on the value of the glass transition temperature [7]. Especially this last observation indicates the relevance of kinetic and non-equilibrium conditions for the glass transition.

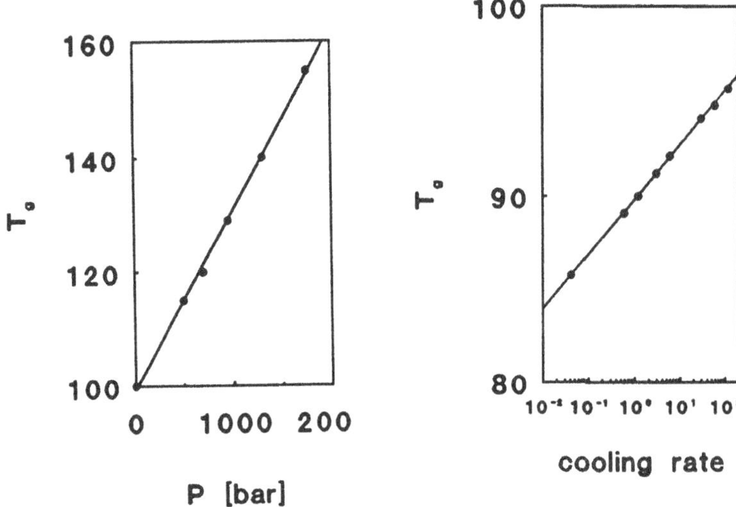

Figure 13. The variation of the glass transition temperature of polystyrene with a) pressure [4] and b) cooling rate [7].

Another aspect of the non-equilibrium conditions becomes apparent in the following experiment. An interested experimentator monitors the volume of the polymer glass (prepared along the first route) as a function of time and observes a continuous densification of the glass. This is one aspect of the physical aging, typical of glassy systems. Figure 14 depicts some more detailed experimental results on the volume relaxation behaviour of PVAc [36]. Polymer samples are jumped a) from different initial temperatures T_i to the same final temperature $T_e > T_i$ and b) from the same initial temperature to different final temperatures $T_e < T_i$.

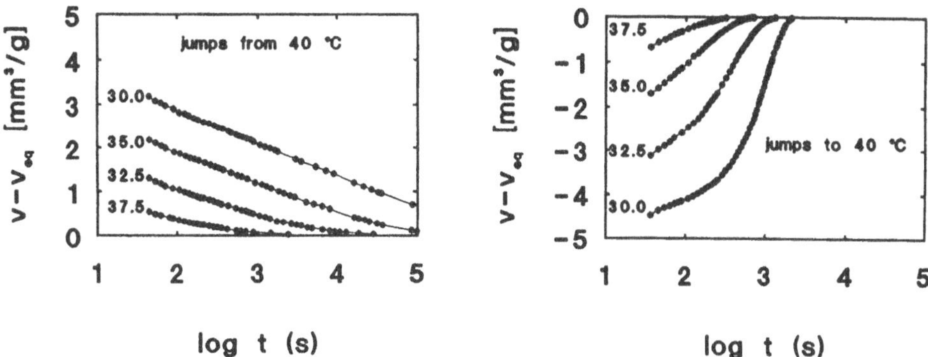

Figure 14. Volume departure from equilibrium vs time after single temperature jumps for polyvinylacetate [36]. •: experimental data, lines are drawn for convenience. a) jumps from 40.0°C to indicated temperatures. b) jumps from indicated temperatures to 40.0°C.

3.2. METHOD OF REDUCED VARIABLES

If linear behaviour is observed plots of $(v-v_f)/(v_i-v_f)$ at different temperature jumps ΔT should produce a single master curve by shifting the different relaxation curves along the time axis to a particular curve chosen for reference [37]. For the data in figure 14 this can only be done in a small interval of ΔT and for larger values of ΔT significant deviations are noticeable. Moreover the deviation from the master curve are different for upward and downward temperature jumps. For quenches the volume contraction becomes progressively slower whereas for $\Delta T > 0$ initially volume dilatation occur very slowly but becomes progressively faster.

In order to understand the auto-retardation and -acceleration effects more information on the molecular processes responsible for these phenomena must be presented. In a good approximation it may be assumed that the molecular processes responsible for the volume relaxation are also important in other viscoelastic properties [37,38]. Most viscoelastic results are obtained from experiments in a limited temperature and time or frequency regime. It is common practise to compose from these results a master curve, i.e. the viscoelastic response as a function of frequency (time) at a fixed temperature. The time-temperature superposition principle states that a viscoelastic response obtained at time t_o and temperature T_o will be observed also at a different temperature T_1 but at another time t_1 which is related to t_o at T_o by the shift factor a_T [37]

$$t_1 = a_T . t_o$$

The principle of reduced variables affords a valuable simplification in separating the time (frequency) and temperature effects on many viscoelastic properties e.g. dynamic moduli, compliancies, viscosities etc. Depending on the specific viscoelastic function considered the exact shifting or reducing procedures can be different but in all cases a master curve and a temperature shift factor is obtained. Details concerning these reduction procedures are clearly described and discussed in a monograph by Ferry [37]. Furthermore the shift factor a_T obtained from the reducing procedure depends on temperature which can be understood in terms of the free volume concept [37]. According to this hypothesis the segmental mobility in the dense (liquid) state is restricted by neighbouring segments or molecules, i.e. the mobility is mainly determined by the proportion of free volume available to the segments or molecules. This principle was first mentioned by Batchinsky [39] and later on presented by Doolittle [40] in an empirical expression, which describes accurately the viscosity of low molecular mass liquids. Theoretical arguments, supporting the form of the empirical Doolittle equation, were presented by Cohen and Turnbull [41]. Following Doolittle's approach the relation between a_T and free volume fraction f reads

$$\log(a_T) = B(1/f - 1/f_0) \tag{15}$$

where B is a proportionality constant, f and f_0 are the free volume fractions at temperature T and T_0 respectively

If certain assumptions are made concerning the dependence of the free volume on temperature the Williams Landel and Ferry equation, relating the shift factor to temperature, can be obtained [42]

$$\log(a_T) = -c_0^1(T-T_0)/(c_0^2 + T - T_0) \tag{16}$$

with T_0 the reference temperature and c_0^1 and c_0^2 the WLF constants.

Eq. (16) gives an accurate analytical representation of the experimental shift factors derived from different viscoelastic properties for many different polymers, small molecules organic and inorganic glasses. Furthermore the shift factors derived from different properties are often in good agreement. The WLF equation is only valid for temperatures $T_g<T<T_g+100$. It is more correct to express the limited validity in terms of the free volume. If the amount of free volume becomes too large the mobility of molecules is in addition determined by intrinsic molecular factors.

So far the discussion of the concepts of time-temperature superposition and free volume was focused on the temperature dependence of the viscoelastic properties. In an analogous manner the influence of pressure on these properties can also be expressed in terms of master curves at a selected pressure and a pressure dependent shift factor a_P. The combined effect of pressure and temperature can be expressed in a shift factor $a_{T,P}$ [37]. Although the results of the influence of pressure are less abundant (for obvious experimental reasons) the results available have shown that use of reduced variables is equally successful in incorporating pressure. Figure 15 represents the effect of pressure and temperature on the shear viscosity of polystyrene [43]. As shown in figure 16a, these data can be reduced to a single master curve at a choosen reference pressure and temperature. The corresponding shift factor $a_{T,P}$ is depicted in figure 16b. These data are succesfully described by the extended WLF equation [43]

$$\log(a_{T,P}) = \frac{-19.32(T-T_s)}{A_2+T-T_s} \tag{17}$$

$$\text{with } T_s = D_2+D_3.P,\ A_2 = D_4+D_3.P$$
$$D_2 = 393.5K,\ D_3 = 3.05.10^{-7}K.Pa^{-1}, D_4 = 75.44K$$

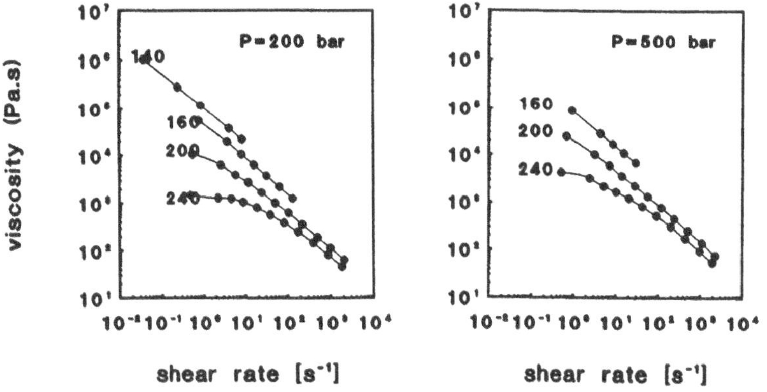

Figure 15. Shear viscosity of polystyrene at indicated temperatures [K] vs shear rate for different pressure at a) 200 bar and b) 500 bar. [43]

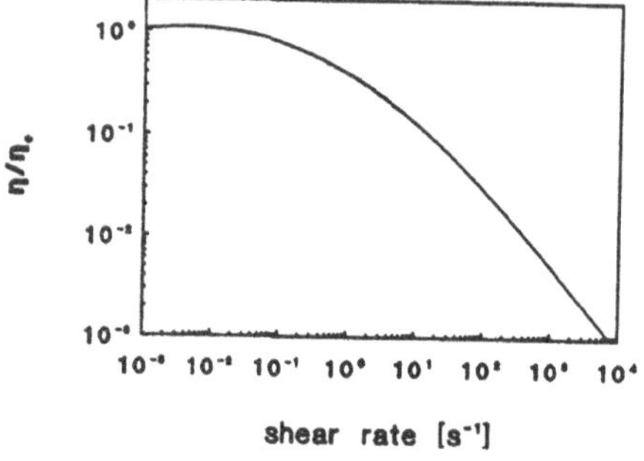

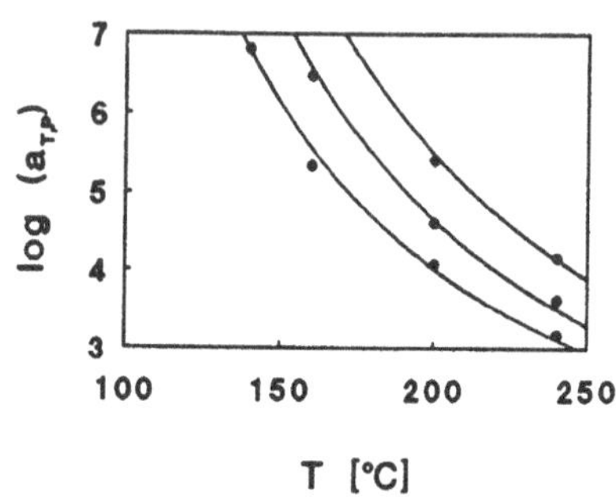

Figure 16. a) Master curve of relative shear viscosity η/η_o vs shear rate at T = 200°C and P = 1 bar [43]. b) Corresponding shift factor $a_{T,P}$ vs temperature at indicated presures. Lines are calculated according to Eq (17). (η_o is the zero shear rate viscosity)

3.3. Physical Ageing Re-examined

It is now possible to understand the non-linear behaviour of the volume relaxation experiments depicted in figure 14. According to eqs. (16) the experiments should be characterised by a constant shift factor and a simple shift along the time axis. However, in these experiments the direct relation between molecular mobility and free volume becomes apparent. The volume relaxation observed in an upward and downward temperature jump is accompanied by a destruction and creation of free volume respectively. Hence the more basic relation between mobility and free volume should be considered instead of the WLF equation. If the change in molecular mobility is taken into account the non linear behaviour of the different relaxation curves can be explained.

The volume relaxation following isothermal pressure jumps is analogous to the temperature jump experiments considered previously. Figure 17 shows the volume relaxation behaviour for pressure jumps from different P_1 to atmospheric pressure [44,45,48]. All these data can be explained by the free volume concept.

Polymer glasses prepared at atmospheric pressure show physical ageing, i.e. a densification in time. High density glasses can be prepared under high pressure conditions. One can consider then an appropriate formation history so that physical ageing can be reduced or circumvented completely. A tentative scheme is the following: the polymer is jumped from the initial temperature T_i = 298K to a temperature T_1 (indicated in figure 18). The polymer is kept at this temperature until the density is equal to the equilibrium density at another temperature T_f.

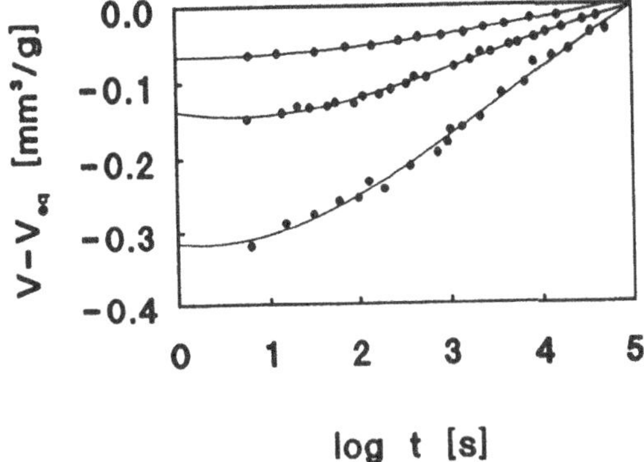

Figure 17. Volume departure from equilibrium vs time after single pressure jumps from indicated initial pressures to atmospheric pressure for polystyrene [44, 45, 48].
•: experimental data, lines are drawn for convenience.

At this moment the temperature is jumped this temperature T_f and the volume is monitored in time. Naively one might hope that no physical ageing will occur since the glass at T_f has attained its equilibrium density. The time response of the glass prepared in this manner is shown in figure 18 [36]. It can be observed that the density does not remain constant in time but at first decreases, reached a minimum and the returns to the equilibrium density. The polymer appears to possess a memory for its formation history.

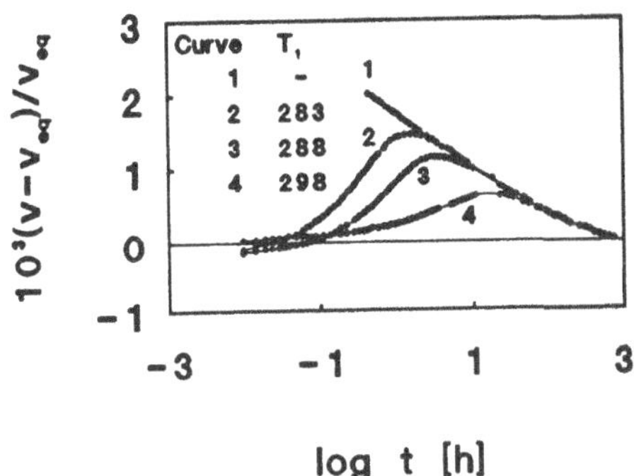

Figure 18. The effect of combined temperature jumps on the volume relaxation behavior [36]. The sample is jumped to an initial temperature T_i to indicated temperatures T_1. When the density of the sample is equal to the equilibrium density at temperature T_2 the sample is jumped to this temperature. The volume relaxation behavior following this last jump is followed is time.

A similar effect can be observed with pressure if e.g. the following formation path is followed: cool the polymer at high pressure to a temperature T_2. Pressure and temperature are chosen so that the density of the fresh sample is equal to the (equilibrium) density of the sample prepared at atmospheric pressure at T_2. The pressure is released and the subsequent time response of the density is followed. Also in this case a memory effect is observed [47]. From these experiments it is clear that the density is not the only factor determining the characteristic behaviour of glasses. In the following contribution these phenomena will be explained.

4. Thermodynamic Properties at High Frequencies

Although transient and dynamic oscillatory experiments are related to each other by Fourier transforms and thus contain the same time-dependent information, experimentally it is often more convenient to perform dynamic oscillatory experiments instead [37]. In these experiments the actual deformations are normally small and at the same time a large constant hydrostatic pressure can be applied. Therefore these experiments are extremely suitable in studying the influence of pressure on the linear viscoelastic behaviour.
The most extensive direct measurement of dynamic bulk compression, the frequency analog of the equilibrium compressibility β_T, has been performed for polyvinylacetate in the frequency range 50 to 1000Hz and for pressures from 1 to 1000 atm [48].
In analyzing these data the significant pressure and temperature dependence of the limiting bulk compliancies at high and low frequencies must be taken into account. From these results the pressure and temperature dependencies at a given frequency can be derived.

In more recent years a new technique to measure the longitudinal bulk compliance M has become available. Photon correlation spectroscopy presents an elegant, non-destructive technique to measure this quantity as a function of pressure and temperature over a considerable time (or frequency) range [49-51].
Free volume also plays a role in other physical properties e.g. the complex dielectric constant [52,53], the T_1 spin relaxation time in NMR experiments [54,55], diffusion of small probe molecules in polymers [56,57].

5. Conclusions

The combination of pressure and temperature result in a much richer dependence of many physical properties then the variation of e.g. temperature alone. In the case of single constituents the influence of pressure on solid-liquid and solid-solid transitions has been discussed. Furthermore, recent results in poly(4-methyl-pentene) have drawn the attention to the possibility of pressure induced amorphisation and at high pressures to crystallisation upon heating and vice versa. This unexpected and peculiar behaviour is currently under intensive investigation and waits for an explanation.
Also intriguing phase behaviour is found in mixtures (solutions and blends) in the amorphous dense liquid state without the interference of the crystalline state. This pressure dependence of

the liquid-liquid miscibility has its influence if one is considering the polymerisation in solution or in the bulk. The most examined example is probably the synthesis of linear low density polyethylene from ethylene at pressures much higher then vapour-liquid critical pressure. Also the processing of polymers blends and solution can be influenced significantly by pressure. A possible explanation for the observed behaviour is presented in the following contribution.

Finally, the influence of pressure on the glass transition temperature and the glassy state itself was discussed. In particular, the importance of the formation route and the occurrence of volume relaxation was demonstrated. The responsible molecular mechanism for the glass transition leads to the study of the time dependence or high frequency behaviour of thermodynamic properties. In this discussion the time-temperature/pressure superposition principle clearly displays the influence of the different variables, i.e time and temperature.

REFERENCES

1. Gibbs, J.W. (1948) *Collected Works*, Yale University Press, New Haven, Vol. 1.
2. Denbigh, K. (1987) *The Principles of Chemical Equilibrium*, Cambridge Univ. Press.
3. Olabisi, O.; Simha, R. (1975) *Macromolecules*, **8**, 206.
4. Quach, A., Simha, R. (1971) *J. Appl. Phys.*, **42**, 4592.
5. Hellewege, K.; Knappe, W.; Lehman, P.(1963) *Kolloid-Z.*, **183**, 110.
6. Haldon, R.A; Simha, R. (1968) *Macromolecules*,**1**, 340.
7. Greiner, R.; Schwarzl, xx. (1984) *Rheologica Acta*, **23**, 378.
8. Rastogi, S.; Hikosaka, M.; Kawabata, H.; Keller, A. (1991) *Macromolecules*, **24**, 6384.
9. Wunderlich, B. (1980) *Macromolecular Physics*, Vol. 3 , Academic Press, NY.
10. Mishima, O.; Calvert, L.D.; Whalley; E. (1984) *Nature*, **310**, 393.
11. Basset, D.C., Turner, B. (1972) *Nature*, **240**, 146.
12. Wunderlich, B.; Moller, M.; Grebowitcz, J.; Bauer, H. (1988) *Adv. Polym. Sci.*, 87.
13. Keller, A. (1983) *Stucture-Property Relationships of Polymeric Solids*, Edt. Hiltner, A., Plenum Press.
14. Bhateja, S.K.; Pae, X.D. (1975) J. *Macrom. Sci., Rev. Macromol. Chem.*, **C13**, 72.
15. Hikosaka, M.; Tsukijima, K.; Rastogi, S.; Keller, A. (1992) *Polymer*, **33**, 2502.
16. Hirakawa, S.; Takemura, T. (1969) *Jpn. J. Appl. Phys.*, **8**, 635.
17. Rastogi, S.; Newman, M.; Keller, A. (1992) Nature, **353**, 353.
18. Koningsveld, R.; Stockmayer, W.H.; Nies, E. *Equilibrium Thermodynamics of Polymer Systems, Vol. I*, Oxford Univ. Press, to be published.
19. Scholte, Th.G. (1972) *J. Polym. Sci., C*, **39**, 281.
20. Sanchez, I.C. (1988) *Encyclop. Phys. Sci Techn., Vol 11*, 1.
21. Lacombe, R.U.; Sanchez, I.C. (1976) *J. Chem. Phys.*, **80**, 2568.
22. Prigogine, I., Defay, R. (1954) *Chemical Thermodynamics*, Longmans and Green,

London.
23. Koningsveld, R.; Staverman, A.J. (1967) *J. Polym. Sci., C*, **16**, 1775.
24. Gordon, M.; Chermin, H.A.G.; Koningsveld, R. (1969) *Macromolecules*, **2**, 207.
25. Zeman, L.; Patterson, D. (1972) *J. Phys. Chem.*, **76**, 1214.
26. Wolf, B.A., Geerissen, H. (1981) *Coll. Pol. Sci.*, **59**, 1214.
27. Saeki, S., Kuwahara, N., Nakata, M., Kaneko, M. (1975) *Polymer*, **16**, 445.
28. Zeman, L.; Biros, J.; Delmas, G.; Patterson, D. (1977) *J. Chem. Phys.* **76**, 1206.
29. Borchard, W. et. al. unpubished results.
30. Rostami, S.; Walsh, D.J. (1984) *Macromolecules*, **17**, 315.
31. Suzuki, Y., Miyamoto, Y., Miyaji, H., Asai, K., *J. Pol. Sci.* (1982) Pol. Lett. Ed., **20**, 563.
32. Sanchez, I.C.; Lacombe, R.U. (1976) *J. Chem. Phys.*, **60**, 2352.
33. Van Opstal, L.; Nies, E.; Koningsveld, R., to be published.
34. Simha, R. (1977) *Macromolecules*, **10**, 905.
35. Haward, R.N. (1973) *The Physics of Glassy Polymers*, Appl. Sci. Publ., London.
36. Kovacs, A. J.(1963) *Fortschr. Hochpolym.-Forsch.*, **3**, 394.
37. Ferry, J.D. (1980) *Viscoelastic Properties of Polymers*, John Wiley, NY.
38. Robertson, R. E. (1979) *J. Polym. Sci., Polym. Phys.*, **17**, 597.
39. Batschinsky, A.J. (1913) *J. Phys. Chem.*, **84**, 644.
40. Doolittle, A.K. (1951) *J. Appl. Phys*, **22**, 1471.
41. Cohen, M.H.; TurnBull, D. (1959) *J. Chem. Phys.*, **31**, 1164.
42. Williams, M.I.; Landel, R.F.; Ferry, J.D. (1955) *J. A. Chem. Soc.*, **77**, 3701.
43. Kadijk, S.E.; Brule, A.A. van den, to be published.
44. Goldbach, G.; Rehage, G. (1967) *Rheol. Acta*, **6**, 30.
45. Goldbach, G.; Rehage, G. (1967) *J. Polym. Sci., C*, **16**, 2289.
46. Tribone, J.J; O'Reilly, J.M.; Greener, J. (1989) *J. Polym. Sci., Pol. Phys.*, **27**, 837.
47. Robertson, R. E.; Simha, R.; Curro, J. G. (1985) *Macromolecules*, **18**, 2239.
48. McKinney, J.E.; Belcher, H.V. (1963) *J. Res. Nat. Bur. Stand.*, **A67**, 43.
49. Patterson, G.D.; Stevens, J.R.; Carrol, P.J. (1982) *J. Chem. Phys.*, **77**, 622.
50. Tribone, J.J.; Jamieson, A.M.; Simha, R. (1984) *J. Polym. Sci., Polym. Symp.*, **71**, 231.
51. Fytas, G. (1989), *Macromolecules*, **22**, 211.
52. Wong, C-P; Schrag, J.L.; Ferry, J.D. (1970) *J. Polym. Sci., A2*, **8**, 911.
53. Samara, G.A. (1992) *J. Polym. Sci., Polym. Phys.*, **30**, 669.
54. Sasabe, H.; Saito, S. (1968), *J. Polym. Sci., A2*, **6**, 1401.
55. Laupetre, F.; Virlet, J.; Boyle, J.P. (1985) *Macromolecules*, **18**, 1846.
56. Fujita, H. (1961) *Adv. Polym. Sci.*, **3**, 1.

MOLECULAR MODELING OF THE INFLUENCE OF PRESSURE ON FLUID AND GLASSY STATE BEHAVIOUR

Erik Nies and Servaas Vleeshouwers
Department of Polymer Technology
Eindhoven University
P.O.Box 513
5600 MB Eindhoven
The Netherlands

ABSTRACT. A model of the dense disordered state, pertinent to chain and small molecule fluids is discussed. It rests on a quasi-lattice or cell model, with additional configurational disorder provided by vacancies. In the theory two parameters define the intermolecular interactions. In polyatomic systems a third parameter ensues, quantifying molecular modes of motion (i.e. soft vibrations and rotations) which are perturbed by the surroundings. In mixtures these parameters are explicit functions of composition and of self and cross interactions.

The theory is successfully applied to describe the equation of state behaviour of the separate constituents and their mixtures. Typically, the experimental data are described within the experimental uncertainty of the measuring technique.
Subsequently, the miscibility behaviour of polymer solutions and mixtures is discussed employing the hole theory. Equation of state contributions are essential to understand e.g. the appearance of lower critical miscibility (LCM) phase behaviour in polymer solutions and mixtures. Furthermore, the intriguing influence of pressure on the upper critical miscibility (UCM) phase behaviour for polymer solutions is predicted successfully. Some comments concerning the pressure dependence of LCM phase behaviour of polymer mixtures are presented.

The equilibrium theory is complemented with a stochastic formalism. This combination allows to discuss the influence of formation parameters on the glassy state. For instance, the dependence of the glass transition temperature on cooling rate and pressure is predicted. Also the equation of state of the resulting glasses is predicted obtained and compares favourably with the experimental data.
The presented formalism opens the way to discuss the dependence of the ultimate properties of materials, obtained along a processing route, on the non-equilibrium conditions experienced during processing.

R. Winter and J. Jonas (eds.), High Pressure Chemistry, Biochemistry and Materials Science, 225–246.

1. Introduction

A proper understanding of the phase behaviour of single components, polymer solutions and mixtures in the melt and the glassy state is of interest scientifically as well as technologically. Important thermodynamic properties are e.g. the equation of state, the miscibility behaviour and physical ageing. The starting point to deal with these questions is a theoretical description of the dense disordered state, pertinent to chain and small molecule fluids. It rests on a quasi-lattice or cell model, with additional configurational disorder provided by a fraction h of vacant sites or holes. These so called hole theories were initiated by Simha and Somcynsky [1] and subsequently applied and refined by different authors [2-8]. In what follows we first review the essentials of the theory for single and multi-constituent systems under single phase conditions. Subsequently, the theory is applied to the equation of state of polymer melts, organic solvents and mixtures of these components. It is shown that the theory quantitatively describes experimental data in the dense state, i.e. densities pertinent to liquids.

Before extending the theory to the glassy state the application to miscibility behaviour is considered. With the pioneering work of Flory [9], Huggins [10], Staverman [11] a.o. a theoretical understanding of upper critical (UCM) phase behaviour, i.e. demixing with decreasing temperature, was obtained. For non-polar systems UCM phase behaviour results due to combinatorial entropy contributions, which at high temperatures lead to miscibility, and unfavourable energetic interactions which at low temperatures cause the system to demix. The understanding of lower critical (LCM) phase behaviour, i.e demixing with increasing temperature, emerged 30 years later with the advent of equation of state theories [12-15]. This at the same time offers the possibility to rationalize the influence of pressure on the miscibility behaviour. With the presented statistical thermodynamic theory it is shown that also the miscibility behaviour and its pressure dependence is predicted successfully.

The theoretical understanding of the equation of state and the glass transition is presently one of the central problems in condensed matter properties. In this work the glass transition is considered to be a consequence of a gradual freeze in [16-18], resulting in molecular relaxation times large in comparison with observation times.
The present thermodynamic equilibrium theory applied is complemented with a stochastic formalism [19,20] to describe and evaluate the time dependent changes in the structure function h [21-25]. This allows us to consider the properties of the polymer glass as a result of the formation history which is retraceable to the equilibrium polymer liquid [24,25].

2. Equilibrium theory: Helmholtz free energy and equation of state.

2.1. Single Constituent

Each unit (segment or molecule) is subjected to thermal motions in the field of its neighbours. This field is computed employing a mean field approximation which places these neighbours onto average positions defined by the sites of a lattice. In order to enhance the configurational disorder, a fraction h of vacant sites is introduced. It may be viewed as an excess free volume function over that prevailing in the fully occupied lattice [1].

The free volume function enters through three factors into the partition function, hence also into the Helmholtz free energy A and all thermodynamic properties derivable from this. These three factors are

a) A combinatory factor arising from the mixing of empty and occupied sites.
b) A free volume term characteristic of cell theories which is computed by approximating the cell potential by a square well.
c) A lattice energy contribution, again characteristic of cell theories.

Thus the Helmholtz free energy can be viewed as a function of temperature T, volume V and free volume h

$$A(V,T) = A[V, T, h(V,T)] \tag{1}$$

We note that the resulting expressions for the contributions (a-c) are functions of the configurational disorder, i.e.the free volume function h. In recent investigations, it has been demonstrated that the external contact fraction q is more appropriate to describe the influence of the configurational disorder [4-8]. This external contact fraction q is related to the free volume function, according to

$$q=\frac{(1-\alpha)y}{(1-\alpha y)} \tag{2}$$

with $\alpha = (2/z)(1-1/s)$; $y = (1-h)$; z the lattice coordination number and s the number of lattice sites occupied by a molecule of molar mass M. Introducing the further abbreviations A = 1.011, B = 1.2045 and indicating reduced variables by tildes, the Helmholtz free energy is given by

$$\begin{aligned} A/NsRT = {} & \frac{\ln(y)}{s} + \frac{(1-y)\ln(1-y)}{y} - \frac{(1-\alpha y)\ln(1-\alpha y)}{\gamma y} \\ & -c_s\ln(v^*\tilde{\omega}(1-\eta)^3) + \frac{c_s q}{2\tilde{T}}(A\tilde{\omega}^{-4}-2B\tilde{\omega}^{-2}) \end{aligned} \tag{3}$$

with $\tilde{\omega}=y\tilde{V}$, $\eta=2^{-1/6}y(1-\alpha)/((1-\alpha y)\tilde{\omega}^{1/3})$ the volume of a lattice site or cell.

From the Helmholtz free energy the pressure equation can be obtained according to

$$p=-\left(\frac{\partial A}{\partial V}\right)_{N,T,y}-\left(\frac{\partial A}{\partial y}\right)_{N,T,V}\left(\frac{\partial y}{\partial V}\right)_{N,T} \tag{4}$$

The temperature and volume or pressure dependence at equilibrium of the characteristic structure function q (or h) is uniquely determined by the minimization of the configurational free energy [1,4]

$$\left(\frac{\partial A}{\partial y}\right)_{N,T,V} = 0 \tag{5}$$

Performing the necessary algebra, we obtain the two coupled equations [4]

$$\frac{\tilde{P}\tilde{V}}{\tilde{T}} = \frac{1}{(1-\eta)} + \frac{2(1-\alpha)y}{(1-\alpha y)\tilde{T}\tilde{\omega}^2}\left[\frac{A}{\tilde{\omega}^2} - B\right] \tag{6}$$

and

$$1 - \frac{1}{s} + \frac{1}{y}\ln(1-y) - \frac{1}{\gamma y}\ln(1-\alpha y) - \frac{\alpha}{\gamma} =$$
$$c_s\left[\frac{(3\eta-1+\alpha y)}{(1-\eta)(1-\alpha y)} + \frac{(1-\alpha)y}{2\tilde{T}(1-\alpha y)^2\tilde{\omega}^2}\left\{2B - \frac{3A}{\tilde{\omega}^2} + 4\alpha y\left(\frac{A}{\tilde{\omega}^2} - B\right)\right\}\right] \tag{7}$$

with $\tilde{P} = P/P^*$, $\tilde{V} = V/V^*$, $\tilde{T} = T/T^*$.

The characteristic scaling parameters of pressure, volume and temperature denoted by P*, V* and T* are combinations of the inter segmental attraction energy ϵ^* and segmental repulsion volume υ^* for a segment of mass M_s and the extra parameters $c_s = c/s$, where 3c represents the number of external or volume dependent modes in the s-mer. The scaling parameters are defined by the following relations

$$P^* = z(1-\alpha)\epsilon^*/\upsilon^*;\ V^* = s\upsilon^*;\ T^* = z(1-\alpha)\epsilon^*/(c_sR) \tag{8}$$

The importance of the c-parameter was first recognized by Prigogine, Trappeniers and Mathot [26]. It is to be taken as a measure of the perturbation of internal rotations of the chain in addition to motions of the chain as a whole, in the dense medium. A comparison between theory and experimental equation of state data serves to estimate the scaling parameters (P*, T*, V*) or alternatively the molecular parameters (ϵ^*, v*, c_s). The theory quantitatively describes the PVT data for solvents, polymers including engineering plastics and their mixtures at densities typical of the liquid state at low and high pressures [2,4,26]. Examples of this are given in the **Applications** section.

In the further application of the theory to the glassy state, one needs to consider the fluctuations in free volume δh_{eq} which can be computed from [21-25]

$$\left\langle (\delta h_{eq})^2 \right\rangle = \left\langle (h - \langle h \rangle_{eq})^2 \right\rangle = kT \left[\left(\frac{\partial^2 A}{\partial y^2} \right)_{N,T,V} \right]^{-1} \tag{9}$$

2.2. Mixtures

The generalization to multi-component systems has been given [3,4] by adopting a simplification introduced by Prigogine et al. in their theory of mixtures, based on the simple cell model [26]. That is, the repulsion volume v^* is replaced by an average, instead of dealing explicitly with the problem of packing differently sized spheres. This approximation requires the restriction to nearly equal sized segments (not necessarily equal sized chemical repeat units). For a binary system of components a and b we have the following defining relation for the mean interaction parameters

$$\langle \epsilon^* \rangle \langle v^* \rangle^m = \epsilon_{aa}^* v_{aa}^{*m} q_a^2 + 2\epsilon_{ab}^* v_{ab}^{*m} q_a q_b + \epsilon_{bb}^* v_{bb}^{*m} q_b^2; \; m = 2,4 \tag{10}$$

where q_i represents the contact fraction of component i.

Eq. (10) ensures that the scaled eqs. 6 and 7 retain their formal validity with all parameters now explicit functions of composition and of self and cross interactions [3,6]. Moreover, averages $\langle c \rangle = \Sigma c_i/\phi_i$; $\langle s \rangle = \Sigma s_i/\phi_i$ enter where ϕ_i [$= s_iN_i/(s_aN_a+s_bN_b)$] is the segment fraction of component i. The scaling parameters (eqs. 8) are then redefined in terms of these compositional averages.
Furthermore, extra contributions from mixing the two chemically distinct components enter the Helmholtz free energy

$$\begin{aligned} A/N\langle s \rangle RT &= \frac{\phi_a \ln(\phi_a)}{s_a} + \frac{\phi_b \ln(\phi_b)}{s_b} \\ &+ \frac{\ln(y)}{\langle s \rangle} + \frac{(1-y)\ln(1-y)}{y} - \frac{(1-\langle \alpha \rangle y)\ln(1-\langle \alpha \rangle y)}{\gamma y} \\ &- \langle c_s \rangle \ln(\langle v^* \rangle \tilde{\omega}(1-\eta)^3) + \frac{\langle c_s \rangle q}{2\tilde{T}}(A\tilde{\omega}^{-4} - 2B\tilde{\omega}^{-2}) \end{aligned} \tag{11}$$

Thus far, random mixing was implicitly assumed between molecules and vacancies, as well as between molecules of different species. An evaluation of the effect of non-randomness, based on the quasi-chemical approach of Guggenheim, has recently been given [6-8].

3. Equilibrium Theory: Phase behaviour.

3.1. Spinodal and critical conditions

The extra mixing terms in the free energy expression (11) are particularly pertinent for the discussion of phase behaviour. The phase stability of mixtures at given pressure P and temperature T is governed by the Gibbs free energy G related to the Helmholtz free energy A and the equation of state

$$G = A + PV \tag{12}$$

In the P-V-T-composition diagram of a binary mixture of components A and B spinodal and critical conditions can be defined and are related to the second and third compositional derivatives of the Gibbs free energy respectively [28]. The spinodal is defined by

$$J_{sp} = \partial^2 G/\partial\phi_b^2 = 0 \tag{13}$$

The critical condition, which is also a spinodal point, is given by

$$J_{sp} = 0 \text{ and } J_{cr} = \partial^3 G/\partial\phi_b^3 = 0 \tag{14}$$

For the hole theories considered here, the detailed expression resulting from the application of Eqs. (13) and (14) have been presented elsewhere [4,29].

3.2. Critical temperature: influence of pressure

The change of the critical temperature with pressure $(\partial T/\partial P)_c$ can be related to the compositional curvature of volume and enthalpy of the mixture [30]

$$(\partial T/\partial P)_c = (\partial^2 V/\partial\phi_b^2)_c \,/\, (\partial^2 H/\partial\phi_b 2)_c) \tag{15}$$

where the subscript c indicates that the derivatives have to be evaluated at the critical temperature and composition for a given pressure.

The curvature of the enthalpy of mixing is directly related to the type of phase behaviour. For LCM and UCM phase behaviour one finds [30]

$$\text{LCM: } (\partial^2 H/\partial\phi_b^2) > 0 \tag{16a}$$
$$\text{UCM: } (\partial^2 H/\partial\phi_b^2) < 0 \tag{16b}$$

Consequently, the slope of the critical temperature with pressure is directly related to the curvature of the volume of mixing as a function of composition. A comparison of the theoretical calculations with experimental results on phase behaviour and its pressure dependence will be presented in the section **Applications**.

4. Non-Equilibrium Theory: Glassy State and Glass Transition.

4.1. THE DYNAMICS OF RELAXATION PHENOMENA: STOCHASTIC THEORY.

Consider a change of temperature and/or pressure in the system from an initial to a final state. The thermodynamic properties change accordingly. On a microscopic level this change in macroscopic properties is caused by rearrangements in the polymer chains. It was suggested by Robertson [19] that the local free volume is an appropriate parameter to describe the local segmental mobility and thus the rate of rearrangements. The system is thought to be subdivided into small volume elements, each with its (local) free volume. The sample is thus characterised by a (time dependent) free volume distribution. For mathematical convenience it is assumed that this distribution is a set of n discrete levels $\{w_i(t), i=1,n\}$

$$w(t) = \{w_i(t),\ i=1,n\} = \left[w_1(t)\ .\ w_i(t)\ .\ w_n(t)\right] \tag{17}$$

with the local free volume in level w_i equal to h_i .

Changes in the occupation of the levels occur by transitions between the different levels and can be regarded as a stochastic process known as a Markov chain. The transition probability P_{ij} that the system changes from state i to state j in the time (t+s) is given by

$$P_{ij}(t+s) = \sum_{k=1}^{n} P_{ik}(t)P_{kj}(s), \qquad 1 \le i,j,k \le n \tag{18}$$

The time derivative of this equation with respect to s yields a set of coupled differential equations which can be written in matrix form

$$\dot{P}(t) = P(t)\cdot A \tag{19}$$

with P(t) the transition probability

$$P(t) = \begin{bmatrix} P_{11}(t) & P_{12}(t) & . & P_{1n}(t) \\ P_{21}(t) & . & . & P_{2n}(t) \\ . & . & . & . \\ P_{n1}(t) & . & . & P_{nn}(t) \end{bmatrix} \tag{20}$$

In an infinitesimal time interval (s→0), only transitions between adjacent levels can occur and it can be shown that the generator matrix A becomes tridiagonal

$$A=\begin{bmatrix} -\lambda_{1,2} & \lambda_{1,2} & 0 & . & . & . & 0 \\ \lambda_{2,1} & -\lambda_{2,1}-\lambda_{2,3} & \lambda_{2,3} & 0 & . & . & . \\ 0 & \lambda_{3,2} & -\lambda_{3,2}-\lambda_{3,4} & \lambda_{3,4} & 0 & . & 0 \\ . & . & . & . & . & . & \lambda_{n-1,n} \\ 0 & . & . & . & 0 & \lambda_{n,n-1} & -\lambda_{n,n-1} \end{bmatrix} \qquad (21)$$

where $\lambda_{i,i-1}$ and $\lambda_{i,i+1}$ are the downward and upward transition rates from state i to state (i-1) and state (i+1)

The formal solution of eq (19) is given by

$$P(t)=P(0)e^{At} \qquad (22)$$

where P(0) is the initial transition probability matrix, in all applications set equal to the identity matrix I.

Equation (22) can be solved by an eigenvalue analysis described in many textbooks [20]. Once the transition probability is known, the calculation of the occupation levels is straightforward

$$w(t+s) = w(t)\cdot P(s) \qquad (23)$$

Equation (23) defines the time evolution of the occupation of the free volume levels. It can be shown that at sufficiently long times the Markov chain evolves to a stationary distribution $\{\xi_i, i=1,n\}$

$$\lim_{t\to\infty} w(t) = [\xi_1 \ . \ \xi_i \ . \ \xi_n] \qquad (24)$$

In this contribution the stationary distribution is identified with the structure function h and its distribution defined in the HH theory [4,24,25]. For sufficiently small fluctuations it can be shown that the fluctuations have a Gaussian distribution fully determined by $<h>_{eq}$ and $<(\delta h_{eq})^2>$.

4.2. TRANSITION RATES: KINETICS

In order to use the stochastic theory the upward and downward transition rates $\lambda_{i,i+1}$ and $\lambda_{i+1,i}$, $1 \leq i \leq (n-1)$, respectively, must be defined. The functional dependence of the local transition rates on regional free volume h_i is assumed to be identical with the dependence of the global mobility μ on the equilibrium overall free volume $<h>_{eq}$. The global mobility of a system with free volume h can be expressed e.g. by a Williams-Landel-Ferry (WLF) type of equation [31]. On a local level then this reads

$$\mu_i \propto \exp\left[2.303\frac{c_1^{\prime}(h_i-f_o^{\prime})}{c_2^{\prime}+h_i-f_o^{\prime}}\right] \tag{25}$$

For the local transition rates the following expressions are used [19-25]

$$\lambda_{i,i-1} = \frac{R}{\beta^2}\sqrt{\frac{\xi_{i-1}}{\xi_i}}\;\mu_i$$
$$\lambda_{i,i+1} = \frac{R}{\beta^2}\sqrt{\frac{\xi_{i+1}}{\xi_i}}\;\mu_i \tag{26}$$

where R (s^{-1}) contains a characteristic rate and a compensation for differences in global and local mobilities.

4.3. THE STOCHASTIC SIMULATION

The numerical simulation procedure will be illustrated for a cooling experiment. A polymer melt, initially, at time t, at temperature T and pressure p, is cooled at a rate T' K/s. At a certain temperature the glass transition temperature will be reached and upon further cooling a polymer glass is formed. The stochastic simulation of this real time experiment proceeds along the following scheme [25]

1. Calculate the equilibrium properties (V, $<h>_{eq}$, $<(\delta h_{eq})^2>$) at T and p corresponding to a time $t+\Delta t$.
2. Calculate the equilibrium distribution $\{\xi_i, i=1,n\}$ at $t+\Delta t$.
3. Compute the transition rates using the equilibrium distribution and the actual distribution $\{w_i(t), i=1,n\}$ at time t.
4. Determine the transition probabilities P(t) (eq 8).
5. Update the actual distribution $\{w_i(t+\Delta t), i=1,n\}$ (eq 9) at time $t+\Delta t$.

6. Increment the time with Δt, and repeat steps 1-6.

A similar experiment can be done by e.g. specifying a pressurizing rate p' or by a simultaneous change in T and p. Furthermore T' and p' can be changed at will during the simulation. E.g. at a given T and p in the glassy state one can set T'=0 and p'=0 and from that time on monitor the physical aging. It is thus possible to simulate the complete formation history of a polymer glass and study the influence of these parameters on the resulting thermal properties. During the simulation the actual distribution w(t) and the calculated average free volume $\langle h(t)\rangle_{ac}$ can differ from the equilibrium values. At sufficiently high temperatures the actual and equilibrium values (and of course also the complete distributions) coincide; thus the system is equilibrated. At a certain temperature the actual distribution w(t) and also the actual average free volume $\langle h(t)\rangle_{ac}$ commence to deviate from the equilibrium condition. Upon further cooling the difference between actual and equilibrium states increases. As depicted in e.g. Figure 9 a transition temperature T_t can be obtained as the intercept of the extrapolated 'glassy' part of the curve with the equilibrium line. This transition temperature is now identified with the glass transition temperature T_g in the real experiment. Furthermore the p,h,T data obtained from a specific simulation experiment can be used to compute the thermodynamic properties [25].

5. Applications

5.1. Equation of state of polymer, solvent and solution.

Extensive experimentation on and analysis of polymer melts has taken place [32]. Whereas Simha and colleagues were concerned with vinyl-type compounds, Zoller and colleagues included in their studies also various engineering plastics, compatible polymer blends and copolymers. The outcome was a very satisfactory agreement between experiment and theory. An illustration is given in Figure 1 for the equation of state data for polystyrene investigated over pressures up to 2 kbar [33]. The results displayed in Figure 1 are typical for the many polymer systems investigated. The characteristic parameters for some constituents are summarized in Table 1.

TABLE 1. Molecular characteristic parameters for some constituents.

Polymer	P^* [bar]	V^* [cm^3g^{-1}]	T^* [K]
Polystyrene	6797	0.9468	9458
Polyvinylacetate	8955	0.8063	7989
Polycarbonate	10004	0.7968	7928
Cyclohexane	8389	1.1143	5247

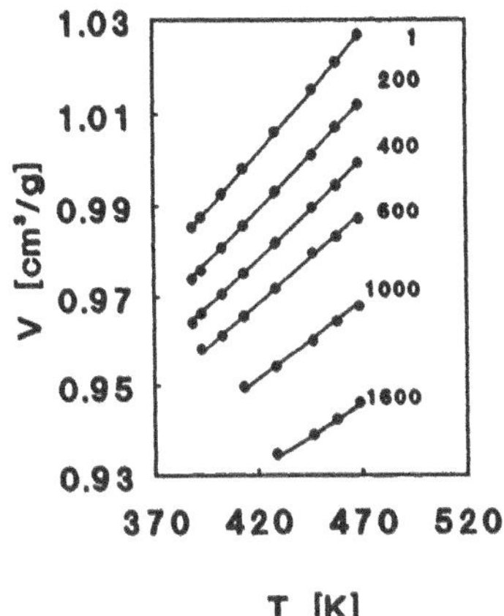

Figure 1. Equation of state data for polystyrene [32]. Lines are drawn according to the hole theory.

The predictions of the theory have been examined also for low molecular mass compounds. Normal paraffins, for example, offer an opportunity to proceed to much higher pressures for the present purpose than in polymer melts. Again satisfactory comparisons have ensued for both homogeneous fluids and their mixtures [3]. For the subsequent discussion of phase equilibria organic solvents are of primary interest. Benzene, carbon tetrachloride and cyclohexane have been investigated [2,3,34,35], as well as mixtures of the latter two [3].
Similar results have been obtained for polymer solutions [35]. Of special interest in connection with its phase behaviour, discussed below, is the pair polystyrene-cyclohexane. Specific volumes at several compositions, molar masses and temperatures have been determined [36]. The requisite scaling volumes and temperatures of the mixtures were derived from phase data. The maximum deviation between predicted specific volume and measurement is below 1% which is gratifying indeed. However, we should also note at this point that the reverse procedure i.e. the prediction of phase relations from equation of state data, is not satisfactory for a quantitative analysis of the subtle compositional and pressure effects to be considered in what follows.

5.2. Miscibility behaviour of polymer solutions.

In the previous paragraph the pure component parameters are obtained from the equation of state behaviour of the constituents. The cross interactional parameters ϵ^*_{ab} and v^*_{ab} are calculated from the critical point (T_c and w_{2c}) for the system CH/PS3 (see Table 2) [5]. With the molecular parameters fixed all configurational thermodynamic properties, e.g. critical conditions, can be predicted for different polymer samples in cyclohexane.

TABLE 2. Critical coordinates for different polystyrene samples in cyclohexane. Molar masses are in kg/mol, compositions are expressed in mass fraction PS and temperatures in Kelvin.

code	M_w	T_c	w_{2c}	T_c	w_{2c}
PS1	51	288.85	0.146	292.10	0.121
PS2	166	296.60	0.099	297.88	0.081
PS3	520	301.15	0.064	301.15	0.064
PS4	600	301.34	--	--	--

In Table 2 it can be observed that the chain length dependence of the critical coordinates is not predicted accurately . The agreement between experiment and theory can be improved if allowance is made for a small change in the cross interactional parameters with chain length [37]. For the present purpose the discrepancies are accepted and we focus on the predictions concerning LCM phase behaviour and the influence of pressure on miscibility. It has been shown previously that the predicted LCM demixing is in reasonable agreement with experimental data. Therefore with the present hole theory a semi-quantitative prediction of the entropy driven LCM demixing is obtained, using parameter values extracted from the enthalpy driven UCMT phase behaviour.

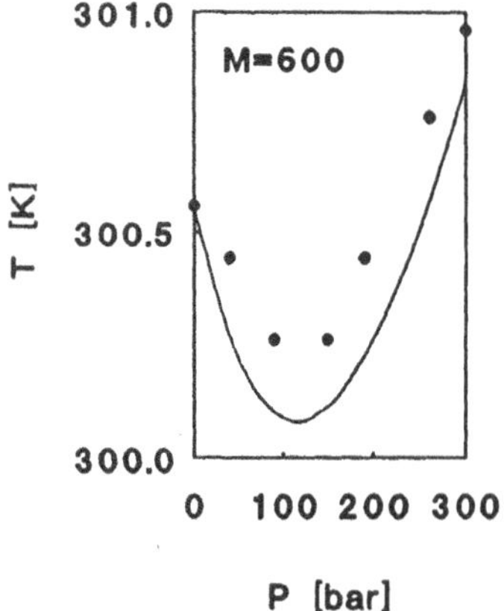

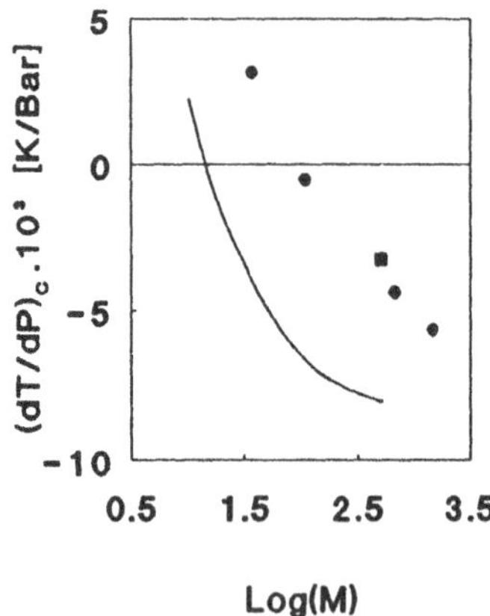

Figure 2. a. Critical temperature as a function of pressure for PS4/CH [38]. b. Initial slope $(\partial T/\partial P)_c$ as a function of molar mass [39].

In Figure 2a the influence of pressure on the critical temperature of the system PS4/CH is shown [38]. Initially the slope of the critical temperature with pressure $(\partial T/\partial P)_c$ is negative but at sufficiently high pressure the slope becomes zero and finally positive. Furthermore Saeki et. al. measured the pressure dependence of the critical temperature for different molar masses in a pressure range where changes in slope with pressure, as shown in figure 2a, are not observed [39]. In Figure 2b the initial slope $(\partial T/\partial P)_c$ is negative for high molar masses in agreement with the data of Wolf and Geerisen. However if the molar mass is decreased sufficiently a positive slope $(\partial T/\partial P)_c$ is observed. In Figure 2 the solid line is the predicted pressure dependence according to the hole theory. Although the agreement with experiment is not quantitative, especially in Figure 2b, the observed agreement with experiment is very gratifying, remembering that no adjustable parameters are involved since they have already been determined as stated previously. According to Eqs. 15 and 16 the change in slope with molar mass and pressure is directly related to the compositional curvature of the volume of mixing. In Figure 3 the predicted excess volume ΔV is plotted as a function of pressure at constant molar mass. As a function of molar mass at atmospheric pressure a similar composition dependence is observed. For sufficiently high molar masses and at moderate pressures the excess volume is positively curved for all polymer concentrations. Consequently a negative slope $(\partial T/\partial P)_c$ is predicted. Upon applying pressure or decreasing the molar mass of the polymer, locally at low polymer concentrations a negative curvature of the excess volume ensues. For high pressures and small molar masses the region extends over a larger polymer concentration range. Depending on the location of the critical composition in a region of positive, zero or negative curvature, the slope $(\partial T/\partial P)_c$ is negative, zero or positive respectively. Unpublished results of Wolf et. al. show that the experimental and theoretical composition dependencies of the excess volume for different molar masses are in qualitative agreement [40]. In view of the agreement observed in Figure 2b only a qualitative agreement can be expected. However the main features of the dependence of the excess volume on composition and chain length are reproduced. Thus a consistent thermodynamic prediction of the complicated pressure dependence is obtained.

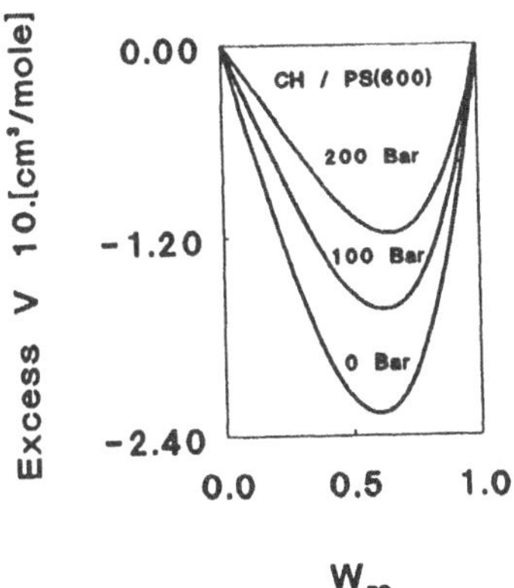

Figure 3. Predicted excess volume of mixing at indicated pressures and corresponding temperatures.

5.3. INFLUENCE OF PRESSURE: CELL AND FREE SITE VOLUMES

In the hole theory the excess volume can be viewed as the result of two volume changes upon mixing, i.e a volume change ΔY due to a change in the fraction of free lattice sites and a volume change $\Delta\Omega$ due to the change of the volume per lattice site (the cell volume)

$$\Delta V = \Delta Y + \Delta\Omega \tag{27}$$

In Figure 4 the different contributions to the total volume change ΔV are shown for three different pressures. The volume change due to the change in the fraction of free lattice sites is very sensitive to pressure. The changes in excess cell volume $\Delta\Omega$ however are practically pressure independent. Upon pressurizing the densification of the liquid related to ΔY is reduced. Consequently, for sufficiently high pressures the excess cell volume becomes relatively more important and cause a local change in curvature of the total excess volume ΔV, also shown in Figure 3. Furthermore the excess cell volume $\Delta\Omega$ is practically independent of the molar mass. Only the occupied fraction of segments ΔY is sensitive to changes in molar mass. Consequently for sufficiently low polymer molar masses the cell volume changes $\Delta\Omega$, relative to ΔY, become more important and cause the total excess volume ΔV to change curvature.

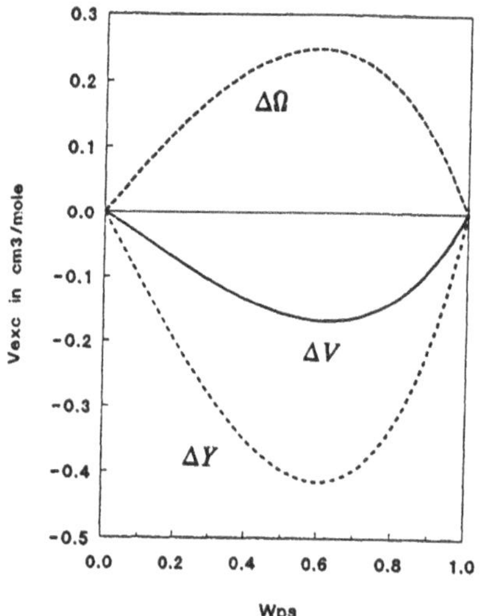

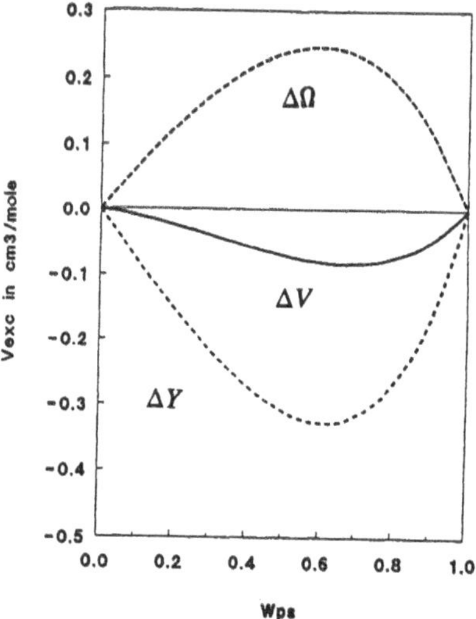

Figure 4. Predicted excess volume of mixing ΔV, free site volume ΔY and cell volume $\Delta\Omega$ at 1 and 300 bar.

5.4. INFLUENCE OF CELL AND FREE LATTICE SITE VOLUMES ON LCMT PHASE BEHAVIOUR.

In polymer solutions showing LCM phase behaviour differences in lattice free site volume are large compared to the changes in cell volume. Therefore, the contribution of the excess cell volume is not able to manifest itself. Consequently in polymer solutions the LCM miscibility gap will be raised with pressure. For polymer mixtures however volume changes due to change in the fraction of free lattice sites are much smaller compared to those in polymer solutions. In this case cell volume effects can become noticeable and, eventually, the LCM miscibility gap may lower in temperature with pressure. Furthermore the two different contributions to the excess volume may cause a complicated pressure dependence of the spinodal condition in the temperature-composition diagram. It is worth mentioning that a complicated pressure dependence on cloud-point curves has been observed experimentally [41].

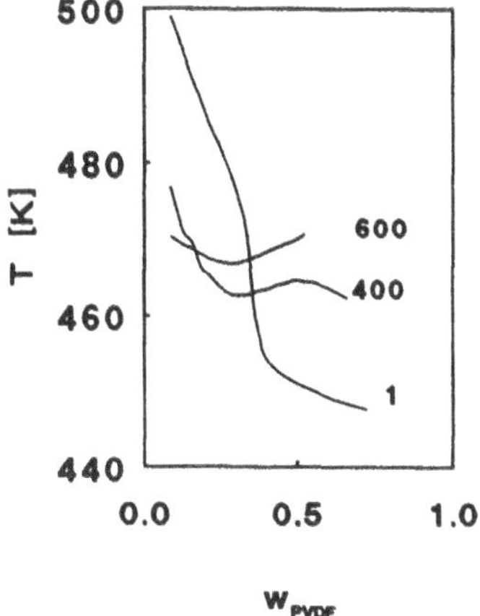

Figure 5. The temperature-composition phase diagram (cloud point curves) for poly(ethylacrylate)/ poly(vinylidenefluoride) for three indicated pressures [bar] [41]. The composition is expressed in mass fraction poly(vinylidenefluoride) w_{PVDF}.

6. Non-Equilibrium: The Glassy State

6.1. Estimation of simulation parameters.

The stochastic theory will be applied to the thermal behaviour of polyvinylacetate (PVAC). With the scaling parameters (Table 1), $<h>_{eq}$ and $<(\delta h_{eq})^2>$ can be calculated. Subsequently the full h-distribution $\{\xi_i\}$ is calculated, using a Gaussian approximation [25]. The parameters c_1', c_2' and f_o'in Eq. (25), describing the correlation between global mobility and free volume are obtained from experimental viscosimetric shift factors, are summarized in Table 3. The description of the shift factors with these parameters, is shown in Figure 6 by the full line.

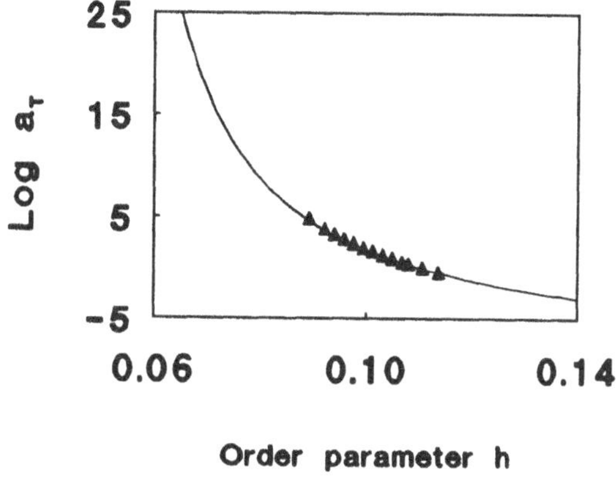

Figure 6. Experimental dynamic mechanical shift factors versus order parameter h (▲) for PVAC [42]; WLF fit, lines.

The parameter R can be estimated from different kinds of experiments. In this application, following Robertson, Simha and Curro [21], we use the relaxation in a jump experiment [42] to obtain a value for R [24,25]. The best description of these experiments (shown in Figure 7) yields the value for the constant R given in Table 3. The set of parameters, summarized in Table 3, is now fixed and we are ready to explore other dynamic simulations.

Table 3. Parameters for PVAC, used in simulations.

Parameter	value
c_1'	9.0
c_2'	0.061
f_o'	0.11
R (s^{-1})	1.

6.2. Simulation results

In Figure 7 the computed volume relaxation behaviour brought about by single temperature jumps is compared to experimental results [24,25]. The observed agreement is typical for the present implementation of the stochastic theory as observed earlier [21]. The agreement can be improved by adopting a temperature dependent R and more refined functions correlating the macroscopic mobility and free volume [21-23,43].

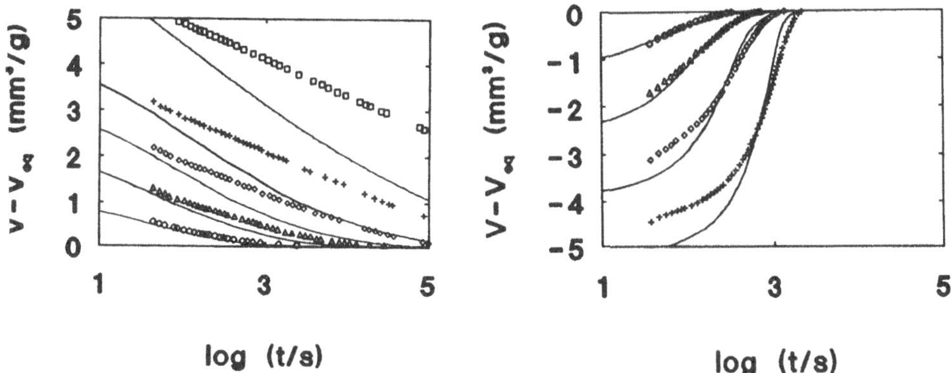

Figure 7. Volume departure from equilibrium versus time after single temperature jumps for PVAC. Symbols, experimental data (Kovacs [43]): (a) jumps from 40°C to 37.5°C (○), 35°C (△), 32.5°C (◇), 30°C (+) and 25°C (□); (b) jumps to 40°C (b) from 37.5°C (○), 35°C (△), 32.5°C (◇) and 30°C (+). Solid lines, calculated.

In Figure 8 the simulated and experimental glass transition temperatures at ambient pressure are presented as a function of cooling rate [24,25]. In the simulations the glass transition temperature is identified with the intercept of extrapolated glass and the liquid like volume traces. An important observation is that the best R value derived from the jump experiment gives an accurate prediction of the glass transition temperature T_g at atmospheric pressure.

This suggests that also the volumetric glass transition temperature, which is more commonly available than volume relaxation experiments, can be used to estimate the parameter R, which enables prediction of the volume relaxation behaviour for a given polymer under different conditions.

The influence of pressure on the formation of the glassy state is shown in Figure 9. The simulated formation conditions are chosen to mimic the experimental formation history (p and cooling rate T') [44]. The dependence of volume on temperature under isobaric cooling is depicted [25]. The agreement for the 0.1MPa curve is excellent: maximum deviations between experimental and theoretical specific volumes ΔV do not exceed $2 \cdot 10^{-3}$ cm^3/g. At higher pressures the predicted specific volumes are systematically too large but the maximum deviation ΔV remains less than $8 \cdot 10^{-3}$cm^3/g at p = 80MPa. Values for the thermal expansion coefficient α_g and the glass transition temperature T_g extracted from the simulation results compare favorably with experimental data. However, with increasing pressure systematic deviations occur between theory and experiment which can be attributed to a too small predicted value of the pressure dependence of the glass transition temperature (dT_g/dp) [25].

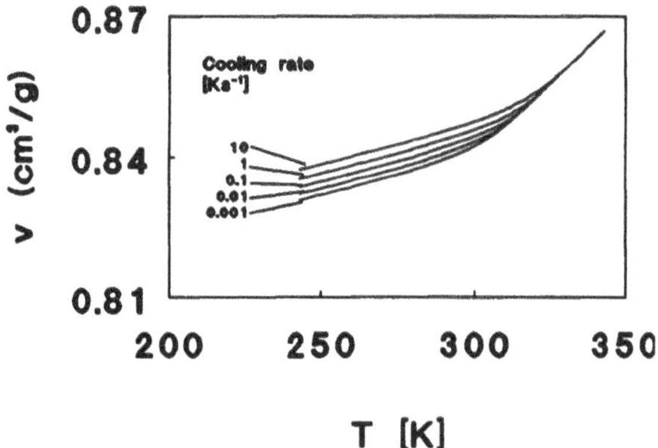

Figure 8. Calculated volume versus temperature for indicated cooling rates for PVAC.

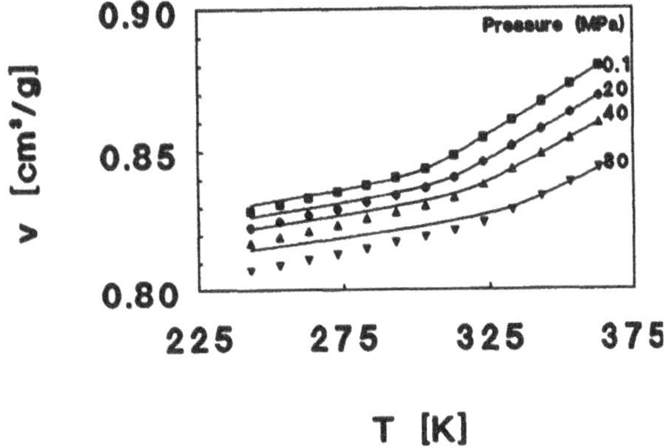

Figure 9. Volume versus temperature for indicated formation pressures for PVAC. Solid lines, calculated isobars; symbols, experimental [45].

The simulated pVT behaviour of a polymer glass formed by cooling at 0.1MPa and by subsequently pressurizing is shown in Figure 10a together with experimental data [44]. Once more the experimental formation history was reproduced accurately [25]. In this case the maximum deviation between experimental and theoretical volume ΔV is smaller than $3 \cdot 10^{-3}$ cm^3/g. For a polymer glass formed by cooling a high pressure polymer melt (p=80 MPa) and by subsequently depressurizing, the results are depicted in Figure 10b, together with experimental data [44]. In this case the deviations ΔV are larger ($\Delta V < 8 \cdot 10^{-3}$ cm^3/g) and can be mainly attributed to the overestimated pressure dependence of the glass transition temperature [25].

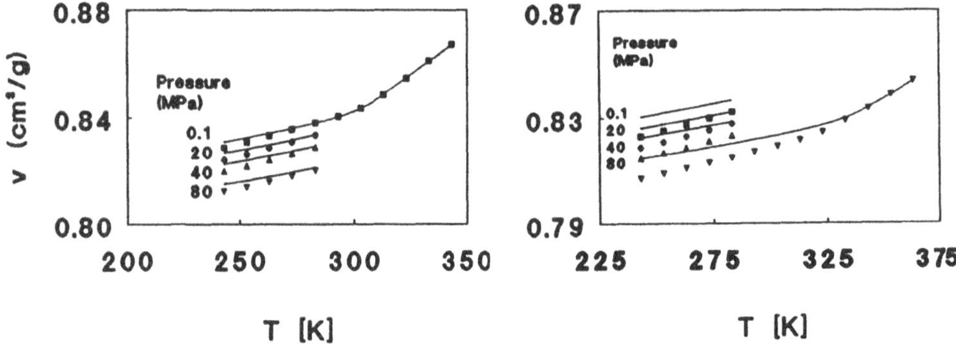

Figure 10. Volume versus temperature for a PVAC glass formed by cooling at a) 0.1 MPa and b) 80 MPa. Solid lines, calculated isobars; symbols, experimental [45].

7. Conclusions

The equilibrium theory outlined yields quantitatively successful equations of state for dense disordered assemblies of low and high molar mass constituents, and their mixtures.
With a limited set of parameters a quantitative description of PVT data is obtained at densities typical of the liquid state at ambient and elevated pressures. This excellent description of the equation of state data also contributes to the accurate prediction of e.g. the complicated influence of pressure found in miscibility behaviour. Furthermore, the molecular mechanism of this complex pressure influence is exposed and is related to the subtle interplay between repulsive and attractive contributions to the total excess volume of the solutions and mixtures.

The equation of state behaviour of glasses, the glass transition temperature and its dependence on formation history is predicted from equilibrium pVT-data, equilibrium dynamic mechanical shift factors and a single non-equilibrium volume relaxation experiment. Simulations at ambient pressure show excellent agreement with experimental data. Discrepancies between theory and experiment at high pressures are almost entirely related to the dependence of T_g on the glass formation pressure [25]. The use of shift factor data, obtained at ambient pressure, to obtain a relation between mobility and free volume, is probably limited to low pressures.

Discrepancies at high pressures, as shown in this contribution, suggest that mobility at high pressure is not only described by the order parameter h, which is here used as a measure for free volume.

The only non-equilibrium data used for the predictions are volume data of a single relaxation experiment, to estimate a value for R. As is shown in this contribution the parameter R can also be estimated from different experiments, e.g the location of the ambient glass transition temperature. Also other properties (e.g. dynamic bulk compressibility, enthalpy relaxation, dynamic light scattering) are being simulated and investigated for PVAC and other polymers as well (polycarbonate, polystyrene, polymethylmethacrylate, etc.).

REFERENCES

1. Simha, R., Somcynsky, T. (1969) *Macromolecules*, **2**, 341.
2. Jain, R.K., Simha, R. (1980) *J. Chem. Phys.*, **72**, 4909.
3. Jain, R.K., Simha, R. (1980) *Macromolecules*, **13**, 1501.
4. Nies. E., Stroeks, A.(1990) *Macromolecules*, **23**, 4088.
5. Stroeks, A., Nies, E. (1990) *Macromolecules*, **23**, 4092.
6. Nies, E., Xie, H., submitted.
7. Xie, H.; Nies, E., submitted.
8 Nies, E.; Xie, H. (1992) *Makromol.Chem., Macromol. Symp.*, **58**, 227.
9. Flory, P.J., (1941) *J. Chem. Phys.*, **9**, 660.
10. Huggins, M.L., (1941) *J. Chem. Phys.*, **9**, 440.
11. Staverman, A.J. (1937) *Recl. Trav. Chim. Pays-Bas*, **56**, 885.
12. Prigogine, I., Bellemans, A., Mathot, V. (1957) *The Molecular Theory Solutions*, North-Holland Publishing Co., Amsterdam.
13. Flory, P.J., Orwoll, R.A., Vrij, A. (1964) *J. Am. Chem. Soc.*, **86**, 3567.
14. Kleintjens, L.A., Koningsveld, R.(1980) *Coll. Polym. Sci.*, **258**, 711.
15. Sanchez, I.C., Lacombe, R.U. (1976) *J. Phys. Chem.*, **80**, 2352.
16. Jackle, J. (1986) *Rep. Prog. Phys.*, **49**, 171.
17. Kirkpatrick, T. R.; Thiramalai, D.; Wolynes, P. G.(1989) *Phys. Rev. A*, **40**, 1045.
18. Donth, E. (1991) *J. Non-Cryst. Solids*, **131-133**, 204.
19. Robertson, R. E. (1979) *J. Polym. Sci., Polym. Phys.*, **17**, 597.
20. Goodman, R. (1988) *Introduction to Stochastic Models*; Benjamin/Cummings: Menlo Park, p 127.
21. Robertson, R. E.; Simha, R.; Curro, J. G. (1984) *Macromolecules*, **17**, 911.
22. Robertson, R. E.; Simha, R.; Curro, J. G. (1985) *Macromolecules*, **18**, 2239.
23. Robertson, R. E.; Simha, R.; Curro, J. G. (1988) *Macromolecules*, **21**, 3216.
24. Vleeshouwers, S; Nies, E. (1991) *Polymer Comm.*, **32**, 418.
25. Vleeshouwers, S; Nies, E., *Macromolecules*, in press.
26. Prigogine, I., Trappeniers, N., Mathot, V. (1953) *Discussions Faraday Soc.*, **15**, 93.
27. Zoller, P. (1978) *J. Polym. Sci, Pol. Phys.*, **16**, 1471.
28. Gibbs, J.W. (1948) *Collected Works*, Yale University Press, New Haven, Vol. 1.

29. Nies,E.; Stroeks, A.; Simha, R.; Jain, R.K. (1990) *J.Coll.Pol.Sci*, **268**, 731.
30. Prigogine, I., Defay, R. (1954) *Chemical Thermodynamics*, Longmans and Green, London.
31. Williams, M.I.; Landel, R.F.; Ferry, J.D. (1955) *J. A. Chem. Soc.*, **77**, 3701.
32. Zoller, P. (1989) *Polymer Handbook*, Eds. Brandrup, J., Immergut, E.H., John Wiley, NY, VI/475.
33. Quach, A., Simha, R. (1971) *J. Appl. Phys.*, **42**, 4592.
34. Jain, R.K., Simha, R. (1979) *J. Chem. Phys.*, **70**, 2792.
35. Simha, R., Jain, R.K. (1985) *J. Colloid Polym. Sci.*, **263**, 905.
36. Wolf, B.A., private communication.
37. Stroeks, A., Nies, E. (1988) *Pol. Eng. Sci.*, **28**, 1347.
38. Wolf, B.A., Geerissen, H. (1981) *Coll. Pol. Sci.*, **59**, 1214.
39. Saeki, S., Kuwahara, N., Nakata, M., Kaneko, M. (1975) *Polymer*, **16**, 445.
40. Wolf, B.A., Private communication.
41. Suzuki, Y., Miyamoto, Y., Miyaji, H., Asai, K., *J. Pol. Sci.* (1982) Pol. Lett. Ed., **20**, 563.
42. Williams, M. L.; Ferry, J. D. (1954) *J. Colloid Sci.*, **9**, 479.
43. Kovacs, A. J.(1963) *Fortschr. Hochpolym.-Forsch.*, **3**, 394.
44. Robertson, R. E. (1985) *Macromolecules*, *8*, 953.
45. McKinney, J. E.; Goldstein, M. (1974) *J. Res. Nat. Bur. Stand.-A. Phys. Chem.* , **78A**, 331.

COMBUSTION AND DIFFUSION FLAMES IN SUPERCRITICAL AQUEOUS FLUIDS AT HIGH PRESSURES

E.U. FRANCK
Institut für Physikalische Chemie
Universität Karlsruhe
Kaiserstraße 12
7500 Karlsruhe
Germany

ABSTRACT. Supercritical water is in wide regions of pressures and temperatures completely miscible with CO_2, CH_4, C_2H_6, N_2, O_2, H_2 and other nonpolar gases and with higher alkanes. Homogeneous, single phase, 70 - 30 % water-methane mixtures, for example, can be prepared at 400 ^oC and above with pressures up to 2000 bar. A slow injection of oxygen creates a steady diffusion flame. UV-spectra from the emission of OH-radicals have been determined and used to derive rotational temperatures, which can be above 3000 K. Flames with supercritical water-hydrogen and other water-alkane mixtures were also obtained. The decrease of spontaneous ignition temperatures with pressure is discussed. A computational procedure permits the prediction of properties of one-dimensional diffusion flames to 3000 bar with realistic equations of state and transport functions. Experiments at temperatures below flame formation were performed. Pyrolysis, hydrolysis and oxydation in the dense hydrothermal fluid can be combined to dispose of toxic wastes. Model reactions have been investigated.

R. Winter and J. Jonas (eds.), High Pressure Chemistry, Biochemistry and Materials Science, 247–263.

I. INTRODUCTION

The miscibility of high pressure supercritical water with other, nonpolar, gases opens up a possibility to study combustion in very dense fluids and to generate high pressure diffusion flames. In binary aqueous systems the two-phase region is confined at high temperatures by a critical curve, which begins at the critical point of pure water (374 °C, 221 bar) and extends into the three-dimensional temperature-pressure-composition space towards higher pressures as a high temperature envelope of the two-phase region. Fig. 1 shows a number of experimentally determined critical curves for binary systems, relevant for combustion. In most of the supercritical aqueous fluids discussed here, water is a diluent for fuel and oxygen. To some extent this has technical reasons. It is also of basic interest, however, to study the possible interference of the dense water with combustion. In addition there is interest in supercritical aqueous combustion for geochemistry and for new acceptable methods for toxic waste disposal. The generation of flames and their UV-spectra and temperature analysis will be described. Combustion products will be shown briefly, and computational results for high pressure, one-dimensional diffusion flames will be presented.

II. APPARATUS

To investigate flames and combustion in - to some extent - corrosive fluids up to 500 °C and pressures of 1000 bar and higher, special reaction cells have to be designed. One type of cell, used in this work, is described in detail elsewhere (7). It is made of a high strength nickel-base superalloy[x] and of cylindrical shape with 250 mm length and 80 mm o.d. and 30 mm i.d. It is mounted horizontally. The internal volume is 30 cm^3. Closures at both ends contain sapphire windows to observe the cell contents. Inlets in the center part permit introduction of thermocouples and capillaries for injection and extraction of fluid. Electric heating jackets provide the desired temperatures.

The reactor is filled, for example, with a homogeneous fluid mixture of 70 mol% water and 30 mol% methane at 450 °C and 1000 bar (see Fig. 1). Then

---

x) ATS 340, No. 24949, DEW, Thyssen, Krefeld.

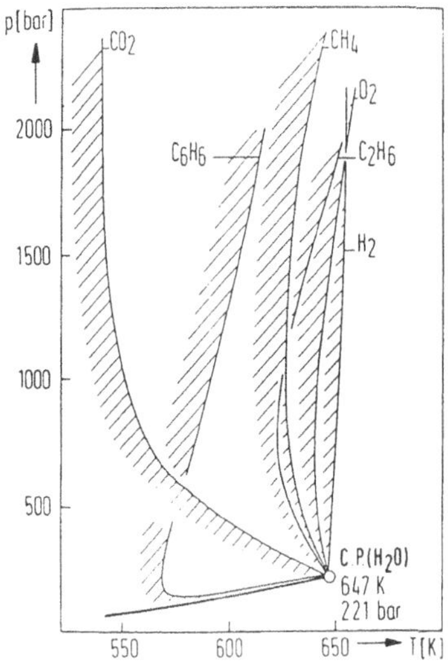

Figure 1: Critical curves of binary aqueous systems with indicated non-polar partners. ——: Vapour pressure of pure water. C.P.: Critical point. /////: Two-phase region. CO_2(1), CH_4(2), C_6H_6(3), C_2H_6(4), O_2(5), H_2(6).

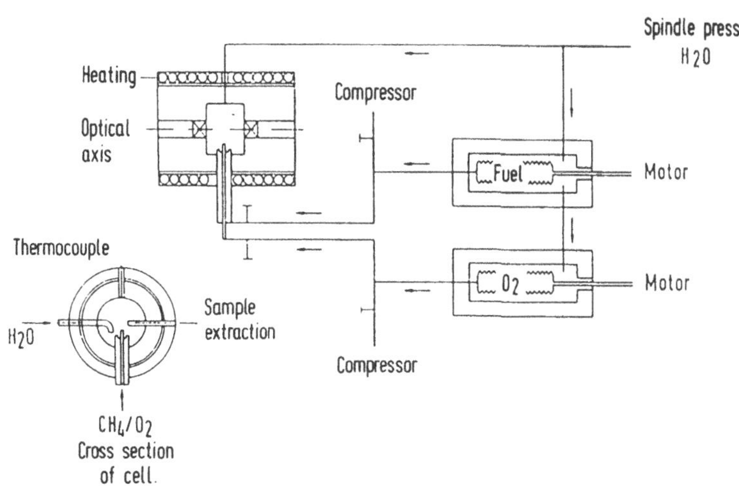

Figure 2: Schematic description of high pressure oxidation apparatus with reaction cell and feed autoclaves for fuel and oxygen.

a slow flow of oxygen has to be injected from below into the sample through a small nozzle. The oxygen flow should be steady and with a rate of a few mm^3 per second. This could be achieved with special high pressure pumps. In this work a different principle was used, shown schematically in Fig. 2. Two "feed autoclaves" contain stainless steel bellows, filled with the pressurized gases. The space outside the bellows is filled with water of the same pressure. With special motors and gears the bellows can be compressed slowly and the respective gas, for example oxygen, is injected into the reation cell. An equivalent part of the cell contents is transferred into the region outside the bellows of one of the feed autoclaves. Thus the fluid performs an almost circular motion until the bellows can not be compressed further. After 30 to 60 minutes, refilling of the feed autoclave is necessary with additional equipment not shown here (8)(9). Steadily burning flames on top of the reaction cell nozzle can be observed through the sapphire windows with a stereo microscope. Video camera and monitor serve to record the phenomena. The gas nozzle is of thin stainless steel, circular and 0.5 mm wide. Other types of nozzles have been used, for example slits (11) with 2 mm length and 0.1 mm width or two concentric tubes of 1.5 and 0.5 i.d. respectively. These can serve to inject two gas flows in parallel.

Other reaction vessels have also been used: To perform combustion by slow counterflow diffusion of fuel and oxygen in dense supercritical water a cylindrical, vertical autoclave was designed. It had a length of 550 mm and 50 and 15 mm o.d. and i.d. Water and gaseous partners could be injected from top and bottom. Samples are taken from central outlets (9)(10). The autoclave material is the same nickel-base alloy as above. There are no windows. With electric heating jackets and a series of thermocouples temperature uniformity or small temperature gradients can be achieved.

Another spherical cell was used for ignition and explosion experiments. Two cylindric superalloy blocks of 80 mm o.d., have flat ends. These two flat surfaces have each a hemi-spherical cavity of 35 mm diameter. Both blocks are mounted within a 100 t hydraulic press, which presses the two parts together to make a complete hollow sphere. Sealing is achieved with an outer circular V-groove with a stainless steel double-delta ring. Closing and opening of the cell with the hydraulic press is quick and easy. It holds pressures well above 2000 bar many times. Ignition of dense gas mixtures can be initiated by heating from outside or by suitable designs from the center.

III. FLAME PHENOMENA

If oxygen is injected as described with flow rates of a few $mm^3 \cdot s^{-1}$ into a supercritical homogeneous mixture of 70 mole% water and 30 mole% methane at 1000 bar and 400 ^oC or higher, spontaneous ignition of a steadily burning diffusion flame is observed. Fig. 3 shows such a conical diffusion flame of about 3 mm height and 0.5 mm diameter at the base, seen through the sapphire windows of 8 mm diameter. No stimulation for ignition, for example with a spark, was necessary. Typically such a flame can be maintained for 30 to 45 minutes with the present arrangements.

Hydrothermal methane-oxygen flames of this kind have been produced at 300, 600, 1000 and 2000 bar at nearly constant ignition rate and environment temperature. The height of the flame cones increases with pressure from about 1 to 5 mm, which is a consequence of the decreasing effective diffusion coefficient. Such an effective coefficient D can be approximately estimated from flame height and injection rate according to Burke and Schumann (12) and Jost (13). For 1000 bar and an assumed average temperature of 900 K one obtains a value of $D \approx \cdot 10^{-3}$ $cm^{2} \cdot sec^{-1}$ (7), which, although highly conjectural, does not entirely disagree with existing knowledge of high pressure transport processes in gases (14).

Similar flames could be produced, if the methane content in the supercritical aqueous phase was reduced to 10 % or increased to 50 %. Oxygen could be replaced by compressed air. The role of oxygen and methane can be exchanged, with methane injected into a dense aqueous oxygen phase. It is interesting to observe analogous flames, where the water has been replaced by argon and even helium at 1000 bar. Surprisingly, size and shape of these flames are similar to those in the aqueous environment, indicating that the interaction between flame and the surrounding fluid is rather non-specific, a fact, which needs further investigation. It is possible also to replace methane by hydrogen with, and even without, added dense water. Again, the general appearance of the flames at 1000 bar does not change substantially. Heavier alkanes, to pentane, have also been used as aqueous mixtures. From propane a growing formation of soot flakes is observed.

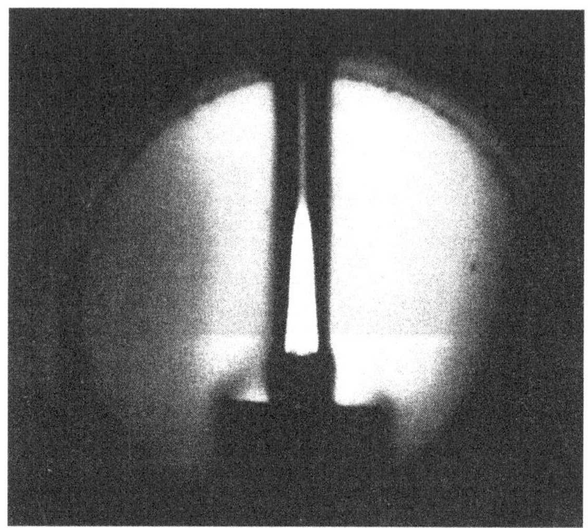

Figure 3: Diffusion flame with oxygen injection from below into supercritical mixture of 70 % water and 30 % methane at 500 °C and 1000 bar. Oxygen injection rate: 3 mm^3/s. Flame height: 3 mm. Observation through sapphire windows.

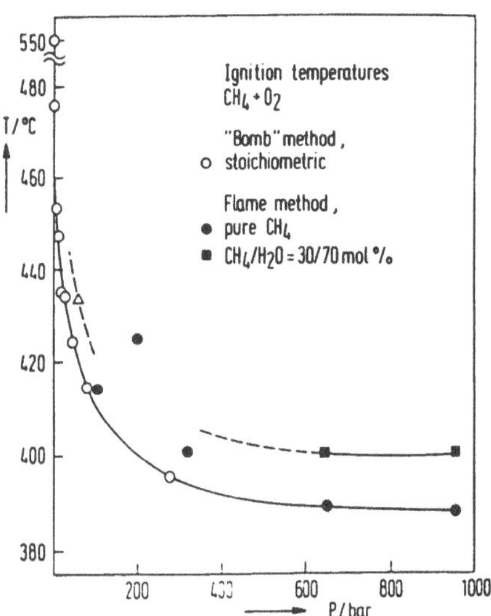

Figure 4: Spontaneous ignition temperatures of stoichiometric CH_4-O_2-mixtures in dependence of pressure.

Flames produced with the slit nozzle of 0.1 mm width and 2 mm length (see above) were still about 0.5 mm thick, but had a flat surface, which is desirable to develop procedures of laser diagnostics with the present flame equipment (11)(15). It has so far not been possible to create premixed high pressure flames. The dense fuel-oxygen mixtures, when introduced through capillaries into the thick-walled hot autoclave, ignited prematurely.

IV. IGNITION

The unexpected observation of low temperatures of spontaneous flame ignition stimulated related experiments. Such spontaneous ignition of nearly stoichiometric methane - oxygen mixtures at 1 bar appears between 500 and 600 ^{o}C, depending on the technical conditions (16). Increase of pressure lowers this temperature. Experiments were made here to pressures beyond the limits reported so far in the literature. Fig. 4 shows some of the results. The white points indicate data obtained with dry, stoichiometric CH_4-O_2-mixtures filled into the window-equipped reaction vessel at room temperature and moderate pressures. Subsequently the temperature is increased with about 10 K per minute. The pressure rises accordingly. Temperature and pressure at a sudden rapid increase are recorded as ignition values. Pressure is measured with a strain gauge transducer. For higher pressures, up to 1000 bar, a flame method was used: Either pure methane or oxygen was filled into the reaction cell and brought to a certain temperature and pressure. Oxygen or fuel are then slowly injected. At favourable conditions a flame appears. Ignition temperatures and pressures obtained in this way (black points in Fig. 4) continue the previous curve reasonably well. Thus pressures to 600 bar reduce the ignition temperature to about 390 C. If, instead of pure methane, a 70/30 mole% supercritical water-methane mixture is used, the flame appearance increases not more than about 20 ^{o}C, which is remarkable, considering the high quenching capability of the dense water.

With the spherical vessel (see above), filled with methane - air - mixtures of stoichiometric CH_4-O_2-relation thermal ignition was observed already at 400 bar and 382 ^{o}C. Central ignition could also be studied with this vessel. Time dependence of pressure increase up to 2500 bar

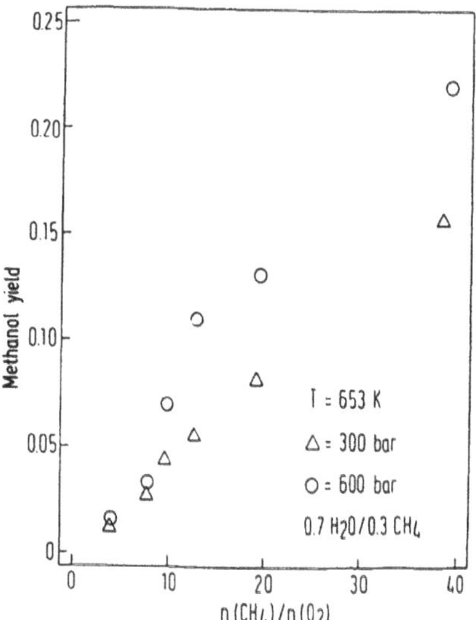

Figure 5: Methanol yield from reacted methane at 380 °C at two different pressures from flameless hydrothermal oxidation.

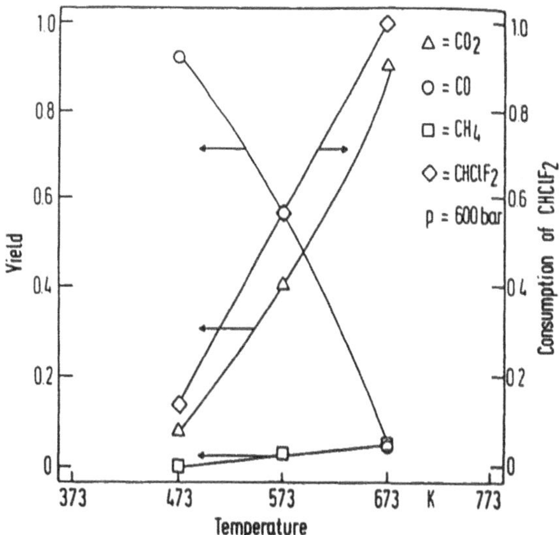

Figure 6: Yield of CO_2, CO, CH_4 and consumption of $CHClF_2$ at 600 bar by hydrolysis and pyrolysis only, without added oxygen, after 30 minutes.

was also investigated with strain gauge and piezoelectric transducers. Typical pressure rising periods are around 50 msec. Details will be reported elsewhere (17).

V. COMBUSTION PRODUCTS

Because temperatures of the supercritical hydrothermal methane flames are well over 2000 ^oC and the flame size is of the order of mm, there must be extremely steep temperature gradients at the flame surfaces. One might expect considerable output of partly oxidized species like methanol, formaldehyd and others, because diffusion through the thermal gradient is fast and the quenching power of the surrounding aqueous phase is high. This presumption was examined. It was not possible, however, to bring an extraction capillary close to the flame because of the high temperatures. The complete fluid content of the reaction cell had to extracted after certain periods and transferred into analytic arrangements. Volumetric measurements and chromatography with suitable columns were used. Methanol and comparable compounds were indeed found in small amounts but the combustion mainly proceeded to CO_2, CO and H_2O.

If the combustion was performed at temperatures below flame ignition, the methanol yield was higher. Only a limited temperature region was available, however, if the sample was to remain homogeneous, because of the phase separation occuring below the critical curves (see Fig. 1). In Fig. 5 yield of methanol, that is the fraction of reacted methane converted to methanol, is plotted as a function of the CH_4 / O_2-ratio at 380 ^oC for two different pressures. High methanol yield is favoured by high methane-oxygen ratio and high pressure. If the H_2O/ CH_4 ratio is raised to 0.9 / 0.1, a methanol yield of 0.30 can be obtained. Details will appear elsewhere (9) (10). A pressure increase to 1000 bar did not further raise the methanol yield. Tentative experiments with added possible catalysts like cobalt compounds or platinum did not have a substantial positive effect on the methanol yield.

It has been suggested, to destroy toxic wastes in dense, supercritical aqueous phases without flame formation (18)(19). A number of test experiments have been performed in the present work. Monochlordifluoro-

-methane, $CHClF_2$, was chosen as a test compound. It turns out, that at high enough temperature and pressure this compound decomposes completely by pyrolysis and hydrolysis even without oxidation by added oxygen or air. Fig. 6 demonstrates the hydrolysis after 30 min reaction time as a function of temperature. The left hand ordinate gives the relative yield of CO_2, CO, and CH_4. On the right side the degree of decomposition of $CHClF_2$ is shown. At 400 ^{o} C after 30min the destruction is nearly complete. The addition of oxygen would considerably accelerate the process.

VI. FLAME SPECTRA AND TEMPERATURES

It is particularly interesting to know the temperature of the supercritical hydrothermal flames. From a number of qualitative observations it is clear, that the temperature must be very high. Thus thermocouples or resistance wires can not be used. The small flame size would also prohibit such methods. Pyrometry was impossible because the emission spectra in the visible range contained a high proportion of chemilumilescence and little black body radiation - even with the presence of soot particles. The application of Na-D- line reversal or emission-absorption methods requires the introduction of sodium into the flame. Instead of pure water very dilute aqueous solutions of NaCl or Na-acetate were used to prepare the hydrothermal flames. Above 200 ^{o} C, however, the inner surfaces of the sapphire windows (Al_2O_3) became increasingly etched and obscured the vision. To a small extent this is observed already with pure dense supercritical water, which rules out the use of OH line-reversal. The possibilities of laser-diagnostic methods, for example LIF, normal Raman and CARS are being investigated at present (15). Difficulties have to be overcome, however, which are to some extent caused by the high density difference between flame and environment. "Adiabatic" flame temperatures based on thermochemistry with combination of a substance balance and a heat balance can be performed in a well established manner (see for example Gaydon and Wolfhard (16)). Under conditions of very high pressure, however, "real" equations of state and realistic transport functions must be included, which will in general reduce the resulting flame temperatures. This has been attempted for one-dimensional diffusion flames (see below).

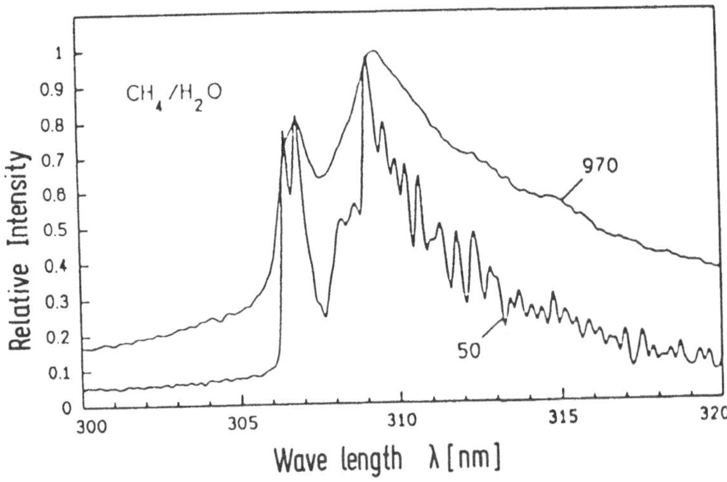

Figure 7: 50 bar and 970 bar emission spectrum of OH-radicals in CH_4/H_2O-oxygen flame at 470 °C environment temperature. CH_4/H_2O-ratio = 0.3 / 0.7. $A^2\Sigma$-$X^2\Pi$ transition between 0-0 vibration levels.

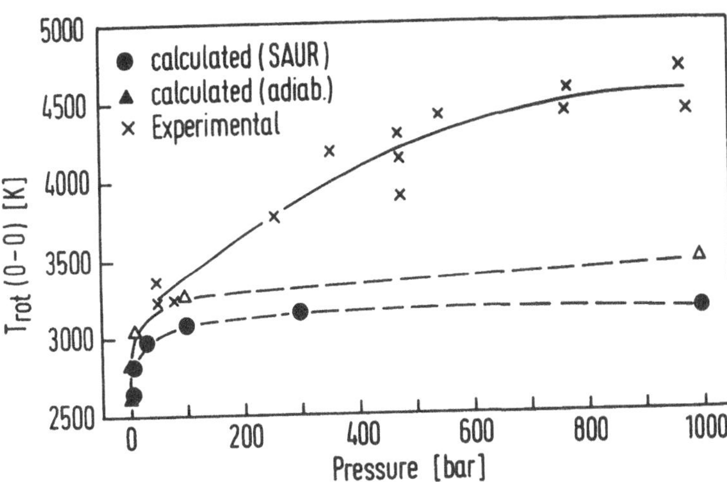

Figure 8: Comparison between experimentally derived (0-0)-OH-rotational temperatures of CH_4/H_2O-O_2 flames with temperatures calculated with "real" thermodynamic and transport functions.

For the present work a determination of the rotational temperature of OH-radicals was undertaken (11). The emission of the $A^2\Sigma - X^2\Pi$ transition between the 0-0-vibration levels near 310 nm is intense, and from its rotational structure a temperature can be derived with a suitable fitting process. The fit is complicated by the considerable pressures broadening of the band elements. This is shown by Fig. 7 for pressure of 50 and 970 bar. Special assumptions concerning the line- widths of the band elements have to be made. The fitting process is based on a procedure developed elsewhere for solid state combustion (20)(21). One example of rotational temperatures in dependence of pressure derived from fitting of spectra like those in Fig. 7, for CH_4 / O_2 flames burning in supercritical water, is shown in Fig. 8. It is obvious, that the experimentally derived OH-rotational temperatures increase to much higher values as the temperatures obtained from adiabatic or "real" calculations with realistic equations of state and transport functions. Similar behaviour was observed with hydrogen, burning in dense, supercritical water. Lower rotational temperatures and better agreements between experiment and calculation were obtained for methane and hydrogen burning in argon to 1000 bar. Further discussion of the high rotational temperatures has probably to await results of laser diagnostics. Very high degrees of OH-rotational excitation have been observed with other flames, although at much lower pressures (16). The effect of pressure can perhaps be explained in terms of a highly exothermic chemical process, forming the OH radicals in the electronically excited state and at the same time with a rotational energy distribution initially equivalent to a very high temperature.

VII. COMPUTATION OF DIFFUSION FLAMES

Parallel to the flame experiments described above, the computational simulation of one-dimensional laminar counterflow diffusion flames to 3000 bar was attemped. The purpose of such computations with simplified conditions was to recognize the general character of flame variations if the interparticle distance is continuously decreased from high "gas-like" to low values where almost constant particle contact is possible. The influence of realistic equations of state and high density transport properties should be investigated. This work was only possible because modelling of methane and other flames with one-dimensional geometry at one bar have been accomplished by Warnatz et al. (22)-(25). The program

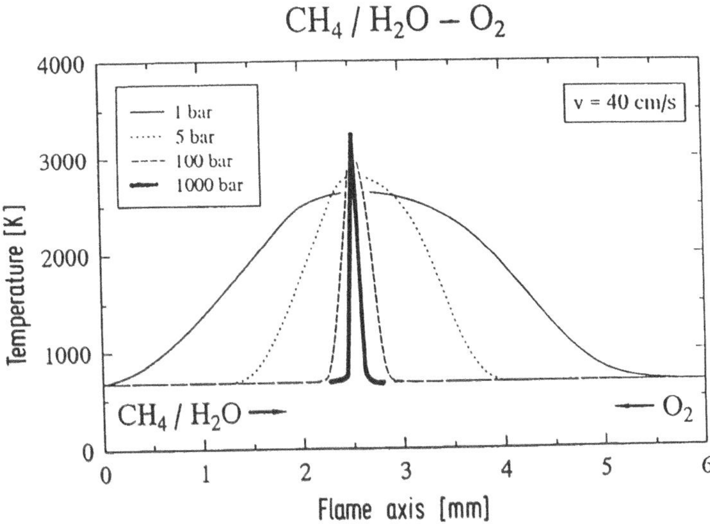

Figure 9: Calculated variation of the temperature flame profile with pressure for a CH_4/H_2O-O_2 flame at constant flow rate.

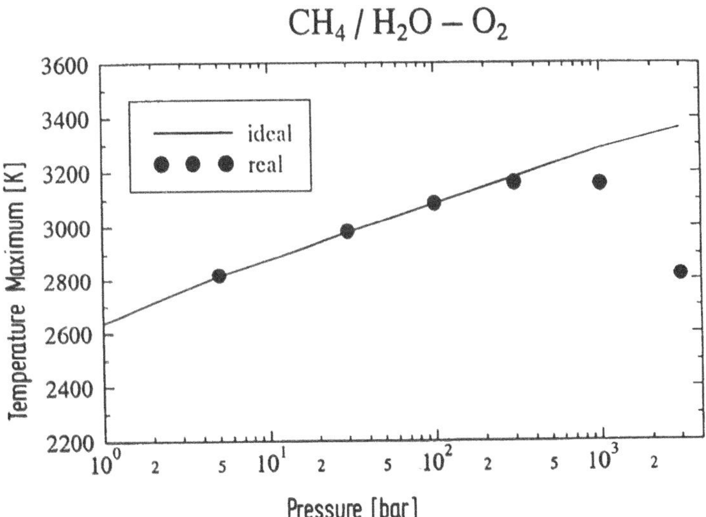

Figure 10: Calculated CH_4/H_2O-O_2 flame front maximum temperatures in dependence of pressure. ●●●: calculated with "real functions" (see text).

packages of these groups had been available for the present work. As a high pressure equation of state a rational equation with repulsive and attractive terms developed recently (26) - (28) was used, which depends on a Carnahan-Starling type of repulsion. At the high flame temperatures the particle repulsion is particularly important. For the high density transport processes a modified Enskog equation was used which takes into account collision transfer of momentum and heat (14). Details are described elsewhere (29((30). Flame profiles along the flame axis, concentration and reaction rate profiles and maximum flame temperatures as functions of pressure have been calculated. The increasing pressures created serious problems connected with the stiffness of the system of differential equations.

Only a few examples are shown. Fig. 9 gives the calculated flame temperatures between 1 and 1000 bar of pressure for constant gas flow rates along the axis perpendicular to the flame front. The fuel is a 70/30 mixture of methane and water and the counterflow gas is oxygen. Increasing pressure creates a very narrow and steep flame profile. The maximum temperature increases. The profiles for methane - argon flames are very similar. Fig. 10 shows the pressure dependence of maximum flame temperatures, calculated as indicated above in two ways: First, simple ideal gas laws are applied. Second, "real" behaviour with equations of state and transport processes is assumed. The difference is not very pronounced. The reason is, that because of the very high temperature within the flame region the deviation from ideality is not very pronounced. It is, however, indicated, that the maximum flame temperature passes through a pressure maximum around 500 bar. This became even more evident with calculations for H_2 / Ar - O_2 flames, which were tentatively extended to 10.000 bar. After a maximum at 500 bar of 3300 K the temperature decreased to 1200 K at 10.000 bar. These calculations use a number of simplifying assumptions. Neverless, it appears to be evident, that pressure increase to several kbar does not raise maximum flame temperatures substantially. Effective diffusion coefficients and thus diffusion controlled reaction rates are decreased. The compressibility of the reactive mixture is also decreased, while thermal conductivity and heat loss are augmented.

VIII. OUTLOOK

Combustion and flames in high density fluids appear to be worth further investigation. Very high reaction rates with liquid-like density can be combined and unusual excitation reached. In the field of toxic waste destruction and possibly chemical synthesis, novel developments can be foreseen.

LITERATURE

1. Tödheide, K.; Franck, E.U.; (1963); Z. Physikal. Chemie, N.F., **37**, 387.
2. Welsch, H.; (1964) "The systems Xe-H_2O and CH_4-H_2O at high pressures and temperatures", Thesis, Inst. of Physical. Chem. Karlsruhe University.
3. Alwani, Z.; Schneider, G.M.; (1967) Ber. Bunsenges. Phys. Chem. **71**, 633.
4. Danneil, A.; Tödheide, K.; Franck, E.U.; (1967), Chemie-Ing.Technik, **39**, 816.
5. Japas, M.L.; Franck, E.U.; (1985), Ber.Bunsenges.Phys.Chem. **89**, 1268.
6. Seward, T.M.; Franck, E.U.; (1981), Ber.Bunsenges.Phys.Chem. **85**, 2.
7. Schilling, W.; Franck, E.U.; (1988), Ber.Bunsenges.Phys.Chem. **92**, 631.
8. Schilling, W.; (1989); "Combustion and Diffusion Flames at Pressures to 1000 bar", Thesis, Inst. of Physical Chem., Karlsruhe University.
9. Hirth, Th.; (1992); "Pyrolysis, Hydrolysis, and Oxidation in Supercritical Water at Pressures to 1000 bar", Thesis, Inst. of Physical Chem., Univ. of Karlsruhe.
10. Hirth, Th.; Franck, E.U.; (1993); Ber.Bunsenges.Phys.Chem. **97**, to appear.
11. Pohsner, G.; (1992) "Spectra and Temperatures of Diffusion Flames in Supercritical Water to 1000 bar"; Thesis, Inst. of Physical Chem., Univ. of Karlsruhe.
12. Burke, S.P.; Schumann, T.E.W.; (1928), Ind. Eng. Chem. **20**, 998.
13. Jost, W.; (1939) "Explosions- und Verbrennungsvorgänge in Gasen", Springer-Verlag, Berlin.
14. Hirschfelder, J.O.; Curtiss, C.F.; Bird, R.B.; (1965); "Molecular Theory of Gases and Liquids", Wiley, N.Y.
15. Wolfrum, J.; (1992), Personal communication, Experimental work in progress. Inst. of Physical Chem., Univ. of Heidelberg.
16. Gaydon, A.G.; Wolfhard, H.G.; (1979). "Flames, their Structure, Radiation and Temperature", Chapman and Hall, London.
17. Steinle, J.; (1993); "Ignition in Highly Compressed Gases", Thesis, Inst. of Physical Chem.; Univ. of Karlsruhe.
18. Modell, M.; Gandet, G.; Simson, M.; Hong, G.; Bilmann, K.; (1982), Solid Wastes Management, **8**, 26.

19. Shaw, R.W.; Brill, T.B.; Clifford, A.A.; Eckert, C.A.; Franck, E.U.; (1991); Chem. Eng. News, December 23.
20. Eisenreich, N.; Schneider, H.; (1986); "Temperatur-Bestimmung von Treibstoff-Flammen durch Berechnung der OH(0-0)-Bande", Fraunhofer-Institut, ICT, Bericht 10186, 7507 Pfinztal-Berghausen, Germany.
21. Eisenreich, N.; Liehmann, W.; (1986); "Zeitaufgelöste Spektroskopie von Verbrennungsvorgängen"; Bericht 12/86, Fraunhofer-Institut, ICT, 7507 Pfinztal-Berghausen, Germany.
22. Warnatz, J.; (1983); Ber.Bunsenges.Phys.Chemie, **87**, 1008.
23. Dreier, T.; Lange, B.; Wolfrum, J.; Zahn M.; Behrendt, F.; Warnatz, J.; (1986); Ber.Bunsenges. Phys. Chem., **90**, 1010.
24. Maas, U.; Warnatz, J.; (1991); VDI-Berichte, **922,** 181:
25. Diessel, E.; Dreier, T.; Lange, B.; Wolfrum, J.; Behrendt, F.; Warnatz, J.; (1992); Ber. Bunsenges. Phys. Chem.; **90**; 579.
26. Christoforakos, M.; Franck, E.U.; (1986); Ber. Bunsenges.Phys.Chem.; **90**, 780.
27. Heilig, M.; Franck, E.U.; (1989); Ber.Bunsenges.Phys.Chem.; **93**, 898.
28. Heilig, M.; Franck, E.U.; (1990); Ber.Bunsenges.Phys.Chem.; **94**; 27.
29. Saur, A.M. (1992); "Computation of Counter-Flow-Diffusion-Flames to Pressures of 3000 bar", Thesis, Inst. of Phys. Chemistry, Univ. of Karlsruhe.
30. Saur, A.M.; Franck, E.U.; (1993); Ber.Bunsenges.Phys.Chem. **97;** to appear.

RAMAN SPECTROSCOPY OF AMMONIA MONOHYDRATE TO 13.5 GPA

A. KOUMVAKALIS* and M. NICOL
University of California
Los Angeles, CA 90024-1569
USA

ABSTRACT. Raman spectra of $NH_3 \cdot H_2O$ and $ND_3 \cdot D_2O$ have been studied from 120 to 450 K between 0 to 13.5 GPa in diamond-anvil high pressure cells. Only one solid and the liquid phase were found. The melting line was determined to about 400 K at 6 GPa. The ν_1 and site-split ν_3 N-H modes at 3300, 3370, 3395, and 3400 cm^{-1} do not shift with pressure, suggesting that the N-H ··· O hydrogen bonds are weak. The O-H ··· N stretch shifts by -31.5 cm^{-1} GPa^{-1} in $NH_3 \cdot H_2O$ and -25.5 cm^{-1} GPa^{-1} in $ND_3 \cdot D_2O$. Two other very weak features unaffected by pressure are assigned to O-H ··· O stretches along the water chains. These spectra imply that $NH_3 \cdot H_2O_{(s)}$ is incompressible along the water chains and compresses by decreasing separations between these chains and the bridging NH_3 molecules. Raman spectra of the external modes are very weak and shift much less with pressure (by 3 to 7 cm^{-1} GPa^{-1}) than might be expected.

1. Introduction

NH_3 and H_2O are thought to be major constituents of Neptune, Uranus, and some jovian and saturnian satellites.[1-5] The cosmic abundance of NH_3 is estimated to be about 18% that of H_2O,[4] although surface analysis of Uranus shows significantly lower NH_3 concentrations.[6] Observations of apparently high H_2O:NH_3 abundances at this and other planetary surfaces raise the question: can NH_3 be sequestered inside these icy bodies as NH_3-rich solids? Experimental data on the phase diagram of the NH_3-H_2O system and properties of the ammonia hydrates are needed to analyze this possibility.

This study was designed to determine spectra or other properties which distinguish phases of $NH_3 \cdot H_2O$ from ice and other ammonia hydrates at high pressure. The $NH_3 \cdot H_2O$ phases stable between 100 and 450 K at pressures to 13.5 GPa and phase lines between these phases also were determined. Parallel studies of the NH_3-H_2O system used these data [7] to show that water ices and $NH_3 \cdot H_2O$ are the principal solids for $(NH_3)_x(H_2O)_{1-x}$ compositions with $x \leq 0.5$, although $NH_3 \cdot 2H_2O$ is stable under some P-T-X conditions. A summary of these results was reported elsewhere.[8]

R. Winter and J. Jonas (eds.), High Pressure Chemistry, Biochemistry and Materials Science, 265–274.

Before this study began, thermodynamic data, ir spectra and several other properties of the NH_3 -H_2O system at ambient pressure were known,[9-19] and the structure of the solid at -95 and -160 °C was known to belong to the $P2_12_12_1$ (D_2^4) space group with a unit cell of dimensions, 0.451 x 0.5587 x 0.970 nm containing 8 molecules.[20] The H_2O molecules are hydrogen bonded into chains running parallel to the a axis with 2 molecules from each of 2 chains in the unit cell. These chains are cross-linked by NH_3 molecules, each tightly held to a water chain by O-H···N and N-H···O bonds. Each H_2O is coordinated to six molecules with two 0.276-nm bonds along the O-H···O-H···O chain, one 0.2775-nm O-H···N bond, and three N-H···O bonds to the second lone-pair of electrons of 0.321, 0.326, and 0.329 nm. Each NH_3 is four coordinate. The H_2O and NH_3 occupy sites of C_1 symmetries.

2. Experimental

Raman spectra were obtained of samples in a Holzapfel diamond-anvil cell.[21] The $NH_3 \cdot H_2O$ was made in two ways: as solids at low T and ambient P and as liquids at ambient T and slightly elevated P. In the early part of this study, $NH_3 \cdot H_2O$ was fractionally crystallized from 30 mole-% aqueous NH_3, and some of the precipitate was loaded directly into a cooled diamond-anvil cell. Acid titration of these samples showed that they were slightly rich in H_2O. Some water-rich mother-liquor probably was not removed. Other samples were loaded by immersing the cell in a liquid solution of H_2O and NH_3 with precisely measured compositions in an autoclave. Both techniques are described in detail elsewhere.[7, 22] Heat-treated Inconel gaskets were used. After these gaskets were found to corrode at high T, gaskets were gold plated to retard corrosion. Pressures were determined from ruby luminescence with Yen's calibration.[23] Instrumental uncertainties contribute $\pm$0.1-GPa to the precision of P. At other than ambient T, the cell was sealed in a cryostat modified from Noack and Holzapfel's design.[24] The sample temperature was measured with a PT100 platinum resistor or a copper-constantan thermocouple. Raman spectra were excited by an Ar-ion laser at 515, 448 or 477 nm and were collected with a SPEX 0.25-m single monochromator with a Princeton Instruments OMA or a SPEX 1400-II 0.75-m double monochromator with a cooled ITT FW120 photomultiplier and photon-counting electronics. Shorter excitation wavelengths yielded higher signal-to-background ratios and reduced interference by ruby and diamond luminescence.

The onset of melting was detected visually as a sudden clearing of the polycrystalline samples with obvious reliefs and cracks while the sample was slowly (2 K/min) heated under constant load or pressure was released isothermally. When melting was detected, the sample T and the ruby luminescence spectrum were measured to determine the melting point. The onset of freezing was similarly detected by the first appearance of solid during slow cooling under constant load. Melting and crystalization were rapid and apparently congruent. Any P and T where solid and liquid were observed to coexist was assumed to be on the melting line.

In melting experiments with the samples loaded as solid (water-rich) ammonia monohydrate in unplated Inconel gaskets, variations of the initial compositions and possible compositional changes because of reactions between the hot sample and Inconel affected the melting data. In the worst cases, mixtures of $NH_3 \cdot H_2O$ and ice VII were observed; the melting points were very close to the melting line of ice VII; and Raman spectra showed high ice content. Such observations were excluded from the data reported here. After liquid loading and gold-plated gaskets was used, the melting data were reexamined. The presence of liquid and solid $NH_3 \cdot H_2O$ and the absence of ices were confirmed by Raman spectroscopy and by visual observations.

3. Results

3.1. RAMAN SPECTRA OF THE INTERNAL MODES

The hydrogen-bonded network structure of $NH_3 \cdot H_2O$ suggests that the internal mode frequencies should be close to those of pure NH_3 or H_2O. The numbers and degeneracies of these modes are less certain. Factor group analysis for the D_2 group of the x-ray structure indicates that a nondegenerate modes of NH_3 and H_2O yields $A+B_1+B_2+B_3$ modes and a degenerate NH_3 mode yields $2A+2B_1+2B_2+2B_3$ modes. All of these modes may be Raman active. X-ray analysis, however, is insensitive to H-atom positions, so the actual symmetry of the structure might be lower than D_2^4 if H atoms in different chains are disordered and are not symmetry related. [7] In interpreting the ir spectrum of $NH_3 \cdot H_2O$, Bertie and Shehata [9] suggested a third possibility. Because almost all of the molecules lie in bc planes at a/4 and 3a/4, the monohydrate may have spectroscopic D_{2h} pseudosymmetry. Thus, molecular modes symmetric with respect to an a/4 or 3a/4 pseudomirror plane would have $A_g+B_{3g}+B_{1u}+B_{2u}$ characters; modes antisymmetric to such planes would have $B_g+B_{2g}+A_u+B_{3u}$ characters. Any gerade mode may be Raman active.

In general, Raman spectra of $NH_3 \cdot H_2O$ were dominated by NH_3 modes. The spectrum of liquid $NH_3 \cdot H_2O$ in the stretching region has 3 bands, at 3400 cm^{-1} (depolarized) and 3227 and 3309 cm^{-1} (polarized) which are assigned as the ν_3, $2\nu_4$, and ν_1 of ammonia. Water vibrations only contribute a broad background to the spectrum as shown in Fig. 1. In solid $NH_3 \cdot H_2O$, the Fermi resonance $2\nu_4$ band almost completely disappears and can be seen only in a few spectra as a weak shoulder. The ν_3 band splits into three components, and a broad band, assigned as the O-H···O bond of water, appears near 2950 cm^{-1}; see Fig. 1. Bertie and Shehata assigned in splitting of the ν_3 vibration in the infrared spectrum to B_1, B_2, and B_3 components, the only infrared-active components for the D_2 factor group. Because D_2 modes of any character may be Raman active and the crystals were not specifically oriented with respect to the polarization and propagation directions of the laser excitation and scattering directions, assignment of these modes is not so unambiguous.

Fig. 1. Raman spectra of $NH_3 \cdot H_2O$ at ambient T. Upper trace: fluid at 2.77 GPa; lower trace: solid at 5.2 GPa. Intensities (I, arbitrary units) have not been corrected for the spectral responses of the spectrometer and detector.

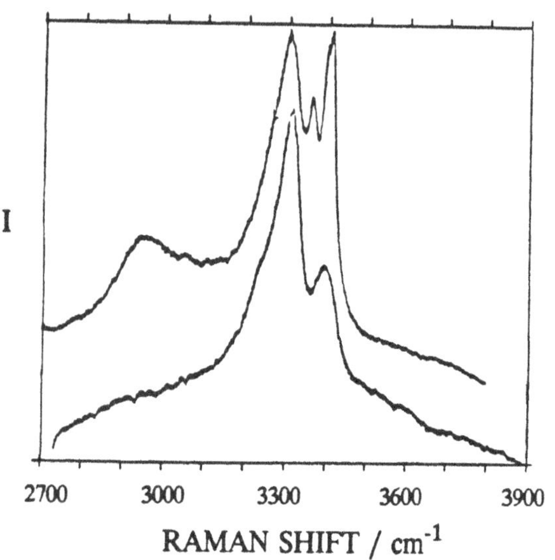

Similar features occur in spectra of $ND_3 \cdot D_2O$ between 2300 and 2550 cm^{-1}. This presents a problem for diamond-anvil cells; the second-order Raman spectrum of diamond overlaps the spectrum of the sample. Fig. 2 shows how, in favorable cases, the second-order diamond spectrum was subtracted from the $ND_3 \cdot D_2O$ spectrum. The upper trace is the spectrum of the $ND_3 \cdot D_2O$ sample in the diamond-anvil cell collected with the OMA, and the middle trace shows the second-order diamond spectrum collected with the OMA. The lower trace in Fig. 2 results by digitally subtracting a fraction of the diamond spectrum from the upper trace until the sharp edge near 2700 cm^{-1} virtually disappears. The vibrational frequencies of $ND_3 \cdot D_2O$ and the ratios of the frequencies of $NH_3 \cdot H_2O$ and $ND_3 \cdot D_2O$ listed in Table 1 were determined from such spectra. For a harmonic vibration, the ratio of $\nu_H : \nu_D$ should be 1.38 for stretching modes and between 1.30 and 1.35 for a bending-stretching combination; the assignments are in reasonable agreement with these values.

Raman spectra of the bending fundamentals were relatively weak; these modes were detected for only a few samples. Good spectra of $NH_3 \cdot H_2O$ in the bending region also were collected at ambient P for the liquid at ambient T and for the solid at 77 K. Lines at 1600 and 1635 cm^{-1} were assigned to the site-split ν_4 vibration of NH_3. Another band near 1100 cm^{-1} was assigned as the ν_2 band of H_2O by comparison with ir data;[8] the first-order diamond spectrum obscures this band at high pressures. The splitting for the weak bending band of $ND_3 \cdot D_2O$ detected near 1170 cm^{-1} could not be resolved.

3.2. PRESSURE DEPENDENCES OF THE INTERNAL MODES

Raman spectra of the stretching modes were collected to 8.8 GPa for $NH_3 \cdot H_2O$ and to 8.2 GPa for $ND_3 \cdot D_2O$. The frequencies of the N-H or N-D modes vary at most

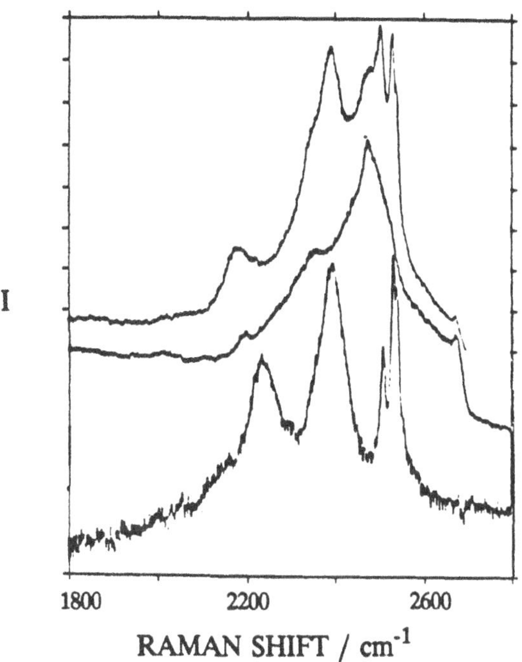

Fig. 2. Raman spectrum of $ND_3 \cdot D_2O$ (lower trace) determined by subtracting from the spectrum of the sample in the diamond cell (upper trace) a fraction of the second-order Raman spectrum of the diamond anvil (middle trace) to remove the sharp edge of the second-order diamond spectrum near 2700 cm^{-1}.

only slightly; linear regression best fits to the available data are given in Table I. The frequencies of the O-H···O and O-D···O stretching modes decrease strongly with increasing pressure by -31.5 and -25.5 cm^{-1} GPa^{-1}, respectively. Two other very weak features detected near 3060 and 3110 cm^{-1} in some $NH_3 \cdot H_2O$ spectra do not shift with pressure within an uncertainty of about 5 cm^{-1}. Within the uncertainties of the measurements, the frequencies of the bending modes were independent of pressure.

3.3. RAMAN SPECTRA OF THE EXTERNAL MODES

Raman spectra in the external mode region of $NH_3 \cdot H_2O$ were collected at ambient T from 3 to 11.8 GPa, at 7.5 GPa and 157 K, and at 13.5 GPa and 181 K. Five bands were clearly resolved between 100 and 230 cm^{-1}. Above 230 cm^{-1}, very weak monohydrate bands were distinguished from background only in a few cases. The pressure dependences of these modes are shown in Fig. 3; the lines are drawn to guide the eye. The frequencies of the lower 3 lines increase by 3 cm^{-1} GPa^{-1}, and the next 2 higher frequency lines increase by about 7 cm^{-1} GPa^{-1}. External mode spectra of $ND_3 \cdot D_2O$ were not collected.

3.4. MELTING AND PHASE STABILITY

All of the melting data determined in these experiments are summarized in Fig. 4. Melting and freezing transitions of individual samples often were sharp, indicating that melting was congruent. In some samples, liquid and solid were observed to

Table I. Positions, pressure dependences, and assignments of the internal modes of the Raman spectra of $NH_3 \cdot H_2O$ and $ND_3 \cdot D_2O$. ν_0's are values extrapolated to ambient pressure.

Raman Frequencies and Pressure Dependences				Ratios	Assignments
$NH_3 \cdot H_2O$		$ND_3 \cdot D_2O$			
ν_0 cm^{-1}	$\partial\nu/\partial P$ cm^{-1}/GPa	$\nu 0$ cm^{-1}	$\partial\nu/\partial P$ cm^{-1}/GPa	$\nu_{OH} : \nu_{OD}$	
1600	(0)	1170		1.37	ν_4 NH_3
1635	(0)				
3130.5	-31.5	2338.7	-25.5	1.34	ν_1 O-H···N
3236					$2\nu_4$ NH_3
3306.1	-0.1	2388.2	0.6	1.38	ν_1 NH_3
3368.0	1.5	2507.6	-0.8	1.34	ν_3 NH_3
3395.2	0.4	2523.9	0.8	1.35	
3397.7	1.4	2527.9	1.8	1.34	

coexist over a larger range of P and T. Hysteresis of about 20 K often was observed between the melting points on heating and freezing points on cooling under constant applied load. Although the general trend of the melting line is obvious, the individual melting observations are highly scattered. The solid curve corresponds to our best estimate of the melting curve of stoichiometric $NH_3 \cdot H_2O$ and is consistent with subsequent work by Boone and Nicol.[25]

Several experimental factors which may contribute to the scatter of the melting data, besides trivial rate effects and possible differences between the temperatures of the samples and the thermometers for a particular cell. This diamond-anvil cell was not temperature compensated and was heated or cooled under constant load conditions. Even when only one phase was present, the pressure of the sample usually increased during heating. Differences between the densities of the solid and fluid and the finite strengths of the cell and gaskets may cause the pressure on the sample to change when the sample melts or freezes under constant load. Thus, any sample would not completely melt at a constant T and P. Melting in off-composition (water-rich) samples prepared by loading cold solid $NH_3 \cdot H_2O$ would be expected to begin at lower T than the congruent melting temperature of the stoichiometric monohydrate. Only one solid phase of $NH_3 \cdot H_2O$ was detected over the range of conditions studied

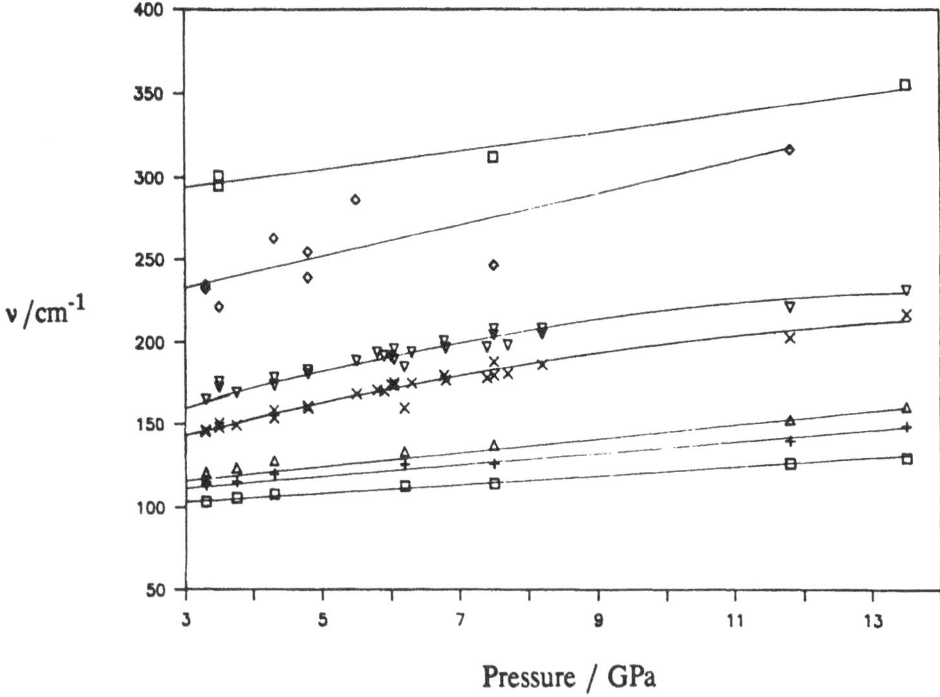

Fig. 3. Pressure dependences of several external modes of the Raman spectrum of $NH_3 \cdot H_2O$ at 295 K. Symbols distinguished different bands; the curves are drawn to aid the eye.

here, 120 to 450 K at pressures to 13.5 GPa! This aspect of $NH_3 \cdot H_2O$ is very different from ice, which occurs in many stable structures under similar conditions.

4. Discussion

The pressure dependences of most of the the internal and external frequencies of $NH_3 \cdot H_2O$ are surprisingly small and may be interpreted in terms of structural changes. The very slight, if any, shifts of the sharp N-H ν_1 and ν_3 bands of NH_3 suggest that the NH_3 molecules are essentially unaffected by their nearest neighbors. We infer that hydrogen bonds from the NH_3 hydrogens to H_2O molecules are weak. The features in the Raman spectrum assigned to the O-H···O vibrations within the H_2O chains are weak and do not shift. We infer that these strong intrachain hydrogen bonds are relatively incompressible. Only the stretch of the hydrogen bonds from H_2O to lone pairs of electrons on NH_3 molecules is the strongly affected by pressure. $\partial\nu / \partial P$ for this mode equals -31.5 cm^{-1} GPa^{-1}, which is smaller than that for

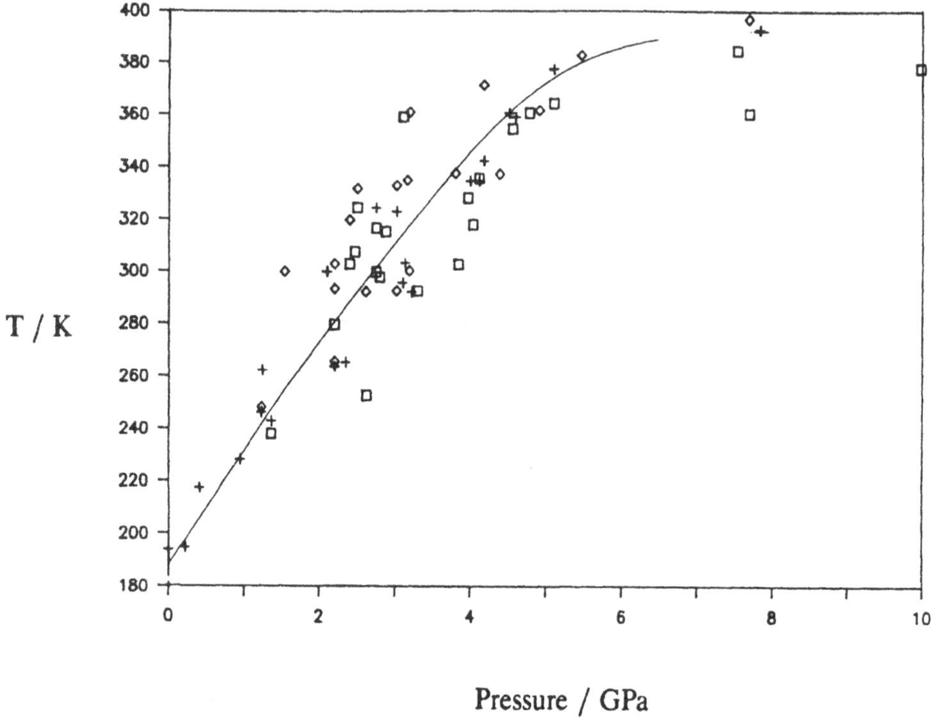

Fig. 4. Phase observations near the melting line of $NH_3 \cdot H_2O$ observed in this study. Symbols denote where solid (□), fluid (◇), or coexisting phases (+) were observed. The solid curve is our best estimate of the melting curve.

the O-H···O hydrogen bonds in ice VI, -72 cm^{-1} GPa^{-1} [26], and about the same as for ice VII, -25.2 cm^{-1} GPa^{-1}.[27] These negative shifts suggest that these X-H···A hydrogen bonds are strong and change strength significantly with P, the attraction between the approaching H and A atoms weakening and lengthening the X-H bond.

None of the lattice vibrations were assigned, although some sense of the assignments can be gained by comparison with ammonia hemihydrate, $NH_3 \cdot 2H_2O$, where bands below 280 cm^{-1} are assigned as pure translations and librations occur above 380 cm^{-1}. [7 and references therein] Two ir bands of $NH_3 \cdot 2H_2O$ at 243 and 267 cm^{-1} were assigned to the intermolecular components of either O-H···N or O-H···O hydrogen bonds. The weaker features above 230 cm^{-1} in the Raman spectra might be due to these vibrations. The external modes shift at 3 to 7 cm^{-1} GPa^{-1}, which is less than the 10 to 30 cm^{-1} GPa^{-1} expected for external modes.[28] This results suggests that these bands belong to phonon motions in directions which are weakly affected by pressure; that is, motions along or parallel to the water chains.

5. Summary

Raman spectra and visual observations were used to delineate the phase diagram of $NH_3 \cdot H_2O$ from 120 to 450 K to 13.5 GPa. The melting curve was determined from 0.1 MPa to more than 6 GPa. Only one solid phase and the liquid were detected. The Raman spectrum of the stretching modes of liquid $NH_3 \cdot H_2O$ consists of 3 bands, at 3227, 3309, and 3395 cm^{-1}, which are assigned to the $2\nu_4$, ν_1, and ν_3 vibrations of NH_3. In solid $NH_3 \cdot H_2O$, site effects and intermolecular coupling split the ν_3 band into 3 components at 3300, 3360, 3395, and 3405 cm^{-1}. Another Raman band of the solid at 2950 cm^{-1} is assigned to the O-H stretch of water, hydrogen bonded to the ammonia molecules which bridge between the chains of water molecules. The frequency of this vibrational mode decreases with increasing pressure by -31.5 cm^{-1} GPa^{-1} in $NH_3 \cdot H_2O$ and by -25.5 cm^{-1} GPa^{-1} in $ND_3 \cdot D_2O$. No other internal mode shows any marked pressure dependence. The Raman spectra of the external modes are weak. The five most prominent bands shift by 3 to 7 cm^{-1}/GPa. These probably are translational modes. Because their shifts are smaller than expected, they probably correspond to phonon motions along the incompressible water chains. These spectra are interpreted in terms of very weak N-H···O bonds; strong, short O-H···O bonds along the water chains which are not greatly perturbed by compression; and strong O-H···N hydrogen bonds which are changed by compression of the solid. The implication is that solid ammonia monohydrate is relatively incompressible along the water chains and is compressed by moving the water chains closer to the bridging ammonia molecules and, therefore, to each other.

Acknowledgments

It is a pleasure to acknowledge assistance provided by S. Boone, H. Cynn, and M.L. Johnson and support provided by NSF-DMR 90-22076.

Notes and References

* Present address: Rocketdyne Division, Rockwell International, Canoga Park, CA 90103.

1. R. Smoluchowski, The Moon and Planets **28** (1983) 137.
2. R.H. Brown, D.P. Crushank, and D. Morrison, **22** (1982) 423.
3. S. Gulkis, E.T. Olson, and M.J. Klein, Science **22** (1983) 454.
4. D.J. Stevenson, in T. Gehrels & M.S. Matthews, eds., *Saturn* (University of Arizona Press, Tucson, 1984), 671.
5. D.J. Stevenson, Nature **298** (1982) 142.
6. S. Gulkis, M.A. Janssen, and E.T. Olsen, Icarus **34** (1978) 10.
7. See, for example, H. Cynn et al. in *Proc. 19th Lunar and Planet. Sci. Conf.* (Lunar and Planetary Inst., Houston, 1989), 433.

8. M. Nicol et al., in A.K. Singh, ed., *Recent Trends in High Pressure Research* (Oxford and IBH Publishing, New Delhi, 1992), 223.
9. J.E. Bertie and M.R. Shehata, J. Chem. Phys. **83** (1985) 1449.
10. J.E. Bertie and M.M. Morrison, J. Chem. Phys. **73** (1980) 4832.
11. J.E. Bertie and M.M. Morrison, J. Chem. Phys. **74** (1981) 4351.
12. R.R. Rupert, J. Amer. Chem. Soc. **31** (1909) 866.
12. R.R. Rupert, J. Amer. Chem. Soc. **32** (1910) 748.
14. R. Smits and S. Postma, Z. Anorg. Chem. **71** (1911) 250.
15. S. Postma, Rec. Trav. Chim. Pays-Bas, **39** (1920) 515.
16. L.D. Elliot, J. Phys. Chem. **28** (1924) 887.
17. D.L. Hildenbrand and W.F. Giauque, J. Amer. Chem. Soc. **75** (1953) 2811.
18. I.L. Clifford and E. Hunter, J. Chem. Phys. **37** (1933) 101.
19. O.S. Binbrek and A. Anderson, Chem. Phys. Lett. **15** (1971) 421.
20. I. Olovsson and D.H. Templeton, Acta Crystallogr. **12** (1959) 827.
21. K.R. Hirsch and W.B. Holzapfel, Rev. Sci. Instr. **52** (1981) 52.
22. A. Koumvakalis, Ph.D. Dissertation (University of California, Los Angeles, 1988).
23. J. Yen and M. Nicol, J. Appl. Phys. (In press).
24. R.A. Noack and W.B. Holzapfel, in *High-Pressure Science and Technology* (Plenum, New York, 1979), pp. 748.
25. S. Boone and M. Nicol, in *Proc. Lunar and Planetary Sci., Vol. 21* (Lunar and Planetary Inst., Houston, 1991), 603.
26. G.E. Walrafen et al., J. Chem. Phys. **77** (1982) 2166.
27. J.F. Bertran, J. Mol. Structr. **95** (1982) 9.
28. W.F. Sherman and G.R. Wilkinson, in R.J.H. Clark and R.E. Hester, eds., *Adv. Infrared and Raman Spectra* (Heyden, London, 1980), **6** 158.

THE HIGH PRESSURE VISCOSITY OF SIMPLE AND POLYMERIC FLUIDS

H.E. King, Jr.; Richard L. Cook; and Chris A. Herbst
Exxon Research and Engineering Co.
Annandale, NJ 08801
USA

ABSTRACT. Three new devices to measure viscosity at high pressure are described: a rolling-ball diamond cell viscometer, a centrifugal-force diamond cell viscometer, and a dynamic light scattering (DLS) device. These provide high pressure measurements over more than a 10 decade range in viscosity. The combined use of the rolling-ball viscometer and DLS has allowed us to examine the response of polymers in solution to high pressures. We find that many *good* solvent systems are insensitive to pressure; however, one system, polyethylene oxide in water, exhibits a pressure-driven collapse of the polymer chain. This is attributed to the effect of pressure on hydrogen bonding. The viscosity rise associated with the pressure-driven glass transition has also been investigated. Glycerol has been studied to a viscosity of over 10^9 cP, corresponding to a pressure of approximately 3 GPa. We find that free-volume theory is an accurate model for this data.

1. INTRODUCTION

Study of the viscosity of a fluid at high pressures provides data of both theoretical and practical importance. The viscosity of dense fluids is far from well understood, in part, because it is a complex function of inelastic, multiple-scattering processes. An important unifying concept, however, is the density scaling behavior of transport which has been both empirically observed[1] and theoretically[2] derived. This concept first arose through the study of thermal effects, however high-pressure data provide a much more stringent test of such a scaling concept. One can access a much wider density range than through temperature variation, and the complexity of changing thermal energy is eliminated by making isothermal measurements. However, there exist few such data taken with high precision over wide pressure ranges because of the difficulty in making such measurements.

R. Winter and J. Jonas (eds.), High Pressure Chemistry, Biochemistry and Materials Science, 275–290.

The practical applications of high pressure viscosity data are numerous. For example: one can estimate diffusion times of constituents involved in high-pressure fluid reactions; the limits of hydrostaticity for any given liquid can be established; the rate of flow for polymer liquids into high-pressure injection molds can be established; etc. Perhaps one of the most direct practical applications of high pressure viscosity data is in lubrication studies. In fluid lubricants pressures in excess of 3 GPa are possible,[3] and the viscosity rise with pressure plays a critical role in determining the fluid film thickness in such a lubricated contact, hence the wear and friction.

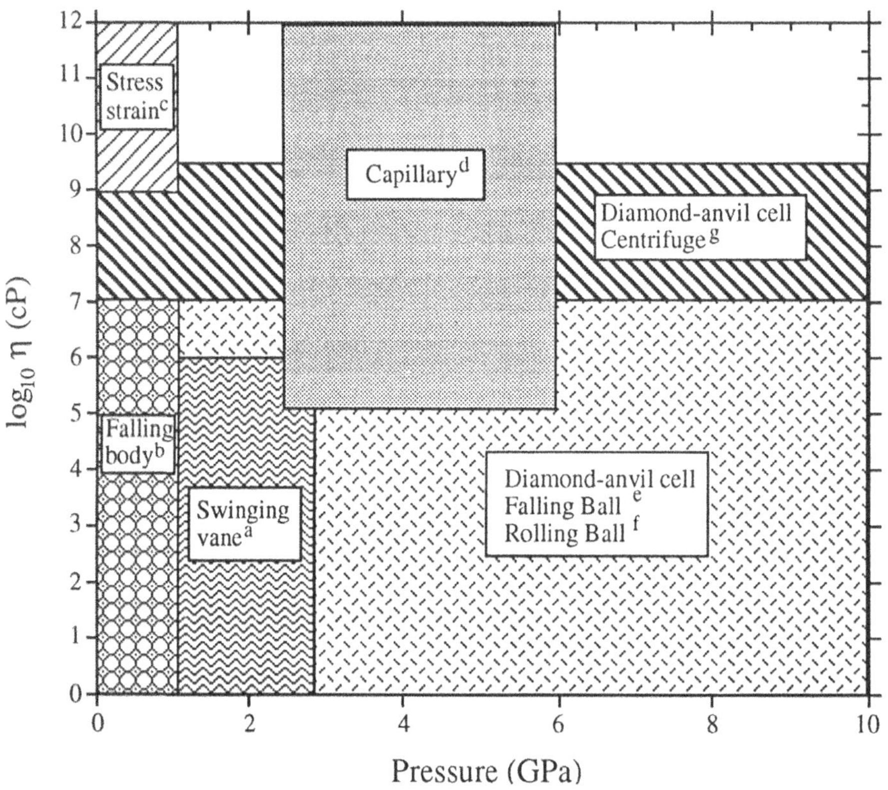

Fig. 1 Chart showing the range of various high pressure viscometers. The centrifugal force viscometer has extended the available range of viscosity to 4 x 10^9 cP and the pressure range as high as that of any known liquid. [a]Bridgman (1949),[4] [b]Bair and Winer (1982)[5] [c]the stress-strain apparatus developed by Bair and Winer (1982)[5] provides an estimate of viscosity under very high shear stress, [d]Barnett and Bosco(1969),[6] [e]Piermarini *et al.* (1978)[7] and Fujishiro *et al.* (1986),[8] [f]King *et al.* (1992),[9] [g]this work.

In addition to these direct uses of high pressure viscosity one can also envision an important secondary use—as a probe of polymer geometry. Through the use of Flory-Huggins theory[10] one has a simple relationship between viscosity and polymer chain dimensions in dilute solutions. As will be demonstrated below, the variation of viscosity with pressure for a given polymer then allows one to extract the chain dimensions at any pressure. Thus, the response of polymer-solvent interactions to density changes can be probed. This is thus a probe of the volume dependence of the free energy which provides an important contrast to the thermal studies which mainly examine the entropic part of the free energy. The subtle balance between entropic and energetic contributions determine the phase behavior and solubility of polymers. For example, this plays an important role in determining the configuration of biopolymers, such as proteins, in solution.

2. EXPERIMENTAL TECHNIQUES

An appreciation for the experimental requirements of a high pressure viscometer can be gained by considering the general attributes of fluids. Viscosity variation with pressure (or temperature) follows an approximately exponential dependence; consequently, the viscometer must be applicable over many orders of magnitude of viscosity. The usable pressure range must also be wide. Among different fluids there is a wide range in behavior; for example, the glass transition pressure at room temperature ranges from a few hundred MPa to nearly 15 GPa. Thus one needs a pressure device that can cover both the low and high pressure range.

There have been many devices used in measuring high pressure viscosity (see King *et al* .[9] for a review), and a selection of these are shown in Fig. 1. The diamond cell viscometer excels in both of the requirements mentioned above. As will now be described, its proper use can also give quite accurate results (a few percent error); thus allowing one to precisely define the pressure-viscosity dependence over a much wider range than previously accessible.

2.1 The Rolling Ball Diamond Cell Viscometer

The rolling-ball viscometer (more fully described in King *et al* .[9]) is shown in Fig. 2. Measurements are made by recording the velocity of the falling ball inside the pressure chamber. The pressure chamber consists of a 200-500 μm diameter hole drilled in a 250 μm thick piece of stainless steel which is compressed between two flat diamond surfaces. The ball, a nickel alloy about 50 μm in diameter, rolls down the surface of one of the diamonds. A video camera relays the image of the ball to a video dimension analyzer (VDA) which returns a voltage proportional to the displacement of the ball. A computer makes periodic readings of the VDA output giving displacement as a function of time. The slope of the line thus obtained yields the velocity of the ball which can be converted into viscosity, η, by a modified Stokes equation

$$\eta = \gamma\left[\frac{2\,R_s^2\,g\cos(\theta)(\rho_s - \rho_f)}{9\,V}\right] \tag{1}$$

where R_s = sphere radius, g = acceleration of gravity, ρ_s = sphere density (7.6 g/ml), ρ_f = fluid density, θ = tilt angle of cell (0° = vertical), V = velocity of the ball, and γ = wall correction. All of the above parameters are experimentally measurable with γ being an instrument constant dependent only on tilt angle. We determine γ for each experiment by taking the ratio of the observed ambient pressure velocity in the diamond cell to the Stokes velocity obtained from an independent measurement of the ambient pressure viscosity in a Contraves Low Shear 30 viscometer. Experiments have shown that γ defined in this manner is then pressure independent. Pressure is measured in the usual manner using the ruby fluorescence peak except that we use a neon reference system to increase the accuracy at low pressures.[11] The fluid density term ρ_f in Eqn. 1 varies with pressure and is calculated from the Tait equation of state. Using our present Merrill-Bassett pressure clamp we can access pressures of from 0.2 to 10 GPa and viscosities from 10^{-1} to 10^7cP have been measured. The resulting viscosity data have a precision of ≈4%. Data taken on materials with known viscosity vs. pressure behavior have established the accuracy of the measurements.[9]

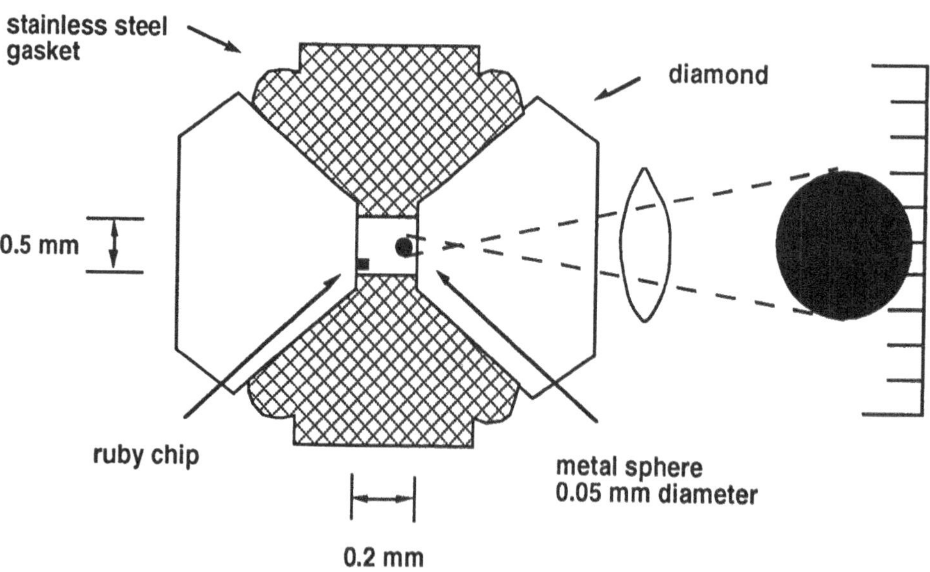

Fig. 2 The components of the rolling-ball diamond-cell viscometer are shown except for the Merrill-Bassett type clamp. The metal sphere moves under the influence of gravity and its velocity is determined through a video system. A small tilt angle (not shown) maintains a constant trajectory for the sphere.

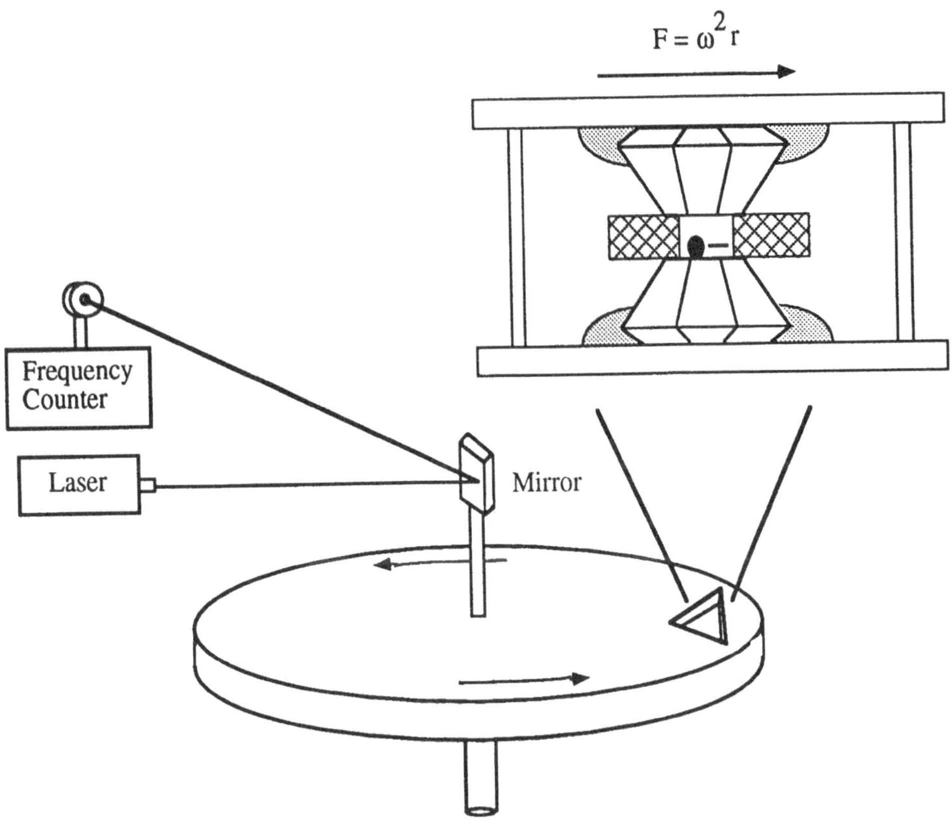

Fig. 3 Schematic diagram of centrifugal force viscometer. The ≈50 μm diameter nickel ball moves along the horizontal diamond surface being driven by centrifugal force. The rotation of a mirror on the shaft of the rotor deflects a laser beam across a detector producing a signal whose frequency gives the speed of the rotor. Our current centrifuge is able to generate an acceleration of 1700g.

2.2 The Centrifugal Force Diamond Cell Viscometer

To measure higher viscosities, we have introduced a modification to the rolling-ball viscometer. The viscosity range of the rolling-ball technique is limited by the speed of the rolling ball. For example, at a viscosity of 10^7 cP, the 50 μm diameter ball rolls at a rate of only 2 μm/day; this displacement is of a marginal magnitude to accurately measure. To overcome this limitation, we have developed a novel method in which we use centrifugal force instead of gravity to drive the ball in the diamond-anvil cell viscometer. By applying a maximum acceleration of 1700 g to the ball we are able to measure viscosities up to 4×10^9 cP while maintaining an acceptable uncertainty of ≈8%.

A schematic of the centrifugal force viscometer is given in Fig. 3. A bench-top centrifuge was fitted with a specially designed rotor to accommodate the diamond cell. The hole in the rotor which accepts the triangular diamond-anvil cell contains a flat surface perpendicular to the rotor's radius; thus the cell may be placed such that force is in one of three cell directions and positioned so that the ball will always travel near the center of the cell, thus avoiding unwanted wall effects.[9] At the beginning of a measurement, the position of the ball relative to the edge of the gasket is measured. The cell is then placed in the centrifuge and accelerated for a specified time, and the ball's position is measured again after the run to determine the total displacement. Because the force applied to the sphere is proportional to ω^2, it is critical to know ω as a function of time. The velocity varies, of course, during acceleration and deceleration but also during the "constant velocity" portion of the run. All these variations are measured by reflecting a laser beam off a mirror attached to the shaft of the rotor and recording the optical pulses using a photo diode. A computer records the measured frequency and total time of the measurement and then integrates over $\omega^2 t$. This is then used in addition to the total sphere displacement (as discussed below), to calculate viscosity through Eqn. 2:

$$\eta = \gamma_{cent}\left[\frac{2\,R_s^2\,(\rho_s - \rho_f)\,r\,\omega^2 t}{9\,(x_f - x_i)}\right] \tag{2}$$

where x_i = initial position of ball, x_f = final position of ball, r = distance of cell from center of rotation = 7.37 cm, and γ_{cent} = wall correction for the centrifugal force measurements. We determine γ_{cent} in a manner similar to that at ambient pressure; however, here the viscosity from a rolling sphere experiment is used as the reference value. The overlap region between the rolling-sphere and centrifugal-force, $\eta \approx 10^5$ to 10^7 cP is used. A constant value for γ_{cent} throughout the experiment is essential for high-precision data, and recent data confirm this to be the case.[12] Runs on the centrifuge varied from 3 minutes at a speed of 800 RPM (acceleration = 50g) for $\eta \approx 10^5$ cP to 1 hour at 4600 RPM (1700g) for $\eta \geq 10^9$ cP. The most critical experimental parameter in making accurate viscosity measurements is maintaining a constant temperature. Heating of the bench-top centrifuge on runs at high speed of longer than 5 minutes is significant so the centrifuge is water cooled when long runs are required. This cooling is able to maintain the temperature to within 2 °C of room temperature (i.e. 22.5 ± 2.0 °C).

2.3 Dynamic Light Scattering

Dynamic Light Scattering (DLS) or photon correlation spectroscopy is a widely used probe in investigating the properties of simple liquids and polymers[13]. Through measurements of the temporal variation of a laser speckle pattern one probes the dynamics on a time scale (sub-MHz) not easily accessible with other light scattering probes. Thus, this is a useful technique for measuring diffusion times of small particles (<1μm) and molecules. Combining DLS with the high pressures attainable in the diamond anvil cell is a potentially important new area of research. For example, one can measure the diffusion time for particles of known diameter in a fluid and extract the fluid viscosity. Similarly, the diameter of polymer molecules at any given pressure can be deduced from their diffusion times in a known viscosity fluid. Previous uses of DLS at high pressure were limited to pressures below 0.5 GPa.[14]

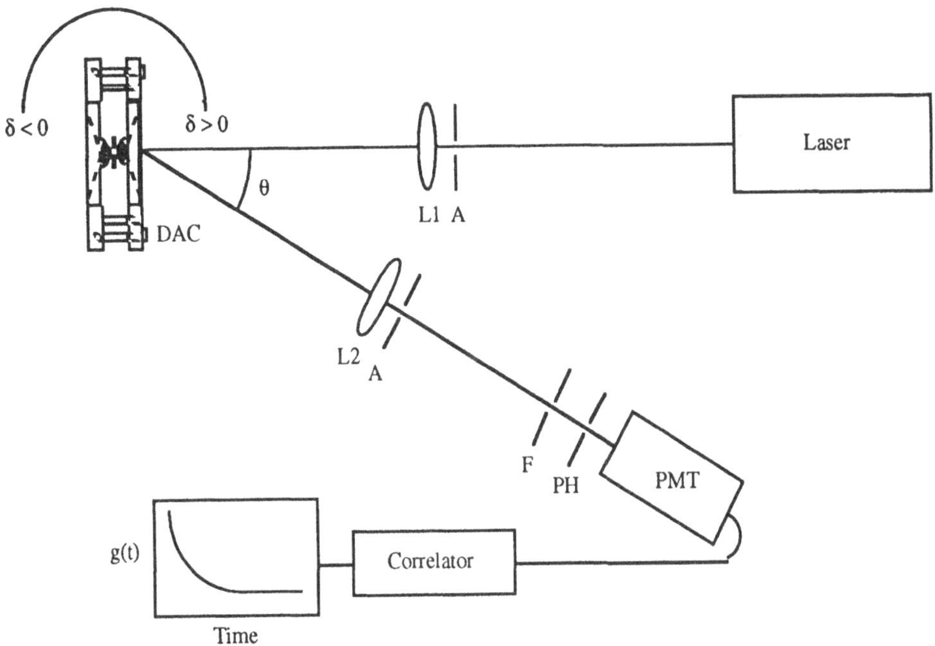

Fig. 4 Optical arrangement used in the DLS experiment, where A is an aperture, L_1 and L_2 are the incident and collection lenses, F is a line filter and PH is a pinhole. The scattering angle was $\theta = 170°$ and the diamond anvil cell (DAC) was rotated at an angle δ.

A schematic of the experiment is shown in Fig 4. Incident radiation is supplied by an argon ion laser, and the scattered light is collected at a scattering angle of 170°. The output from the photo multiplier tube goes into a Brookhaven Instruments Model BI-2030 AT digital correlator. The samples are contained within a Merrill-Bassett type diamond anvil cell in which the backing plates have been slotted to give wide angular access ports (100° full-angle opening). The gasket holes used are large enough (500-750 μm) so that even if the hole diameter slightly decreased under pressure, as is often observed, the incident focused beam diameter will still be much smaller than the gasket diameter. This is important because it is essential to position the cell so that the incident beam is not focused onto either the ruby chip or edges of the metal gasket.

A significant difficulty in DLS measurements with the diamond cell comes from the small size of the pressure chamber. Not only does this lead to a weak scattered intensity, but scattering from the sample is combined with partly coherent scattering from the diamond/air and diamond/sample interfaces thus complicating data interpretation. This heterodyne scattering component can bias the measured diffusion coefficients, but it can be minimized with proper cell placement and good spatial filtering[15].

In a dynamic light scattering experiment one measures the intensity autocorrelation function of the scattered light, which can be related to the exponential relaxation time for a dilute solution of monodisperse particles

$$|g^{(1)}(\tau)| = \exp(-\Gamma\tau) \qquad (3)$$

where Γ is the decay rate or line width. This line width is related to the diffusion coefficient by $\Gamma = Dq^2$ where $q = (4\pi n/\lambda)\sin(\theta/2)$ is the magnitude of the scattering vector, n is the index of refraction, λ is the wavelength of the laser and θ is the scattering angle. One then applies the Stokes-Einstein equation:

$$D = k_BT / 3\pi\eta d, \qquad (4)$$

where D is the translational diffusion coefficient, k_B is Boltzmann's constant, T is the absolute temperature, η is the viscosity and d is the particle diameter, in calculating the particle diameter or fluid viscosity. We recently demonstrated the accuracy of this technique when used for a sample in a diamond anvil cell by measuring the viscosity of methanol to 3.0 GPa.[15]

3. RESULTS

3.1 Polymeric Fluids

Dilute polymer solutions represent one of the most basic and well understood forms of polymeric fluids. The polymer concentration in such fluids is below the overlap concentration ($c^* = [\eta]$, where $[\eta]$ is the intrinsic viscosity); hence the behavior is characterized by that of individual polymer chains. These chains are solvated by a simple monomeric solvent (typically a simple hydrocarbon or aqueous solvent), and solvent quality can range from *good* to *poor*. In a *good* solvent the polymer chain is expanded beyond its simple random-walk radius due to a preferred interaction of the chain with the solvent over that of the chain with itself. At the θ point, the random-walk radius is attained for a *poor* solvent. Typically a *good* solvent becomes *poor* at some temperature, and a further temperature change will then cause the polymer chain to collapse leading to phase separation. Because the behavior of these fluids is characterized by individual chain behavior, these systems are interesting ones in which to explore the effects of pressure. High pressure can be expected to alter the relative importance of entropic and enthalpic parts of the free energy, and correspondingly the radius of individual chains. The effects of pressure upon the behavior of such fluids has not been widely studied, but some of the pioneering efforts in this area has come from the group at Mainz.[16]

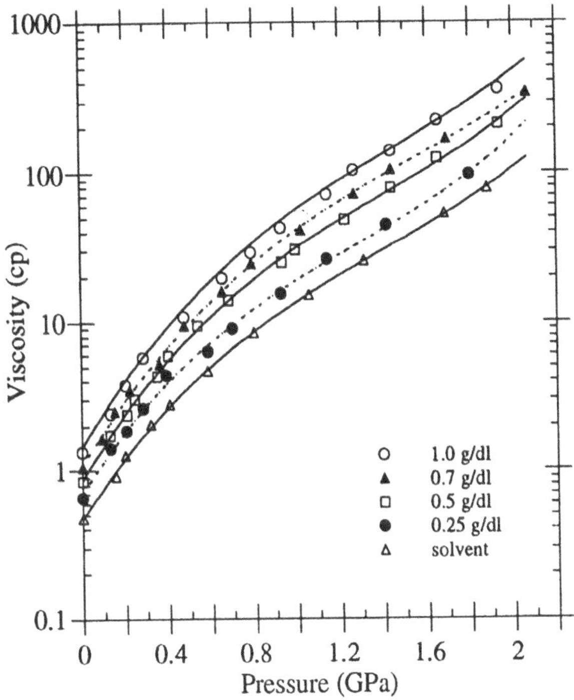

Fig. 5 Viscosity - pressure relationship for PS solutions in THF. The approximately exponential dependence of viscosity on pressure for THF (bottom curve) is typical simple fluid behavior. The polymer viscosity rises faster with pressure than the solvent as discussed in the text.

Fig. 5 shows the pressure dependence of viscosity for four concentrations of 300,000 M_W polystyrene(PS) in tetrahydrofuran (THF), a good solvent. As can be seen in comparison with THF itself, also shown in Fig. 5, each of the polymer solutions has a higher ambient pressure viscosity and a more pronounced pressure dependence. Note how the curves diverge from the THF data and from one another as pressure increases. The ambient-pressure behavior follows the well known Huggins equation:

$$\eta_{sp} = [\eta]\, c + k_H\, [\eta]^2\, c^2 + \dots \qquad (5)$$

where c is the polymer concentration, k_H is the Huggins interaction parameter and $[\eta]$ is the intrinsic viscosity which is defined as:

$$[\eta] \equiv \lim_{c \to 0} \left(\frac{\eta_{sp}}{c} \right) \qquad (6)$$

The viscosity in this case is normalized to that of the solvent, $\eta_{sp} = \eta/\eta_{THF} - 1$. One can also apply this equation at any given pressure, but to do so requires that the concentration, measured in this case as mass per unit volume, be clearly defined. In Fig. 6 is shown the application of Eqn. 5 at ambient pressure and two elevated pressures. Although these three data sets apparently give different solutions to the Huggins equation, one finds that making the assumption that the concentration at elevated pressure is increased in proportion to increased solvent density causes all the data to fall on the ambient-pressure Huggins equation solution. The implications of this are important because we can directly relate [η] to the polymer chain dimensions through the Flory equation[10]:

$$[\eta] = \Phi' \langle s^2 \rangle^{3/2} M^{-1} \quad (7)$$

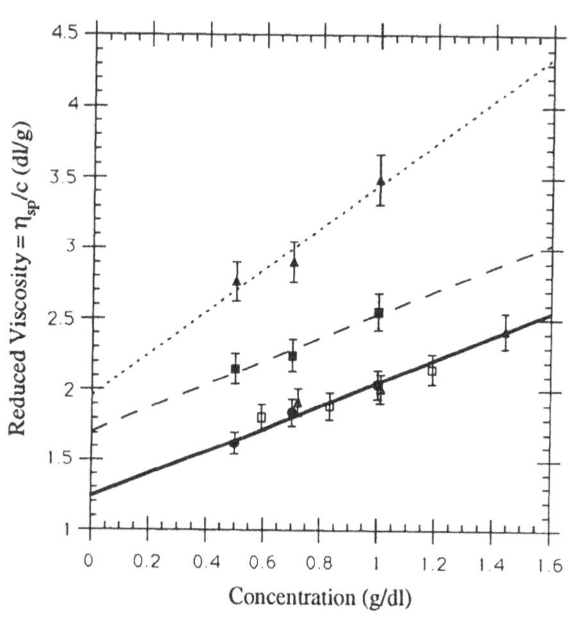

Fig. 6 Huggins plot of PS in THF showing how the concentration correction affects [η] calculation. The solid symbols are uncorrected (i.e. mass/mass conc.) and the open symbols are corrected (i.e. mass/volume conc.). The circles, squares, and triangles are at 0, 0.4, and 1.8 GPa respectively. Note that uncorrected data gives an [η] that increases with pressure (dashed lines). The same data, once corrected, falls on the ambient pressure (solid) line (see text).

where $\Phi' = 6^{3/2}\,\Phi \approx 3.7 \times 10^{22}$ dL mol^{-1} cm^{-3}, Φ = Flory constant, $<s^2>^{1/2}$ = radius of gyration, and M = molecular weight. Clearly these viscosity results suggest that [η] and, consequently the polymer chain dimensions, are invariant (to within experimental accuracy ≈10%) with increasing pressure. Results on a variety of dilute solutions suggest this is a general phenomena.[17]

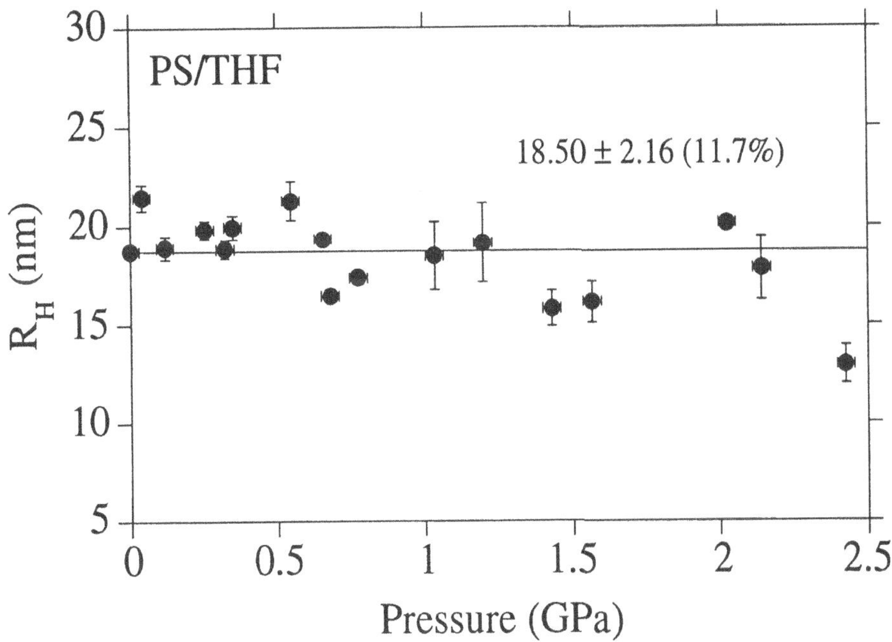

Fig. 7 Hydrodynamic radius for 300,000 M_w polystyrene in tetrahydrofuran at a concentration of 1 g/dL measured by dynamic light scattering in a diamond anvil cell. These represent the highest pressure measurements of this sort

The surprising nature of this result suggested the importance of further study by DLS to directly examine the polymer hydrodynamic radius as a function of pressure. The results for 300,000 M_w PS in THF are shown in Fig. 7. This is the first time that a polymer chain dimension has been measured to such high pressures, and studies of other systems give similar results.[14] These results, of course, are in good agreement with our viscosity measurements. The thermodynamic implication of these findings is that polymer solutions are more ideal than previously believed.[17] However, the microscopic interactions leading to this behavior are not yet understood.

Hydrogen bonded systems can be expected to exhibit quite different behavior. It is well established that hydrogen bonding in water significantly declines at high pressure.[18] We have recently examined dilute solutions of polyethylene oxide (PEO)

in water (a good solvent at ambient pressure) where hydrogen bonding is known to play a significant role in the solvation of the PEO chain. We find that the viscosity of the solution decreases with increasing pressure. Applying an analysis similar to that above and then using [η] to calculate the polymer chain dimension leads to the result in Fig. 8. The radius, R_η, decreases from the swollen value typical of a *good* solvent to the collapsed value typical of a *poor* (or θ solvent). A further pressure increase causes a phase separation. We can associate this progressive change in solvent quality with the loss of hydrogen bonding. Hydrogen bonding plays a pivotal role in stabilizing the water solubility of PEO; for example, the structurally similar polypropylene oxide has relatively weaker hydrogen bonding and thus is insoluble in water.[19] Understanding the configuration of macromolecules in water based solutions is fundamental to many technologies as well as to much of biology. The complexity of this problem is apparent when one considers that the overall configuration is typically a balancing act between

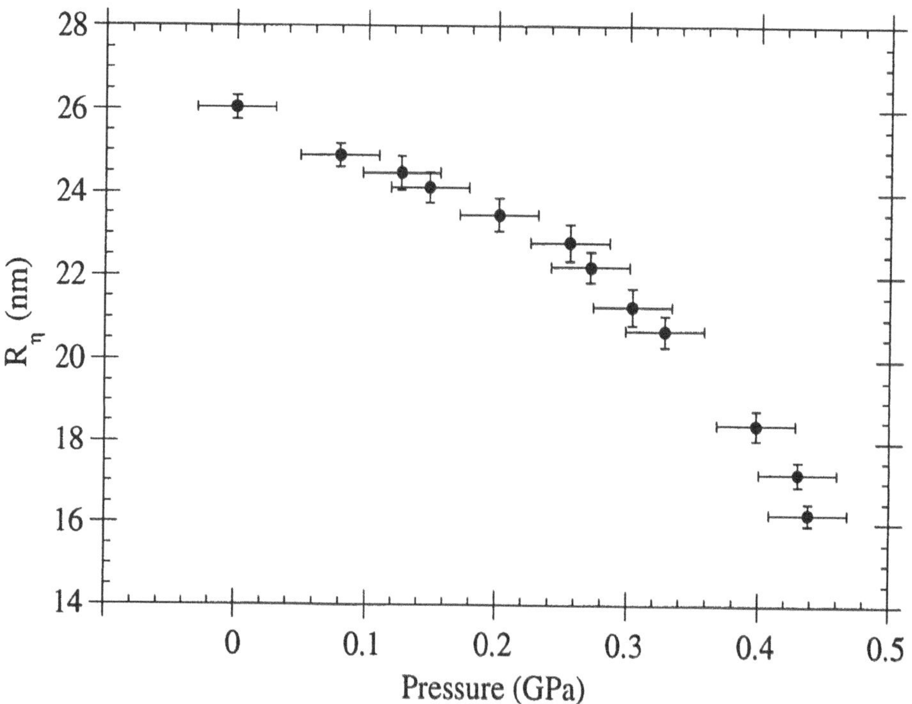

Fig. 8 Radius of gyration for polyethylene oxide in water calculated from viscometric data. The decrease in radius with pressure shows a contraction of the polymer chain from the expanded, good solvent configuration to the contracted, theta solvent conformation. Intrinsic viscosity is calculated from the Huggins equation assuming a Huggin's coefficient of 0.3. The radius, R_η is calculate from the Flory relationship: $[\eta] = \Phi' R_\eta^3 M^{-1}$ where $\Phi' = 3.7 \times 10^{22}$ dL mol^{-1} cm^{-3} and $R_\eta = R_g^{2/3} R_h^{1/3}$.

several forces, i.e. electrostatic forces, van der Waals interactions, hydrogen bonding, and hydrophobic interactions. High pressure studies such as these where one can tune the relative importance of these forces provide a means to further our understanding of such systems.

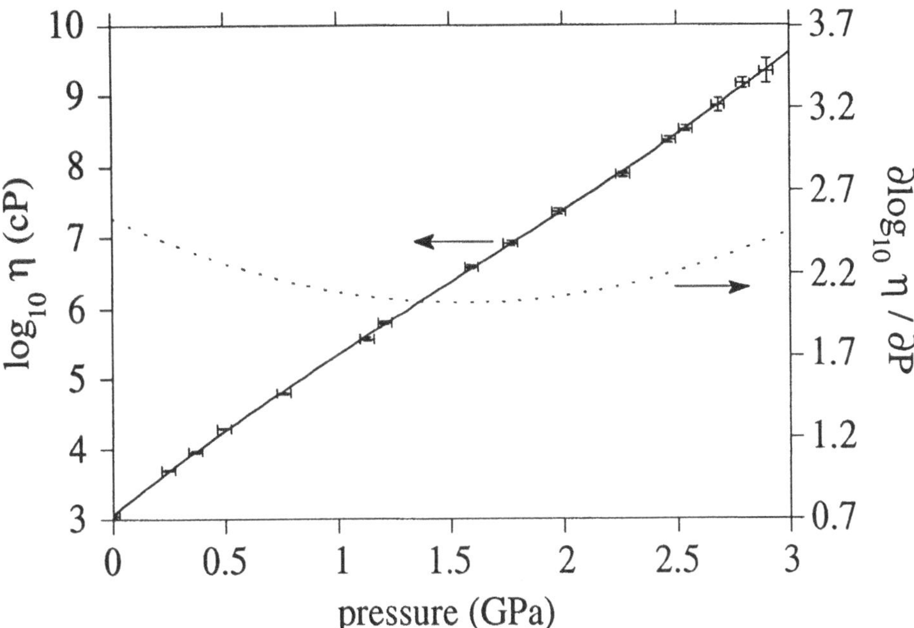

Fig. 9 Pressure dependence of the viscosity of glycerol measured with the rolling ball and centrifugal force techniques and the pressure derivative of the viscosity. Each data point represents the mean of multiple measurements and the calculated standard deviations are shown. These data sets overlap between pressures of 1 - 1.5 GPa, and agree to within 5%. The pressure derivative is calculated from the smoothed viscosity data; the minimum near 1.5 GPa is in good agreement with the analytical theory described in the text.

3.2 Pressure Induced Glass Formation in Simple Fluids

The general effect of elevated pressure on fluids is to increase their viscosity and for many fluids this increase is continuous leading to glass formation. (A convenient and widely used approximation for fluid viscosity at the glass transition is $\approx 10^{15}$ cP, although this number can vary substantially.) Glass formation and glassy behavior is important to nearly all areas of science and technology; consequently it is a rigorously studied phenomena. Although variable temperature investigations are common in this field, an isothermal, purely density-driven approach to the glassy state is of great interest[1,2] The experimental challenge is to measure the high pressure viscosity over several decades with sufficient accuracy to allow a differentiation of various models. The importance of this is borne out by the finding that extrapolation of temperature dependent viscosity data taken over only a few decades in viscosity are not reliable.[20] Thus, investigation into the nature of the viscosity-pressure relationship requires high accuracy data spanning many decades. Data taken using the new centrifugal force viscometer can address this need.

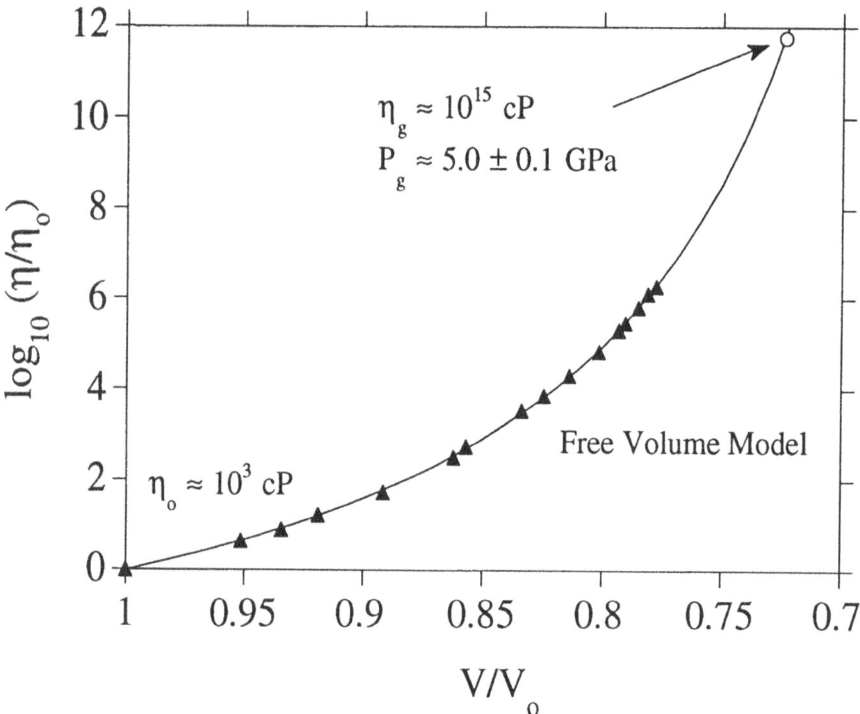

Fig. 10 Reduced viscosity (η_0 = viscosity at ambient pressure) plotted versus reduced volume (V_0 = ambient pressure volume). The free volume equation models the data over the entire compression range. Extrapolation gives an estimate for the glass transition pressure (5.0 ± 0.1 GPa) in reasonable agreement with that determined from stress gradient measurements (4.4 ± 0.5 GPa).

One of the most widely studied glass forming liquids is glycerol; however, its high pressure viscosity has only been investigated over a two decade range.[21] Our viscosity measurements were made on a sample of Aldrich spectrophotometric grade glycerol (99.5+%) which was dried over a molecular sieve. The water content was less that 0.2%. Fig. 9 shows the high pressure rolling ball and centrifugal force viscosity measurements made at 22.5 ± 1.0 °C. These data extend over a pressure range of nearly 3 GPa and encompass a range of nearly seven orders of magnitude in viscosity. The sigmoidal shape of the high pressure log η versus pressure curve is less pronounced for glycerol than for other glass formers,[6] but $\partial \log\eta/\partial P$ exhibits a clear minimum; this locates the inflection point of log η near 1.5 GPa, in a viscosity range consistent with results for other liquids.

We find that over the entire viscosity range our data for $\eta(P)$ can be accurately modeled using a free volume equation,[1,22] as shown in Fig. 10. This equation has the form $\eta = A\exp[BV_\infty/(V-V_\infty)]$, where $V-V_\infty$ is the free volume of the liquid and V is the pressure-dependent volume, obtained by fitting P-V data[23] with the Tait equation of state. There are several important features about the resulting free-volume fit. First, one can extrapolate this function to a viscosity of 10^{15} cP and from this obtain an estimate of the glass transition pressure, P_g = 5.0±0.1 GPa. This is in good agreement with our determination of P_g = 4.4±0.5 GPa by ruby fluorescence measurements of stress gradients. Second, the value of V_∞ appears to be physically significant. We find that the volume we obtain, 75.5 ± 0.4 Å^3, is in excellent agreement with the effective hard sphere volume, 74.5 Å^3, determined through a modified Carnahan-Starling—van der Waals equation of state.[24] Finally, we find that the use of this model allows one to analytically determine the origin of the inflection point in log $\eta(P)$. The volume at the inflection point is given by $V_I/V_0 = V_\infty/V_0 + 2/(K_0' + 1)$ where V_0 is the volume at ambient pressure and K_0' is the pressure derivative of the bulk modulus. This calculation gives P_I = 1.4 GPa for glycerol, which compares well with the value of 1.5 GPa found in Fig. 7. As can be seen, non-linearity in the P-V curve, i.e. K_0', completely determines the existence of the inflection point within a physically accessible pressure regime, i.e. $V_I/V_0 < 1.0$. The features described here for glycerol are quite general and similar results have been found for over 25 other, chemically dissimilar compounds.[12,25]

4. CONCLUSIONS

The ability to determine high pressure viscosities over a ten decade range is now possible. Furthermore, the accuracy is sufficient that many subtle features such as the viscosity changes accompanying polymer radius of gyration changes are resolvable. In conjunction with dynamic light scattering studies in the diamond anvil cell, this now opens the possibility of fully utilizing high pressure studies in the investigation of complex fluids. Such fluids can, in general, order on several different length scales, and the response of these fluids to high pressure can give insight into the forces controlling them. In general these "soft condensed matter" phases are the result of a subtle balance of entropy and energy.[26] An additional area of importance is the study of glassy behavior. With access to such a wide viscosity and pressure range, the density scaling relationships, suggested from variable temperature data, can, for the first time, be fully tested.

5. ACKNOWLEDGMENTS

This work has benefited from discussions with many people and among them are: Eric Herbolzheimer, Dennis Peiffer, and Dudley Herschbach.

6. REFERENCES

[1] Doolittle, A. K. *J. Appl. Phys.* **1951**, *22*, 1471.
[2] Cohen, M. H.; Turnbull, D. *J. Chem. Phys.* **1959**, *31*, 1164.
[3] *CRC Handbook of Lubrication*; Booser, R. E., Ed.; CRC Press, Inc.: Boca Raton, FL, 1984.
[4] Bridgman, P. W. *Proc. Am. Acad. Arts. Sci.* **1949**, *77*, 117.
[5] Bair, S.; Winer, W. O. *Trans. ASME-J. Lub. Tech.* **1982**, *104*, 357.
[6] Barnett, J. D.; Bosco, C. D. *J. Appl. Phys.* **1969**, *40*, 3144.
[7] Piermarini, G. J.; Forman, R. A.; Block, S. *Rev. Sci. Instrum.* **1978**, *49*, 1061.
[8] Fujishiro, I.; Nakamura, Y.; Matsuhiro, S. *Bull. Jap. Soc. Mech. Eng.* **1986**, *29*, 1280.
[9] King, H. E., Jr.; Herbolzheimer, E.; Cook, R. L. *J. Appl. Phys.* **1992**, *71*, 2071.
[10] Flory, P. J. *Principles of Polymer Chemistry*; Cornell University Press: Ithica, NY, 1953.
[11] King, H. E., Jr.; Prewitt, C. T. *Rev. Sci. Instrum.* **1980**, *51*, 1037.
[12] Cook, R. L.; Herbst, C. A.; King, H. E., Jr. *J. Phys. Chem.* **1993**, submitted.
[13] Pecora, R. *Dynamic Light Scattering*; Plenum Press: New York, 1985, pp 420.
[14] Herbst, C.; King, H. E., Jr. **1992**, in preparation.
[15] Herbst, C.; King, H. E., Jr.; Gao, Z.; Ou-Yang, H. D. *J. Appl. Phys.* **1992**, *72*, 838.
[16] Schott, N.; Will, B.; Wolf, B. A. *Makromol. Chem.* **1988**, *189*, 2067.
[17] Cook, R. L.; King, H. E., Jr.; Peiffer, D. G. *Macromolecules* **1992**, *25*, 2928.
[18] Jonas, J.; DeFries, T.; Wilbur, D. J. *J. Chem. Phys.* **1976**, *65*, 582.
[19] Kjellander, R.; Florin, E. *J. Chem. Soc., Faraday Trans. 1* **1981**, *77*, 2053.
[20] Laughlin, W. T.; Uhlmann, D. R. *J. Phys. Chem.* **1972**, *76*, 2317.
[21] Bridgman, P. W. *Proc. Am. Acad. Arts. Sci.* **1926**, *61*, 57.
[22] Turnbull, D.; Cohen, M. H. *J. Chem. Phys.* **1970**, *52*, 3038.
[23] Bridgman, P. W. *Proc. Am. Acad. Arts. Sci.* **1932**, *67*, 1.
[24] Ben-Amotz, D.; Herschbach, D. R. *J. Phys. Chem.* **1990**, *94*, 1038.
[25] Herbst, C. A.; Cook, R. L.; King, H. E., Jr. **1992**, submitted for publication.
[26] Pincus, P., personal communication, 1992.

MOLECULAR DYNAMICS SIMULATION OF DENSE CYCLOHEXANE IN POROUS SILICA

T. W. Zerda and A. Brodka
Texas Christian University
P.O. Box 32915
Fort Worth, TX 76129
USA

ABSTRACT

Molecular dynamics simulations at temperature 313 K and pressures up to 4 kbar for six-center Lennard-Jones models of C_6H_{12} in porous medium are reported. Thermodynamic, structural and dynamic properties of molecules confined to the pores are compared with the bulk fluids. It is observed that translational diffusion and reorientational motion of molecules in the pores are much slower than in the bulk phase. The liquid-plastic phase transition of cyclohexane in the pores is shifted toward higher pressures and depends on the pore size.

1. INTRODUCTION

Recently, we used molecular dynamics (MD) simulation method to investigate thermodynamic and transport properties of cyclohexane and SF_6 in small pores as a function of temperature and pore diameter [1-4]. It was found that MD calculations realistically predict liquid - solid phase transition both in the bulk form and inside small pores. In small pores supercooling and depression of the transition point in comparison to the bulk phase was observed. These results matched quite well experimental data for cyclohexane in porous silica [5,8].

In this paper we report a study on the pressure induced liquid-plastic phase transition of cyclohexane in modeled porous medium using MD simulation. Molecular dynamics calculations were performed for cyclohexane in the pure phase and in two cavities of diameters of about 30 Å and 50 Å. We calculated diffusion coefficients, rotational relaxation times and correlation times of the angular velocities.

2. SIMULATION METHOD

The molecular dynamics simulation of pure cyclohexane was identical as in our previous paper [3]. We considered a set of 108 rigid molecules in a cubic box and applied periodic boundary conditions. The cyclohexane molecule in the chair form was modeled by six LJ sites placed at the centers-of-mass of CH_2 groups.

To study dynamics of cyclohexane in porous silica we adopted the cavity model described in Ref. [4]: a rigid silica cluster of 12 Å diameter is placed at the corners of the cubic box filled with cyclohexane; the space between the clusters represents a cavity of size

R. Winter and J. Jonas (eds.), High Pressure Chemistry, Biochemistry and Materials Science, 291–297.

D_c. The silica cluster was cut from a bulk amorphous structure, and it was represented by LJ centers placed at oxygen atoms. We considered two systems: a small system with 95 cyclohexane molecules in 30 Å pore, and a large system with 243 molecules in the 50 Å pore. Our calculations were carried out for T=313.4 K and experimental densities of pure cyclohexane obtained for pressures up to 4 kbar [5]. Because the diameter of the cavity, D_c, depends on the box length, L,

$$D_c = \sqrt{3}\,L - d_c \tag{1}$$

where d_c is the diameter of silica cluster, it decreases from 34 Å to 31 Å for the small system and from 50 Å to 46 Å for the large one for pressures raising from 1 bar to 4 kbar. The potential parameters for oxygens were ε_{OO}/k_B=230 K and σ_{OO}=3.0 Å [1].

To calculate the forces we considered neighboring LJ sites inside a sphere of radius $3\sigma_{CC}$ and shifted force potential. The equations of motions were solved using the predictor-corrector method [9], fifth-order for translations and fourth-order for rotations (quaternions and angular velocities). The calculations were carried out for time-step of 5 fs. The initial configuration of the cyclohexane molecules was an fcc structure, characteristic for the plastic phase, and random orientations were assigned. Initial translational and angular velocities were chosen to be consistent with the required temperature, and the resultant momentum of the whole system was zero. During 2000 time-steps of the equilibration run velocities were scaled to obtain the required temperature, and then the system relaxed through the next 8000 time-steps. Coordinates and velocities of molecules were stored at intervals of ten time-steps, and quantities such as temperature, pressure and energies were calculated.

3. RESULTS AND DISCUSSION

3.1. Structural properties

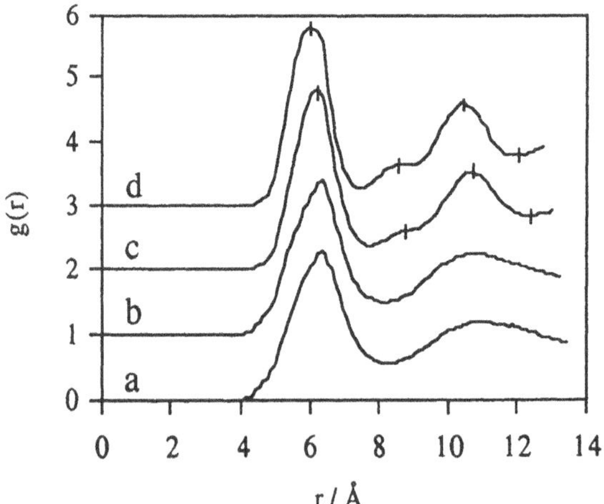

Fig. 1. Radial distribution function for pure cyclohexane for pressures a - 0.001, b - 0.6, c - 0.7, d - 3 kbars. The short vertical lines denote coordination zones for an ideal fcc lattice.

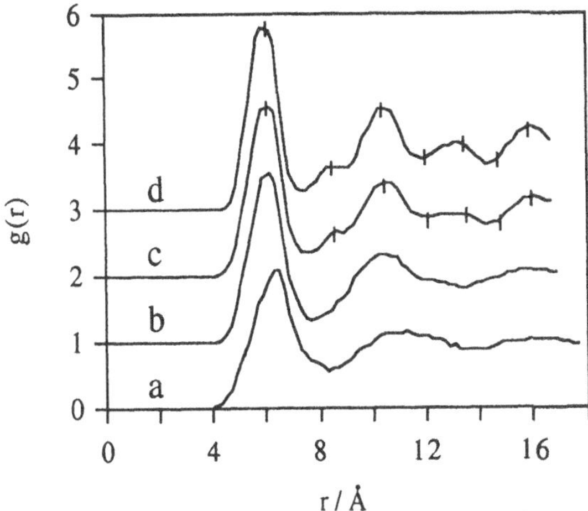

Fig. 2. Radial distribution function for C_6H_{12} in the 50 Å pore. a - 0.001, b - 2.5, c - 3, d - 4 kbar. Vertical lines denote coordinatione zones for fcc lattice.

Examples of the pair radial distribution function for pure cyclohexane are presented in figure 1. Up to the pressure of 0.6 kbar, the radial distribution functions show features typical for the liquid state. For higher pressures they show fcc structure. These results are in good agreement with experimental studies of Wisotzki and Wurflinger [5] who observed a phase transition between 0.6 kbar and 0.7 kbar.

Radial distribution functions, g(r), were calculated for the large system (D_c~50 Å) and they are presented in figure 2. The fcc structure is observed for pressures higher than in the case of bulk cyclohexane. For small cavity (D_c~30 Å) the shapes of the radial distribution function are characteristic for the liquid state for the whole density range under consideration, and we do not present them here.

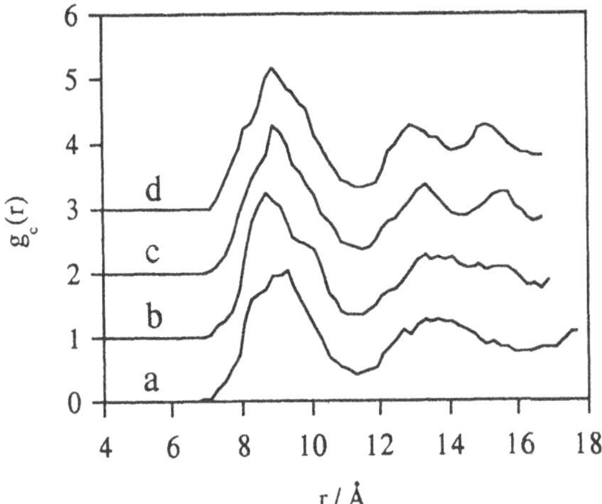

Fig. 3. Density of C_6H_{12} as a function of distance r from the center of silica cluster. Symbols as in Fig.2.

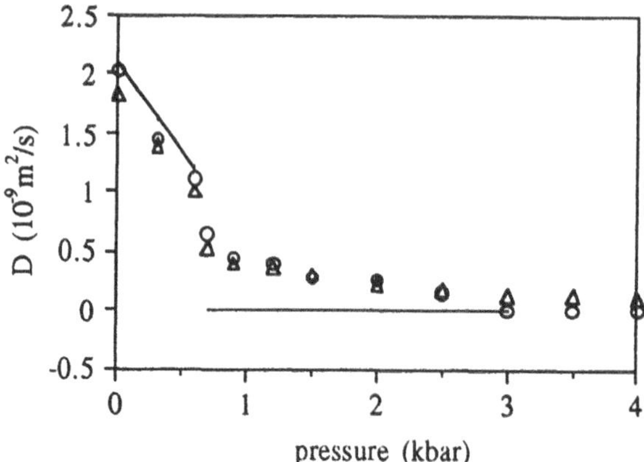

Fig. 4. Diffusion coeffcient as a function of pressure. Solid line - pure cyclohexane, cirles - cylohexane in the 50 Å pore, triangles - in the 30 Å pore.

From the positions of the centers of mass we found the distribution of cyclohexane around the silica cluster. Fig. 3 shows the densities of C_6H_{12}, $g_c(r)$, for the large cavity. One observes a layer about 5 Å thick near the silica surface. At higher densities the distribution function $g_c(r)$ shows succeeding layers which are not observed at low densities. We traced the positions of molecules during the simulation and we divided the molecules into "surface" molecules which were localized near the silica surface and "center" molecules which moved inside the inner part of the cavity. For the three lowest densities a few molecules exchanged their positions between the contact layer and inner part of the cavity, and those molecules were not counted either as the surface or as the center molecules. For the highest densities surface molecules were well localized in the monolayer.

3.2. Dynamic properties

Positions, orientations and velocities of molecules recorded during the simulations were used to calculate time dependent correlation functions. The mean square displacement, $\Delta r^2(t)$, and the velocity correlation functions were calculated for translational motion. We calculated the correlation functions $G_{20}(t)$ which describe reorientations of the anisotropic

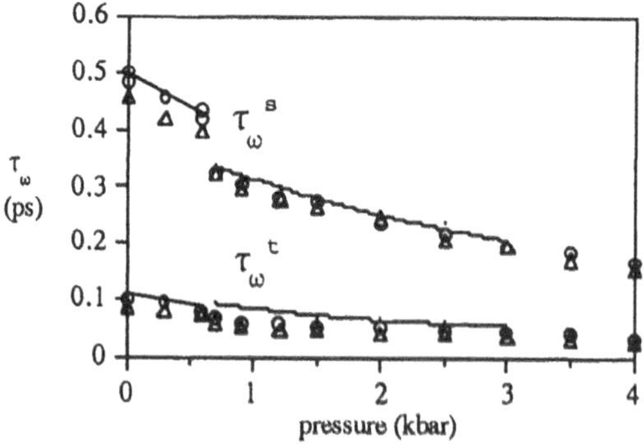

Fig. 5. Pressure dependence of the angular velocity correlation times for spinning, $\tau_\omega{}^s$, and tumbling motions, $\tau_\omega{}^t$. Symbols as in figure 4.

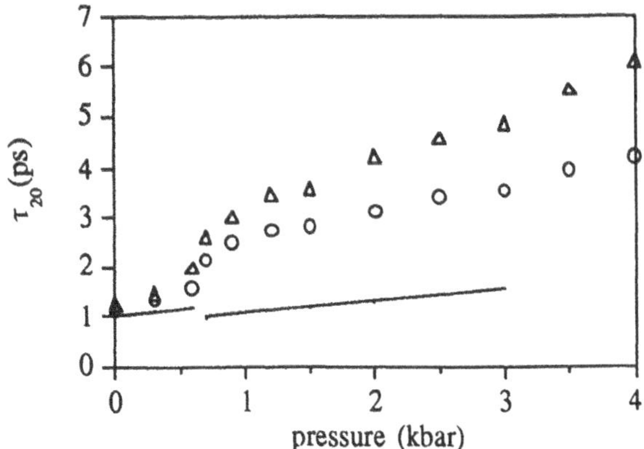

Fig. 6. Pressure dependence of rotational relaxation time τ_{20}. Symbols as in Fig. 4.

part of Raman tensor for A_{1g} mode of C_6H_{12}, and the angular velocity correlation functions for the spinning and tumbling motions, $G_\omega{}^s(t)$ and $G_\omega{}^t(t)$. The shapes of the correlation functions are very similar to those presented previously [3,4], and we limit the following discussion to the diffusion coefficient and rotational and angular velocity correlation times. Diffusion coefficients were obtained from the slopes of the mean square displacements. Relaxation times were obtained by integration of the corresponding correlation funtions.

For bulk cyclohexane a sharp drop of the diffusion coefficient indicates the liquid-plastic phase transition for a pressure value somewhere between 0.6 and 0.7 kbar [2]. The diffusion coefficients in the liquid state are slightly larger, less than 10%, than the experimental values obtained by Jonas et al. [10]. This discrepancy may be a result of a simplified structure of the cyclohexane molecule adopted in this study. Molecules in the cavity of 50 Å diameter behave like a liquid up to density corresponding to the pressure of 2.5 kbar and just above this value their translational motion is frozen. For the small cavity, D_c~30 Å, even at the highest density, a small diffusion of molecules is still observed. We also observed that mobility of the surface molecules is lower than that of the center molecules. The observed differences result from different neighborhoods of the molecules.

The relaxation time τ_{20} describes reorientations of the C_3 axis of cyclohexane, and it is related to the angular velocity correlation time for tumbling motion, $\tau_\omega{}^t$. For bulk cyclohexane the tumbling motion in the solid phase is slightly less restricted that in the liquid phase. In the liquid state collisions between molecules, which result from translational motion, contribute to hindrances in the tumbling motion, whereas in the plastic phase this contribution decreases due to the ordered and frozen positions of molecules.

This conclusion is confirmed by a decrease of the time $\tau_\omega{}^t$ and an increase of the relaxation time τ_{20} across the transition point, see figures 5 and 6. Higher density combined with relative large mobility of molecules lead to greater hindrances of the tumbling motion. For molecules in the cavity additional restrictions are result of interactions with silica. We observed that reorientation of the center molecules depends on the pore diameter, whereas rotational relaxation of the surface molecules is independent of the cavity size. Because, the number of the center molecules which are not influenced at all, or influenced only slightly by the surface, increases with cavity diameter, one observes diminishing of the relaxation times τ_{20} with cavity size.

Figure 5 shows that the freedom of the spinning motion is much greater than that of the tumbling motion. Confinement of the molecules in silica cavity does not change the density dependence of the angular velocity correlation time $\tau_\omega{}^s$. For the center molecules the values of the time $\tau_\omega{}^s$ are almost the same as for pure cyclohexane. However, for the surface molecules those times are shorter than for the bulk liquid. This means that the restrictions in the spinning motion originate from the interactions with silica. Similarly as in our previous study [4], the times $\tau_\omega{}^s$ do not change with the cavity diameter what suggests that the immediate neighborhood is essential for hindrances of the spinning motion.

4. CONCLUSIONS

MD simulations of pure cyclohexane properly predict the liquid-plastic phase transition. In the plastic phase, for densities corresponding to pressure 0.7 kbar and higher molecular translations are frozen but rotational freedom of molecules is still observed. A transition to the solid state makes the spinning motion more difficult, whereas the tumbling motion becomes less restricted.

Confinement of molecules in the cavity shifts the transition point toward higher densities. For the large cavity, D_c~50 Å, the transition point was found for densities corresponding to pressures between 2.5 kbar and 3 kbar. In the case of the small cavity, D_c~30 Å, a transition to the plastic phase was not observed, at least for pressures up to 4 kbar. For the small cavity the surface interactions greatly affect dynamics of most molecules and prevent the transition to the solid phase. Silica surface potential plays a crucial role in determining dynamics of molecules in the pores. Since the molecules do not leave the surface layer their motion can be treated as a 2-D motion. Reorientational and translational motions in the monolayer are highly restricted and for pores of small diameters where the monolayer accounts for most of the trapped molecules, the system behaves as a supercooled liquid. Geometrical restrictions such as pore shape and interconnectivity of pores in real sol-gel glass may additionally affect behaviour of molecules.

ACKNOWLEDGMENT

This study was supported by AFOSR grant No. 90-0165.

REFERENCES

1. Brodka, A., and Zerda, T. W. (1991) 'Molecular dynamics of SF_6 in porous silica' J. Chem. Phys., 95, 3710.
2. Brodka, A. and Zerda, T. W. (1992) 'Molecular dynamics simulation of reorientational motion of SF_6 in porous sol-gel glass' J. Non-Crystal. Sol. 139, 215.
3. Brodka, A., and Zerda, T.W. (1992) 'Molecular dynamics simulation of liquid-solid phase transition of cyclohexane' J. Chem. Phys., in press.
4. Brodka, A., and Zerda, T.W. (1992) 'Molecular dynamics simulation of liquid-solid phase transition of cyclohexane in porous silica' J. Chem. Phys., in press.
5. Wisotzki, K.D., and Wurflinger, A. (1982) 'PVT data for liquid and solid cyclohexane, cyclohexanone, and cyclopentanol up to 3000 bar' J. Phys. Chem. Solids, **43**, 13.
6. Mu, R. and Malhotra, V. M. (1991) 'Effects of surface and physical confinement on the phase transitions of cyclohexane in porous silica' Phys. Rev. B44, 4296.
7. Dore, J.C., Dunn, M., Hasebe, T., and Strange, J.H. (1989) 'Orientationally disordered crystals in porous silica: cyclohexane' Colloids and Surfaces, 36, 199.

8. Jackson, C.L., and McKenna, G.B. (1990) 'The melting behavior of organic materials confined in porous solids' J. Chem. Phys., 93, 9002.
9. Gear, C.W. (1971) 'Numerical initial value problems in ordinary differential equations' Prentice-Hall, London.
10. Jonas, J., Hasha, D., and Huang, S.G. (1980) 'Density effects on transport properties in liquid cyclohexane' J. Phys. Chem., 84, 109.

HIGH PRESSURE NUCLEAR QUADRUPOLE RESONANCE STUDIES OF HYDROGEN BONDED SYSTEMS.

M. MAĆKOWIAK and M. ZDANOWSKA-FRĄCZEK
Institute of Molecular Physics
Polish Academy of Sciences
Smoluchowskiego 17
60-179 Poznań, Poland.

ABSTRACT. *Pressure induced variations of the proton localisation in the hydrogen bond are reflected by changes of the NQR frequency observed also for nuclei located outside the hydrogen bond. A variational correlated ground state wavefunction theory is applied to investigate the high pressure deformation of hydrogen-bond potential. The evolution of the hydrogen bond potential with increasing pressure and the effects of deuterium substitution on the quantum-fluctuation -driven phenomena are discussed. High pressure NQR studies were performed for KDP-like hydrogen bonded ferroelectrics and several deuterated and undeuterated crystals. The correlation between the magnitude of geometric isotope effect and the NQR-frequency isotope shift as well as the value of pressure coefficient have been found and explained on the basis of ground-state variational theory in terms of microscopic parameters of a compressed single hydrogen bond. It is shown that the value of the pressure coefficient of the NQR frequency is related to the degree of proton transfer in hydrogen bonded complexes. The utility and significance of the high pressure NQR techniques as well as some technical aspects of the high pressure radiospectroscopic studies are discussed and demonstrated on the basis of some selected results.*

1. Introduction.

Nuclear Quadrupole Resonance (NQR) spectroscopy represents a well established and powerful tool for studying the dynamic structure of matter. A great majority of the NQR studies have used temperature as the only experimental variable, while pressure was left constant, usually at atmospheric pressure. The use of pressure will provide another dimension over which to investigate the molecular system. In some cases pressure is found to be a complementary variable to temperature, in other cases pressure is the essential variable. In particular, the high pressure technique enables one to separate the effects resulting from a change in crystal volume from those which are exclusively temperature dependent and creates the possibility of

R. Winter and J. Jonas (eds.), High Pressure Chemistry, Biochemistry and Materials Science, 299–308.

modification of classical or quantum molecular motions due to a variation of potential barrier determining the nature of a particular interaction. High pressure can also stimulate some phase transitions.

Many crystalline materials contain pairs of oxygen atoms connected by hydrogen bonds. By lowering the temperature some of these substances undergo order-disorder phase transformations to states of spontaneously broken symmetry characterized by long-range proton order. The effects of high pressure upon hydrogen-bonded solids have been the subject of recent interest [1-5]. Under pressure the character of the bistable potential in which a proton moves along the bond can be significantly altered. Various hydrogen-bonded systems in crystals are quite delicately balanced and small changes in external parameters can cause large changes in their physical properties. Variations of the proton localisation within the hydrogen bond are reflected in changes of the electric field gradient and, thereby, in changes of the NQR frequency. Application of a relatively small hydrostatic pressure (of the order of several hundred MPa) causes considerable changes in the NQR frequency observed also for nuclei located outside the hydrogen bond. Up to now pressure studies have been carried out for only a few hydrogen-bonded systems, mainly for ferroelectric crystals of the KDA-type. Our recent NQR pressure studies of hydrogen bonded systems have shown that the sign and magnitude of the pressure coefficient of the NQR frequency are related to the degree of proton transfer in the hydrogen bond [6,7].

The purpose of the present paper is to apply the ground-state variational theory for predicting and interpreting a range of phenomena induced by high pressure in terms of microscopic parameters of a compressed single hydrogen bond. Another application is to the interpretation of pressure dependences of the nuclear quadrupole resonance frequencies observed in several deuterated and undeuterated crystals with hydrogen bonds of different degree of proton transfer [8]. On the basis of the obtained results the correlation between the geometric isotope effect, the NQR-frequency pressure coefficient, and isotope shifts is discussed for various hydrogen bonded crystals with different H-bond length.

2. Experimental.

The nuclear quadrupole resonance frequency measurements were carried out using a pulse spectrometer equipped with an automatic frequency sweep. The hydrostatic pressures were generated by means of a Unipress, three-stage, helium-gas compressor. Pressures were measured to an accuracy of 2 MPa with a manganin gauge located close to the high-pressure vessel. The pressure cell was machined from beryllium-copper alloy and heat treated to a Rockwell hardness of C-39 to C-41. The pressure vessel was suspended in a cryostat, and nitrogen gas was used as the cryogenic medium. The high-pressure electric feedthrough serving to introduce the electric rf wires connecting the measuring coil and spectrometer was made of corundum monocrystal polished to optical accuracy [9]. The rf feedthrough with a 3 mm thick corundum plate as a main sealing element has the best exploitation parameters. The beryllium-copper surfaces of pressure chamber, between which this plate is placed, are also polished to an optical smoothnesss. The tightness of this kind of plug results from the principle of the unsupported area, which plays an essential role in the high pressure technique. The mechanical strength of corundum determines the upper limit of the applied pressure. This kind of the rf feedthrough is especially useful for working with a gas pressure transmitting medium at low temperatures and at very high pulses (there is no electric breakdown) but at frequency lower than 100 MHz.

3. Results and discussion.

3.1. NQR STUDY OF HIGH-PRESSURE DEFORMATION OF THE HYDROGEN BOND POTENTIAL.

Recent development of the ground-state variational wave-function theory of hydrogen bonded crystals allows one to study the quantum aspect of the proton tunneling driven high pressure phenomena and pressure deformation of the hydrogen bond potential. In the previous papers [4,5] we applied the ground-state correlated wave-function theory for interpreting high-pressure induced phenomena in hydrogen bonded materials. This internally consistent model that simultaneously incorporates proton interaction energies and two-state quantum mechanical tunneling is directly applicable to the interpretation of the high-pressure experimental data. With increasing pressure the proton tunneling frequency increases. Such fluctuations tend to destroy the long-range proton ordered state. The competition between proton tunneling and proton-proton interactions may lead to a phase transition of the order-disorder type.

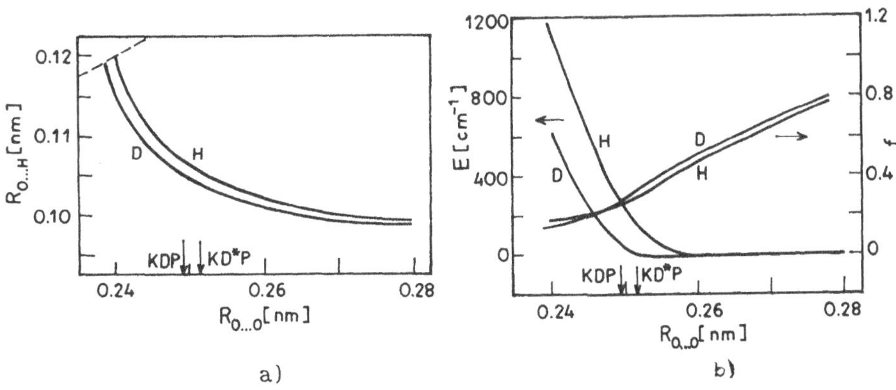

Fig. 1. a) Theoretical dependence of the proton/deuteron-oxygen distance R_{OH} on the hydrogen-bond length $R_{O...O}$ as calculated from the ground-state variational-correlated wave-function theory.
b) Tunnel splitting (in cm^{-1}) and proton interaction energy (in arbitrary units) plotted as a function of oxygen-oxygen separation $R_{O...O}$. The effect of isotope substitution of proton (H) for deuteron (D) is shown.

On the basis of the double Morse potential model for the hydrogen bond [1], we have examined the correlation between oxygen-oxygen distance and oxygen-hydrogen distance in the bond and studied an isotope effect on it. Results of calculations are presented in Fig. 1a . It is assumed that the relationship between R(O...O) and R(OH) is the same whether deduced for systems of differing chemical compounds at one pressure, or for a given system in which changes in bond distances are induced by changes in pressure.

The variation of tunnel splitting and proton-proton interaction energies with oxygen separation (or equivalently pressure) for deuterated and undeuterated hydrogen bonds is shown in Fig. 1b . As seen from this figure, R(O...O) dependences of tunel splitting and proton-proton interaction are not simple. This fact should be reflected on the pressure behavior of both quantities depending on the bond length of the hydrogen bond involved as well as the mass of the tunneling particle.

The evolution of the hydrogen bond potential plotted for some selected pressures is shown in Fig. 2 . With increasing pressure the character of the bistable potential that a proton moves in along the bond is significantly altered. The separation between the two equivalent minima and the barrier separating them are greatly reduced. This leads to an enhancement of the resonance splitting of the proton ground vibrational state and a more rapid tunneling of the proton from one side of the bond to the other. Increased quantum fluctuations lead to the pressure induced order-disorder phase transition in hydrogen-bonded ferroelectrics.

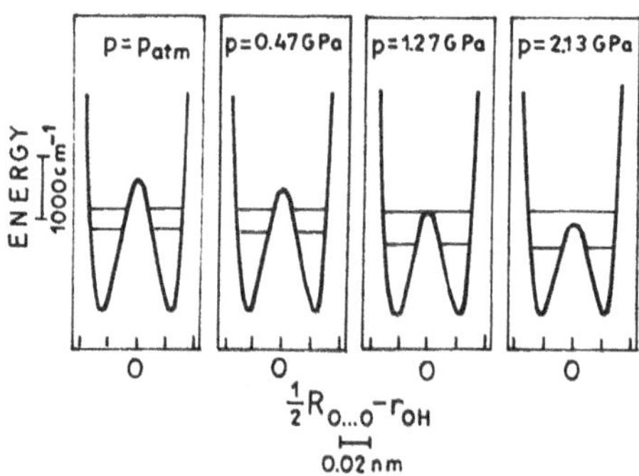

Fig. 2. Pressure dependence of the hydrogen-bond potential in KDP.

The ferroelectric state disappears if the tendency to disorder increases (as a result of an enhanced frequency of tunneling through a narrower and lower barrier) compared to the tendency to order in the dipole field. It is interesting to point out, that even in the so called "quantum paraelectric state", i.e. under pressure higher than 1.71 GPa (when the ferroelectricity in KDP vanishes), the hydrogen-bond potential in KDP is still bistable. The hydrogen bond is expected to symmetrize at much higher pressure. Therefore the pressure induced disappearing of the ferroelectric state is not connected with the hydrogen-bond symmetrization. The pressure scale in Fig. 2 was determined using the value of the single hydrogen-bond compressibility coefficient $\delta R(O...O) / \delta p = -0.0015$ nm/ GPa.

The temperature dependence of the NQR frequency of the As nuclei in KDA under various pressure conditions is shown in Fig.3 . The increase of pressure lowers both the NQR frequency and the phase transition temperature. The decrease of the Curie temperature under the influence of pressure is caused by an increase of the tunneling frequency of the protons which leads to a higher disorder in relation to the ordering influence of dipolar field.

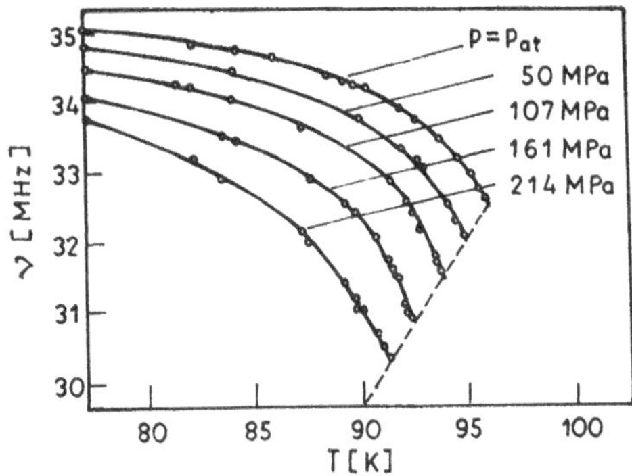

Fig.3. KH_2AsO_4. $\nu(^{75}As)$ vs. T. Ferroelectric phase transition at 96 K at the atmospheric pressure. NQR signal vanishes above the transition point; broken line indicates the location of the phase transition point under applied pressure.

3.2. PRESSURE DEPENDENCE OF THE PROTON TRANSFER EQUILIBRIUM IN HYDROGEN BONDED COMPLEXES AND FERROELECTRICS.

Changes in the electron density distribution in both the donor and acceptor molecules are one of the consequences of hydrogen bond formation. The NQR spectroscopy, being particularly sensitive to subtle changes in electron density, is a valuable tool in studies of hydrogen bonds in solids. A proper selection of proton donor and acceptor enebles one to modify the position of proton inside the bridge which reveals itself as a change in the NQR frequency. Studies by numerous authors [6,7] showed a direct relation between the NQR frequency and the degree of proton transfer in the hydrogen bonds. The literature lacks systematic studies of the high pressure effect on the parameters of the NQR spectrum for hydrogen bonds of different degree of proton transfer. Complexes of pentachlorophenol with nitrogen bases make a model molecular system for such a study and an appropriate selection of the base strength makes it possible to modify properties of the hydrogen bond. Thus, both weak hydrogen bonds of the covalent type

A-H...B as well as strong ionic bonds with transferred protons A...H-B can be studied within one class of compounds (Fig.4). The intermediate range, in which there is the 50 % proton transfer, seems to be particularly interesting since in this range one can expect some critical phenomena.

In this experiment a correlation between a sign and magnitude of the pressure coefficient of the NQR frequency and the degree of proton transfer within the hydrogen bond has been found. The dependence of the pressure coefficient of Cl NQR frequency $\delta\nu/\delta p$ determined for the lowest resonance frequency on pKa of base is shown in Fig. 5 . The pressure coefficient of the NQR frequency is negative for weak hydrogen bonds and changes insignificantly with the increasing pKa. At a pKa about six the critical increase of the absolute value of $\delta\nu/\delta p$ by one order of magnitude is observed. It reaches an extreme value for complexes in which the degree of proton transfer amounts to about 50% . Further increase of pKa gives an increase of the pressure coefficient from large negative to positive values. For complexes of A(-)...H-B(+) type the pressure coefficient has a positive value.

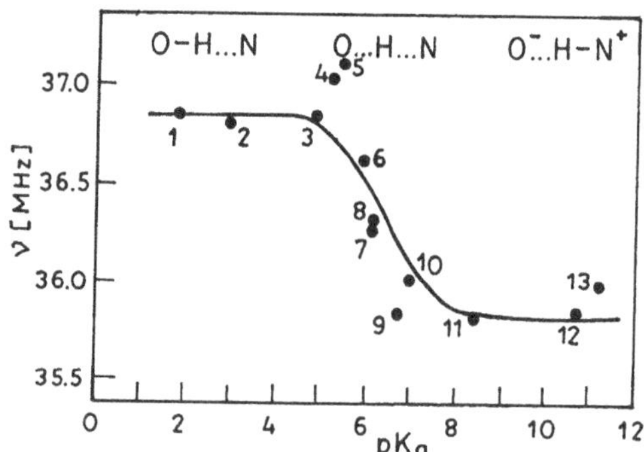

Fig.4. Dependence of ^{35}Cl NQR frequency for complexes of pentachlorophenol on basicity of nitrogen bases (T=77K) : 1 - 4-cyanopyridine, 2 - 3-bromopyridine, 3 - quinoline, 4 - pyridine, 5 - isoquinoline, 6 - 2-methylpyridine, 7 - 4-ethylpyridine, 8 - 4-methylpyridine, 9 - 2,4-dimethylpyridine, 10 - imidazole, 11 - morpholine, 12 - triethylamine, 13 - piperidine.

The proton position along the hydrogen bond is strictly related to its length. The relation between the distance of electronegative atoms and oxygen-proton distance as well as the proton tunneling effect has been theoretically explained by the ground-state variational theory (Fig.1). According to this theory the transition from the long hydrogen bond with a double minimum potential to the short bond with a single minimum potential is very sharp and is of a critical character.

Obviously, the expression "symmetric bond with a single-well potential" should be interpreted dynamically, since a direct differentiation between a single- and double-well potential is impossible when the tunneling frequency is too high.

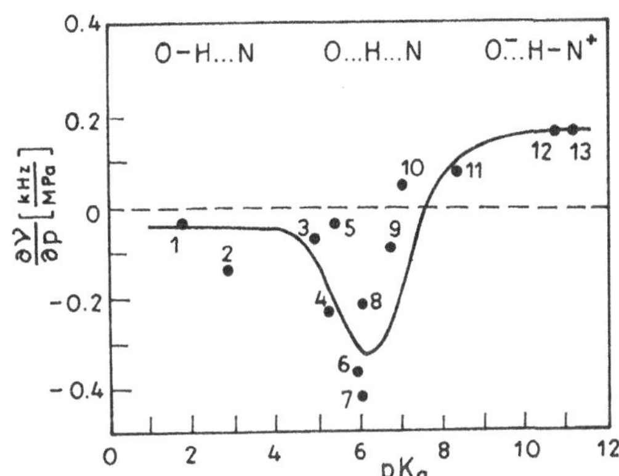

Fig.5. Dependence of pressure coefficient of ^{35}Cl NQR frequency for complexes of pentachlorophenol on pKa of nitrogen bases at 77 K. Notation as in Fig.4.

The different signs of the pressure coefficients of the NQR frequency for both forms of hydrogen bonds, covalent bonds and ionic bonds with proton transfer, may be explained again by the ground-state variational theory. The applied pressure shortens the bond, shifts the position of the proton towards the center of the bond, and therefore causes a change in dipole moment and in the EFG. The pressure induced change of the electric field gradient is proportional to $\delta R_{(O-H)}/\delta R_{(O...N)}$. Thus one can expect a change in the sign of $\delta\nu/\delta p$ going from covalent O-H...N (negative value of $\delta R_{(O-H)}/\delta R_{(O...N)}$; the proton moves away from the oxygen nucleus) to ionic $O^{(-)}...H-N^{(+)}$ bonds (positive value of $\delta R_{(O...H)}/\delta R_{(O...N)}$; the proton moves towards the oxygen nucleus).

The pressure induced variations of the electronic structure of the hydrogen bonded systems were monitored at the opposite ends of the hydrogen bond in the pentachlorophenol-2-chlor-pyridine complex (Fig.6). The existence of such dimer provides unique possibility to observe the variations of the electronic structure of the hydrogen bonded system by Cl NQR spectrum at the opposite ends of the dimer. With increasing pressure the proton moves towards the center of the bond. Therefore, the pressure coefficient of the NQR frequency is negative for the monitoring chlorine nucleus at the one side of the hydrogen bond and positive for the chlorine at the opposite side.

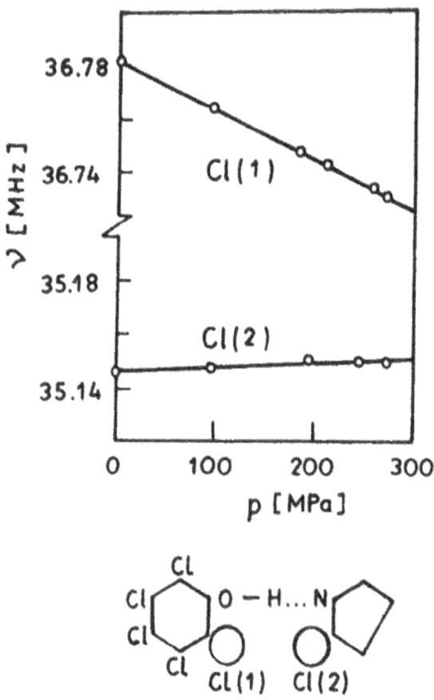

Fig.6. Pressure dependence of the ^{35}Cl NQR frequency in pentachlorophenol-2-chlor-pyridine complex monitored at the opposite ends of the asymmetric hydrogen bond (T=77 K).

Similar effects have been observed in the ferroelectric phase of ammonium hydrogen bis-chloroacetate $NH_4H(ClCH_2COO)_2$ (abbreviated as AHCA). The crystal belongs to an important group of hydrogen-bonded ferroelectrics. The AHCA crystal is composed of ammonium ions and chloroacetate anions in which the two chloroacetate radicals are kept together by a short hydrogen bond (0.2457 nm). In the paraelectric phase (above 120 K) the chloroacetate radicals are crystallographically equivalent and are connected with a hydrogen bond across the center of symmetry. In the ferroelectric phase the dimeric anion loses its center of symmetry and the chlorine sites at the ends of the anion become crystallographically inequivalent. The direction of spontaneous polarization coincides with that of the hydrogen bond direction. The presence of a short asymmetric hydrogen bond and the effect of deuteration (deuteration shifts the transition temperature to 130 K) suggest an order-disorder type of transition. Our NQR pressure studies have shown that the proton motion in hydrogen bridges plays a primary role in the ferroelectricity of AHCA crystal [10,11]. Figure 7 shows the temperature dependence of the NQR frequency of the Cl nuclei in AHCA under isobaric conditions at two selected pressures: p_{at} and p = 200 MPa. An increase in pressure lowers the NQR line splitting and the point of its disappearance is shifted downward with the slope dT_c/dp = -0.0195 deg/MPa. The separation between the NQR lines is proportional to the spontaneous polarization which can be considered as the order parameter.

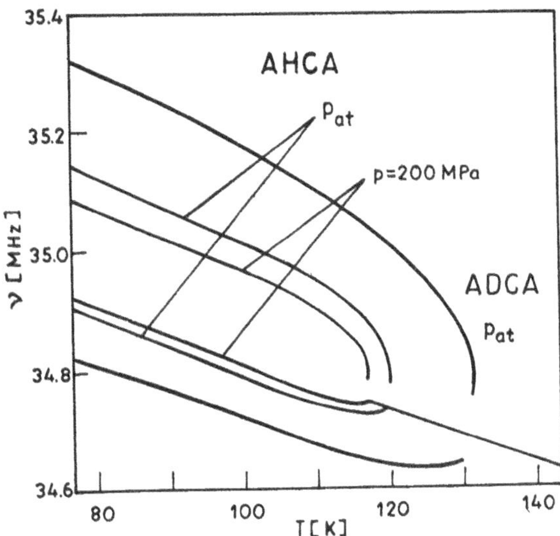

Fig.7. Temperature dependence of the ^{35}Cl NQR frequency in AHCA crystal under isobaric conditions at two selected pressures. The effect of deuteration on the NQR spectral parameters and the phase transition is also shown.

The isotope and pressure effects create a further evidence of the decisive role of the hydrogen bonds in the ferroelectric phase transition in AHCA. On the basis of the pressure dependence of the Cl NQR in AHCA one can suggest that, similar to KDA, the phase transition is initiated by ordering of the proton motion in the hydrogen bonds. It is noteworthy that high pressure acts in a way contrary to that of deuteration. Deuteration shifts T_c upwards and enhances the NQR line splitting (Fig. 7). The high pressure acts in a way to destroy the ferroelectric phase in AHCA. Substitution of deuterium for hydrogen leads to an elongation of hydrogen bonds and changes in the electric field gradient, but this effect is very strongly dependent on the bond length involved.

Hence the results of the present work support the predictions of the ground-state-variational wave function theory of hydrogen bonds and allow one to describe the observed phenomena in an internally consistent way in terms of microscopic parameters of a compressed single hydrogen bond.

Acknowledgement.

This research was partially supported by the Foundation for Polish Science Development under grant PONT No. 1/43/92.

References.

[1] Matsushita, E. and Matsubara, T. (1982) 'Note on isotope effect in hydrogen bonded crystals', Prog. Theor. Phys. 67, 1-18.

[2] Schweizer, K.S. and Stillinger, F.H. (1984) 'Phase transitions induced by proton tunneling in hydrogen-bonded crystals. Ground state theory', Phys. Rev. B29, 350-360.

[3] Stillinger, F.H. and Schweizer, K.S. (1983) 'Ice under pressure: transition to symmetrical hydrogen bonds', J. Phys. Chem. 87, 4281-4288.

[4] Maćkowiak, M. (1987) 'Isotope effect on high-pressure behavior of hydrogen-bonded crystals', Physica 145B, 320-328.

[5] Maćkowiak, M. (1989) 'NQR study of high-pressure deformation of the hydrogen bond potential', J. Mol. Struct. 192, 189-198.

[6] Maćkowiak, M., Kozioł, P. and Stankowski, J. (1986) 'Pressure dependence of the proton transfer equilibrium in hydrogen bonded complexes', Z. Naturforsch. 41a, 225-229.

[7] Kozioł, P., Maćkowiak, M., Stankowski, J. and Jadżyn, J. (1985) 'High pressure NQR studies of hydrogen bonds for complexes of pentachlorophenol with nitrogen bases', J. Mol. Struct. 131, 147-158.

[8] Maćkowiak, M. and Kozioł, P. (1988) 'Effect of pressure on the symmetric hydrogen bond in $(CCl_3COO)_2HK$', phys. stat. sol. 108(a), 739-745.

[9] Maćkowiak, M., Stankowski, J. and Zdanowska, M. (1979) 'NQR study of hindered rotation in trans-1,2-dichloroethane under high pressure', J. Magn. Reson. 33, 41-49.

[10] Zdanowska-Frączek, M. and Lipiński, E. (1983) 'The influence of hydrostatic pressure on ferroelectric $NH_4H(ClCH_2COO)_2$', J. Magn. Reson. 55, 1-11.

[11] Zdanowska-Frączek, M. (1986) 'The proton motion effect in the ferroelectric phase of $NH_4H(ClCH_2COO)_2$', Z. Naturforsch. 41a, 286-289.

HIGH PRESSURE MECHANISTIC STUDIES OF INORGANIC AND ORGANOMETALLIC SYSTEMS

RUDI VAN ELDIK
Institute for Inorganic Chemistry
University of Witten/Herdecke
Stockumer Straße 10
5810 Witten
Germany

ABSTRACT. A general introduction on the application of high pressure techniques in kinetic studies of chemical reactions in solution is presented. Some experimental advances in recent years are reviewed. A systematic treatment of the effect of pressure on typical processes such as solvent and ligand exchange, ligand substitution, electron-transfer, and addition reactions in inorganic and organometallic systems is given. These include thermal-, photochemical- and radiation-induced processes. Reaction volume profiles for such processes are presented where possible and the nature of the reaction pathway is discussed. Finally, some applications of high pressure techniques in synthetic organometallic chemistry are covered.

1. Introduction

Within the context of this NATO ASI it is my intention to demonstrate how high pressure kinetic techniques are and have been used in mechanistic studies of chemical reactions in solution. We have been working in this area for almost 15 years and have concentrated on studies dealing with inorganic, organometallic and bioinorganic systems. Numerous reviews have appeared in recent years, and readers are advised to consult these for further information [1-6]. Similar studies have been performed on organic systems and readers are referred to the contributions by G. Jenner in this volume [3,7].

The fundamental idea behind this work is that when dealing with mechanistic studies in general, it is essential to investigate the effect of as many chemical (concentration, pH, solvent, ionic strength) and physical (temperature, pressure) variables as possible on the observed kinetic behaviour of the reaction under study, in order to obtain as much indirect information as possible on the nature of the underlying reaction mechanism. Only then can the suggested mechanism approach the real one. Almost all chemical reactions exhibit a characteristic pressure dependence [1,3,7] over a moderate pressure range of a few hundred MPa, or a few kbar. This pressure dependence is used to calculate a volume of activation, $\Delta V^{\#}$, via the relationship in (1), according

$$\left(\frac{\partial \ln k}{\partial P}\right)_T = -\frac{\Delta V^{\#}}{RT} \qquad (1)$$

to which a process that is accelerated by pressure exhibits a negative $\Delta V^{\#}$ value, i.e. a drop in

R. Winter and J. Jonas (eds.), High Pressure Chemistry, Biochemistry and Materials Science, 309–328.

volume in going to the transition state, and vice versa for the deceleration of a reaction by pressure. The value of $\Delta V^{\#}$, extrapolated to ambient pressure in the case of a non-linear dependence of lnk on pressure, is combined with partial molar volume data for the reactant and product species, or with the reaction volume of the overall reaction, to construct a volume profile that describes the volume changes that occur along the reaction coordinate. Such volume profiles, of which many will be presented in this report, greatly assist the assignment of the intimate mechanism on the basis of the location of the transition state in reference to that of the reactant and product states. Thus the reaction mechanism is interpreted in terms of specific volume changes along the reaction coordinate, which consist of intrinsic and solvational contributions that result from changes in bond lengths and angles, and changes in solvent electrostriction, respectively. Much of the achieved advances result from the development of instrumentation to study slow and fast reactions, including flow systems and relaxation techniques, at pressures up to 300 MPa. The techniques involve stopped-flow, T-jump, P-jump, NMR, ESR, flash-photolysis and pulse-radiolysis instrumentation [1,4,8,9].

The examples presented in this contribution cover various types of reactions in inorganic (coordination) and organometallic chemistry, viz. ligand exchange and substitution, metal-carbon bond formation, addition and elimination, electron-transfer reactions. Typical examples dealing with reactions in bioinorganic systems are presented in the following contribution.

2. Substitution Reactions

In this section we will focus an various types of thermal substitution reactions including solvent and ligand exchange processes, ligand substitution, complex formation, aquation/solvolysis and hydrolysis reactions. Substitution reactions have been the topic of many mechanistic investigations because of the fundamental importance of such reactions in many chemical, biological and catalytic processes. For a general substitution reaction (2), where X is the leaving group and Y the entering ligand, there are three simple pathways: (i) the dissociative (D) process, with an

$$ML_nX + Y \rightarrow ML_nY + X \tag{2}$$

intermediate of lower coordination number; (ii) the associative (A) process, with an intermediate of higher coordination number; (iii) the interchange (I) process, in which no intermediate of lower or higher coordination number is involved, that can be more associative (I_a) or more dissociative (I_d) in nature depending on whether bond formation or bond breakage, respectively, is more important. In general the volume of activation ($\Delta V^{\#}$) should differ significantly for such processes, and during the last two decades it has become a well established criterion to assess the mechanism [1,3-5].

In the case of symmetrical solvent or ligand exchange reactions, $\Delta V^{\#}$ will be a direct measure of the degree of bond formation/bond breakage in the transition state, and a continuous spectrum of transition states can be envisaged as illustrated in Figure 1 [4,5]. The experimental data reported for such reactions [4,10] clearly demonstrate the sensitivity of $\Delta V^{\#}$ towards the size of the metal center and the coordinated ligands that tune the intimate nature of the exchange process. In this respect the $\Delta V^{\#}$ data for solvent exchange reactions of first row high spin trivalent transition metal ions exhibit a definite trend across the series from more negative for the large to more positive values for the smaller cations, i.e. a changeover from A to I_d or D [5,11]. A similar changeover in mechanism was found for solvent exchange on high spin first row divalent

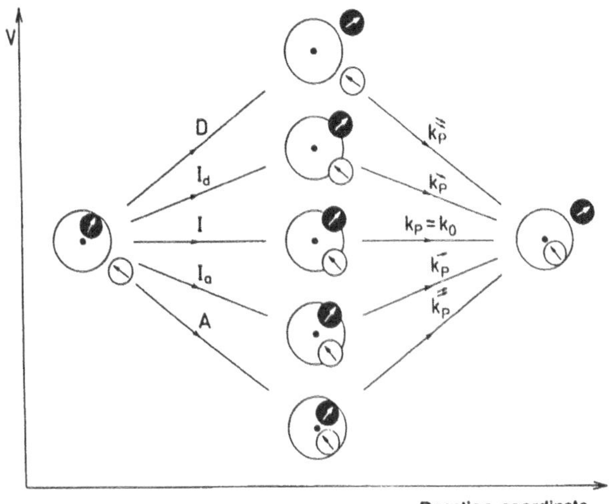

Figure 1. Volume profiles for solvent exchange reactions

transition metal ions [5], for which both the size of the metal ion and the bulkiness of the coordinated solvent molecules play a crucial role. Recently a mechanistic changeover was reported for solvent exchange on tetrasolventoberyllium(II), from associative (A) for the smallest solvent H_2O to dissociative (D) for the bulky solvent tetramethylurea and dimethylpropyleneurea [12]. Such mechanistic changeovers have also attracted the attention of theoreticians in an effort to account for the observed effects [13].

The sensitivity of $\Delta V^{\#}$ data can also be employed to resolve the mechanism of induced solvent exchange reactions as in the case of monohydroxo complexes for which typical data are summarized in Table 1 [4]. The higher reactivity of the hydroxo complexes is accompanied by a more dissociative character for water exchange, i.e. a more positive $\Delta V^{\#}$ [14]. Similar rate enhancements observed for acetonitrile exchange in going from $Ru(\eta^6\text{-}C_6H_6)(CH_3CN)_3^{2+}$ to $Ru(\eta^5\text{-}C_5H_5)(CH_3CN)_3^{+}$ are also accompanied by a significant increase in $\Delta V^{\#}$ and a changeover in mechanism from interchange (I) to dissociative (D), respectively [15]. The introduction of polyaminocarboxylate chelates such as o-phenylenediamine-N,N,N'N'-tetraacetate in the coordination sphere of Fe(III), results in the formation of a seven-coordinate species in which the coordinated solvent molecule (H_2O or DMF) undergoes an enhanced exchange reaction with the bulk solvent and is characterized by positive $\Delta V^{\#}$ values in line with an I_d mechanism [16,17].

In general, the above described trends observed in $\Delta V^{\#}$ data for solvent exchange reactions, also hold for ligand exchange reactions, and in particular for complex formation reactions. The latter reactions are non-symmetrical ligand substitution processes for which $\Delta\overline{V} \neq 0$, and the construction of volume profiles has greatly assisted the elucidation of the substitution mechanism [3,5]. Again a clear mechanistic changeover was observed for complex formation reactions of divalent first-row transition-metal elements from I_a to I_d along the series [10], in agreement with that found for the corresponding solvent exchange reactions. A typical example of a volume profile recently reported for the complex formation of V^{3+} with NCS^- in DMSO is given in

TABLE 1. Rate constants and activation parameters for water exchange on some hexaaqua and monohydroxypentaaqua metal ions

M^{3+}	k^{298} s^{-1}	k_{OH}/k	$\Delta H^{\#}$ kJ mol^{-1}	$\Delta S^{\#}$ J K^{-1} mol^{-1}	$\Delta V^{\#}$ cm^3 mol^{-1}	pK_a	Mechanism
Ga^{3+}	4.0×10^2	275	67.1	+30.1	+5.0	≈3.9	I_d
$Ga(OH)^{2+}$	1.1×10^5		58.9	-	+6.2		I_d
Fe^{3+}	1.6×10^2	750	64.0	+12.1	-5.4	2.9	I_a
$Fe(OH)^{2+}$	1.2×10^5		42.4	+5.3	+7.0		I_d
Cr^{3+}	2.4×10^{-6}	75	108.6	+11.6	-9.6	4.1	I_a
$Cr(OH)^{2+}$	1.8×10^{-4}		110.0	+55.6	+2.7		I
Ru^{3+}	3.5×10^{-6}	170	89.8	-48.2	-8.3	2.7	I_a
$Ru(OH)^{2+}$	5.9×10^{-4}		95.8	+14.9	+0.9		I
Rh^{3+}	2.2×10^{-9}	19100	131.2	+29.3	-4.2	3.5	I_a
$Rh(OH)^{2+}$	4.2×10^{-5}		103.0	-	+1.3		I

Figure 2 [18]. In this case both the solvent exchange and complex formation reactions follow an I_a mechanism as seen from the more compact transition state. Specific labilization effects observed for solvent exchange reactions also show up in complex formation reactions and control

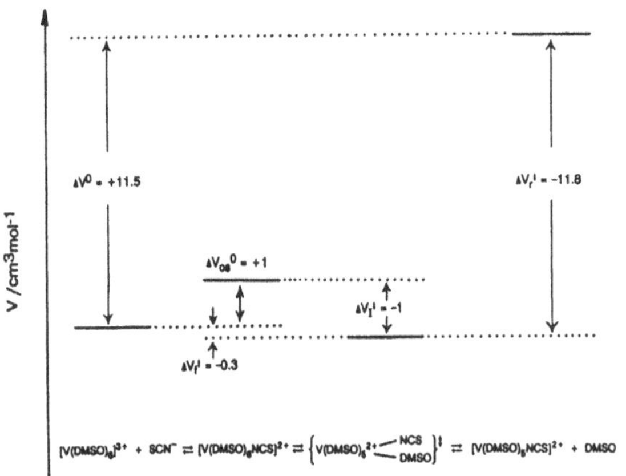

Figure 2. Volume profile for the reaction between $V(DMSO)_6{}^{3+}$ and NCS^-

the nature of the substitution mechanism. For instance, complex formation of Fe^{3+} with desferrioxamine B and acethydroxamic acid follows an I_a mechanism for $Fe(H_2O)_6{}^{3+}$, but an I_d mechanism for $Fe(H_2O)_5OH^{2+}$ [19]. In fact the nature of the complex formation mechanism is strongly controlled by the nature of the other ligands present in the coordination sphere, which can cause a gradual changeover in mechanism during the stepwise complexation of acethydroxamic acid [19]. Thus such measurements coupled to the construction of volume profiles increasingly assist the classification of complex formation reactions [20].

Aquation and solvolysis reactions of octahedral complexes also exhibit very characteristic $\Delta V^{\#}$ values, and the mechanism is once again controlled by the nature of the metal center and the bulkiness of the coordinated ligands [3,5]. For instance, the aquation reactions of series of complexes of the type $M(NH_3)_5L^{3+}$ exhibit very consistent $\Delta V^{\#}$ values that favour the operation of an I_a mechanism for M = Cr(III), compared to an I_d mechanism for M = Co(III) [21]. A similar trend was observed for the solvolysis reactions of complexes of the type $M(NH_3)_5$-$(OSO_2CF_3)^{2+}$ in MeCN and MeOH for which $\Delta V^{\#}$ steadily decreases along the series Co(III) > Rh(III) > Cr(III), in agreement with a gradual changeover from I_d to I_a [22]. There are quite a number of cases reported in the literature where aquation and solvolysis reactions of Fe(II)-diimine complexes are characterized by very large positive $\Delta V^{\#}$ values that favour the operation of a D mechanism [23,24]. Similarly, complex formation and aquation reactions of pentacyano-ferrate(II) also exhibit significantly positive volumes of activation, ca. +18 cm^3mol^{-1}, which strongly support a D mechanism [25-27]. Base hydrolysis reactions of complexes of the type Co-$(NH_3)_5X^{(3-n)+}$ exhibit $\Delta V^{\#}$ values between +19 and +43 cm^3mol^{-1}, for 30 different complexes, depending on the nature of the leaving group X^{n-}, especially its charge [5,28]. A detailed volume profile analysis clearly demonstrated the dissociative nature of such processes which occur via the formation of the conjugate base species. Similar studies on the base hydrolysis of related Cr(III) complexes revealed $\Delta V^{\#}$ values that are significantly smaller than in the case of the Co(III) complexes, which led to the suggestion that the conjugate base species in the case of the Cr(III) complexes may undergo an interchange type of substitution reaction [29,30]. At this point it should be noted that the interpretation of $\Delta V^{\#}$ data may not be that straightforward especially in cases where an overall volume profile cannot be constructed. In addition, the detailed interpretation may require a further refinement of the suggested ligand substitution mechanisms as discussed in more detail elsewhere [31-33].

Ligand substitution reactions of a series of organometallic complexes, mainly metal-carbonyl species, have also been studied using high pressure kinetic techniques [5,6]. For instance, replacement of coordinated CO on $M(CO)_4$phen by $P(OMe)_3$ is characterized by a $\Delta V^{\#}$ value of +13.8 cm^3mol^{-1} for M = Cr, compared to -21 cm^3mol^{-1} for M = Mo, which clearly demonstrates the changeover in mechanism from D to A in going from the smaller to the larger metal center [34]. The corresponding volume profile for the Cr system is given in Figure 3, from which the volume increase in going to the transition state for the forward and reverse reaction can be clearly seen. Such measurements could greatly assist the longstanding uncertainty about the meaning of the two-term rate law usually observed for such reactions, which can be due to parallel or reversible processes [35]. The size of the central metal atom and the steric hindrance on the coordinated chelate also determine the nature of the chelate-ring replacement mechanism of complexes of the type $M(CO)_4(S\text{-}S)$, which could be resolved with the aid of $\Delta V^{\#}$ measurements [36]. In comparison to the data reported in Figure 3, associative substitution reactions of metal carbonyl complexes are characterized by significantly negative $\Delta V^{\#}$ values [37].

Our interest in the mechanistic behaviour of solvento metal carbonyl complexes, due to their potential role in catalytic cycles, led to a series of high pressure studies involving the displacement of coordinated solvent molecules. In most cases such species are short-lived and can only be studied by flash-photolysis. However, in the case of THF as solvent, complexes of the type

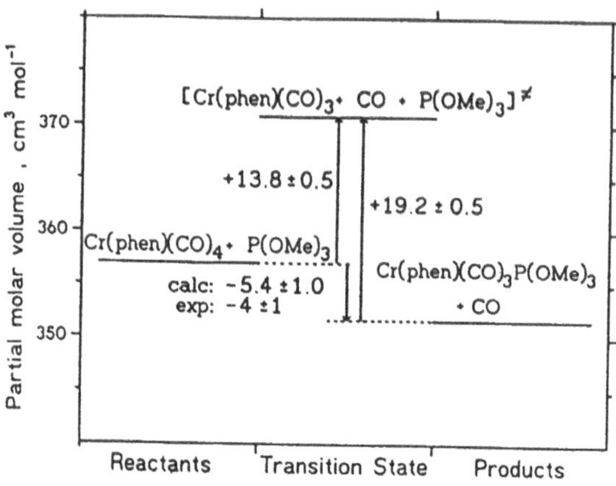

Figure 3. Volume profile for the reaction
$Cr(phen)(CO)_4 + P(OMe)_3 \rightarrow Cr(phen)(CO)_3P(OMe)_3 + CO$

$M(CO)_5THF$ can be prepared in solution, and their substitution behaviour can be studied using stopped-flow techniques [38]. A systematic variation of M and the entering nucleophile resulted in the data summarized in Table 2, from which it follows that $\Delta V^{\#}$ becomes more negative in

TABLE 2. Summary of rate and activation parameters for the reaction

$$M(CO)_5THF + L \xrightarrow{k} M(CO)_5L + THF$$

M	Ligand	k at 25 °C $M^{-1}sec^{-1}$	$\Delta H^{\#}$ kJ mol^{-1}	$\Delta S^{\#}$ J $K^{-1}mol^{-1}$	$\Delta V^{\#}$ cm^3mol^{-1}
Cr	piperidine	1.53	45 ± 2	-90 ± 8	-2.2 ± 0.6
	$P(C_6H_5)_3$	0.36	78 ± 1	+8 ± 8	-1.9 ± 1.0
	$P(OC_2H_5)_3$	0.37	53 ± 1	-74 ± 2	-3.6 ± 0.7
Mo	piperidine	15.2	60 ± 2	-21 ± 8	-3.6 ± 1.2
	$P(C_6H_5)_3$	2.5	65 ± 4	-17 ± 14	-8.3 ± 1.0
	$P(OC_2H_5)_3$	1.7	55 ± 2	-55 ± 8	-5.8 ± 0.9
W	piperidine	0.75	38 ± 6	-122 ± 20	-4.4 ± 0.5
	$P(C_6H_5)_3$	0.071	50 ± 3	-100 ± 11	-12.2 ± 0.4
	$P(OC_2H_5)_3$	0.31	49 ± 3	-89 ± 10	-14.9 ± 1.0

going to the larger metal center (Cr < Mo < W) and larger entering nucleophile. Thus bond formation is more predominant for these systems, and a gradual changeover to a more associative substitution mechanism from Cr to W, as outlined in (3), was suggested [38]. Application of

$$\begin{array}{ccccc} & M(CO)_5THF & \underset{k_{-1}}{\overset{k_1}{\rightleftharpoons}} & M(CO)_5 + THF & \\ k_3 & \downarrow +L, -THF & & \downarrow +L & k_2 \\ \text{"A"} & \longrightarrow & M(CO)_5L & \longleftarrow & \text{"D"} \end{array} \qquad (3)$$

$$k_{obs} = \{k_1k_2/k_{-1} + k_3\}[L]$$

high pressure flash-photolysis techniques have enabled a systematic study of the substitution behaviour of intermediate species of the type $M(CO)_5S$, produced via flash photolysis of $M(CO)_6$ in a coordinating solvent S. The effect of pressure on a series of such reactions [39,40] has resulted in the data summarized in Table 3, from which a similar trend as for the THF complexes can be seen. Similar studies on the displacement of coordinated solvent molecules via ring-closure of P-olefins for cis-$(CO)_4W(S)(PPh_2(CH_2)_nCH=CH_2)$ (n = 1 to 4, S = chlorobenzene)

TABLE 3. Summary of $\Delta V^{\#}$ data for solvent displacement reactions of the type $M(CO)_5S + L \rightarrow M(CO)_5L + S$

S	L	$\Delta V^{\#}$, cm^3mol^{-1}		
		M = Cr	M = Mo	M = W
fluorobenzene	1-hexene	+9.4 ± 0.7	+5.8 ± 0.8	+2.5 ± 0.2
toluene		+10.8 ± 0.7	+3.2 ± 0.3	
benzene		+10.9 ± 1.0		
chlorobenzene		+5.4 ± 0.4	+3.2 ± 0.3	+0.4 ± 0.3
n-heptane		+6.2 ± 0.2	+2.2 ± 0.3	+2.7 ± 0.4
fluorobenzene	piperidine	+6.1 ± 0.3		
toluene		+4.8 ± 1.4		
benzene		+4.2 ± 0.3		
chlorobenzene		+0.2 ± 0.2		
n-heptane		+1.4 ± 0.4		

resulted in $\Delta V^{\#}$ values of +7.7, +5.1, + 10.7 and +10.5 cm^3mol^{-1} for n = 1 to 4, respectively [41]. These values indicate that chelate ring-closure follows an I_d mechanism for n = 1,2 and a D mechanism for n =3,4, the difference being related to the pre-association of the olefin moiety with the metal center in the case of the shorter chainlengths.

When the attacking nucleophile is a bidentate ligand, flash photolysis of $M(CO)_6$ results in the reaction sequence outlined in (4), in which ring-closure (k_2) now involves CO displacement as compared to solvent displacement discussed above. The $\Delta V^{\#}$ data reported for such reactions [42-44] clearly demonstrate that the larger metal centers (Mo and W) tend to ring-close in an associative way and exhibit significantly negative $\Delta V^{\#}$ values. The smaller Cr center must loose CO

$$\mathrm{M(CO)_6} \xrightarrow{h\upsilon} \mathrm{M(CO)_5 + CO}$$

$$\mathrm{M(CO)_5 + S} \xrightarrow{\text{fast}} \mathrm{M(CO)_5S}$$

$$\mathrm{M(CO)_5S + N\text{-}N} \xrightarrow{k_1} \mathrm{M(CO)_5N\text{-}N + S} \tag{4}$$

$$\mathrm{M(CO)_5N\text{-}N} \xrightarrow{k_2} \mathrm{M(CO)_4(N\text{-}N)} + \mathrm{CO}$$

prior to ring-closure, unless there is no bulkiness on the entering ligand that prevents an associative ring-closure reaction.

Ligand substitution reactions of electronically excited organometallic complexes also exhibit very characteristic pressure dependences [45]. For instance, photosubstitution reactions of the type outlined in (5) are all accompanied by significantly positive $\Delta V^{\#}$ values, which support the

$$\mathrm{M(CO)_6 + L} \xrightarrow[313\ \text{nm}]{h\upsilon} \mathrm{M(CO)_5L + CO} \tag{5}$$

(M = Cr, Mo, W; L = piperidine, pyridine, acetonitrile)

operation of a dissociative mechanism [46]. The photosubstitution of CO in $M(CO)_4$phen (M = Cr, Mo, W) was investigated, since the photoactivity of the lower lying MLCT states has been a controversial issue in the literature [47]. It was suggested that excitation of the MLCT state is followed by thermal back population of the higher energy LF state from which a dissociative substitution reaction occurs. On the other hand, it was argued that the MLCT states themselves are photoactive and could undergo substitution in an associative way. The pressure dependence of the photosubstitution quantum yield was measured as a function of irradiation wavelength in order to populate the different excited states. Indeed, the quantum yields measured for LF and MLCT excitation revealed very different pressure dependencies, and the corresponding volumes of activation are summarized in Table 4. The results clearly demonstrate that in the Mo and W complexes, MLCT and LF photosubstitution occur according to associative and dissociative mechanisms, respectively. For the smaller Cr complex, the associative MLCT path does not seem to be possible and even this reaction has to follow a dissociatively activated process, presumably of the interchange (I_d) type. This example nicely demonstrates the value of pressure as a key kinetic parameter to distinguish between associative and dissociative photosubstitution mechanisms. High pressure techniques were recently also applied to the study of excited state processes [48,49], the photoreactions of $(\eta^5\text{-}C_5H_5)Fe(CO)_2(COCH_3)$ [50], and the MLCT photochemistry of $(CO)_5ReMn(CO)_3$(α-diimine) complexes [51].

TABLE 4. Quantum yields and volumes of activation for the reaction in toluene

$$M(CO)_4phen + PEt_3 \xrightarrow{h\upsilon} M(CO)_3(PEt_3)phen + CO$$

M	λ_{irr} nm	$\Delta V^{\#}$(MLCT) cm^3mol^{-1}	$\Delta V^{\#}$(LF)
Cr	366		+9.6 ± 1.6
	546	+2.7 ± 0.3	
Mo	366		+6.0 ± 0.2
	546	-13.3 ± 1.2	
W	366		+8.2 ± 0.5
	54	-12.0 ± 0.7	

3. Formation and Cleavage of Metal-Carbon σ Bonds

The mechanisms of inorganic and organometallic free radical reactions, such as the formation of metal-carbon bonds, can conveniently be studied by using pulse-radiolysis techniques. Such processes are closely related to substitution reactions, although they usually involve a formal change in the oxidation state of the metal center due to the covalent nature of the metal-carbon σ bond. In order to investigate the pressure dependence of such processes it was necessary to design special windows for the high pressure cell that allows the penetration of 2 MeV and 5 MeV electrons [52]. In the first study, the reaction volume for reaction (6) was found to be -19.7 cm^3mol^{-1}, whereas $\Delta V^{\#}$ for the forward and reverse reactions equalled +4.7 and +24.4 cm^3mol^1, respectively [53]. These values indicate that metal-carbon bond formation is signifi-

$$Ni^{II}(cycalm) + \cdot CH_3 + H_2O \rightleftharpoons Ni^{III}(cyclam)(CH_3)H_2O^{2+} \quad (6)$$

cantly assisted by pressure since the reverse homolysis reaction is strongly inhibited by pressure. In a similar way it could be shown that the formation of a cobalt-carbon σ bond is assisted by pressure in the reaction of $Co^{II}(nta)(H_2O)_2^-$ with $\cdot CH_3$ to produce $Co^{III}(nta)(H_2O)(CH_3)^-$, where nta = nitrilotriacetate [54]. This is due to an overall reaction volume of -16.4 ± 1.6 cm^3mol^{-1} (determined from the pressure dependence of the equilibrium constant), which results from a small positive volume of activation for the forward bond formation reaction and a large positive volume of activation for the reverse homolysis reaction. The volume profile in Figure 4 clearly indicates the significantly higher partial molar volume of the transition state, which is interpreted in terms of an I_d mechanism, i.e. the reaction is controlled by solvent exchange on Co(nta)-$(H_2O)_2^-$. The large volume collapse following the transition state is ascribed to metal-carbon bond formation accompanied by oxidation of Co^{II} to Co^{III}.

The above example demonstrates that metal-carbon bond formation can reveal mechanistic information on solvent exchange processes. For this reason we studied the effect of pressure on the formation of a series of organo-chromium(III) species, in an effort to obtain mechanistic information on solvent exchange and complex formation reactions of $Cr(H_2O)_6^{2+}$ [55]. This reaction was studied for 10 different aliphatic radicals $\cdot$R, which all exhibited very similar bond

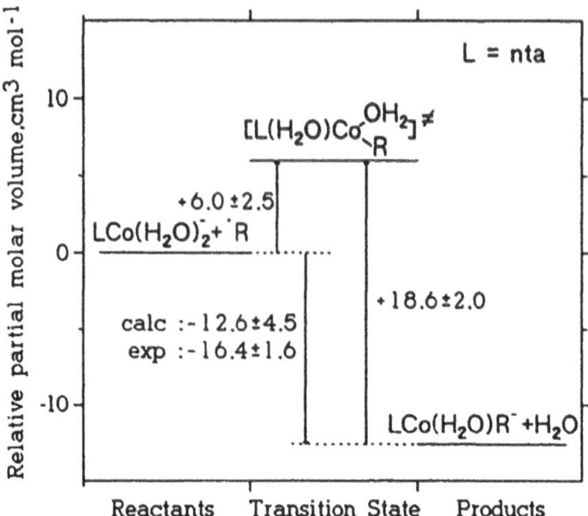

Figure 4. Volume profile for the reaction
$Co(nta)(H_2O)_2^- + \cdot CH_3 \rightarrow Co(nta)(H_2O)(CH_3)^- + H_2O$

formation rate constants and small positive volumes of activation with an average value of +4.3 ± 1.0 cm^3mol^{-1}, independent of the nature of R. This constant value and the fact that the bond formation rate constants are all very similar and close to that for water exchange on $Cr(H_2O)_6^{2+}$,

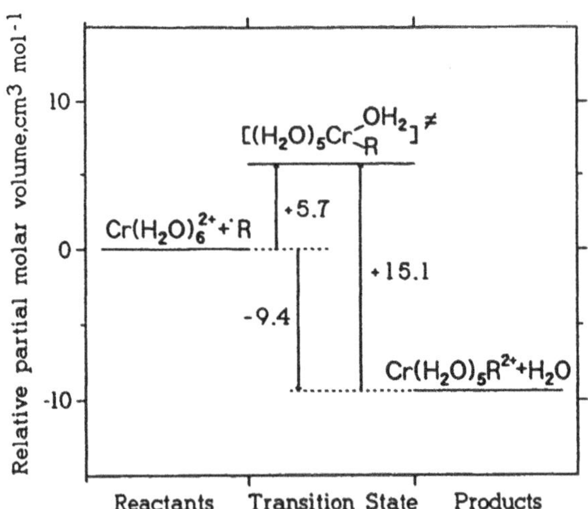

Figure 5. Volume profile for the reaction
$Cr(H_2O)_6^{2+} + \cdot C(CH_3)_2OH \rightarrow Cr(H_2O)_5C(CH_3)_2OH^{2+} + H_2O$

suggested that metal-carbon bond formation is controlled by solvent exchange on $Cr(H_2O)_6^{2+}$, which follows an I_d mechanism. This process is presumably enhanced by Jahn-Teller distortion in $Cr(H_2O)_6^{2+}$, similar to that for $Cu(H_2O)_6^{2+}$ [56]. Combining these results with that of an earlier homolysis study, resulted in the volume profile given in Figure 5, which clearly demonstrates the dissociative nature of the complex formation process. In addition, the large volume collapse following the transition state must be due to the intrinsic volume decrease due to Cr-R bond formation and the contraction due to Cr^{II}-R → Cr^{III}-R^-.

High pressure kinetic techniques have also been applied to the spontaneous and acetate-catalyzed heterolysis reactions of Cr-R^{2+} species [57]. The reported $\Delta V^{\#}$ data clearly indicate that coordinated acetate causes a trans-labilization which results in a dissociative substitution of R by the solvent. Similar effects have been observed for a series of anionic species [58]. Balt and co-workers [59-61] have studied the effect of pressure on the homolytic cleavage of the cobalt-carbon bond in a series of alkyl cobalamins. These reactions are in general characterized by large positive volumes of activation.

4. Electron-Transfer Reactions

It has been reported in recent years [1,3,5,62] that many electron-transfer reactions of inorganic and organometallic systems exhibit characteristic $\Delta V^{\#}$ values. In the case of symmetrical outer-sphere reactions, the $\Delta V^{\#}$ data can be accounted for on the basis of the Marcus-Hush-Stranks theoretical treatment, for which significant contributions have come from Swaddle and co-workers [63-66]. These calculations revealed information on the nature of the transition state for the electron transfer process in terms of its adiabatic or non-adiabatic character, and in general showed a good agreement with the experimental values. A typical example is the self exchange reactions of $Fe(H_2O)_6^{3+/2+}$ and $Fe(H_2O)_5OH^{2+}/Fe(H_2O)_6^{2+}$ for which $\Delta V^{\#}$ has the values -11.1 and +0.8 cm^3mol^{-1}, respectively [64]. These values can be accounted for essentially quantitatively on the basis of an adiabatic outer-sphere mechanism for the former and a hydroxide-bridged inner-sphere mechanism for the latter reaction. Wherland and co-workers [67-69] have studied the effect of pressure on a series of outer-sphere electron-transfer reactions involving a series of cobalt clathrochelates and cyclopentadienide complexes of $Ru^{II/IV}$ and $Os^{II/IV}$.

In our own work we investigated a series of non-symmetrical outer-sphere reactions between complexes of the type $Co(NH_3)_4(NH_2R)X^{(3-n)+}$ and $Fe(CN)_6^{4-}$ [70,71]. The reported $\Delta V^{\#}$ values for electron-transfer within the precursor ion-pair intermediate are: +26.5 (R = H, X^{n-} = H_2O), +29.8 (R = H, X^{n-} = py), +34.4 (R = H, X^{n-} = DMSO), +18.8 (R = H, X^{n-} = N_3^-), +25.9 (R = H, X^{n-} = Cl^-), +25.1 (R = CH_3, X^{n-} = Cl^-), +31.3 (R = i-C_4H_9, X^{n-} = Cl^-) cm^3mol^{-1}. Theoretical calculations indicate that the transition state lies approximately halfway between the reactant and product states on a volume basis along the reaction coordinate. These reactions are accompanied by a substantial decrease in electrostriction due to charge neutralization, Co^{III} → Co^{II} and $Fe(CN)_6^{4-}$ → $Fe(CN)_6^{3-}$, such that the observed $\Delta V^{\#}$ can largely be accounted for in this way. More recently we used high-pressure pulse-radiolysis techniques to determine $\Delta V^{\#}$ for the oxidation of phenanthroline-copper(I) by molecular oxygen [72], and stopped-flow techniques to determine $\Delta V^{\#}$ for the oxidation of benzenediols by $Fe(CN)_4bpy^-$ and $IrCl_6^{2-}$ [73,74], and for the reduction of a hexaamminecobalt(III) complex by V(II) [75].

5. Addition, Insertion and Elimination Reactions

In general oxidative addition reactions are characterized by large negative $\Delta V^{\#}$ values, partially due to bond formation and partially due to charge creation, i.e. increasing electrostriction, in the transition state [1,76,77]. Similarly, cyclo-addition reactions of $Fe(CO)_3HTE$ (HTE = heptatrienone) with tetracyanoethane are characterized by volumes of activation between -29 and -33 cm^3mol^{-1} [78].

A recent study of [2+2] cyclo-addition reactions (7) on the coordinated ligand of pentacarbo-

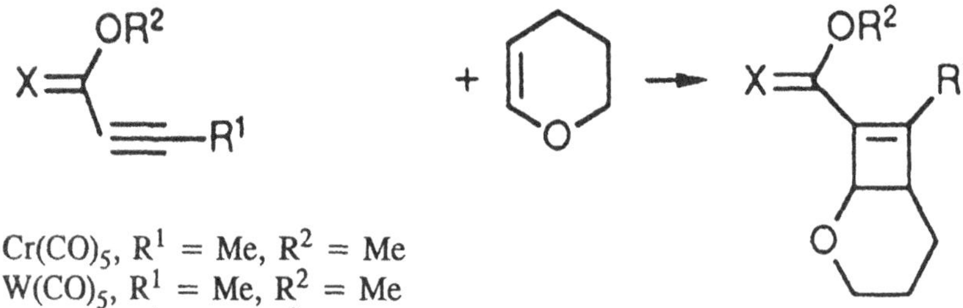

X = $Cr(CO)_5$, R^1 = Me, R^2 = Me
X = $W(CO)_5$, R^1 = Me, R^2 = Me
X = $W(CO)_5$, R^1 = Ph, R^2 = Et

(7)

nyl carbene complexes of Cr and W, revealed a significant acceleration by pressure (average $\Delta V^{\#}$ of -16 $\pm$ 1 cm^3mol^{-1}) and almost no dependence on the polarity of the solvent [79].

These data suggest that the reaction follows a nonpolar concerted, synchronous one-step mechanism. The observed pressure acceleration is very similar to that found for the insertion of dipropylcyanamide and 1-(diethylamino)propyne into the metal-carbene bond of pentacarbonyl(methoxyphenylcarbene)chromium and -tungsten, for which $\Delta V^{\#}$ varies between -17 and -25 cm^3mol^{-1} [80].

Addition reactions of α,β-unsaturated Fischer carbene complexes (8) exhibit $\Delta V^{\#}$ values

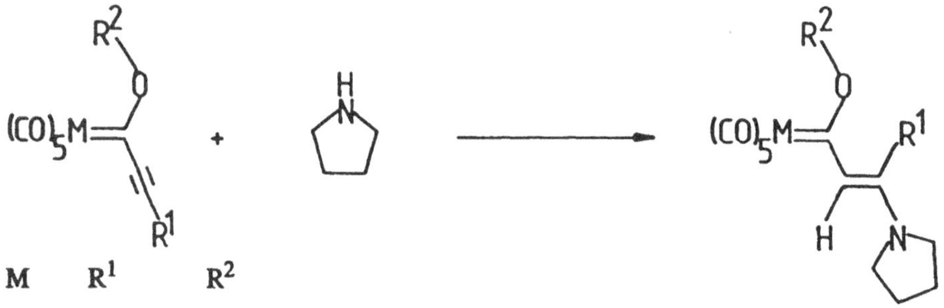

(8)

M	R^1	R^2
Cr	Ph	Et
Mo	Ph	Et
W	Ph	Et
W	Me	Me

between -15 and -17 cm^3mol^{-1}, and are characterized by extremely low $\Delta V^{\#}$ and very negative $\Delta S^{\#}$ values [81]. A systematic solvent dependence study revealed a 50 fold decrease in the addition rate constant in going from acetonitrile to n-heptane as solvent, accompanied by a significant decrease in $\Delta V^{\#}$ (to more negative values).

These data exhibit a good correlation with the solvent parameter q_P (the pressure derivative of q) as shown in Figure 6, from which it follows that the intercept ($\Delta V^{\#}_{intr}$) has a value of approx. -14 cm^3mol^{-1}.

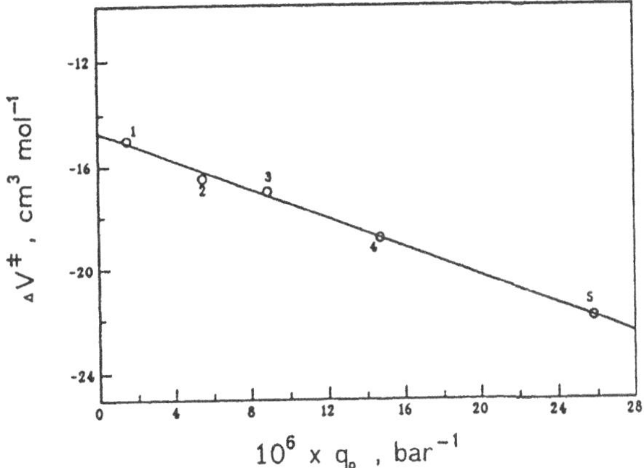

Figure 6. Plot of $\Delta V^{\#}$ versus q_P for reaction (8) with M = W, R^1 = Ph and R^2 = Et. Solvents: acetonitrile (1), 1,2-dichlorobenzene (2), chlorobenzene (3), benzene (4), n-heptane (5).

The absence of a meaningful kinetic isotope effect and the above reported results indicate that the addition reaction (8) follows a two step process with a polar transition state leading to a zwitterionic intermediate.

A similar study for the addition of a series of p-substituted anilines to a Fischer carbene complex as shown in (9), resulted in the data summarized in Table 5 [82]. It can be seen from the data that $\Delta V^{\#}$ is significantly more negative than that reported above for the corresponding reaction with pyrrolidine. Furthermore, the second-order rate constants increase with increasing basicity of the amine (expressed as pK_a value in H_2O as solvent), which is accompanied by a

$(CO)_5Cr{=}$ ... + p-$X{-}C_6H_4{-}NH_2$ $\xrightarrow{MeCN}$ ... (9)

X = CN, CH_3CO, Cl, F, H, CH_3, CH_3O

TABLE 5. Summary of rate and activation parameters for the addition of p-substituted anilines to Fischer carbenes

p-substituent	$pK_a(H_2O)$	k x 10^2 $M^{-1}s^{-1}$	$\Delta H^{\#}$ kJ mol^{-1}	$\Delta S^{\#}$ J $K^{-1}mol^{-1}$	$\Delta V^{\#}$ cm^3mol^{-1}
CN	1.74	0.69	29 ± 3	-190 ± 9	-27.9 ± 0.6
CH_3CO	2.75	2.8	29 ± 2	-179 ± 8	-26.6 ± 0.5
Cl	3.99	21	26 ± 2	-169 ± 6	-24.5 ± 0.4
H	4.63	73	24 ± 1	-167 ± 5	-22.2 ± 0.8
F	4.65	65	23 ± 1	-172 ± 4	-24.6 ± 0.9
CH_3	5.08	167	24 ± 2	-159 ± 8	-21.1 ± 1.0
CH_3O	5.36	497	18 ± 2	-170 ± 6	-21.1 ± 1.0

significant decrease in $\Delta H^{\#}$ and an increase in $\Delta V^{\#}$ (to more positive values) with increasing basicity. These trends clearly demonstrate that the slower reactions occur via a late transition state for which $\Delta V^{\#}$ will be close to the reaction volume $\Delta\overline{V}$, whereas the faster reactions will occur via an early transition state, i.e. less negative $\Delta V^{\#}$. These trends as well as the absence of a significant kinetic isotope effect suggest that a similar mechanism as suggested for the addition of pyrrolidine is operative.

The effect of pressure on the addition of a series of substituted pyridines to tricarbonyl(1,5-η-dienyl)iron(II) (dienyl = C_6H_7, 2-MeOC_6H_6 and C_7H_9) was investigated in an effort to obtain further mechanistic insight into such addition reactions, a typical example is given in reaction (10) [83-85]. The reactions exhibited no significant pressure dependence for the addition and

$$[(CH_2)_n\text{-dienyl-}Fe(CO)_3]BF_4 + N\text{-pyridine-}Et \longrightarrow [Et\text{-pyridine-}N\text{-}(CH_2)_n\text{-dienyl-}Fe(CO)_3]BF_4 \qquad (10)$$

n	A
1	H
1	OMe
2	H

dissociation of 4-cyanopyridine [83], but revealed significantly negative and positive $\Delta V^{\#}$ values for these reactions with 4-formylpyridine, respectively [85]. In the case of 4-ethylpyridine, the addition reaction is characterized by a slightly negative $\Delta V^{\#}$ value [84]. These results demonstrate that amine basicity controls the location of the transition state along the reaction coordinate during the addition reaction. The mechanism can therefore vary from a pure interchange of bonds on the coordinated dienyl ligand (i.e. an almost zero $\Delta V^{\#}$) to a highly associative bond formation process (negative $\Delta V^{\#}$), depending on the basicity of the attacking nucleophile.

By way of comparison, elimination and reductive elimination reactions are expected to show significantly positive volumes of activation. For instance, reductive elimination of H_2 from $H_3Ru_3(\mu_3\text{-COMe})(CO)_9$ is characterized by a $\Delta V^{\#}$ of +20 ± 2 cm^3mol^{-1} [86]. The reverse hydrogenation of $HRu_3(\mu\text{-COMe})(CO)_{10}$ exhibits a $\Delta V^{\#}$ of +9.6 ± 0.6 cm^3mol^{-1}, which is consistent with a CO dissociation mechanism. Hydride migration from bridging to terminal coordination modes can result in a significant volume increase as reflected by the $\Delta V^{\#}$ value of $+4.1 \pm 0.3$ cm^3mol^{-1} recently reported for this process in $(\mu\text{-H})_2Ru_3(\mu_3\text{-CHCO}_2Me)(CO)_9$ [87].

6. Synthetic Applications

The characteristic pressure dependencies of the typical reactions discussed in the previous sections create the possibility to employ the acceleration or deceleration by pressure in synthetic applications either in a direct way by controlling the product distribution, or in an indirect way by tuning the reactivity of the catalytic center in the case of homogeneous catalysis involving transition metal complexes. This has been done for many organic systems, and to some extent also for organometallic systems in recent years [3,7,88]. In general, a $\Delta V^{\#}$ value of -10 cm^3mol^{-1} corresponds to an increase in rate constant by a factor of two at a moderate pressure of 100 MPa (i.e. 1 kbar). Some reactions show significantly more negative $\Delta V^{\#}$ values, such that much larger accelerations, especially at higher pressures (up to 10 kbar), can be expected. Furthermore, in cases where parallel reaction paths exist that exhibit significantly different $\Delta V^{\#}$ values, it is in principle possible to tune the product distribution.

In the case of addition and insertion reactions to coordinated ligands of metal carbene complexes discussed in the previous section, these reactions can all be accelerated by pressure. Various techniques can be employed to remove the metal carbonyl fragment in order to obtain the organic product, which can easily be converted into an ether, aldehyde or ester, depending on the method employed. Alternatively, the application of an excess CO at a moderate pressure will convert the metal carbonyl fragment to the $M(CO)_6$ species and produce the pure organic fragment. In this way the metal carbonyl fragment will be available to reinitiate the overall insertion/addition process.

High pressure techniques could be employed to prepare [4+2] cyclo-adducts via the Diels-Alder reaction between Danishefsky's diene and butyl glyoxylate as shown in reaction (11), for which some typical results are summarized in Table 6 [89]. In the case of heat sensitive organotin compounds, good yields of cyclo-addition products could be obtained under high pressure conditions, see reaction (12), whereas reactions at ambient pressure and high temperature resulted in low yields and reverse diastereoselectivity due to isomerization of the reactant dienophiles [90].

Application of pressure also had a significant affect on various substitution and rearrangement reactions, viz. the silylation of tertiary alcohols [91] and the synthesis of α-silylated esters and lactones [92].

Me_3SiO … OMe + H–C(=O)–CO_2Bu^n ⟶ Me_3SiO … CO_2Bu^n … O … OMe (cis) + Me_3SiO … CO_2Bu^n … O … OMe (trans) (11)

TABLE 6. Effect of pressure on [4+2] cyclo-addition reactions of dienes to butyl glyoxylate

Diene 3-R	Reaction conditions	Yield	Isomer ratio cis:trans
$OSiMe_3$	C_6H_6, 80 °C, 0.1 MPa, 20 h	40 %	1 : 1
$OSiMe_3$	Et_2O, 25 °C, 1000 MPa, 24 h	80%	5 : 1
$OSiMe_2Bu^t$	C_6H_6, 80 °C, 0.1 MPa, 15 h	30 %	4 : 1
$OSiMe_2Bu^t$	Et_2O, 25 °C, 1000 MPa, 24 h	85 %	10 : 1
$OSiMe_2Bu^t$	Et_2O, 25 °C, 0.1 MPa, 1 % $Eu(fod)_3$, 48 h	75 %	7 : 3

(12)

R^1=H, R^2=R^3=Me	Z:E=70:30	0.1 MPa, 180 °C, 30 %	cis: trans = 19 : 81
	Z:E=80:20	2300 MPa, 70°C, 80 %	cis: trans = 90 : 10
R^1=OMe, R^2=R^3=H	Z:E=75:25	0.1 MPa, 180 °C, 42 %	cis: trans = 22 : 78
	Z:E=75:25	2000 MPa, 50 °C, 80 %	cis: trans = 70 : 30

7. Conclusions

The results discussed in this contribution for the different types of reactions of inorganic and organometallic systems have clearly demonstrated that the additional physical parameter pressure can add a decisive dimension to mechanistic studies of such systems. The possibility to correlate the volume of activation with partial molar volumes of the reactant and product species, or with the volume of activation for the reverse reaction and the overall reaction volume, has led to the construction of reaction volume profiles that enable us to visualize the chemical process in terms of volume changes along the reaction coordinate. With the data presently available, intrinsic volume changes are in general well understood and can be interpreted with confidence. This is, however, not the case for volume changes arising from solvational effects for which significantly less data are available and the interpretation is more speculative. The quoted examples have also clearly demonstrated that $\Delta V^{\#}$ data can reveal unique information on the mechanism of chemical reactions that are induced by flash photolysis or pulse radiolysis.

It is important to realize that the presented interpretation of the $\Delta V^{\#}$ data is based on a simplified version of the transition state theory, which has its limitations and restrictions, and various modifications are being considered. These include the introduction of stochastic and transport models to account for the back-flux over the activated complex barrier in dense media.

The fact that the rate-determining step of a particular process exhibits a characteristic pressure dependence, creates the possibility to tune the reactivity of particular reactions via the application of moderate pressures up to 1000 MPa (i.e. 10 kbar). This can lead to the selective synthesis of particular reaction products in cases where the product distribution proves to be pressure dependent. Some of the examples presented here exhibit a large pressure sensitivity, meaning that even the application of moderate pressures could be extremely appropriate in synthetic processes. In addition, $\Delta V^{\#}$ data and the revealed mechanistic information should be of predictive value in the optimization of industrial processes. In this way more moderate experimental conditions could be reached, and products could be synthesized that are not possible at ambient pressure.

Acknowledgements

The author gratefully acknowledges the very stimulating collaboration with numerous graduate students, post docs and visiting scientists, their names appear on the work cited from this laboratory. Financial support from the Deutsche Forschungsgemeinschaft, Volkswagen-Stiftung, Fonds der Chemischen Industrie, Max-Buchner-Forschungsstiftung, NATO Scientific Affairs Division and the German-Israeli Foundation is greatly appreciated. The work discussed in this contribution was partly performed in collaboration with G.R. Dobson (Denton, USA), P.C. Ford (Santa Barbara, USA), J.B. Keister (Buffalo, USA), W.D. Wulff (Chicago, USA), J.G. Leipoldt (Bloemfontein, SA), D. Meyerstein/H. Cohen (Beer-Sheva, Israel), D.J. Stufkens (Amsterdam, NL), and H. Fischer (Konstanz, FRG).

8. References

[1]. van Eldik, R. (Ed.) (1986), Inorganic High Pressure Chemistry: Kinetics and Mechanisms, Elsevier, Amsterdam.
[2]. van Eldik, R. and Jonas, J. (Eds.) (1987), High Pressure Chemistry and Biochemistry, D. Reidel, Dordrecht, Series C 197.
[3]. van Eldik, R., Asano, T. and le Noble, W.J. (1989), Chem. Rev., 89, 549.
[4]. Akitt, J.W. and Merbach, A.E. (1990), NMR Basic Principles and Progress, 24, 189.
[5]. van Eldik, R. and Merbach, A.E. (1992), Comments Inorg. Chem., 12, 341.
[6]. van Eldik, R. (1992), in Perspectives in Coordination Chemistry, Williams, A.F., Floriani, C. and Merbach. A.E. (Eds.), VHCA Basel, VCH Weinheim, 55.
[7]. le Noble, W.J. (Ed.) (1989), Organic High Pressure Chemistry, Elsevier, Amsterdam.
[8]. Kotowski, M. and van Eldik, R. (1989), Coord. Chem. Rev., 93, 19.
[9]. Wishart, J.F. and van Eldik, R. (1992), Rev. Sci. Instrum., 63, 3224.
[10]. Merbach, A.E. (1987), Pure Appl. Chem., 59, 161.
[11]. Dellavia, I., Helm, L. and Merbach, A.E. (1992), Inorg. Chem., 31, 2230.
[12]. Pittet, P.-A., Elbaze, G., Helm, L. and Merbach, A.E. (1990), Inorg. Chem., 29, 1936.
[13]. Kang, S.-K., Lam, B., Albright, T.A. and O'Brien, J.F. (1991), New J. Chem., 15, 757.
[14]. Laurenczy, G., Rapaport, I., Zbinden, D. and Merbach, A.E. (1991), Magn. Res. Chem., 29, 545.
[15]. Luginbühl, W., Zbinden, P., Pittet, P.A., Armbruster, T., Bürgi, H.-B., Merbach, A.E. and Ludi, A. (1991), Inorg. Chem., 30, 2350.

[16]. Mizuno, M., Funahashi, S., Nakasuka, N. and Tanaka, M. (1991), Inorg. Chem., 30, 1550.
[17]. Mizuno, M., Funahashi, S., Nakasuka, N. and Tanaka, M. (1991), Bull. Chem. Soc. Jpn., 64, 1988.
[18]. Dellavia, I., Sauvageat, P.-Y., Helm, L., Ducommun. Y. and Merbach, A.E. (1992), Inorg. Chem., 31, 792.
[19]. Birus, M. and van Eldik, R. (1991), Inorg. Chem., 30, 4559.
[20]. Laurenczy, G., Bugnon, P. and Merbach, A.E. (1992), Inorg. Chim. Acta, 198-200, 159.
[21]. Curtis, N.J., Lawrance, G.A. and van Eldik, R. (1989), Inorg. Chem., 28, 329.
[22]. Suvachittanont, S. and van Eldik, R. (1989), Inorg. Chem., 28, 3660.
[23]. Burgess, J., Galema, S.A. and Hubbard, C.D. (1991), Polyhedron, 10, 703.
[24]. Blandamer, M.J., Burgess, J., Fawcett, J., Guardado, P., Hubbard, C.D., Nuttall, S., Prouse, L.J.S., Radulovic, S. and Russell, D.R. (1992), Inorg. Chem., 31, 1383.
[25]. Bal Reddy, K. and van Eldik, R. (1991), Inorg. Chem., 30, 596.
[26]. Alsheri, S. and Burgess, J. (1991), Inorg. Chim. Acta, 181, 153.
[27]. Stochel, G., Chatlas, J., Martinez, P. and van Eldik, R. (1992), Inorg. Chem. in press.
[28]. Kitamura, Y., Lawrance, G.A. and van Eldik, R. (1989), Inorg. Chem., 28, 333.
[29]. Guardado, P., Lawrance, G.A. and van Eldik, R. (1989), Inorg. Chem., 28, 976.
[30]. House, D.A., Bal Reddy, K. and van Eldik, R. (1991), Inorg. Chim. Acta, 186, 5.
[31]. Lay, P.A. (1991), Coord. Chem. Rev., 110, 213.
[32]. Lay, P.A. (1991), Comments Inorg. Chem., 11, 235.
[33]. Swaddle, T.W. (1991), Comments Inorg. Chem., 12, 237.
[34]. Schneider, K.J. and van Eldik, R. (1990), Organometallics, 9, 1235.
[35]. Schneider, K.J. and van Eldik, R. (1990), Organometallics, 9, 92.
[36]. Awad, H.H., Dobson, C.B., Dobson, G.R., Leipoldt, J.G., Schneider, K., van Eldik, R. and Wood, H.E. (1989), Inorg. Chem. 28, 1654.
[37]. Vest, P., Anhaus, J., Bajaj, H.C. and van Eldik, R. (1991), Organometallics, 10, 818.
[38]. Wieland, S., van Eldik, R. (1991), Organometallics, 10, 3110.
[39]. Zhang, S., Dobson, G.R., Bajaj, H.C., Zang, V. and van Eldik, R. (1990), Inorg. Chem., 29, 3477.
[40]. Zhang, S., Zang, V., Bajaj, H.C., Dobson, G.R. and van Eldik, R. (1990), J. Organomet. Chem., 397, 279.
[41]. Zang, V., Zhang, S., Dobson, C.B., Dobson, G.R. and van Eldik, R. (1992), Organometallics, 11, 1154.
[42]. Bal Reddy, K. and van Eldik, R. (1990), Organometallics, 9, 1418.
[43]. Zhang, S., Zang, V. Dobson, G.R. and van Eldik, R. (1991), Inorg. Chem., 30, 355.
[44]. Bal Reddy, K., Hoffmann, R., Konya, G., van Eldik, R. and Eyring, E.M. (1992), Organometallics, 11, 2319.
[45]. Wieland, S. and van Eldik, R. (1990), Coord. Chem. Rev., 97, 155.
[46]. Wieland, S. and van Eldik, R. (1990), J. Phys. Chem., 94, 5865.
[47]. Wieland, S., Bal Reddy, K. and van Eldik, R. (1990), Organometallics, 9, 1802.
[48]. Ford, P.C. and Crane, D.R. (1991), Coord. Chem. Rev., 11, 153.
[49]. Crane, D.R. and Ford, P.C. (1991), J. Am. Chem. Soc., 113, 8510.
[50]. Ryba, D.W., van Eldik, R. and Ford, P.C. (1992), Organometallics, in press.
[51]. van der Graaf, T., Bal Reddy, K., van Eldik, R., Langford, C.H., Rossenaar, B.D., Stufkens, D.J. and Vlcek Jr, A (1992), Inorg. Chem., in press.
[52]. Wishart, J.F. and van Eldik, R. (1992), Rev. Sci. Instrum., 63, 3224.

[53]. van Eldik, R., Cohen, H., Meshulam, A. and Meyerstein, D. (1990), Inorg. Chem., 29, 4156.
[54]. van Eldik, R., Cohen, H. and Meyerstein, D. (1991), Angew. Chem. Int. Ed. Eng., 30, 1158.
[55]. van Eldik, R., Gaede, W., Cohen, H. and Meyerstein, D. (1992), Inorg. Chem., 31, 3695.
[56]. Powell, D.H., Helm, L. and Merbach, A.E. (1991), J. Chem. Phys., 95, 9258.
[57]. Cohen, H., Gaede, W., Gerhard, A., Meyerstein, D. and van Eldik, R. (1992), Inorg. Chem., 31, 3805.
[58]. Gaede, W., van Eldik, R., Cohen, H. and Meyerstein, D., unpublished results.
[59]. de Bolster, M.W.G. and Kranenburg, R.A.C. (1991), Inorg. Chim. Acta, 183, 119.
[60]. Gamelkoorn, H.J., de Bolster, M.W.G. and Balt, S. (1992), Recl. Trav. Chim. Pays-Bas, 111, 178.
[61]. Gerards, L.E.H., de Bolster, M.W.G. and Balt, S. (1992), Inorg. Chim. Acta, 192, 287.
[62]. van Eldik, R. (1991), High Pressure Research, 6, 251.
[63]. Jolley, W.H., Stranks, D.R. and Swaddle, T.W. (1990), Inorg. Chem., 29, 385.
[64]. Jolley, W.H., Stranks, D.R. and Swaddle, T.W. (1990), Inorg. Chem., 29, 1948.
[65]. Doine, H. and Swaddle, T.W. (1991), Inorg. Chem., 30, 1858.
[66]. Jolley, W.H., Stranks, D.R. and Swaddle, T.W. (1992), Inorg. Chem., 31, 507.
[67]. Murguia, M.A. and Wherland, S. (1991), Inorg. Chem., 30, 139.
[68]. Anderson, K.A. and Wherland, S. (1991), Inorg. Chem., 30, 624.
[69]. Anderson, K.A., Kirchner, K., Dodgen, H.W., Hunt, J.P. and Wherland, S. (1992), Inorg. Chem., 31, 2605.
[70]. Krack, I. and van Eldik, R. (1986), Inorg. Chem., 25, 1743.
[71]. Krack, I. and van Eldik, R. (1990), Inorg. Chem., 29, 1705.
[72]. Goldstein, S., Czapski, G., van Eldik, R., Cohen, H. and Meyerstein, D. (1991), J. Phys. Chem., 95, 1282.
[73]. Hubbard, C.D., Bajaj, H.C., van Eldik, R., Burgess, J. and Blundell, N.J. (1991), Inorg. Chim. Acta., 183, 7.
[74]. Hubbard, C.D., Gerhard, A. and van Eldik, R. (1991), Inorg. Chem., 30, 5023.
[75]. Bänsch, B., Martinez, P. and van Eldik, R. (1992), J. Phys. Chem., 96, 234.
[76]. Leipoldt, J.G., Steynberg, E.C. and van Eldik, R. (1987), Inorg. Chem., 26, 3068.
[77]. Venter, J.A., Leipoldt, J.G. and van Eldik, R. (1991), Inorg. Chem., 30, 2207.
[78]. Hallinan, N., McArdle, P., Burgess, J. and Guardado, P. (1987), J. Organomet. Chem., 333, 77.
[79]. Pipoh, R., van Eldik, R., Wang, S.L.B. and Wulff, W.D. (1992), Organometallics, 11, 490.
[80]. Schneider, K.J., Neubrand, A., van Eldik, R. and Fischer, H. (1992), Organometallics, 11, 267.
[81]. Pipoh, R., van Eldik, R. and Henkel, G., Organometallics, submitted for publication.
[82]. Pipoh, R. and van Eldik, R., Organometallics, submitted for publication.
[83]. Odiaka, T. and van Eldik, R. (1992), J. Organomet. Chem., 425, 89.
[84]. Odiaka, T. and van Eldik, R. (1992), J. Chem. Soc., Dalton Trans, 2215.
[85]. Odiaka, T. and van Eldik, R. (1992), J. Organomet. Chem., in press.
[86]. Anhaus, J., Bajaj, H.C., van Eldik, R., Nevinger, L.R. and Keister, J.B. (1989), Organometallics, 8, 2903.

[87]. Keister, J.B., Frey, U., Zbinden, D. and Merbach, A.E. (1991), Organometallics, 10, 1497.
[88]. Jurczak, J. and Baranowski, B. (Eds.) (1989), High Pressure Chemical Synthesis, Elsevier, Amsterdam.
[89]. Jurczak, J., Golebiowski, A. and Rahm, A. (1986), Tetrahedron Lett., 27, 853.
[90]. Jurczak, J. and Rahm, A., Chapter 11 in ref. [88].
[91]. Dauben, W.G., Gerdes, J.M. and Look, G.C. (1986), Synthesis, 532.
[92]. Yamamoto, Y., Maruyama, K. and Matsumoto, K. (1984), Organometallics, 3, 1583.

BIOINORGANIC KINETICS AT ELEVATED PRESSURE. APPLICATION OF STOP-PED-FLOW, T-JUMP, FLASH-PHOTOLYSIS AND PULSE-RADIOLYSIS TECHNIQUES

RUDI VAN ELDIK
Institute for Inorganic Chemistry
University of Witten/Herdecke
Stockumer Straße 10
5810 Witten
Germany

ABSTRACT. In this contribution the application of various high pressure kinetic techniques in mechanistic studies of bioinorganic systems are presented. These systems include the antitumor activity of platinum metal complexes, the substitution behaviour of cobalamin (vitamin B_{12}), the transport of small molecules such as O_2 and CO by biological carriers, and long-distance electron-transfer processes in cytochrome c. Volume profiles for such reactions are discussed where available, and the mechanistic interpretation of the observed pressure dependencies are presented. Present and future goals are discussed in order to demonstrate the possible application of high pressure kinetic techniques.

1. Introduction

In the preceding contribution it was demonstrated how high pressure kinetic techniques can be used in the elucidation of substitution, electron transfer, and addition/elimination reaction mechanisms of inorganic and organomettalic systems. In particular it was emphasized how such data can be combined with partial molar volume and reaction volume data to construct a volume profile for the reaction under investigation, according to which the chemical process can be analysed in terms of volume changes along the reaction coordinate. This has been done for numerous systems [1-5] in the area of inorganic and organometallic solution phase reactions, but untill recently very little was done in the area of bioinorganic chemistry. The latter area concerns the inorganic aspects of biochemical reactions, especially those in which transition metal ions and complexes play a vital role in the control of the biochemical process.

In this contribution the available information will be reviewed for a variety of substitution, ligation and electron-transfer reactions that are of biological importance and that can be classified as bioinorganic systems. Where possible the construction and detailed interpretation of volume profiles will be presented, and the mechanistic information revealed by the pressure dependence of these processes will be discussed. Readers are referred to the preceding chapter for more information on the general principles involved and instrumentation used in such investigations.

R. Winter and J. Jonas (eds.), High Pressure Chemistry, Biochemistry and Materials Science, 329–344.

2. Substitution Reactions

There are a number of important biological processes in which ligand substitution reactions on transition metal centers play an important role in the control over the overall process. Two examples will be treated here.

The first deals with the antitumor activity of platinum metal complexes. There is presently a significant interest in the interaction of cis-Pt^{II}(diamine) and cis-Pd^{II}(diamine) complexes with DNA and its constituents in an effort to improve the understanding of the antitumor activity of such complexes and their use in chemotherapy [6]. It is generally accepted that substitution reactions involving the metal complexes and the DNA moieties play an important role in such processes [6]. Many studies deal with the structural identification of reaction products using NMR and X-ray techniques, such that a good understanding of the bonding modes has been achieved. Considerably less is known about the reactivity, i.e. kinetics and mechanism, of the bonding process. In our work we have focused on the substitution behaviour of such platinum metal complexes and their interaction with typical nucleic bases, nucleosides and 5'-nucleotides [7-17].

In principle, square planar complexes of d^8 metal ions undergo substitution reactions according to an associative mechanism. Furthermore, such reactions are usually characterized by two-term rate laws, from which indirect information on the importance of the solvolysis pathway can be obtained [7]. In fact, the formation of solvento complexes is in many cases the rate-determining step, since such species are usually very labile and undergo rapid substitution reactions to produce the final reaction product. A fundamental question that has often been raised in the literature concerns the possibility to force such substitution reactions to proceed via a dissociative mechanism by increasing the steric hindrance on the non-exchanging ligands. A series of studies on diethylenetriamine (dien) and substituted dien complexes of the type $Pd(R_5dien)Cl^+$, revealed that increasing steric hindrance can slow down the substitution rate constant by up to six orders of magnitude, but it does not affect the associative nature of the substitution mechanism [7,11-13,18]. This was also confirmed by significantly negative $\Delta V^{\#}$ data for solvent exchange reactions on complexes of the type $Pd(R_5dien)H_2O^{2+}$ [12,13]. Typical volume profiles for two non-symmetrical substitution reactions are given in Figure 1, from which it follows that the transition state has a significantly lower partial molar volume than either the reactant or product species, which demonstrates the associative nature of the substitution process. Recent reports have, however, indicated that a changeover in mechanism from associative to dissociative can be achieved via electronic effects, for instance via trans-labilization by a metal-carbon bond to induce a dissociative ligand exchange process [19,20]. On the other hand, the majority of solvent exchange, ligand exchange and complex formation reactions of Pt(II) and Pd(II) complexes recently reported in the literature [21-23], are all characterized by significantly negative volumes of activation in line with an associative substitution mechanism.

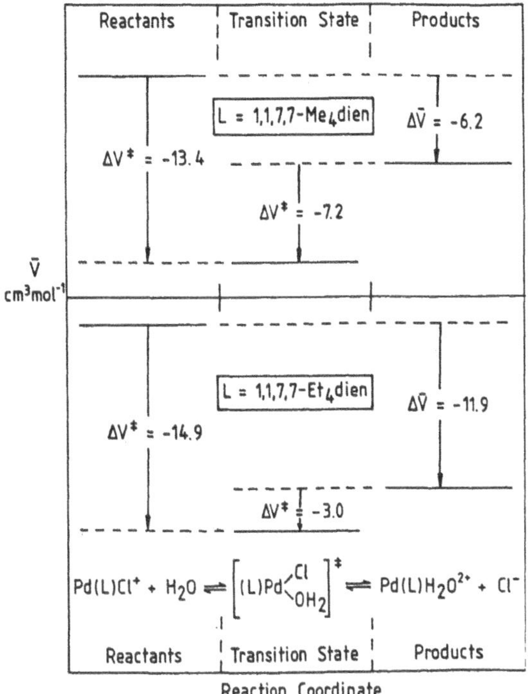

Figure 1. Volume profiles for the reaction
$Pd(L)Cl^+ + H_2O \rightarrow Pd(L)H_2O^{2+} + Cl^-$

In terms of the antitumor activity of cis-$Pt(NH_3)_2Cl_2$, it is well known that the spontaneous aquation reactions lead to a series of aqua and hydroxo complexes, of which the aqua species are the most reactive ones [24,25]. The reaction of cis-$Pt(NH_3)_2(H_2O)_2{}^{2+}$ with 2'-deoxy-5'-guanosinemonophosphate, the nucleotide that shows preference binding in the DNA sequence [6], is characterized by significantly negative $\Delta V^{\#}$ values for the formation of the 1:1 and 1:2 substitution product [26], which demonstrates the associative nature of this process. In subsequent work in our group, significantly more reactive complexes of the type $Pd(R_4en)(H_2O)_2{}^{2+}$ were employed as model species to investigate their interaction with the purine nucleosides adenosine and inosine [15], and with purine 5'-nucleosides monophosphates [16]. The reported activation volumes are once again significantly negative, between -4 and -11 cm^3mol^{-1}, and underline the associative nature of the ligand substitution reactions. Thus high pressure kinetic measurements can also contribute to our understanding of the fundamental substitution reactions that occur during the binding of the antitumor agent to DNA.

The second example deals with the substitution reactions of cobalamins (vitamin B_{12}), which has attracted considerable attention from kineticists (see ref. [27] and literature cited therein). In these systems the usually kinetic inert Co(III) ion is labilized considerably by the corrin ring, and there has been some disagreement in the literature concerning the mechanism of these substitution reactions. Substitution reactions of Co(III) complexes usually follow an I_d mechanism [1], such that labilization by the corrin ring could cause a changeover to a D mechanism [28]. A series of detailed studies were performed on the substitution mechanism of the axially coordinated H_2O molecule in aquacobalamin, hereafter referred to as B_{12}-H_2O^+ [27,29-36], in an effort to resolve the I_d or D nature of the mechanism. The available $\Delta V^\#$ data for the complex formation and reverse aquation reactions of aquacobalamin are summarized in Table 1. The data indicate that with the exception of the reaction with $Fe(CN)_6^{4-}$, all the $\Delta V^\#(k_f)$ values are in line with an I_d mechanism in which B_{12}-H_2O^+ bond cleavage causes the major increase in volume. The deviation in the case of $Fe(CN)_6^{4-}$ can be accounted for in terms of ion-pair formation that should be accompanied by a significant decrease in electrostriction and increase in volume. The results for the neutral, -1 and -2 charged ligands are very similar and also indicate that the substitution is controlled by the breakage of the Co-H_2O bond. There is a larger scatter in the data for $\Delta V^\#(k_d)$, since here the leaving group (L^{n-}) must have a significant effect on this value.

TABLE 1. Summary of $\Delta V^\#$ data for the reaction

$$B_{12}\text{-}H_2O^+ + L^{n-} \underset{k_d}{\overset{k_f}{\rightleftharpoons}} B_{12}\text{-}L^{(1-n)+} + H_2O$$

L^{n-}	$\Delta V^\#(k_f)$	$\Delta V^\#(k_d)$ cm^3mol^{-1}	$\Delta\overline{V}$ [a]	Ref.
$Fe^{II}(CN)_5NO^{2-}$	$+8.9 \pm 0.5$			[32]
$Fe^{II}(CN)_5H_2O^{2-}$	$+8.2 \pm 0.8$			[32]
$Fe^{II}(CN)_6^{4-}$	$+16.2 \pm 1.2$			[32]
$S_2O_3^{2-}$	$+6.0 \pm 0.6$			[35]
I^-	$+5.5 \pm 0.8$	$+11.5 \pm 1.6$	-5.8 ± 2.3	[36]
N_3^-	$+6.9 \pm 0.2$			[32]
	$+6.4 \pm 0.1$			[31]
HN_3	$+5.5 \pm 0.3$			[31]
C_5H_5N	$+8.7 \pm 1.2$	$+16.9 \pm 0.8$	-8.2 ± 2.0	[33]
$SC(NH_2)_2$	$+3.6 \pm 0.5$	$+6.3 \pm 2.5$	-2.7 ± 3.0	[35]
	$+9.1 \pm 0.9$	$+6.7 \pm 0.3$	$+2.4 \pm 1.2$	[34]
$SC(NHMe)_2$	$+6.7 \pm 1.0$	$+10.0 \pm 0.8$	-3.3 ± 1.8	[34]

[a] $\Delta\overline{V} = \Delta V^\#(k_f) - \Delta V^\#(k_d)$

For the reaction with pyridine [33], the observed rate constants reached a limiting value at high pyridine concentration and this was interpreted as evidence for a D mechanism. However, in later work [27,34] it was shown that this curvature must be due to a precursor formation step in terms of an I_d mechanism. Furthermore, the non-linear concentration dependence observed for the reaction with pyridine enabled the kinetic separation of the precursor formation constant and the rate-determining interchange rate constant, such that a detailed volume profile for the overall

process (Figure 2) could be constructed [34]. This example clearly demonstrates the importance

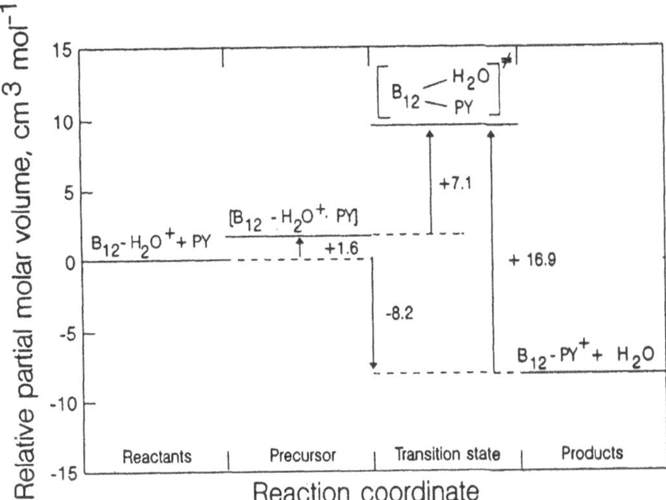

Figure 2. Volume profile for the reaction
$B_{12}\text{-}H_2O^+ + py \rightleftharpoons B_{12}\text{-}py^+ + H_2O$
according to an I_d mechanism (py = pyridine)

of considering all available kinetic and other information in order to assign a mechanism that is as close as possible to the real one. Furthermore, the cis-labilization caused by the corrin ring is not enough to cause a changeover from an I_d to a D substitution process as in the case of porphyrin ligands [28]. The labilization effect is however still very impressive if one considers that normal Co(III) complexes react on a hour time-scale at 60 °C, compared to the B_{12} system that reacts on a second time-scale at 25 °C.

3. Ligation Reactions

The mechanistic understanding of the binding of small molecules such as O_2, CO and NO to ferrous hemes and hemoproteins has attracted significant attention from scientists all over the world in recent years. Model heme complexes are usually employed to improve our understanding of the reactions of the corresponding proteins. It is especially the mechanism of the binding and release process that controls the overall transport of such molecules in biological systems. Application of picosecond and nanosecond flash photolysis techniques [37] has suggested that in the case of simple model hemes the germinate pair [Fe L] exists as a single kinetic intermediate as shown in reaction (1). In the case of proteins, the germinate pair exists in two

$$\text{Fe-L} \rightleftharpoons [\text{Fe L}] \rightleftharpoons \text{Fe} + \text{L} \qquad (1)$$

configurations, a fast-reacting "contact pair" and a slow-reacting "separated pair" as shown in

$$\text{Fe-L} \rightleftharpoons \underset{\text{contact pair}}{[\text{Fe} \quad \text{L}]} \rightleftharpoons \underset{\text{separated pair}}{[\text{Fe} \,||\, \text{L}]} \rightleftharpoons \text{Fe} + \text{L} \qquad (2)$$

reaction (2) [38]. In the separated pair, the ligand wanders through the heme pocket before forming the contact pair that is the precursor to bond formation. In an effort to obtain more mechanistic information on the various steps in reactions (1) and (2), a detailed pressure dependence study was performed on the binding kinetics of small neutral molecules to ferrous hemes and hemoproteins [39].

Two model heme systems were studied, viz. protoheme dimethyl ester (PHDME) and monochelated protoheme (MCPH), using two different photolysis techniques outlined in (3). In the first the axial ligand was photodissociated and the bimolecular recombination to form the original

$$\text{L-Fe-L} \underset{k}{\overset{h\nu}{\rightleftharpoons}} \text{Fe-L} + \text{L}$$

$$\text{OC-Fe-B} \underset{+CO}{\overset{h\nu,\ -CO}{\rightleftharpoons}} \text{Fe-B} \underset{-L}{\overset{+L}{\rightleftharpoons}} \text{L-Fe-B} \qquad (3)$$

complex was monitored. In the second the Gibson technique was used during which CO is photodissociated and the produced five-coordinate intermediate picks up the attacking nucleophile L. Typical $\Delta V^{\#}$ data for the addition of a series of neutral ligands are summarized in Table 2. The data show a clear correlation between $\Delta V^{\#}$ and k_{on}, which can be explained in terms of a

TABLE 2. $\Delta V^{\#}$ data for the bimolecular addition of various neutral ligands to five-coordinate ferrous model heme complexes in toluene as solvent [39]

Heme complex	L	k_{on}(25°C) $M^{-1}s^{-1}$	$\Delta V^{\#}$ cm^3mol^{-1}
MCPH	CO	1.1×10^7	-19.3 ± 0.4
MCPH	O_2	1.0×10^8	-11.3 ± 1.0
(MeNC)PHDME	MeNC	3.9×10^8	$+11.6 \pm 0.8$
(t-BuNC)PHDME	t-BuNC	2.5×10^8	$+9.9 \pm 1.0$
(1-MeIm)PHDME	1-MeIm	1.5×10^8	$+10.9 \pm 3.1$

change in the rate-limiting step of reaction (1) for which different factors will dominate $\Delta V^{\#}$. For the slower reactions recombination is rate-limiting and results in a negative volume of activation due to bond formation. For the faster reactions the processes become diffusion controlled in toluene, and are slowed down by increasing pressure since the viscosity of toluene shows a large pressure dependence, viz. $\Delta V^{\#}_{vis} = +22$ cm^3mol^{-1}.

In a subsequent study [40], the reaction of CO with MCPH was studied as a function of pressure in a very viscous medium, viz. 90/10 (v/v) mineral oil/toluene solution. A typical set of results, reproduced in Figure 3, clearly shows a changeover in rate-determining step from bond formation to diffusion-controlled on increasing the pressure to 300 MPa. A least square

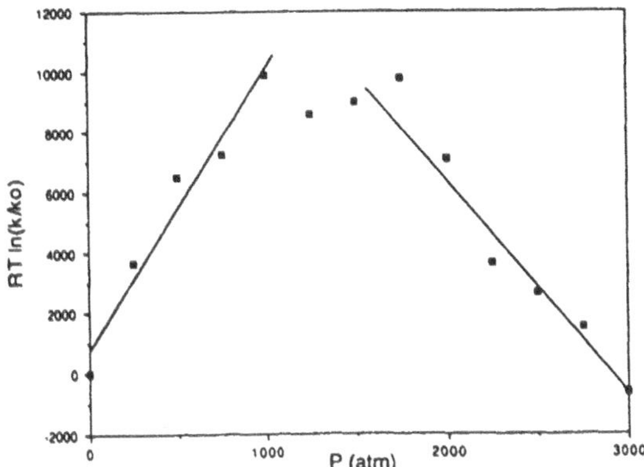

Figure 3. Plot of RT ln (k_{obs}/k^o) versus pressure for the reaction of CO with the intermediate formed by nanosecond laser flash photolysis of MCPH-CO [40].

analysis of the data in the lower pressure range results in a $\Delta V^{\#}$ value of -9.6 cm^3mol^{-1}, whereas the data in the higher pressure range result in $\Delta V^{\#}$ = +7.1 cm^3mol^{-1}. These results clearly demonstrate how the different steps in reaction (1) can become rate-limiting as a function of the viscosity of the medium and the applied pressure.

Similar techniques were applied to study the effect of pressure on the bimolecular association rate constant for the reaction of sperm whale myoglobin with a series of neutral ligands in water as solvent [39]. The results are summarized in Table 3, from which it follows that only the

TABLE 3. $\Delta V^{\#}$ data for the bimolecular addition of various ligands to deoxymyoglobin in aqueous buffer

L	k_{on}(25 °C) $M^{-1}s^{-1}$	$\Delta V^{\#}$ cm^3mol^{-1}
CO	5.2 x 10^5	-10.0 ± 0.8
O_2	2.5 x 10^7	+5.2 ± 0.5 [a]
	1.3 x 10^7	+7.8 ± 1.3 [b]
MeNC	1.4 x 10^5	+8.8 ± 1.0
t-BuNC	2.1 x 10^3	+9.3 ± 0.3

[a] Data obtained using T-jump technique [41], see following Discussion
[b] Data reported in ref. [42]

reaction with CO is characterized by a negative $\Delta V^{\#}$ value, in line with a bond formation process. The positive $\Delta V^{\#}$ data for the other ligands in Table 3 are ascribed to the entering of the ligand into the protein, which will be accompanied by significant desolvation and presumably conformational changes on the protein chain. Thus in terms of the 4-step mechanism in (2), bond formation seems to be rate-determining for the binding of CO, whereas the entrance of the ligand into the protein pocket seems to be rate-determining for the binding of the remaining ligands. The effect of pressure on the escape of the ligand from the protein-separated pair, i.e. [Mb| |L] → Mb + L, resulted in significantly positive volumes of activation, viz. +11.7 ± 0.1, +12.6 ± 1.7 and +9.1 ± 3.5 cm^3mol^{-1} for L = CO, O_2 and MeNC, respectively [39]. These positive numbers are consistent with the notion of a "gate" that operates in both directions, and reflect both small conformational changes in the protein and the solvation of the exiting ligand.

The large differences observed above for the binding of O_2 and CO to deoxymyoglobin (Mb) motivated us to perform detailed volume profile analyses for these reactions [41,43]. In these studies the binding of O_2 and CO was studied using a T-jump technique, whereas the release of O_2 and CO was studied using a stopped-flow technique in which the O_2 and CO were rapidly removed from an equilibrated solution in a chemical way. Furthermore, the pressure dependence of the overall equilibrium constant was studied via UV-VIS spectrophotometry. These data could now be employed to construct a volume profile for the reaction of myoglobin with O_2 and CO, which have been combined in Figure 4 [43]. The volume profile for the binding of O_2 is characterized by a significant volume increase in going from the reactant to the transition state (compare data reported in Table 3), followed by a significant volume collapse in going to the product state. The observed volume increase has been ascribed to rate-determining movement of

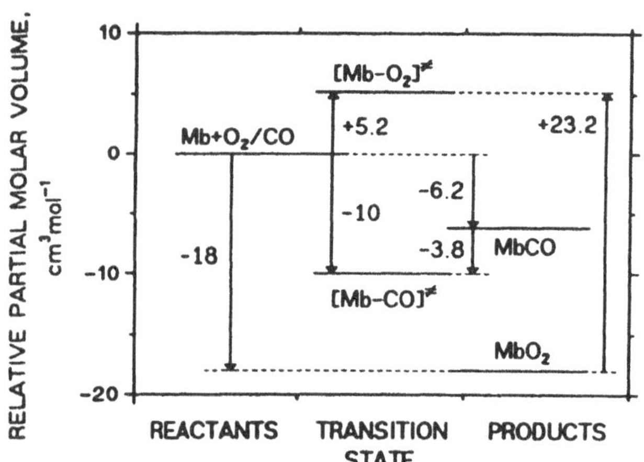

Figure 4. Comparison of the volume profiles for the reactions $Mb + O_2 \rightleftharpoons MbO_2$ and $Mb + CO \rightleftharpoons MbCO$

O_2 through the protein to the heme pocket, which may involve significant hydrogen bonding with the distal histidine as well as desolvation [41]. This step is followed by rapid bond formation with the Fe(II) center, during which the change in spin from high to low, the movement of the Fe(II) center into the porphyrin plane, and associated conformational changes will account for the drastic volume collapse. The overall reaction volume of -18 cm^3mol^{-1} demonstrates the large volume collapse caused by the binding of O_2. The volume profile for the binding of CO shows a significant volume decrease on going from the reactant to the transition state, which has been ascribed to rate-determining bond formation [39,43]. Surprisingly, the reverse bond cleavage reaction is accompanied by a volume decrease, which may be related to the significantly different bonding mode of CO compared to O_2. It is known that the porphyrin iron binds O_2 under an angle of 115° and exhibits a hydrogen bond to histidine E7, which results in a sterically favoured orientation of the oxygen molecule in the porphyrin pocket. In contrast, CO does not show any hydrogen bonding, and its favoured linear bonding geometry is not possible due to histidine E7. This causes the heme pocket to widen significantly as compared to deoxymyoglobin due to the steric tension. It follows that Fe-CO bond breakage will be accompanied by a decrease in steric tension and a slight volume collapse due to reorganization of the protein pocket as CO leaves the iron coordination site. The significantly different bonding mode of CO must also account for the much smaller absolute reaction volume observed in this case, viz. -6 compared to -18 cm^3mol^{-1} in the case of the binding of oxygen [43].

It must be realized that the interpretation of the overall volume profile is rather complicated since we are dealing with a 4-stage mechanism as outlined in (2). Ideal would be to have the pressure dependence of each step and to construct a volume profile subdivided into volume contributions coming from the different steps, as recently done for the complex formation reactions of Fe(III) with hydroxamic acid [44]. It should also be noted that such measurements and their interpretation were recently critized by Frauenfelder and co-workers [45] on the basis of low-temperature high-pressure investigations at < 160 °K in 75 % glycerol solutions. These experimental conditions are so irrelevant to the study of biological processes that occur at ambient temperature in aqueous solution, that the validity of their criticism remains questionnable. The $\Delta V^{\#}$ data reported in this contribution have also been found by other investigators, and also seem to hold for other biological binding processes of oxygen and related molecules [46,47].

4. Electron-Transfer Reactions

The application of high pressure kinetic techniques in the mechanistic study of electron-transfer reactions in inorganic and organometallic systems has been discussed to some extent in the preceding chapter of this volume. It was demonstrated that electron-transfer reactions in general exhibit characteristic volumes of activation that can be accounted for in terms of existing theories. In this section we will discuss two systems which involve electron-transfer reactions that are of biological significance.

The first system deals with the oxidation of L-ascorbic acid (vitamin c) by different Fe(III) complexes. In the case of $Fe(CN)_6^{3-}$, the oxidation reaction can only proceed according to an outer-sphere mechanism for which the overall reaction scheme is presented in (4) [48]. In this

$$
\begin{array}{lll}
H_2A & \rightleftharpoons HA^- + H^+ & , K_1 \\
HA^- & \rightleftharpoons A^{2-} + H^+ & , K_2 \\
H_2A + Fe(CN)_6^{3-} & \rightarrow Fe(CN)_6^{4-} + H_2A\cdot^+ & , k_a \\
HA^- + Fe(CN)_6^{3-} & \rightarrow Fe(CN)_6^{4-} + HA\cdot & , k_b \\
A^{2-} + Fe(CN)_6^{3-} & \rightarrow Fe(CN)_6^{4-} + A\cdot^- & , k_c \\
H_2A\cdot^+/HA\cdot/A\cdot^- + Fe(CN)_6^{3-} & \rightarrow Fe(CN)_6^{4-} + A + 2H^+ & , \text{fast}
\end{array}
\quad (4)
$$

reaction sequence L-ascorbic acid (H_2A) is oxidized to L-dehydroascorbic acid (A) via the radicals $H_2A\cdot^+$, $HA\cdot$ and $A\cdot^-$. The rate of the process strongly depends on pH, since the latter controls the actual ascorbic acid/ascorbate species (H_2A, HA^- or A^{2-}) that participate in the rate-determining step, $K_1 = 3.2 \times 10^{-4}$ and $K_2 = 1.3 \times 10^{-11}$ M at 25 °C. It is possible to select the experimental conditions in such a way that mainly the reaction with H_2A or the reaction with HA^- is observed kinetically. Under these conditions the process is fairly pH independent and a detailed temperature and pressure dependence study could be performed. The observed rate and activation parameters are reported in Table 4, from which the large rate acceleration by increasing pH and pressure can clearly be seen. This increase in k is accompanied by a significant

TABLE 4. Rate and activation parameters for the oxidation of L-ascorbic acid by $Fe(CN)_6^{3-}$

pH	k at 25 °C $M^{-1}s^{-1}$	$\Delta H^{\#}$ kJ mol^{-1}	$\Delta S^{\#}$ J $K^{-1}mol^{-1}$	$\Delta V^{\#}$ cm^3mol^{-1}
0.30 [a]	0.58	34.7 ± 0.9	-133 ± 3	-16.6 ± 0.5
5.25	842	20.8 ± 0.8	-119 ± 3	-16.3 ± 0.4

[a] Oxidation of H_2A, $k = k_a$ in reaction (4)
[b] Oxidation of HA^-, $k = k_b >> k_a$ in reaction (4)

decrease in $\Delta H^{\#}$, but no significant change in $\Delta V^{\#}$. Theoretical calculations based on the Marcus-Hush theories (see ref. [48] for details), resulted in a $\Delta V^{\#}$ value that is ca. 6 cm^3mol^{-1} more positive than the experimental values. This difference can easily be compensated for if it is assumed that the electron transfer reaction is not fully adiabatic and a correction for a non-adiabatic contribution towards $\Delta V^{\#}$ is made.

For die oxidation of L-ascorbic acid by aquated Fe(III), both inner-sphere and outer-sphere electron transfer reactions are possible since we are dealing with a labile metal center. The question therefore arises whether this redox process is substitution or electron-transfer controlled [49]. In acidic medium H_2A is the main ascorbic acid species, whereas $Fe(H_2O)_6^{3+}$ and $Fe(H_2O)_5OH^{2+}$ are the oxidants. A detailed temperature and pressure dependence study as a function of $[H^+]$ enabled the determination of the rate and activation parameters for the reaction steps indicated in (5), for which the data are summarized in Table 5 along with solvent exchange

$$\begin{aligned} Fe(H_2O)_6^{3+} + H_2A &\rightarrow Fe(H_2O)_6^{2+} + HA\cdot + H^+ , \; k_a \\ Fe(H_2O)_5OH^{2+} + H_2A &\rightarrow Fe(H_2O)_6^{2+} + HA\cdot , \; k_b \\ Fe(H_2O)_6^{3+} + HA\cdot &\rightarrow Fe(H_2O)_6^{2+} + A + H^+ , \; \text{fast} \end{aligned} \quad (5)$$

data for these complexes [50]. Very surprising is the observation that the second order rate constants for the electron-transfer and solvent exchange reactions are indeed very similar, which could suggest that electron-transfer is substitution controlled. However, a more detailed analysis indicates that the theoretical electron-transfer rate constant shows a good agreement for the oxidation by $Fe(H_2O)_6^{3+}$, but a significant deviation for the oxidation by $Fe(H_2O)_5OH^{2+}$. This suggests that only the former reaction follows an outer-sphere mechanism. The values of $\Delta H^{\#}$ and $\Delta S^{\#}$ are also very similar for the two processes, but there are remarkable differences in

TABLE 5. Rate and activation parameters for the oxidation of L-ascorbic acid by aquated Fe(III) and for solvent exchange on aquated Fe(III)

Reaction	k_{exp} at 25 °C $M^{-1}s^{-1}$	k_{calc} [a] $M^{-1}s^{-1}$	$\Delta H^{\#}$ kJ mol^{-1}	$\Delta S^{\#}$ J K^{-1}mol^{-1}	$\Delta V^{\#}$ cm^{3}mol^{-1}
$Fe(H_2O)_6^{3+} + H_2A$	3.4 [b]	1.1	72 ± 3	+7 ± 11	+14 ± 2
$Fe(H_2O)_5OH^{2+} + H_2A$	1730 [c]	40	57 ± 2	+9 ± 5	+4.6 ± 0.7
$Fe(H_2O)_6^{3+} + H_2O$	2.9		64 ± 2	+12 ± 7	-5.4 ± 0.4
$Fe(H_2O)_5OH^{2+} + H_2O$	2162		42 ± 1	+5 ± 4	+7.0 ± 0.3

[a] calculated from the Marcus cross relationship
[b] $k_{exp} = k_a$ in reaction (5)
[c] $k_{exp} = k_b$ in reaction (5)

$\Delta V^{\#}$. First of all, $\Delta V^{\#}$ for the oxidation by $Fe(H_2O)_6^{3+}$ is ca. 20 cm^{3}mol^{-1} more positive than $\Delta V^{\#}$ for solvent exchange on this complex, indicating that substitution cannot be the rate-controlling step. On the other hand, the $\Delta V^{\#}$ data for the oxidation by $Fe(H_2O)_5OH^{2+}$ and solvent exchange on the latter complex, are indeed very similar. It follows that the more labile $Fe(H_2O)_5OH^{2+}$ species rapidly binds H_2A, which is followed by a fast electron-transfer step in terms of an inner-sphere mechanism. On the contrary, the less labile $Fe(H_2O)_6^{3+}$ species presumably undergoes outer-sphere electron-transfer prior to a possible ligand substitution process. This example clearly demonstrates the usefulness of combined temperature and pressure dependence studies in order to elucidate the intimate nature of the rate-determining step. The $\Delta V^{\#}$ data reported for the outer-sphere electron-transfer reactions in Tables 4 and 5 are mainly associated with volume changes on the Fe(III) center, viz. a volume collapse due to the reduction of $Fe(CN)_6^{3-}$ to $Fe(CN)_6^{4-}$ and a volume increase due to the reduction of $Fe(H_2O)_6^{3+}$ to $Fe(H_2O)_6^{2+}$, respectively [48,49].

In the remainder of this section we will focus on the effect of pressure on long-distance electron-transfer reactions in cytochrome c. Such processes have received significant attention from several laboratories, investigating long-range electron-transfer processes in inorganic systems and in proteins [51], in recent years. We have performed a detailed study on the effect of pressure on some intramolecular electron-transfer reactions in ruthenated cytochrome c species, as well as on some intermolecular electron-transfer reactions between such species [52,53], using pulse-radiolysis and stopped-flow techniques. The intramolecular electron-transfer reactions in horse heart $(NH_3)_5Ru^{II}$-His 33 and candida krusei $(NH_3)_5Ru^{II}$-His 39 undergo significant acceleration on increasing pressure with corresponding $\Delta V^{\#}$ values of -17.7 ± 0.9 and -18.3 ± 0.7 cm^{3}mol^{-1}, respectively. The intermolecular process between $Ru(NH_3)_6^{2+}$ and hh cyt c exhibits a similar pressure acceleration for which $\Delta V^{\#}$ is -15.6 ± 0.6 cm^{3}mol^{-1} [52]. At this point there seems to be no correlation between the electron-transfer distance and the value of $\Delta V^{\#}$, but further investigations are presently underway. A qualitative interpretation of the quoted $\Delta V^{\#}$ values suggests that they mainly result from volume changes due to electrostriction on the ruthenium center, during the oxidation of Ru(II) to Ru(III) on the surface of the protein. This becomes clear from a comparison of the available data summarized in Table 6. The intramolecular electron-transfer from cyt c to Ru(III) exhibits the opposite pressure dependence than for the related reverse process reported above. Our data for the intermolecular electron-transfer reaction between hh cyt c^{III} and $Ru(NH_3)_6^{2+}$ is very similar to that reported for the reaction with $Co^{II}(phen)_3^{2+}$ [54]. Again the reverse reaction exhibits the opposite trend in the latter case. When using the hexacyano complex of Fe(II/III) as redox partner [54], the observed $\Delta V^{\#}$ and $\Delta\overline{V}$ values go in the opposite direction since reduction of $Fe(CN)_6^{3-}$ will be accompanied by a volume collapse, whereas the reverse reaction should show a volume increase. Finally, we recently measured the effect of pressure on the oxidation and reduction of hh cyt c with $Ru(NH_3)_5isn^{2+/3+}$ [53], which exhibits very similar absolute $\Delta V^{\#}$ values (see Table 6). These data can be combined to construct a volume profile for the overall reaction (Figure 5), from which it follows that the transition state is practically halfway between the reactant and product states on a volume basis.

TABLE 6. Summary of rate and activation parameters for long distance electron-transfer reactions

Reaction [a]	k_{298}	$\Delta V^{\#}$ cm^3mol^{-1}	$\Delta\overline{V}$ cm^3mol^{-1}	Ref.
A_5Ru^{II}-hh^{III}	39 s^{-1}	-17.7		[52]
A_5Ru^{II}-ck^{III}	87 s^{-1}	-18.3		[52]
hh^{II}-$A_4Ru^{III}isn$	400 s^{-1}	+4.0		[53]
ck^{II}-$A_4Ru^{III}isn$	220 s^{-1}	+3.4		[53]
A_6Ru^{II} + hh^{III}	6.3 x 10^4 $M^{-1}s^{-1}$	-15.6		[52]
hh^{II} + $A_5Ru^{III}isn$	1.1 x 10^5 $M^{-1}s^{-1}$	+16	+33 [b]	[53]
$A_5Ru^{II}isn$ + hh^{III}	1.5 x 10^3 $M^{-1}s^{-1}$	-17		[53]
hh^{II} + $Co^{III}(phen)_3^{3+}$	1.9 x 10^3 $M^{-1}s^{-1}$	+8.5	+20	[54]
$Co^{II}(phen)_3^{2+}$ + hh^{III}	-	-11.5 [b]		[54]
hh^{II} + $Fe^{III}(CN)_6^{3-}$	3.0 x $10^6 M^{-1}s^{-1}$	-24 [b]	-37	[54]
$Fe^{II}(CN)_6^{4-}$ + hh^{III}	1.2 x $10^4 M^{-1}s^{-1}$	+13		[54]

[a] Abbreviations: A = NH_3, isn = isonicatinamide, phen = 1,10-phenanthroline
[b] Calculated from the relationship $\Delta\overline{V}$ = $\Delta V^{\#}$ (forward reaction) - $\Delta V^{\#}$ (back reaction)

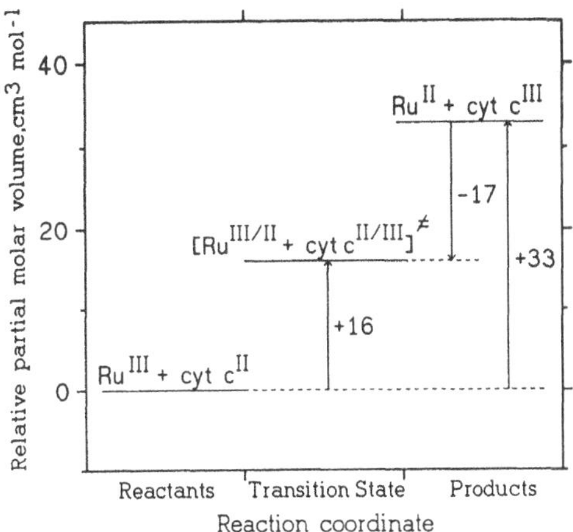

Figure 5. Volume profile for the reaction [53]
$Ru^{III}A_5isn^{3+}$ + hh cyt c^{II} ⇀ $Ru^{II}A_5isn^{2+}$ + hh cyt c^{III}

The above discussed data exhibit interesting trends and motivate further work along these lines on model as well as protein systems, to contribute towards a better understanding of the mechanism of long-range electron-transfer processes.

5. Conclusions

The examples discussed in the previous sections demonstrate that high-pressure kinetic data can contribute to a detailed understanding of the mechanism of chemical processes that are of biological significance. The complex nature of such processes naturally complicates the interpretation of the observed effects, and creates extraordinary challenges for innovative research efforts. When dealing with mechanistic analysis of such systems it is important to investigate as many as possible chemical and physical parameters, only then will a realistic description of the system be possible. The combination of a wide range of rapid kinetic techniques, as illustrated by some of the examples, allows an understanding of the processes that occur in different time domains. It is in this respect that especially the application of pulsed-laser flash-photolysis and pulse-radiolysis techniques have revealed new and promising information. Further developments to understand fundamental processes at a molecular level, by employing higher resolution and faster spectroscopic techniques at elevated pressure, are urgently needed. Only then will a meaningful contribution to the mechanistic clarification of complex biochemical processes be possible. A correlation of structural, energetic and volume data may assist the development of theoretical models that can describe the mechanistic behaviour of such processes.

Acknowledgements

The author gratefully acknowledges the very stimulating interaction with numerous graduate students, post docs and visiting scientists, their names appear on the work cited from this laboratory. Financial support from the Deutsche Forschungsgemeinschaft, Volkswagen-Stiftung, Fonds der Chemischen Industrie, Max-Buchner-Forschungsstiftung, NATO Scientific Affairs Division and the German-Israeli Foundation is greatly appreciated. The work discussed in this contribution was partly performed in collaboration with D. Magde (San Diego, USA), A.E. Merbach (Lausanne, Switzerland), E.L.J. Breet/J.J. Pienaar (Potchefstroom, SA), J.F. Wishart (Brookhaven National Laboratory, USA) and S.S. Isied (New Brunswick, USA).

6. References

[1] van Eldik, R. Asano, T. and le Noble, W.J. (1989), Chem. Rev., 89, 549.

[2] Akitt, J.W. and Merbach, A.E. (1990), NMR Basic Principles and Progress, 24, 189.

[3] van Eldik, R. and Merbach, A.E. (1992), Comments Inorg. Chem., 12, 341.

[4] van Eldik, R. (1992), in Perspectives in Coordination Chemistry, Williams, A.F., Floriani, C. and Merbach, A.E. (Eds.), VHCA Basel, VCH Weinheim, 55.

[5] van Eldik, R. (1992), Organometallics in Organic Synthesis Vol. 4, Vieweg Verlag, in press.

[6] For recent reviews on this topic see: Sherman, S.E. and Lippard, S.J. (1987), Chem. Rev., 87, 1153; Sundquist, W.I. and Lippard, S.J. (1990), Coord. Chem. Rev., 100, 293; Keppler, B.K. (1990), New. J. Chem., 14, 389; Keppler, B.K., Berger, M.R. Klenner, Th. and Heim, M.E. (1990), Adv. Drug. Res., 19, 243; Reedijk, J. (1987), Pure Appl. Chem., 59, 181; Reedijk, J., Fichtinger-Schepman, A.M.J., van Oosterom, A.T. and van de Putte, P. (1987), Struct. Bonding, 67, 53; Umapathy, P. (1989), Coord. Chem. Rev., 95, 129.

[7] Kotowski, M. and van Eldik, R. (1986) in Inorganic High Pressure Chemistry: Kinetics and Mechanisms, van Eldik, R. (Ed.), Elsevier, Amsterdam, Chapter 4.

[8] Kotowski, M. and van Eldik, R. (1986), Inorg. Chem., 25, 3896.

[9] Breet, E.L.J. and van Eldik, R. (1987), Inorg. Chem., 26, 2517.

[10] Breet, E.L.J. and van Eldik, R. (1987), Inorg. Chem., 26, 4264.

[11] Pienaar, J.J., Kotowski, M. and van Eldik, R. (1989), Inorg. Chem., 28, 373.

[12] Berger, J., Kotowski, M., van Eldik, R., Frey, U., Helm, L. and Merbach, A.E. (1989), Inorg. Chem., 28, 3759.

[13] Helm, L., Merbach, A.E. Kotowski, M. and van Eldik, R. (1989), High Pressure Research, 2, 49.

[14] Hohmann, H., Hellquist, B. and van Eldik, R. (1991), Inorg. Chim. Acta, 188, 25.

[15] Hohmann, H., Hellquist, B. and van Eldik, R. (1992), Inorg. Chem., 31, 345.

[16] Hohmann, H., Hellquist, B. and van Eldik, R. (1992), Inorg. Chem., 31, 1090.

[17] Shoukry, M., Hohmann, H. and van Eldik, R. (1992), Inorg. Chim. Acta, 198-200, 187.

[18] van Eldik, R. (1992), Pure Appl. Chem., 64, 1439.

[19] Romeo, R. (1990), Comments Inorg. Chem., 11, 21.

[20] Frey, U., Helm, L., Merbach, A.E. and Romeo, R. (1989), J. Am. Chem. Soc., 111, 8161.

[21] Hallinan, N., Besancon, V., Forster, M., Elbaze, G., Ducommun., Y. and Merbach, A.E. (1991), Inorg. Chem., 30, 1112.
[22] Frey, U., Elmroth, S., Moullet, B., Elding, L.I. and Merbach, A.E. (1991), Inorg.Chem., 30, 5033.
[23] Elmroth, S., Burgarcic, Z. and Elding. L.I. (1992), Inorg. Chem., 31, 3551.
[24] Miller, S.E. and House, D.A. (1989), Inorg. Chim. Acta, 161, 131 and 166, 189.
[25] Miller, S.E. and House, D.A. (1990), Inorg. Chim. Acta, 173, 53.
[26] Evans, D.J., Green, M. and van Eldik, R. (1987), Inorg. Chim. Acta, 128, 27.
[27] Marques, H.M. Bradley, J.C. and Campbell, L.A. (1992), J. Chem. Soc., Dalton Trans., 2019.
[28] Leipoldt, J.G., van Eldik, R. and Kelm, H. (1983), Inorg. Chem., 22, 4146.
[29] Marques, H.M. (1991), J. Chem. Soc., Dalton Trans., 339.
[30] Marques, H.M. (1991), J. Chem. Soc., Dalton Trans., 1437.
[31] Marques, H.M., Breet, E.L.J. and Prinsloo, F.F. (1991), J. Chem. Soc., Dalton Trans., 2941.
[32] Stochel, G., van Eldik, R., Kunkely, H. and Vogler, A. (1989), Inorg. Chem., 28, 4314.
[33] Stochel, G. and van Eldik, R. (1990), Inorg. Chem., 29, 2075.
[34] Meier, M. and van Eldik, R., Inorg. Chem., submitted for publication.
[35] van Herk, A.M. (1986), Ph.D. Thesis, Free University of Amsterdam, The Netherlands.
[36] Hassinoff, B.B. (1974), Can. J. Chem., 52, 910.
[37] Traylor, T.G., Magde, D., Taube, D.J. and Jongeward, K. (1987), J. Am. Chem.. Soc., 109, 5864.
[38] Jongeward, K.A., Magde, D., Taube, D.J., Marstero, J.C., Traylor, T.G. and Sharma, V.S. (1988), J. Am. Chem. Soc., 110, 380.
[39] Taube, D.J., Projahn, H.-D., van Eldik, R., Magde, D. and Traylor, T.G. (1990), J. Am. Chem. Soc., 112, 6880.
[40] Traylor, T.G., Luo, J. Simon, J.A. and Ford, P.C. (1992), J. Am. Chem. Soc., 114, 4340.
[41] Projahn, H.-D., Dreher, C. and van Eldik, R. (1990), J. Am. Chem. Soc., 112, 17.
[42] Hassinoff, B.B. (1974), Biochemistry, 13, 311.
[43] Projahn, H.-D. and van Eldik, R. (1992), Inorg. Chem., 30, 3288.
[44] Birus, M. and van Eldik, R. (1991), Inorg. Chem., 30, 4559.
[45] Frauenfelder, H., Alberding, N.A., Ansari, A., Braunstein, D., Cowen, B.R., Hong, M.K., Iben, I.E.T., Johnson, J.B., Luck, S., Marden, M.C., Mourant, J.R., Ormos, P., Reinisch, L., Scholl, R., Schulte, A., Shyamsunder, E., Sorensen, L.B., Steinbach, P.J., Xie, A., Young, R.D. and Yue, K.T. (1990), J. Phys. Chem., 94, 1024.
[46] Adachi, S. and Morishima, I. (1989), J. Biol. Chem., 264, 19896.
[47] Balny, C. and Travers, F. (1989), Biophys. Chem., 33, 237.
[48] Bänsch, B., Martinez, P., Zuluaga, J., Uribe, D. and van Eldik, R. (1991), Z. Phys. Chem., 170, 59.
[49] Bänsch, B., Martinez, P., Uribe, D., Zuluaga, J. and van Eldik, R. (1991), Inorg. Chem., 30, 4555.
[50] Swaddle, T.W. and Merbach, A.E. (1981), Inorg. Chem., 20, 4212.
[51] For recent reviews on this topic see: Isied, S.S. (1991), in Metals in Biological Systems, Sigel, H. and Sigel, A. (Eds.), Marcel Dekker, Inc., New York, 27, 1; Isied, S.S. (1991) in ACS Advances in Chemistry Series, Bolton, J.R., Mataga, N. and McLendon, G. (Eds.), American Chemical Society, Washington D.C., 228, 229; Long-Range Electron

Transfer in Biology, Structure and Bonding (1991), Vol 75.; Winkler, J.R. and Gray, H.B. (1992), Chem. Rev., 92, 369; Isied, S.S., Ogawa, M.Y. and Wishart, J.F. (1992), Chem. Rev., 92, 381.

[52] Wishart, J.F., van Eldik, R., Sun, J., Su, C. and Isied, S.S. (1992), Inorg. Chem., 31, 3986.

[53] Bänsch, B., Meier, M., Martinez, P., van Eldik, R., Su, C. and Wishart, J.F., prepared for publication.

[54] Heremans, K., Bormans, M., Snauwaert, J. and Vandersypen, H. (1982), Faraday Disc., Chem. Soc., 74, 343.

HIGH PRESSURE KINETIC EFFECTS AS MECHANISTIC PROBES IN ORGANIC CHEMISTRY

G. JENNER
Laboratoire de Piézochimie Organique, EHICS
B.P. 296
67008 Strasbourg
France

ABSTRACT. The high pressure kinetic methodology is a useful tool for the determination and the assessment of mechanisms in organic chemistry. The volume of activation is an indicator of the position of the transition state and, consequently, of the type of mechanism. If accurately determined, it enables the location of polarized, congested or strained transition states and the detection of fine effects (return mechanism in radical decompositions, π-participation in solvolyses, secondary orbital interactions in pericyclic processes, angular hydrogen transfer in ene reactions...). Selected examples are presented : isomerization reactions (aromatization of (Dewar) benzenes, isomerization of azobenzenes), thermal decomposition reactions (thermolysis of radical inhibitors and β-lactones), substitution and elimination reactions, addition of keto compounds to multiple bonds, pericyclic reactions (sigmatropic shifts, cycloaddition and ene reactions).

1. Introduction

Pressure and volume are interrelated, since organic reactions are run at constant temperature. In fact, it is necessary to consider the volume occupied by an organic molecule -the Van der Waals volume- and the space available to the molecule, the difference is the empty interstitial or void space that cannot be occupied.

Considering organic molecules submitted to pressure, compression induces densification, liquefaction -providing appropriate critical parameters-, solidification and, at sufficiently high pressure, metalation. A commonly view to visualize the pressure effect on molecules in solution takes into account molecular deformation (squeezing, twisting, flattening) resulting in a change of molecular behavior which may be different from the conformation prevailing at ambient pressure. This notion is actually wrong : the effects of pressure on reactions *in solution result basically from equilibrium shifts due to volume differences and the resulting energy term PΔV*.

Let us imagine two groups of reacting molecules A and B confined in a closed system on which pressure is gradually applied. In a first stage, an increase in pressure raises the concentration of A and B simply by reducing the volume. This is the concentration effect, very important so far as gases are concerned. A further increase in pressure densifies the liquid phase. Consequently, the space available for the formation of the transition state is restricted in such way that the overwhelming effect is the pressure effect on the reaction rate. This is called the kinetic effect which is the principal one to be considered for organic reactions in solutions.

R. Winter and J. Jonas (eds.), High Pressure Chemistry, Biochemistry and Materials Science, 345–366.

1.1. THE VOLUME OF ACTIVATION

The kinetic effect is described by the transition state theory [1,2]

reactants ----------------> transition state ----------------> products

From a thermodynamic point of view, the reaction occurs only if $\Delta G<0$. The condition is necessary but not sufficient, as it depends also on the height of the energy barrier. The classical thermodynamics tells only the equilibrium position between reactants and products, but does not provide information on the passage from reactants to products, hence neglects the importance of the mechanism that pictures the pathway followed by the reaction.

A chemical reaction is defined both by its energy and volume profile. Fig. 1 considers a one-step condensation reaction -meaning that one transition state is involved-. The reaction volume $\overline{\Delta V}$ is given by $\overline{\Delta V} = \overline{V}_f - \overline{V}_i$ ($\overline{V}_i$, $\overline{V}_f$: partial molar volume of initial and final state respectively). The rate constant k is expressed as :

$$k = \chi \frac{k_B T}{h} \exp(-\Delta G^{\neq}/RT)$$

χ : probability parameter
k_B : Boltzmann constant
h : Planck constant

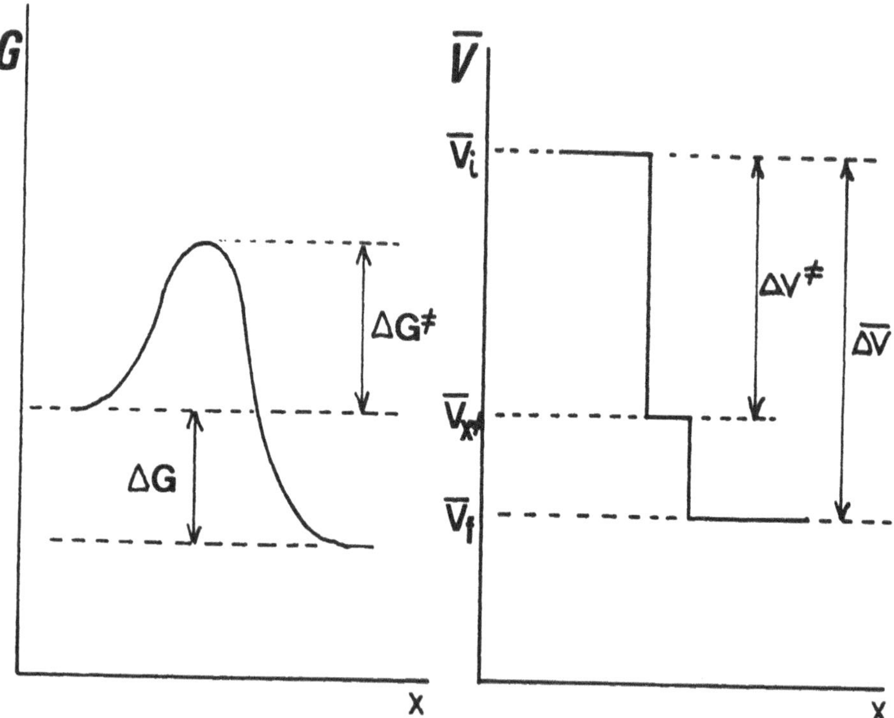

Figure 1. Energy and volume profile of a one-step condensation reaction (x is the reaction coordinate).

The pressure effect on rate constant k is expressed as :

$$\frac{\partial Lnk}{\partial P} = -\frac{\partial \Delta G^{\neq}}{RT\partial P} = -\frac{\Delta V^{\neq}}{RT} = -\frac{(\overline{V}_{X\neq}-\overline{V}_i)}{RT}$$

This is the basic relationship in organic piezochemistry. $\Delta V^{\neq}$ is the volume of activation and represents stricto sensu the difference between the partial molar volumes of the transition state and the initial state. It is essential to determine the position of the transition state on the reaction axis, as the mechanistic pathway is mostly deduced from the knowledge of transition state properties. In addition, the magnitude of $\Delta V^{\neq}$ informs immediately on the yield that can be expected when running the reaction at a given pressure. If the rate-determining step (or transition state) is close to the products, the pressure kinetic effect is usually remarkable. If it is on the reactant side, this effect is small (when no other volume effect is involved - see below).

$\overline{V}_i$ and $\overline{V}_f$ can be routinely determined from density measurements extrapolated to infinite dilution. $\overline{V}_{X\neq}$ is not directly accessible, it can only be obtained from the knowledge of $\overline{V}_i$ and $\Delta V^{\neq}$. The determination of the volume of activation proceeds from the relation reported above by plotting k-values against pressure in a relatively narrow pressure range, usually up to 100-200 MPa. Number of calculation procedures to determine $\Delta V^{\neq}$ with variable accuracy have been proposed [3,4].

1.2. COMPONENTS OF THE VOLUME OF ACTIVATION

As defined above, the meaning of $\Delta V^{\neq}$ is apparently simple. However, many other volume effects may operate : solvation, formation of charge transfer complexes, creation of hydrogen bonds, existence of dispersion forces... The most important volume contribution, in addition to volume changes arising from simple geometrical modifications in bond length and bond angle due to reaction, results from the generation (or its reverse) of ionic charges. This effect is called electrostriction. The development of charges or polarized species during the reaction increases the strength of interactions between reactant and solvent, thereby resulting in a negative volume contribution which affects the structure of the transition state and which may be very large. As a consequence, the selection of the solvent is of great importance for the pressure kinetic study of such reactions : Menshutkin reaction [5], [2+2] cycloadditions [6] taken as examples in Table 1.

TABLE 1. Solvent dependence of the volume of activation of ionogenic reactions

Solvent	ε_{20}	$\Delta V^{\neq}$ ($cm^3.mol^{-1}$)	
		Menshutkin reaction [5] methyl iodide + pyridine	[2+2] cycloaddition[6] TCNE + butylvinyl ether
carbon tetrachloride	2.24	-37.5	-50
benzene	2.28	-35.1	-43
chlorobenzene	5.70	-30.7	--
dichloromethane	9.10	--	-37
acetone	21.10	--	-35
acetonitrile	32.10	--	-29
methanol	32.60	-26.6	--
nitrobenzene	34.80	-21.3	--

Another type of volume contribution has been examined in recent years [7]. Hindered reactions were shown to be more sensitive to pressure than unhindered ones. They were related to the conceptual steric volume of activation $\Delta V_{\sigma}^{\neq}$ making allowance for the additional pressure kinetic effect. More generally, $\Delta V_{\sigma}^{\neq}$ concerns not only steric hindrance, but all kind of strain (ring, torsion, crowding ...) [7]. In all reactions examined, a more congested reaction is featured by a later transition state (Table 2).

TABLE 2. Effect of the bulkiness of alkyl groups on the volume of activation in radical copolymerizations of maleic anhydride with gem-disubstituted alkenes $CH_2 = C(R_1)(R_2)$

R_1	R_2	$-\Delta V^{\neq}$ ($cm^3.mol^{-1}$)	$-\Delta V_{\sigma}^{\neq}$ ($cm^3.mol^{-1}$)
CH_3	C_3H_7	31	17
CH_3	iC_3H_7	33	19
CH_3	tC_4H_9	~60	~45
CH_3	$CH_2tC_4H_9$	46	32
CH_3	C_6H_5	35	21
C_6H_5	C_6H_5	~50	~35

It is understandable that a careful analysis of all volume effects that can affect a given reaction must be made for the correct determination of its volume profile. Besides, there are some pressure induced physical effects which should not be dismissed :

- *freezing point*

As pressure raises, the melting point of organic compounds increases, the physical state of the medium must be checked. Such operation is rarely necessary in kinetic studies, as pressures do not exceed 200-300 MPa. An important exception, however, is benzene who freezes at 306.6 K at 100 MPa.

- *solubility*

Liquid solubilities of solids and gases are closely related to the pressure effect on melting points. Thus, it would be unwise to use saturated or nearly saturated solutions of solids under pressure. In any case, kinetics should be investigated with dilute solutions.

- *viscosity*

The viscosity of liquids is one of the most pressure sensitive physical properties as it increases exponentially with pressure. This dramatic effect has possible consequences for diffusion processes. An illustrative example is the termination step of radical polymerization which is diffusion-controlled under pressure. Instead to be accelerated by pressure -it is a bimolecular reaction-, it is actually retarded, which explains the high molecular weight of polymers obtained under pressure.

1.3. VALIDITY OF THE METHOD

The viscosity effect was argued against the validity of pressure kinetic studies as a mechanistic tool [8,9]. Actually, there is only one known example of an organic reaction for which there is an extremum in rate at high pressure : the displacement of bromide ion from ethyl bromide by the eugenoxide ion in isopropanol-eugenol mixtures. This reaction which is accelerated over most of the pressure range, drops precipitously in rate when the pressure exceeds 2000 MPa [10]. Nevertheless, in the low pressure range required for kinetic measurements, viscosity effects do not play a significant role to invalidate the transition state theory.

A second argument against the mechanistic pressure criterion comes from the fact that most part of the shrinkage in condensation reactions would result from the reaction itself [11]. However, this assumption was vigorously refuted by Klärner on the basis of comparison of the packing coefficients between pericyclic and acyclic transition states, giving good confidence in the consistency of the pressure kinetic methodology [12]. In addition, the above argument of the absence of correlation between $\Delta V^{\neq}$ and the structure of transition state ignores the hundred $\Delta V^{\neq}$- values determined for pericyclic processes which can widely vary according to the route followed (see below). Innumerable data are now available supporting the conjecture that concerted cycloadditions should have smaller transition states than the diradical stepwise analogues.

A former article reported on the interpretation of volumes of activation in a very general way [10], whereas the kinetic effect of pressure as a mechanistic probe has been perceived for a long time [13-20]. The volume of activation has also been applied to mechanistic studies in coordination chemistry [21].

The knowledge of $\Delta V^{\neq}$ and $\overline{\Delta V}$, as stated above, enables the positioning of the transition state in the volume profile. Their ratio θ is of highest interest, as it gives an indication of the progression of the transition state depending on the mechanism followed. The method is now highlighted with selected examples taken from earlier and recent articles [22].

2. Volumes of activation and mechanistic studies

2.1. ISOMERIZATIONS, REARRANGEMENTS AND RELATED REACTIONS

2.1.1. *Intramolecular Rearrangement of Amides.* The volume of activation of cyclohexane inversion shows a low negative value (-1.9 $cm^3.mol^{-1}$) indicating obviously that only C-C bond rotation is involved [23]. The situation differs for the intramolecular rearrangement of formamides (Table 3) [24].

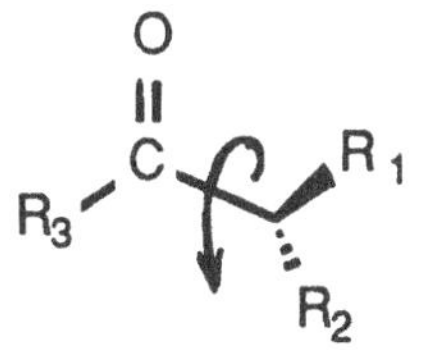

The positive $\Delta V^{\neq}$-values can be accommodated on the basis of steric requirements of the rotating group. Unsubstituted amides show the lowest values whereas rotation about the C-N bond in N,N-dimethylamides is more pressure decelerated. The observed $\Delta V^{\neq}$-values are also indicative of conformational and electrostriction effects, depending on the nature of the alkyl and aryl groups.

TABLE 3. Activation values for amide rotation

R_1	R_2	R_3	$\Delta G^{\neq}$ (kJ. mol^{-1})	$\Delta V^{\neq}$ (cm^3.mol^{-1})
H	H	H	68	2.3
H	H	Me	71	2.2
Me	Me	Me	75	10.3
Me	Me	Ph	62	8.6
Et	Et	Ph	61	7.7
Pr^i	Pr^i	Ph	57	5.4
Me	Pr^i	Me	72	10.6

2.1.2. *Aromatization of (Dewar) Benzenes.* The aromatization of hexamethyl (Dewar) benzene shows a surprisingly negative value for $\Delta V^{\neq}$ (-12 cm^3.mol^{-1}) against all expectation. The volume expansion from the cleavage of the 1,4-bond in simple (Dewar) benzene is effectively in accord with a positive activation volume (5 cm^3.mol^{-1}) [25]. The result can directly be related to the induction of considerable steric interactions between the six methyl groups in the planar molecule, hexamethylbenzene, in agreement with the concept of $\Delta V_{\sigma}^{\neq}$ underlined in the introduction.

Me Me Me Me Me Me → Me Me Me Me Me Me

2.1.3. *Isomerization of N=N and C=N Compounds.* The mechanism of cis-azoalkanes isomerization was the subject of great debate (Fig. 2). It is related to their thermal decomposition (see 2.2.1). It was suggested that at least two mechanisms can be envisioned : inversion of a R group through a nitrogen atom (i.e. semilinearization) and alternative rotation about the N=N bond [26] (Fig. 2). Intermediate diazenyl radicals could also be responsible for the isomerization process as inferred by $\Delta V^{\neq}$-values [27]. The values measured for the azonorbornane reaction are $\Delta V^{\neq} = +6.1$ cm^3.mol^{-1} [26]. Similar values were found in the isomerization of other cis-diazenes [28]. It was concluded that the reaction proceeds via a non-radical inversion mechanism with a transition state lying midway along the reaction coordinate [26-28].

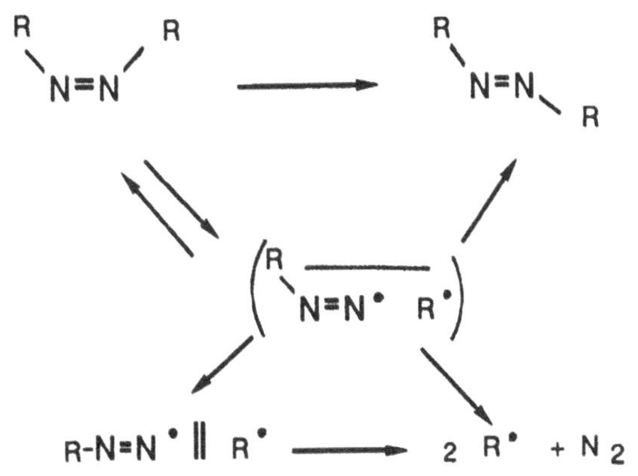

Figure 2. Possible mechanisms for isomerization of cis-azoalkanes.

Via a similar approach, Asano's group determined pressure effects on Z-E isomerization of various azobenzenes [29]. Two opposing mechanisms may be conceivable : rotation about the N=N bond (involving a rotational transition state) or flip-flop inversion of one of the N atoms (leading to an inversion transition state). The essential difference between the two mechanisms is that the rotational mechanism proceeds via a dipolar transition state exemplified by a deep minimum in the volume profile (Fig. 3).

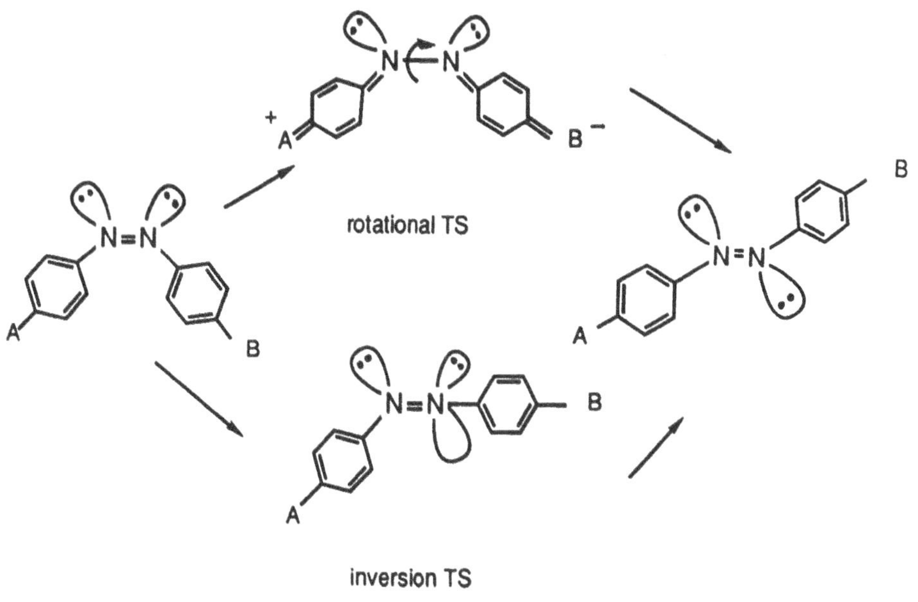

Figure 3. Mechanistic pathways in the isomerization of azobenzenes.

It was found that the mechanism depends on the nature of A and B. With push-pull substituents, both mechanisms may compete to various degrees, as revealed by pressure kinetic studies [30]. This is particularly evidenced in the isomerization of trans-4-dimethylamino-4'-nitroazobenzene (Table 4) : in low polar solvents, inversion is predominant whereas it changes to rotation in the most polar solvents [30]. It should be pointed out that, controversely, in case A = NMe_2, the isomerization was claimed to proceed only via an inversion transition state [31].

For other azobenzenes, the situation greatly differs according to substituent A. In the case of nitroazobenzene, the rotational mechanism cannot compete with the inversion mechanism. When A is a methoxy group, the effect is only modest. For A = NEt_2 and B = H there is a significant solvent effect on $\Delta V^{\neq}$ (cf. results in hexane and methanol), suggesting that the lowering of the rotational transition state can be realized by the strong electron-donor dialkylamino group without the assistance from a pulling group providing a suitable solvent.

TABLE 4. Solvent dependence of $\Delta V^{\neq}$ in Z-E isomerization of azobenzenes[a]

Solvent	$\Delta V^{\neq}$ ($cm^3.mol^{-1}$)			
	A = H	A = NMe_2	A = OMe	A = NEt_2
Hexane	-3.67	-3.0	-1.18	-1.68
benzene		-22.0	-2.64	-1.73
dichloroethane			-6.44	
methanol	-3.48	-27.6	-4.63	-17.55

a) B is NO_2 except when A is NEt_2, T is 20-25°C except when A is NMe_2 (40°C)

A recent study emphasizes the pressure effect on the degenerate isomerization of hexafluoroacetone anils [32]. These reactions involve isomerization about a C=N bond and are featured by low $\Delta V^{\neq}$-values. The results are explained by assuming two different inversion states.

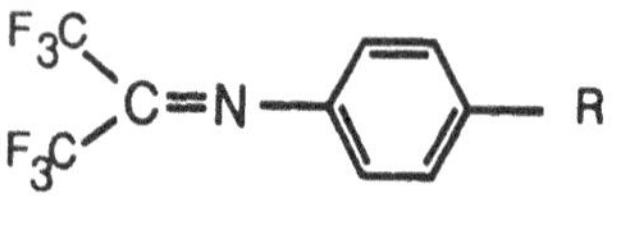

$(F_3C)_2C{=}N{-}C_6H_4{-}R$

R = OMe $\Delta V^{\neq}$ = + 3.7 $cm^3.mol^{-1}$

R = NMe_2 $\Delta V^{\neq}$ = + 0.5 $cm^3.mol^{-1}$

2.2. THERMAL DECOMPOSITION REACTIONS

2.2.1. *Thermolysis of radical initiators.* Homolytic scission reactions are characterized by $\Delta V^{\neq}$-values which can largely vary. The general mechanism of decomposition of initiators must make allowance for recombination of the initial geminate radicals (return), for diffusion of the radical pair and for the possibility of cage reactions [17].

$$\text{initiator (A-B)} \underset{k_{-i}}{\overset{k_i}{\rightleftharpoons}} \underset{\text{(radical pair)}}{\overline{A^\circ B^\circ}} \begin{cases} \xrightarrow{k_c} \text{cage reactions} \\ \xrightarrow{k_d} \text{separative diffusion} \end{cases}$$

with $$k_{obs} = k_i \left[\frac{k_d + k_c}{k_{-i} + k_d + k_c} \right]$$

and $$\Delta V^{\neq}_{obs} = \Delta V_i^{\neq} + RT \frac{\partial}{\partial P} \,\mathrm{Ln} \left[1 + k_{-i} / \left(k_d + k_c \right) \right]$$

Selected $\Delta V^{\neq}$-values are listed in Table 5.

TABLE 5. Volume of activation for decomposition of initiators[a]

Initiator	$\Delta V^{\neq}$ ($cm^3.mol^{-1}$)	Mechanism
$Me_2C(CN)$-N=N-$C(CN)Me_2$	+4	concerted
PhCOO-$OCMe_3$	+4	concerted
Me_2C-C_6H_4-N=N-C_6H_4-CMe_2	+5	concerted
(Z) EtCH=C(CH_2Et)COO-$OCMe_3$	+9	mixed or stepwise
MeC_5H_4-N=N-C_5H_4Me	+16	stepwise
adamantyl-N=N-adamantyl	+17	stepwise
Ph_3C-N=N-$C_6H_4NO_2$	+18 to +21	stepwise
Ph_3C-N=N-Ph	+16 to +19	stepwise

a) Taken from ref. 17, 28,33.

Neuman concluded that small $\Delta V^{\neq}$-values (0-5 $cm^3.mol^{-1}$) are consistent with concerted cleavage and large ones indicate stepwise pathways involving one-bond scission and formation of intermediates that can return to the starting initiator. As an example, decomposition of phenylazotriphenylmethane yields a phenyldiazenyl radical that recombines with the ge minate trityl radical to regenerate the initiator in competition with separate diffusion and β-scission [33].

2.2.2. *Miscellaneous thermolyses.* The large $\Delta V^{\neq}$-values (+31 $cm^3.mol^{-1}$) found in the thermolysis of liquid bibenzyl at 400°C was interpreted in the light of cage effects [34].

The thermolysis of the β-lactone L yielded also a high $\Delta V^{\neq}$-value but in the negative way ($-28\ cm^3.mol^{-1}$) in spite of the large positive reaction volume ($+52\ cm^3.mol^{-1}$) [35]. The value is indicative of a stepwise pathway via a highly dipolar transition state (Fig. 4).

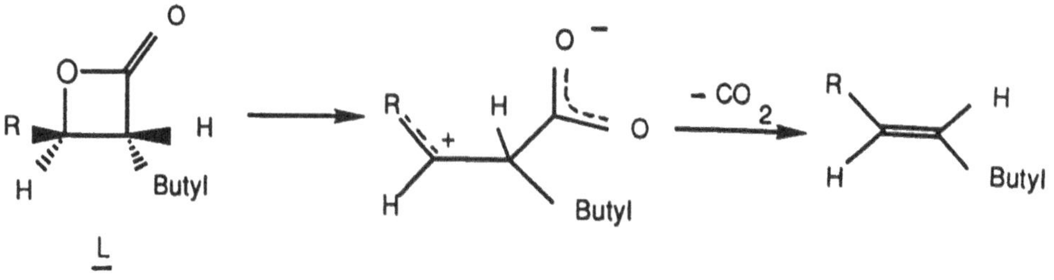

Figure 4. Thermolysis of β-lactone L

2.3. SUBSTITUTION AND ELIMINATION REACTIONS

These reactions are extremely common in organic chemistry. They proceed via a wide spectrum of mechanisms. The example below illustrates the general scheme.

$$R\text{-}CH_2\text{-}CH_2\text{-}X \xrightarrow{OH^-} RCH_2\text{-}CH_2OH + X^- \quad \text{substitution}$$

$$R\text{-}CH_2\text{-}CH_2\text{-}X \xrightarrow{OH^-} RCH{=}CH_2 + H_2O + X^- \quad \text{elimination}$$

2.3.1. *Elimination reactions.* They are characterized either by the unimolecular mechanism (E1) or by the bimolecular mechanism (E2) or by more complex mechanisms (E1cB, E2C). They offer similarities i.e. in basic media, eliminations proceed either via an E1, E2 or E1cB pathway. The latter is a E1-type elimination with formation of a carbanion as intermediate.

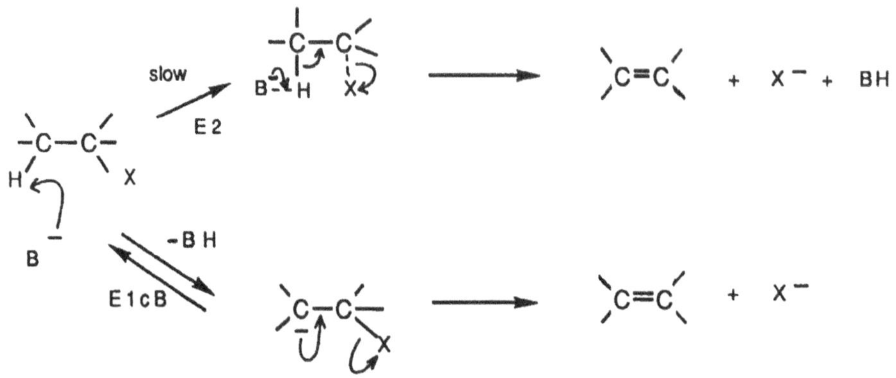

The pressure effect was examined on second-order elimination reactions of halides and other neutral substrates with anionic bases [36]. E2 reactions should display a negative volume of activation as the bond formation (B-H) and the partial bond cleavage (C-X)

occur in concert. E1cB mechanisms should be characterized by a positive $\Delta V^{\neq}$-value as a reflection of the volume increase arising from the cleavage of the C-X bond whereas there is no volume variation in the equilibrium process involving proton transfer. These assumptions are sustained by the experimental $\Delta V^{\neq}$-values for reactions for which the mechanism is well established [36] (Table 6).

TABLE 6. Volume of activation for some elimination reactions

substrate	$\Delta V^{\neq}$ ($cm^3.mol^{-1}$)	Mechanism
2-Bromobutane	-10	E2
1-Chloro-2-phenylethane	-12	E2
(Z)-β-bromostyrene	-6	E2
Diacetone alcohol	+6	E1cB
Mandelonitrile	+12	E1cB
Chloroform	+16	E1cB

2.3.2. *Substitution reactions.* The pendant of elimination reactions are substitution reactions which occur via a diversity of mechanisms. The most popular in aliphatic nucleophilic substitution are the S_N1 and S_N2 mechanisms.

$$Z^- + R_1R_2R_3C{-}X \xrightarrow[S_N2]{slow} [Z\cdots C(R_1)(R_2)(R_3)\cdots X] \longrightarrow Z{-}CR_1R_2R_3 + X^-$$

$$R_1R_2R_3C{-}X \xrightarrow[S_N1]{slow} X^- \; R_1R_2R_3C^+ \xrightarrow{Z=H_2O} R_1R_2R_3C{-}O^+H_2 \underset{}{\overset{-H^+}{\rightleftharpoons}} R_1R_2R_3C{-}OH$$

The sign of $\Delta V^{\neq}$ cannot operate the dichotomy between both mechanisms since S_N1 and S_N2 are both accelerated by pressure. However, the magnitude of $\Delta V^{\neq}$ is different : the ionogenic S_N1 reaction is normally more accelerated by pressure than the S_N2 process. In the alcoholysis of 2,2,2-triphenylethyl-p-toluene sulfonate in the presence of sodium alkoxide, both S_N1 and S_N2 processes compete, with an enhanced beneficial effect for the first mechanistic type [37].

Solvolysis reactions are strongly accelerated by pressure, when π-participation is operative in the process. Compounds 1 and 2 were submitted to alcoholysis under pressure [38]. The respective $\Delta V^{\neq}$-values were determined -13.3 and -24.0 $cm^3.mol^{-1}$ (S_N1 processes).

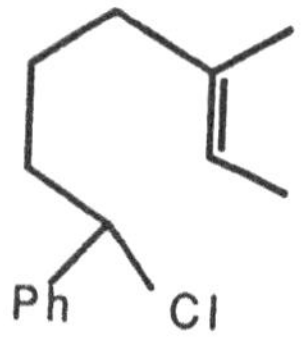

π-participation supposes the formation of an extra bond between the charged center and the olefinic bridge with an additional volume contraction. However, the significant $\Delta V^{\neq}$-difference observed between both solvolyses is much larger than expected. This was ascribed to conformational changes required for the prealignment of double bonds leading to extended π-participation (in possible relation to a steric factor $\Delta V_{\sigma}^{\neq}$).

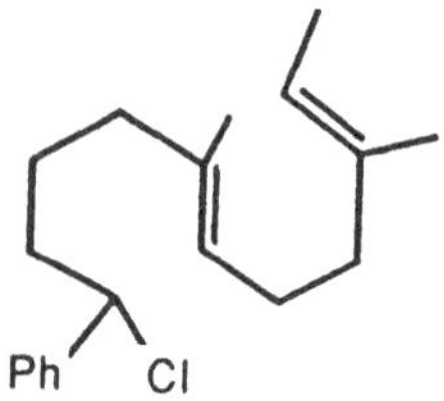

The base-catalyzed hydrolysis of esters is mainly featured by the BAC2 mechanism:

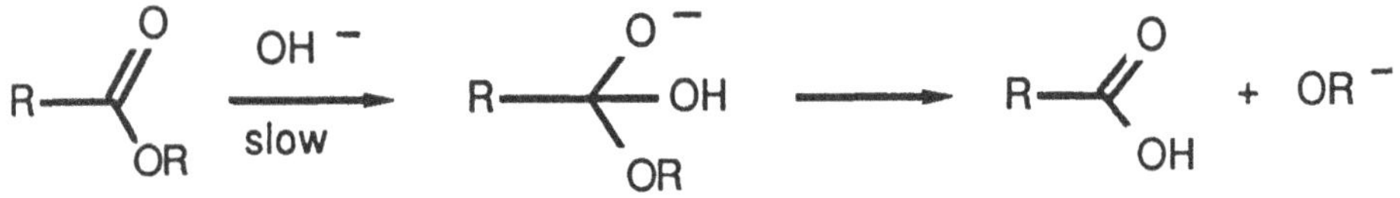

The volume of activation is usually negative if such mechanism operates [39]. However, in one case an eliminative E1cB mechanism could be detected (Fig. 5) through the volume of activation [40] (Table 7).

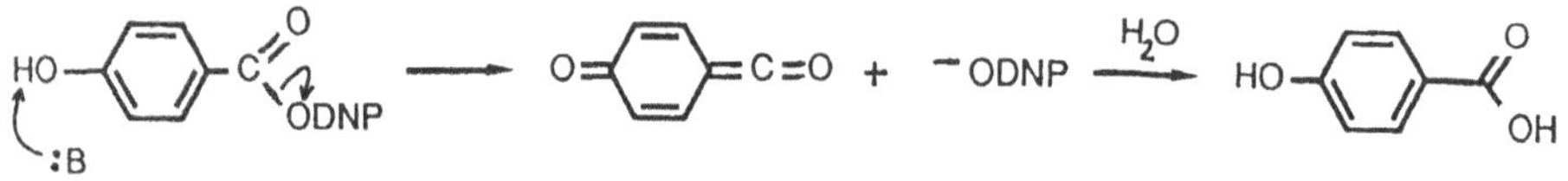

Figure 5. E1cB mechanism in the hydrolysis of 2,4-dinitrophenyl-4-hydroxybenzoate.

TABLE 7. Volumes of activation for ester hydrolyses

Ester[a]	$\Delta V^{\neq}$ ($cm^3.mol^{-1}$)	Mechanism
(ONP) p-nitrophenylacetate	-9	$B_{AC}2$
(PNP) p-nitrophenylacetate	-20	$B_{AC}2$
(DNP) p-methoxybenzoate	-19	$B_{AC}2$
(DNP) m-hydroxybenzoate	-15	$B_{AC}2$
(DNP) p-hydroxybenzoate	+12	E1cB

a) ONP, PNP, DNP : ortho-, para-, di- (nitrophenyl).

2.4 ADDITION TO MULTIPLE BONDS

The addition of keto compounds to multiple bonds in the presence of a base is called the Michael reaction. In the presence of a tertiary amine, the reaction is diverted along an α-substitution pathway.

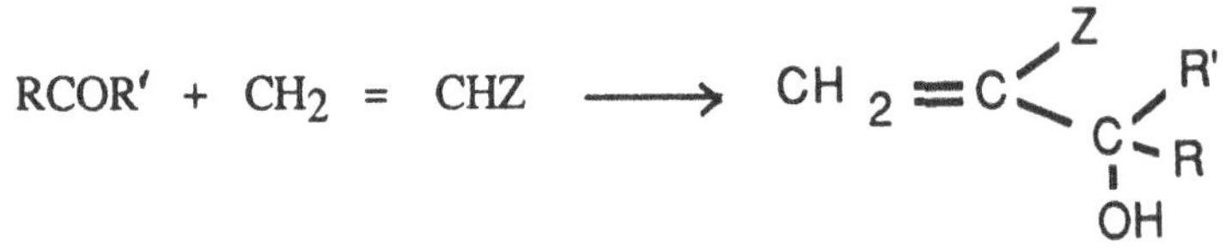

$$RCOR' + CH_2 = CHZ \longrightarrow CH_2{=}C(Z){-}C(R')(R)OH$$

This reaction was found to exhibit an extraordinarily sensitivity to pressure [41] with an apparent major shrinkage of the reacting species along the reaction coordinate. As an example, in the addition of acetaldehyde to acrylonitrile, the volume of activation has the astonishing value of ~ -79 $cm^3.mol^{-1}$, one of the highest absolute $\Delta V^{\neq}$-value ever observed [42]. The result was explained by assuming the volume of activation as the sum of two volume contributions characterizing the following steps (Fig. 6).

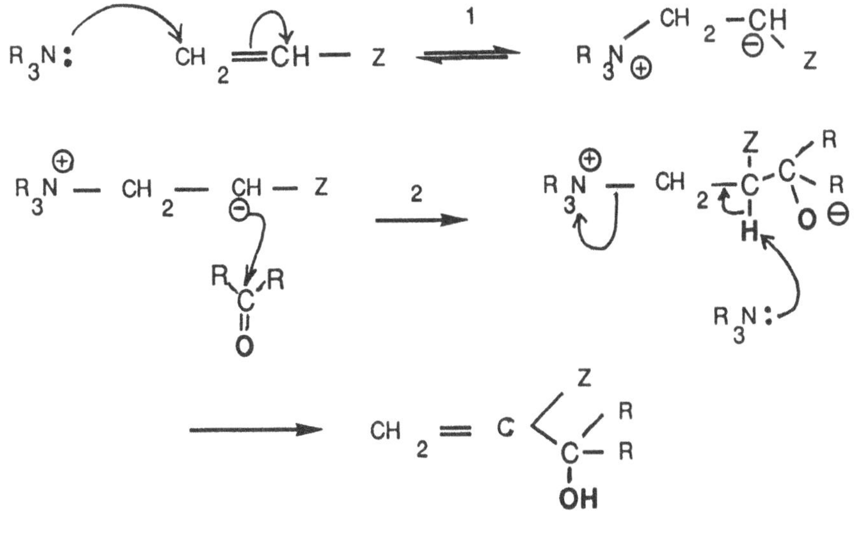

$$R_3N: + CH_2{=}CH{-}Z \overset{1}{\rightleftharpoons} R_3N^{\oplus}{-}CH_2{-}CH^{\ominus}{-}Z$$

$$R_3N^{\oplus}{-}CH_2{-}CH^{\ominus}{-}Z + R_2C{=}O \xrightarrow{2} R_3N^{\oplus}{-}CH_2{-}C(Z)(H){-}C(R)(R)O^{\ominus} \quad (+ R_3N:)$$

$$\longrightarrow CH_2{=}C(Z){-}C(R)(R)OH$$

Figure 6. Proposed mechanism for the α-substitution pathway.

$$\Delta V^{\neq}_{obs} = \Delta V_1 + \Delta V_2^{\neq}$$

Step 1 is an equilibrium whereas step 2 is rate-determining. ΔV_1 is assumed to amount -40 to -50 $cm^3.mol^{-1}$ and $\Delta V_2^{\neq}$ about -20 to -30 $cm^3.mol^{-1}$ [42].

2.5. PERICYCLIC REACTIONS

This family involves reactions occurring via cyclic or pseudocyclic transitionstates. They are characterized by a large spectrum of mechanisms from a rate-limiting formation of intermediates, such as biradicals, to full concertedness. Three kinds of pericyclic reactions are examined here. More exhaustive literature is provided in ref. [20].

2.5.1. *Sigmatropic rearrangements.* The volume of activation in Cope and Claisen rearrangements is consistent with concerted processes (-10 to -20 $cm^3.mol^{-1}$) [20]. In one case, a biradical pathway could be evidenced [43] : the $\Delta V^{\neq}$-value for the [1,4] sigmatropic rearrangement of 2-alkoxypyridine N-oxides is -30 $cm^3.mol^{-1}$ when R is $PhCH_2$ and +10 $cm^3.mol^{-1}$ when R is Ph_2CH (Fig. 7).

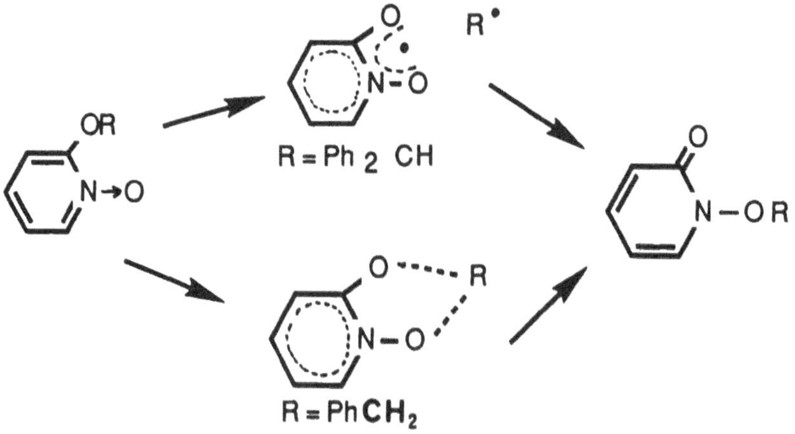

Figure 7. [1,4] shift in 2-alkoxypyridine N-oxides

Degenerate Cope rearrangements present $\Delta V^{\neq}$--values ranging from -0.5 to -26 cm^3. mol^{-1} indicating variable transition state structures from cyclic to loose biradical-like TS [44]. The [1,7] sigmatropic hydrogen shift in the interconversion of previtamin D_3 and vitamin D_3 is concerted ($\Delta V^{\neq}$ = -5.1 $cm^3.mol^{-1}$) [45], as is also the [1,9] sigmatropy of 2-(benzyloxy)-3-bromotropone ($\Delta V^{\neq}$ = -11 $cm^3.mol^{-1}$) [46].

An apparently related reaction is the rearrangement of O-silylated ketene acetals during which silicon migrates from oxygen to α-carbon. With the help of the high pressure methodology, the process was found to occur intermolecularly at variance with the previously suggested intramolecular route [47] (Fig. 8).

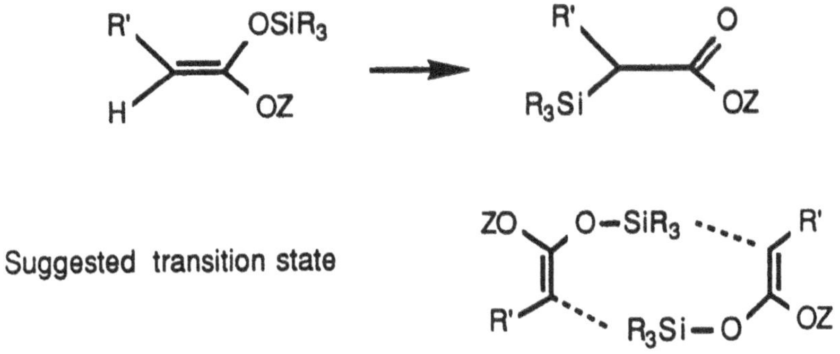

Suggested transition state

Figure 8. Intermolecular rearrangement of O-silylated ketene acetals.

2.5.2. *Cycloaddition reactions.* A paradigm of pericyclic processes is the Diels-Alder reaction for which impressive mechanistic arguments have been mustered to operate mechanistic dichotomy. The high pressure kinetic methodology was applied to many reactions. The following example is a parangon of [4+2] cycloadditions involving dienes and dienophiles with normal electron demand. For the addition of acrylonitrile to 2-methylfuran, the volume of activation was determined for the forward reaction [48] and its reverse [49] (Table 8).

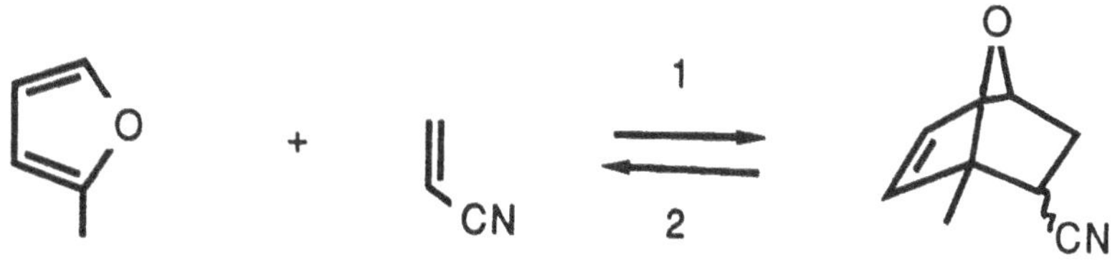

TABLE 8. Volume data for the condensation of 2-methylfuran and acrylonitrile

Volumes \ Reaction	1	2
$\Delta V^{\neq}$ ($cm^3.mol^{-1}$)	-30.3	-2.0
$\overline{\Delta V}$ ($cm^3.mol^{-1}$)	-28.7	+28.7
$\lvert\theta\rvert$ a)	1.06	0.07

a) θ is the ratio $\Delta V^{\neq} : \overline{\Delta V}$

The $\Delta V^{\neq}$-value for the forward reaction is fairly negative, in harmony with a quasicyclic and tight transition state, so far as the volume profile is concerned. In virtue of the principle of microreversibility the transition state in the reverse reaction should be early. This is actually observed and points to a perfect concertedness of the pericyclic process. In addition, a very interesting deduction can be made in observing that $\lvert\Delta V^{\neq}\rvert > \lvert\overline{\Delta V}\rvert$ for reaction 1. This is in possible relation with the existence of secondary orbital interactions contributing to enhance the stability of the transition state making it more rigid [50]. The slightly negative $\Delta V^{\neq}$-value for the reverse reaction is unique to date. It means that this pericyclic thermolysis process is accelerated by pressure, a rather unexpected result since it is solvent unsensitive. The example reveals the fruitfulness of the $\Delta V^{\neq}$-criterion and its notable mechanistic prevalence over other kinetic and thermodynamic arguments such as activation enthalpy and entropy.

Diels-Alder reactions involving neutral electron demand substituents are highly interesting to investigate since the donor-acceptor properties of the addends are less favourable to concertedness with possible occurence of stepwise mechanisms : while dimerization of isoprene (θ = 0.88) seems to proceed via a concerted route, 2,3-dimethylbutadiene (θ = 0.73) and cyclopentadiene (θ = 0.72) cyclodimerize probably via mixed pathways [20].

The high pressure cyclodimerization of 1,3-cyclohexadiene yields five products. The [4+2] endo-2 and threo-4 are formed by concerted mechanisms ($\Delta V^{\neq} = -28$ and -32 $cm^3.mol^{-1}$ respectively), but the [2+2] cycloadducts are visibly formed in two steps (-18 and -22 $cm^3.mol^{-1}$) as is also, more surprisingly, the [4+2] exo-2 adduct [51]. The thermal dimerization of the highly substituted and, therefore crowded, 1,1,2,2,3,3-hexamethyl-4,5-bis(methylene)cyclopentane, proceeds via two parallel [4+2] and [4+4] stepwise processes as indicated by low $\Delta V^{\neq}$-values all of the same order of magnitude (-15 $cm^3.mol^{-1}$) [52] (Fig. 9).

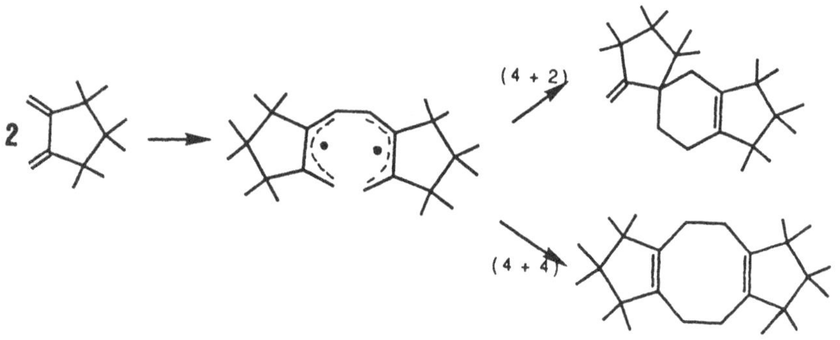

Figure 9. Cyclodimerization of hexamethyl bis (methylene) cyclopentane.

A closely related cycloaddition is the homo-Diels-Alder reaction. The condensation of norbornadiene and acetylenic dienophiles occurs via a concerted pathway ($\theta \sim 0.85$-0.90) [53]. The homo-Diels-Alder reaction of homofuran with olefins was examined under pressure [54]. Two mechanisms can be envisioned : a concerted $[(\pi^2+\sigma^2)+\pi^2]$ cycloaddition and an electrocyclic ring opening followed by a 1,3-dipolar cycloaddition.

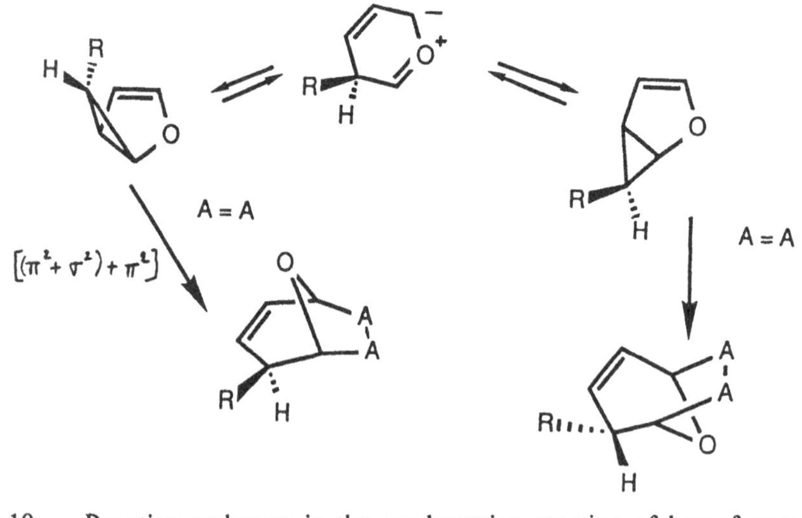

Figure 10. Reaction pathways in the condensation reaction of homofurans and fumaronitrile (A=A)

An elegant study based on kinetic measurements involving homofuran and derived methyl epimers at high pressures (up to 700 MPa) could clarify the mechanistic and stereochemical course of the reaction (Fig. 10). An increase in pressure was beneficial to the bimolecular addition compared to the unimolecular ring opening.

The pressure study of the cycloaddition of phenylhalocarbenes to alkenes led to $\Delta V^{\neq}$-values ranging from -10 to -18 $cm^3.mol^{-1}$. They support a mechanism involving either reversible formation of carbene-alkene complex or an early one- or two bond transition state [55] (Fig. 11).

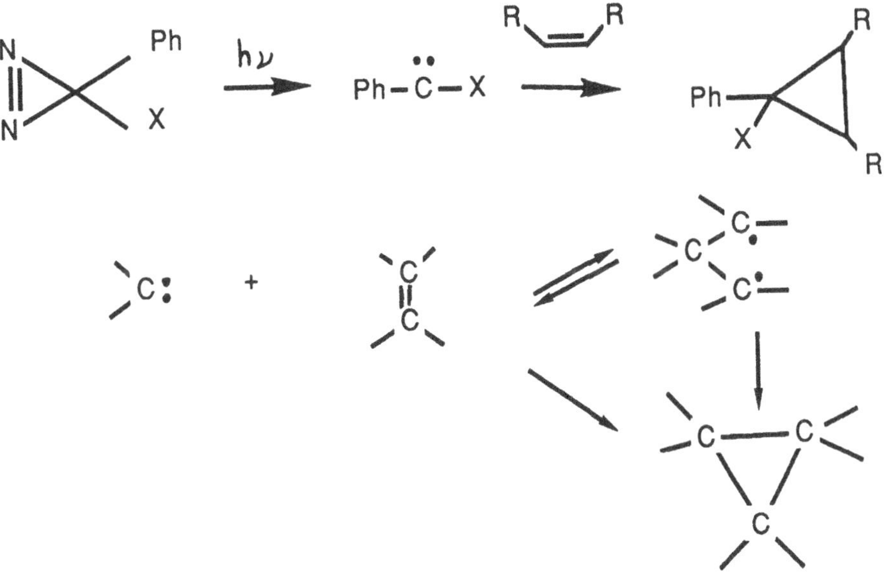

Figure 11. Addition of phenylhalocarbenes to alkenes.

2.5.3 *Ene reactions.* The ene reaction can be paralleled with the Diels-Alder cycloaddition despite the lack of symmetrical transition state. The reaction involves hydrogen transfer from Z to B and formation of a X-A bond as below :

A wide variety of mechanisms has been found or suggested (diradical, concerted, stepwise with three-membered or zwitterion or [2+2] cyclic intermediates). Pressure kinetic studies are a valuable tool to distinguish between concerted and stepwise mechanisms according to the type of hydrogen transfer (Z--H--B) [20] (Table 9).

TABLE 9. Mechanisms for ene reaction.

Ene	Enophile	Type of transfer Z -- H -- B	θ	Mechanism[b)]
1-Hexene	dimethylmesoxalate	C -- H -- O	1.05	Cc
2-Ethylbutene	" " "	C -- H -- O[a)]	1.05	Cc
2-Ethylbutene	" " "	C -- H -- O[a)]	1.12	Cc
β-Pinene	diethylmesoxalate	C -- H -- O	1.06	Cc
β-Pinene	diMeacetylenedicarboxylate	C -- H -- C	1.12	Cc
Allylbenzene	diethylazodicarboxylate	C -- H -- N	0.75	S(?)
Cyclopentene	" " "	C -- H -- N	0.68	S
Cyclohexene	" " "	C -- H -- N	0.58	S
1,4-Cyclohexadiene	" " "	C -- H -- N	0.88	Cc
Acetone	diethylmesoxalate	O -- H -- O	0.88	Cc

a) (Z) and (E) addition respectively

b) Cc : concerted S : stepwise

C--H--O and C--H--C hydrogen transfer processes are concerted. Interestingly, the θ-values are all higher than unity, indicative of a strained transition state and accommodating the view that it is more "compressed" than the final state. This can be interpreted on the basis of a non-linear hydrogen transfer process with an angular transition state [56] (Fig. 12).

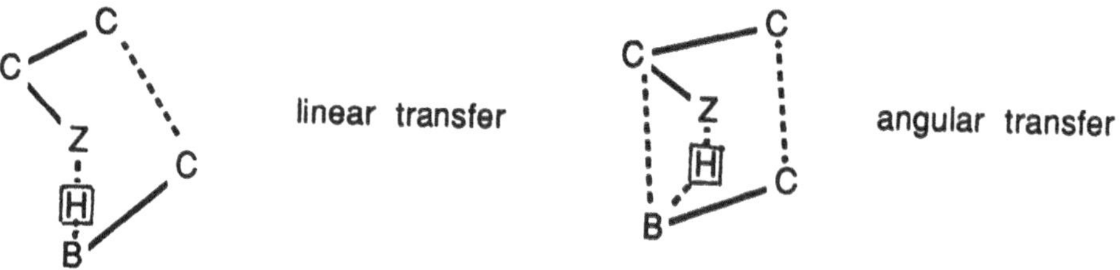

Figure 12. Linear and angular transfer in ene reactions.

Contrastingly, most examined (C--H--N) ene processes follow a sequential mechanism either via a diradical pathway or a possible three-membered intermediate [57]. The notable exception is the 1,4-cyclohexadiene reaction [58]. Comparison between the cyclohexene and the 1,4-cyclohexadiene ene additions ($\Delta V^{\neq}$ = -19.7 and -30 $cm^3.mol^{-1}$ respectively) suggests two different mechanisms explained by the greater flexibility and the spatial conformation behavior of cyclohexene vs the rigidity of 1,4-cyclohexadiene -a quasi planar molecule- for which the hydrogen atom is transferred orthogonally. A recent paper has reviewed the volume profiles for the ene reaction on the basis of linear free energy relationships [59].

3. Conclusion :

As abundantly shown in this chapter, high pressure kinetic effects permit assessment of mechanisms for organic reactions through the volume of activation. Obviously, $\Delta V^{\neq}$ is mechanistically interpretable only if the reaction profile is carefully analyzed (examination of solvent and steric effects, comparison of $\Delta V^{\neq}$ with the reaction volume ΔV). Very often, the volume of activation is an additional argument to support mechanisms advocated by other lines of evidence. In some cases, $\Delta V^{\neq}$ is even more conclusive than any other parameter. More detailed information can also be gained when $\Delta V^{\neq}$ is determined with accuracy.

In conclusion, the method is useful for the kineticist to elaborate mechanistic models more expeditiously.

REFERENCES :

[1] Le Noble, W.J. (1980) 'The volume profile of chemical reactions', Rev. Phys. Chem. Jpn 50, 207-216.

[2] Jenner, G. (1984) 'Analyse du profil volumique des réactions péricycliques à l'aide de la piézochimie', Bull. Soc. Chim. France II, 275-284.

[3] Lohmüller, R. Macdonald, D.D. Mackinnon, M. and Hyne, J.B. (1978) 'The volume of activation of benzyl chloride hydrolysis and its pressure dependence', Can. J. Chem. 56, 1739-1745.

[4] Asano, T. and Okada, T. (1984) 'New simple functions to describe kinetic and thermodynamic effects of pressure', J. Phys. Chem. 88, 238-243

[5] Hartmann, H. Brauer, H. Kelm, H. and Rinck, G. (1968) 'Zum Einfluss des Lösungsmittel auf chemische Reaktionen in flüssiger Phase', Zeit. Phys. Chem. (N.F.) 61, 53-62.

[6] Fleischmann, F.K. and Kelm, H. (1973) 'Cycloaddition reaction betwenn tetracyanoethylene and n-butyl vinyl ether', Tetrahedron Lett., 3773-3776.

[7] Jenner, G. (1985) 'The pressure effect on strained transition states', J. Chem. Soc. Faraday Trans. 1, 81, 2437-2460.

[8] Firestone, R.A. and Vitale, M.A. (1981) 'Vibrational activity. Acceleration of bond-making reactions by solvent viscosity as a model for enzymic catalysis', J. Org. Chem. 46, 2160-2164.

[9] Firestone, R.A. and Saffar, S.G. (1983) 'Effect of solvent density on the rate of a Diels-Alder reaction', J. Org. Chem. 48, 4783-4784.

[10] Kelm, H. and Palmer, D. (1978) 'Determination and interpretation of volumes of activation', in H. Kelm (ed.), High Pressure Chemistry, Reidel Publishing Company, Dordrecht, pp.281-309.

[11] Firestone, R.A. and Smith, G.M. (1989) 'The roles of changes in bonding vs. packing fraction in the pressure-induced acceleration of the Diels-Alder reaction', Chem. Ber. 122, 1089-1094.
[12] Klärner, F.G; Ruster, V. Zimny, B. and Hochstrate, D. (1991) 'The effect of pressure on pericyclic reactions', High Press. Res. 7, 133-135.
[13] Hamann, S.D. (1964) 'High pressure chemistry', Ann. Rev. Phys. Chem. 15, 349-370.
[14] Whalley, E. (1964) 'Use of volumes of activation for determining reaction mechanisms', Adv. Phys. Org. Chem. 2, 93-162.
[15] Le Noble, W.J. (1967) 'Reactions in solutions under pressure', J. Chem. Educ. 44, 729-739.
[16] Eckert, C.A. (1972) 'High pressure kinetics in solution', Ann. Rev. Phys. Chem. 23, 239-264.
[17] Neumann, R.C. (1972) 'Pressure effects as mechanistic probes of organic radical reactions', Acc. Chem. Res. 5, 381-387.
[18] Jenner, G. (1975) 'High pressure kinetic investigations in organic and macromolecular chemistry', Angew. Chem. Int. Ed. 14, 137-143.
[19] Le Noble, W.J. and Kelm, H. (1980) 'Chemistry in compressed solutions', Angew. Chem. Int Ed. 19, 841-856.
[20] Jenner, G. (1989) 'Pericyclic reactions', in W.J. le Noble (ed.), Organic High Pressure Chemistry, Elsevier, Amsterdam, pp.143-203.
[21] Kotowski, M. and van Eldik, R. (1989) 'Application of high pressure kinetic techniques to mechanistic studies in coordination chemistry', Coordin. Chem. Rev. 93, 19-57.
[22] Values of volumes of activation are presented in a voluminous article. Van Eldik, R. Asano, T. and le Noble, W.J. (1989) 'Activation and reaction volumes in solution', Chem. Rev. 89, 549-688.
[23] Lüdemann, H. Rauchschwalbe, R. and Lang, E. (1977) 'Determination of the activation volume of simple molecular rearrangements by ^{1}H-NMR spectroscopy at high pressures", Angew. Chem. Int. Ed. 16, 331-332.
[24] Hauer, J. Völkel, G. and Lüdemann, H.D. (1980) 'High pressure NMR. $\Delta V^{\neq}$ and $\Delta G^{\neq}$ for the rotation of the amide group', J. Chem. Res. (S), 16-17.
[25] Le Noble, W.J. Brower, K.R. Brower, C. and Chang S. (1982) 'Pressure effects on the rates of aromatization of hexamethyl Dewar benzene, J. Am. Chem. Soc. 104, 3150-3152.
[26] Van Eldik, R. Kelm, H. Schmittel, M. and Rüchardt, C. (1985) 'Concerning the mechanism of isomerization of cis-azoalkanes', J. Org. Chem. 50, 2998-3000.
[27] Neumann, R.C. and Binegar, G.A. (1983) 'Cis-azoalkanes. Mechanisms of scission and isomerization', J. Am. Chem. Soc. 105, 134-135.
[28] Neumann, R.C. Berge, C.T. Binegar, G.A. Adam, W. and Nishizawa, Y. (1990) 'Cis-diazenes. Pressure effects on their thermal deazatization and isomerization reactions', J. Org. Chem. 55, 4564-4568.
[29] Asano, T. (1989) 'Stereoisomerizations, in W.J. le Noble (ed.), Organic High Pressure Chemistry, Elsevier, Amsterdam, pp.123-142.
[30] Asano, T. and Okada, T. (1984) 'Thermal Z-E isomerization of azobenzenes', J. Org. Chem. 49, 4387-4391.
[31] Nishimura, N. Tanaka, T. and Sueishi, Y. (1985) 'Evidence of an inversion mechanism for the thermal cis-trans isomerization of push-pull azobenzenes', J. Chem. Soc. Chem. Comm., 903-904.

[32] Swaddle, T.W. Doine, H. Kinrade, S.D. Sera, A. Asano, T. and Okada, T. (1990) 'Hith pressure 19FNMR study of the degenerate isomerization of hexafluoroacetone anils', J. Am. Chem. Soc. 112, 2378-2382.

[33] Neumann, R.C. and Lockyer, G.D. (1983) 'One-bond azo initiators. Thermal decomposition of substituted (phenylazo) triphenylmethanes', J. Am. Chem. Soc. 105, 3982-3987.

[34] Brower, K.R. (1980) 'Kinetics and pressure effect in the thermolysis of liquid bibenzyl', J. Org. Chem. 45, 1004-1008.

[35] Isaacs, N.S. and Laila, A.H. (1983) 'The volume profile for thermolysis of a beta-lactone', Tetrahedron Lett. 24, 2897-2900.

[36] Brower, K.R. Muhsin, M. and Brower, H.E. (1976) 'Activation volume as a measure of E2-E1cB character in elimination reactions', J. Am. Chem. Soc. 98, 779-782.

[37] Lee, K. and Okamoto, Y. (1976) 'Reaction of 2,2,2-Triphenylethyl-p toluene sulfonate in alcohol solutions under high pressure', J. Org. Chem. 41, 1552-1555.

[38] Ho, N.H. and le Noble, W.J. (1989) 'Pressure effects in a solvolysis involving extended π participation', J. Org. Chem. 54, 2018-2021.

[39] Isaacs, N.S. and Najem, T. (1986) 'Characterisation of $B_{AC}2$ and E1cB routes of ester hydrolysis by measurement of activation volumes', Physica 139&140B, 701-704.

[40] Isaacs, N.S. and Najem, T. (1984) 'Evidence from activation volumes for an elimination mechanism in the hydrolysis of a hydroxybenzoate', J. Chem. Soc. Chem. Comm., 1361-1362.

[41] Hill, J.S. and Isaacs, N.S; (1988) 'Nucleophile-catalysed addition of aldehydes and ketones to acrylic compounds', J. Chem. Res., (S) 330-331, (M) 2641-2676.

[42] Hill, J.S. and Isaacs, N.S. (1986) 'Functionalisation of the α position of acrylate systems by the addition of carbonyl compounds', Tetrahedron Lett. 27, 5007-5010.

[43] Le Noble, W.J. and Daka, M.R; (1978) 'Effect of pressure an concerted and stepwise sigmatropic shifts', J. Am. Chem. Soc. 100, 5961-5962.

[44] Schulman, E.M; Merbach, A.E. Turin, M. Wedinger, R. and le Noble, W.J. (1983) 'Effect of pressure on the rates of sigmatropic shifts', J. Am. Chem. Soc. 105, 3988-3991.

[45] Dauben, W.G. Kowalczyk, B.A. and Funhoff, D.J. (1988) 'Interconversion of previtamin D_3 and vitamin D_3 at high pressure', Tetrahedron Lett. 29, 3021-3024.

[46] Sugiyama, S. Mori, A. and Takeshita, H. (1987) 'High pressure analysis of [1,9] sigmatropy of 2-(benzyloxy)-3 bromotropone', Chem. Lett., 1247-1250.

[47] Yamamoto, Y. Maruyama, K. and Matsumoto, K. (1984) 'A new procedure for the synthesis of α-silylated esters and lactones. Migration of silicon from oxygen to carbon at high pressure', Organomet. 3, 1583-1585.

[48] Rimmelin, J. Jenner, G. and Rimmelin, P. (1978) 'Influence de la pression dans la synthèse diénique du système oxa-7-bicyclo [2.2.0] heptane à partir du méthyl-2 furanne', Bull. Soc. Chim. France II, 461-464.

[49] Jenner, G. Papadopoulos, M. and Rimmelin, J. (1983) 'Activation parameters and location of the transition state in the retro-Diels-Alder reaction of a 7-oxabicyclo [2.2.1] hept-5-ene derivative, J. Org. Chem. 48, 748-749.

[50] Jenner, G. (1991) 'Volume profiles of pericyclic reactions involving rigid transition states. A pressure study', New J. Chem. 15, 897-899.

[51] Klärner, F.G. Dogan, B.M. Ermer, O. von Doering, W. and Cohen, W.P. (1986) 'Mechanism of the thermal 1,3-cyclohexadiene dimerization', Angew. Chem. Int. Ed. 25, 108-110.

[52] Baran, J. Mayr, H. Ruster, V. and Klärner, F.G. (1989) 'Stepwise [4-2] and [4+4] cyclodimerizations of 1,1,2,2,3,3 hexamethyl-4,5 bis (methylene) cyclopentane', J. Org. Chem. 54, 5016-5019.

[53] Jenner G. and Papadopoulos, M. (1982) 'The pressure effect in the homo-Diels-Alder reaction', Tetrahedron Lett. 23, 4333-4336.

[54] Klärner, F.G. and Schröer, D. (1989) 'Organische Reaktionen unter hohem Druck : zum Mechanismus der homo-Diels-Alder-Reaktion von Homofuran', Chem. Ber. 122, 179-185.

[55] Turro, N.J. Okamoto, M. Gould, I.R. Moss, R.A. Lawrynowicz, W. and Hadel, L.P. (1987) 'Volumes of activation for the cycloaddition reactions of phenylcarbenes to alkenes', J. Am. Chem. Soc. 109, 4973-4976.

[56] Jenner, G. and Papadopoulos, M. (1983) 'Concertedness in thermal intermolecular ene reactions of dimethyl mesoxalate and substituted alkenes', J. Org. Chem. 47, 4201- 4204.

[57] Jenner, G. Ben Salem, R. El'yanov, B and Gonikberg, E.M. (1989) 'Piezochemical interpretation of (C--H--X) hydrogen transfer in ene reactions', J. Chem. Soc. Perkin Trans II, 1671-1675.

[58] Jenner, G. and Ben Salem, R. (1990) 'Anatomy of ene and Diels-Alder reactions between cyclohexadienes and azodicarboxylates', J. Chem. Soc. Perkin Trans 2, 1961-1964.

[59] El'yanov, B. Gonikberg, E.M. and Jenner, G. (1992) 'Linear free energy relationship for the pressure dependence of rate constants in ene reactions', J. Chem. Soc. Perkin II, 137-142.

THE FUTURE OF HIGH PRESSURE ORGANIC SYNTHESIS

G. JENNER
Laboratoire de Piézochimie Organique,
European Higher Institute of Chemistry
B.P. 296
67008 Strasbourg
France

ABSTRACT. From the substantiation that, when compared to other activation processes, piezoactivation is hardly used as an organic synthetic strategy, the merits of the high pressure methodology are emphasized : synthesis of thermal fragile molecules, synthesis of molecules sensitive to Lewis acid catalysts, synthesis of strained or bulky structures, enhancement of chemo- and stereoselectivity, asymmetric induction.

Perspectives of high pressure organic synthesis are outlined. The prospective fields of interest encompass the synthesis of strained polycyclic molecules, the synthetic aspect of host-guest molecules with unorthodox topology (macrocyclic fused compounds, cryptands), the synthesis of natural products (peptides, sugars, total synthesis of potential pharmacological compounds), organometallic chemistry, the synthesis of polycyclic thio compounds.

Coupling of high pressure with ancillary activation methods should widen the application field and give consideration to the methodology as a pertinent and valid synthetic strategy.

1. The apparent marginality of piezosynthesis

In comparison with other physical methods for the activation of organic molecules, pressure has been employed for decades. Yet, high pressure equipment has found seldom room in the laboratory environment. Some reasons are obvious, other can only be perceived : heavy and costly devices (in fact, not true), small available volumes, chemical ostracism against non-glassy devices, limited applications, difficulty of transposition from laboratory to pilot plants.

In the 1910-1920 period, the achievement of the ammoniac synthesis via the Haber process under 20 MPa and the Claude alternative under 100 MPa were considered as a good starting-board for rapid growing. The success of the low-density polyethylene process under 200-300 MPa in the mid-thirties encouraged hopes for further developments. However, there was a lack of determining findings in the area until the discovery of the diamond synthesis by General Electric in the 1950's under very high pressures and temperatures. More recently, superhard boron nitride has been commercialized, but once again, there has been no major high pressure application in organic chemistry, however, with the notable exception of the oxo process which requires rather moderate pressures (5-20 MPa).

R. Winter and J. Jonas (eds.), High Pressure Chemistry, Biochemistry and Materials Science, 367–392.

In view of this clear-cut introduction, the future looks discouraging. In fact, every time the organic chemist needs piezoactivation -high pressure activation-, he tries to escape buying the equipement by seeking other more conventional methods. The following examples are revealing.

1.1. CHEMICAL ACTIVATION VIA CATALYSIS

The Diels-Alder reaction between multiring aromatics and maleic anhydride, which does not proceed under ambient pressure, gives fair yields under 1000 MPa at 100° C [1]. However, use of lower temperatures at 0.1 MPa leads to nearly quantitative yields in the presence of Nafion-H® as catalyst [2]. In the same manner, the Diels-Alder approach to podocarpane diterpenes can be achieved either under 600 MPa pressure or via $SnCl_4$ catalysis, the latter method being even more effective [3] !

Furans were long considered as reluctant dienes to enter cycloaddition with monoactivated dienophiles. Dauben demonstrated the ease of Diels-Alder reactions involving furans under high pressure [4]. However, soon after, it was reported that the same reactions could be achieved at normal pressure in the presence of catalysts such as Cu(I) and Cu(II) compounds [5,6], ZnI_2 [7], doped clays [8].

Sluggish [4+2] dimerizations are accelerated by high pressure [9,10]. However, catalyzed cyclodimerizations at atmospheric pressure proceed much easier : butadiene is dimerized in the presence of zeolites [11] and cyclohexadiene by cation-radicals [12]. The homo-Diels-Alder reaction involving norbornadiene is fairly accelerated by pressure [13]. However, at normal pressure catalysis by transition metal catalysts derivatives [14,15], leads to excellent yields of homo-cycloadducts.

1.2. PHYSICAL ACTIVATION

1.2.1. *Thermal activation.* This is evidently the first reflex in synthetic chemistry. Nevertheless, in that case, high pressure is generally far superior to just heating [16]. in the following reaction the yield at 0.1 MPa is 74 % at 100° C, whereas it reaches 90 % at 20° C under 1200 MPa [17].

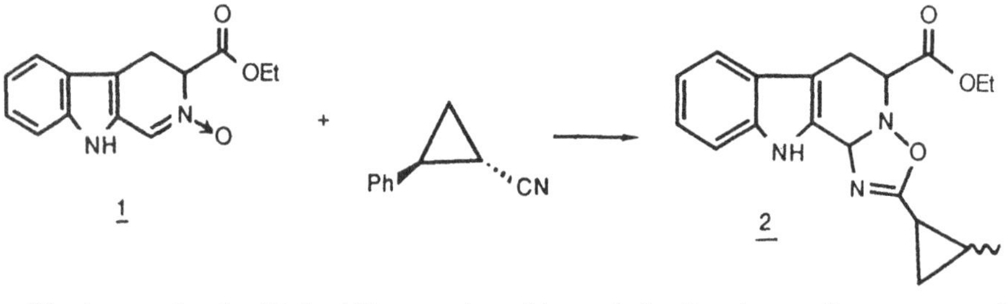

O OEt N O NH 1 + Ph CN → O OEt N O NH N 2

The intramolecular Diels-Alder reaction of long-chain dienyl-pyrrolinones proceeds either by heating to 120° C for 17 h (10 % yield) or better under 1200 MPa at 40° C (30 %) [18]. Another example related to the reaction shown above emphasizes the advantage of using high pressure vs. heating [19] (Table 1) (cycloaddition of **1** and R-C≡N).

TABLE 1. Cycloaddition of nitrone **1** and nitriles

R	Conditions	Yield of **2** %
COOEt	50° C, 1200 MPa, 28 h	75
	80° C, 0.1 MPa, 7 days	26
Ph	50° C, 1200 MPa, 2 days	56
	80° C, 0.1 MPa, 7 days	11
Me	50° C, 1200 MPa, 3 days	87
	80° C, 0.1 MPa, 6 days	12

1.2.2. *Ultrasonication.* Since the effects of ultrasonic waves are often compared to pressure effects due to cavitation phenomena generating local high pressures and temperatures, sonochemistry has been proposed in place of piezochemistry [20]. Table 2 refers to examples apparently sustaining this proposal [21] and confirmed later in the synthesis of abietanoid o-quinones [22] (Table 2).

TABLE 2. Cycloadditions[a] involving o-quinone **3**

Diene	Yield of adducts % Thermal reaction[b]	Ultrasound[c]	Hith pressure[d]
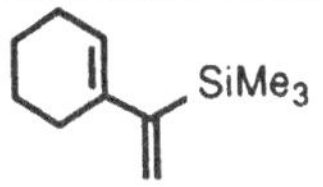 1-(1-cyclohexenyl)-1-(trimethylsilyl)ethene ($SiMe_3$)	61	57	61
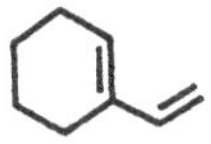 1-vinylcyclohexene	40	65	67
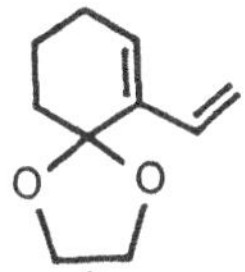 vinylcyclohexene ethylene ketal		65	75
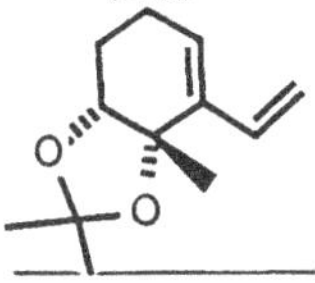 vinylcyclohexene acetonide		76	73

a : 3 is [structure of o-quinone 3]

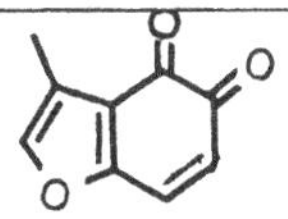

b : Reflux in methanol, 16 h
c : Neat, 45° C, 2 h
d : 1000-1100 MPa, 20° C, 2 h in toluene or methanol

1.2.3. *Microwaves*. Microwave heating is a relatively modern technique to promote chemical synthesis though the intimate causes have not been fully clarified [21]. [4+2] cycloadditions seem to be excellent candidates in this respect [23].

1.2.4. *Solvent "internal pressure"*. Recently a strange correlation has been established between pressure and solvents with high internal pressure [24]. A dramatic rate acceleration of some Diels-Alder reactions occurs when the reactions are carried out at normal pressure in 5 M lithium perchlorate-diethyl ether (LPDE). An outstanding example is the synthesis of cantharidine involving as a first step the Diels-Alder reaction of furan and 2,5-dihydrothiophene-3,4-dicarboxylic anhydride. The reaction was successfully achieved under high pressure up to 100 % yield [25]. Using LPDE as reaction medium, a 70 % yield could be obtained at ambient pressure and temperature [24]. However, it was shown since that the effect of LPDE is to act like Lewis acid catalysis and is not connected with any internal pressure effect [26]. As a consequence reactions that are insensitive to Lewis acid catalysis are not promoted in LPDE solutions.

These examples seemingly discourage use of high pressure strategies in effecting organic syntheses. Additional gegen-arguments can even be put forward. As a physical parameter, pressure affects the reaction only kinetically, it is unable to alter the electronic properties of the reaction partners -at least in the pressure range considered in this paper, up to 2000 MPa-. This means that the driving parameter in high pressure reactions is the volume of activation $\Delta V^{\neq}$. Accordingly, reactions with low $\Delta V^{\neq}$-values (early transition states) are hardly promoted by pressure. In other words, if two reactions are competing, the reaction featured by the more negative $\Delta V^{\neq}$-value is also the most accelerated. Consequently, some desired structures cannot be obtained at high pressures whereas catalysts not only influence yields, but also chemo-, stereo- and regioselectivities. The cycloaddition of norbornadiene and dimethylacetylene dicarboxylate yields two products **4** and **5** when it is catalyzed by $RuH_2\,CO(PPh_3)_3$, whereas the thermal reaction yields only the homo-Diels-Alder adduct **4**. At high pressure, in the catalytic reaction, **5** is dramatically disfavoured [27] (Table 3 : E is CO_2Me).

E——E E E E E

4 5

TABLE 3. Cycloaddition of norbornadiene and dimethylacetylene dicarboxylate

Catalyst	Pressure (MPa)	Time (h)	Conversion %	Selectivity % 4	5
none	0.1	96	85	100	0
none	300	16	100	100	0
$RuH_2CO(PPh_3)_3$	0.1	96	100	31	69
$RuH_2CO(PPh_3)_3$	300	16	100	87	13

[a] : Benzene, 80° C

This introduction presents high pressure as a marginal activation process. In fact, this strategy has not rivalled other current synthetic methodologies to date. This is apparently reflected by the low growing rate of paper edition reporting high pressure studies (Fig. 1).

What is it actually and what can be expected for the future ?

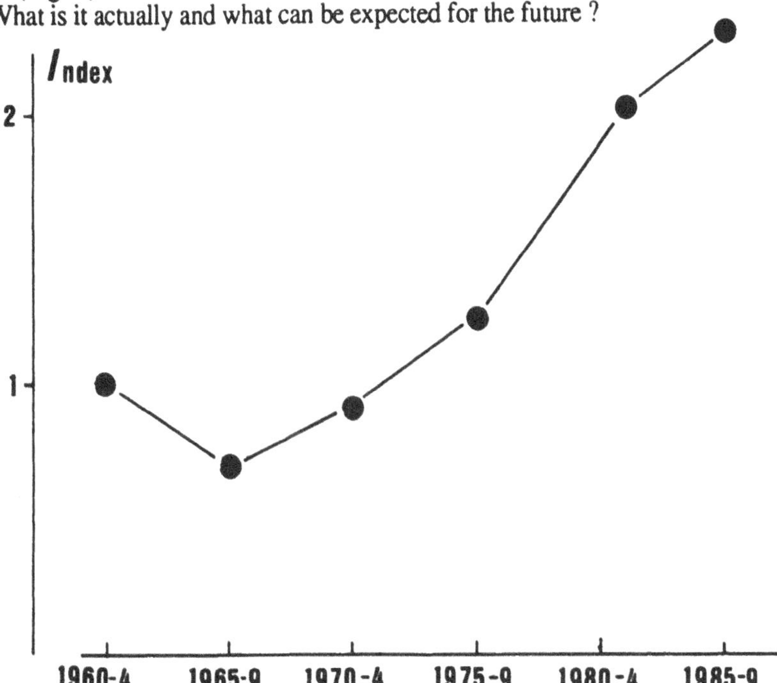

Figure 1. Edition of high pressure papers in organic chemistry (polymer chemistry not included) (Basis 1 for the period : 1960-1964).

2. The merits of piezosynthesis

First of all, it must be stressed out that it is totally incorrect to draw a parallelism between pressure and sonication effects on one side and pressure and Lewis acid catalysis on the other hand. In actual fact, the reactions presented above appear to be limiting cases. There exist numerous illustrative examples highlighted below, in which neither high temperature nor Lewis acids are able to bring the reactions to observable rates.

a) p-Benzoquinone adds to chiral 2,4-pentadienoic acid derivatives only under high pressure [28]. In the same way, with application of high pressure (1700 MPa), the [4+2] addition of crotonates to appropriate dienes affords cycloadducts in 40-75 % yield, which are intermediates in the synthesis of yohimbine-type indole alkaloids [29]. The interesting total synthesis of jatropholones was made possible by using 500 MPa pressure ; the adduct **6** was obtained in 80 % yield (0 % at 0.1 MPa) [30].

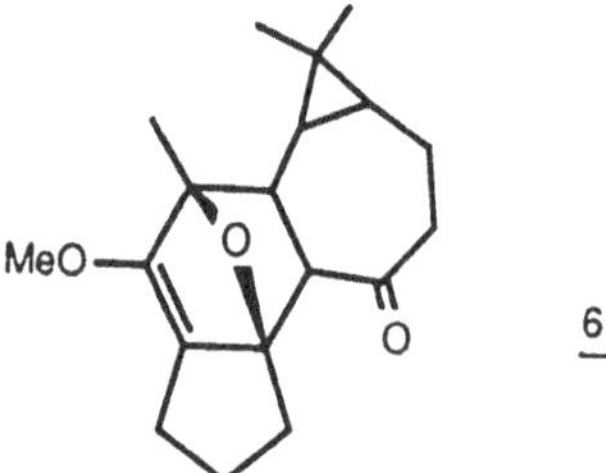

b) Intramolecular Diels-Alder reactions involving furans were examined under different conditions [31,32]. Only high pressure was shown to lead to high yield of **7**.

HP

$(CH_2)n$ ⇌ $(CH_2)n$ **7**

c) [4+2] cycloaddition of electrophilic 1,3-disulfonyl-2-pyridones with vinyl ethers proceeds regio- and stereoselectively yielding bicyclic lactams **8** under high pressure [33]. Lewis acids are uneffective and elevated temperatures induce polymerization of the dienophile.

RO_2S-N ... SO_2Tol + OR' ⟶ RO_2S, N, O, SO_2Tol, OR' **8**

In addition, every time cycloadducts can be obtained either under thermal, or catalytic or high pressure conditions, the latter proves generally to be a more powerful synthetic way as products are obtained in better yields or in greater purity or with a more appropriate stoechiometry. As an example, cycloaddition of cyclopropenes to conjugated dienes proceeds under pressure in high yield with 1:1 diene-dienophile stoechiometry whereas thermal reactions require large excesses of diene [34].

X + Y ⟶ X, Y

Recently, when comparing Lewis acid and high pressure methodologies in Diels-Alder additions of N-alkyl-N-sulfinyl dienophiles **9** and **11**, it was found that high pressure led to the best yields for the intra- as well as for the intermolecular reaction [35] (Table 4).

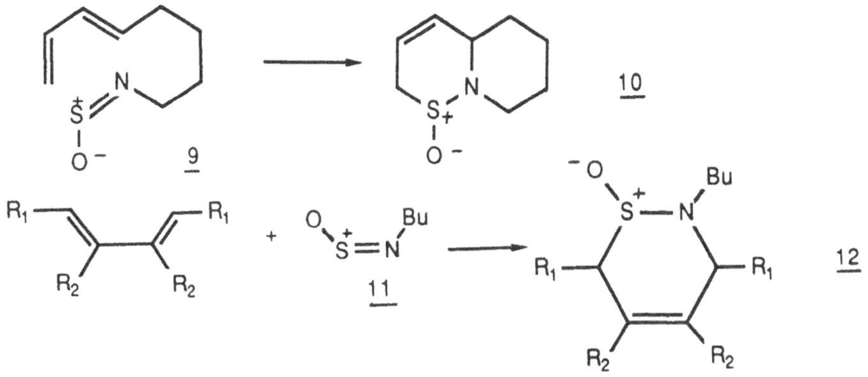

TABLE 4. Intra- and intermolecular cycloaddition of **9** and **11**

Conditions	Yield of **10** %	Yield of **12** % R^1=H, R^2=Me	R^1=Me, R^2=H
$TiCl_4$, -78° C		4	27
$SnCl_4$, -78° C		55	47
BF_3, -78° C	40	68	54
1200 MPa, 20° C	82	96	80

It is obvious that addition reactions deserve the major piezochemical interest. The selection of reactions can be accounted for specific reasons. To this purpose, it is desirable to recall the fundamental pressure relationship connecting the rate constant with the volume of activation :

$$\partial \mathrm{Ln}k / \partial P = -\Delta V^{\neq} / RT$$

The relation tells that in the pressure range at the laboratory scale (500-2000 MPa), the quasi-unique effect is of kinetic order. Reactions featured with a large volume shrinkage deserve the greatest attention to this respect. This should be outlined although the assertion has been highlighted in most high pressure books and reviews [36-39]. Table 5 shows approximative amplification kinetic ratios.

TABLE 5. Calculated rate constant ratios at 25° C

P (MPa) $\Delta V^{\neq}$ ($cm^3.mol^{-1}$)	10	20	30	40	50
500	3.3	11	37	122	405
750	5	25	128	643	3253
1000	7	52	375	2702	1.96×10^4
1500	13	172	2264	3×10^4	3.9×10^5

2.1. SYNTHESIS OF HEAT-SENSITIVE MOLECULES

The pressure parameter is of high value in the synthesis of thermal fragile molecules. If a reaction needs high temperature conditions to occur, then it can be expected to proceed with relative ease at lower temperature under high pressure. The rate acceleration is sometimes so dramatic that it is worthwhile to take advantage to decrease temperature as much as possible in order to shift the reaction in the desired direction by impeding or even, suppressing inverse or side reactions. Corollarly, lowering temperature has a beneficial effect on the stabilization of heat-sensitive molecules. A nice example is provided by Gladysz experiments [40]. The thermal extrusion of small stable molecules such as CO, CO_2, N_2 from neutral molecules is retarded by pressure which is known to decelerate fragmentation reactions [41].

The temperature decrease ($\Delta\theta = T_1 - T_2$) necessary to preserve isokinetic conditions can be estimated from the relation [42] :

$$T_1 / T_2 = 1 + R\,(\Phi/\Delta E)\,\Delta V^{\neq}$$

Φ is a complex function of pressure and ΔE is the activation energy at 0.1 MPa. Accordingly, $\Delta\theta$ depends not only on pressure, but also on the activation energy. From the relation it is clear that the more negative is $\Delta V^{\neq}$, the more affected is $\Delta\theta$. As an example, the [4+2] addition of acrylonitrile to 2-methylfuran has following activation parameters [43] : $\Delta V^{\neq} = -31\ cm^3.mol^{-1}$ at 40° C, $\Delta E = 75.2\ kJ.mol^{-1}$. Replacing these values in the relation yields $\Delta\theta \sim 80°$ C. Consequently, the reaction can be carried out at ambient temperature leading to high yields whereas no adduct is formed at 100° C due to its extreme thermal instability with reversal to starting materials. Numerous furan, thiophene and, even, pyrrole adducts have been prepared via the high pressure route [36, 38, 44]. An interesting case is provided by the [4+2] addition of simple 1,4-benzoquinones which takes place at room temperature only at 2000 MPa pressure [45].

The synthesis of bis-oxabicyclo compounds is notoriously difficult under ambient pressure, as the eventually formed adducts are highly unstable. While 2-pyrone is reported to react only with maleic anhydride and N-phenylmaleimid, other less reactive dienophiles lead to 1,3-cyclohexadienes and Diels-Alder products derived from the intermediate diene.

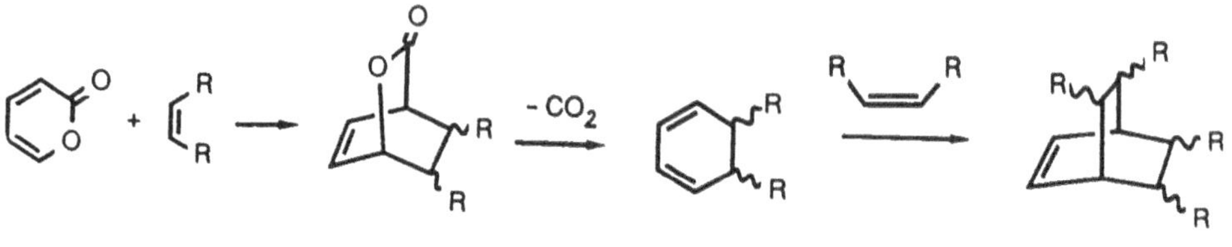

Operating at high pressure (500-1000 MPa) enables to stop the reaction at the 3-oxa-2-bicyclo [2.2.2] oct-7-ene stage [46]. The high pressure route has found a remarkable application in asymmetric Diels-Alder additions of esters of 3-hydroxypyran-2-one yielding useful intermediates for pharmacological purposes [47].

In the same area, strained olefins such as cyclopropene ketals react smoothly with pyrones (620 MPa). The adducts extrude CO_2 upon release of pressure, except with the parent pyrone , for which an unstable adduct can be recovered at ambient pressure [48].

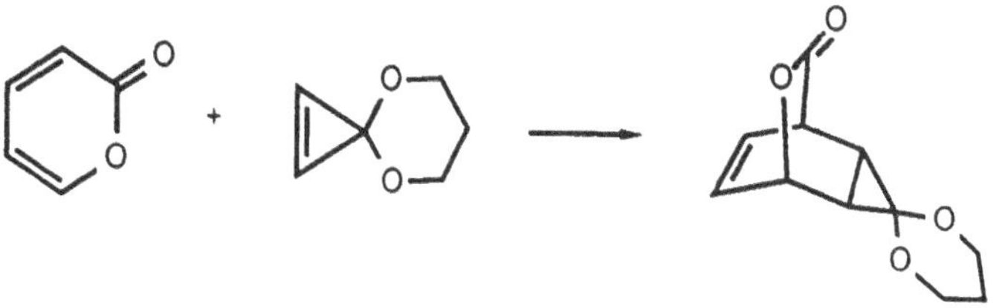

Parabenzoquinones are interesting dienophiles. However, [4+2] cycloadducts are usually obtained in low yields due to their propensity to aromatize under reaction conditions. The use of 1500 MPa pressure at 20° C alleviates this difficulty [49]. The aromatic character of pyrroles does not incline them to enter [4+2] addition. High pressure and low temperature are a mean to prepare cycloadducts [50].

Other reactions include [2+2] addition of toluene-4-sulfonyl isocyanate to glycals [51]. The cycloadducts formed are stable only under pressure and prone to revert to starting glycals upon recovering. The [3+2] dipolar cycloaddition of nitrones with vinyl ethers is fairly promoted by pressure [52]. A dramatic example is provided by the reaction:

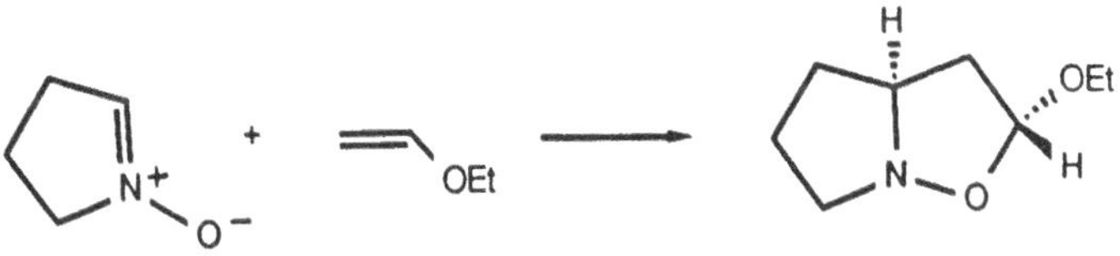

No cycloadduct can be obtained at 80° C and 0.1 MPa. Under a pressure as low as 200 MPa at 50° C, 83 % cycloadduct is formed in a very stereoselective way. Another illustrative report in the field considers the high pressure 1,3-dipolar cycloaddition of nitronic esters to steroidal dipolarophiles [53]. The thermolability of the starting 1,3-dipolarophiles precludes reaction at 0.1 MPa. The cycloadduct is produced with an insignificant yield, which is raised to 65-80 % under 1400 MPa.

Azide-electron deficient olefin dipolar cycloadditions afford triazolines in good yields under high pressure at room temperature. Heating is formally precluded due to triazoline instability [54].

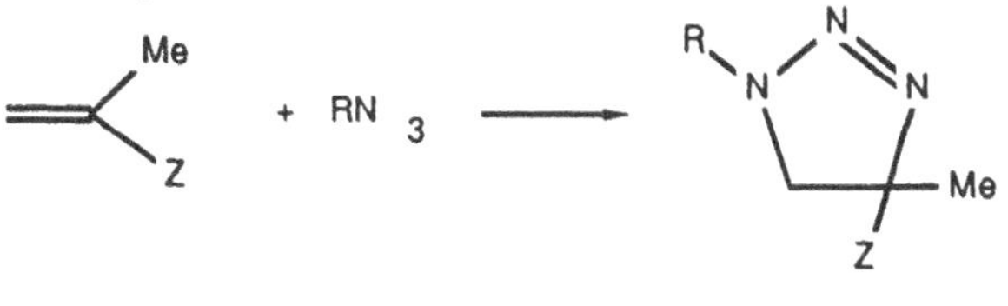

2.2. REACTIONS SENSITIVE TO LEWIS ACID CATALYSTS

The pressure parameter is of high value in reactions sensitive to conventional Lewis acid catalysts. Lewis acids can change the course of a reaction involving sensitive substrates (isomerization, polymerization...). [2+2] addition of isocyanates to glycals cannot proceed under Lewis acid conditions, which cause dimerization of glycals. The cycloadducts are obtained in good yields under pressure [51]. Though cycloaddition of trichloroacetyl isocyanate to a sugar aldehyde (rhamnal) proceeds under normal pressure conditions in the presence of protonic catalysts, the adducts are unstable whereas high pressure readily induces crystallization of cycloadducts [55].

In the crucial step of the synthesis of the antibiotic aklavinone, it was demonstrated that high pressure (1700 MPa, 65° C, 48 h) was the best method to synthesize the adduct (75 % yield), as the enone and the diene were substrates sensitive to Lewis acids [56].

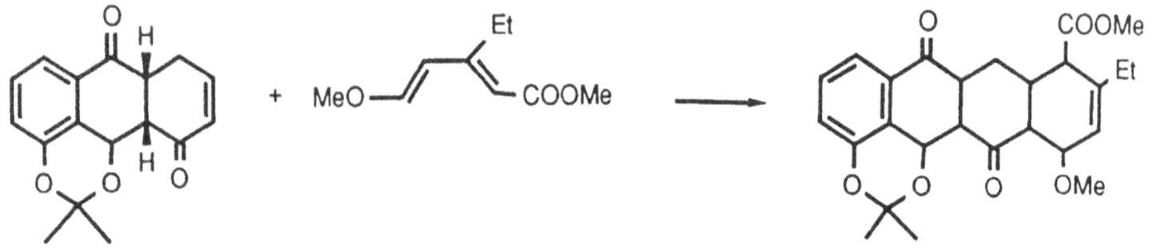

The total synthesis of bruceantin, an antitumor diterpenoid, involves as a key step the conjugate addition of a O-silylated ketene acetal to an activated enone [57]. Lewis acids cause degradation of the starting materials whereas the cycloadduct is formed quantitatively under 1500 MPa. The acid catalyzed intramolecular cyclization of unsaturated carbonyl compounds is not general as the produced sensitive alcohols might readily dehydrate. A method based on the use of silica gel as a mild Lewis acid at high pressure is generally applicable to a wide variety of unsaturated carbonyl systems [58].

An interesting property of the quadricyclane molecule is its mode of addition ($\pi^2+\sigma^2+\sigma^2$) to yield strained polyadducts (bishomodiene reaction). Quadricyclane is converted into its valence isomer -norbornadiene- in the presence of Lewis acids. High pressure is the favourite method to synthesize new quadricyclane adducts [59].

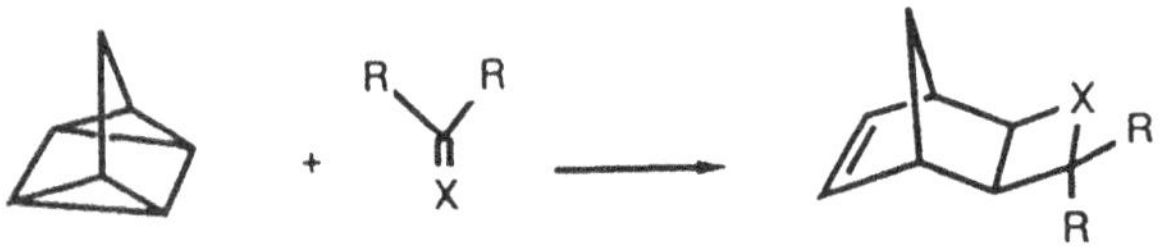

Annelated oxetans (X = O) [60] and thietans (X = S) [61] are easily produced in this way. In particular, carbon disulfide reacts under 900 MPa at 25° C to yield **13**.

A sequential double ($\pi^2+\sigma^2+\sigma^2$) process involving quadricyclane leads to a new polycyclic adduct. The bishomodiene reaction of two molecules of quadricyclane with one molecule of methyl propynoate yields 96 % of the bisadduct **14** under 1000 MPa at 100 ° C (E is COOMe) [62].

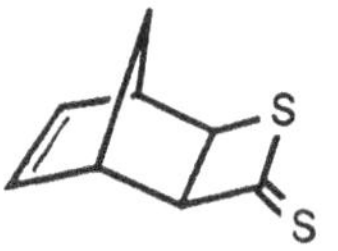

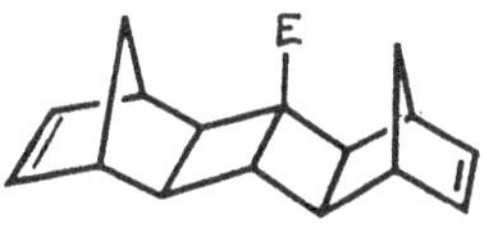

2.3. SYNTHESIS OF STRAINED AND BULKY MOLECULES

The pressure parameter is of high value for the synthesis of strained and crowded molecules via sterically demanding reactions. Some thirty years ago, the late Prof. Gonikberg found a strong pressure acceleration in Menshutkin reactions when comparing rate constants in the N-alkylation of crowded pyridines vs. unhindered ones [63]. Since the size of substituents involves geometric concepts and is reflected by the volume occupied by the substituents, it is reasonable to take into account the pressure parameter. In fact, steric manifestations should have an effect on the volume of activation [64]. The demonstration could be made that steric demands of substituents and/or enhanced strain energy are measurable as components of the volume of activation [65]. Unless other effects act in the opposite way, steric congestion produces late transition states, justifying the use of high pressure to overcome steric hindrance. The total volume of activation may be expressed as :

$$\Delta V^{\neq} = \Delta V_0^{\neq} + \Sigma\Delta V_\sigma^{\neq}$$

$\Delta V_0^{\neq}$, $\Delta V_\sigma^{\neq}$: volume of activation relative to uncrowded standard reaction and to strain or steric hindrance respectively.

The additional volume $\Delta V_\sigma^{\neq}$ is a dynamic term. When the steric demand increases, the transition state shifts downwards towards the product along the reaction axis. An important consequence of this concept is its application to organic synthesis. Intrusion of strain relief and steric hindrance induces a stronger pressure acceleration. Some relevant investigations in this field encompass following systems.

2.3.1. *Strained systems.* Cyclopropenes are very strained molecules. Under 800 MPa, at room temperature, the gem-dimethylsubstituted cyclopropene **15** dimerizes to yield a bicyclobutane **16** in 62 % yield [66]. Under ambient pressure, the yield is only 3 %.

15 CO_2Me → CO_2Me $MeCO_2$ 16

[2.2] paracyclophane **17** is a twisted molecule exhibiting high strain. High pressure makes the molecule amenable to react with maleic derivatives according to a Diels-Alder reaction [67].

17 + (O, X, O) →

Other strained systems were synthesized by methods utilizing high pressure [68].

2.3.2. *Congested systems.* The Menshutkin reaction offers the most straightforward example. Quaternary salts of sterically hindered bis-quinolizidine monolactams such as α-isolupanine and oxosparteine are formed in quantitative yield with the application of 2000 MPa pressure at 20° C [69]. The Menshutkin reaction has been used for the high pressure synthesis of cryptands [70]. Particularly remarkable is the quaternization of chiral diazacoronands with bis (iodoethyl) ethers via a double Menshutkin reaction, leading to excellent yields of the bisammonium salt under 1000 MPa, especially with increasing complexity of the diazacoronand. Numerous cryptands were synthesized in this way [71].

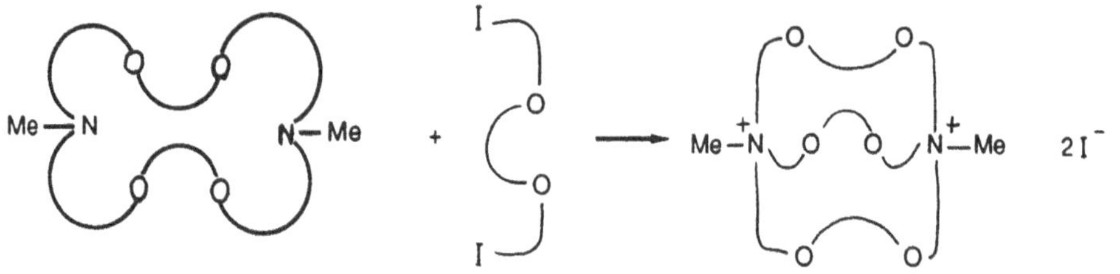

Some difficult reductions of hindered ketones were successfully carried out under high pressure with tributyltin hydride. Table 6 portrays prominent examples [72].

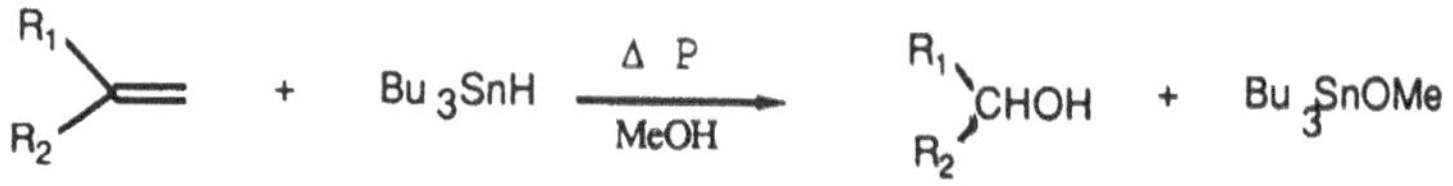

TABLE 6. High pressure reduction of ketones with Bu_3SnH (55° C, 24 h)

R_1	R_2	Yield of alcohols % 0.1 MPa	Yield of alcohols % 1000 MPa
Me	Bu^t	18	64
$Prop^i$	CH_2Ph	15	58
$Prop^i$	$Prop^i$	4	62
$Prop^i$	CH(Me)Ph	2	63
Et	CH(Me)Ph	12	68
Bu^t	CH(Me)Ph	0	57

The high pressure methodology is also rewarding in the synthesis of hindered peptides. Peptides containing an α,α-disubstituted glycine part are synthesized by four-component condensation under 900 MPa [73].

2.4. SELECTIVITY

The pressure parameter is a mean to improve selectivity.

2.4.1. *Chemoselectivy.* The chemoselectivity of a reaction carried out under pressure depends on the volume of activation. At ordinary pressure, many reactions involving unsaturated substrates are affected by parallel polymerization which limits the yield of the desired product. The condensation of unsaturated carbonyl compounds and enol derivatives is an illustrative exemple [74]. At atmospheric pressure, the Diels-Alder reaction yields dihydropyrans in appreciable yield only above 150° C, but polymerization, retro-Diels-Alder reaction and rearrangement to open-chain products occur as well. Combination of high pressure and low temperature leads selectively to the targeted cycloadducts.

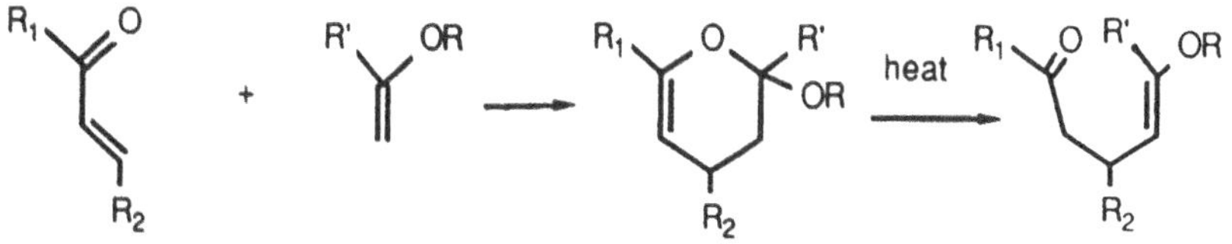

Cycloheptatriene reacts with dienophiles across the norcaradiene structure yielding [4+2] or [6+2] adducts and can undergo ene reactions. Under high pressure, [6+2] adducts are never formed due to the non-concerted nature of this process. With diazene esters, the ene and the [4+2] addition reaction should be equally favoured on the basis of their similar volumes of activation. Nevertheless, the equilibrium cycloheptatriene $\rightleftharpoons$ norcaradiene -a 6π electrocyclic reaction- is promoted by pressure, thus leading besides the ene product **18** to the unprecedented [4+2] norcaradiene adduct **19** [75].

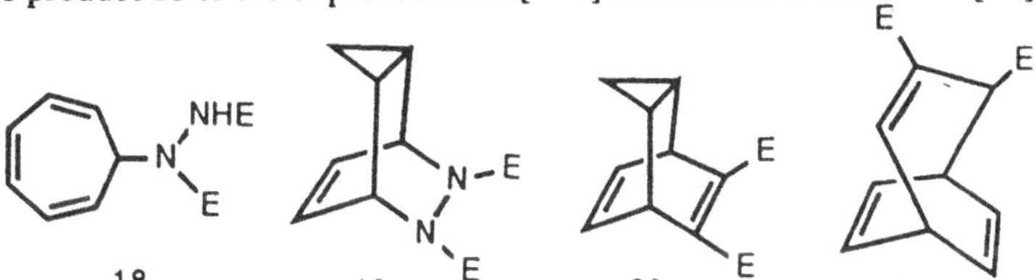

Acetylenic bonds add to cycloheptatriene yielding **20** and **21**. Under 900 MPa, only the norcaradiene adduct **20** is formed. Under high pressure conditions, methyl propynoate ($E = CO_2Me$) adds to cycloheptatriene yielding four products (**22** - **25**) all exhibiting the norcaradiene structure.

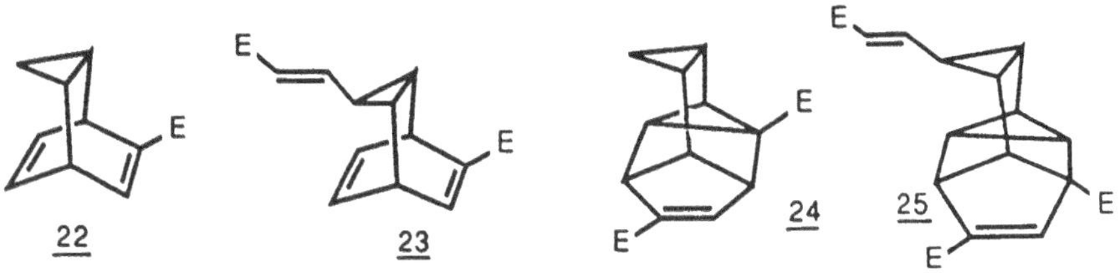

22 is the normal [4+2] adduct, **23** is formed according to a sequential process via an initial ene reaction followed by [4+2] addition of the acetylenic bond, **24** and **25** are resulting homo-Diels-Alder adducts. These examples show that high pressure promotes the selective formation of cycloheptatriene adducts retaining the norcaradiene structure.

Cyclopropylketones are chemoselectively and nearly quantitatively reduced to cyclopropylalcohols by Bu_3SnH under 1400 MPa [72], whereas only ring-opened products are obtained at ambient pressure.

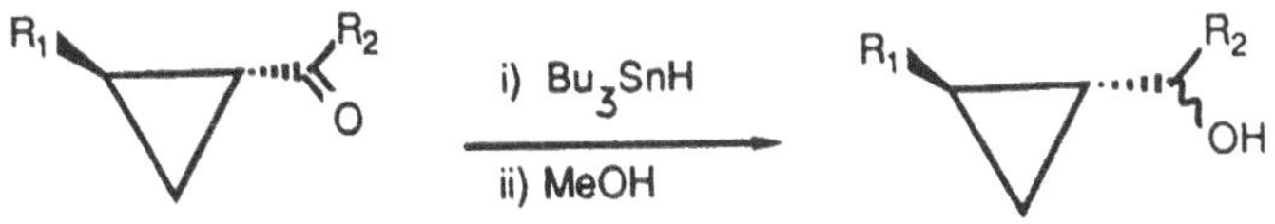

i) Bu_3SnH
ii) MeOH

Cleavage of esters occurs chemo- and regioselectively under pressure in the presence of tertiary amines [76].

2.4.2. *Stereoselectivity and optical activity.* High pressure may improve stereoselectivily. The addition of butenolides to electron-rich dienes under high pressure conditions leads to the expected cycloadducts in a high regio- and stereoselective way [77]. A significant pressure-induced diastereoselectivity is observed in Diels-Alder reactions involving vinyl ethers and enaminoketones [78].

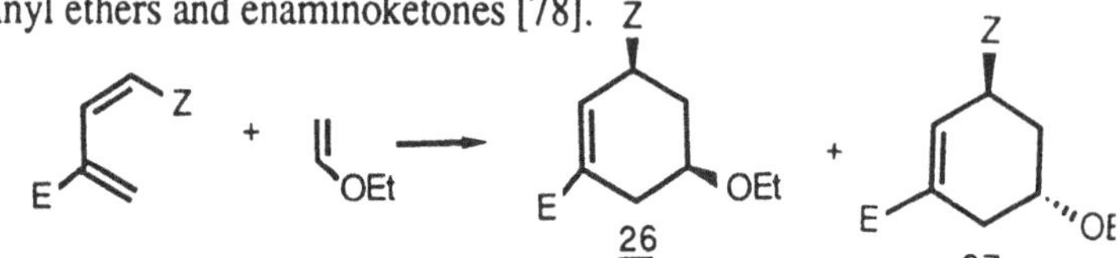

If Z = (phthalimido) and E = CCl_3 , **26** : **27** is 1.67 : 1.00 at 0.1 MPa
13.67 : 1.0 at 600 MPa

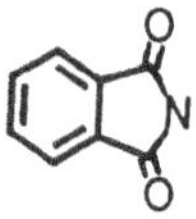

The pressure effect on diastereoselectivity is highlighted in the aldol reaction between silyl enol ethers and aldehydes (Mukaiyama reaction) under neutral conditions [79]. There is a strong pressure effect, as the erythro : threo ratio increases from 0.3 to 3 when pressure is varied from 0.1 to 1200 MPa. This means that while at low pressure a chair transition state is preferred, high pressure induces a crossover to a boat transition state.

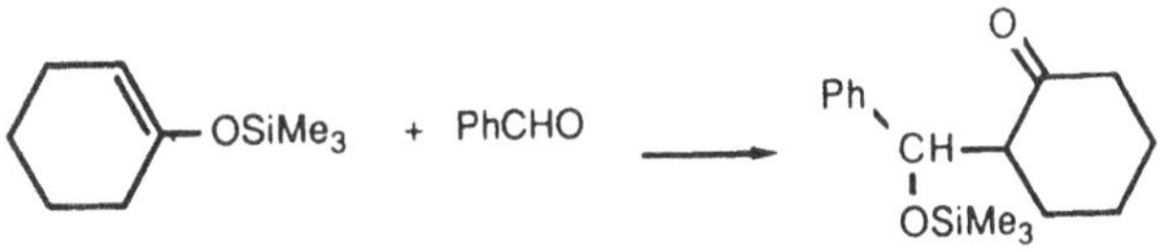

The high pressure methodology represents a way to give access to optically pure synthons. A nice exemple is the enantioselective synthesis of β-aminoesters via [80] :

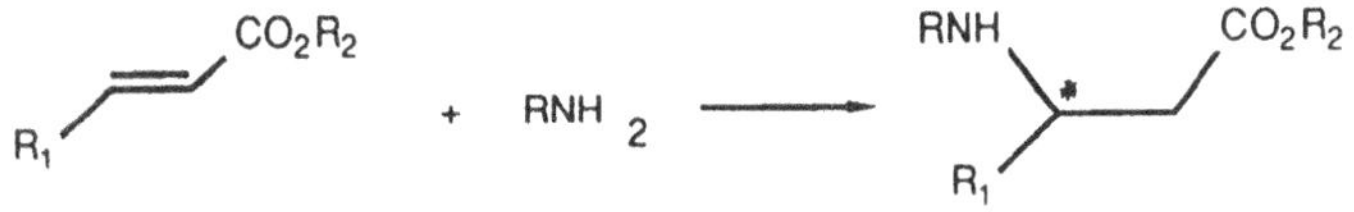

Asymmetric induction is obtained in the high pressure reaction of 2,3-O-iso-propylidene-D-glyceraldehyde with 1-methoxybutadiene [81]. The effect on asymmetric induction is very perceptible : under high pressure the transition state leading to adduct **28** is preferred. The study shows that ultra high pressure (2000 MPa) and use of solvents with high freezing points are very favourable on the direction of asymmetric induction [82].

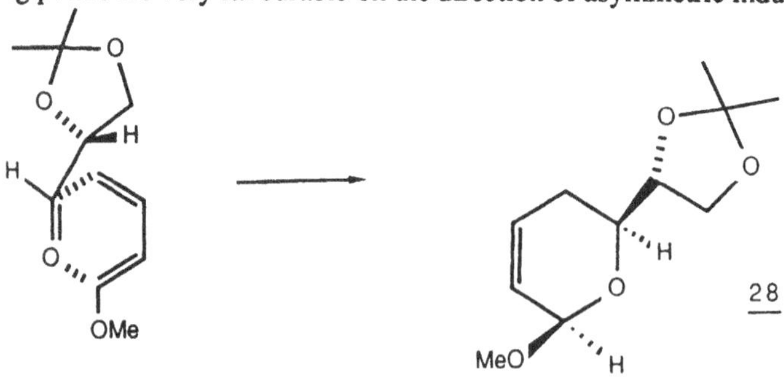

3. The future of organic piezochemistry

3.1. WHICH REACTIONS ?

The advantage of using high pressure in organic chemistry is directly related to the magnitude of $\Delta V^{\neq}$. It must be evidently negative with the highest absolute value. Such situation is encountered in addition reactions, particularly if they involve encumbered transition states or electrostriction such as in Menshutkin reactions. The development of electric changes during the reaction course has not yet been evoked in this chapter, as it is not frequently met. For an emerging example, the addition of carbonyl compounds to acrylic derivatives can be mentionned [83]. In this reaction, $\Delta V^{\neq} \sim -70\ cm^3.mol^{-1}$, one of the largest $\Delta V^{\neq}$-values ever determined. Concerning steric hindrance, recent studies encompass the synthesis of angularly trifluoromethylsubstituted tricyclic compounds impossible to make at ambient pressure [84].

The latest developments which should give adherence to the high pressure methodology can be envisionned in the following respective fields.

3.1.1. *Synthesis of strained polycyclic molecules.* Polycyclic molecules are attracting increased interest due to their fascinating complex structures. As pressure acts as a stabilizing factor in preserving the strained structure, numerous exotic molecules can be synthesized in that way [68].

3.1.2. *Synthesis of fused molecules.* Aromatic macrocyclic molecules with fixed topology were synthesized in recent times by Stoddart and Kohnke [85]. Molecular belts and collars are formed by lateral fusion of adjacent carbon-carbon double bonds. One of the key step of the laborious synthetic approach relies upon cascade Diels-Alder reactions involving the bisdiene **29** with its four exocyclic tetramethylene fragments and the bisdienophile **30** to yield **31**. Subsequent consecutive [4+2] cycloadditions involving **29** and **31** and their resulting adduct with **30** eventually afford **32**. The last step occurs only under high pressure (1200MPa) yielding **32** in 36 % yield.

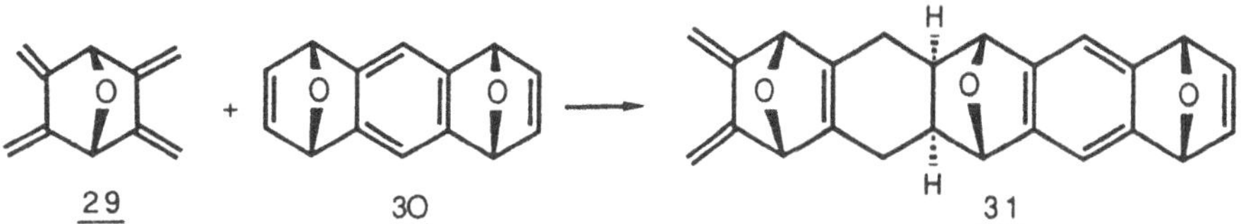

29 + 30 ⟶ 31

29 + 31 —(1000 MPa)⟶ 32

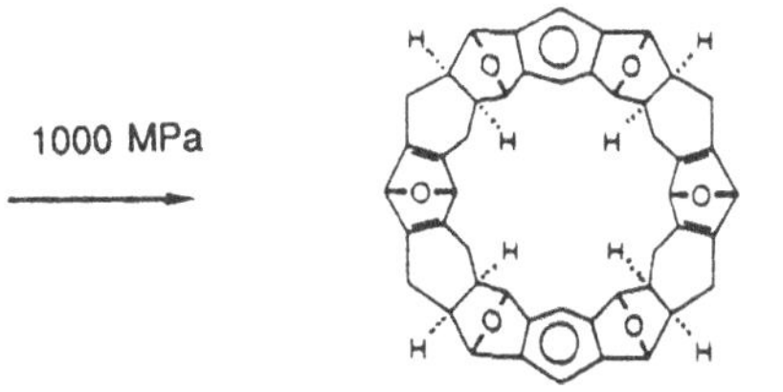

Other new macrocyclic fused compounds with unusual topology can be synthesized under pressure using similar approaches : stereoregular oligomers of **32** [86] and trinacrene , a cage compound with barrel-like structure [85]. Such rigid materials may be candidates for the synthesis of organic like-zeolites, molecular columns and molecular nets. Some other interesting properties such as molecular recognition may open a wide application field.

3.1.3. *Synthesis of functionalized crown ethers.* The chemistry of host-guest molecules expands rapidly. The high pressure methodology in the synthesis of cryptands is particularly suitable when sensitive groups must be introduced into the architecture of the cryptand. Crown ether compounds incorporating heterodonor groups exhibit interesting guest selectivities in cation binding and transport processes. Arylation of double-armed diaza-crown ethers by S_NAr reactions is successfully effected under high pressure [87]. Such functionalized crown ethers show high binding and transport selectivity for Ag+ ion.

3.1.4. *Synthesis of natural or pseudo-natural compounds.*

a) Peptides. A peptide can be formed by aminolysis of an acylated intermediate derived from an ester substrate or by transpeptidation. Aminolysis reactions are readily achieved under high pressure, i.e. aminolysis of lactones to hydroxyamides occurs with fair to excellent yields [88]. Lower pressures (150-200 MPa) promote peptide formation in coupling reactions catalyzed by carboxypeptidases [89]. A more general study considered several protease-catalyzed peptide formation [90]. The important conclusion was that the effect of pressure is exerted through molecular recognition.

Peptide coupling between non-activated aminoacid alkyl esters and aminoacid alkali salts yields dipeptides in fair yield under 700-1100 MPa [91].

R_2 / R_1NH / CO_2Me + R_3 / H_2N / CO_2Na —i) Δ P, ii) H_3O^+→ R_2 / R_1NH / O / NH / CO_2H / R_3

Conformational mobility of peptide backbones is influenced by the presence of bulky α,α-disubstituted glycines. The synthesis of such hindered peptides can be achieved via the Ugi reaction. In this frame, high pressure offers a way to synthesize tripeptides possessing glycine fragments with very bulky groups [73].

b) Carbohydrates and related compounds. Substituted-5,6-dihydro-2H-pyrans are versatile synthons for the stereoselective total synthesis of sugars, antibiotics and pheromones [38]. As an illustrative example, the chiral adduct **28** produced under high pressure (see above) [81] is the starting material for the synthesis of 4-deoxyheptoses **33**, 2-deoxyriboses **34** and purpurosaminide **35** [38].

Pure diastereoisomerically cycloadducts with 100 % asymmetric induction were obtained in 70 % yield in the [4+2] addition of 1-methoxybutadiene and a galactopyranose-6-ulose [92]. With sugar aldehydes as dienophiles high degree of asymmetric induction is reached under 2000 MPa [93]. It should be pointed out that in all cases, there is preference for transition states leading to cis-adducts due to their more compact structure vhich is obviously thermodynamically favoured by pressure.

There is still a vast application field as exemplified by recent articles reporting Diels-Alder reactions involving sugars [94] and condensation reactions such as glycosylations [95].

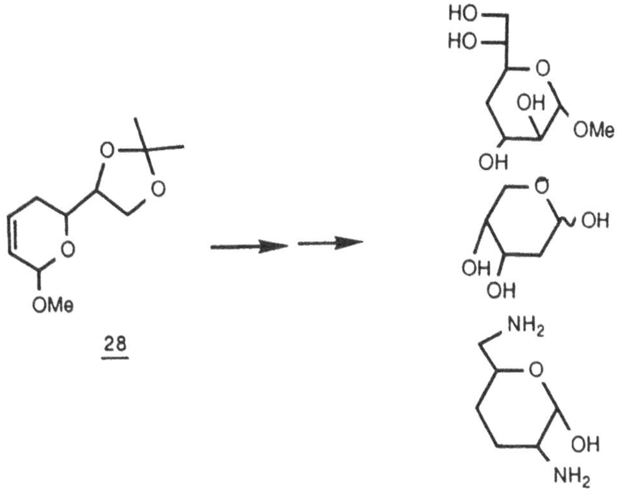

c) Other natural products. The total synthesis of natural products is meeting with general interest because of the challenging synthetic problems and the applicability as drugs or pheromones. High pressure Diels-Alder reactions were the key step in the synthesis of quinoline-quinones (sreptonigrone) [96], diterpenes (taxodione, royleanone) used as antitumor agents [97], abietanoid diterpenes (tanshinol family) used in the chinese traditional medecine [98].

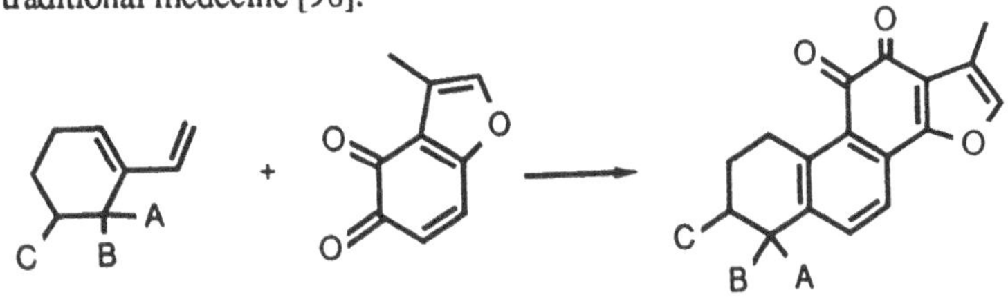

In the synthesis of solenopsin B (a piperidine alkaloid), the key step was the high pressure conversion of an alcohol into a phenyl sufide [99].

3.1.5. *Organometallic reactions*. This field has been approached from the high pressure standpoint only recently. Stereoselective allylation of aldehydes with allylic stannates takes place under very mild conditions under 1000 MPa [100].

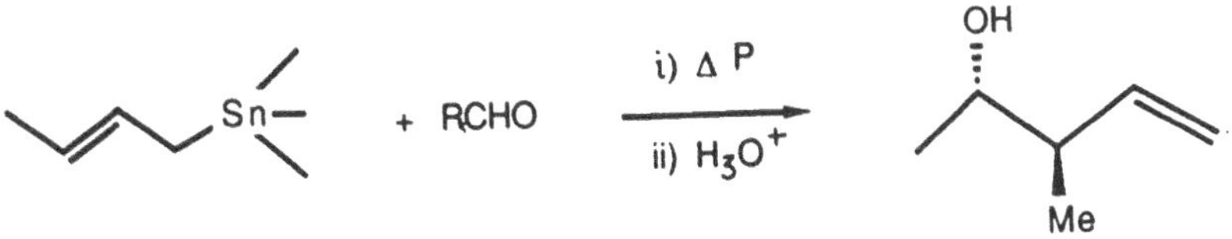

Hexamethylditin add to alkynes under high pressure [101]. The products can be coupled with benzyl bromide leading to a new vinyltin synthon. Both reactions are catalyzed by a Pd(0) compound at room temperature.

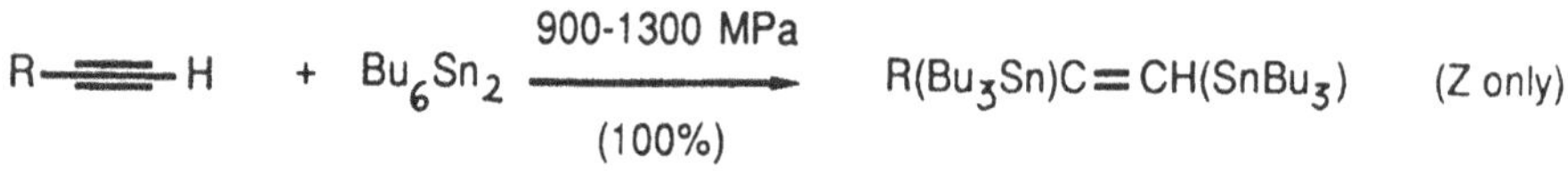

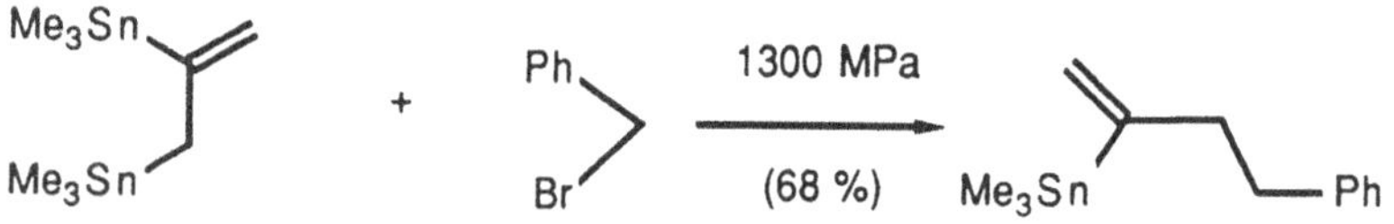

$$R{-}C{\equiv}C{-}H \;+\; Bu_6Sn_2 \xrightarrow[(100\%)]{900\text{-}1300\ MPa} R(Bu_3Sn)C{=}CH(SnBu_3) \quad (Z\ only)$$

In the same way, Pd-catalyzed cross-coupling reaction between aryl- or vinyltins and chloroformates proceed smoothly (52-96 %) under 1000 MPa [102].

O-Silylated ketene acetals are isomerized into α-silylated esters and lactones in high yields under 1000 MPa [79]. The reaction is typical of silicon migration from oxygen to carbon.

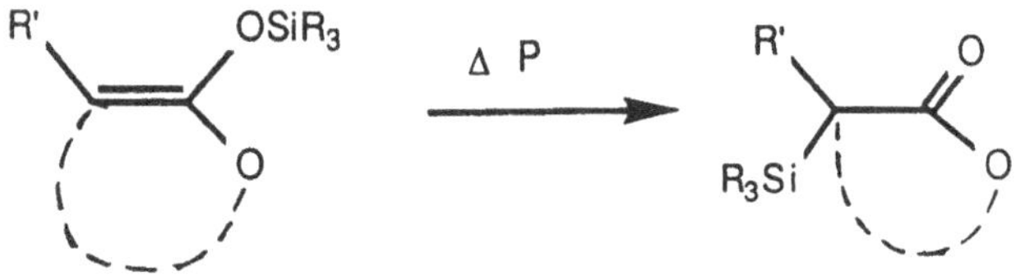

The high pressure methodology appears as a promising way to synthesize 7,3-disilabicyclo [2.2.2] octa-2,5-diene derivatives (1000 MPa, 100° C) which are precursors of tetramethyldisilene [103].

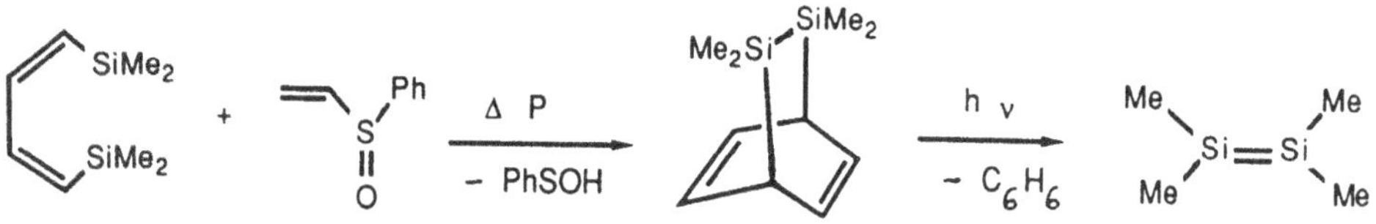

Manganacycles are formed in excellent yields by the sequential insertion of carbon monoxide and alkenes into alkylmanganese pentacarbonyl complexes at 600 MPa [104]. These compounds serve as useful intermediates for the synthesis of carbonyl compounds.

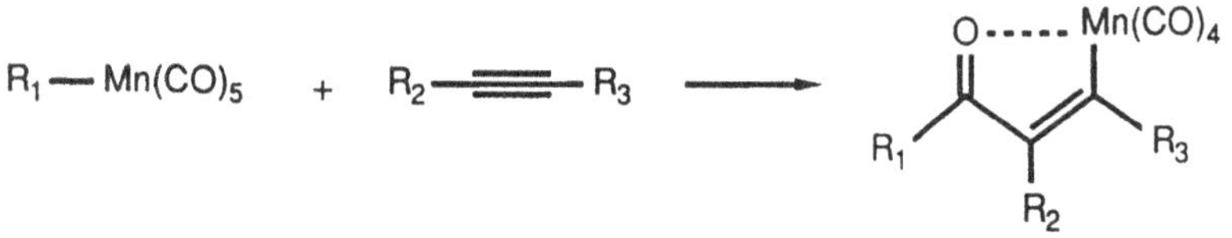

$$R_1{-}Mn(CO)_5 \;+\; R_2{-}C{\equiv}C{-}R_3 \longrightarrow$$

3.1.6. *Synthesis of thiacyclic compounds.* Carbon disulfide reacts under high pressure with various unsaturated substrates to yield thietanes [61], tetrathiofulvalenes from acetylenic esters [105]. The latter compounds are good candidates for conducting organic materials after complexation by tetracyanoquinodimethane. Recent studies reporting high pressure CS_2 or COS reactions embody the synthesis of 1,3-dithiolan-2-ones from oxiranes [106], of 1,3-thiadiazines [107], 1,2-dithiazoles [108] and 1,3,5-thiadiazin-4-ones [109] from dialkylcyanamides. Cycloadditions involving N-sulfinyl dienophiles are promoted by high pressure yielding dihydro-thiazine-1-oxides [110].

3.2. IN WHICH WAY ?

At the beginning of this chapter several activation methods were compared. Three activation processes only have been employed to date in conjunction with pressure : thermal and photonic activation as physical methods and catalysis as chemical activation.

3.2.1. *Physical activation.* Temperature is the easiest accessible parameter. All high pressure vessels designed for liquid phase reactions are able to withstand to fairly high thermal stresses corresponding to temperatures in the range 100-200° C. When the effects of pressure and temperature are convergent, good results can be expected.

The investigation of photochemical effects under pressure requires highly specialized techniques. There has been little work to date. One of the reasons is that not much is known about the pressure dependence of the fundamental steps in photochemistry [111]. An earlier article reported that the pressure-induced acceleration rate in the photodimerization of a naphthoate was much less than in the corresponding thermal reaction [112]. However, a recent work reported a remarkable pressure-induced diastereoselectivity in the photodimerization of 1,3-cyclohexadiene [113]. $\Delta\Delta V^{\neq}$ could reach 11 $cm^3.mol^{-1}$, depending on the triplet energy of the sensitizer and the solvent used.

3.2.2. *Catalysis.* Catalysts have already been utilized in compressed solutions. As agitation devices cannot be generally provided in high pressure vessels -at least for pressures > 500 MPa-, only homogeneous catalysis can be taken into account.

Pressure can be somewhat lowered when lanthanide catalysts -weak Lewis acids- are used in Diels-Alder reactions involving 1-methoxybutadiene [114] (Table 7). Though the yields in the catalyzed reaction are lower than those obtained at 2000 MPa, calculation of hypothetical yields for the thermal reaction at 1000 MPa -the pressure under which the catalyzed reaction takes place- would give values as low as 1-3 %.

OMe + R_1 R_2 O → R_1 R_2 O OMe

TABLE 7. Catalyzed Diels-Alder reaction

R^1	R^2	Yields % no catalyst[a]	Yields % $Eu(fod)_3$[b]
Me	CO_2Me	85	81
H	Me	62	15
H	Ph	80	12
H	furyl	73	17

a : 2000 MPa, 50-65° C b : 1000 MPa, 50° C, catalyst (1 %)

The high pressure Eu(fod)$_3$-catalytic method was applied to the total synthesis of purpurosamine B [115]. A more active catalyst is $EtAlCl_2$ which was used in the regioselective synthesis of angularly methylated octalones at 1200 MPa [116].

4. Conclusion

The high pressure methodology is doubtless of utmost interest for innumerable reactions involving sluggish and reluctant partners. Basic reaction properties such as selectivity can be positively affected by pressure. In the author's opinion, high pressure should be coupled with other physical and chemical activation methods in order to increase the operation field and accordingly, to remove the long-standing restriction to the universal application of this chemistry. As outlined above, the technique should be incorporated into the network of synthetic strategies in the field of natural products, host-guest molecules and exotic chemistries such as the rapidly expanding buckminsterfullerene chemistry [117].

REFERENCES

[1] Jones, W.H.; Mangold, D. and Plieninger, H. (1962) 'Diensynthesen unter hohem Druck', Tetrahedron 18, 267-272.
[2] Olah, G.A.; Meidar, D. and Fung, A.P. (1979) 'Catalysis of Diels-Alder reactions by Nafion-H', Synthesis, 270-271.
[3] Engler, T.A. and Naganathan, S. (1986) 'An efficient benzoquinone based Diels-Alder approach to podocarpane diterpenes', Tetrahedron Lett. 27, 1015-1018.
[4] Dauben, W.G. and Krabbenhoft, H.O. (1976) 'Organic reactions at high pressure. Cycloadditions with furans', J. Am. Chem. Soc. 98, 1992-1993.
[5] Vieira, E. and Vogel, P. (1982) 'Copper (I) and copper (II)-catalyzed Diels-Alder additions of α-substituted acrylonitrile to furan', Helv. Chim. Acta 65, 1700-1704.
[6] Moore, J.A. and Partain, E.M. (1983) 'Catalyzed addition of furan with acrylic monomers', J. Org. Chem. 48, 1105-1106.
[7] Brion, F. (1982) 'On the Lewis-acid catalyzed Diels-Alder reaction of furan', Tetrahedron Lett. 23, 5299-5302.
[8] Laszlo, P. and Luchetti, J. (1984) 'Easy formation of Diels-Alder cycloadducts between furans and α,β-unsaturated aldehydes and ketones at normal pressure', Tetrahedron Lett. 25, 4387-4388.
[9] Jenner, G. and Rimmelin, J. (1974) 'Etude des réactions péricycliques sous pression. Dimérisation thermique de l'isoprène', Tetrahedron 30, 3081-3085.
[10] Klärner, F.G.; Dogan, B.M.; Ermer, O.; von Doering, W. and Cohen, M.P. (1986) 'Mechanism of the thermal 1,3-cyclohexadiene dimerization', Angew. Chem. Int. Ed. Engl. 25, 108-110.
[11] Dessau, R.M. (1986) 'Catalysis of Diels-Alder reactions by zeolites', J. Chem. Soc., Chem. Comm., 1167-1168.
[12] Bauld, N.L. (1989) 'Cation radical cycloadditions and related sigmatropic reactions', Tetrahedron 45, 5307-5363.
[13] Jenner, G. and Papadopoulos, M. (1982) 'The pressure effect on the homo-Diels-Alder reaction', Tetrahedron Lett., 4333-4337.

[14] Tom Dieck, H.; Mallien, M. and Diercks, R. (1989) 'Diazadiene-iron(0) catalyzed synthesis of tetracyclononenes from norbornadiene and internal alkynes', J. Mol. Catal. 51, 53-58.

[15] Lautens M., T. and Crudden, C.M. (1989) 'Scope of the cobalt-catalyzed [2+2+2] homo-Diels-Alder reaction', Organomet. 8, 2733-2735.

[16] Lee, J.; Tang, J. and Snyder, J.K. (1987) 'Preparation and dienophilicity of 3-methyl-4,5-benzofurandione', Tetrahedron Lett. 28, 3427-3430.

[17] Place, R.; Hermkens, P.H.; Smits, J.M.; Nivard, R.J. and Ottenheijm, H.C. (1987) 'Employment of nitriles in the stereoselective cycloaddition to nitrones', J. Org. Chem. 52, 1047-1051.

[18] Harkin, S.A. and Thomas, E.J; (1983) 'Macrocycle formation using intramoleculare Diels-Alder reactions', Tetrahedron Lett. 24, 5535-5538.

[19] Hermkens, P.H.; van Maarseveen, J.H.; Kruse, C.G. and Scheeren, H.W. (1988) '1,3-Dipolar cycloaddition of nitrones with nitriles', Tetrahedron 44, 6491-6504.

[20] Einhorn, C.; Einhorn, J. and Luche, J.L. (1989) 'The use of ultrasonic waves in synthetic organic chemistry', Synthesis, 787-813.

[21] Lee, J. and Snyder, J.K. (1989) 'Ultrasound-promoted Diels-Alder reactions' J. Am. Chem. Soc. 111, 1522-1524.

[22] Lee, J. and Snyder, J.K. (1990) 'Ultrasound-promoted cycloadditions in the synthesis of salvia miltiorrhyza abietanoid o-quinones', J. Org. Chem. 55, 4995-5008.

[23] Stambouli, A.; Chastrette, M. and Soufiaoui, M. (1991) 'Réactions de cycloaddition [4+2] sous micro-ondes des dérivés du glyoxal', Tetrahedron Lett. 32, 1723-1724.

[24] Grieco, P.A.; Nunes, J.J. and Gaul, M.D. (1990) 'Dramatic rate accelerations of Diels-Alder reactions in 5 M LPDE : the cantharidin problem reexamined', J. Am. Chem. Soc. 11, 4595-4596.

[25] Dauben, W.G.; Kessel, C.R. and Takemura, K.H. (1980) 'Simple, efficient total synthesis of cantharidin via a high pressure Diels-Alder reaction', J. Am. Chem. Soc. 102, 6893-6894.

[26] Forman, M.A. and Dailey, W.P. (1991) 'The LPDE rate acceleration of the Diels-Alder reaction : Lewis acid catalysis by lithium ion', J. Am. Chem. Soc. 113, 2761-2762.

[27] Papadopoulos, M.; Ben Salem, R. and Jenner, G. (1990) 'High pressure effect on the chemoselectivity of catalyzed [2+2] and [2+2+2] addition reactions involving acetylenic esters', High Press. Res. 5, 644-646.

[28] Dauben, W.G. and Bunce, R.A. (1982) "Asymmetric induction in the Diels-Alder reaction of p-benzoquinone with chiral 2,4-pentadienoic acid derivatives', Tetrahedron Lett. 23, 4875-4878.

[29] Ferroud, C. Revial, G. and d'Angelo, J. (1985) 'Crotonates as dienophiles in high pressure induced Diels-Alder cycloadditions', Tetrahedron Lett. 26, 3981-3984.

[30] Smith, A.B.; Liverton, N.J.; Hrib, N.J.; Sivaramakhrishnan, H. and Winzenberg, K. (1986) 'Total synthesis of (+) Jatropholones A and B. Exploitation of the high pressure technique', J. Am. Chem. Soc. 108, 3040-3048.

[31] Keay, B.A. and Dibble, P.W. (1989) 'High pressure intramolecular Diels-Alder reactions of the furan diene', Tetrahedron Lett. 30, 1045-1046.

[32] Harwood, L.M.; Ishikawa, T.; Phillips, H. and Watkin, D. (1991) 'Synthetic approaches to phorbols via the intramolecular Diels-Alder reaction of furans', J. Chem. Soc., Chem. Comm., 527-530.

[33] Posner, G.H. and Switzer, C. (1987) 'Diels-Alder cycloadditions using electrophilic sulfonylpyridones', J. Org. Chem. 52, 1642-1644.

[34] Rigby, J.H. and Kierkus, P.C. (1989) 'Elaboration of fused gem-dimethylcyclopropane systems via cyclopropene cycloaddition', J. Am. Chem. Soc. 111, 4125-4126.

[35] Bell, S.I. and Weinreb, S.W. (1988) 'Lewis acid and high pressure promoted Diels-Alder cycloadditions of N-alkyl-N-sulfinyl dienophiles', Tetrahedron Lett. 29, 4233-4236.

[36] Matsumoto, K.; Sera, A. and Uchida, T; (1985) 'Organic synthesis under high pressure', Synthesis, 1-26 and 999-1027.

[37] Dauben, W.G.; Gerdes, J.M. and Look, G.C. (1986) 'Organic reactions at high pressure. The effect of pressure change upon reaction rates of bimolecular processes', Synthesis, 532-535.

[38] Jurczak, J. and Baranowski, B. (Eds) (1989) 'High pressure chemical synthesis', Elsevier, Amsterdam.

[39] Matsumoto, K. and Acheson, R.M. (Eds) (1991) 'Organic synthesis at high pressures', Wiley, New York.

[40] Gladysz, J.A.; Lee, S.J.; Tomasello, J.A. and Yu, Y.S. (1977) 'High pressure cycloadditions of pyrones', J. Org. Chem. 42, 4170-4172.

[41] To date, there is only one known example of a pressure accelerated decomposition reaction (free of electrostriction effects) : Jenner, G.; Papadopoulos, M. and Rimmelin, J. (1983) 'Activation parameters and location of the transition state in the retro-Diels-Alder reaction of 7-oxabicyclo [2.2.1] hept-5-ene', J. Org. Chem. 48, 748-749.

[42] Jenner, G. (1979) Synthèse organique sous hautes pressions statiques', New J. Chem. 3, 329-334.

[43] Rimmelin, J.; Jenner, G. and Rimmelin, P. (1978) 'Influence de la pression dans la synthèse diénique du système oxa-7 bicyclo [2.2.1] heptane', Bull. Soc. Chim. France II, 461-464.

[44] Matsumoto, K.; Uchida, T. and Acheson, M. (1981) 'The synthesis and reactions of heterocycles under high pressures', Heterocycles 16, 1367-1387.

[45] Jurczak, J.; Kozluk, T.; Filipek, S. and Eugster, C.H. (1983) '[2+4] cycloadditions under high pressure. First realization of a Diels-Alder addition of furan with simple 1,4-benzoquinones', Helv. Chim. Acta 66, 222-225.

[46] Pfaff, E. and Plieninger, H. (1982) 'Synthese einer Reihe von 3-oxo-2-oxabicyclo [2.2.2] oct-7-en Derivaten', Chem. Ber. 115, 1967-1981.

[47] Prapansiri, V. and Thornton, E.R. (1991) 'Asymmetric Diels-Alder reactions. Inverse electron demand reactions of esters of 3-hydroxypyran-2-one with vinyl ethers', Tetrahedron Lett. 32, 3147-3150.

[48] Boger, D.L. and Brotherton, C.E. (1986) 'Thermal reactions of cyclopropenone ketals', J. Am. Chem. Soc. 108, 6695-6713.

[49] Dauben, W.G. and Baker, W.R. (1982) 'The high pressure Diels-Alder reaction of p-benzoquinone with dienic esters', Tetrahedron Lett. 23, 2611-2614.

[50] Drew, M.G.; George, A.V.; Isaacs, N.S. and Rzepa, H.S. (1985) 'High pressure synthesis of some derivatives of 7-azabicyclo [2.2.1] heptane', J. Chem. Soc. Perkin Trans. 1, 1277- 1284.

[51] Chmielewski, M.; Kałuza, Z.; Bełziecki, C.; Sałanski, P.; Jurczak, J. and Adamowicz, H. (1985) 'High pressure (2+2) cycloaddition of toluene-4-sulfonyl iscyanate to glycals', Tetrahedron 41, 2441-2449.
[52] Dicken, C.M. and De Shong, P. (1982) 'Reactions at high pressure. [3+2] dipolar cycloaddition of nitrones with vinyl ethers', J. Org. Chem. 47, 2047-2051.
[53] Kamernitzki, A.V.; Levina, J.S.; Mortikova, E.M. and El'yanov, B.S. (1975) 'High pressure induced 1,3-cycloaddition of nitronic esters to the steroidal dipolarophiles', Tetrahedron Lett., 3235-3238.
[54] Anderson, G.T.; Henry, J.R. and Weinreb, S.M. (1991) 'High pressure induced 1,3-dipolar cycloadditions of azides with electron-deficient olefins', J. Org. Chem. 56, 6946-6948.
[55] Chmielewski, M.; Kaluza, Z.; Mostowicz, D.; Bełzecki, C.; Baranowska E.; Jacobsen, J.P.; Sałanski, P. and Jurczak, J. (1985) 'High and atmospheric pressure cycloaddition of trichloroacetyl isocyanate to 3,4-di-O-acetyl-L-rhamnal', Tetrahedron 43, 4555-4563.
[56] Guingant, A. and d'Angelo, J. (1986) 'Approach to the aklavinone series through a new high pressure induced cycloaddition strategy', Tetrahedron Lett. 27, 3729-2732.
[57] Bunce, R.A.; Schlecht, M.F.; Dauben, W.G. and Heathcok, C.H. (1983) 'Conjugate additions of O-silylated ketene acetals to activated enones' Tetrahedron Lett. 24, 4943-4946.
[58] Dauben, W.G. and Hendricks, R.T. (1992) 'The acid-catalyzed cyclizations of unsaturated carbonyl compounds utilizing silical gel at high pressure', Tetrahedron Lett. 33, 603-606.
[59] Papadopoulos, M. and Jenner, G. (1983) 'Pressure effect and estimation of concertedness in some (2+2+2) cycloadditions of quadricyclane', New J. Chem. 7, 463-464.
[60] Papadopoulos, M.; Jost, R. and Jenner, G. (1983) 'Synthesis of annelated oxetans via cycloaddition of quadricyclane under thermal and high pressure conditions', J. Chem. Soc., Chem. Comm., 221-222.
[61] Jenner, G. and Papadopoulos, M. (1985) 'Synthèse de thiétannes par cycloaddition du quadricyclane avec des liaisons thiocarbonyles', Tetrahedron Lett. 26, 725-726.
[62] Jenner, G. (1987) 'High pressure sequential ($\pi^2+\pi^2+\pi^2$) and ($\pi^2+\sigma^2+\sigma^2$) cycloaddition reactions', Tetrahedron Lett. 28, 3927-3930.
[63] Gonikberg, M.G. (1963) 'Study of steric effects in chemical reactions by means of high pressure', Russ. J. Phys. Chem. Engl. Trans. 37, 248-255.
[64] Jenner, G. and Kellou, M. (1981) 'Etude de l'effet stérique en chimie organique au moyen du volume d'activation', Tetrahedron 37, 1153-1160.
[65] Jenner G. (1985) 'The pressure effect on strained transition states', J. Chem. Soc., Faraday Trans. 1, 81, 2437-2460.
[66] Franck-Neumann, M.; Miesch, M.; Barth, F. and Jenner, G. (1989) 'De l'instabilité des cyclopropènes à double liaison fonctionnalisée monosubstituée', Bull. Soc. Chim. France, 661-666.
[67] Matsumoto, K. (1985) 'Diels-Alder reactions of [2.2] paracyclophane', Chem. Lett., 1681-1682.
[68] Jenner, G. (1991) 'High pressure synthesis of strained polycyclic molecules', in R. Pucci and G. Piccitto (eds), Molecular Systems under High Pressure, North-Holland Publishers. Amsterdam. pp. 361-380.

[69] Jurczak, J.; Tkacz, M. and Majchrzak-Kuczynska, U. (1983) 'Formation of quaternary salts of sterically hindered bis-quinolizidine monolactams under high pressure', Synthesis, 920.
[70] Jurczak, J. and Pietraszkiewicz, M. (1986) 'High pressure synthesis of cryptand and complexing behaviour of chiral cryptands', Top. Curr. Chem. 130, 185-200.
[71] Jurczak, J.; Kasprzyk, S.; Mąkosza, M.; Ostaszewski, R. and Sałanski, P. (1990) 'Recent advances in the organic synthesis at high pressure. Effect of pressure on the macrocyclization reaction', High Press. Res. 5, 641-643.
[72] Degueil-Castaing, M.; Rahm, A. and Dahan, N. (1986) 'Reduction of ketones by tributyltin hydride. The effect of pressure on steric hindrance', J. Org. Chem. 51, 1672-1676.
[73] Yamada, T.; Hanagi, Y.; Omote, Y.; Mizayawa, T.; Kuwata, S.; Sugiura, M. and Matsumoto, K. (1990) 'Four-component condensation (Ugi reaction) at high pressure', J. Chem. Soc., Chem. Comm., 1640-1641.
[74] Jenner, G.; Abdi-Oskoui, H. and Rimmelin, J. (1979) 'Influence de la pression dans les condensations entre dérivés carbonylés insaturés et éthers vinyliques', Bull. Soc. Chim. France, 33-39.
[75] Jenner, G. and Papadopoulos, M. (1986) 'High pressure addition reactions involving cycloheptatriene-norcaradiene equilibrium', J. Org. Chem. 51, 585-588.
[76] Yamamoto, Y.; Furuta, T.; Matsuo, J. and Kurata, T. (1991) 'Cleavage of esters under nearly neutral conditions at high pressure', J. Org. Chem. 56, 5737-5738.
[77] Ortuño, R.M.; Guingant, A. and d'Angelo, J. (1988) 'High pressure induced Diels-Alder cycloadditions of butenolides to electron-rich dienes', Tetrahedron Lett. 29, 6989-6992.
[78] Buback, M.; Trost, W.; Hübsch, T.; Voss, E. and Tietze, L.F. (1989) 'Diastereoselectivity and kinetics of intermolecular hetero Diels-Alder reactions under high pressure', Chem. Ber. 122, 1179-1186.
[79] Yamamoto, Y.; Maruyama, K. and Matsumoto, K. (1983) 'Aldol reaction of silyl enol ethers with aldehydes under neutral conditions', J. Am. Chem. Soc. 105, 6963-6965.
[80] d'Angelo, J. and Maddaluno, J. (1986) 'Enantioselective synthesis of β-amino esters through high pressure addition of amines to α,β-ethylenic esters', J. Am. Chem. Soc. 108, 8112-8114.
[81] Jurczak, J.; Bauer, T.; Filipek, S.; Tkacz, M. and Zygo, K. (1983) 'Asymmetric induction in the high pressure cycloaddition of isopropylidene-D-glyceraldehyde to 1-methoxybuta-1,3-diene', J. Chem. Soc., Chem. Comm., 540-542.
[82] Jurczak, J. and Bauer, T. (1986) 'Stereochemistry of Diels-Alder reaction at high pressure. Influence of pressure on asymmetric induction', Tetrahedron 42, 5045-5052.
[83] Hill, J.S. and Isaacs, N.S. (1986) 'Functionalization of the a-position of acrylate systems by the addition of carbonyl compounds', Tetrahedron Lett. 27, 5007-5010.
[84] Bégué, J.P.; Bonnet-Delpon, D.; Lequeux, T.; d'Angelo, J. and Guingant, A. (1992) 'First direct access to angularly tricyclic compounds', Synlett, 146-148.
[85] Stoddart, J.F. (1989) 'The making of molecular belts and collars', J. Incl. Phenom. Mol. Recogn. Chem. 7, 227-245.

[86] Ashton, P.R.; Isaacs, N.S.; Kohnke, F.H.; Mathias, J.P. and Stoddart, J.F. (1989) 'Stereoregular oligomerization by repetitive Diels-Alder reaction', Angew. Chem. Int. Ed. Engl. 28, 1258-1261.
[87] Tsukube, H.; Minatogawa, H.; Munakata, M.; Toda, M. and Matsumoto, K. (1992) 'High pressure functionalization of diaza-crown ethers', J. Org. Chem. 57, 542-547.
[88] Matsumoto, K.; Hashimoto, S.; Uchida, T.; Okamoto, T. and Otani, S. (1989) 'High pressure aminolysis of lactones to hydroxyamides', Bull. Chem. Soc. Jpn 62, 3138-3142.
[89] Kunugi, S.; Tanabe, K.; Fukuda, M.; Makimoto, S. and Taniguchi, Y. (1987) 'The effect of pressure on protease-catalyzed peptide formation', J. Chem. Soc., Chem. Comm., 1335-1336.
[90] Kunugi, S.; Tanabe, K.; Yamashita, K.; Morikawa, Y.; Ito, T.; Kondoh, T.; Hirata, K. and Nomura, A. (1989) 'Protease-catalyzed peptide formation under high pressure', Bull. Chem. Soc. Jpn 62, 514-518.
[91] Gante, J.; Kalthof, U.; Klärner, F.G. and Weber, T. (1990) 'Peptide synthesis under high pressure', Angew. Chem. Int. Ed. Engl. 29, 1025-1026.
[92] Jurczak, J.; Bauer, T. and Jarosz, S. (1984) 'High pressure stereospecific cycloaddition of a galactopyranose-6-ulose to 1-methoxybutadiene', Tetrahedron Lett. 25, 4809-4812.
[93] Jurczak, J.; Bauer, T. and Jarosz, S. (1986) 'Asymmetric induction in the reaction of 1-methoxybutadiene with sugar aldehydes', Tetrahedron 42, 6477-6486.
[94] Gołebiowski, A. and Jurczak, J. (1991) 'High pressure [4+2] cycloaddition of 1-methoxybutadiene to threoninals', Tetrahedron 47, 1037-1044.
[95] Sasaki, M.; Gama, Y.; Yasumoto, M. and Ishigami, Y. (1990) 'Glycosylation reaction under high pressure', Tetrahedron Lett. 31, 6549-6552.
[96] Boger, D.L. and Kasper, A. (1989) 'A general solution to implementing the 4π participation of 1-azabutadienes in Diels-Alder reactions', J. Am. Chem. Soc. 111, 1517-1519.
[97] Engler, T.A.; Sampath, U.; Naganathan, S.; Van der Velde, D. and Takusagawa, F. (1989) 'A new general synthetic approach to diterpenes', J. Org. Chem. 54, 5712-5727.
[98] Haiza, M.; Lee, J. and Snyder, J.K. (1990) 'Asymmetric syntheses of abietanoid o-quinones', J. Org. Chem. 55, 5008-5013.
[99] Kotsuki, H.; Kusumi, T.; Inoue, M.; Ushio, Y. and Ochi, M. (1991) 'Synthesis of solenopsin B', Tetrahedron Lett. 32, 4159-4162.
[100] Yamamoto, Y.; Maruyama, K. and Matsumoto (1984) 'Organometallic high pressure reactions', J. Chem. Soc., Chem. Comm., 489-490.
[101] Mitchell, T.N.; Dornseifer, N.M. and Rahm, A. (1991) 'Palladium catalyzed reactions of organotins', High Press. Res. 7, 165-167.
[102] Jousseaume, B.; Jurczak, J.; Sałanski, P. and Verlhac, J.B. (1991) 'Pd-catalyed alkoxycarbonylation of vinyl and arylorganotins under high pressure', High Press. Res. 7, 162-164.
[103] Sekiguchi, A.; Maruki, I.; Ebata, K.; Kabuto, C. and Sakurai, H. (1991) 'High pressure synthesis, structure and novel photochemical reactions of a disilabicyclodiene', J. Chem. Soc., Chem. Comm., 341-343.
[104] DeShong, P.; Sidler, D.R.; Rybczynski, P.J.; Slough, G.A. and Rheingold, A.L. (1988) 'A general method for the preparation of carbonyl compounds and

butenolides from organomanganese pentacarbonyl complexes', J. Am. Chem. Soc. 110, 2575-2585.

[105] Rice, J.E. and Okamoto, Y. (1981) 'One-step synthesis of tetrathiofulvalenes under high pressure', J. Org. Chem. 46, 446-447.

[106] Taguchi, Y.; Yasumoto, M.; Shibuya, I. and Suhara, Y. (1989) 'The synthesis of dithiolan-2-ones on the reaction of oxiranes with carbon disulfide under high pressure', Bull. Chem. Soc. Jpn 62, 474-478.

[107] Tsuchiya, T.; Yasumoto, M.; Shibuya, I. and Goto, M. (1990) 'Novel preparation of 1,3,5-thiadiazine derivatives by cyclocondensation of carbon disulphide under high pressure', J. Chem. Soc. Perkin Trans. 1, 1218-1219.

[108] Tsuchiya, T.; Yasumoto, M. and Shibuya, I. (1989) 'Cycloaddition of dialkylaminodithiazole-3-thione with dialkylcyanamide under high pressure', Chem. Lett., 1357-1360.

[109] Yasumoto, M.; Tsuchiya, T.; Taguchi, Y.; Shibuya, I., Yonemoto, K. and Goto, M. (1991) 'Novel single step preparation of 1,3,5-thiadiazin-4-ones', Chem. Lett., 1229-1232.

[110] Bell, S.I. and Weinreb, S.M. (1991) 'Metalation of 3,6-dihydrothiazine 1-oxides prepared via Diels-Alder cycloadditions of N-sulfinyl dienophiles', J. Org. Chem. 56, 373-377.

[111] Schmidt, R. and Brauer, H.D. (1988) 'The application of high pressure in photophysics and photochemistry', in W.J. le Noble (ed.), Organic High Pressure Chemistry, Elsevier, Amsterdam, pp. 357-379.

[112] Hamann, S.D.; Linton, M. and Sasse, W.H. (1980) 'High pressure photodimerization of a naphthalene derivative', Aust. J. Chem. 33, 1419-1423.

[113] Chung, W.S.; Turro, N.J.; Mertes, J. and Mattay, J. (1989) 'Pressure-induced diastereoselectivity in photo-induced Diels-Alder reactions', J. Org. Chem. 54, 4881-4887.

[114] Jurczak, J.; Golebiowski, A. and Bauer, T. (1985) 'Lanthanide-catalyzed [4+2] cycloadditions under high pressure', Synthesis, 928-929.

[115] Golebiowski, A.; Jacobsson, U. and Jurczak, J. (1987) 'High pressure approach to the total synthesis of 6-epi-D-purpurosamine B', 43, 3063-3066.

[116] Aben, R.W.; Minuti, L.; Scheeren, H.W. and Tattichi, A. (1991) 'High pressure Lewis acid catalyzed Diels-Alder reaction of 3-methylcyclohex-2-en-1-one', Tetrahedron Lett. 32, 6445-6448.

[117] Wudl, F. (1992) 'The chemical properties of buckminsterfullerene (C_{60}) and the birth and infancy of fulleroids', Acc. Chem. Res. 25, 157-161.

HIGH PRESSURE NMR STUDIES OF CHEMICAL AND BIOCHEMICAL SYSTEMS

Jiri Jonas
Department of Chemistry
School of Chemical Sciences
University of Illinois
Urbana, Illinois 61801
U.S.A.

ABSTRACT. Nuclear magnetic measurements at high pressure provide unique information about the microscopic behavior of chemical and biochemical systems. In view of the high information content of advanced high resolution NMR techniques including 2D and 3D NMR, the ability of performing these experiments at high pressure opens a very promising direction in high pressure research.

This review is organized in the following way. The introductory section outlines the fundamental reasons for performing high pressure experiments on chemical and biochemical systems. The second section discusses the specialized instrumentation needed for high pressure NMR experiments. The following two sections, which deal with specific examples of high pressure NMR studies on chemical and biochemical systems, represent the main part of this review. The material included in these two sections illustrates well how high resolution, high pressure NMR can study a wide spectrum of problems ranging from the dynamics of complex liquids, reaction kinetics in liquid solutions, to experiments on model membranes and studies of pressure induced unfolding of proteins.

1. Introduction

In recent years, there has been a major expansion of high pressure research providing unique information about systems of interest to a wide range of scientific disciplines. Since nuclear magnetic resonance has been applied to a wide spectrum of problems in chemistry, physics and biochemistry, it is not surprising to find that high pressure NMR techniques have also had many applications in these fields of science. Clearly, the high information content of NMR experiments is enhanced by the ability of performing these experiments at high pressure.

It was the aim of a recent monograph [1], in the series on NMR Basic Principles and Progress, to illustrate the wide range of problems of interest to physics, chemical physics, biochemistry, and chemical reaction kinetics, which can be successfully studied by high pressure NMR. In many different ways, this monograph demonstrated the power of modern experimental and theoretical techniques to investigate very complex systems.

Our understanding of the liquid state has lagged far behind our basic knowledge of the solid and gas states. While there exists an ideal crystal lattice model for solids and an ideal gas model for gases, no such simple ideal model is available for liquids. This lag is due to the inherent complexity of liquids: they have neither the rigid geometric structure of solids nor the complete randomness of gases. In recent years, there has been significant progress toward a better

R. Winter and J. Jonas (eds.), High Pressure Chemistry, Biochemistry and Materials Science, 393–441.

understanding of the liquid state based on both theoretical and experimental work. In particular, studies in which pressure has been used as an experimental variable have contributed in a major way to the recent advances in our knowledge of liquids [2].

There are several fundamental reasons for performing experiments on liquids at high pressure:

(i) It is important to realize that changing temperature at constant pressure affects molecular motions in two distinct ways: not only is the average kinetic energy of the molecule changed, but there is always an accompanying change in the average volume available for the motion of the molecule. Only by using both pressure and temperature as experimental variables can one separate the effects of density and temperature on molecular motions. Due to the close packing of molecules in a liquid, even a small change in density can produce a considerable change in the molecular dynamics of the system; therefore, in order to test rigorously a theoretical model of a liquid, or a model of a specific dynamic process in a liquid, one must perform isochoric, isothermal, and isobaric experiments.

(ii) Use of pressure allows one to extend the range of measurements above the boiling point and also permits the study of supercritical dense fluids [3].

(iii) In an analogy to the concept of pressure tuning of energy levels in solids [4], one can pressure tune the intermolecular interactions in liquids [5].

(iv) Studies of re-orientational motions and reaction dynamics require variation of the solvent shear viscosity η. In most studies η is changed by the use of different solvents but in a high-pressure experiment η can be changed by changing pressure. Since introduction of different solvents changes the solute-solvent intermolecular potential the high pressure single solvent experiment represents a much cleaner experiment [6].

The large majority of NMR applications in chemistry and biochemistry deal with liquids in which the NMR lines are narrowed by motional averaging to a natural line width of the order of 0.1 to 1 Hertz. High-resolution NMR spectra of complex molecules in the liquid phase usually exhibit a great deal of structure and yield a wealth of information about the molecule. Therefore, it is not surprising that the multi-nuclear high-resolution Fourier transform NMR spectroscopy at high pressure represents the most promising technique in the studies of the dynamics in liquids [1]. The high information content of the many advanced NMR techniques including 2D-NMR techniques such as NOESY, COSY, and ROESY have yet to be fully explored in the high pressure NMR experiments. Recent advances in superconducting magnet technology have resulted in the development of superconducting magnets capable of attaining a high homogeneity of the magnetic field over the sample volume so that even without sample spinning one can achieve high resolution. It is quite impressive that one is able to achieve a NMR line width of 1.2 Hz for sample diameter of 10 mm at the proton frequency of 300 MHz (H_o = 7.05 Tesla). At the same time, Fourier transform techniques make all these high resolution experiments much easier to be performed at high pressure than was the case with classical CW techniques. The experimental capability of recording high resolution NMR spectra on dilute spin systems opened an exciting new direction of high pressure NMR spectroscopy dealing with pressure effects on biochemical systems.

The rationale for using pressure as an experimental variable in studies of proteins and model membranes has been thoroughly discussed by a number of authors; therefore, I highlight only the most important points. First of all, changing the temperature of a system at atmospheric pressure produces a simultaneous change in thermal energy and volume; therefore, in order to separate the thermal and volume effects, one has to perform high pressure experiments. Second, the environment (e.g. viscosity) of a biomolecule can be changed continuously with pressure without having to change the solvent composition. Third, it is well known that proteins undergo pressure denaturation, but few details are known about this important process or how it is related to thermal or solvent induced denaturation. Fourth, the current investigations of the structure and dynamics of model membranes by changing the temperature of the system could be greatly enhanced by high pressure experiments, since pressure is as important a thermodynamic variable as is temperature. Fifth, in our studies of simple liquids, we found that quite often the volume effects determine the mechanism of a dynamic process, whereas temperature only changes the frequency of the motions without actually affecting the mechanism. Therefore, one can expect to obtain new, fundamental information about various processes from the high pressure NMR experiments.

The effect of pressure on chemical equilibria and reaction rates is described by the following standard equations which define the reaction volume ΔV and activation volume $\Delta V^{\ddagger}$:

$$\Delta V = -(RT\partial \ln K)/\partial P)T \quad \Delta V^{\ddagger} = -(RT\partial \ln k/\partial P)T, \qquad (1)$$

where the other symbols have their usual meaning. Another important relationship that can be obtained from the high pressure experiments is dT_M/dP where T_M denotes the transition temperature between two phases, e.g. for the transition between the LC and gel phase of a phospholipid bilayer. There are many excellent recent reviews [7-15] discussing the wide spectrum of biochemical problems that can be investigated by various high pressure techniques.

In spite of the fact that NMR is one of the most promising spectroscopic tools for the investigation of biochemical systems at ambient conditions, very few high pressure NMR studies on biological molecules have been reported to date. In fact, in their 1983 review of the effect of high pressure on proteins and other biomolecules, Weber and Drickamer [8] emphasized the need for high pressure, high resolution NMR experiments on biological systems because of the high information content of the high resolution NMR techniques. Similarly, Morishima [16] in his 1987 review pointed out the scarcity of high pressure NMR studies on biochemical systems and indicated the need for high resolution, high pressure NMR experiments in this area.

The following high resolution NMR studies of biochemical systems have been reported so far. Wagner [17] investigated the pressure effects on the rotation of the Tyr 35 and Phe-45 aromatic rings in BPTI, at 37° C and a pressure up to 120 MPa. Williams et al. [18,19] studied the effect of pressure on the dynamics and α-helical structure in several homopolypeptides; and Morishima [16,20] investigated the effects of pressure on hemoproteins [20-25]. Spin equilibrium in myoglobin (Mb) and hemoglobin (Hb), hemichrome formation, and motions of the proximal and distal histidines in cyanomet Mb were investigated. The effects of pressure on the ionization of histidine and rotation of amide groups in peptides have been studies by Ludemann and his coworkers [26,27].

In our laboratory we have used high resolution, high pressure NMR spectroscopy to investigate various problems dealing with simple molecular liquids [28-34] and predicted the promise of this high pressure technique in studies of biological systems. However, at the beginning, progress has been quite slow in the application of high pressure NMR to biochemical systems because the technique presents major experimental difficulties and requires the development of specialized equipment. The early studies mentioned above used the capillary techniques introduced by Yamada [35], which have two inherent limitations for the investigation of biochemical systems: a limited pressure range (maximum pressures of 1.5 - 2 kbar) and a very small sample volume - usually a 1 mm wide sample cell. This, of course, markedly decreases the sensitivity and precludes the use of dilute biochemical samples which are very often necessary to prevent artefacts due to aggregation. Only very recently we have succeeded in developing high pressure NMR probes [36] which allow high resolution and high sensitivity experiments on biochemical systems.

This review is organized in the following way. After a discussion of experimental high pressure, high resolution NMR techniques (Sect. 2.0), the main results of several recent NMR studies will discuss the dynamics of complex liquids (Sect. 3.1); and the fundamental studies of dynamical solvent effects on reaction rates (Sect. 3.2). In addition, a discussion of pressure effects on dynamics of liquids in confined geometries (Sect. 3.3) and the effects of pressure on intercalated solids (Sect. 3.4) will illustrate the wide scope of problems that can be investigated by high pressure NMR spectroscopy.

The second main part of this review deals with high resolution NMR studies of biochemical systems, as this area of applications of high pressure NMR is very promising. Specifically, the following studies are discussed. First, the pressure induced unfolding of lysozyme will be reviewed in Sect. 4.1, followed by two examples of high pressure NMR experiments on model membranes. Both the deuteron and phosphorous NMR studies of dipalmitoylphosphatidylcholine bilayers will be briefly discussed in Sects. 4.2 and 4.3.

2. High Pressure, High Resolution NMR Instrumentation

Over the last twenty years, we have built a number of high pressure NMR probes suitable for experiments on gases, liquids and disordered solids. Their design and construction have been discussed in detail in earlier studies [34,37-39] and reviews [28,30,33] covering the work of our laboratory. The very first multinuclear high resolution FT NMR experiment on liquids [40-42] at high pressure was performed in our laboratory in 1971 [40] using an electromagnet. However, the main applications were delayed till the availability of commercial high homogeneity superconducting magnets equipped with the superconducting and room-temperature shims [38]. This Section discusses in a general way the design of the high pressure, high resolution NMR probe.

The schematic drawing of the high pressure, high resolution NMR probe is shown in Fig. 1.

The pressure vessel was made from a non-magnetic titanium alloy (RMI 6Al-2Sn-4Zr-6Mo) which has a tensile strength of 170 kpsi and excellent resistance to corrosion. In our earlier designs, we used the classical Bridgman seal, but we found the C-seals have many advantages for high pressure NMR probes.

Fig. 1 shows the details of the C-seal fitted into closure plug flange which has been polished to 32 rms. When the closure plug is tightened, the C-seal is slightly compressed and better fits to the flange. Additionally, soft plating smears into imperfections of the flanges thus filling the microscopic voids and machining marks. This allows the seal to hold low pressures. As pressure is increased inside the vessel, it tends to "open" the C seal, and then the seal is held tighter against the

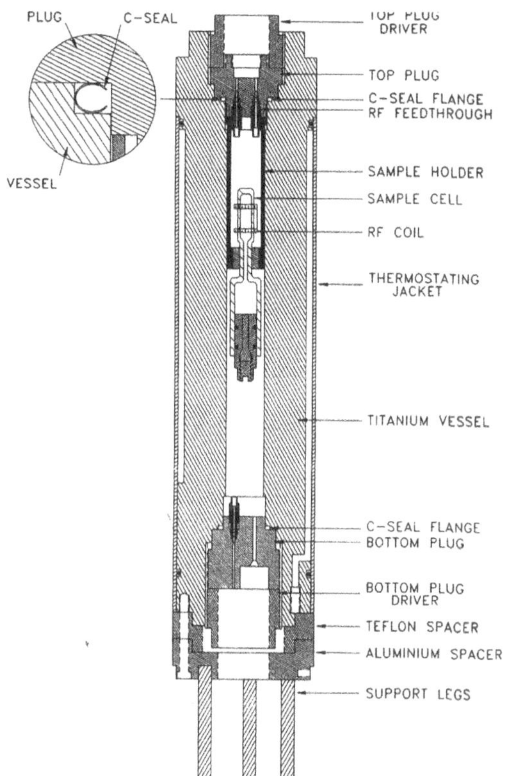

Fig. 1. Schematic drawing of the high pressure, high resolution NMR probe operating at proton frequency of 300 MHz. Pressure range 1 bar to 7 kbar. Temperature range -20° C to 100° C.

flanges at higher pressures. The metallic C-seals (EG&G, Inc.) used in our laboratory were made of Inconel X-750 and were heat treated over a long aging cycle. The plating of the C-seal depends upon its intended use. We used the golden coated C-seals in our high temperature, high pressure studies [37], which allowed us to go up to temperatures as high as 700° C at pressures up to 2 kbar which would not be possible with the lead based sealings (melting point 327° C). For lower temperatures we are using lead plated C-seals. We found the C-seals performance in all our applications to be very good. This type of sealing was working both with liquid and gaseous pressurizing mediums. It is very easy to use and can be re-used many times. In our 300 MHz system the same set of C-seals has been used for over 100 high pressure (up to 5 kbar) runs with opening and closing of the vessel between each experiment without any observable decrease in sealing performance.

Fig. 2 shows the sample holder and the details of the closure plug with RF leads.

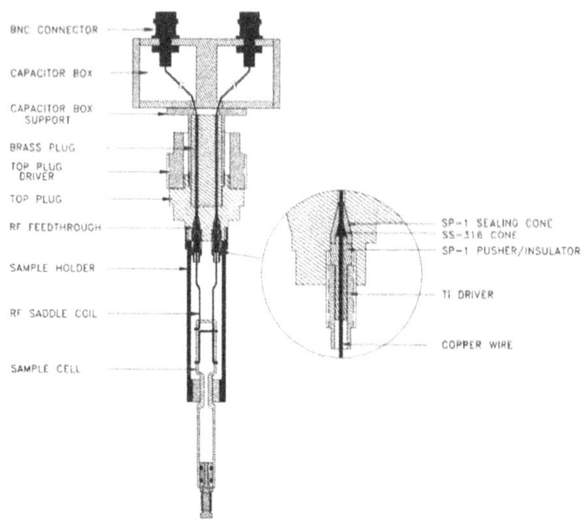

Fig. 2. Schematic drawing of the sample holder and the closure plug.

In this specific high pressure, high resolution NMR probe, we used the conventional design [34] of the high-pressure RF feedthroughs using SS316 cones soldered on copper wire, and the plastic Vespel (SP-1 polyimide, DuPont de Nemours and Company, Wilmington, DE) was used as a sealing cone to provide electrical insulation from the plug body. This probe was used in the 7.05 Tesla (300 MHz proton frequency) superconducting magnet (Oxford Instruments, Inc.), and the resolution for protons was 1.2 Hz for 10 mm sample diameter. The sensitivity (S/N ratio) of this probe is comparable to ambient pressure commercial probes of the GN 300 NMR spectrometer (General Electric, Inc.)

The type as well as the shape of the sample cell is related to the type of the experiment performed. In general, we use two types of high pressure sample cells which differ in the way the pressure is transmitted to the sample. One is a piston based design, while the other uses a bellows system. Both systems are shown in Fig. 3. In the piston design, the part of the sample cell which is in the RF coil region is made of 10 mm NMR tube which is then connected via capillary tubing to precision bore tubing that contains a piston, which is made of teflon, with a plunger pushed by a driver that allows for tighter contact with the glass surface. For a good sealing in order to prevent sample contamination, two rubber O-rings (Parker Seal Group) were used on the piston.

In the sample cell which uses bellows a glass (Pyrex 7740) or quartz to metal (SS-316) seal (Quartz Scientific, Inc., Palo Alo, Calif.) connects the actual 10 mm glass sample chamber to the SS-316 tubing. Stainless steel bellows (Mechanized Science Seals, Inc., Los Angeles, Calif.) accommodate the volume change of the liquid due to compression.

A different shape of sample cell shown in the center (b) of Fig. 3 was used for our ^{2}H NMR studies in powder samples of pyridine intercalated $CdPS_3$ [43] and in membrane [44] studies. The "pipe" like shaped sample cell is shown in Fig. 3 and the whole system together with sample holder B is presented in Fig. 4. The sample cell itself is made using a thick-walled 5 mm NMR tube (Wilmad Glass Co.) which has been attached through a fine pyrex tube to a glass-to-metal seal when bellows are used or to a precision bore tubing when piston system is used. The powdered sample occupies the horizontal NMR tube region of the sample cell and is kept in place by gravity. This also allows the use of a solenoid coil which has higher sensitivity than the saddle coil used in the other sample cells which is very important in ^2H NMR powder studies where high power, short pulses are required.

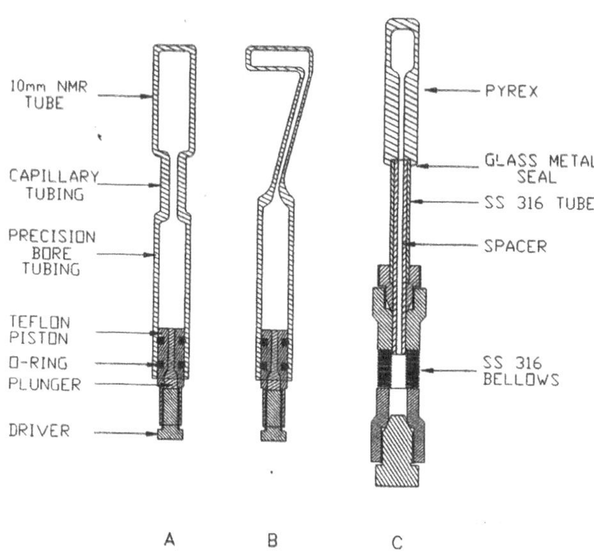

Fig. 3. Schematic drawings of the sample cells used for the high pressure NMR experiments.

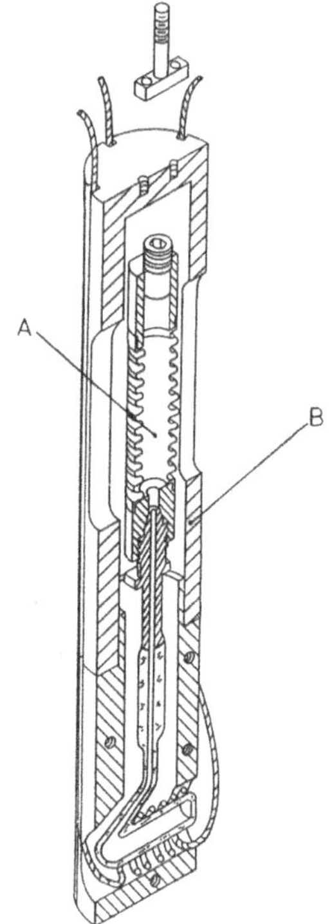

Fig. 4. Schematic drawing of the sample holder used for experiments on intercalates [43] and model membranes [44].

3. Applications of High Pressure NMR Spectroscopy to Chemical Systems

3.1 TRANSPORT AND RELAXATION IN COMPLEX LIQUIDS

In contrast to the number of experimental and theoretical studies of the dynamics in low viscosity (η < 0.01P) liquids, including our own high pressure studies [2], very few investigations dealing with the dynamics in complex viscous liquids (η > 1P) have been reported [45]. In particular, the high pressure experiments offer a great promise because of the ability of varying viscosity of complex liquids over many orders of magnitude should provide an advantage over nonviscous liquids in testing theories. In the following Sects. 3.11 and 3.12 we discuss the results of the NMR studies of self-diffusion [46,47] and high resolution natural abundance ^{13}C relaxation [47] in complex liquids of 2-ethylhexylbenzoate (EHB) and 2-ethylhexylcyclohexanoate (EHC) which structural formulas are given in Fig. 5.

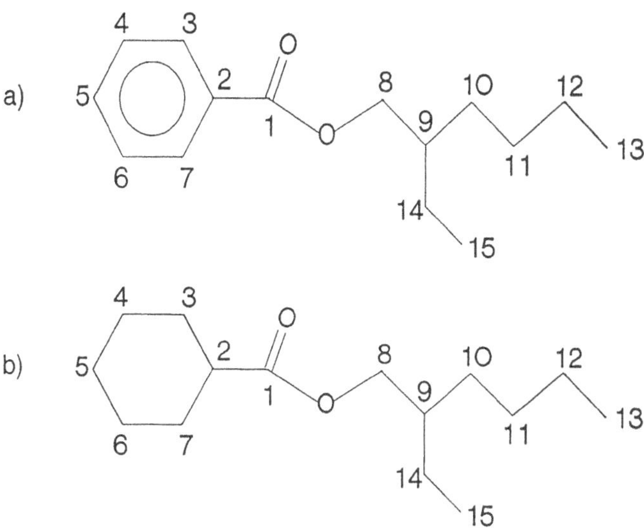

Fig. 5. Structural formulas of 2-ethylhexylbenzoate (EHB) (a) and 2-ethylhexylcyclohexanoate (EHC) (b).

Selection of EHB for this study was motivated by a lack of understanding of the relationship between the molecular properties and bulk fluid properties of elasto-hydrodynamic (ehd) lubricants [48], which operate under conditions of high pressure. In this respect EHB has been chosen as a model synthetic hydrocarbon based ehd lubricant, its molecular structure being complex enough to represent a real ehd fluid, while still being simple enough to allow detailed investigation of its molecular dynamics. Realization of the need for improving our understanding of the relationship between molecular structure of a complex liquid and its bulk fluid properties led to the study which investigated the molecular dynamics in liquid EHC. It was of interest to compare the various properties of EHC with those of EHB in order to determine the specific effect of replacing the benzene ring in EHB with a saturated, flexible cyclohexane ring in EHC.

It is important to emphasize that a range of 5 orders of magnitude in self-diffusion and fluidity were covered in our experiments on EHB and EHC. This wide change in the fluidity provides a unique opportunity to investigate the relationship between the molecular level quantities, e.g. reorientational correlation times and the fluidity of the medium.

3.11 Self-Diffusion in EHB and EHC Liquids and the Validity of the Stokes-Einstein Equation at the Molecular Level. The self-diffusion coefficients, densities and shear viscosities of liquid EHB and EHC [46,47] were measured as a function of pressure from 1 bar to 4.5 kbar within the temperature range from -20° C to +80° C. Diffusivities greater than 3×10^{-8} $cm^2 s^{-1}$ were measured by using the spin-echo Bessel analysis method [49] and diffusivities less than 3×10^{-8} $cm^2 s^{-1}$ were obtained by measuring the H_1 field dependence of the rotating frame proton spin lattice relaxation time [50] according to the method developed by Burnett and Harmon [51]

The self-diffusion coefficients of EHC and EHB measured as a function of pressure and temperature exhibit a behavior consistent with that observed for shear viscosity [46,47]. As it is shown in Fig. 6, EHC diffuses faster than EHB at -20° C whereas at 80° C their diffusion coefficients are nearly identical. The

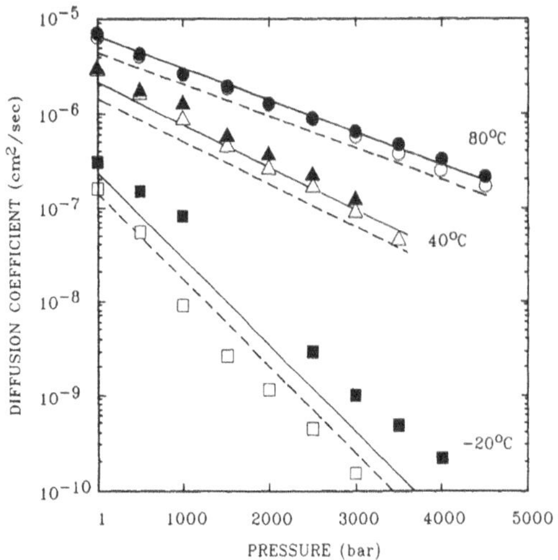

Fig. 6. Self-diffusion coefficients of liquid EHB (open symbols) and EHC (full symbols) as a function of pressure in the temperature range -20° C to 80° C. The lines represent the predictions from the experimental shear viscosity using the Stokes-Einstein equation in the slip boundary limit (full line) and the stick boundary limit (dashed line).

different diffusion behavior of EHB and EHC must be due to the fact that in EHB there is a possibility of conjugation of the phenyl with the ester group which results in a hindered rotation about the C-C bond between the phenyl and the carbonyl group. Several resonant structures exist for EHB, and particularly at lower temperatures the rotation about the C-C bond will be slow, so that the C_6H_5COO part of the molecule represents a relatively rigid planar structure. Since no such conjugation can occur in EHC, this explains the observed differences in diffusion, viscosity, and compressibility behavior. This interpretation is also supported by the experimental T_1 data discussed later in Sect. 3.12.

The availability of a wide range of diffusion coefficients and viscosities enables us to provide a test of the Stokes-Einstein equation at the molecular level for EHB and EHC. This hydrodynamic equation relates the diffusion coefficient (D) to viscosity (η) and molecular radius (a) according to

$$D = kT/(C\pi a\eta) \qquad (2)$$

where C is the constant equal to 4 in the slipping boundary limit and 6 in the sticking boundary limit. Equation (2) is derived for a macroscopic sphere moving in a continuum, and theoretically thus should only apply to solute-solvent systems where the solute is large compared with the solvent; in this case the stick condition (C=6) should apply. It has been suggested that the slip condition (C=4) may be approached when the solute and solvent are of similar size (i.e.), self-diffusion. Interestingly, several studies [2] have shown that Eq. (2) proves a reasonable estimate of self-diffusion for a number of simple molecules such as cyclohexane, methylcyclohexane, and benzene. Our data provide the opportunity to test the relationship over a much wider range of viscosity than these previous studies. The experimental results obtained showed that the Stokes-Einstein equation generally provides a good description of diffusion [46,47] of EHB and EHC over a 5 order of magnitude data range, while the constant C varied only between 2.6 to 5.2 as shown in Table 1. In view of the range of viscosities and diffusion coefficients, this result is quite remarkable. Figure 6 compares the experimental pressure dependence of the self-diffusion coefficient for liquid EHB and EHB the theoretical predictions from viscosity using Eq. (2) in the slipping and the sticking boundary conditions.

One can take advantage of the finding that the Stokes-Einstein equation is valid over many orders of magnitude changes in viscosity even for complex liquids. The conventional techniques for measuring shear viscosity in highly viscous liquids ($\eta > 10P$) at high pressure are experimentally very difficult. They lack accuracy and reproducibility, and in addition, are very time consuming. Therefore, it is easier to measure the self-diffusion coefficients of highly viscous liquids at high pressure by NMR and then calculate viscosity η from the Stokes-Einstein equation. Particularly for workers in the applied field of lubrication, it is sufficient to know whether viscosity is, e.g., 10^2 or 10^3 poise at some specific thermodynamic state. The NMR approach provides a very good estimate of the viscosity and its behavior at extreme conditions of high pressure and high temperature.

3.12 High Resolution Natural Abundance ^{13}C NMR Relaxation Study of Liquid EHB and EHC. Both experimental [52] and theoretical [53] studies have demonstrated the value of natural abundance ^{13}C NMR relaxation experiments to yield a detailed information about the motional dynamics of complex liquids. So far, however, all such ^{13}C NMR studies have been performed only as a function of temperature at atmospheric pressure. Taking advantage of our recent development of NMR instrumentation [36] which permits high resolution, high sensitivity NMR experiments on liquids at high pressure, we decided to measure natural abundance ^{13}C NMR spin-lattice relaxation times, T_1, and Nuclear Overhauser Enhancement (NOE) in liquid EHC and EHB as a function of

Table 1. Stokes-Einstein Constants for EHB and EHC

	-20° C		40° C		80° C	
P,MPa	EHB	EHC	EHB	EHC	EHB	EHC
0.1	2.6	3.4	3.5	2.9	4.2	3.9
50	2.3	2.8	3.5	2.9	4.2	4.3
100	3.5	2.6	3.7	2.5	4.6	4.2
150	3.0	--	3.9	3.1	4.3	3.6
200	--	--	3.5	3.0	4.4	4.3
250	3.9	4.2	3.0	3.1	4.4	4.4
300	4.4	5.3	3.2	3.7	4.4	3.6
350	3.6	5.0	3.6	--	4.5	3.2
400	3.3	5.2	--	--	4.6	3.4
450	3.9	5.8	--	--	4.6	4.1

pressure from 1 bar to 5 kbar within the temperature range from -20 to + 80° C. The general expressions relating ^{13}C T_1 and NOE to the spectral density functions for the intramolecular dipolar coupling mechanism are

$$\frac{1}{NT_1} = \frac{1}{10}\frac{\hbar^2\gamma_C^2\gamma_H^2}{r_{CH}^6}[J(\omega_H - \omega_C) + 3J(\omega_C) + 6J(\omega_H + \omega_C)] \tag{3}$$

$$NOE = 1 + \frac{\gamma_H}{\gamma_C}\left[\frac{6J(\omega_H + \omega_C) - J(\omega_H - \omega_C)}{J(\omega_H - \omega_C) + 3J(\omega_C) + 6J(\omega_H + \omega_C)}\right] \tag{4}$$

where N is the number of directly attached protons, $J(\omega)$ is the spectral density function, and ω_H and ω_C are the proton and carbon resonant frequencies, respectively. The particular form of $J(\omega)$ depends on the model for molecular reorientation. Because of asymmetric shape and high degree of internal mobility in EHC and EHB, the form of $J(\omega)$ needed to describe the relaxation is complex.

Figure 7, which gives the natural abundance ^{13}C NMR spectra of liquid EHB obtained at 80° C and 5 kbar pressure, illustrates the excellent resolution

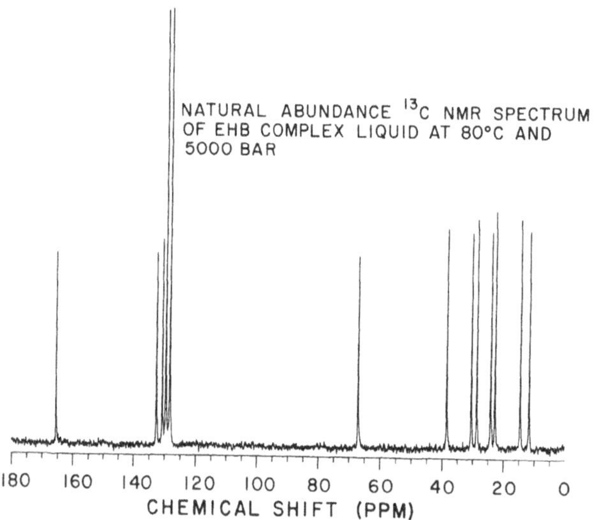

Fig. 7. High resolution, natural abundance ^{13}C NMR spectrum of EHB at 80° C and 5 kbar.

obtainable even under high pressure conditions without sample spinning. This high resolution permits the ^{13}C T_1 measurements for each individual carbon in EHC and EHB and makes it possible to probe directly the overall and internal motions in these complex liquids. The fact that our experiments cover both the motionally narrowed regime and the slow motion regime as the viscosity changes extend over five orders of magnitude enabled us to test rigorously the various theoretical models proposed to describe the dynamics of complex molecules of asymmetric shape and high flexibility.

However, in this overview, we mention only the general trend in the T_1 values for two selected carbon atoms in EHC and EHB. The pressure dependence of the ^{13}C relaxation rates, $1/NT_1$, for selected carbons 9 (methine) and 13 (methyl) in EHB and EHB at 80° C is plotted in Fig. 8. Inspection of Fig. 8 shows that for the methine chain carbon 9 the $1/NT_1$ values for EHB are higher than the corresponding values for EHC, reflecting the higher mobility of the cyclohexyl ring. It is interesting to find that $1/NT_1$ for EHB are larger than those for EHC even at 80° C in spite of the fact that the viscosities of EHB and EHC are nearly the same at this temperature. It is not surprising that the side chain methyl group has nearly identical $1/NT_1$ values both for EHC and EHB due to the C_{3v} symmetry of the end CH_3 group and the high flexibility of the chain. Figure 9 was included to show that at -20° C the motion of both EHB and EHC falls into the slow motion regime for all carbons with the exception of methyl carbons 13 and 15 (for carbon numbering, see Fig. 5).

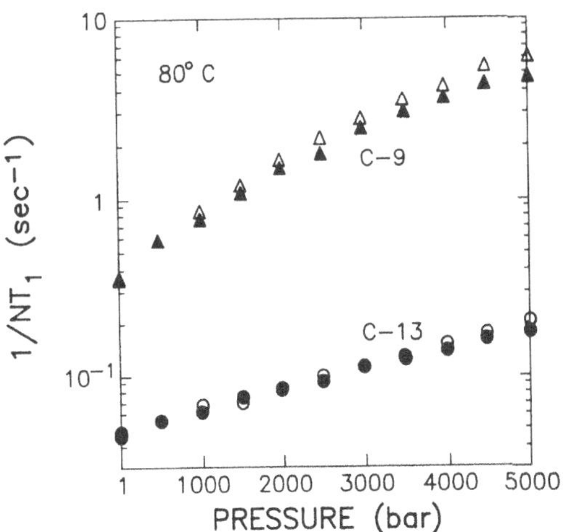

Fig. 8. Pressure dependence of the ^{13}C NMR relaxation rate, $1/NT_1$, for the methine carbon 9 and the methyl carbon 13 in liquid EHB (open symbols) and EHC (full symbols) at 80° C. (Motionally narrowed regime)

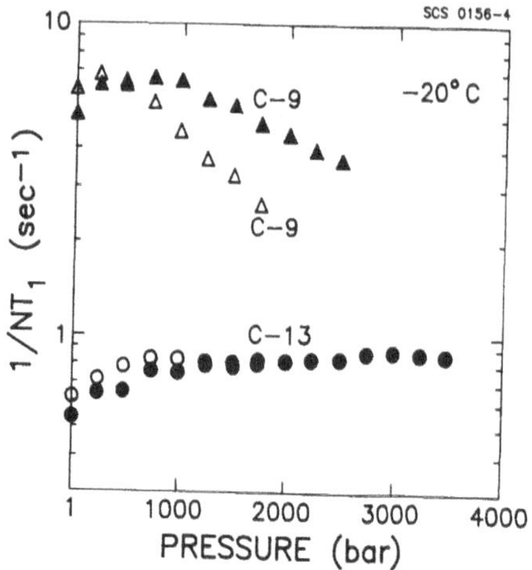

Fig. 9. Pressure dependence of the ^{13}C NMR relaxation rate, $(1/NT_1)$, for the methine carbon 9 and the methyl carbon 13 in liquid EHB (open symbols) and EHC (full symbols) at -20° C.

As an illustration, we present Figure 10 which gives the NOE values for the methine carbon 9 at 80° C and -20° C. The NOE behavior of all other carbons

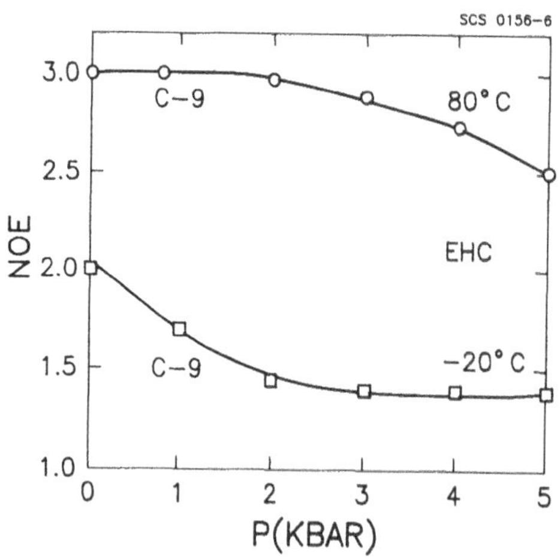

Fig. 10. Pressure dependence of the ^{13}C NOE for the methine carbon 9 in liquid EHC at 80° C and -20° C.

with the exception of the end methyl group carbons 13 and 15 shows a similar trend with pressure and temperature. The NOE's values determined for each individual carbon at 80° C indicate that at least for this temperature the extreme narrowing condition is value (ω_C and ω_H)τ << 1. Only at the highest pressures at 80° C we observe a decrease of the NOE values from 3.0 to 2.5. In contrast, the NOE values are lower than 2.0 for all pressures at -20° C indicating clearly the slow motion regime. Of course, this is true for all carbons, again with the exception of the end methyl groups 13 and 15 which rotate relatively freely even at -20° C and high pressures.

For an approximate evaluation of the relative mobility of the individual carbon atoms, we may consider the value of $R = \tau_{min}/\tau_D$ where τ_{min} is the correlation time, calculated from the T_1 minimum at lower temperature, where $\omega\tau_{min}$ = 0.6158, and τ_D is the correlation time calculated from the Debye equation. For example, at -20° C for EHB the R values for individual carbons exhibit the following progression R(5) ≌ > R (3,7) ≌ R (4,6) > R (10,14) > R [11,12,13,15].

The experimental data also suggest that the ring and the whole molecule is undergoing anisotropic reorientation, and the aliphatic chain is undergoing multiple internal rotation about each C-C bond. According to the model for multiple internal rotation [54], our observation that the plot of NT_1 vs carbon number, starting with the methine carbon 9 shows a linear dependence is indicative of equal rotational diffusion constants for all bonds in the chain except for the terminal methyl group.

A theoretical analysis of the T_1 and NOE data shows that simple models involving overall and internal rotation, anisotropic reorientations with multiple internal rotations, do not reproduce the experimental data. It appears that one has to use a Cole-Davidson distribution [55] of correlation times. This approach has successfully been used in our laboratory to interpret the temperature and pressure dependence of deuterium relaxation times in selectively deuterated glycerols [56]. Availability of T_1 and NOE values for fluidity changes of five orders of magnitude offers a unique opportunity not only to test rigorously various theoretical models for dynamics of complex liquids, but for the first time, these experiments will allow one to characterize the density and temperature effects on the molecular dynamics in complex liquids.

3.2 HIGH PRESSURE NMR STUDIES OF THE KRAMERS TURNOVER FOR ISOMERIZATION REACTIONS IN LIQUID SOLUTIONS

According to theoretical models [57,58] describing the dynamical solvent effects on reaction rates in liquid solutions, the reaction coordinate is coupled to the solvent, enabling the system to gain sufficient energy to cross the barrier, lose energy, and become trapped into the product well. In absence of electrostatic interactions, this coupling is produced by collisions between the solvent and solute molecules. In contrast to classical transition state theories for isomerization reactions, the stochastic models propose a dependence of the transmission coefficient κ upon so-called "collision frequency" α which reflects the actual coupling of the reaction coordinate to the surrounding medium. According to theoretical models, the transmission coefficient κ is found to be a strong nonmonotonic function of α with two different limits. Activation due to collision rate is limiting and κ is proportional to α for the energy-controlled regime at low collision frequencies. At high collision frequencies in the diffusive regime, particles which have crossed but not yet cleared the barrier may suffer collisions and recross the barrier. The reaction in this limit is said to be diffusion controlled and the rate is inversely proportional to α. Between these two regimes there is a nonmonotonic transition, Kramers turnover (crossover). In our laboratory we observed such turnover behavior for the chair-chair inversion of cyclohexane in several solvents [59]. Hochstrasser et al. [60] have reported Kramers turnover region in the isomerization of trans-stilbene by using high pressure gaseous ethane. Troe et al. [61] have also observed Kramers turnover for trans-stilbene combining experiments performed both in liquid and in gas at high pressure. However, the great majority of systems studied in dense liquid media show the high friction behavior.

It was not surprising that our results for isomerization of cyclohexane were met with a considerable degree of skepticism as the prevalent view was that the rapid internal energy flow would prevent observation of the low friction regime in dense liquid solvents. However, in their multidimensional molecular dynamics calculations, Chandler et al. [62] have reported that observation of the inertial behavior depends strongly on the relative strength between the intermolecular coupling and the intramolecular coupling. They concluded that the RRKM picture of unimolecular kinetics does not describe cyclohexane isomerization in liquid solutions as the energy exchange between the molecular and the stochastic bath occurs with similar ease as the energy rearrangement among intramolecular modes.

Clearly, the inefficient internal energy flow is the key ingredient in order to observe the energy controlled regime or the Kramers turnover for a system in dense solvent fluids. Therefore, we turned our attention to systems where we may expect an inefficient internal energy flow [63]. The theoretical predictions by Marcus et al. [64] and by Uzer and Hynes [65] about the heavy metal atom bottleneck effect on internal energy redistribution in a molecule provided the main motivation for our experiments dealing with Rh complexes in liquid solutions. Experimental studies on several organometallic compounds [66] have also suggested that a heavy metal atom acted as a barrier which reduced the rate of intramolecular vibrational energy transfer between the liquids; however, this effect was not observed for another system [67].

In our recent studies [68,69] we investigated the effects of temperature and pressure on the internal rotation rate of coordinated ethylene in π-cyclopentadienyldiethylenerhodium and π-cyclopentadienylethylenetetrafluoroethylene-rhodium in several liquid solvents. Figure 11 shows the molecular structure of the of the Rh complexes. It was found that the rotation of coordinated ethylene is initially accelerated by pressure, reaches a maximum and then decreases at high pressure. The experimental data, as interpreted in terms of stochastic models of isomerization reactions, indicate a Kramers turnover for the pressure dependence of the rotation of coordinated ethylene in the Rh complexes. The observation of the energy controlled regime in this system may be the consequence of the so-called heavy metal atom bottleneck effect which reduces the intramolecular energy transfer within the molecule. Table 2 summarizes the results of our experimental studies dealing with the dynamical solvent effects on reaction rates in dense liquid solvents.

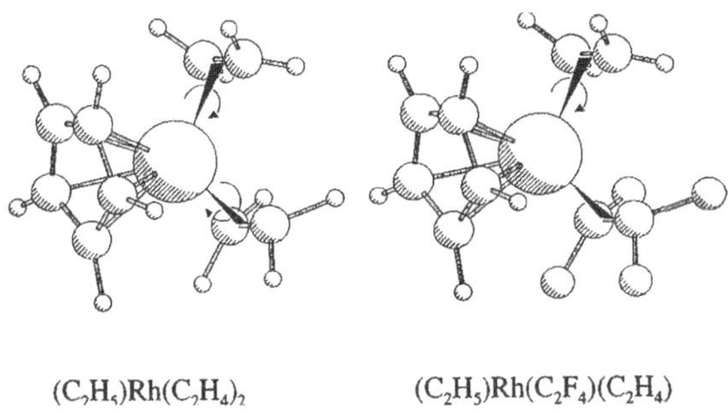

Fig. 11. Molecular structure of π-cyclopentadienyldiethylenerhodium and and π-cyclopentadienylethylenetetrafluoroethylenerhodium

Table 2. Summary of Recent Results

System	Process	Result	Ref.
Cyclohexane	Conformational Isomerization	Kramers Turnover	[59,70]
1,1-difluorocyclohexane	Conformational Isomerization	?	[71]
N,N-Dimethyltrichloroacetamide	Hindered Rotation	Diffusive Regime	[72]
π-cyclopentadienyldiethylene-rhodium	Ethylene Rotation	Inertial Regime	[68]
π-cyclopentadienylethylenetetra-fluoroethylenerhodium	Ethylene Rotation	Kramers Turnover	[69]

In all our studies we used high-pressure, high resolution NMR techniques to investigate the dynamical solvent effects on the reaction rates in liquid solutions. It is appropriate to comment on the importance of using pressure as an experimental variable in the studies of dynamical solvent effects on reaction rates, as the collision frequency which reflects the coupling of the reaction coordinate to the medium can be related through simple hydrodynamic arguments to shear viscosity η. The collision frequency α in different solvents is given by the coefficient of friction ζ

$$\alpha = \zeta/m \tag{5}$$

and the molecular mass m of the solute. An estimate for ζ can be obtained by applying the Stokes law

$$\zeta = c\pi\eta(\sigma/2) \tag{6}$$

where σ is the hard-core diameter and the c is equal to 4 in the slipping boundary limit, whereas it is 6 for the sticking boundary limit. Both theoretical and experimental studies show that the slipping boundary conditions are appropriate for cases discussed in our experiments. Therefore, the collision frequency is given by

$$\alpha = (2\pi/m)\,\eta\sigma. \tag{7}$$

In most studies the shear viscosity η is changed by the use of different solvents but in the high pressure experiment, viscosity can be varied by changing pressure. One has to realize that viscosity represents only an approximative measure of the degree of coupling of the reaction coordinate to the reaction medium and consequently by changing solvents one may influence the reaction rate by different molecular shape, size, or strength the intermolecular interactions of the solvent molecule used. Therefore, different solvents of the same shear viscosity may not have the same effect on the reaction rate measured. Clearly, using the same solvent and changing its viscosity by pressure represents a much cleaner experiment.

As most of the studies of simple isomerization reactions involve laser spectroscopic techniques, one should also comment on the relative merits of the NMR technique. There are several advantages of using NMR to study isomerization and hindered rotation in liquid solutions. The system chosen can be very simple, and the molecule can be studied in its ground state. For example, the chair-chair isomerization of cyclohexane is a relatively simple process which can be characterized by two degrees of freedom.

However, there are some disadvantages connected with the use of the NMR line shape technique [73] to calculate the experimental rates. For this technique to be applicable, motions must fall within a narrow time scale, and the restricted range of measurable rates leads to a reletively large error in determined activation parameters. This inherent weakness of the NMR lineshape analysis approach has been overcome in our laboratory using the NMR rotating frame technique [70] to measure rates. For example, for cyclohexane the highest measurable rate by the NMR lineshape technique is about 5.10^3 s^{-1}, whereas the NMR rotating frame method allows one to measure rates up to 5.10^5 s^{-1}.

In our study [70] we presented the results of pressure effects on the conformational chair to chair isomerization rate of cyclohexane. Rotating frame spin-lattice relaxation time $T_{1\rho}$ measurements were made to determine these rates and combined with previous results [59] using the line shape fitting analysis technique. The data indicates that this system is in the inertial to turnover regions in the Kramers' model. By increasing the temperature we were able to shift the system further into the inertial regime by lowering the viscosity which is proportional to the collisional frequency. At this temperature, the interconversion of cyclohexane is too fast to be measured by the line shape technique, therefore, we used the $T_{1\rho}$ method to measure isomerization rates and their pressure dependence.

For large amplitude isomerization motion, and for E_0 > RT the rate constant has been related to the viscosity by the following expression:

$$k = F(\eta) \exp(-E_0/RT), \tag{8}$$

where $F(\eta)$ is a function of the solvent shear viscosity η and E_0 is the internal barrier height. According to this formula, the energy barrier effects can be separated from the frictional effects of the solvent. The accurate activation energy should be determined by the temperature variation of the rate at fixed viscosity. If different isoviscosity plots yield the same value of E_0 then the activation barrier would have to be pressure independent for a given solvent.

The isoviscosity plots of ln k vs 1/T for cyclohexane in CS_2 solvent are shown in Figure 12. The points on the lines were extracted from the pressure fits of viscosity

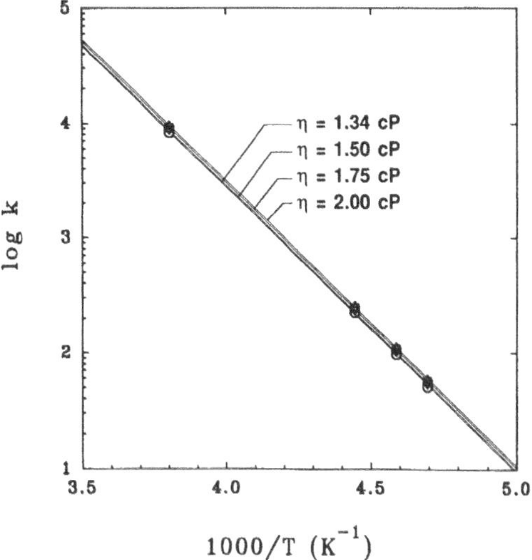

Fig. 12. The isoviscosity plots of ln k vs 1/T for cyclohexane in CS2; (0) - η = 1.34 cP, (∇) - η = 1.50 cP, ($\square$) - η = 1.75 cP, (Δ) - η = 2.0 cP.

and rate constants. The parallel relationship among the four lines for viscosities ranging from 1.34 to 2.0 cP clearly indicates that the internal barrier height is not sensitive to the solvent viscosity. From the isoviscosity plots we get the activation energy which is 11.28 ± 0.01 kcal/mol for all four viscosities. High accuracy of the isoviscosity data, reflected by the value of correlation coefficients close to one, is remarkable. Since these plots have identical slopes we have ruled out the possibility that the barrier height is a function of pressure in the range studied. The pressure induced decrease in the energy of activation should be about 0.25 kcal/mol in order to account for the observed increase of the reaction rate over a 5 kbar pressure range. Such a significant decrease of the activation energy would be detectable from our isoviscosity plots. However, they remain constant within ± 0.01 kcal/mol. Therefore, the pressure induced acceleration of the isomerization rate of cyclohexane is due to collisional frequency changes characteristic for molecular systems in the inertial regime.

A practical way to discuss the isomerization dynamics is provided by the relationship between the transmission coefficient κ and the solvent viscosity.

Figure 13 shows the dependence of the normalized transmission coefficient upon viscosity as generated from the experimental data for cyclohexane

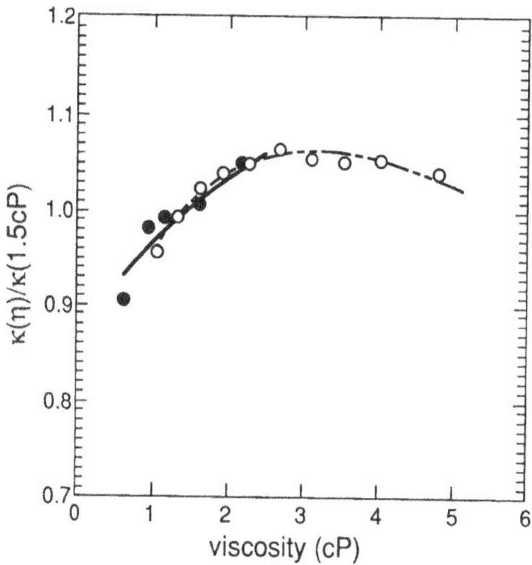

Fig. 13. The normalized transmission coefficient as a function of viscosity for cyclohexane isomerization in CS_2 solvent. (●) - NMR rotating frame relaxation data at 263 K, (○) - NMR line shape data at 225 K. The lines are the best fits to the stochastic model (solid line -- T = 263 K, broken line -- T = 225 K).

isomerization in CS_2 solvent. As in our previous work [59], the value of $\Delta V^{\ddagger}_{TST}$ = - 1.5 cm^3 mol^{-1} has been assumed. We emphasize that in Figure 13 the two lines are experimental $\kappa(\eta)/\kappa$ (1.5 cP) vs η plots for two different temperatures. The plots indicate the strong correlation between the reduced transmission coefficient and solvent viscosity. Of particular importance is the fact that the transmission coefficient increases with η by about 20% and that the turnover occurs between 2.5 and 3.0 cP. The results obtained by the NMR line shape analysis [59] at 225 K are also shown for comparison. The data taken at 263 K using the NMR rotating frame relaxation technique clearly indicate that cyclohexane is in the inertial regime of Kramers' model.

In summary, our recent study [70] reported the results of the NMR rotating frame relaxation study of the isomerization rate of cyclohexane in carbon disulfide at various temperatures and pressures. It is the first time that the $T_{1\rho}$ technique under high pressure has been applied for the chemical exchange measurements. By combining the $T_{1\rho}$ and NMR line shape techniques and generating the isoviscosity plots we were able to show that the barrier height to isomerization is independent of pressure. The isomerization rate constant in the pressure range studied was a monotonic linear increasing function of pressure indicating that cyclohexane in this viscosity, temperature, and pressure range is in the inertial regime of the Kramers' model. The viscosity dependence of the reduced transmission coefficient suggests that the reaction coordinate is weakly coupled to the surrounding medium and the isomerization dynamics shows an energy controlled behavior.

3.3 PRESSURE EFFECTS ON LIQUIDS IN CONFINED GEOMETRIES

In view of its fundamental and technological importance, it is not surprising to find that the problem of liquid behavior in confined geometries continues to attract attention [74,75]. The study discussed in this Section 3.3 represents a continuation of our efforts[76-79] to improve the understanding of the dynamics of molecular liquids confined to porous silica glasses prepared by the sol-gel process.

There were several results of our earlier study [76] which provided the main motivation for these experiments [80]. First of all, we have established the applicability of the two-state fast exchange model to analyze the spin-lattice relaxation time, T_1, data for polar molecular liquids confined to sol-gel prepared porous silica glasses with a wide range of pore diameters. It is important to point out that this allows one to determine the relaxation rate, $(1/T_1)$, for the surface layer liquid. Therefore, it is of interest to find out whether the two-state, fast exchange model is also applicable for analysis of relaxation data for confined liquids at high pressures, as this approach will allow one to investigate for the first time the effects of pressure on the reorientational dynamics of the surface layer liquid.

In addition, our investigation [76] of temperature effects on the reorientational dynamics of pyridine-d_5 confined to porous glasses showed that the Arrhenius-type temperature dependence of the deuteron T_1 yielded activation energies, (E_a), which became smaller in smaller pores, i.e., E_a for bulk liquid pyridine was higher than E_a for the surface layer pyridine. Based on this finding, it would be informative to calculate the activation volumes, $\Delta V^{\ddagger}$, for the relaxation times of confined pyridine at high pressures in order to determine whether the $\Delta V^{\ddagger}$'s are also dependent on the pore size.

However, so far all NMR or other spectroscopic studies on liquids in confined geometries were performed at ambient pressure. The reason for this may be related to the inherent experimental difficulty of this type of high pressure experiment, as one has to avoid the contamination of the confined liquid by the pressure transmitting medium (e.g., dissolved gas) and also prevent mechanical crushing of the porous silica glass (e.g., piston apparatus, diamond anvil cell). Fortunately, we solved the problem of pressure transmitting medium by choosing perfluoropolyether liquid lubricant, Krytox, made by DuPont. For the purposes of our experiments, these polymeric fluids are immiscible with pyridine-d_5 and other molecular liquids, and therefore can be used as pressure transmitting liquids. The main goal of our novel experiments was to determine the effects of pressure on the dynamics of liquids in the surface layer by using the two-state, fast exchange model and compare to the pressure effects observed for bulk liquids. With this goal in mind, the deuteron NMR spin-lattice relaxation times, T_1, in liquid pyridine-d_5 and nitrobenzene-d_5 confined to porous silica glasses with 18.4 Å, 24.1 Å, 32.8 Å, 52.1 Å and 69.4 Å pore radii were measured as a function of pressure up to 5 kbar at 300K [80].

The ^{2}H NMR relaxation time measurements were carried out by using a NMR pulse spectrometer equipped with an Oxford Instruments high resolution 4.2 Tesla superconducting magnet. The experimental procedures were described in detail in our earlier study [76].

According to the two-state, fast-exchange model, the liquid in a pore is assumed to have two distinct phases: a bulk phase, which has the same relaxation properties as the bulk liquid, and a surface-affected phase for which the relaxation rate is greatly enhanced. If diffusion between the two phases is much faster than the relaxation rate, a single exponential decay will be observed. The observed relaxation rate (T_1^{-1}) as a function of pore radius (R) can be approximated by a linear relationship between T_1^{-1} and R^{-1}

$$T_1^{-1} = T_{1b}^{-1} + 2\alpha/R\,(T_{1s}^{-1} - T_{1b}^{-1}), \qquad (9)$$

where T_{1b} is the relaxation time for bulk liquid, T_{1s} is the relaxation time for the liquid in the surface layer, and α is the thickness of the surface layer.

As we discussed in our earlier study [76], the 1 bar NMR relaxation experiments on pyridine-d_5 confined to porous silica glasses were analyzed by the two-state model, and it was important to find that this model also describes well the experimental relaxation data obtained for pressures up to 5 kbar. Figs. 14 and 15 show the deuteron T_1's of liquid pyridine-d_5 and nitrobenzene-d_5, respectively, as a function of pore radius (R^{-1}) in porous silica glasses at pressures from 1 bar to 5 kbar. Clearly, the strong interaction of these wetting liquids with the silica surface represents the main reason for the applicability of the two-state, fast exchange model. As a result, we were able to determine the effect of pressure on the T_{1s}^{-1} and compare it with the pressure dependence of T_{1b}^{-1} for bulk pyridine, and nitrobenzene, respectively.

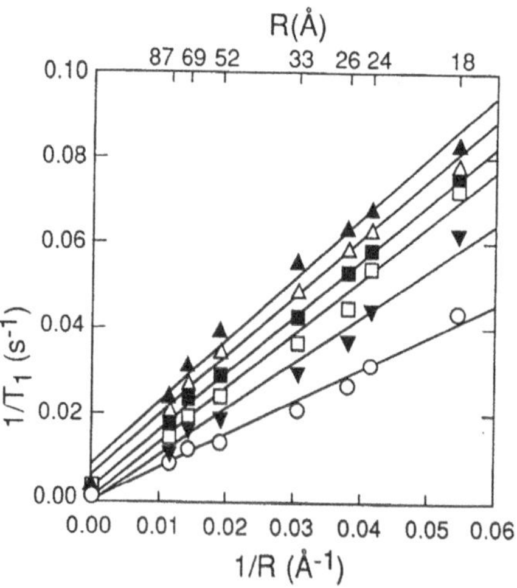

Fig. 14. The ²H spin-lattice relaxation rate (T_1^{-1}) of liquid pyridine-d_5 as a function of pore radius (R^{-1}) in porous sol-gel silica glasses at 300 K (○ 1 bar; ▼ 1 kbar; □ 2 kbar; ■ 3 kbar; Δ 4 kbar; ▲ 5 kbar).

In a first approximation, by assuming isotropic reorientation, the deuteron relaxation rate (T_1^{-1}) due to quadrupolar interactions is given by the following expression

$$(T_1^{-1}) = 3/8\,(e^2qQ/\hbar)\,(1 + \eta^2/3)\,\tau_\theta \qquad (10)$$

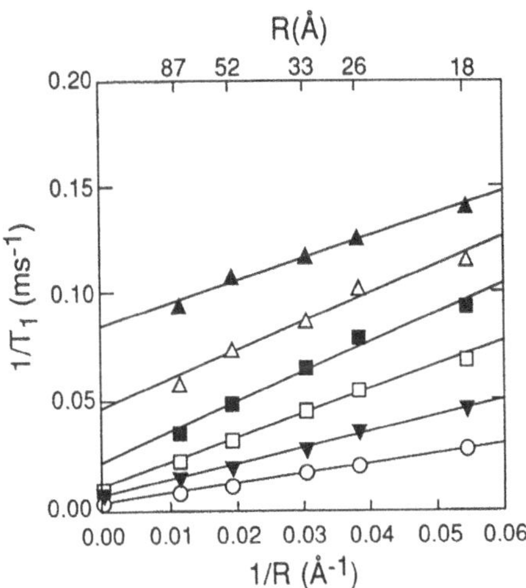

Fig. 15. The ^2H spin-lattice relaxation rate (T_1^{-1}) of liquid nitrobenzene-d_5 as a function of pore radius (R^{-1}) in porous sol-gel silica glasses at 300K (0 1 bar; ▼ 1 kbar; □ 2 kbar; ■ 3 kbar; Δ 4 kbar; ▲ 5 kbar).

where ($e^2qQ/\hbar$) is 2π times the nuclear quadrupole coupling constant in Hz, η is the asymmetry parameter, and τ_θ is the reorientational correlation time. In agreement with our earlier results [6], we find $\tau_{\theta,b}$ = 2.5 psec for bulk pyridine-d_5 at 1 bar and 300 K, and $\tau_{\theta,s}$ = 81.4 psec for pyridine in the surface layer. As expected, the mobility of pyridine appears to be significantly reduced in the surface layer.

Fig. 16 shows the pressure dependence of the reorientational times, $\tau_{\theta,s}$ and $\tau_{\theta,b}$, normalized to the value $\tau_{\theta,b}$ at 1 bar. From Figure 16, one sees that at 1 bar the $\tau_{\theta,s}$ is 32 times larger than $\tau_{\theta,b}$ while at 5 kbar the $\tau_{\theta,s}$ is only 18 times larger than the value of $\tau_{\theta,b}$. In fact, the pressure increase to 5 kbar changes τ_θ by a factor of 3.8 for the bulk liquid but only by a factor of 2.1 for the surface layer liquid. Indeed, one can see that the correlation time of the bulk liquid exhibits a different pressure response than the $\tau_{\theta,s}$ of the surface layer liquid. Fig. 17 compares the effects of pressure on the ratio of the surface layer liquid relaxation rate (T_{1s}^{-1}) to that for the bulk liquid (T_{1b}^{-1}) for both liquid pyridine-d_5 and nitrobenzene-d_5, respectively. As we pointed out earlier [76], nitrobenzene interacts less strongly with the silica surface than pyridine and, therefore, the surface relaxation enhancement (T_{1s}^{-1})/(T_{1b}^{-1}) is much smaller than that observed for pyridine at 1 bar. Similarly, as in the case of pyridine, the increase in pressure leads to the decrease in this ratio)T_{1s}^{-1})/(T_{1b}^{-1}).

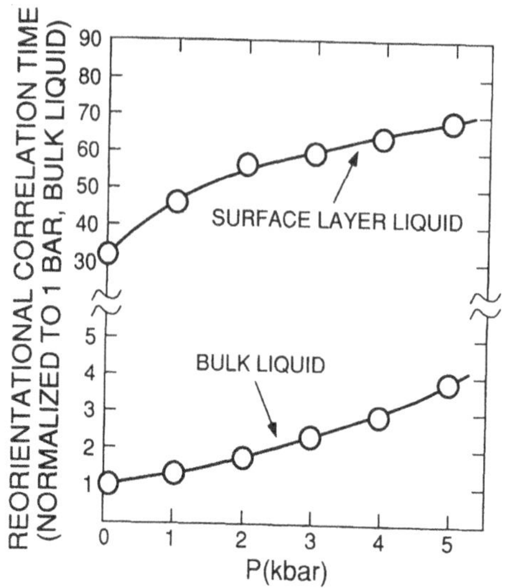

Fig. 16. The effect of pressure on the reorientational correlation times for the bulk liquid pyridine-d_5, and pyridine-d_5 in the surface layer. The τ_θ values are normalized to the τ_θ value of bulk pyridine at 1 bar and 300 K.

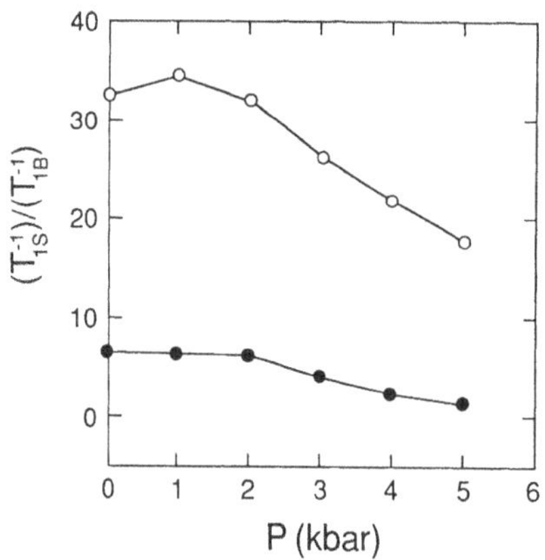

Fig. 17. Pressure dependence of the ratio of surface layer relaxation rate $(1/T_{1s})$ to the bulk liquid relaxation rate $(1/T_{1b})$ for liquid pyridine-d_5 (o) and nitrobenzene-d_5 (●) at 300 K.

These results suggest that the activation volumes ($\Delta V^\ddagger = -RT\,(\partial \ln T_1/\partial P)_T$ for T_1 will be a function of the pore radius. In fact, Table 3, which gives the values of $\Delta V^\ddagger$ determined for both pyridine and nitrobenzene confined to glasses of different

Table 3. The effect of confinement on the activation volumes, $\Delta V^{\ddagger}$, for the deuteron NMR spin-lattice relaxation rate, T_1^{-1}, for liquid pyridine-d_5 and nitrobenzene-d_5 confined to porous silica glasses at 300 K.

Pore Radius	$\Delta V^{\ddagger}$ (cm³/mole)	
R(Å)	Pyridine-d_5	Nitrobenzene-d_5
18	2.8	7.8
26	3.5	8.9
33	4.1	9.8
52	5.0	11.3
87	5.1	12.3
Bulk	6.6	15.6

pore radius, illustrate how $\Delta V^{\ddagger}$ decreases with the decreasing pore size. As a matter of fact $\Delta V^{\ddagger}$ changes from 6.6 cm³/mole for the bulk pyridine to 2.8 cm³/mole for pyridine confined to a glass with 18 Å pore radius. It is interesting to note that the trend of $\Delta V^{\ddagger}$ towards lower value accompanying confinement in smaller pores parallels the behavior of activation energies for confined pyridine which were found to be 2.8 kcal/mole for bulk pyridine and 1.6 kcal/mole for pyridine in 18 Å pore glass. Even a qualitative interpretation of the trend in the activation energies and activation volumes must be approached with caution. First of all, it is well known that for low-viscosity liquids the Arrhenius-type plots of various relaxation times represent a convenient way to plot the data, but the reorientational motion cannot be represented as an activated process. Similarly, the concept of activation volume $\Delta V^{\ddagger}$ has only a phenomenological significance for the low viscosity liquids. Nevertheless, the observed trends in $\Delta V^{\ddagger}$ with confinement are significant for the polar liquids studied, and it is definitively worthwhile to determine the pressure effects on $\Delta V^{\ddagger}$ in chemically modified glasses where the surface hydroxyl groups were replaced by Si-O-Si(CH_3)$_3$ groups which cannot form hydrogen bonds with pyridine and nitrobenzene. There is another compelling reason for a more detailed study of the E_a's and $\Delta V^{\ddagger}$'s for confined liquids, as their changes with confinement as reported in this study contrasts sharply with the increase of the relaxation rate of the surface layer liquid which result indicates slower motions and would suggest an opposite trend. Since we have also found an analogous behavior for $\Delta V^{\ddagger}$'s of $T_{1\rho}$ (rotating frame relaxation times) of nitrobenzene-d_5 confined to porous glasses, we believe that this result merits further investigation for other polar liquids, as it may provide the information needed to characterize better the details of the motional dynamics of confined liquids.

In particular, the experiments on non-wetting, non-polar liquids such as cyclohexane and methylcyclohexane are of special interest, as for these liquids the topological confinement effects dominate over the surface interaction effects, and the effects of reduced-dimensionality of the motions of the confined liquids are readily observable [76]. The described experimental approach which allows investigation of the effects of pressure on the dynamic behavior of surface-layer liquids, may provide a new tool in studies [46,47] of model liquid lubricants at extreme conditions of pressure and temperature. Systematic high pressure NMR studies of various molecular liquids including model lubricants confined to porous silica glasses are in progress.

3.4. HIGH PRESSURE NMR STUDIES OF DISORDERED SOLIDS

As it was discussed in our earlier reviews [30,33], the high pressure NMR techniques used for studies of liquids are readily adaptable (see Sect. 2) to experiments on disordered solids which are relatively compressible even in the range 1 bar to 5 kbar. In this section, the main results of a recent high pressure deuterium solid-state study of the dynamics of pyridine intercalated $CdPS_3$ are reviewed to illustrate the specific information content of such high pressure NMR experiments.

Our recent results on the dynamics of pyridine in several layered compounds provided motivation for the study of the effects of pressure on the motion dynamics of pyridine-d_5 in $CdPS_3$ [43]. In the $CdPS_3$/pyridine system, the solid host layers are separated by two-dimensional van der Waals (VDW) gaps. The pyridine is then intercalated into the gaps and the molecular reorientational motion can be determined through the analysis of the deuterium powder pattern lineshapes. Previously, our ^2H NMR temperature study on the polycrystalline pyridine/$CdPS_3$ system indicated that the pyridine experienced a large-amplitude reorientational diffusion about an in-plane axis perpendicular to the molecular C_2 symmetry axis in addition to small amplitude librational motion (see Fig. 18). We concluded that the molecular was oriented in the VDW gap such that the molecular plane was

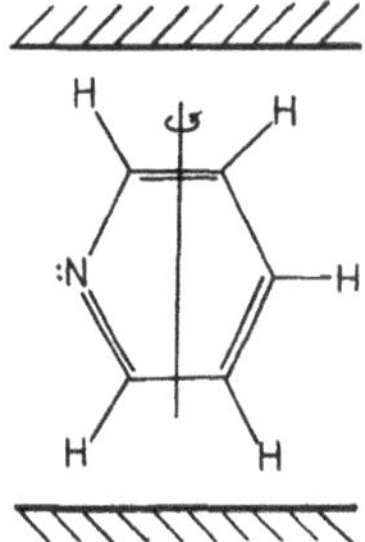

Fig. 18. The orientation of the pyridine in the VDW gap of the $CdPS_3$ host materials and the rotational motion of the intercalated pyridine.

perpendicular to and the C_2 symmetry axis was parallel to the host layers. Both the dynamics and orientation of the pyridine in $CdPS_3$ host were very similar to that seen earlier for $(C_5H_5N)_3\,TaS_2$.

Due to the lamellar nature of the host material, The $CdPS_3$ is compressible only along the c-axis in the pressure range investigated. The application of pressure to the intercalation compound results in a decrease of the VDW gap dimension. In this study we examine the effect of compression on the reorientational motion of the intercalated molecule.

Pressure could effect the intercalated molecular in one of several ways. The compression could cause deintercalation of the pyridine or the stage of the complex (number of host layers between occupied VDW gaps) could be altered. It has been suggested elsewhere (81) that pressure could cause an increase in the stage of the potassium/graphite system without deintercalation. If this was the case in the pyridine/$CdPS_3$ system, the occupied gaps would have a higher density of pyridine after compression. This could conceivably alter the reorientational motion of the pyridine. Even in the absence of a change in staging, pressure might modify the mode of reorientational motion seen for the pyridine. Another possible result of the application of pressure to the compound would be lineshape changes similar to those obtained by decreasing the temperature.

In the high pressure study of pyridine intercalated $CdPS_3$ four isotherms were investigated: 270K, 300K, 330K and 360K. At each temperature, powder pattern lineshapes were collected at pressures ranging from 1 bar to 4500 bar. A typical series of lineshapes obtained as a function of pressure is shown in Fig. 19. At

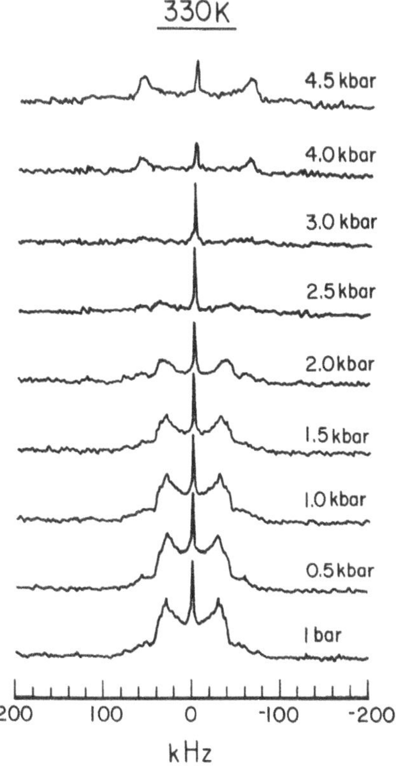

Fig. 19. High-pressure deuterium solid-state NMR lineshapes of $(C_5H_5N)_{0.41}CdPS_3$ obtained by using a solid echo sequence incorporating composite $90°$ pulses. The echo delay used was 35 μs and 8192 scans were averaged using equilibrium delays between 11 s and 8 s to avoid saturation. Apodization with an exponential function was performed providing 1000 Hz of linebroadening for signal-to-noise improvement. The lineshapes were symmetrized in order to increase S/N by zeroing out the quadrature channel; no changes other than an increase in S/N were observed in the frequency domain signal after zeroing when compared with the raw spectra.

ambient pressure and 330K, three motionally narrowed components and an isotropic pyridine component (0 Hz) are observed. The reorientational motion responsible for this powder pattern has been previously determined. The pyridine is undergoing a rapid rotational diffusion of three-fold or higher symmetry about an in-plane axis perpendicular to the molecular C_2 symmetry axis. With increasing pressure, the three motionally narrowed components broaden and decrease in intensity. By 2.5 kbar, little intensity indicates that the system is experiencing irreversible dephasing typically seen in an intermediate exchange regime. This behavior was observed previously for the pyridine/$CdPS_3$ system and was reported in our temperature study on that system. At 3.0 kbar, the rigid component appears; by 4.0 kbar, no large-amplitude reorientational motion remains.

The experimental results were analyzed in detail in our original study [43], and we were able to conclude that the application of pressure does not change the reorientational dynamics of the intercalated molecule in this system. Instead, the motion is suppressed or hindered due to the compression of the host lattice. Also, there was no clear evidence for a change in the stage of the $CdPS_3$/pyridine complex. Increasing the pressure caused lineshape results very similar to those seen with a decrease in temperature. Although our results show similar effects for a pressure increase and temperature decrease, the two methods probe the system in different ways. High-pressure NMR techniques could be used in "VDW gap tuning" experiments. The simulation of different host environments could be achieved through the application of pressure in order to study the effect of a variety of hosts on the dynamics of a particular intercalated molecule or ion.

4. NMR Studies of Biochemical Systems Under High Pressure

4.1 HIGH RESOLUTION NMR STUDY OF THE PRESSURE INDUCED UNFOLDING OF LYSOZYME

Most studies dealing with the denaturation of proteins have been carried out at atmospheric pressure using temperature or the chemical composition of the medium as experimental variables. The interpretation of the results of such experiments is not straightforward since the change of temperature produces simultaneous changes both in volume and thermal energy, and their effects are difficult to separate. In contrast, using pressure as the experimental variable in studies of solutions of proteins allows one to perturb the environment of the protein in a continuous controlled way by changing only intermolecular distances [8].

It is well known that reversible denaturation (unfolding) of a protein in solution can be caused not only by an increase in temperature or by changes in the composition of the medium but also by application of high pressure [82]. So far there have appeared only a few high pressure studies dealing with the denaturation of proteins. In particular, Weber and co-workers [83] used various fluorescence techniques to study the pressure denaturation of lysozyme, and on the basis of their results proposed a plurality of pressure-denaturated forms for this enzyme. Their experiments, which used either intrinsic or extrinsic fluorescence probes followed the unfolding process in a non-specific, global way, characteristic of the various experimental techniques used so far to follow pressure denaturation of proteins. In contrast, high resolution NMR spectroscopy is a powerful technique for probing local changes in proteins. Until our recent work [84], high resolution NMR has not been used to follow the pressure induced denaturation of a protein in solution. Since the process of denaturation is slow on the NMR time scale, the proton resonances from both native and denaturated states can be detected in the NMR spectrum.

Several studies reported in the literature provided the main motivation for our work. First of all, the already mentioned high pressure fluorescence study of lysozyme and lysozyme with bound tri-N-acetylglucosamine (tri-NAG) by Weber and co-workers [83] represents the main reason why we have chosen to investigate lysozyme. Secondly, the pioneering study of Kundrot and Richards [85] determined the crystal structure of hen egg white lysozyme at a hydrostatic pressure of 1 kbar and found a differential compressibility in the different regions of the protein. Third, Dobson and co-workers [86] investigated the thermal denaturation of lysozyme using high resolution NMR by following NMR signals for various residues located in different regions of the protein, and concluded that unfolding can be represented by a cooperative, two-state process.

In our study we used high resolution NMR to follow the effect of pressure on the equilibrium constant for the native and denatured forms of lysozyme by observing the proton resonances of the following residues: His-$15^{\epsilon 1}$, Leu-$17^{\delta 2}$, Trp-$28^{\epsilon 3}$, Cys-64^{α} and Trp-$108^{\epsilon 3}$. Since these residues lie in the regions for which Kundrot and Richards [85] determined different compressibilities, we were interested to find out whether these differences are also reflected in the determined reaction volumes (ΔV) for the individual residues. Dobson and co-workers [86] proposed that thermal denaturation of lysozyme is a cooperative two-state process since all the observed resonances belonging to different residues gave equivalent ΔH and ΔS values for the unfolding process. Therefore, the determination of ΔV's for the same residues may contribute to resolving the fundamental question of how similar or different is pressure denaturation compared to thermal denaturation. We also attempted to determine whether pressure denaturation of lysozyme is a cooperative, two-state process in view of the proposal of a plurality of pressure denaturated forms of lysozyme made by Weber and co-workers [83] on the basis of their high-pressure fluorescence study. Since it has been observed that binding of tri-N-acetyl-glucosamine (tri-NAG), a substrate analog of lysozyme, stabilizes the native form of lysozyme, we have also studied the pressure denaturation of lyusozyme with bound tri-NAG in order to determine the effect of tri-NAG binding on the unfolding process; in particular attempting to detect possible differences in ΔV values obtained for residues located in or close to the active site.

The pressure effects on the high resolution proton NMR spectra of D_2O olutions of egg-hen white lysozyme at pH = 3.9 and a temperature of 68.5° C were measured in the pressure range from 1 bar to 5 kbar. Analogous experiments were performed for lysozyme with bound tri-NAG. The proton residues of interest in this study are representative of different structural regions of lysozyme with sufficiently resolved proton spectra for accurate intensity measurements. His-$15^{\epsilon 1}$ has well resolved native and denatured resonances, while the other residues have resolved native resonances only. His-$15^{\epsilon 1}$, Trp-$28^{\epsilon 3}$ and Trp-$108^{\epsilon 3}$ and Leu-$17^{\delta 2}$ are in α-helix structures, and Cys-64^{α} is in a β-sheet structure.

At this point, it is important to show the 3-D structure of lysozyme in Figure 20 and indicate the specific locations of the residues studied. At this point, a brief summary of the main results of the X-ray study of lysozyme by Kundrot and Richards [85] will help in understanding the subtle differences in the ΔV values obtained for the various residues of lysozyme in the absence of tri-NAG. By analyzing the X-ray data obtained for lysozyme at 1 bar and 1 kbar, Kundrot and Richards found differential compressibilities for the different regions of lysozyme.

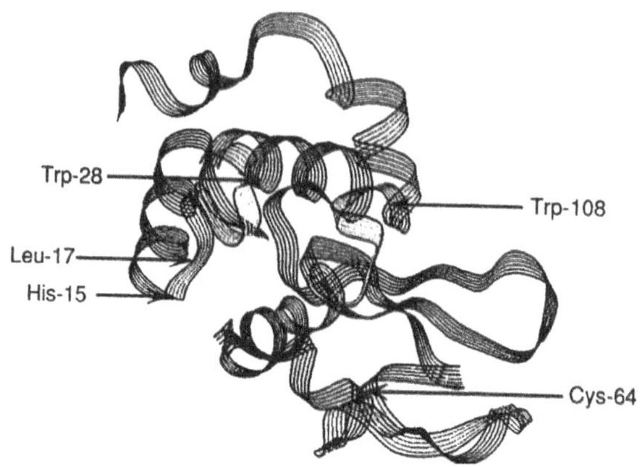

Fig. 20. Schematic ribbon 3-D structure of lysozyme indicating the specific locations of residues studied. (See the text for details).

Domain 1, which contains residues 1 to 39 and residues 89 to 129, exhibited a compressibility of 5.7 x 10^{-3} $kbar^{-1}$, while domain 2, including residues 40 to 88, was essentially incompressible. The interdomain region was also compressible. We note that only one residue Cys-64^{α} studied in the present work is located in the incompressible domain 2. The experimental procedures and equipment were discussed in detail in our original study [84].

Figure 21 shows the ^{1}H NMR spectrum of lysozyme in the region from 10 to 5 ppm obtained at 68.5° C and 3.0 kbar pressure. The spectrum resembles the spectra of native and partially denatured lysozyme obtained by Dobson and Evans [87] at 68° C and 77° C and atmospheric pressure; and illustrates the degree of resolution attainable with the pressure vessel. All chemical shifts are given in ppm relative to the HDO peak. No internal standard was used as the accurate determination of the chemical shifts was not required. The resonance corresponding to His-$15^{\epsilon 1}$ is clearly observed at 8.75 ppm, the resonance for His-$15^{\epsilon 1}$ in the denatured state is at 8.47 ppm, and the resonance for Trp-$108^{\epsilon 3}$, Trp-$28^{\epsilon 3}$, and Cys-64^{α} in the native state are resolved at 6.38, 6.09 and 5.70 ppm, respectively at 3 kbar. The native form resonance for Leu-$17^{\delta 2}$ was resolved at -0.80 ppm. Under native conditions, the integrated areas under the above resolved peaks are identical and represent the RF absorption of a single proton. The small spurious shoulder near the Trp-$28^{\epsilon 3}$ resonance appeared only in two of our seven independent experiments; however, we included these experiments in the calculations which follow. Any error introduced due to the spurious peak was included in the standard deviation indicated as error bars for each pressure. With increasing pressures, the overall spectrum approaches the characteristics of the spectrum for thermally denatured lysozyme, with the resolved native peaks decreasing in intensity. For the residues of interest, only the resonance of denatured His-$15^{\epsilon 1}$ is resolved and increases with pressure. The percentage of the denatured and native state for the His-$15^{\epsilon 1}$ residue was obtained from the integrated areas under the two peaks. For all the other residues the

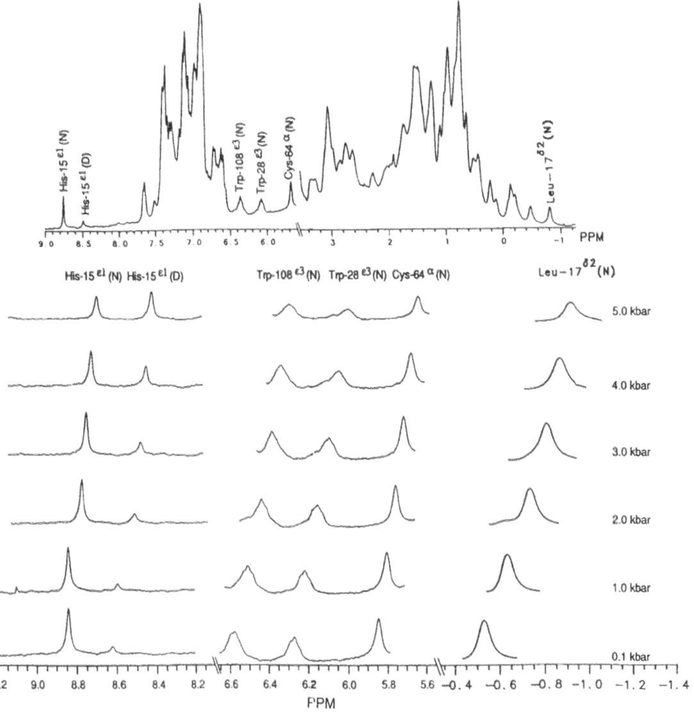

Fig. 21. High resolution proton NMR spectrum of lysozyme at pH 3.9 and 68.5° C at 3.0 kbar. The pressure effects on NMR spectra of residues studied are also shown in the pressure ranged from 0.1 kbar to 5 kbar.

percent of the native form was calculated from the integrated area under the respective native peak by reference to the area under both His-15$^{\epsilon 1}$ peaks taken as 100%.

The calculated percentages of the denatured form for each residue can be plotted as a function of increasing pressure. For each residue the results are given in the absence and in the presence on tri-N-acetyl-glucosamine binding to lysozyme. Each data point was the mean from seven independent experiments.

The experimental data can be plotted in linear form as in K_{eq} versus pressure, and the equilibrium constant for denaturation (K_{eq}) is calculated from the primary data as follows:

$$K_{eq} = \% \text{ denatured form}/\% \text{ native form}$$

The slopes of the lines are related to ΔV, the reaction volume, by the equation [8]

$$p\Delta V = RT \ln K \tag{11}$$

where p is pressure, R is the gas constant, and T is temperature in degrees Kelvin.

Table 4 summarizes the means and standard deviations for the ΔV values calculated from the slopes of the ln K_{eq} versus pressure plots, for all five residues.

Table 4. ΔV values calculated from the pressure dependence of the ln K_{eq} for five amino acid residues of lysozyme, in the presence and absence of tri-N-acetyl-glucosamine (tri-NAG).

	ΔV (cm^3/mol)	
Residue	Without tri-NAG	With tri-NAG
His-15$^{\epsilon 1}$	-11.8 ± 0.5[a]	-12.0 ± 0.3
Leu-17$^{\delta 2}$	-10.2 ± 0.5	-10.6 ± 0.9
Trp-28$^{\epsilon 3}$	-10.8 ± 0.7	-10.4 ± 0.5
Cys-64$^{\alpha}$	- 9.3 ± 0.7	-10.9 ± 0.4
Trp-108$^{\epsilon 3}$	-10.3 ± 0.6	-18.1 ± 1.7

[a]The ΔV values are given as the mean ± standard deviation from seven independent experiments.

In the absence of tri-N-acetylglucosamine the ΔV values range from -9.3 ± 0.7 to -11.8 ± 0.5 cm^3/mol. Differences between the mean values that exceed the standard deviations can be considered statistically significant; therefore, the ΔV values for the three nonpolar residues, Leu-17$^{\delta 2}$, Trp-28$^{\epsilon 3}$, and Trp-108$^{\epsilon 3}$ are essentially equal, while those for His-15$^{\epsilon 1}$ and Cys-64$^{\alpha}$ are larger and smaller, respectively. In the presence of tri-N-acetylglucosamine. Leu-17$^{\delta 2}$, His-15$^{\epsilon 3}$, and Trp-28$^{\epsilon 3}$ have the same ΔV values as in the absence of the substrate analog, whereas Cys-64$^{\alpha}$ and Trp-108$^{\epsilon 3}$ have more negative ΔV values. In fact the ΔV for Trp-108$^{\epsilon 3}$ decreases dramatically from -10.3 ± 0.6 to -18.1 ± 1.7 cm3/mol.

In the absence of tri-NAG the ΔV values are essentially identical -10.2 ∽ 10.8 cm^3/mol for the three nonpolar residues Leu-17$^{\delta 2}$, Trp-28$^{\epsilon 3}$ and Trp-108$^{\epsilon 3}$. Interestingly, ΔV for Cys-64$^{\alpha}$ is somewhat smaller -9.3 ± 0.7 cm3/mol than for the other residues, probably because Cys-64$^{\alpha}$ is located in the incompressible domain 2. In addition, our results indicated that this residue is more resistant to pressure denaturation than the other residues as at 5 kbar the concentration of its denatured form is about ∽42%, whereas for all the other residues of the denaturated form is about ∽53%. One possible origin of this resistance to denaturation may be related to the fact that Cys-64$^{\alpha}$ is a disulfide linked residue in the loop region of lysozyme. We shall comment further on the significance of this difference in the pressure response of this residue when considering the results obtained for lysozyme with bound tri-NAG and its direct relevance to the conclusions by Weber and co-workers [83] in their high pressure fluorescence study of lysozyme in the absence and presence of tri-NAG. His-15$^{\epsilon 1}$ is the only charged residue that is analyzed in the present work, and it may well be that in the native state it is partially protected from D_20 while in the denaturated state electrostriction could lead to a larger decrease in volume when compared to the uncharged residues.

In their experiments on lysozyme using fluorescence techniques Weber and co-workers [83] obtained a ΔV = -19.7 cm^3/mol which is more negative than our average value of -10.5 cm^3/mol. However, this difference is not surprising as our experiments were carried out at pH = 3.9 and 68.5° C, whereas the fluorescence study was performed at pH = 7.6 and a temperature of 23° C. According to Zipp and Kauzmann [82] the Δ V must become less negative with increasing temperature.

In the context of the general question of possible similarities or differences between thermal denaturation and pressure denaturation, it is appropriate to mention the NMR results obtained by Dobson and co-workers [88] at atmospheric pressure who determined the equilibrium constant for the denaturation of lysozyme over the temperature range 70-78° C at pH = 3.8 using the residues His-$15^{\epsilon 1}$, Leu-$17^{\delta 2}$, Trp-$28^{\epsilon 3}$, Trp-$108^{\epsilon 3}$. The fact that a linear temperature dependence was obtained for all the resonances indicates that thermal denaturation can be adequately described by a two-state model. Therefore, our finding that statistically significant although small differences in Δ V values do exist for the same residues investigated suggests that pressure denaturation resulting from changes in intermolecular distances may provide novel information about the details of the folding/unfolding process.

In the presence of tri-NAG the Δ V values are essentially unchanged for the His-$15^{\epsilon 1}$; Leu-$17^{\epsilon 2}$; Trp-$108^{\epsilon 3}$ residues, but we find a more negative value for Cys-64^{α} and much more negative value for Trp-$108^{\epsilon 3}$. Both these residues are near the active site and can be affected by the binding of the substrate analog. It is well known that tri-NAG binds to lysozyme in a cleft at the surface of the enzyme and occupies about half of the cleft. The hydrogen bonds between the tri-NAG and lysozyme are as follows: sugar residue A is hydrogen bonded to Asp-101; residue B also to Asp-101; the sugar residue C is hydrogen bonded to Trp-62, Trp-63 and to the main chain at residues 59 and 107.

Our results show that for Trp-$108^{\epsilon 3}$ the equilibrium is shifted towards the native state when tri-NAG is bound, as expected from the stabilization of lysozyme by bound tri-NAG towards unfolding. Interestingly, the other more remote regions of lysozyme are not affected as revealed by the His-$15^{\epsilon 1}$, Leu-$17^{\delta 2}$ and Trp-$28^{\epsilon 3}$ behavior. With increasing pressure, the equilibrium between the native and denaturated forms of lysozyme, as reflected by the Trp-108 residue, approaches that found for the other residues. It is important to point out that binding of tri-NAG changes the Δ V for Trp-$108^{\epsilon 3}$ from -10.3 ± 0.6 to -18.1 ± 1.7 cm^3/mol. This large change in Δ V is most likely related to the fact that tri-NAG occupies about half of the active site and produces a significant free volume in the immediate vicinity of the Trp-$108^{\epsilon 3}$ residue. It is important to note that Cys-64^{α} also senses the binding of the substrate analog, but in this case denaturation is favored at high pressure. This finding has a direct relevance to the results obtained by Weber and his co-workers [83] who observed two distinct regions of fluorescence change in lysozyme which suggested the existence of more than one pressure-denaturated forms of lysozyme. However, upon addition of tri-NAG to lysozyme the two-step pressure change in fluorescence was replaced by a smooth single step. In spite of the fact that the pressure range in their study was 11 kbar, and different experimental conditions of pH and temperature were used, nevertheless, we see a connection between our results to their findings. As we already mentioned, the Cys-64^{α} residue shows only ~42% denaturation at 5 kbar in contrast to about ~53% found for the other residues. After adding tri-NAG the denaturation reflected by the Cys-64^{α} residue is favored at high pressure, and the Cys-64^{α} residue behaves the same way as the other residues. This may be related to the finding of Weber and co-workers [83] that bound tri-NAG essentially removes the two-step fluorescence change with pressure for lysozyme.

Since the results for Cys-64^{α} were obtained in the limited pressure range up to 5 kbar, we can only conclude that the behavior of Cys-64^{α} in the absence and presence of tri-NAG suggests the possible existence of multistep denaturation in lysozyme and a single step process in lysozyme with bound tri-NAG. To unequivocally prove or disprove the existence of multiple pressure denaturated forms of lysozyme can only be accomplished by performing the high pressure NMR experiments under comparable experimental conditions to those used in the fluorescence study of Weber and co-workers [83]. This will require maximum pressures of 10 kbar. The building of such a high resolution NMR probe which will allow variable temperature measurements in the pressure range from 1 bar to 10 kbar is in progress. The planned experiments should prove highly informative about the unfolding process judging from the small but statistically significant differences in the behavior of the various residues observed in the limited pressure range of 5 kbar in the present study. In addition, the information content of pressure effects on chemical shifts of the residues studied will also be explored.

4.2. DEUTERIUM NMR STUDIES OF CHAIN PERDEUTERATED DIPALMITOYLPHOSPHATIDYLCHOLINE BILAYERS AT HIGH PRESSURES

In our laboratory we have initiated systematic high pressure NMR studies on model phospholipid membranes [36,44,89]. Of particular relevance to the present study are two of our recent high pressure NMR studies [44,89] of perdeuterated 1,2-dipalmitoyl-sn-phosphatidylcholine (DPPC-d_{62}). One series of experiments [44] was performed on pure lipid bilayers of DPPC-d_{62} in the liquid-crystalline (LC) state and on bilayers in the LC state containing the charged form of the local anesthetic tetracaine (TTC). The results showed that the order parameters of all segments of the acyl chains increased with pressure in the LC state. The more highly ordered regions of the chains were affected slightly more than the regions near the methyl ends. The addition of tetracaine increased the disorder of the chains, and pressure reversed the effect of anesthetic on the lipid as seen by reversal of the changes in lineshape and the measured order parameters. The second study [89] concentrated on the pressure effects on the gel states of DPPC-d_{62} and used the high pressure ^{2}H-NMR techniques to detect various gel phases and to generate a pressure-temperature phase diagram for this phospholipid.

The capability to study the gel state of phospholipids by NMR methods was only attained after the major technical advances in solid state NMR techniques in the late 1970s. A most fruitful area of application has been the use of ^{2}H-NMR to study the gel state of model, reconstituted, and native biological membranes [90]. A fairly unified picture of the structure and dynamics of the gel phases of various phospholipids has emerged from these studies. Another ongoing area of research has involved the use of high pressure to induce the various gel phases and study these phases using various physical techniques. High-pressure IR and Raman spectroscopy [91] has been used very productively in this regard. The use of pressure to study the gel state has revealed several new phases which may only exist at very low temperatures at ambient pressure. The majority of high pressure studies of the phase diagram for DPPC focused on the liquid crystal (LC) to gel transition using volumetric [92] and adiabatic compression [93] techniques.

High pressure ^{2}H-NMR has not been used to probe the gel phases of a phospholipid until our study [89]. Gel state spectra were observed at several temperatures and pressures up to 5 kbar. At this point, a few general remarks about the gel state of DPPC are appropriate. In contrast to the liquid-crystalline state, the gel state (referring to several gel phases) is highly ordered. At ambient pressure a sharp transition from the LC to gel state occurs at 41° C (37° C for the perdeuterated lipid) and is readily observable by ^{2}H-NMR as a marked increase in width of the spectrum, in both, perdeuterated and selectively labeled samples. It was once thought that the gel state consisted of motionless acyl chains in the all trans state; however, ^{2}H-NMR spectra clearly refute this. At intermediate temperatures below the LC state from 5° C to 35° C, the characteristic axially symmetric rigid lattice spectrum with sharp shoulders at ±63 kHz is absent, and spectra indicate the presence of motional averaging. The observed lineshapes have been modeled and interpreted in terms of various combinations of long axis rotation, chain fluctuations, and trans/gauche isomerizations. At very low temperatures (<0° C) a rigid lattice spectrum is seen for DPPC indicating absence of motion on the NMR timescale.

The gel state is in reality made up of several phases, each with a particular structure. Most often studied are the $P_{\beta'}$ (pretransition) and $L_{\beta'}$ phases. Also present is a subtransition (L_c) phase which appears upon prolonged refrigeration of the $L_{\beta'}$ phase. No attempt will be made here to review the structure and properties of these phases. Somewhat less well-known is the existence of other low-temperature/high pressure phases. Wong et al. [94] have done extensive work in this area and have shown the existence of at least five pressure induced gel phases which they have names GI through GV corresponding to the increasing pressure of the transition; GI and GII refer to $P_{\beta'}$ and $L_{\beta'}$, respectively, but, the structure and dynamics of GIII, IV and V are not well studied by other methods.

In addition to Wong's studies, small angle neutron diffraction [95,96] has shown the existence of a pressure induced interdigitated phase in DPPC bilayers at pressures above 1 kbar and temperatures above 40° C.

Davis [97] originally demonstrated the changes of perdeuterated DPPC NMR spectral lineshapes associated with temperature changes at ambient pressure in the gel state. However, at that time the rich phase behavior of this lipid was not known, variable pressure studies had not been performed, and a detailed analysis of lineshape changes with changes in conditions had not been attempted.

The gel state of DPPC bilayers shows a variation in lineshapes which depend on the particular pressure and temperature. The types of lineshapes seen in our high pressure investigation of d_{62}-DPPC are depicted in Fig. 22. This figure illustrates the lineshape changes which occur with pressure at four representative temperatures (7° C, 34° C, 41° C, 75° C). A decrease in temperature or increase in pressure will cause a general change in lineshape from the fast motion axially asymmetric powder pattern of the LC state (Fig. 1d, 75° C and 1900 bar) to the almost rigid methyl lattice spectrum at 7° C and 1.6 kbar (Fig. 21a). Throughout most temperatures and pressures, a splitting from approximately ±10 to ±20 kHz is seen for the methyl group. It is also evident that no methylene resonances can be seen in the gel state, but only an unresolved curved lineshape which is uniquely determined by the particular conditions of pressure and temperature.

In order to measure phase transitions in a quantitative way, we used two methods: a direct measurement of the quadrupole splitting of the methyl group or first moment analysis.

Table 5 lists the phase changes which were measured in our study [89] and the best method for the measurement of the change. Methyl splitting and/or first moment analysis were able to demonstrate phase changes in all cases except the P_β' to L_β' (GI to GII) transition, which was determined by measuring the ratio of the central peak intensity to the shoulder intensities. This approach worked because the spectra of the P_β' phase have two components which appear like a mixture of a liquid-like and a more order phase. (In this case, the central peak intensity corresponds to the liquid-like spectral component, which decreases suddenly upon change to the L_β' phase). A sudden change in the magnitude of

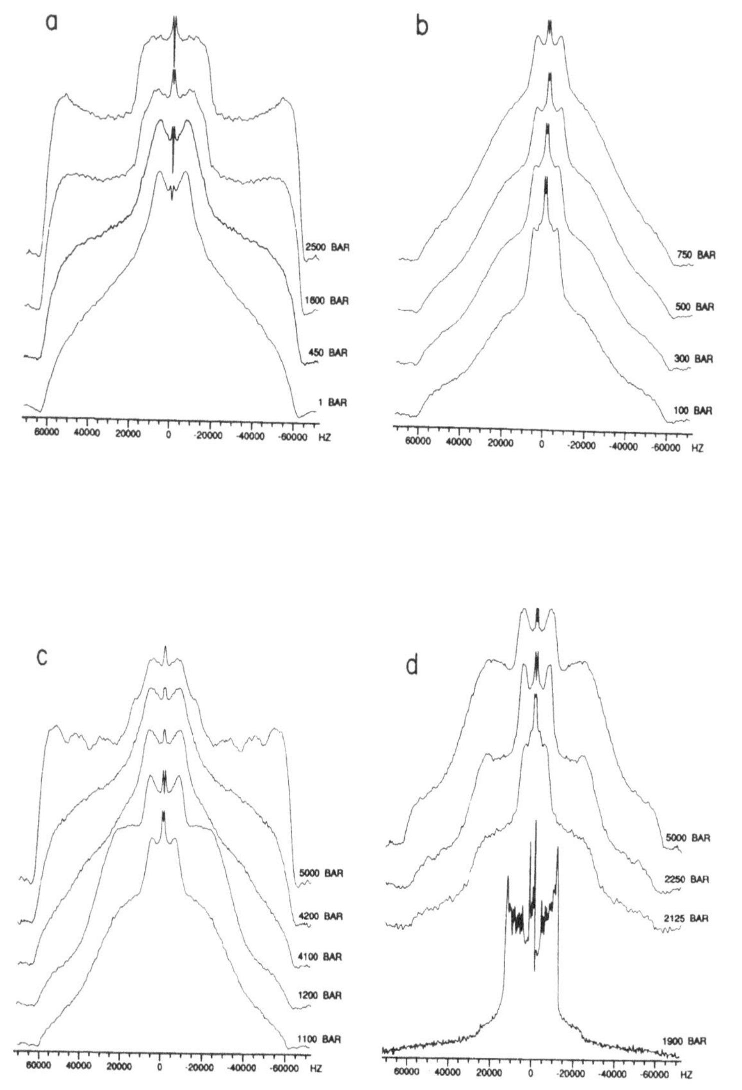

Fig. 22. Representative lineshapes and lineshape changes as a function of pressure in gel states of d_{62}-DPPC at : (a) 7° C; (b) 34° C; (c) 41° C; (d) 75° C. Pressure conditions are noted for each spectrum.

either the methyl splitting or first moment, with a variation of pressure was taken as an indication of a phase change.

Table 5. Measurement of gel phase transitions

Phase transition	Temperature/pressure	Method
GI/GII	34° C/350 bar	Peak height/shoulder height ratio
GII/Gint[a]	41° C/1150 bar	Methyl splitting[b], 1st moment
	75° C/2100 bar	Methyl splitting[b], 1st moment
GII/GIII	7° C/735 bar	Methyl splitting[b], 1st moment
	24° C/1500 bar	Methyl splitting[b], 1 st moment
GIII/GX	7° C/1800 bar	Methyl splitting, 1st moment
	24° C/3200 bar	1st moment
	41° C/4190 bar	Methyl splitting, 1st moment[b]

[a]Gint is the interdigitated gel phase.
[b]This method is the best of the two methods for detecting the transition at this particular PT point.

Figure 23 includes the phase transitions observed by the methyl splitting and first moment analysis as listed in Table 5. The observed phases are labeled according to Wong [98] and Prasad [99] and a few of their data points are included for comparison. The LC-GI, GI/GII, GI/Gint, GI/GII, and GII/GIII transitions were all clearly identified using our ^{2}H-NMR spectral measurements. An additional phase transition not observed by lineshape changes was identified as corresponding to the GII/GIII transition as shown in Fig. 23. A plot of the first moment at 7° C and 24° C does not easily identify this transition (explaining the difficulty in seeing this change from lineshapes); however, measurement of methyl splitting clearly shows a phase change. The phase change from GIII to X phase (Fig. 23) corresponds to the transformation to a rigid lattice type spectrum (Fig. 22a, 1600 bar) and is also detected by first moment analysis at 41° C. This phase change occurs at lower temperatures and higher pressures than the GII - GIII transition, but is not consistent with Wong's GII/GIV phase transition which occurs at higher pressures [98]. We also noticed some interesting lineshape changes at low temperatures and at very high pressures as seen in Fig. 22c. For 5 kbar and 41° C there seems to be a further splitting of the methyl region into two peaks, and the shoulder into three peaks. There is also the double methyl splitting at high pressure coupled with a very high shoulder intensity. At the present time, we have no explanation for this observation.

The use of pressure to study the gel phase has allowed us to observe phase changes that are not seen at ambient temperature. It is clear from these data, that the gel state cannot be thought of as a single phase beyond the $P_{\beta'}$ transition, undergoing gradual lineshape change in ^{2}H spectra with a gradual decrease in motion; rather there are abrupt changes in motion and structure at phase transitions occurring at specific temperatures and pressures.

It would be interesting to expand these high pressure ^{2}H-NMR studies to lipids specifically labeled at several sites. Relaxation studies and lineshape modeling would then allow a detailed investigation of the molecular motions and conformations in the various gel states detected in this study. Work along this direction is in progress in our laboratory.

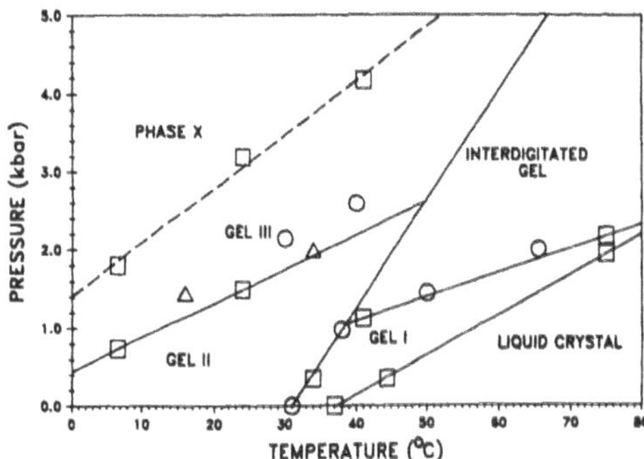

Fig. 23. Pressure/temperature phase diagram of d_{62}-DPPC, representing the transitions measured by the methods listed in Table 5, and by other investigators: (-□-) shows transitions measured in this study (See Table 5); (-△-) GII/GIII transition points measured by Wong et al. [94]; (-0-) transition points measured by Prasad et al. [99]. The previously known phase boundaries are indicated with a solid line. The dashed phase transition line was obtained here by first moment analysis and has not been previously reported.

4.3 HIGH PRESSURE ^{31}P NMR OF STRUCTURE AND DYNAMICS OF PHOSPHOCHOLINE HEADGROUP IN DPPC BILAYERS

For pure phosphatidylcholine bilayers, the orientation of the headgroup has been well characterized by a variety of physical methods showing that headgroups are aligned approximately parallel to the bilayer surface [100]. Studies of the dynamic properties of the headgroups demonstrated that the reorientational motions were reduced by lowering the temperature in the LC state, and that a distinct decrease of motions accompanied the phase transition from the LC to the gel phase [101].

Because only one phosphorus ^{31}P with 100% natural abundance is contained in the phospholipid molecule, ^{31}P NMR has become an important tool to study the phospholipid headgroup structure and dynamics. Many ^{31}P NMR experiments have been carried out at ambient pressure [102] giving information about the orientation, mobility, and interaction of the phospholipid headgroups.

Boulanger et al. [103] have reported that the addition of TTC to phosphatidylcholine dispersions in water produced different effects depending on whether the anesthetic was positively charged (pH 5.5) or uncharged (pH 9.0). The changes in bilayer properties were monitored by ^{2}H and ^{31}P NMR spectroscopy. The fact that the changes observed at low pH in the ^{2}H and ^{31}P NMR spectra of the phospholipid headgroup in the presence of TTC resemble those produced by ions was most relevant to our study. These results also suggested that the headgroup undergoes a conformational change upon interaction with the anesthetic. Scherer and Seelig [104] have carried out thorough and elegant studies on the effects of electric charge on phospholipid headgroups by investigating mixtures of 1-

palmitoyl-2-oleoyl-sn-glycero-3-phosphocholine (POPC) with charged amphiphiles. Deuterium, ^{31}P and ^{14}N NMR spectra were recorded for bilayers with varying concentrations of amphiphiles. They found [104] that the charged amphiphiles have a major effect on the phosphocholine headgroup. The headgroups which are normally aligned parallel to the membrane surface in a pure phospholipid membrane, upon addition of a positively charged amphiphile change in orientation by a movement of the N^+ end of the dipole toward the water phase. This result was generalized for positively charged anesthetics, charged phospholipids and peptides.

The results obtained by Scherer and Seelig [104] in the LC phase were extended by MacDonald et al. [105] to the gel state of 1,2-dimyristoyl-sn-glycero-3-phosphocholine (DMPC) using ^2H NMR spectroscopy of specifically choline-deuterated DMPC. McDonald et al. [105] provided evidence that the choline headgroup of phosphatidylcholine in the gel state also responds to charge changes at the membrane surface; thus, the response of the choline headgroup of phosphatidylcholine to surface charges is qualitatively and quantitatively similar in the LC state and in the gel state of membrane lipids.

In our recent study [106] the ^{31}P NMR lineshapes and ^{31}P spin-lattice relaxation times of pure DPPC bilayers and DPPC bilayers containing charged tetracaine were measured as a function of pressure from 1 bar to 5.1 kbar at 50° C. The goals of the study were as follows: 1) To determine the behavior of the headgroup in the LC phase and the various gel phases accessible at high pressure; 2) To compare the pressure effects on pure DPPC bilayers with those observed for DPPC/tetracaine bilayers; 3) To determine whether the concept of molecular electrometer as introduced by Seelig [104] is also applicable for gel phases induced at high pressures.

Using a temperature of 50° C and the pressure range from 1 bar to 5 kbar in our experiments allows us to determine ^{31}P NMR lineshapes in the liquid crystalline state and various gel states of DPPC. Fig. 24 shows the representative ^{31}P NMR spectra for DPPC multilamellar bilayers at various pressures in the LC phase and the various gel phases induced by high pressure. As it can be seen from Fig. 24, the lineshape distortion caused by proton-phosphorous dipolar interactions was removed by ^1H decoupling. The ^{31}P NMR spectra of DPPC in the LC state, Gel I, the interdigitated gel Gi phase and the Gel II phase show a typical axially symmetric powder lineshape. In contrast, the lineshape becomes axially asymmetric in the high pressure Gel III phase. The asymmetry increases further in the last accessible gel phase Gel X for which a typical rigid lattice axially symmetric powder pattern can be observed. The experimental chemical shift anisotropy values, $\Delta\sigma$, were determined using the standard procedures [100]. For the axially symmetric lineshapes the anisotropic part of the time averaged shielding tensor is defined as

$$\Delta\sigma = \sigma_{\parallel} - \sigma_{\perp} \tag{12}$$

where $\sigma_{11} = \sigma_{22} = \sigma_{\perp}$ and $\sigma_{33} = \sigma_{\parallel}$.

For the axially asymmetric powder pattern the shielding anisotropy was calculated using the following expression

$$(\Delta\sigma)^2 = 3/2\,[(\sigma_{11} - \sigma)^2 + (\sigma_{22} - \sigma)^2 + (\sigma_{33} - \sigma)^2] \tag{13}$$

where $\sigma_{11}, \sigma_{22}, \sigma_{33}$ have their usual meaning and σ denotes the isotropic part of the shielding tensor.

$$\sigma = 1/3\,(\sigma_{11} + \sigma_{22} + \sigma_{33}). \qquad (14)$$

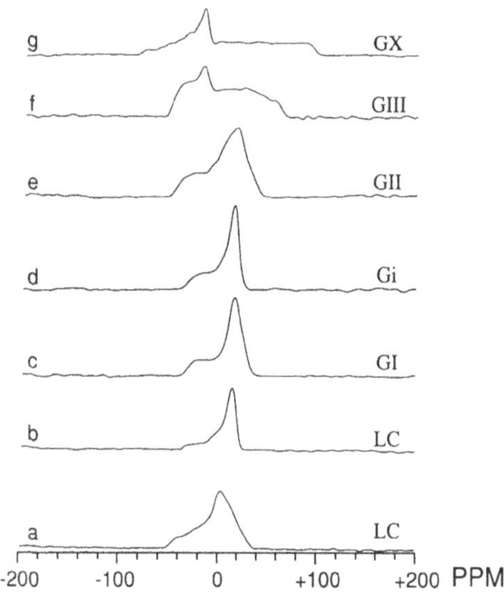

Fig. 24. ^{31}P NMR spectrum obtained using a single 90° pulse without ^{1}H decoupling in pure DPPC bilayers at 50° C and 1 bar (a); and ^{31}P NMR spectra obtained using a fully phase-cycled Hahn echo sequence with inversely gated ^{1}H decoupling in pure DPPC bilayers at 50° C and 1 bar in the LC phase (b); 1 kbar in the gel GI phase (c); 1.75 kbar in the interdigitated Gi gel phase (d); 2.5 kbar in the GII gel phase (e); 3.70 kbar in the GIII gel phase (f); 5.10 kbar in the GX gel phase (g).

The calculated $\Delta\sigma$ values are plotted as a function of pressure in Fig. 25. Only the absolute value of $\Delta\sigma$ can be determined from Eq. (13), but in view of the fact that $\Delta\sigma$ is negative [104] for the LC phase and the temperature-induced gel phases (Gel I, Gel II), we expected that $\Delta\sigma$ will also be negative for the high pressure gel phases of DPPC.

The experimental ^{31}P NMR lineshapes for DPPC bilayers containing tetracaine are shown in Fig. 26 and the calculated $\Delta\sigma$ values are also plotted as a function of pressure in Fig. 25, in order to permit a direct comparison between the $\Delta\sigma$ values obtained for pure DPPC and DPPC/TTC bilayers. In Fig. 26, the small peak near the main peak in the interdigitated gel phase (Fig. 3, c and d) arises from the buffer phosphate ion. According to Brown and Seelig [107], no spectral changes in the ^{31}P lineshapes of DPPC are due to general ionic strength effects, and consequently, the addition of the buffer used in our DPPC/TTC experiment has no effect on the observed ^{31}P lineshapes. Similarly, as for pure DPPC bilayers, the ^{31}P lineshapes shows the characteristic axially symmetric powder patterns for the LC phase and the interdigitated Gi gel phase but changes to an axially asymmetric powder pattern in the high pressure gel phases.

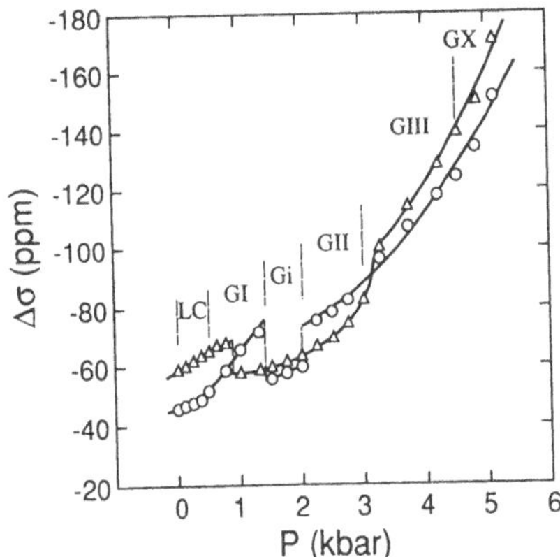

Fig. 25. Pressure dependence of ^{31}P chemical shift anisotropy values for DPPC bilayers (0) and DPPC/TTC bilayers (Δ) at 50° C. The pressure ranges for different phases of pure DPPC are indicated (for detailed discussion, see the text).

The effect of pressure on the measured chemical shift anisotropy values, $\Delta\sigma$, is shown graphically in Fig. 25 both for pure DPPC and DPPC/TTC bilayers. For pure DPPC the absolute value of $|\Delta\sigma|$ increases slightly with pressure in the LC phase in the pressure range from 1 bar to 500 bar. In the Gel I phase the $|\Delta\sigma|$ value continues to increase with pressure until the phase transition Gel I/Gi is reached. As expected, the $|\Delta\sigma|$ values for the interdigitated gel phase of DPPC are smaller than in the other gel phases. An abrupt change to higher $|\Delta\sigma|$ value occurs at the Gi/Gel II phase transition pressure. In the gel phases Gell II, Gel III and Gel X, the $|\Delta\sigma|$ values increase continuously with increasing pressure, and it is evident that there are no distinct changes in the $\Delta\sigma$ vs. pressure plots associated with the expected Gel II/Gel III and Gel III/Gel X phase transitions.

In agreement with literature [103,104] the $|\Delta\sigma|$ value increases from -46 to -59 ppm when charged tetracaine is added to DPPC bilayers in the LC state at 1 bar and 50° C. Again, we find that $|\Delta\sigma|$ increases with pressure up to 0.8 kbar, but then, the $|\Delta\sigma|$ value changes abruptly to a lower value indicating a direct transition to the interdigitated gel phase. As pressure increases, the $|\Delta\sigma|$ value gradually increases in the region between 1 kbar to 3.1 kbar, and finally approaches the value of $\Delta\sigma$ found for pure DPPC bilayers at about 3.1 kbar. It can be noticed that the $\Delta\sigma$ values in this pressure range are lower than those of pure DPPC in Gel I and Gel II phases but greater than those in the interdigitated gel Gi phase of pure DPPC. Finally, in the region between 3.1 and 5.1 kbar the $|\Delta\sigma|$ values increase monotonically and are greater than those of pure DPPC.

The general pressure-temperature phase diagram for DPPC-d_{62} was described in detail in our earlier study (see Sect. 4.2) [89], which also provided the notation for designating the various gel phases. The ^{31}P lineshapes, as shown in Fig. 24 at different pressures agree well with those for the corresponding temperature-induced phases (P_β' and L_β') at ambient pressure with the exception of the

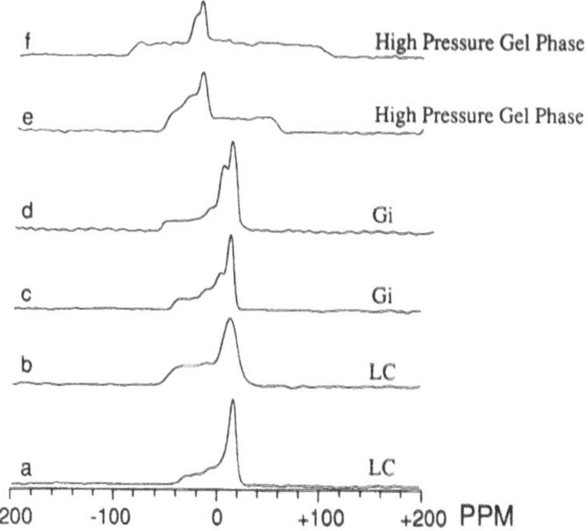

Fig. 26. ^{31}P NMR spectra obtained using a fully phase-cycled Hahn echo sequence with inversely gated ^{1}H decoupling in DPPC/TTC bilayers at 50° C and 1 bar in the LC phase (a); 750 bar in the LC phase (b); 1 kbar in the interdigitated Gi gel phase (c); 2.75 kbar in the interdigitated Gi gel phase (d); 3.25 kbar in the high pressure gel phase (e); 5.1 kbar in the high-pressure gel phase (f) (for details, see the text).

interdigitated gel phase, Gi, which cannot be attained by varying temperature only. The observed $\Delta\sigma$ = -46 ppm at 1 bar and 50° C in the LC phase is in good agreement with that of -47 ppm reported previously by others [104]. The transition from the LC phase to the Gel I phase is accompanied by a reduction of gauche isomers in the acyl chains affecting the available space for the headgroup motions and resulting in an increase of the absolute values of $|\Delta\sigma|$. The abrupt decrease of $|\Delta\sigma|$ that accompanied the Gel I/Gi phase transition provides supporting evidence for the existence of the interdigitated phase, Gi, in which the phospholipid headgroup has more available space and is thus less restricted. In this phase, the faster motion of the headgroup averages the chemical shift anisotropy leading to lower $|\Delta\sigma|$ values. In contrast, in the Gel II phase, there is a further reduction of gauche conformers [108], and the two acyl chains are almost extended which makes the chains tilt with respect to the bilayer normal. As a result, the available headgroup space decreases again, and the $|\Delta\sigma|$ value returns to higher values. In the Gel III phase, the headgroup motion is further restricted, and the $|\Delta\sigma|$ value continues to increase with pressure. In agreement with our results [89] for DPPC-d_{62} the phase change from Gel III to Gel X phase corresponds to a transformation to a rigid lattice spectrum, which shows a typical ^{31}P axially asymmetric powder pattern. However, in contrast to the deuterium lineshapes, there is no discontinuity in $\Delta\sigma$ at the phase transition pressures corresponding to Gel II/Gel III and Gel III/Gel X phase transitions. This may be due to the fact that the phase transitions are driven by the cooperative interactions between the DPPC fatty acid chain segments.

The increase in pressure in all phases is accompanied by an increase in the ^{31}P chemical shift anisotropy $|\Delta\sigma|$, as headgroup motions are slowed down, and the motional averaging is reduced. However, in view of the results obtained by neutron scattering [95,96] and X-ray diffraction [109], an alternative explanation should be considered for the $|\Delta\sigma|$ behavior in the LC phase. Increasing pressure causes a decrease in overall volume available for the DPPC molecules by "squeezing" them together, with an increase in the overall interactions. Since there is an increase in the thickness of the bilayer with pressure, a decrease in the overall cross-sectional area of the phospholipid molecule is needed to explain the decreased volume. The average surface area per headgroup thus decreases. With pressure, the headgroup which is normally oriented parallel to the bilayer surface, may change its orientation, thus contributing to the measured increase in bilayer thickness and leading to higher $|\Delta\sigma|$ values. In the gel phases this explanation is not applicable, as the bilayer thickness decreases with increasing pressure [95,96]. As is well documented in the literature [102], the knowledge of the $^{31}P\,\Delta\sigma$ value along is not sufficient to determine the average orientation of the headgroup; therefore, we attribute the increase in $|\Delta\sigma|$ due to pressure in the various phases of DPPC as a consequence of slower motions.

For the DPPC/TTC bilayers, the dependence of $\Delta\sigma$ upon pressure, as shown in Fig. 25, can be divided into three regions. In the first region, between 1 bar and 0.8 kbar, which corresponds to the the LC phase, the $|\Delta\sigma|$ values for the DPPC/TTC bilayers are greater than those for pure DPPC bilayers, as expected. The $\Delta\sigma$ change due to the charged form of tetracaine will be discussed in more detail later. The distinct break which occurs at about 0.8 kbar represents the phase transition which is different from pure DPPC in three respects. First of all, the critical pressure of the main phase transition is shifted from $\backsim$ 0.5 kbar to 0.8 kbar. This increase in critical pressure upon addition of anesthetics has also been observed in the DPPC/TTC system by differential scanning calorimetry and small-angle neutron scattering [96]. Moreover, it is consistent with the depression of the main phase transition temperature at ambient pressure. Second, the phase transition is accompanied by an abrupt change in the $\Delta\sigma$ value which is not observed in the pure DPPC bilayers. Finally, instead of the observed $|\Delta\sigma|$ increase for pure DPPC, the $|\Delta\sigma|$ value decreased significantly at the phase transition pressure. Structural studies of the DPPC/TTC system by X-ray diffraction showed that under certain conditions incorporation of TTC caused the hydrocarbon chains to interdigitate [110]. In addition, it was found by DSC experiments [111] that the pretransition peak disappeared upon the addition of anesthetics. In the second region, from $\backsim$ 0.8 kbar to 3 kbar, the $|\Delta\sigma|$ values are significantly lower than those of pure DPPC bilayers in the Gel I and Gel II phase, but slightly higher than those in the interdigitated Gi gel phase. Thus, our results indicate that the addition of the charged form of TTC leads to the formation of the interdigitated gel Gi phase directly from the LC phase. In the Gi phase the $|\Delta\sigma|$ values increase with pressure and gradually approach the $\Delta\sigma$ values of pure DPPC in Gel II and Gel III phases. In the third region, from about 3.1 kbar to 5.1 kbar, the $|\Delta\sigma|$ values are greater than those observed for pure DPPC bilayers, similar to the situation for the LC phase. The question of the possible expulsion of anesthetic from the bilayer by high pressure cannot be answered by this study, but the fact remains that the $|\Delta\sigma|$ for the DPPC/TTC system is higher than the corresponding $\Delta\sigma$ values observed for the pure DPPC bilayers, even in the pressure range from 3.1 kbar to 5.1 kbar.

As we indicated in the introduction, one of the objectives of this study was to examine our data in light of the results obtained by Scherer and Seelig [104] and MacDonald et al. [105] who investigated the influence of membrane surface charges upon the conformation of the choline headgroup in various phosphatidylcholines.

Boulanger et al. [103] reported that at low pH the changes in the ^2H and ^{31}P NMR spectra of the phospholipid headgroup in the presence of tetracaine resemble those produced by the presence of ions and suggested that the headgroup undergoes conformational changes upon interaction with the anesthetic. In their detailed study of charge effects on the phospholipid headgroup in phosphatidylcholine bilayers, Scherer and Seelig [104] used various charged amphiphiles and extended their observations also to charged local anesthetics. It is well established that the P-N dipole of the headgroup is approximately parallel to the membrane surface in pure phospholipid membranes, but the addition of charged ions, positively charged amphiphiles, and positively charged local anesthetic moves the N^+ end of the dipole toward the water phase and changes the headgroup orientation.

Scherer and Seelig [104] introduced the following expression to calculate $\Delta\sigma$ after the addition of a positively charged amphiphile which leads to a decrease in the ^{31}P $\Delta\sigma$ value

$$\Delta\sigma = -47.6 - 46.2\, X_B \tag{15}$$

where X_B is the molar ratio of the positively charged amphiphile. It is of interest to use this expression, derived for 1-palmitoyl-2-oleoyl-sn-glycero-3-phosphocholine (POPC) to calculate the $\Delta\sigma$ change for our system. Using this expression to calculate $\Delta\sigma$ at 1 bar and 50° C for DPPC/TTC ($X_B = 0.33$), we obtain $\Delta\sigma = -62.9$ ppm, which compares favorably with the experimental value $\Delta\sigma = -59$ ppm. This encouraging result led us to examine in more detail experimental $\Delta\sigma$ values for the LC phase. Since the above expression was derived for POPC and a different temperature, we modified this expression for our experiments on DPPC to the following form:

$$\Delta\sigma = -46 - 39.4\, X_B. \tag{16}$$

From Fig. 27 we can readily see that within the LC phase the $\Delta\sigma$ values for DPPC/TTC parallel those for the pure DPPC bilayers independently of the pressure. It is interesting to note that equation (16) describes quantitatively the behavior of $\Delta\sigma$ for our system. What it suggests is that the orientation of the headgroup changes upon the addition of the charged anesthetic, and that the difference in $\Delta\sigma$ for the pure DPPC system and DPPC/TTC is not affected by the increasing pressure in the LC phase.

In view of the results obtained by McDonald et al. [105], we turn our attention to the high pressure gel phases, i.e., to the pressure region above ∼ 3.5 kbar. Fig. 28 shows the dependence of $\Delta\sigma$ for DPPC/TTC upon $\Delta\sigma$ for pure DPPC, but in this case there is a small increase in the difference in the $\Delta\sigma$ values with increasing pressure. Nevertheless, the average $\Delta\Delta\sigma$, defined as the difference between $\Delta\sigma$ for DPPC/TTC and that for pure DPPC is -12.4 ppm, while equation (5) predicted a $\Delta\Delta\sigma = -13$ ppm. This result provided supporting evidence for the proposal of McDonald et al. [105] that the choline headgroup of phosphatidyl- choline responds to the surface charge effects even in gel-state lipid membranes.

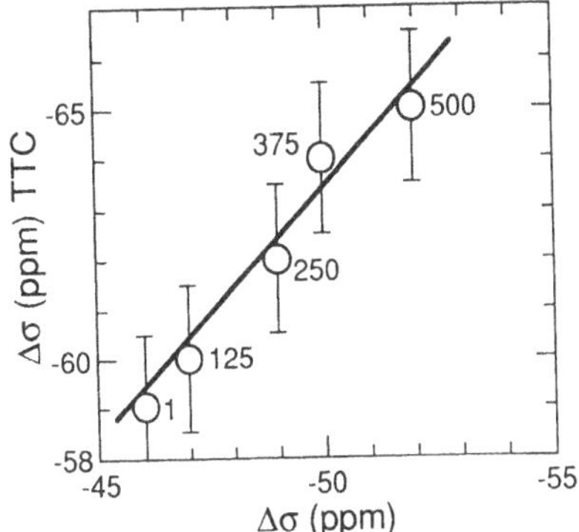

Fig. 27. [31]P chemical shift anisotropy $\Delta\sigma$ for DPPC/TTC as a function of $\Delta\sigma$ for pure DPPC in the liquid-crystalline phase at 50° C. The numbers next to the experimental points denote pressure in bars.

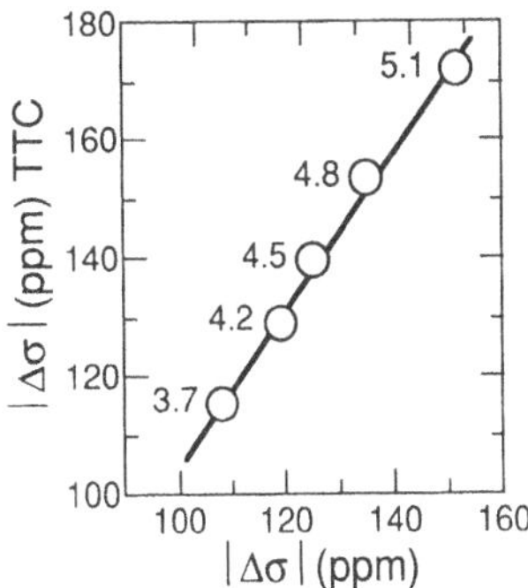

Fig. 28. [31]P chemical shift anisotropy $\Delta\sigma$ for DPPC/TTC as a function of $\Delta\sigma$ for pure DPPC in the high-pressure gel phases at 50° C. The numbers next to experimental points denote pressure in kilobars.

Acknowledgements

This work was supported in part by the National Science Foundation under grants NSF CHE 90-17649 and NSF DMR 89 20538, the National Institutes of Health under grant PHS 5 RO1 GM42452-04, and the Air Force Office for Scientific Research under grant AFOSR 89-0099.

References

[1] (1991) "High Pressure NMR", 'NMR Basic Principles and Progress', P. Diehl, E. Fluck, H. Gunther, R. Kosfeld and J. Seelig (eds.), Guest-Editor: J. Jonas, Springer-Verlag, New York.
[2] Jonas, J. (1984) Acc. Chem. Res. 17, 74-80.
[3] Jonas, J. and Lamb, D. M. (1987) ACS Symp. Ser. 329, 15-28.
[4] Drickamer, H. G. (1987) NATO ASI Ser. C 197, 263-278.

[5] Schindler, W., Zerda, T. W. and Jonas, J. (1984) J. Chem. Phys. 81, 4306-4313.
[6] Peng, X. and Jonas, J. (1990) J. Chem. Phys. 93, 2192-2200.
[7] Heremans, K. (1982) Ann. Rev. Biophys. Bioeng. 11, 1-21.
[8] Weber, G. and Drickamer, H. G. (1983) Quart. Rev. Biophys. 16, 89-112.
[9] Wong, P. T. T. (1987) 'High Pressure Chemistry and Biochemistry', in R. van Eldik and J. Jonas (eds.), NATO ASI, Series C, Vol. 197, D. Reidel, Dordrecht, p. 381-400.
[10] Weber, G. (1987) 'High Pressure Chemistry and Biochemistry', in R. van Eldik and J. Jonas (eds.), NATO ASI, Series C, Vol. 197, D. Reidel, Dordrecht, p. 401-420.
[11] Heremans, K. (1987) 'High Pressure Chemistry and Biochemistry', in R. van Eldik and J. Jonas (eds.), NATO ASI, Series C, Vol. 197, D. Reidel, Dordrecht, p. 421-421.
[12] MacDonald, A. G. (1987) 'Current Perspectives in High Pressure Biology', in H. W. Jannash, R. E. Marquis and R. E. Zimmerman (eds.), Academic, New York, p. 207-223.
[13] Heremans, K. (1987) 'Current Perspectives in High Pressure Biology', in H. W. Jannasch, R. E. Marquis and A. M. Zimmerman (eds.), Academic, New York, p. 225-234.
[14] Weber, G. (1987) 'Current Perspectives in High Pressure Biology', in H, W. Jannasch, R. E. Marquis and A. M. Zimmerman (eds.), Academic, New York, p. 235-244.
[15] Wong, P. T. T. (1987) 'Current Perspectives in High Pressure Biology', in H. W. Jannasch, R. E. Marquis and A. M. Zimmerman (eds.), Academic, New York, p. 287-314.
[16] Morishima, I. (1987) 'Current Perspectives in High Pressure Biology', in H. W. Jannasch, R. E. Marquis and A. M. Zimmerman (eds.), Academic, New York, p. 315-322.
[17] Wagner, G. (1980) FEBS Lett. 112, 280-284.
[18] Williams, R. K., Fyfe, C. A., Epand, R. M., and Bruck, D. (1978) Biochem. 17, 1506-1509.
[19] Williams, R. K., Fyfe, C. A., Bruck, D., and van Vein, L. (1979) Biopolymers 18, 757=763.
[20] Morishima, I., Ogawa, S., and Yamada, H. (1979) J. Amer. Chem. Soc. 101, 7074=7076.
[21] Morishima, I., Ogawa, S., and Yamada, H. (1980) Biochem. 19, 1569-1575.
[22] Morishima, I. and Hara, M. (1982) J. Am. Chem. Soc. 104, 6833-6834.
[23] Morishima, I. and Hara, M. (1983) Biochem. 22, 4102-4107.
[24] Morishima, I. and Hara, M. (1983) J. Biol. Chem. 258, 14428-14432.
[25] Morishima, I. and Hara, M. (1984) Biochem. Biophys. Res. Commun. 121, 229-236.
[26] Hauer, J., Muller, K., Ludemann, H. D., and Jaenicke, R. (1981) FEBS Lett. 135, 135-138.
[27] Hauer, J., Ludemann, H. D., and Jaenicke, R. (1981) Naturwissenshaften 36, 366.
[28] Jonas, J. (1973) Advs. Magn. Reson. 6, 73-139.
[29] Jonas, J. (1975) Ann. Rev. Phys. Chem. 26, 167-190.
[30] Jonas, J. (1978) NATO ASI, Series C, 41, 65-110.
[31] Jonas, J. (1980) Rev. Phys. Chem. Japan 50, 19-35.
[32] Jonas, J. (1982) Science 216, 1179-1184.
[33] Jonas, J. (1987) NATO ASI, Series C, 197, 193-235.

[34] Jonas, J. (1972) Rev. Sci. Instr. 42, 643-649.
[35] Yamada, Y. (1974) Rev. Sci. Instr. 45, 640-642.
[36] Jonas, J., Xie, C. L., Jonas, A., Grandinetti, P. J., Campbell, D., and Driscoll, D. (1988) Proc. Natl. Acad. Sci. USA 85, 4115-4117.
[37] DeFries, T. H. and Jonas, J. (1979) J. Magn. Reson. 35, 111-119.
[38] Jonas, J., Hasha, D. L., Lamb, W. J., Hoffman, G. A., and Eguchi, T. (1981) J. Magn. Reson. 42, 169-172.
[39] Vander Velde, D. G. and Jonas, J. (1987) J. Magn. Reson. 71, 480-484.
[40] Wilbur, D. J. and Jonas, J. (1971) J. Chem. Phys. 55, 5840-5841.
[41] Wilbur, D. J. and Jonas, J. (1973) J. Magn. Reson. 10, 279-289.
[42] Wilbur, D. J. and Jonas, J. (1975) J. Chem. Phys. 62, 2800-2807.
[43] McDaniel, P. L., Liu, G., and Jonas, J. (1989) Physica A 156, 203-211.
[44] Driscoll, D. A., Samarasinghe, S., Adamy, S., Jonas, J., and Jonas, A. (1991) Biochem. 30, 3322-3327.
[45] Artaki, I. and Jonas, J. (1985) J. Chem. Phys. 82, 3360-3370.
[46] Walker, N. A., Lamb, D. M., Adamy, S. T., Jonas, J., and Dare-Edwards, M. P. (1988) J. Phys. Chem. 92, 3675-3679.
[47] Jonas, J., Adamy, S., Grandinetti, P. J., Masuda, Y., Morris, S. J., Campbell, D. M., and Li, Y. (1990) J. Phys. Chem. 94, 1157-1164.
[48] Dawson, D. and Hugginson, G. R. (1977) 'Elastohydrodynamic Lubrication', Pergamon, London.
[49] Lamb, D. M., Grandinetti, P. J., and Jonas, J. (1987) J. Magn. Reson. 72, 532-539.
[50] Walker, N. A., Lamb, D. M., Jonas, J., and Dare-Edwards, M. P. (1987) J. Magn. Reson. 74, 580-583.
[51] Burnett, L. J. and Harmon, J. F. (1972) J. Chem. Phys. 57, 1293-1297.
[52] Lyerly, J. R., Jr., McIntyre, H. M., and Torchia, D. A. (1974) Macromolecules 7, 11-14.
[53] London, R. E., and Avitable, J. (1977) J. Am. Chem. Soc. 99, 7765-7776.
[54] Levine, Y. K., Birdsall, M., Lee, A. G., Metcalfe, J. C., Partington, P., and Roberts, G. C. K. (1974) J. Chem. Phys. 60, 2790-2899.
[55] Davidson, D. W. and Cole, R. M. (1951) J. Chem. Phys. 19, 1484-1490.
[56] Wolfe, M. and Jonas, J. (1979) J. Chem. Phys. 71, 3252-3262.
[57] Skinner, J. L. and Wolynes, P. G. (1978) J. Chem. Phys. 69, 2143-2150.
[58] Montgomery, Jr., J. A., Chandler, D., and Berne, B. J. (1979) J. Chem. Phys. 70, 4056-4066.
[59] Hasha, D. L., Eguchi, T., and Jonas, J. (1982) J. Am. Chem. Soc. 104, 2290-2296.
[60] Lee, M., Holton, G. R., and Hochstrasser, R. M. (1985) Chem. Phys. Lett. 118, 359-363.
[61] Schroeder, J. and Troe, J. (1987) Annu. Rev. Phys. Chem. 38, 163-190.
[62] Kuharski, R. A., Chandler, D., Montgomery, Jr., J. A., Rabbi, F.., and Singer, S. J. (1988) J. Phys. Chem. 92, 3261-3267.
[63] Hartke, B., Manz, J., and Mathis, J. (1989) Chem. Phys. 139, 123-146.
[64] Lopez, V. and Marcus, R. A. (1982) Chem. Phys. Lett. 93, 232-234; Lederman, S. M., Lopez, V., Fairen, V., Voth, G. A., and Marcus, R. A. (1989) Chem. Phys. 139, 171-184.
[65] Uzer, T. and Hynes, J. T. (1986) J. Phys. Chem. 90, 3524-3527.
[66] Rogers, P. J., Selco, J. I., and Rowland, F. S. (1983) Chem. Phys. Lett. 97, 313-316.
[67] Wrigley, S. P. and Rabinovitch, (1984) Chem. Phys. Lett. 98, 386-392.
[68] Xie, C.-L., Campbell, D., and Jonas, J. (1988) J. Chem. Phys. 88, 3396-3398.

[69] Peng, X. and Jonas, J. (1990) J. Chem. Phys. 93, 2192-2200.
[70] Campbell, D. M., Mackowiak, M., and Jonas, J. (1992) J. Chem. Phys. 96, 2717-2723.
[71] Ashcroft, J., Xie, C.-L., and Jonas, J. (1989) J. Chem. Phys. 90, 5386-5391.
[72] Xie, C.-L., Campbell, D., and Jonas, J. (1990) J. Chem. Phys. 92, 3736-3740.
[73] Jonas, J., Allerhand, A., and Gutowsky, H. (1965) J. Chem. Phys. 42, 3396-3399.
[74] (1989) 'Molecular Dynamics in Restricted Geometries', J. Klafter and J. M. Drake (eds.), Wiley, New York.
[75] (1990) 'Dynamics in Small Confining Systems', J. M. Drake, J. Klafter, and R. Kopelman (eds.), Extended Abstracts Proceedings of Symposium of 1990 Fall Meeting of the Materials Research Society, Boston, MA.
[76] Liu, G., Li, Y., and Jonas, (1989) J. Chem. Phys. 90, 5881-5882.
[77] Mackowiak, M., Liu, G., and Jonas, J. (1990) J. Chem. Phys. 93, 2154-2155.
[78] Liu, G., Mackowiak, M., Li, Y., and Jonas, J. (1990) Chem. Phys. 149, 165-171.
[79] Liu, G., Li, Y., and Jonas, J. (1991) J. Chem. Phys. 95, 6892-6901.
[80] Xu, Shu, Zhang, Jing, and Jonas, J. (1992) J. Chem. Phys. 97, 4564-4566.
[81] Bloch, J. M., Katz, H., Moses, D., Cajipe, V. B., and Fischer, J. E. (1985) Phys. Rev. B 31, 6785-6788.
[82] Zipp, A. and Kauzmann, W. (1973) Biochem. 12, 4217-4228.
[83] Li, T. M., Hook, III, J. W., Drickamer, H. G., and Weber, G. (1976) Biochem. 15, 5571-5580.
[84] Samarasinghe, S. D., Campbell, D. M., Jonas, A., and Jonas, J. (1992) Biochem. 31, 7773-7778.
[85] Kundrot, C. E. and Richards, F. M. (1987) J. Mol. Biol. 193, 157-170.
[86] Delepierre, M., Dobson, C. M., and Poulsen, F. M. (1982) Biochem. 21, 4756-4761.
[87] Dobson, C. M., Evans, P. A., and Williamson, K. L. (1984) FEBS Lett. 168, 331-334.
[88] Wedin, R. E., Delepierre, M., Dobson, C. M., and Poulsen, F. M. (1982) Biochem. 21, 1098-1103.
[89] Driscoll, D. A., Jonas, A., and Jonas, J. (1991) Chem. Phys. Lipids 58, 97-104.
[90] Opella, S. J. (1982) Annu. Rev. Phys. Chem. 33, 533-562.
[91] Wong, P. T. T. (1987) 'High Pressure Chemistry and Biochemistry', R. van Eldik and J. Jonas (eds.), D. Reidel, Dordrecht, p. 381-400.
[92] Tosh, R. E., and Collins, P. J. (1986) Biochim. Biophys. Acta 859, 10-14.
[93] Russell, N. D. and Collins, P. J. (1982) J. Chem. Phys. 77, 5766-5770.
[94] Wong, P. T. T., Siminovitch, D. J., and Mantsch, H. H. (1988) Biochim. Biophys. Acta 947, 139-171.
[95] Braganza, L. F. and Worcester, D. L. (1986) Biochem. 25, 2591-2596.
[96] Winter, R. and Pilgrim, W.-C. (1989) Ber. Bunsenges Phys. Chem. 93, 708-717.
[97] Davis, J. H. (1979) Biophys. J. 27, 339-358.
[98] Wong, P. T. T. and Mantsch, H. H. (1985) J. Phys. Chem. 89, 883-886.
[99] Prasad, S. K., Shashidhar, R., Gaber, B. P., and Chandrasekhas, S. C. (1987) Chem. Phys. Lipids 43, 227-235.
[100] Seelig, J., and Seelig A. (1980) Q. Rev. Biophys. 13, 19-61.
[101] Browning, J. L. (1981) Biochem. 20, 7144-7151.
[102] Seelig, J. (1978) Biochim.. Biophys. Acta 505, 105-141.
[103] Boulanger, Y., Schreier, S., Leitch, L. C., and Smith, I. C. P. (1980) Can. J. Biochem. 58, 986-995.

[104] Scherer, P. G., and Seelig, J. (1989) Biochem. 27, 7720-7728.
[105] McDonald, P. M., Leisen, J., and Marassi, F. M. (1991) Biochem. 30, 3558-3566.
[106] Peng, X. and Jonas, J. (1992) Biochem. 31, 6383-6390.
[107] Brown, M. F. and Seelig, J. (1977) Nature 269, 721-723.
[108] Wong, P. T., Siminovitch, D. J., and Mantsch, H. H. (1988) Biochim. Biophys. Acta 947, 139-171.
[109] Stamatoff, J., Guillon, D., Powers, L., Cladis, P., and Madsen, D. (1978) Biochem. Biophys. Res. Commun. 85, 724-728.
[110] McIntosh, T. J., McDaniel, R. V., and Simon, S. S. (1983) Biochim. Biophys. Acta 731, 109-114.

THE BEHAVIOUR OF PROTEINS UNDER PRESSURE

KAREL HEREMANS
Department of Chemistry
Katholieke Universiteit Leuven
B-3001 Leuven
Belgium

ABSTRACT. This paper reviews some recent experimental observations on the behaviour of proteins under hydrostatic pressure. The results are correlated with findings on the influence of parameters such as temperature and pressure on living systems. Some processes are discussed which may be of importance for industrial applications such as high pressure food processing. In many cases a quantitative molecular interpretation is still lacking.

1. Introduction

The occurrence of life on this planet seems to be restricted to rather narrow ranges of temperature and pressure. Laboratory experiments strongly suggest that these conditions are dictated by the dependence of the stability of proteins on these parameters. In this respect it is of interest to note that the antagonistic effect of pressure on temperature induced phenomena, that is characteristic for proteins, has also been observed in bacteria (1). Diagrams that show the effect of temperature and pressure on the viability of bacteria from the deep sea show a strong resemblance to those observed for the reversible and irreversible denaturation of proteins (2,3). It was found that the temperature of maximum tolerance for growth increases with increasing pressure up to about 100 MPa. The antagonistic effect does not continue at higher pressures. Similarly, it has been observed that the thermal stability of turnip yellow mosaic virus is increased under high pressure (4).

In this review, some recent experiments are described in which progres has been made in our understanding of the molecular responses of proteins to pressure for a few selected systems. To organize the discussion it is helpful to consider four structural levels in proteins: The *primary structure* describes the order of the covalently linked amino acids. This level is not affected by pressure. The pressure-induced breaking of a covalent bond in a protein has never been observed in the range of a few Gpa. The *secondary structure*

R. Winter and J. Jonas (eds.), High Pressure Chemistry, Biochemistry and Materials Science, 443–469.

describes the way in which the polypeptide chain forms either intramolecular (to form a α-helix) or intermolecular (to form a ß-sheet) hydrogen bonds. Pressure may have an effect on these structures and their transformations. This has important consequences for pressure induced conformational transitions and for pressure-induced denaturation of proteins in general. The *tertiary structure* describes how the secondary structure domains fold into a three dimensional shape as a consequence of the noncovalent interactions of the amino acids side chains. Pressure does have profound effects on this strucural level but they cannot easily be separated by experimental methods from the changes that place on the secondary structures. There is no doubt that an understanding of the changes of the tertiary structure are fundamental for our understanding of the molecular mechanism of conformational transitions as well protein denaturation which is sometimes regarded as the counterpart of protein folding. The *quaternary structure* of proteins is formed by the interaction of noncovalent bonds between polypeptide subunits. The effect of pressure on the interaction between protein subunits has received considerable attention in the work of Weber and coworkers. This topic is discussed in detail in separate contributions in this volume.

As a general rule, pressure induces reversible effects on proteins below 100-200 MPa. These include protein compressibility, the binding of small molecules, changes in conformation and changes in the interaction of the subunits in multimeric proteins. Irreversible effects may include complete inactivation of enzymes and denaturation of proteins. In general these effects occur above 300 MPa.

The effects of pressure on the activity of enzymes may be the consequence of the effect of pressure on one of the many possible steps that control the activity of an enzyme. The simplest possible scheme for such a reaction may be represented as follows:

$$E + S \rightleftharpoons ES \rightleftharpoons ES^* \rightarrow E + P$$

$nE \rightleftharpoons E_n$ Subunit interactions

$E \rightleftharpoons E^*$ Conformational changes

It may be seen that pressure effects may expected on the substrate enzyme interaction, on the catalytic step itself, on the pressure induced conformational changes that may take place in the protein or on the pressure induced dissociation of the enzyme into subunits (5). A meaningful interpretation of activation volumes of enzymatic reactions requires a separate study of all other possible pressure induced changes that may occur. In the following sections examples will be given of each of these possibilities.

It is clear from the previous considerations that pressure induces effects in proteins which can, at least in principle, be understood from the volume changes that are characteristic for noncovalent interactions. It can be seen from Table I that the strongest effects may be expected for processes in which ionic species are involved. This can be understood from a quantitative point of view from solvent electrostriction models (6). While the formation of new hydrogen bonds seems to be stabilized by pressure, the exchange of an existing hydrogen bond of an amino acid residue with the solvent may show a very small if not a neglegible volume change. Opinions on the effect of pressure on hydrophobic interactions are as divergent as are the opinions on the nature of hydrophobic interactions themselves. The subdivision proposed in Table I suggests that stacking of aromatic rings is stabilized by increasing pressure. Interactions between aliphatic chains cannot easily be studied by spectroscopic techniques and the volume changes are obtained from pressure studies on the micelle formation of surfactants (7).

2. Protein compressibility

Data on the compressibility of proteins have received considerable attention in recent years in view of the correlation between the volume fluctuations of a protein and the isothermal compressibility (8):

$$(\Delta V)^2 = kT\, V\, \beta_T$$

Unfortunately, the isothermal compressibility, β_T, is difficult to obtain experimentally because of the rather low solubility, on a molar basis, of proteins in aqueous solutions. In practice the adiabatic compressibility, β_S, is obtained from the velocity of sound (9). The two compressibilities are related via the well known equations:

$$\beta_S = \beta_T\, (Cv/Cp)$$

$$\beta_S = \beta_T - \alpha^2\, T/d\, Cp$$

For proteins the thermal expansion, α, and the heat capacity, Cp, are not easily accessible either. In practice the adiabatic compressibility is used in theoretical considerations.

The compressibility is defined as the relative change of the molar volume with pressure. The molar volume of a protein consists of three contributions: the constitutive volumes of the atoms, the void volume that results from the imperfect packing of the amino acids in the interior of the protein and the volume decrease that results from the hydration of the peptide units and the amino acid residues. Since the molar volume can be calculated from the constitutive atomic volume of the amino acid residues, one may assume that contributions from the void volume and the hydration volume compensate.

Table I. Reactions volumes (mL/mol) of model reactions for pressure effects on noncovalent interactions in proteins and their reactions (10).

Electrostatic interactions	
$CH_3COOH + H_2O \rightleftharpoons CH_3COO^- + H_3O^+$	- 12
$EtNH_3^+ + H_2O \rightleftharpoons EtNH_2 + H_3O^+$	+ 5.4
$H_2PO_4^- + H_2O \rightleftharpoons HPO_4^{--} + H_3O^+$	- 25
Hydrogen bonding	
Phenol-Et_2O (in hexane)	- 6.1
ε-caprolactam-caprolactam (in CCl_4)	- 1.3
Phenol-phenol (in CCl_4)	- 2.4
Hydrophobic interactions	
- stacking of aromatic rings	
FAD (internal complex)	- 4.6
FMN + AMP	- 4
Charge transfer interaction	- 6.5
- aliphatic systems	
various data sources	+ 1/CH_2

One may expect that secondary structures such as α-helix and ß-sheet have a low compressibility. The compressibility of protein molecules may then reflect the interactions between secondary structure domains. Fluctuations of the protein structure may then be visualized as the movement of secondary structures with respect to each other. This hypothesis may be tested experimentally by measuring the compressibility of mutants of one particular protein. Contributions of the hydration to the compressibility may be obtained by varying the solvent composition. No data are available however.

The studies of Kaminsky and Richards (11) indicate that more subtle mechanisms may be at work. These authors obtained the partial specific volume and the adiabatic compressibility of the oxidized and the reduced form of *E. coli* thioredoxin. Both quantities are higher for the oxidized than for the reduced form. The authors attribute the differences to alterations in the solvent layer. The observations also fit into the observed correlations between the adiabatic compressibility and the partial specific volume as found by Gekko and Hasegawa (12) for a set of 25 proteins.

Sarvazyan (9) has recently reviewed the available information from ultrasonic velocimetry studies performed on biological compounds by his own and by other groups. Kharakoz (13) started a systematic study of the compressibility of amino acids as a function of temperature in order to develop an empirical structure-compressibility relationship. The compressibilities of the atomic groups can be interpreted in terms of hydration and intramolecular interactions. A simple additivity scheme which is applicable to amino acids and other small molecules suggests the possibility to calculate the compressibility of proteins. The development of equipment to measure the ultrasonic velocity as a function of pressure is under development. The method has been used to follow the change in compressibility of the temperature induced transition in chymotrypsinogen (9). As expected the high temperature conformation has a higher compressibility than the low temperature form.

3. Ligand-protein interactions

Pressure effects on the binding of small substrate-like molecules to proteins and enzymes are studied in order to obtain information on the noncovalent interactions in the binding process as such, but also to probe the conformational dynamics of the protein itself. As an example for the probing of the specificity of the interaction, we discuss the binding of small molecules to the serine proteases chymotrypsin and trypsin. The influence of the conformational dynamics of the protein on the ligan binding is very well illustrated in the binding studies of oxygen and carbon monoxide to heme proteins.

A topic which is not directly within the scope of this review but of general interest, is the effect of pressure on model systems for general anesthesia. The observation that pressures of the order of 150 atm can reverse general anesthesia has been a stimulus for research into the mechanism of anesthesia. However there are instances where an additive rather than antagonistic effects are observed (14). Also, firefly luciferase, a model system for the protein theory of general anesthesia, does not show the pressure reversal effect with helium gas (15). Experiments on poly(L-lysine) show that volatile anesthetics such as chloroform and halothane induce a α-helix to ß-sheet transition. This anesthetic-induced transition may be reversed by the application of pressure up to 350 atm (16). Small-angle neutron scattering has been used to study the effect of pressure on the incorporation of the local anesthetic tetracaine into lipids (17). Pressure-reversal of the anesthetic-induced transition was observed, but the effects are highly dependent on the concentration.

3.1 SERINE PROTEASES

The serine proteases chymotrypsin and trypsin have a substrate binding pocket which is open to the solvent. Chymotrypsin is specific for nonpolar amino acids while trypsin is specific for amino acids with a positive charge. The different nature of the binding pocket is clearly reflected in the volume changes that are obtained for the binding of proflavin, a dye molecule that has a positive charge (10).

Chymotrypsin + Proflavin $\Delta V = 0$ mL
Trypsin + Proflavin $\Delta V = +17$ mL

The absence of a volume change for the binding of proflavin to chymotrypsin is puzzling in view of the negative volume changes that are observed for the stacking of aromatic molecules. When the pressure induced conformational change (section 4.1) is taken into account then a negative volume change is calculated (- 7 ml) which is in close agreement with the model systems. The binding of Biebrich Scarlet to chymotrypsin shows a negative volume change which may reflect additional interactions that are formed between the dye and the protein.

Chymotrypsin + Biebrich Scarlet ΔV = - 5 mL

The general conclusion from these examples of binding studies is that one has to make sure that no pressure induced conformational changes take place in the protein which are coupled to the binding equilibrium.

3.2 HEME PROTEINS

The binding of carbonmonoxide to hemes and hemoproteins has been a model system for binding studies for about twenty years. An interesting feature of these binding studies is that the ligand must enter the binding pocket which either forms a channel to the solvent or is only accessible for the ligand via a diffusion mechanism through the protein matrix.

The data obtained by Hasinoff (18) are a good starting point for an understanding of the work that is still going on in a number of laboratories both on model systems and on heme proteins. It can be seen from Table II that the binding of CO with hemes is essentially controlled by diffusion. Recent work by Traylor and coworkers (19) has shown that for certain model compounds it is possible to change the viscosity of the medium (90/10, v/v, mineral oil/toluene) by the application of pressure in such a way that the binding proces is rate limiting at low pressures (negative activation volume) while at higher pressure the diffusion of the reacting species becomes rate limiting (positive activation volume).

It may also be seen from Table II that the activation volume for the binding of CO is negative which indicates that the bond formation step is rate limiting. For other ligands the activation volume for the binding to myoglobin is positive, suggesting that the diffusion of the ligand trough the protein matrix is rate limiting (20). The complete volume profile for the binding of oxygen and carbonmonoxide to sperm whale myoglobin has been obtained by Projahn and van Eldik (21) by a combined use of high pressure stopped-flow and temperature jump relaxation techniques.

Table II. Second order rate constants and activation volumes (mL/mol) for the binding of CO and O_2 to heme and sperm whale myoglobin (18).

	k_2	$\Delta V^{\#}$	rate limiting process
Heme + CO	$3\ 10^7$	+ 2	solvent viscosity
Heme + CO (EgOH/H_2O)	$2\ 10^6$	+ 14	solvent viscosity
Myoglobin + CO	$4\ 10^5$	- 9	binding to heme iron
Myoglobin + O_2	$1\ 10^7$	+ 8	diffusion in protein matrix

$$Mb + CO \rightleftharpoons Mb\text{-}CO \qquad \Delta V = -6.2 \text{ mL}$$

$$Mb + O_2 \rightleftharpoons Mb\text{-}O_2 \qquad \Delta V = -18 \text{ mL}$$

The observed difference in reaction volume is attributed to the difference in binding geometry, hydrogen bonding and conformational changes in the heme pocket.

While the model systems suggest a fairly simple picture of the reaction between CO, O_2 and myoglobin, the binding studies with proteins show a much more complex situation. A high pressure stopped-flow study of the binding and the kinetics of azide and imidazole to sperm whale and horse metmyoglobin showed a systematic difference between the two proteins (22). As indicated in Table III, the data for dog metmyoglobin are quite similar to those of horse metmyoglobin (22,23).

The activation volumes are positive indicating a rate control from the protein matrix. The difference in activation and reaction volumes for these natural mutants strongly suggest a specific role for the protein matrix. The amino acid sequence in the surroundings of the heme suggest that the substitution Arg(45)Lys may be crucial. This is supported by studies on sperm whale mutants which indicate that a salt-bridge between Arg(Lys) and the heme group may be rate limiting step in the binding process (24). In order to exclude the possible contributions from other amino acids substitutions in the vicinity it would of interest to perform the same mutation on human myoglobin (25). It is of interest to note that similar differences between the natural mutants have been observed in laser flash photolysis studies for the binding of O_2 and CO (26).

Recently Adachi et al (27) have studied mutants of human myoglobin. They have concentrated their efforts on the substitution of Leu(29), Leu(72) and Leu(104). These residues are part of hydrophobic clusters on the heme distal and proximal side. Table IV shows the dramatic effect of the Leu(29) mutants on the activation volumes for the binding of CO and O_2.

Table III. Reaction and activation volume (mL/mol) for the binding of azide and imidazole to metmyoglobin (22,23).

	azide		imidazole	
	ΔV	$\Delta V^{\#}$	ΔV	$\Delta V^{\#}$
sperm whale	- 9	+ 4	0	+ 8
horse	- 5	+ 11	+ 10	+ 13
dog	- 5	+ 12	+ 8	+ 12

An interesting feature of the Leu(29) mutants is that the CO stretching mode which shows only one peak in the wild type, shows two peaks for the Ile mutant and three peaks for the Ala mutant. This suggests that the bound CO has three possible conformations which interconvert rapidly. An additional feature of the Ala mutant is that the activation volume is pressure dependent, suggesting a pressure effect on the additional orientations that become available for the CO ligand in the protein by the mutation. It is of particular interest that some of these states may also be obtained in sperm whale myoglobin either by freezing the protein at 100K and by the application of pressure (200 MPa) at this temperature or by first applying pressure and then lowering the temperature (28).

Unno and coworkers (29,30) have studied the effect of pressure on the binding of CO to the R- and T-state of human hemoglobin. In both cases the activation volumes are negative but quite different, - 9 mL for the R- and - 32 mL for the T-state. In addition, the activation volume for the R-state is almost zero at about 100 MPa. This suggest that the R-state is more compressible than the T-state. Experiments on the islotated α- and ß-chains show pressure dependent activation volumes suggesting a turnover from a rate limiting bond formation process to a diffusion process at higher pressure. Nanosecond laser photolysis experiments under pressure support this conclusion. Laser flash photolysis allows one to obtain a volume profile of a reaction down to the nanosecond timescale and one may expect a further refinement of the analysis from picosecond experiments (31). The development of time resolved photoacoustic calorimetry allows one to measure the dynamics of enthalpy as wel as volume changes with a time resolution as fast as 10 ns (32). When a photoexcited molecule transfers its excess energy to the solvent, acoustic waves are produced which can be nalyzed in terms of solvent and conformational contributions of the protein. The technique has been, among others, applied to the study of the binding of CO to myoglobin. Studies on mutant sperm whale myoglobin give evidence for the role of the salt-bridge between Arg(45) and the propionate of the heme group (24). The differences between horse and sperm

Table IV. Activation volumes (mL/mol) for the binding of ligands to human myoglobin and mutants (27).

Human myoglobin	$\Delta V^{\#}(CO)$	$\Delta V^{\#}(O_2)$
Wild type	- 21	3.6
Leu(29)Ile	- 3.7	9.2
Leu(29)Ala	+ 10	13
Leu(72)Ala	- 21	- 3.0
Leu(104)Ala	- 23	0.11

whale myoglobin support the ideas about the role of the salt-bridge in the binding process (33). Studies on carboxyhemoglobin show enthalpy and volume events that occur in less than 100 ns (34). The photoacoustic calorimetry technique is a very promising tool to obtain mechanistic details on photoexcitable reactions.

3.3 SOLVENT EFFECTS

Studies of the effects of solvent composition on the activation parameters are not very numerous. However, in a few cases interesting effects have been reported which have not been interpreted into a unifying concept along with temperature and pressure. One recent study is that by Balny and Travers (35) on the binding of CO to horseradish peroxidase. As can be seen from Table V, the effect of organic solvents on the activation volume is quite dramatic. In contrast, the activation energy is nearly independent of the solvent composition. Other pressure studies from the same group have been discussed by Balny et al (36).

Whereas more systematic studies would be needed before any substantial conclusions should be expressed, one is tempted to consider the correlation of these results with other solvent studies under pressure. Li et al (37) have observed that methanol (up to 30% v/v) decreases the midpoint of the denaturation pressure of the riboflavin binding protein from egg white from 640 to 470 MPa. The volume of denaturation is reduced from - 74 mL to - 56 mL. Bonafe et al (38) have observed a stabilizing effect of glycerol (6 M) on the pressure dissociation into subunits of *G. paulistus* hemoglobin. These authors have interpreted these results in terms of the preferential hydration model, and therefore exclusion of glycerol from the assembly domain, a concept introduced by Timasheff and coworkers. Gekko and Timasheff (39) have reported a stabilizing effect of glycerol on the temperature denaturation of proteins. More systematic studies on the effect of solvents on the pressure

Solvent	$\Delta V^{\#}$
Water	- 24
Ethylene glycol, 40% (v/v)	- 7
Methanol, 50% (v/v)	- 9

Table V. Solvent effects on the activation volume (mL/mol) (20°C) for the binding of CO to horseradish peroxidase (35).

effects of various protein reactions such as pressure induced disassembly and denaturation may be fruitful.

4. Conformational transitions

The native structure of a protein, i.e. the conformation that shows biological activity in solution, is the result of a delicate balance between stabilizing interactions within the chain and destabilizing interactions from the solvent (40). Changes in the solvent composition, such as the presence of sugars, salts or extreme pH conditions, may have a strong influence on the stability of the structure. Since the interactions are largely noncovalent, they are also affected by temperature and pressure. The effect of pressure on enzymes and proteins may be reversible or irreversible. A fundamental difference between pressure and temperature induced processes is that no changes in covalent bonding have been observed in the pressure range studied so far.

4.1 SERINE PROTEASES

Many studies on proteins reveal processes that are connected with transitions in the conformation of the protein. The number of instances where these changes are well defined is rather small. A few cases have been studied in detail with Raman spectroscopy. By determining small changes in the shape of the amide I band it is possible to follow the pressure-induced reversible changes in the secondary structure of chymotrypsin and trypsin (41,42). These changes may be correlated with the open nature of the substrate binding site and the salt-bridge in the vicinity of the binding pocket. In contrast, chymotrypsinogen, DFP-chymotrypsin (a covalently linked inhibitor to the active site of the enzyme) and elastase do not show a pressure-induced conformational change up to 300 MPa.

In chymotrypsin and trypsin a salt-bridge between Ile-16 and Asp-194 stabilizes the active conformation of the enzyme. The drastic reduction in the activity between pH 7 and pH 10 may also be explained by the disruption of the salt-bridge. The changes in conformation which may be followed by X-ray

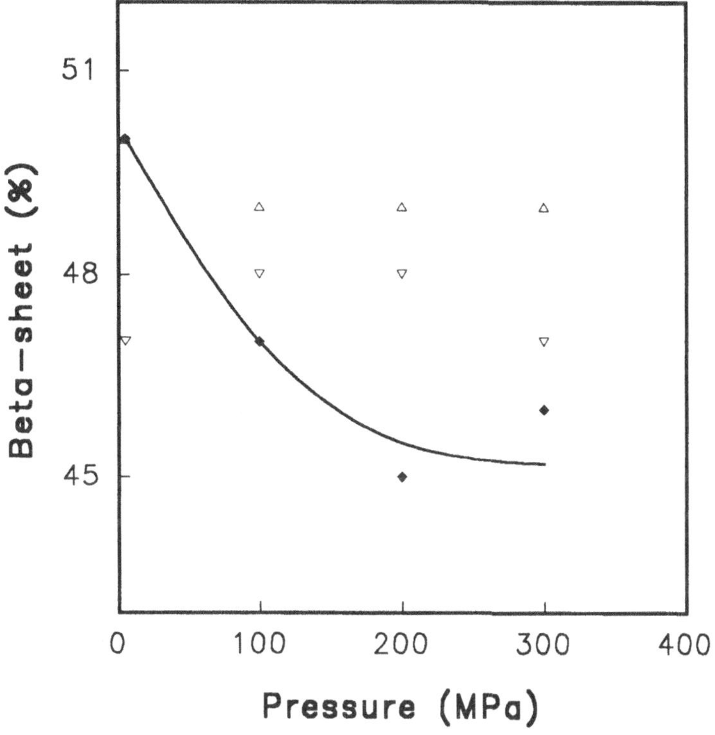

Figure 1. The pressure dependence of the ß-sheet content (%) for chymotrypsin at pH = 7.8 (♦), DFP-chymotrypsin (▵) and chymotrypsinogen (▿). The solid line corresponds to the theoretical pressure dependence of the inactivation with ΔV = - 35 mL/mol. After (42).

diffraction, are completely reversible. Raman spectroscopy up to 300 MPa shows that the Amide I band changes reversibly indicating a change in conformation similar to the pH induced changes. A more detailed analysis of the Amide I band in terms of the contributions of the secondary structure shows a small decrease in the ß structure of the protein (Figure 1).

The reaction volume for the disruption of the salt-bridge is - 35 mL/mol. This may be understood from the fact that the environment of the salt-bridge is hydrophobic and the reaction volume for the formation of ionic species from neutral molecules:

$$R\text{-}COOH + R\text{-}NH_2 = R\text{-}COO^- + R\text{-}NH_3^+ \quad \Delta V = -18 \text{ mL}$$

The consequences of the observed conformational changes of the proteins for substrate or inhibitor binding studies are clear. The volume changes obtained from pressure studies on the binding of ligands contain a contribution of the pressure effect on the conformation of the protein. For chymotrypsin this means that the observed pressure independence of the binding of proflavin (section 3.1) is the result from two compensating effects. For the binding of proflavin to the active conformation of the protein, ΔV = - 7 mL/mol. This is in agreement with expectations from hydrophobic interactions. Similar considerations apply when one obtains experimental activation volumes for reactions catalyzed by chymotrypsin. They contain contributions from the pressure effect on the protein which have nothing to do with the catalytic act per se.

4.2 HEME PROTEINS

Changes in the conformation of hemoproteins may be followed from the resonance Raman spectrum which is essentially the spectrum of the porphyrin moiety. The alkaline isomerization of horse cytochrome c has been followed as a function of pressure. Pressure forces the alkaline conformation back to the normal neutral conformation (43).

Spin transitions in hemoproteins are affected by pressure (44). High pressure favours the low spin state of the heme iron. A single mutation in cytochrome P-450cam has been shown to affect the pressure sensitivity and the binding capacity of camphor (45).

Pressure may also affect electron transfer between proteins presumable via its effect on the solvent reorganization that takes place (46,47). In cytochrome c oxidase, electron transfer between cytochrome a and cytochrome a_3 may also be blocked by the osmotic pressure of the medium. This raises the interesting question as to the relation between the osmotic and the hydrostatic pressure effects on proteins (48). Cytochrome oxidase has the characteristics of an osmotically active system, the exterior of the protein acting as an semipermeable membrane for the interior space. Experiments with hydrostatic and osmotic pressure indicate that there is a channel in the protein through which water has to enter or exit during the catalytic cycle. The idea to open and close channels with osmolytes was first used by Zimmerberg and Parsegian (49) for voltage dependent ionic channels. The activity of cytochrome oxidase may be measured as a function of the glycerol concentration and the activation volume obtained in H_2O is smaller than in D_2O (25 vs. 56 mL) (50). It is clear that these results may also be interpreted in terms of the preferential exclusion of glycerol from the surface of the protein as discussed in section 3.3.

In the future we may expect a number of interesting results from spectroscopic studies on the conformation of proteins. Raman as wel as infrared spectroscopy are particularly promising in this respect. The diamond anvil cell which needs a very small amount of biological substance offers great expectations (51,52). Le Tilly and coworkers have shown recently how infrared spectroscopy may be used to study small but delicate influences of the medium on the hydrogen bonding of helices in myoglobin (53). The recent developments of the NMR technique also offer many new opportunities although a larger amount of substance is needed (54).

5. Protein-protein interactions

Protein-protein interactions play a fundamental role in many biological processes and high pressure studies have contributed substantially to the understanding of the factors that control these interactions (55).

Several authors studied various aspects of these processes. Of particular interest is the fact that most of these studies have been performed with optical techniques such as absorption or fluorescence spectroscopy. Examples of these studies are discussed in detail in the contributions by Silva and Weber in this volume. However, many enzymes do not show any optical changes so that a sampling of the activity without a release of the pressure would be welcome. This approach has been used for the study of phosphofructokinase from *E. coli* with a newly developed instrument (56). This tetrameric enzyme dissociates reversibly into inactive dimers and monomers above 80 MPa. In the presence of D-fructose-6-phosphate, a ligand which bridges the intersubunit interface at the active site, a massive stabilization against pressure dissociation is observed (57).

The interaction between cytochrome c and cytochrome b_5 has been studied with high-pressure techniques and site-directed mutagenesis to probe the contributions from charged amino acid side chains (58). More than a dozen of mutants have been made of cytochrome b_5 and the reaction volume determined. The effect of the removal of a total of four salt-bridges by mutations of charged amino acids into their uncharged amides, may be seen from Table VI. It is interesting to note that the reaction volume changes by 82 ml/mol for the removal of four charged residues. This suggests that two third of the total reaction volume is due to electrostriction. This is also in agreement with the proposal that the solvation for the disruption of a salt-bridge amounts to about 30 ml/mol (42). The 40 ml/mol observed for the quadruple mutant may be explained by the interaction of the solvent with non-polar residues that are buried in the interface of the complex. The observation that the ΔG value changes by only 14%, indicates that electrostatics does not contribute significantly to the stability of the complex formation.

Another important aspect of the interactions between proteins has been highlighted by the work of Weber and his coworkers (59). Using fluorescence techniques it was shown that in protein dimers the rate of exchange of subunits under pressure approaches the rate of dissociation of the dimers. In tetramers, the situation is different. The rate of subunit exchange is between 5-10 times slower than the dissociation rate. The dissociation of the dimers exhibits the characteristic of the classical stochastic chemical equilibria. The dissociation of tetramers, like those of the more complex protein aggregates such as the

TABLE VI. Effect of removal of salt-bridges on the ΔV (mL/mol) and the ΔG (kcal/mol) for the interaction between cytochrome b_5 and cytochrome c (58).

ΔSB	Protein	ΔG	ΔV
0	Wild type	8.69	122
1	Wild type DME	8.12	80
2	E48Q, DME	8.00	60
3	E44Q, E48Q, DME	7.60	52
4	E44Q, E48Q, D60N, DME	7.50	40

ΔSB: number of salt-bridges removed, DME: dimethylester heme, E44Q: glutamic acid 44 replaced by glutamine, D60N: aspartic acid 60 replaced by asparagine.

extracellular hemoglobin of *G. paulistus* (60), is described by deterministic mechanical equilibria of macroscopic bodies. In extreme cases the pressure-induced dissociation of these systems is entirely concentration-independent (60). The effect of the solvent (glycerol/water) on the pressure dependence of the subunit interaction in this hemoglobin has already been mentioned in section 3.3.

Studies on the monomer-dimer equilibrium of Arc repressor, have shown that the pressure-induced monomer is much more compact than a monomer formed by urea-induced dissociation (61). The structure of the pressure-induced monomer has many features of the molten globule state. This state, intermediate between the folded tertiary structure and the unfolded denatured state, has retained a high degree of secondary structure while fluorescence studies show that several hydrophobic side chains are exposed to the solvent. It would of considerable interest to see whether the induction of the molten globule state by high pressure is a general phenomenon in proteins.

The effect of pressure on the dissociation of enzyme subunits may also account for the inactivation of membrane bound proteins. Heremans and Wuytack (62) have shown that the Ca^{2+}-ATPase from sarcoplasmic reticulum is inactivated by pressure and that the pressure at which this effect occurs is strongly dependent on the temperature, suggesting a role for the lipids in the inactivation of the enzyme. Verjovski-Almeida et al (63) have shown that the inactivation may be explained by the pressure-induced dissociation of the subunits of the enzyme.

The interaction of proteins with non-protein materials, such as surfaces, may be another tool to understand the behaviour of proteins. The strong enhancement of the Raman Scattering intensity of molecules when adsorbed on metal colloids such as silver, gold or copper, may become an important tool to study the interaction of proteins with metal surfaces (64). The effect is known as Surface Enhanced (Resonance) Raman scattering (SE(R)RS). More generally, it opens new possibilities for the study of the colloidal state of matter.

Studies on the effect of pressure on SERS have been impeded in the past by decreasing intensities under pressure on silver as well as gold sols. We have encountered similar pressure induced intensity reductions in SERRS studies on iron phenanthroline and cytochrome c adsorbed on silver sols prepared with the citrate reduction method. In all cases the reduction in intensity is only partially reversible (65).

However, a pressure induced intensity increase of the SERS spectrum of iron phenanthroline, phtalazine and cetylpyridinium chloride adsorbed on silver colloids prepared by the reduction of silver nitrate with white phosphorus as first obtained by Faraday (66). For iron phenanthroline adsorbed on silver Faraday colloids, a pressure of 100 MPa applied for three hours increases the intensity ca three times. When adsorbed on citrate colloids under similar conditions, the SERRS spectrum is reduced by almost 70%. Pressurization of the colloids in the absence of the adsorbate, has no effect on the intensity of the SERRS spectra. These experiments, though preliminar in nature, demonstrate the feasibility of the technique to study pressure induced effects in proteins adsorbed on metal surfaces.

6. Protein denaturation

The effects of pressure which have been discussed up to now are all reversible. The application of higher pressures may induce irreversible effects which may include complete inactivation of enzymes and denaturation of proteins. In general these effects occur above 300 MPa. However, there are exceptions which may be of particular help in our general understanding of the mechanism of denaturation of proteins. The process of denaturation may be studied by a number of techniques. Recently, some new techniques have been introduced which reveal knew aspects of the denaturation process. Masson and coworkers (67) used the electrophoresis technique to follow pressure induced subunit interactions and ligand binding. Jonas and coworkers developed NMR to study the pressure induced changes in lipids and proteins (54,68). Wong has introduced optical techniques such as Raman (69) and infrared spectroscopy (51,70). These techniques allow the analysis not only of the conformation of the polypeptide backbone but also the interaction of the side chains with the solvent and other macromolecules. It is interesting to note that the pressure at which denaturation takes place as measured by vibrational spectroscopy, corresponds closely with the one found from absorption and fluorescence techniques (71). In view of the higher concentration that is needed, the effect of the aggregation process becomes clearly visible, making the overall process irreversible. The introduction of the diamond anvill cell has revolutionized this field.

The combined use of the diamond anvil cell and vibrational spectroscopy such as Fourier transform infrared, Raman scattering and possibly Fourier transform Raman scattering proves to have distinct advantages for the study of pressure-induced changes in biomolecules as well as in living tissues. Raman scattering has the advantage that experiments can be performed in water and that virtually no sample preparation is needed. The main disadvantage is possible interference of fluorescent impurities in protein samples. Whereas this can be overcome in some cases by photobleaching, in many cases infrared spectroscopy is the only solution. Colored samples may be conveniently studied under resonance conditions. The Raman scattering is then considerable enhanced because of the coupling of the scattering with electronic transitions. The advantages of infrared absorption are also clear. There is no interference from fluorescence and there is no contribution from colored centers in proteins such as in heme proteins. Minor disadvantages are the need to work in D_2O because of the strong absorption of water in the protein absorption region. Sample preparation is needed to find the convenient range of absorption. The advantages of infrared and Raman scattering are, at least in principle, combined in the Fourier transform Raman technique. However, at present there are no pressure studies on biomolecules reported in the literature.

A more detailed study of the differences between pressure- and temperature-induced changes in the denaturation of proteins and the gels that are formed, would be particularly relevant to the application of pressure techniques in the food industry. This may be demonstrated by the studies on the white of an egg. At the beginning of this century, Bridgman showed that the appearance of the pressure-induced coagulum of egg white is quite different from the coagulum induced by temperature. He also observed that the effect of temperature "seems to be such that the ease of (the pressure-induced coagulation) increases at low temperatures, contrary to what one might expect" (72). It is now clear that this is the consequence of the unique behaviour of proteins (2,3). The denaturation temperature of proteins increases with increasing pressure up to about 100 MPa (Figure 2). When the pressure is increased, the temperature of denaturation increases. The pressure at which the protein denatures below room temperature is usually lower than that at room temperature. This phenomenon of re-entrant transitions which results in elliptic phase diagrams, is also observed in liquid crystals (73). Besides its observation in proteins, there is also evidence that it occurs in polysaccharides such as starch (74) and in lipids (75). It has also been observed in the destruction of bacteriophages (76) and bacteria (77).

One possible explanation for the re-entrant phase behaviour of proteins comes from a model which has been proposed to explain the melting-curve maxima that are observed at high pressure in the case of metals, intermetallic compounds and inorganic salts. It is assumed that two species exist in

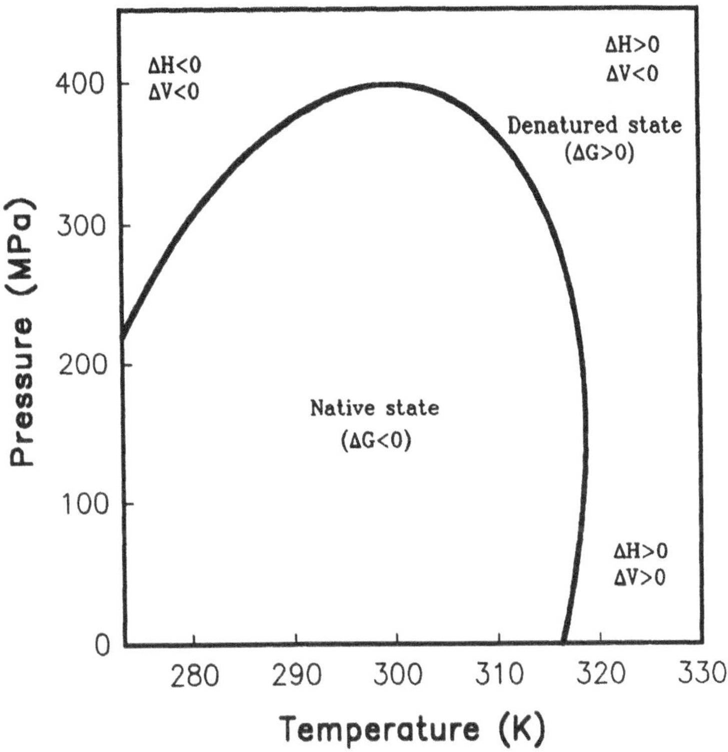

Figure 2. Typical temperature-pressure phase diagram of proteins. At low pressure, pressure stabilizes the protein against heat denaturation. At room temperature pressure-induced denaturation takes place at higher pressures than at low temperature. After (2,3).

equilibrium in the liquid state which may be shifted by the action of temperature and pressure (78). If we transfer this model to protein denaturation, this means that the denatured state of the protein exists in two conformations that can be changed by temperature and pressure.

The characterization with Fourier transform infrared and Raman spectroscopy of pressure-induced changes in food materials, should it make possible to correlate the macroscopic appearance of processed foodstuffs and the microscopic conformation of biological macromolecules present in food. Among other factors to be studied are the stabilizing effects of various conditions such as pH, the presence of sucrose, lipids, etc. These studies are relevant to the effect of the solvent composition on the phase diagram of proteins. Except for the study of the effect of pH on the stability of metmyoglobin (79) and the studies of Carrier and coworkers (80) on the influence of lipids on the pressure induced transitions in polypeptides, this field is still open for fundamental studies. The interaction of proteins with nucleic acids is another important topic which has barely been touched (81).

What is the highest pressure that a protein can resist? This question is not only of fundamental importance but has also practical implications. There are reports in the literature that indicate that some proteins may resist pressures over 1000 MPa while still retaining their biological activity. One of the approaches towards an understanding of this resistance is to study pressure effects on secondary structures in model systems such as Poly(L-lysine). Such studies reveal that the unordered polypeptide undergoes a reversible pressure-induced change to the α-helix structure at about 900 MPa (82). It is of interest to note that lipids may have a stabilizing effect on pressure-induced changes in the structure of these polypeptides (80). The α-helical structure of poly-lysine remains intact up to 1.2 GPa at neutral pH when bound to the lipid DPPG. When free in solution, at pH>11 and 4°C, poly-lysine forms a α-helix which is stable up to 2 GPa. The same compound forms, at pH > 11 and 50°C, a ß-sheet, which transforms into a α-helix at ca. 200 MPa. At neutral pH, the ß-sheet of the peptide, when bound to the lipid DMPA remains intact up to 1.9 GPa. In our laboratory we have observed stabilization effects of lipids and detergents on the conformation of hydrophobic peptides, such as gramicidin, incorporated into nonaqueous phases (83).

While pressure-induced protein denaturation has been a topic of research for many years, it is clear that there is little agreement on the definition of the denatured state. When making extrapolations from studies on model systems to proteins, care should be taken. Vibrational spectroscopy reveals that secondary structures such as α-helix and ß-sheet may be discernable up to very high pressures, but it is also clear that in the denatured state these structures disappear (52,70). Studies on a number of proteins which differ in size and composition in secondary structures are going on in our laboratory. The analysis of the amide I' band may be used to study pressure induced changes up to about 2 GPa. The variation in the bandwidth of the amide I' band also contains useful information on the changes that take place. The denaturation pressures for lysozyme and chymotrypsinogen, as shwon in Figure 3., correspond to those observed with fluorescence techniques (71). The most remarkable observation is that for a few proteins (data not shown) the changes seem to be reversible. However, in most cases either a precipitate (e.g. for lysozyme) or a gel-like structure (e.g. for chymotrypsinogen) is formed. Whereas infrared spectroscopy gives mainly information on the backbone of the protein, the information content in terms of secondary structure is rich. NMR studies, up to say 1 GPa, may prove to be a possible tool that gives more spatial information on amino acid side chains. Such studies may be technically feasible, although not easy, in the future.

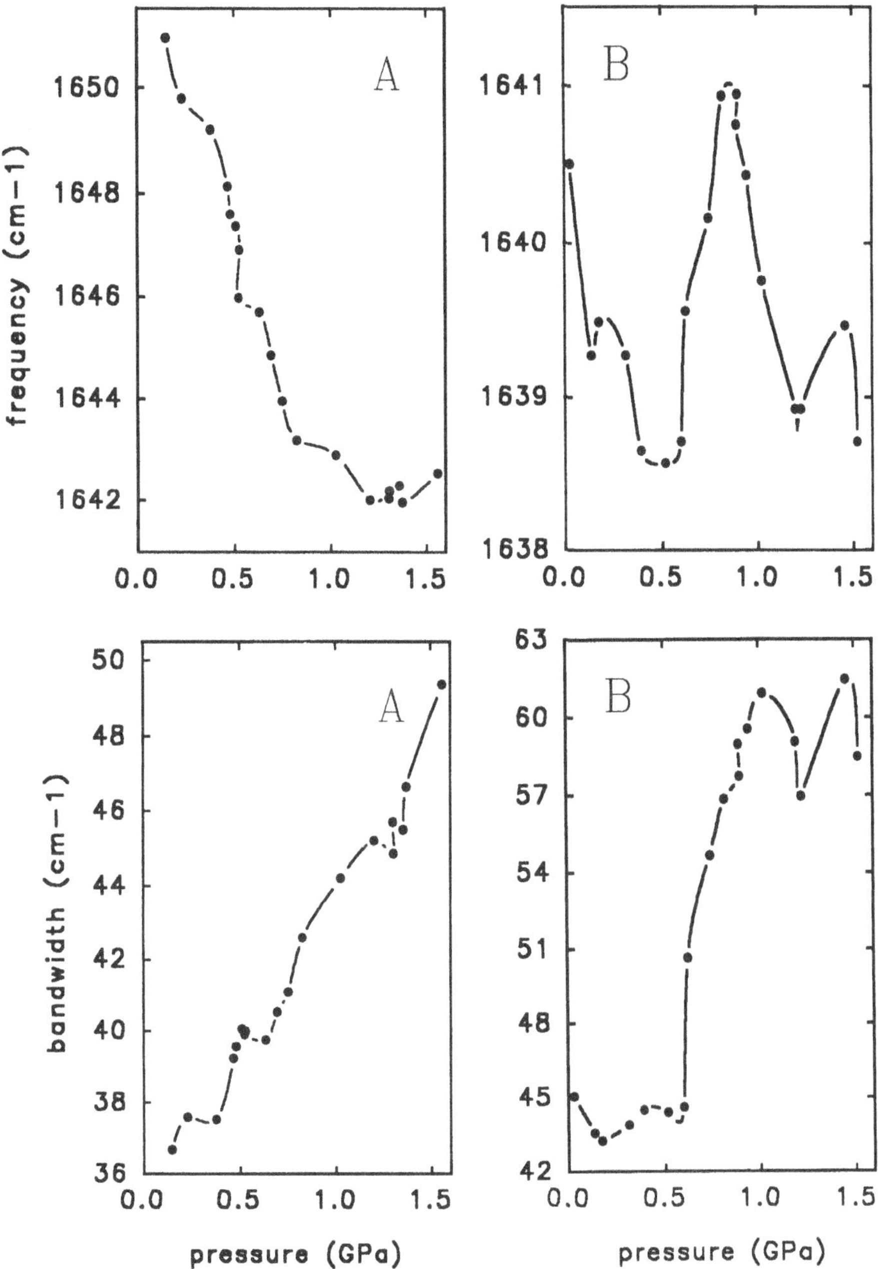

Figure 3. Pressure dependence of the amide I' band frequency maximum (top) and bandwidth (bottom). Data for lysozyme (A) and chymotrypsinogen (B). The pressure-induced denaturation is irreversible in both cases (52).

7. Pressure-induced gel formation

Studies of the effect of pressure on the gel state of matter are not numerous despite its important role in many areas of science. Studies on gels may also reveal important aspects of biological macromolecules in view of the functional similarities (84). From a practical point of view, one may mention the important aspects related to the formation of the gel state as a consequence of high pressure treatment of food materials. However, an extensive review on the structural and mechanical properties of biopolymer gels does not contain a single reference to pressure induced gels (85).

Gels may be formed from proteins or polysaccharides. Basically two types of gels may be discerned. The first is formed starting from random coil polypeptides such as is the case for gelatin and polysaccharides. These systems have the characteristics of conventional polymer solutions. A second type of gel is formed starting from globular or rod-like proteins which, after a partial unfolding, form a network by the aggregation of the globular particles. These gels have properties common with colloidal and emulsion systems.

Early work on the pressure effect on the sol-gel transition in gelatin, polyvinyl alcohol and methyl cellulose concentrated on the role of noncovalent bonding in the process (86). Recent work by Gekko and coworkers concentrated on the thermodynamic aspects of gel formation in gelatin (87), collagen (88), agarose (89) and ovalbumin (90). It is of considerable practical interest that temperature-induced gels from ovalbumin, soy protein and the gels from carrageenan are destabilized by high pressure, whereas gels from agarose and gelatin are stabilized.

Recent work on pressure-set gels of egg white has indicated that they show a higher gel strength than those induced by temperature. On the other hand they are softer than the temperature-induced gels (91). While the molecular details of the gel formation are not known in detail, infrared and Raman spectroscopy studies reveal an increasing contribution from intermolecular ß-sheet in the heat-set gels (92). Work on the pressure-induced gel formation of chymotrypsinogen suggest a similar contribution from ß-sheet (70,93). Unpublished work in our laboratory suggests that similar mechanisms are taking place in other proteins (52).

An important aspect for the food industry is that the natural color and flavour of the pressure induced gels, in contrats to the temperature-induced gels, is kept intact. The difference in mechanical properties of temperature-induced and pressure-induced gels of food proteins (egg white, crude actomysosin, soy protein) has been studied in some detail by Hayashi and coworkers (91,94). In general the pressure-induced gels are glossy and soft, the hardness is increased and the adhesiveness decreased with increasing pressure, they have a larger extensibility and are not fractured by a high stress. In addition, the original color and flavor is preserved. For egg white and egg yolk it was found that vitamins were not destroyed by high pressure treatment.

8. Living systems

The antagonistic effect of pressure on temperature induced phenomena that is characteristic for proteins (section 6) has also been observed in bacteria by Yayanos (1). Diagrams that show the effect of temperature and pressure on the viability of bacteria from the deep sea show a strong resemblance to those observed for the reversible and irreversible denaturation of proteins as shown in Figure 2. It was found that the temperature of maximum tolerance for growth increases with increasing pressure up to about 100 MPa. The antagonistic effect does not continue above 100 MPa.

Studies on the adaptation of living organisms to the depths of the oceans have shown that the terms deep and high pressure begin to apply at depths of about 500 m or less (95). Of particular interest are the studies which concentrate on the characterization of molecules isolated from deep sea organisms (96).

The molecular basis for pressure adaptation remains unclear. The same is true for the survival of organisms in other extreme conditions such as pH, salt etc (40). A few studies suggest however that organisms can be protected agains pressure damage by solvent such as deuterium oxide and dimethylsulfoxide as well by heat shock (97). If the mechanism that is proposed (section 3.3) proves to be correct - the solvent protection of the proteins against denaturation - then this may be a general survival mechanism in nature. Along the same lines, the extreme stability of bacterial spores against pressure is of interest (98). The extreme low activity of water may account for this stability. It is clear that more experimental work, specifically on the effect of solvent conditions on the pressure sensitivity of proteins and organisms, is needed before this conclusion can be accepted as being of general applicability.

In recent years vibrational spectroscopy has been shown to be a useful tool in the study of cells and cellular components (99,100,101). Molecular details on the resistance of cells and organisms against pressure processing may therefore be expected from spectroscopic studies. This also applies to the destabilizing effect of pH and the stabilizing effect of polyalcohols against high pressure treatment of microorganisms. Then, we may be able to figure out why life does not like to flourish under conditions were high temperature and high pressure are superimposed.

9. Industrial applications

Applications of high pressure technology for the processing of food has a long history. In the United States, Hite and coworkers (102) have made extensive investigations on the possibility to preserve fruit and vegetables by killing microorganisms with high pressure treatment of about 500 Mpa for 30 minutes. The experiments by Bridgman on egg white showed that pressure treatment resulted in coagulation of the white with an appearance like that of a boiled egg (72). This suggested that microorganisms are killed by the action of pressure on the proteins. The observation that protein denaturation profiles (2,3) may be correlated with the survival of bacteria (1) strongly suggest that this is the primary mechanism.

On the other hand there remains the distinct possibility that pressure may act upon the colloidal constituents of biological material. This possibility has been suggested a long time ago (103) but has not been further explored as extensively as the protein denaturation hypothesis. The collodial hypothesis was put forward after the observation that the pressure required to kill bacteria is inversely proportional to their complexity. Such studies have been performed by many groups but in particular in France by Macheboeuf and Basset (104). Studies on proteins, enzymes, immunoglobulins and the survival of bacteria, spores and viruses were reported.

In Japan the tradition in high pressure research on biological systems started with the paper of Kiyama and Yanagimoto (105) followed by the fundamental work of Suzuki (2). It took about another 30 years before the trend setting paper of Hayashi appeared (106). In 1989 a Japanese R&D association was founded to stimulate cooperation between industry, universities and government. In 1990 several products were put on the market: strawberry jam, grape-fruit juice, etc. The society organized meetings to discuss progress in the field. The papers were presented and published in Japanese. The recent developments, including the developments in Japan, have been discussed at the First European meeting on High Pressure and Biotechnology (107). The first research project has now been set up on a european scale with the participation of twelve research teams from industry as well from universites to study basic aspects of high pressure food processing. Future developments will decide whether *Bridgmanization*, as the process may be called in honour of the man who made the field of high pressure since the beginning of this century, will be a contribution as vital to the quality of food as the process of *Pasteurization* is now.

However, the use of pressure as an alternative to temperature treatment, has brought about the need for fundamental studies on the pressure-temperature behaviour of macromolecular food constituents such as proteins, lipids and polysaccharides (106,108,109). Pressure treatment of food materials, like temperature treatment, involves the inactivation of microorganisms and enzymes. In most cases this is not only a thermodynamic but a kinetic problem as well. In addition, the macroscopic appearance of foodstuffs is of primary importance. In most cases information is lacking as to the correlation between macroscopic appearance of processed foodstuffs and the molecular conformation of the constituents as a function of temperature and pressure. The mechanisms of protein gelation and the sol-gel behaviour of polysaccharides is far from being understood. Of considerable attraction is the fact the pressure induced gels of foodstuffs preserve their natural color and flavour compared to temperature induced gels.

Acknowledgment

The research performed in our laboratory has been supported over the years by the Belgian National Science Foundation and the Research Fund of our university.

References

1. Yayanos, A.A. (1986) Proc. Nat. Acad. Sci. U.S.A., **83**, 9543-9546.
2. Suzuki, K. (1960) Rev. Phys. Chem. Japan., **29**, 91-98.
3. Hawley, S.A. (1971) Biochemistry, **10**, 2436-2442.
4. Goldbeck, A., Lechner, M.D., Witz, J., Nordmeier, E. and Ibel, K. (1991) Eur. J. Biophys., **20**, 151-156.
5. Morild, E. (1981) Adv. Protein Chem., **34**, 93-166.
6. Hamann, S.D. (1988) in le Noble, W.J., (Ed.), Organic High Pressure Chemistry, Elsevier, Amsterdam, pp. 50-99.
7. Taniguchi, Y. and Suzuki, K. (1983) J. Phys. Chem., **87**, 5185-5193.
8. Cooper, A. (1976) Proc. Nat. Acad. Sci. U.S.A., **73**, 2740-2741.
9. Sarvazyan, A.P. (1991) Annu. Rev. Biophys. Biophys. Chem., **20**, 321-342.
10. Heremans, K. (1988) in le Noble, W.J., (Ed.), Organic High Pressure Chemistry, Elsevier, Amsterdam, pp. 380-425.
11. Kaminsky, S.M. and Richards, F.M. (1992) Protein Science, **1**, 22-30.
12. Gekko, K. and Hasegawa, Y. (1986) Biochemistry, **25**, 6563-6571.
13. Kharakoz, D.P. (1991) J. Phys. Chem., **95**, 5634-5642.
14. Tarasiuk, A., Grossmann, Y. and Kendig, J.J. (1991) Br. J. Anesthesia., **66**, 88-96.
15. Moss, G.W.J., Lieb, W.R. and Franks, N.P. (1991) Biophys. J., **60**, 1309-1314.
16. Chiou, J.S., Tatara, T., Sawamura, S., Kaminoh, Y., Kamaya, H., Shibata, A. and Ueda, I. (1992) Biochim. Biophys. Acta., **1119**, 211-217.
17. Winter, R., Christmann, M.-H., Böttner, M., Thiyagarajan, P. and Heenan, R.K. (1991) Ber. Bunseng. Phys. Chem., **95**, 811-820.
18. Hasinoff, B. B. (1974) Biochemistry, **13**, 3111-3117.
19. Traylor, T.G., Luo, J., Simon, J.A. and Ford, P.C. (1992) J. Am. Chem. Soc., **114**, 4340-4345.
20. Taube, D.J., Projahn, H.-D., van Eldik, R., Magde, D. and Taylor, T.G. (1990) J. Am. Chem. Soc., **112**, 6880-6886.
21. Projahn, H.-D. and van Eldik, R. (1991) Inorg. Chem., **30**, 3288-3293.
22. Heremans, K. and Bormans, M. (1986) Physica, **139 & 140B**, 870-874.
23. Bormans, M., Jans, P., Selleslag I. and Heremans, K. Unpublished data.

24. Westrick, J.A., Peters, K.S., Ropp, J.D. and Sligar, S.G. (1990) Biochemistry, **29**, 6741-6746.
25. Adachi, S., Sunohara, N., Ishimori, K. and Morishima, I. (1992) J. Biol. Chem., **267**, 12614-12621.
26. Adachi, S. and Morishima, I. (1989) J. Biol. Chem., **264**, 18896-18901.
27. Adachi, S., Sunohara, N., Ishimori, K. and Morishima, I. (1992) J. Biol. Chem., **267**, 12614-12621.
28. Frauenfelder, H., Alberding, N.A., Anasari, A., Braunstein, D., Cowen, B.R., Hong, M.K., Iben, I.E.T., Johnson, J.B., Luck, S., Marden, M.C., Mourant, J.R., Ormos, P., Reinisch, L., Scholl, R., Schulte, A., Shyamsunder, E., Sorensen, L.B., Steinbach, P.J., Xie, A., Young, R.D. and Yue, K.T. (1990) J. Phys. Chem., **94**, 1024-1037.
29. Unno, M., Ishimori, K. and Morishima, I. (1990) Biochemistry, **29**, 10199-10205.
30. Unno, M., Ishimori, K., Morishima, I., Nakayama T. and Hamanoue, K. (1991) Biochemistry, **30**, 10679-10685.
31. Traylor, T.G., Magde, D., Taube, D.J., Jongeward, K.A., Bandyopadhyay, D., Luo, J. and Walda, K.N. (1992) J. Am. Chem. Soc., **114**, 417-429.
32. Peters, K.S., Watson, T. and Marr, K. (1991) Annu. Rev. Biophys. Biophys. Chem., **20**, 343-362.
33. Westrick, J.A. and Peters, K.S. (1990) Biophys. Chem., **37**, 73-79.
34. Peters, K.S., Watson, T. and Logan, T. (1992) J. Am. Chem. Soc., **114**, 4276-4278.
35. Balny, C. and Travers, F. (1989) Biophys. Chem., **33**, 237-244.
36. Balny, C., Masson, P. and Travers, F. (1989) High Pressure Research., **2**, 1-28.
37. Li, T.M., Hook, J.W., Drickamer, H.G. and Weber, G. (1976) Biochemistry, **15**, 3205-3211.
38. Bonafe, C.F.S., Villas-Boas, M., Suarez, M.C. and Silva, J.L. (1991) J. Biol. Chem., **266**, 13210-13216.
39. a) Gekko, K. and Timasheff, S.N. (1981) Biochemistry, **40**, 4667-4676. b) Gekko, K. and Timasheff, S.N. (1981) Biochemistry, **40**, 4677-4686.
40. Jaenicke, R. (1991) Eur. J. Biochem., **202**, 715-728.
41. Heremans, L. and Heremans, K. (1989) J. Mol. Structure, **214**, 305-314.
42. Heremans, L. and Heremans, K. (1989) Biochem. Biophys. Acta, **999**, 192-197.
43. Vanhecke, F. and Heremans, K. Unpublished data.
44. Heremans, K. (1987) in van Eldik, R. and Jonas, J. (Eds.) High Pressure Chemistry and Biochemistry, Reidel, Dordrecht, pp. 421-445.
45. Di Primo, C., Hui Bon Hoa, G., Douzou P. and Sligar, S. (1990) Eur. J. Biochem., **193**, 383-386.
46. Heremans, K. Bormans, M., Snauwaert, J. and Vandersypen, H. (1982) Faraday Disc. Chem. Soc., **74**, 343-348.

47. Heiber-Langer, I., Hooper, A.B. and Balny, C. (1992) Biophys. Chem., **43**, 265-277.
48. Kornblatt, J.A. and Hui Bon Hoa, G. (1990) Biochemistry, **29**, 9370-9376.
49. Zimmerberg, J. and Parsegian, V.A. (1986) Nature, **323**, 36-39.
50. Kornblatt, J.A. and Kornblatt, M.J. (1992) Biochim. Biophys. Acta, **1099**, 182-184.
51. Wong, P.T.T. (1991) Can. J. Chem., **69**, 1699-1704.
52. Goossens, K., Smeller, L. and Heremans, K. Unpublished data.
53. Le Tilly, V., Sire, O., Alpert, B. and Wong, P.T.T. (1992) Eur. J. Biochem., **205**, 1061-1065.
54. Jonas, J. (1990) Ber. Bunseng. Phys. Chem., **94**, 307-315.
55. Weber, G. (1992) Protein Interactions, Chapman and Hall, New York.
56. Hui Bon Hoa, G., Hamel, G., Else, A., Weill, G. and Hervé, G. (1990) Anal. Biochem., **187**, 258-261.
57. Deville-Bonne, D. and Else, A.J. (1991) Eur. J. Biochem., **200**, 747-750.
58. Rodgers, K.K. and Sligar, S.G. (1991) J. Mol. Biol., **221**, 1453-1460.
59. Erijman, L. and Weber, G. (1991) Biochemistry, **30**, 1595-1599.
60. Silva, J.L., Villas-Boas, M., Bonafe, C.F.S. and Meirelles, N.C. (1989) J. Biol. Chem., **264**, 15863-15868.
61. Silva, J.L., Silveira, C.F., Corriera Jr, A. and Pontes, L. (1992) J. Mol. Biol., **223**, 545-555.
62. Heremans, K. and Wuytack F. (1980) Fed. Europ. Biochem Soc. Letters, **117**, 161-163.
63. Verjovski-Almeida, S., Kurtenbach, E., Amorim, A.F. and Weber, G. (1986) J. Biol. Chem., **261**, 9872-9878.
64. Cotton, Th.M., Kim, J-H. and Chumanov, G.D. (1991) J. Raman Spectrosc., **22**, 729-742.
65. Vanhecke, F. and Heremans, K. (1990) High Pressure Research, **5**, 746-748.
66. Janssen, E., Smeller, L. and Heremans, K. (1992) in ref. 107, pp. 431-433.
67. Masson, P., Arciero, D.M., Hooper, A.B. and Balny, C. (1990) Electrophoresis, **11**, 128-133.
68. Jonas, J. (1992) in ref. 107, pp. 123-127.
69. Heremans, K. and Wong, P.T.T. (1985) Chem. Phys. Lett., **118**, 101-104.
70. Wong, P.T.T. and Heremans, K. (1988) Biochem. Biophys. Acta., **956**, 1-9.
71. a) Li, T.M., Hook, J.W., Drickamer, H.G. and Weber, G. (1976) Biochemistry, **15**, 5571-5580. b) Chryssomallis, G.S., Torgerson, P.M., Drickamer, H.G. and Weber, G. (1981) Biochemistry, **20**, 3955-3959.
72. Bridgman, P.W. (1914) J. Biol. Chem., **19**, 511-512.
73. Klug, D.D. and Whalley, E. (1979) J. Chem. Phys., **71**, 1874-1877.
74. Thevelein, J.M., Van Assche, J.A., Heremans, K. and Gerlsma, S.Y. (1981) Carbohydrate Research, **93**, 304-307.
75. Winter, R. and Pilgrim, W.-C. (1989) Ber. Bunseng. Phys. Chem., **93**, 708-717.
76. Gross, P. and Ludwig, H. (1992) in ref. 107, pp. 57-59.

77. Sonoike, K., Setoyama, T., Kuma, Y. and Kobayashi, S. (1992) in ref. 107, pp. 297-301.
78. Rapoport, E. (1967) J. Chem. Phys., **46**, 2891-2895.
79. Zipp, A. and Kauzmann, W. (1973) Biochemistry, **12**, 4217-4228.
80. Carrier, D., Mantsch, H.H. and Wong, P.T.T. (1990) Biochemistry, **29**, 254-258.
81. Macgregor, R.B., Jr (1992) Biochim. Biophys. Acta, **1129**, 303-308.
82. Carrier, D., Mantsch, H.H. and Wong, P.T.T. (1990) Biopolymers, **29**, 837-844.
83. Smeller, L., Goossens, K. and Heremans, K. Unpublished data.
84. Douzou, P. (1987) Proc. Nat. Acad. Sci. U.S.A., **84**, 6741-6744.
85. Clark, A.H. and Ross-Murphy, S.B. (1987) Adv. Polymer Science, **83**, 57-192.
86. Suzuki, K., Taniguchi, Y. and Enomoto, T. (1972) Bull. Chem. Soc. Japan, **45**, 336-338.
87. Gekko, K. and Fukamizu, M. (1991) Int. J. Biol. Macromol., **13**, 295-300.
88. Gekko, K. and Koga, S. (1983) Agric. Biol. Chem., **47**, 1027-1033.
89. Gekko, K. and Fukamizu, M. (1991) Agric. Biol. Chem., **55**, 2427-2428.
90. Gekko, K. (1992) in ref. 107, pp. 105-113.
91. Okamoto, M., Kawamura, Y. and Hayashi, R. (1990) Agric. Biol. Chem., **54**, 183-189.
92. Clark, A.H., Saunderson, D.H.P. and Suggett, A. (1981) Int. J. Peptide Protein Res., **17**, 353-364.
93. Ismail, A.A., Mantsch, H.H. and Wong, P.T.T. (1992) Biochim. Biophys. Acta, **1121**, 183-188.
94. Hayashi, R., Kawamura, Y., Nakasa, T. and Okinaka, O. (1989) Agric. Biol. Chem., **53**, 2935-2939.
95. Somero, G.N. (1992) Ann. Rev. of Physiol., **54**, 557-577.
96. Gaill, F., Wiedemann, H., Mann, K., Kühn, K., Timpl, R. and Engel, J. (1991) J. Mol. Biol., **221**, 209-223.
97. Komatsu, Y., Obuchi, K., Iwahashi, H., Kaul, S.C., Ishimura, M., Fahy, G.M. and Rall, W.F. (1991) Biochim. Biophys. Research. Comm., **174**, 1141-1147.
98. Butz, P., Ries, J., Traugott, U., Weber, H. and Ludwig, H. (1990) Pharm. Ind., **52**, 487-491.
99. Rigas, B., Morgello, S., Goldman, I.S. and Wong, P.T.T. (1990) Proc. Nat. Acad. Sci. USA., **87**, 8140-8144.
100. Wong, P.T.T., Wong, R.K., Caputo, T.A., Godwin, T.A. and Rigas, B. (1991) Proc. Nat. Acad. Sci. USA, **88**, 10988-10992.
101. Wong, P.T.T. (1992) in ref. 107, pp. 33-36.
102. Hite, B.H., Giddings, N.J. and Weakly, C.E. (1914) West. Va. Univ. Agr. Expt. Sta. Bull., **146**, 1-67.
103. Wilson, R. and Poulter, T.C. (1929) in Bridgman, P.W. (1946) Rev. Mod Phys., **18**, 1-93.

104. Macheboeuf, M.A. and Basset, J., references in Johnson, F.H., Eyring, H. and Stover, B.J. (1954) The Theory of Rate Processes in Biology and Medicine. New York, Wiley.
105. Kiyama, R. and Yanagimoto, T. (1951) Rev. Phys. Chem. Japan, 21, 41-43.
106. Hayashi, R. (1989) In Spiess, W.E.L. and Schubert, H. (Eds.) Engineering and Food, Vol 2, Elsevier, London, pp. 815-826.
107. Balny, C., Hayashi, R., Heremans, K. and Masson, P. (Eds.) (1992) High Pressure and Biotechnology, INSERM/Libbey, France.
108. Farr, D. (1990) Trends in food science and technology, July, 14-16.
109. Hoover, D.G., Metrick, C., Papineau, A.M., Farkas, D.F. and Knorr, D. (1989) Food Technology, March, 99-107.

PRESSURE DISSOCIATION OF THE SMALLER OLIGOMERS: DIMERS AND TETRAMERS

GREGORIO WEBER
School of Chemical Sciences
University of Illinois at Urbana-Champaign
Urbana, Illinois 61801 USA

ABSTRACT: The dissociating effects of pressure upon the oligomeric proteins have been qualitatively observed in large and complex aggregates and also in homogeneous dimers and tetramers, but only the latter permit, on account of their simple stoichiometry, a determination of the thermodynamic parameters. Dimers behave under pressure as simple homogeneous systems with a well defined association volume and unique free energy of association while tetramers appear to be heterogeneous populations as regards these two properties. Interconversion of the members of the tetramer population is temperature-dependent with an energy of activation of 20 kcal mol^{-1}. Tetramers and most dimers show a loss in the free energy of association, and time-dependent recovery when the subunits are separated. This is attributed to a conformational drift, that occurs when intersubunit contacts are replaced by water-subunit contacts. The conformational drift on dissociation results in a change in equilibrium constant with the degree of dissociation in certain dimers, and is responsible for the long-observed cold inactivation of enzymes.

IMPORTANCE OF PRESSURE PERTURBATION IN STUDIES OF OLIGOMERS

The specific association of independent polypeptide chains is one of the most important determinants of the functional properties of organisms, but our knowledge of the energetics and dynamics of protein associations is still very primitive. In large measure this has been due to the lack of appropriate methods of investigation of the energetics of protein association. For this purpose one requires methods of perturbation of the association which would not, by themselves, significantly alter the properties of the separated chains. Until recently we did not possess any such methods: The component chains of the oligomeric proteins

R. Winter and J. Jonas (eds.), High Pressure Chemistry, Biochemistry and Materials Science, 471–487.

could be separated only by changes in pH, large increases in ionic strength or addition of detergents, but these agents are known to produce changes in the conformations of the polypeptide chains that cannot be distinguished from those that result from the separation of the chains. This situation has been completely changed by the observation that many indefinite protein aggregates showed dissociation after a moderate rise in hydrostatic pressure, in some cases as small as the one or two hundred bars that produce, in the ultracentrifuge, the change in the sedimentation velocity with the rotor speed, a phenomenon of common observation with oligomeric proteins. Moreover, a series of observations carried out during the decade of the seventies, have indicated that hydrostatic pressures of 5 kilobars and higher are necessary to produce gross changes in the conformations of proteins made up of a single peptide chain by a process designated as "pressure denaturation" [1].

The disaggregating influence of hydrostatic pressure became apparent in the course of studies of large, indefinite aggregates like tubulin, actin, hemoglobin S or the capsid of tobacco-mosaic virus, all cases that because of the unknown, and variable, stoichiometry do not lend themselves to the thermodynamic or kinetic analysis that is feasible for the small, limited aggregates of identical particles. In what follows I shall address myself, primarily, to the observations of homogeneous dimers, trimers and tetramers, which we have investigated in our laboratory in the last ten years.

METHODS OF INVESTIGATION

The methods of investigation of the state of aggregation of oligomers fall naturally into two categories: On one hand there are direct methods, that permit the derivation of the degree of dissociation by an estimation of the average size (molecular weight or molecular volume) of the particles. On the other hand there are indirect methods, that rely on some detectable difference, usually spectroscopic but potentially of many other kinds as well, between the intact aggregate and the constitutive particles.

Among the direct methods, which should be the most reliable, some are rendered ineffective because they require protein concentrations much larger than those at which the concentration-dependent dissociation of dimers and tetramers is of possible observation. This is the case of ultracentrifugation in which the effects of hydrostatic pressure on the dissociation is only recognized as a perturbation of the sedimentation velocity of the aggregate, rather than as an observable equilibrium between this and the constitutive subunits. Light scattering is limited to those cases in which a large particle splits into a number of small ones, as happens with the multimers hemocyanin, erythrocruorin or the virus capsids. Fluorescence polarization of covalent conjugates with a fluorophore of appropriate lifetime has proven invaluable for the observations of the dissociation of aggregates of 2 to 4 particles, because in these cases the change in average particle volume with dissociation falls in the range in which dependence of the fluorescence polarization on the volume is largest [2].

Electrophoresis under hydrostatic pressure was introduced by Hawley [3], who used it to demonstrate the separation of seemingly native and denatured chymotrypsinogen at low temperature. Paladini *et al.*[4] employed acrylamide gel electrophoresis to demonstrate the dissociation of tryptophan synthase under a pressure of 1 kbar, and more recently an improved electrophoresis apparatus of Paladini *et al.*[5] has been shown capable of considerably higher resolution, by taking advantage of the low freezing point of water under pressure, which permits operation down to temperatures as low as -20 ^{o}C. It seems certain that this method may greatly enlarge the possibilities of demonstration of conformationally different dissociated species, as recently observed by Erijman *et al.* in a study of *Rhodobacter* Rubisco [6].

Indirect methods of measurement of aggregation depend always on assumptions of the relations of particular properties to the degree of dissociation of the system, and are therefore not always the most reliable, but are usually the most sensitive. Thus, dissociation of several dimers and tetramers, are accompanied by shifts in the average spectral displacement of the fluorescence of over 1,000 cm^{-1}. As it is possible to measure the center of mass of the fluorescence with a precision of about 20 cm^{-1} one could in principle detect differences in dissociation of about 2%, of which direct methods are certainly not capable.

A specific advantage of the spectroscopic methods is that they often permit strong conclusions about the microscopic changes that accompany the dissociation. Thus, the usual red-shift of the tryptophan fluorescence on dissociation is indicative of a change in polarity of the tryptophan environment, and a change as large as 1,000 cm^{-1} can hardly be due to anything else than the access of water to the vicinity of the tryptophan. A more detailed description of the particular microscopic events accompanying protein compression is promised by the application of nuclear magnetic resonance under pressure, pioneered by Jonas and collaborators [7].

THE THERMODYNAMIC PARAMETERS OF DISSOCIATION

Both direct and indirect methods require for their thermodynamic interpretation the conversion of the quantities under observation into corresponding degrees of dissociation of the system. In a plot of the degree of dissociation α against the applied pressure p, plateau values are commonly observed at the lowest and highest pressures. They are interpreted as corresponding to $\alpha \to 0$ and $\alpha \to 1$ respectively, and by an appropriate interpolation one can determine the degrees of dissociation corresponding to intermediate pressures. Ideally one should be able to obtain from the plot of α against p the dissociation constant K(p) as a function of pressure, and therefore, by extrapolation to $p \ll 1$, the free energy of association of the aggregate at atmospheric pressure, $\Delta G(0)$. From the slope of the plot of ln[K(p)] against p the standard change in volume ΔV upon association of the subunits to form the aggregate may be derived. Although the latter quantity is expected to vary with pressure because of the difference in compressibility of the aggregate and subunits, we shall consider ΔV as constant, and pressure independent, on account of the notorious incompressibility of proteins [8].

The general relation of the dissociation constant under pressure K(p) to the applied pressure is

$$K(p)=K(0)\ \exp(p\Delta V/RT) \qquad (1)$$

where K(0) refers to the dissociation constant under atmospheric pressure. The use of the *dissociation* constant and the change in volume with *association* permits us to dispense with the confusing minus sign in the exponential, which would appear if K and ΔV referred *both* to either association or dissociation. In principle the thermodynamic equilibrium of an aggregate particle and its n identical subunits is dependent upon the n-1 power of the concentration C, according to the relation

$$K(0)\exp(p\Delta V/RT)= n^n\alpha^n C^{n-1}/(1-\alpha) \qquad (2)$$

and a plot of the pressure against $\ln[\alpha^n/(1-\alpha)]$ has slope $\Delta V/RT$ and intercept $K(0)/(n^n C^{n-1})$. Because of the method used in its determination, ΔV computed from the slope, under conditions of constant concentration and variable pressure, will be called here ΔV_p. Alternatively we can determine ΔV under conditions of constant dissociation and variable protein concentration: If two solutions of protein of concentrations C_1 and C_2 are subjected to pressures p_1 and p_2 that respectively produce the same dissociation, equation (2) indicates that

$$\Delta V_C = RT\ \ln[(C_2/C_1)^{n-1}]/(p_2-p_1) \qquad (3)$$

where the subscript C in ΔV_C indicates the procedure, change in the pressures of equal dissociation at different concentrations, to distinguish this from ΔV_P, obtained by means of equation (2). Evidently $\Delta V_C=\Delta V_P$ if the thermodynamic relation (1) results from a dynamic equilibrium between aggregate and subunits. A number of dimer proteins studied, and the allophycocyanine trimer satisfy this condition within the experimental error (app. 10%) of the determinations of ΔV. On the other hand a very large discrepancy is observed in tetramers [9]: ΔV_C is three to four times larger than ΔV_P or equivalently the displacement p_2-p_1 is smaller by the same factor than the displacement expected from ΔV_P. An even more glaring discrepancy occurs with multimer particles like hemocyanin or erythrocruorin, in which a concentration dependence is barely discernible or even altogether absent, and this is also true for the capsids of Brome mosaic virus and R17 phage [10]. The only simple explanation that we were able to formulate to explain the discrepancies of ΔV_P and ΔV_C, is that we are dealing with a population of particles of different free energies of aggregation on account of intrinsic conformational differences, or differences in the total volume of association, or both [9]. At intermediate degrees of dissociation the applied pressure selects for dissociation those multimeric aggregates with free energies of association $\Delta G_i < p_i \Delta V_i$ while the remaining aggregates are not split at all at that pressure. We expect that the sharpness of the selection by pressure is dependent upon the rate of interconversion of the members of the population, and one would not have much trouble in visualizing the difficulty, perhaps the impossibility, for a particle made of 180 subunits of changing its free energy of association, determined by chance at the moment of its aggregation, if it requires collective motions of many subunits for any further significant change. More difficult to explain is the observation of large differences in ΔV_P and ΔV_C in tetramers like lactate dehydrogenase, glyceraldehyde phosphate dehydrogenase or glycogen phosphorylase.

ENERGY TRANSFER METHODS OF STUDY OF THE EQUILIBRIUM

It has been possible to obtain more direct information about the character of the equilibrium between subunits and tetramer employing a method that follows the exchange of subunits among the aggregates by permanently attaching covalent fluorescent labels to some of the subunits. The method can be applied in two different ways: energy transfer between identical fluorophores (homotransfer) or different fluorophores (heterotransfer). In the heterotransfer method two identical protein solutions are labeled with two different fluorophores, such that the absorption spectrum of one, the acceptor overlaps as completely as possible the fluorescence spectrum of the other, the donor. After mixing

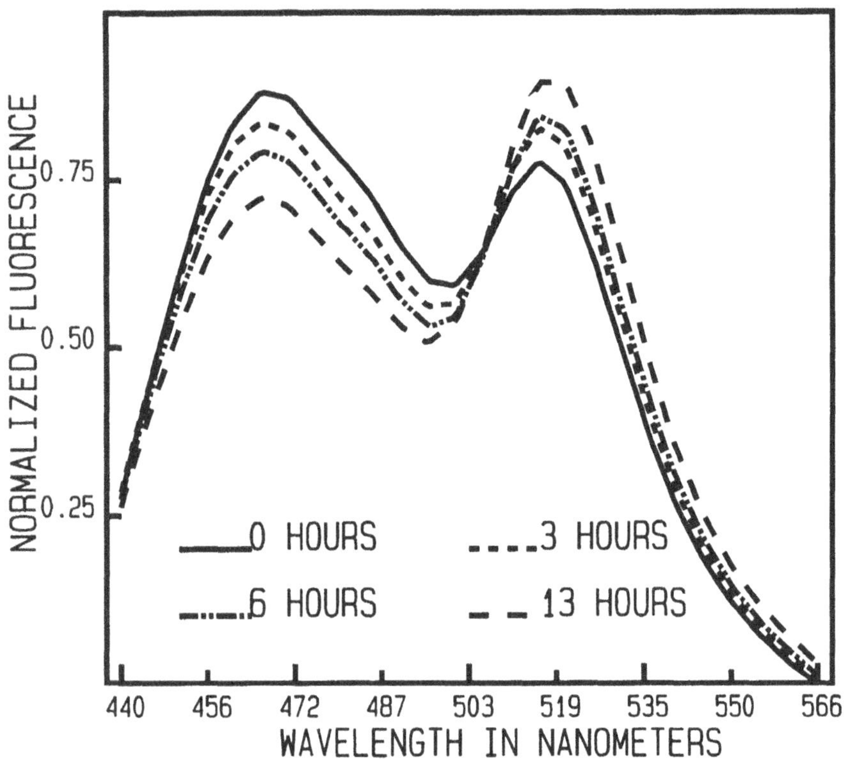

Figure 1. Increase in electronic energy transfer from coumarin labeled subunits (emission maximum at 465 nm) to fluorescein labeled subunits (maximum at 518 nm), at 0° C, on formation of mixed-labeled tetramers of glyceraldehyde phosphate dehydrogenase. From reference [13]

---

the solutions the fluorescence is excited with a wavelength preferentially absorbed by the donor and the progressive time-dependent changes in the fluorescence spectrum are followed. Immediately after mixing the spectrum of the donor predominates over that of acceptor, but as subunits are progressively exchanged and more tetramers contain subunits labeled with both donor and acceptor the spectrum of the latter increases relatively to the former. After a time long in comparison to that of a cycle of dissociation and association a final spectral equilibrium is reached. The characteristic time for subunit exchange, t_x is directly given as the time necessary for the completion of a fraction 1-e of the total spectral change. Figure 1, from Erijman and Weber [11], shows the progressive changes in the spectrum after mixing two solutions of glyceraldehyde phosphate

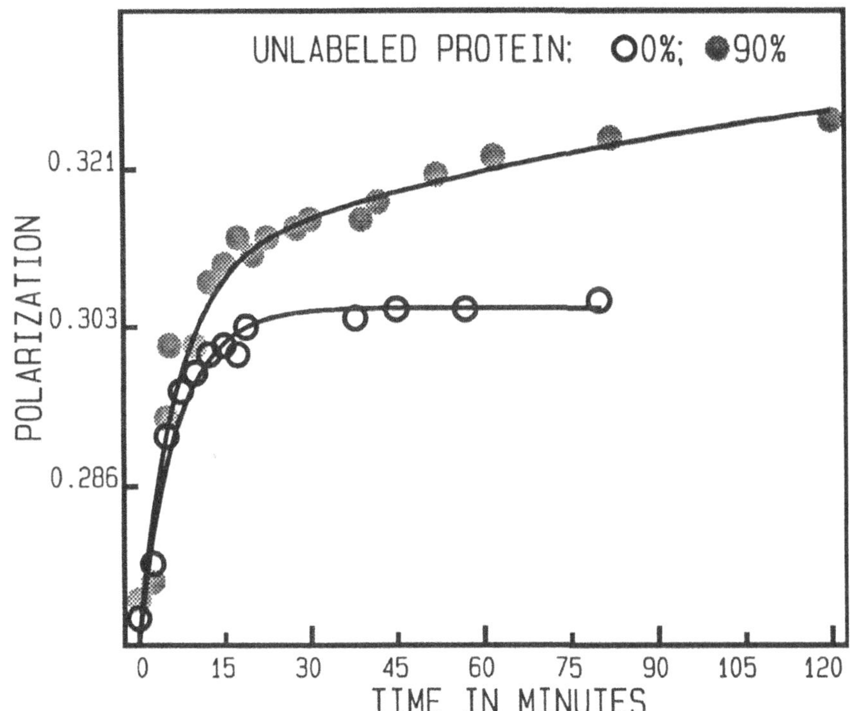

Figure 2. Increase in fluorescence polarization, at 0 °C, of rabbit muscle glycogen phosphorylase multiply labeled with fluorescein, showing the large difference between the rate of pressure dissociation and of subsequent subunit exchange of the undissociated tetramers. From reference [12]

---

dehydrogenase, respectively labeled with coumarin and fluorescein, at 0 °C, after rapidly raising the pressure from atmospheric to 1.4 kbar, at which half of the tetramers are dissociated. Previous observations [9] indicate that dissociation at this temperature is completed in a few minutes, but the spectral changes that depend upon the exchange of subunits owing to the dynamic equilibrium established between monomers and tetramers, takes several hours for its completion. In opposition, several dimers show no appreciable difference in the times required for subunit exchange and dissociation. In the homotransfer method a tetramer molecule is labeled with fluorescein in most of its subunits. As a result the emitted radiation is partially depolarized because of possible energy transfer from the excited fluorophore to a neighboring one preceding the emission. If the labeled solution is mixed with an excess of unlabeled protein, subunit exchange will be revealed by the progressive increase in the polarization of the radiation as the exchanged labels find themselves forming part of tetramers that contain no other labelled subunits [11] . The maximum polarization will be reached when the subunits, labelled and unlabeled, form aggregates that are randomly mixed. Usually the starting labeled solution contains a label distribution that is unknown, except for the average number of labels per aggregate, which is usually significantly greater than 1. However, computation shows that the characteristic time for the attainment of the final label distribution involves a number of cycles of association and dissociation that does not depend on the initial label distribution or even on the degree of dissociation of the solution. t_x thus determined will be exactly equal to the characteristic time for subunit exchange of intact protein if labeled and unlabeled subunits do not differ in the rates of dissociation and association that determine the equilibrium. Use of this method with glycogen phosphorylase is shown in figure 2 [12]. Application of a pressure of 1.4 kbar, results in the dissociation of ca. 50% of the subunits which is shown by a rapid increase in polarization that remains at a stable value independent of time. It correspond to the appearance of isolated labeled monomers that remain separated from each other by distances (several hundred Å), large enough to make impossible the transfer of electronic energy between them. The characteristic time t_d for this rise is simply the reciprocal of the rate of dissociation under pressure. Previous addition of a tenfold excess of unlabeled protein produces precisely the same rapid rise, but this is now followed by a much slower rise in polarization that corresponds to the formation of tetramers with a single labelled subunit, and exhibiting therefore polarization similar to the monomers. The characteristic time of the secondary rise in polarization t_x is the reciprocal of the rate of exchange of subunits among the tetramers.

The relations of t_d and t_x determine the two extreme, opposite cases shown in figure 3. On the one hand if $t_x \approx t_d$ the exchange of subunits is fast enough to generate mixed tetramers within the time of dissociation and the maximum scrambling, and polarization of the fluorescence, is reached with characteristic time t_d. In this case one cannot make any possible distinction between members of the tetramer population: they are all demonstrably equivalent and the equilibrium brought about by the application of pressure is a *stochastic* equilibrium. The very opposite is the case in which $t_x >> t_d$. If t_x->infinity we have the limiting case in which the rise in pressure splits a fixed fraction of the molecular population, and the remainder stays unsplit for as

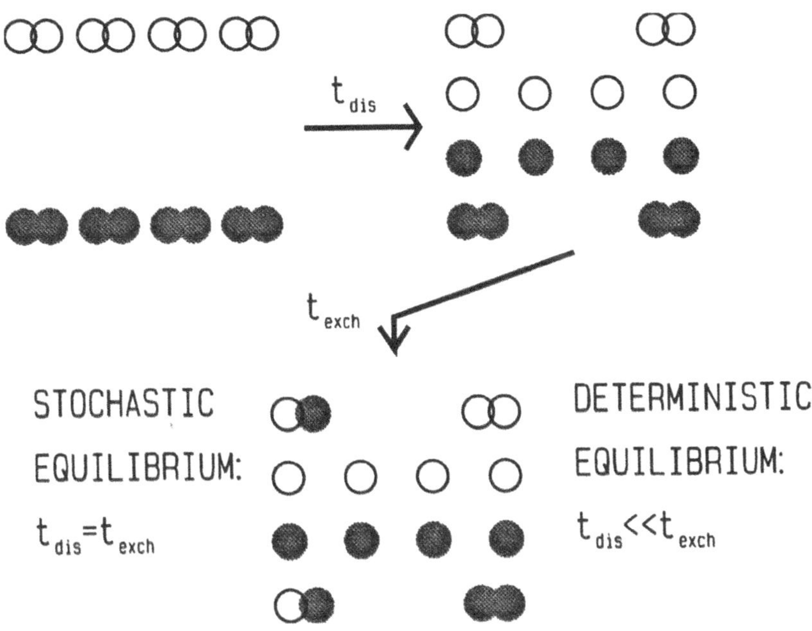

Figure 3 Schematics of the transition from stochastic to deterministic equilibrium, as arising with the increase in the time for subunit exchange over that of aggregate lifetime.

long as we care to carry out the experiment. This situation virtually applies to the case of icosahedral viruses: The particles split by the pressure reassemble into non-infective units on decompression, and the particles that were not split preserve the original infectivity. If t_x is finite, though still much longer than t_d, as it happens in the tetramers at low temperature we have an intermediate case in which the exchange of the property of pressure sensitivity among the particles is slow enough so that we can no longer speak of a stochastic equilibrium. All these cases in which $t_x >> t_d$ can be designated as *deterministic* equilibria [11] . This designation reminds us that these molecular aggregates behave like a collection of macroscopic bodies, in which the time to reach equilibrium is wholly independent of the time during which such equilibrium may persist. Figure 3 shows schematically the origin of the distinction of stochastic and deterministic equilibria.

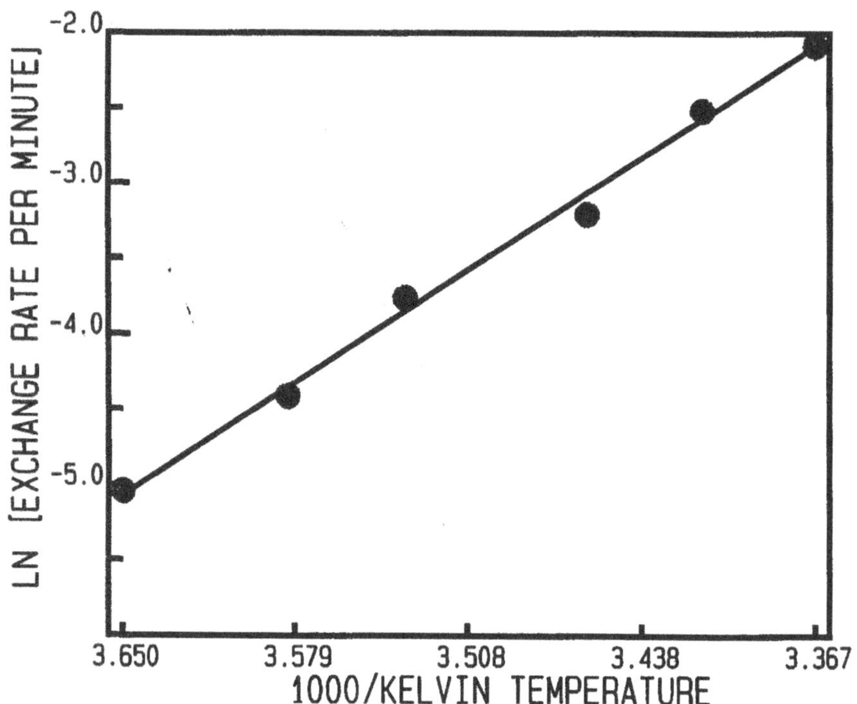

Figure 4 Showing the increase in the rate of exchange of subunits among the tetramers of lactate dehydrogenase with temperature, at atmospheric pressure. From reference [13]

The study of the subunit exchange at atmospheric pressure permits also to determine the influence of the temperature upon this process. From observations on the effects of two successive applications of pressure Erijman and Weber [11] concluded that the differences in free energy of subunit association with temperature and pressure are largely dependent on changes of the rate of dissociation, while the rates of association may be considered invariant, at least to a first approximation. It then follows that if the rate of subunit exchange depended upon the rate of dissociation it would become slower as the temperature increases. However as indicated in figure 4 the opposite is observed [13]: The subunit exchange in lactate dehydrogenase increases with increasing temperature. At 25 °C t_x= 7.4 min approximately twice the rate of dissociation and at 1 °C t_x= 150 min. Thus the effects of a temperature increase are opposite on the rates of dissociation and subunit exchange: The former decreases while the latter increases. In lactate dehydrogenase the increase in the rate of subunit exchange corresponds to an energy of activation of 20 kcal mol^{-1} and we assign this to the energy of activation for the exchange of members of the tetramer population with different free energies of association.

SIGNIFICANCE OF THE CHANGES IN VOLUME ON ASSOCIATION

In dimers and trimers, in which the equilibrium is purely stochastic, one may unequivocally compute a volume change upon dissociation. We may take the β_2 dimer of tryptophan synthase as an example; the dimers of hexokinase and *Rhodobacter* Rubisco differ from it only in details. The separation of two cubic subunits of molecular weight 42 kD involves creation of a surface of contact with the solvent of 1,400 A^2 and the associated decrease in volume (150 ml mol^{-1}) corresponds to 300 A^3 per molecule. Thus, on replacing the neighbor subunit by solvent the decrease in average intermolecular separation is smaller than 0.25 Å. Statistically we expect all protein interfaces to have equivalent properties and show a similar average shortening of bond distances upon dissociation in all aggregates. It may be noted here that the determination of distances between contacting groups at the two sides of an intersubunit boundary, starting with the X-ray crystallographic map, cannot be expected to be accurate enough for an appropriate calculation of volume change on dissociation, although nuclear magnetic resonance may be able to provide, at least for some of the contacting groups, the necessary information.

Although the equilibria involved in the pressure dissociation of dimers is typically stochastic and therefore very different from the deterministic equilibria of hemocyanin or the virus capsids, the range of pressures in which dissociation occurs is very similar in both cases. If we calculate the pressure span dp covered in the change of dissociation from 10% to 90% by means of equation (2) we obtain

$$\delta p= (n+1)\ (RT/\Delta V)\ \ln 9 \qquad (4)$$

As we expect a similar contribution δV to the volume from each of the particles we set $\Delta V=n\delta V$. From the observations in the several dimers δV=50 to 100 ml mol^{-1} and thus $\delta V/RT$=2 to

4 at room temperature. Then setting $(n+1)/n=1$, equation 4 gives δp= 0.7 to 1.4 kbar in good agreement with the observations of multimers. In spite of the independence on particle concentration the pressure of mid-dissociation $p_{1/2}$, falls in the multimers also between 1 and 2 kbar. The pressure must exert its effect homogeneously on all subunit interfaces and similar results may be expected in multimers and the simpler aggregates if compression at the subunit interface is the determinant factor in the pressure dissociation. The demonstration of this latter property will be discussed in my second lecture.

EFFECTS OF SUBUNIT SEPARATION.

We need to distinguish the effects that follow directly the application of pressure, both on the aggregate and subunits, from those that are dependent on the separation of the chains, and this distinction is not always easy. There is no direct test that can yield unequivocally the part played by each of these two causes, and one has to rely upon the comparison of various cases, and the consistency and generality of the arguments that we offer in favor of either of the alternative possibilities to reach a decision as to their relative merit.

In a minority of the dimers (e.g. yeast enolase, ARC protein) the complete reversibility of the effects of applied pressure are shown by the exact correspondence of the degrees of dissociation observed at any given pressure whether this is reached by raising it from atmospheric or decreasing it from the pressure of complete dissociation. In other dimers (β_2 tryptophan synthase, hexokinase, rubisco), as well as in tetramers, the dissociation profile observed on raising the pressure is systematically displaced towards higher pressure in relation to the profile that follows pressure decrease. This difference must necessarily arise from a loss of affinity of the chains for each other when they become separated, and that we have attributed to a *conformational drift* [14]. The conformational drift is best explained as following the substitution of atomic contacts with the neighboring subunits by those with the solvent. Most of the intersubunit contacts are from weak van der Waals interactions, mainly dispersion forces between mutually induced dipoles, which in the isolated subunit are replaced by forces of permanent dipole-induced dipole character with the water molecules. From the analysis of the thermodynamics of the association of dimers, trimers and tetramers, that will be discussed in my second lecture, I have concluded that the intersubunit bonds have average energies of 1 to 2 kcal mol^{-1} while the water-intersubunit bonds are nearly three times higher, 4 to 4.5 kcal mol^{-1}.

This difference is quite sufficient to generate considerable changes in conformation that will be, in general, in the direction of spoiling the affinity of the subunits for each other. Optical spectroscopy, and more recently magnetic resonance, have provided ample evidence of the existence of conformational differences between aggregate and subunits [14,15].

The loss of intersubunit affinity is not permanent, and in each of the cases mentioned above it reverses in times of minutes to days, to generate aggregates with precisely the same pressure-dissociation characteristics of the original. Thus the difference between compression and decompression curves may best be described as a *hysteresis*, or memory effect. It is noteworthy that in the cases in which no apparent hysteresis is observable like enolase or ARC, there is optical, and in the latter case also magnetic resonance, evidence of changes in conformation of the separated subunits [14,15]. The difference between the cases that show clear hysteresis and those that do not must be due to the rapid recovery of subunit affinity on reassociation, a recovery that needs to be fast in relation to the time of establishment of the equilibrium in these latter cases.

DIMER-MONOMER EQUILIBRIA ON DILUTION

Both enolase and ARC have dissociation constants of the order of 10^{-7} which make it possible, employing the fluorescence methods, to follow the dissociation by simple dilution. In both these cases an important anomaly is observed [15,16]: The logarithmic span of the concentration between $\alpha=0.1$ and $\alpha=0.9$ is 1.4-1.7 units, nearly one half of the 2.86 units expected for an equilibrium with a fixed free energy of association. The diminished span indicates unequivocally a loss of affinity that increases with the degree of dissociation of the dimer and which amounts to a difference of some 1.5 kcal mol^{-1} between the dissociation constants observed at $\alpha<0.1$ and at $\alpha>0.9$. The origin and significance of this loss has been extensively discussed [16].

It is evident that the loss of free energy of association upon simple dilution is another manifestation of the conformation drift of the separated subunits, and proves that in the pressure dissociation of aggregates the direct pressure effects cannot have more than a very secondary importance in its production. More interesting still are two general conclusions that may be reached form these observations: In the first place it is obvious that the dissociation equilibria of proteins are maintained by a series of cyclical changes that are virtually unidirectional; dissociation is followed by conformational drift and loss of affinity, and association is followed by the opposite processes and one cannot divide the equilibrium into individual steps, each of them balanced by an equal opposing process [17]. In other words *the principle of detailed balance* is evidently inapplicable here, and demonstrably it has not the generality that, without sufficient experimental proof, has been attributed to it. Its general acceptance in the past has been due to the fact that cases of sufficient complexity were neither experimentally studied nor theoretically envisioned. It is also a consequence of the experimental observations that the dissociation of protein aggregates furnish clear examples that typical thermodynamic equilibria of sufficient complexity cannot be profitably described without consideration of the time dependency and the order of the various elementary processes responsible for them. At equilibrium itself there is an *arrow of time*, a possibility that has been consistently denied, I believe for the very same reasons that have lead to a consensus acceptance of the generality of detailed balance.

TEMPERATURE EFFECTS AND THE COLD INACTIVATION OF ENZYMES

The phenomena of conformational drift and hysteresis are the more marked the lower the temperature at which the cycle of compression and decompression is carried out and the recovery of the properties at atmospheric pressure is correspondingly delayed. Ruan and Weber [9] showed in glyceraldehyde phosphate dehydrogenase that if the tetramers formed on decompression are kept at close to 0° C the recovery of subunit affinity, as well as of enzymic activity, may be indefinitely postponed. However, recovery of both is complete after 2 hours at 20 $^{\circ}$C. In many proteins storage at cold temperature for days to weeks is followed by a progressive loss of enzyme activity and this cold inactivation appears to be the same as is obtained in a much shorter time by dissociation under pressure followed by reassociation on decompression. Therefore we have proposed that cold inactivation is also a manifestation of the conformational drift [14], and that both low temperature and pressure induce the same continuing processes:

dissociation->conformational drift->reassociation->recovery

As commented above the energy transfer method permits to demonstrate that exchange of subunits takes place with increasing ease as the temperature is raised. Therefore the cold inactivation is not directly related to the increased probability of dissociation but to the delayed recovery of the cumulative effects of the conformational drift at the lower temperatures. We also note that the observations of shortened logarithmic span of the concentration dependence of the dissociation imply rapid recovery of the conformation upon reassociation, which would be inconsistent with the observation of hysteresis upon pressure dissociation. In accordance with this notion hysteresis after pressure dissociation is completely absent in both enolase and ARC protein.

CONCLUSIONS

The observations on the dissociation of oligomers have revealed several important features that had not been recognized in the equilibria of small molecular complexes and which still demand investigation. They include the conformational drift of the particles after dissociation by pressure or dilution, the hysteresis and persistence of conformational drift after reassociation and the existence of thermodynamic equilibria owing to deterministic rather than stochastic molecular properties. Of these features one of the most important is the increase in complexity of the subunit interactions on passing from a dimer, or trimer to a tetramer. This is sufficient to generate heterogeneous molecular populations of which the members have rates of interconversion that appear to span several orders of magnitude. The origin of such an effect is as yet unclear and may lead us into an appreciation of some hitherto unrecognized factors of the origin of biological specificity.

REFERENCES

1. The state of the subject as to 1987 is described in Weber G.(1987) 'Dissociation of oligomeric proteins by hydrostatic pressure' in *High Pressure Chemistry and Biochemistry*, R.van Eldyk & J.Jonas (eds) Dordrecht, pp401-420.
   Heremans K.H. (1982) 'High pressure effects on proteins and other biomolecules', Annu.Rev.Biophys.Bioeng, **11**, 1-21.
   Weber G. & Drickamer H.G. (1983) 'The effects of high pressure upon proteins and other biomolecules', Quart. Rev.Biophys. **16**, 89-112.

2. Weber G.(1992) *Protein Interactions*, Chapman & Hall, New York & London,Chapter XII pp186-197

3. Hawley S.A.(1973) 'Electrophoretic separation of conformational states of alpha chymotrypsinogen A',

Biochem.Biophys. Acta **317**,236-239
Hawley S.A. & Mitchell R.M. (1975) 'An electrophoretic study of reversible denaturation; chymotrypsinogen A at high pressures', Biochemistry **14**,

4. Paladini A.A., Silva J.L., & Weber G. (1987) Slab gel electrophoresis of proteins under high hydrostatic pressure', Analyt.Biochem. **161**,358-364.

5. Paladini A.A., & Erijman L. (1992) 'Improved flat bed gel electrophoresis at high pressure', In preparation.

6. Erijman L., Paladini A.A., Lorimer G.H. & Weber G. (1992), 'Plurality of protein conformations of Rubisco probed by high-pressure electrophoresis', submitted for publication.

7. Samarasinghe S.D.,Campbell D.M. Jonas A. & Jonas J. (1992) 'High resolution NMR study of the pressure-induced denaturation of lysozyme', Biochemistry **31**, 7773-7778.
Jonas J. (1992) This volume.

8. Gavish B., Gratton E. & Hardy J,C. (1983) 'Adiabatic compressibility of globular proteins', Proc.Natl. Acad. Sci.USA, **80**,750-754.
Gekko K. & Hasegawa Y. (1986) 'Compressibility-structure relations of globular proteins', Biochemistry **25**, 6563-6571.

9. Ruan K.C. & Weber G. (1989) 'Hystersis and conformational drift of pressure-dissociated glyceraldehyde phosphate dehydrogenase', Biochemistry **28**, 2144-2153

10. Bonafe C.S.F. Villas Boas M., Suarez M.S. & Silva J.L. (1991) 'Reassembly of a large multisubunit protein promoted by non-protein factors', J.Biol.Chem. **266**, 13210-13216
Silva J.L, Villas Boas M., Bonafe C.S.F., & Meirelles N.C. (1989) 'Anomalous pressure dissociation of large protein aggregates', J.Biol.Chem. **264**,15863-15868.
Silva J.L. & Weber G. (1988) 'Pressure-induced dissociation of Brome Mosaic Virus', J.Mol.Biol. **199**,149-161
Da Paoian A.T., Oliveira A.C.,Gaspar L.C., Silva J.L. & Weber G. (1992) 'Reversible pressure dissociation of R17 bacteriophage. The physical individuality of virus particles', Submitted for publication.

11. Erijman L. & Weber G. (1991) 'Oligomeric protein associations: The transition from stochastic to

deterministic equilibrium', Biochemistry **30**,1595-1599

12. Ruan K.-C. & Weber G. (1992) 'Pressure-induced dissociation of glycogen phosphorylase A from rabbit muscle' Submitted for publication

13. Erijman L. & Weber G. (1992) 'Use of sensitized fluorescence in the study of the exchange of the subunits in protein aggregates', Photochem.Photobiol. in press.

14. Weber G.(1986) 'Phenomenological description of the association of protein subunits subjected to conformational drift. Effects of dilution and of hydrostatic pressure', Biochemistry, **25**, 3626-3631
King L. & Weber G. (1986) 'Conformational drift of dissociated lactate dehydrogenase', Biochemistry **25**, 3632-3637.
King L. & Weber G. (1986) Conformational drift and cryoinactivation of lactate dehydrogenase Biochemistry **25**, 3637-3640.
Silva J.L., Miles E.W. & Weber G. (1986) 'Pressure dissociation and conformational drift of the β dimer of tryptophan synthase', Biochemistry **25**,5780-5786

15. Silva J.L., Silveira C.F., Correia A. Jr. & Pontes L. (1992) 'Dissociation of a native dimer to a molten globule monomer', J.Mol.Biol. **223**,545-555.
Jonas J., this volume.

16. Xu J.-G. & Weber G. (1982) 'Dynamics and time-averaged chemical potential of proteins: Importance in oligomer associations', Proc.Natl.Acad.Sci.USA **79**, 5268-5271.
Weber G. (1989) 'Dynamics of oligomeric proteins', J.Mol. Liquids **42**,255-268.

17. Weber G. (1992) *Protein Interactions* Chapman and Hall, New York & London, Chapter XIV.

RELATIONS OF BOND ENERGIES AND ENTROPY WITH VOLUME, PRESSURE AND TEMPERATURE IN PROTEIN AGGREGATES

GREGORIO WEBER
School of Chemical Sciences
University of Illinois at Urbana-Champaign
Urbana, Illinois 61801 USA

ABSTRACT. A procedure is described to compute the average energies of the protein-protein and protein-water bonds that determine the equilibrium between a protein aggregate and its subunits, starting from the standard enthalpy and entropy changes of subunit association. In four dimers, one trimer and two tetramers the data are unequivocal in deciding that, in contrast to previously held notions, the entropy change that drives the association arises from the conversion of the strong protein-water bonds into much weaker protein-protein bonds. The effects of temperature and pressure on the energy of the bonds may be computed by introducing the intermolecular potentials typical of apolar and dipolar interactions, and by specification of the proportion of each type that charcaterizes the protein-protein and protein-water bonds. The dissociating effect of pressure is shown to depend upon the differential compressibility of the bonds which preferentially destabilizes the more apolar protein-protein bonds.

ENTHALPY AND ENTROPY OF SUBUNIT ASSOCIATIONS

A study of the pressure dissociation of four dimers, a trimer and two tetramers at a series of temperatures have provided us with good data on the contribution of enthalpy and entropy to the free energy of association and these results are shown in Table 1. The general reaction of subunit association can be formulated as

$$2 \ \text{P-W} \longrightarrow \text{P-P} + \text{W-W} \qquad (1)$$

where P-W represents the species in which the separated subunit interfaces contact the water, P-P the aggregate and W-W the water contacts that follow the disappearance of the

R. Winter and J. Jonas (eds.), High Pressure Chemistry, Biochemistry and Materials Science, 489–509.

water-subunit interfaces. Such reaction involves two separate balances of the enthalpy and entropy contributions:

$$\Delta H = H(P\text{-}P) + H(W\text{-}W) = 2H(P\text{-}W)$$
$$T.\Delta S = T[S(P\text{-}P) + S(W\text{-}W) - 2S(P\text{-}W)] \qquad (2)$$

where H and S are the enthalpies and entropies associated to the formation of the contacts in equation (1) and ΔH and $T.\Delta S$ the enthalpy and entropy changes that are experimentally determined.

Without need of analysis we observe in Table 1 the decisive importance of the increased entropy in the stability of the aggregates: In six out of the seven cases shown on the table the enthalpy change is positive so that the balance of the energy of the bonds favor the dissociation of the aggregates. In the seventh case, that of glyceraldehyde phosphate dehydrogenase, while both enthalpy and entropy change favor association, the entropy change is predominant, and necessary in ensuring the stability of the aggregate at the low physiological concentrations.

TABLE 1. Enthalpy (ΔH) and Entropy ($T.\Delta S$) contributions, in kcal mol^{-1}, to the free energy of association of oligomers at 1 ^{o}C.

Protein	Subunit Number	Subunit mass in KDa	Subunit Bonds	ΔH	T.ΔS
Yeast Hexo Kinase [13]	2	43	141	17	38
E. Coli[#] Tryptophan Synthase [12]	2	43	141	17.7	28.4
Rhodobacter Rubisco[*] [14]	2	55	166	13.9	26.8
Phosphorylase A Dimer [16]	2	95	239	4	17
Allophyco-cyanine [17]	3	33	118	42	65
GAPDH [15]	4	34	120	-14.1	17.8
Phosphorylase A Tetramer [16]	4	95	239	33	66

[#] At 4 ^{o}C, [*] At 15 ^{o}C. Reference numbers in square brackets. Bond numbers per subunit interface are assumed the same for solvent-solvent, protein-protein, and protein-solvent contacts.

We note that if we consider the enthalpies and entropies appearing in (2) as independent quantities we would not be able to determine the six of them from the values of the overall energy and entropy contributions to the Gibbs free energy of subunit association. However, the problem is greatly simplified if we consider that there are necessary relations between E_{P-W}, E_{P-P} and E_{W-W}, respectively the energies of the protein-solvent, protein-protein and solvent-solvent bonds, and the entropies determined by these same bonds at the respective interfaces, P-W, P-P and W-W. In that case the six unknowns are reduced to three, and if we add to it that the water-water association energy E_{W-W} is exactly known we would have reduced the unknowns to two, for the determination of which we have the two experimental data ΔH and $T\Delta S$.

The question is therefore whether we can postulate unique and general relations between the bond energies and the changes in entropy associated to the isothermal formation of M bonds of a given type. To examine this possibility we must clearly define what we understand by entropy, and how it comes to be uniquely associated to the energy of the bonds.

This first task, the clear definition of the entropy, is an indispensable one, which has been generally eschewed by the many who have written about the influence of entropy in the formation of molecular complexes, including protein complexes. It is the more necessary when we ponder on the remark of Truesdell that "..no informed reader will presume that any one author that uses the word entropy means the same as any other author does" [1].

I shall adopt the following definition of entropy:

Entropy is the part of the energy of a body that is manifested in the heat exchanges with other bodies, it is in fact the energy equipartitioned among the active degrees of freedom of the particles, atoms or molecules, that form the bodies. If two bodies at the same temperature are brought in intimate contact no changes in energy or entropy can occur, unless changes in the bonds holding the particles i.e. chemical reaction occurs at the same time.

The above definition of entropy equates it with $T.\Delta S$ rather than with ΔS and this definition facilitates the conception of entropy necessary to understand the origin of entropy-driven reactions. A reading of the literature on protein interactions, as well as on other areas of physical chemistry, gives the distinct impression that by considering ΔS, rather than $T\Delta S$, the integration of the entropy as part of the total internal energy is obscured and the possibility of a simple explanation for many molecular phenomena is thereby lost. Additionally I note that we never measure entropies (ΔS), although we may calculate them, and what we extract from the experimental data is not ΔS but always $T.\Delta S$.

From this definition of entropy it follows immediately that "bonds" , in the more general sense of this term, are the causes that limit the entropy, and that bond energy and entropy are partially exchangeable and complementary forms of the internal energy of a body. It is not possible to *isothermally* decrease or increase the entropy of a body without producing a corresponding change in the energy of the bonds that link the particles. Two simple rules govern the isothermal energy exchanges between bond energies and entropy: Firstly the change in entropy dS must follow the Clausius rule dS=q/dT where q is the heat *isothermally* absorbed. If a source of entropy is postulated as being associated to a chemical reaction it must be capable of resulting in an isothermal change of heat content of the appropriate magnitude. Secondly, the equipartitioned heat responsible for the entropy fluctuates randomly among the particles producing corresponding changes in the bonds. The changes in entropy arises from the changes in the stationary distributions of bonds in reactants and products, and follows the Boltzmann rule: dS=R ln δN where δN is the change in the number of complexions or dispositions of the bonds as a result of the isothermal reaction. It necessarily involves the computation of the change in the number of complexions associated to the change in bond energies. It is not possible -although it is regularly attempted in the literature- to propose the difference in energy of *unique* bond dispositions of products and reactants as the origin of entropy differences. The entropy of a single disposition exists no more than the temperature of a single molecule.

ENTROPY AND BOND STRENGTH.

With these preliminaries we can now attempt to determine the energies of the protein-protein and protein-solvent bonds starting from the experimental data of protein dissociation. The first step is the introduction of a general relation between the bond energies and the associated entropies. Consider M *identical* interactions holding an interface like those between the water and protein molecules that appear in equation (1). At any given time there is a single probability p of finding anyone of these interactions broken, and therefore a complementary probability 1-p of live interaction. An individual interface characterized by the number j is held together by that number of bonds, and a collection of those interfaces make up a heterogeneous population described by a distribution f_j of bonds. To construct this distribution we require a rule that specifies the dynamics of the population, that is a rule governing the probabilities of the transitions $j \Rightarrow j+1$ and $j \Rightarrow j-1$, that respectively increase and decrease by one the number of bonds present. These rules must ensure an indispensable further property of the distribution: It must be a *regenerative* distribution, in the sense that any arbitrarily chosen set isolated from it should generate the original equilibrium distribution by a dynamic process that operates according to the stated transition probabilities. Besides, the changes in the fraction of active bonds must follow the law of mass action. Both criteria are fulfilled by the rule that we adopt: The transition probability of decrease in the number of bonds by one, j⇒j-1, is the product of the number of live bonds, j, and the *a priori* probability, p that any one of them will be broken

$$W(j \Rightarrow j-1) = p.j \tag{3}$$

Similarly the probability of increase in the number of bonds depends only upon the number (M-j) of unmade bonds and the *a priori* probability of making one more of such bonds, 1-p:

$$W(j \Rightarrow j+1) = (M-j).(1-p) \tag{4}$$

The probability p of bond breakage is related to the absolute temperature T by the equation p=exp(-E/RT) where E is the energy attributed to the bond. Starting with any arbitrary distribution of values of f_j we can compute the final equilibrium distribution of bonds by iterative application of the relations (3) and (4), and a demonstration is given in figure 1. It is shown by this figure, and by many similar computations, that the equilibrium distribution coincides with the normal distribution, with the analytical formulation

$$f_j = \binom{M}{J} (1-p)^J \, p^{M-J} \; ; \qquad 0 \leq j \leq M \tag{5}$$

where $\binom{M}{J}$ is the binomial coefficient or number of combinations of M objects taken j at a time, including also j=0. It is more relevant to analyze the effect of changes in p through those of E/RT, the energy of the individual bond in thermal energy units. Figure 2 shows the total energy E_t and the entropy S of the distribution as a function of E/RT. The energy is given by the relation

$$E_t = E \sum_{j=0}^{M} j \, f_j \tag{6}$$

In calculating the entropy we use the classical Boltzmann relation

$$S = k \ln N \tag{7}$$

where N is the number of complexions. Taking into account that we often deal with a small number of conformations of unequal probability [2] the molar entropy of the distribution is

$$S= R \ln \left(\sum_{j=0}^{M} \binom{M}{J} f_j\right) \tag{8}$$

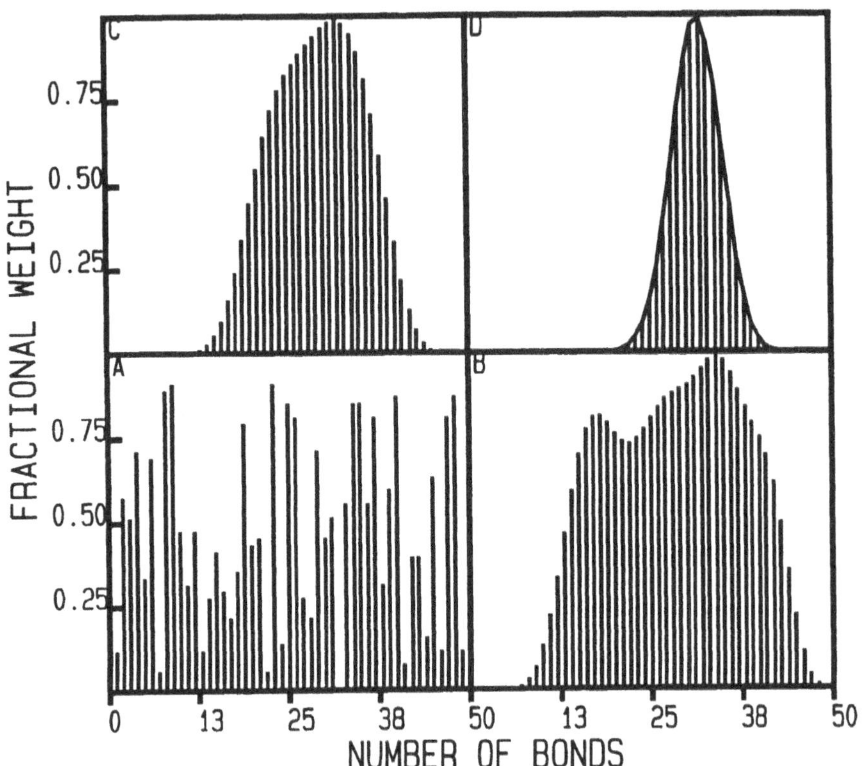

Figure 1A..1D. Iterative generation of the equilibrium distribution of bonds by application of equations (3) and (4), starting from a set of random fractional values (A), with M=50, E/RT=1. B: After 100 iterations, C: after 200, D:1330 iterations. The envelope in D is the distribution calculated by eq.(5).

As depicted in figure 2 the energy approaches the expected value E_t=ME(1-p), increasing therefore monotonically with the number of bonds and their individual value. On the other hand the entropy has an evident maximum for p=1/2, *i.e.* E/RT=ln 2. For values of E/RT smaller than ln 2 it drops rapidly and for values greater than ln 2 it does so much more slowly. Simple inspection of figure 2 shows that the sign and magnitude of the entropy change in a chemical reaction is largely determined by the enthalpy of the weaker bonds. Also, that free energy changes that only involve bonds of considerable strength are accompanied by small, often negligible changes in entropy. It is important to acknowledge the limitations of the treatment of the

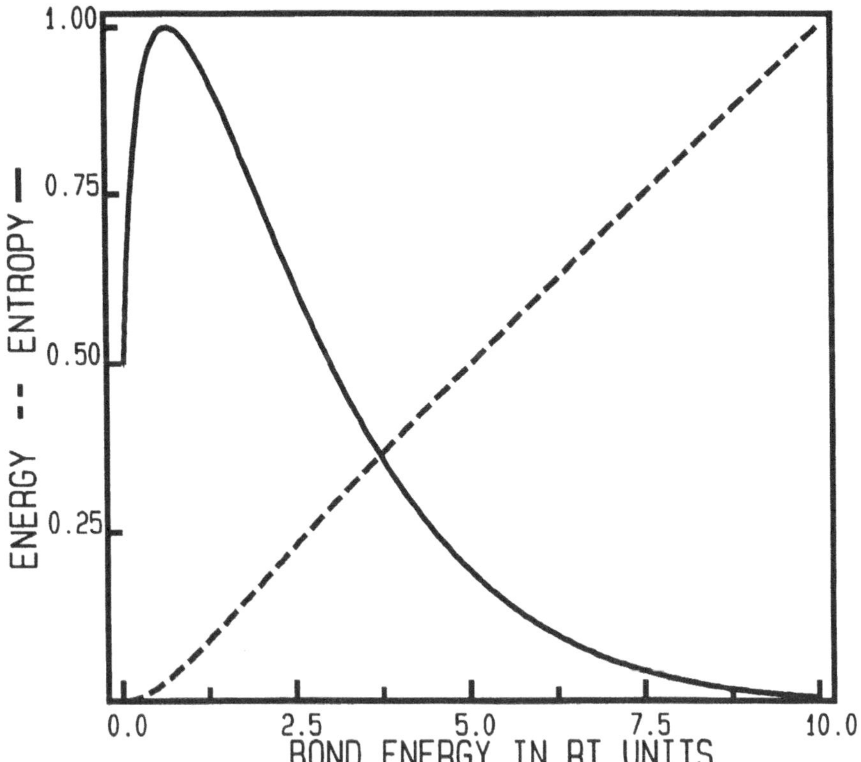

Figure 2. Dependence of entropy and enthalpy for bond distributions that follow equation (5), as a function of E/RT. Enthalpy is plotted as decreasing, entropy as increasing, upwards. Ordinate scales are normalized to the corresponding maxima.

---

relations of entropy and bond energy that I have just offered: In the first place I have considered all bonds to be equivalent in strength, which is evidently not the case for both P-P and P-W interactions. However, though the bonds will cover a spectrum of energies, the average bond energy deduced by considering them equivalent cannot significantly differ from the actual weighted average that would be obtained if we proceeded to introduce the various energies and their relative prevalence. A second limitation is significant: An important premise in our derivation of the statistical relation of bond energy and entropy is that bonds are considered either as made or broken, while in reality bonds depend on interaction energies that vary continuously with distance. However, there is no reason to believe that the relations of energy and entropy would be

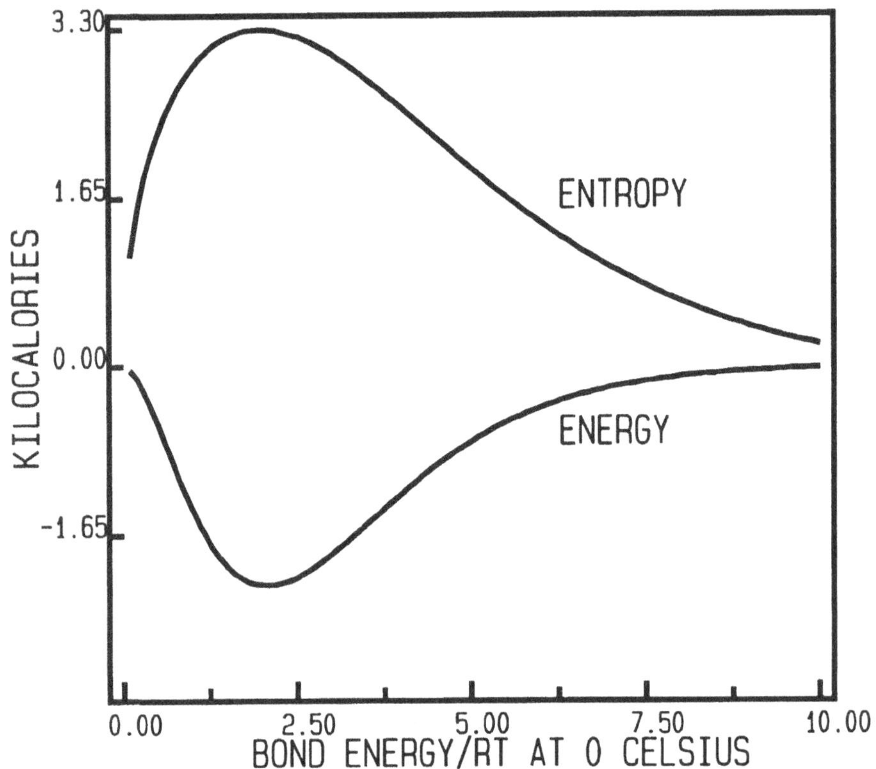

Figure 3. Energy and entropy changes of a set of 100 identical bonds in the interval 0 to 20 °C plotted against bond energy in thermal units at 0 °C. Entropy-enthalpy compensation is restricted to bonds with E less than ≅7 RT. Entropy change is plotted upwards, energy change downwards.

---

appreciably different in either circumstance, because both are subject to the constraints of energy conservation which determines the fraction of lost bond energy that appears as increased entropy in either case.

Figure 2 shows clearly the origin of the often-observed enthalpy-entropy compensation: Decrease in bond energy is necessarily followed by increase in the number of complexions in the population and, in opposition, increase in bond energy diminishes the number of possible complexions. Figure 3 displays the possibilities and limitations of this compensation more clearly by a plot of the changes in energy and entropy of a population of 100 equal bonds, over an interval of 20 ^{o}C, as a function of the bond energy expressed in thermal units. It is evident that the magnitude of the compensation is important only in the case of low energy bonds, in practice those bonds of energy less than 5-7 RT. The energy-enthalpy compensation that follows a moderate increase in temperature was first noted in enzyme catalyzed reactions by biochemists, and from the above discussion it follows that it is necessarily associated with molecules possessing a large number of weak bonds. Thus, it points unmistakably to proteins as having the necessary characteristics for the observation of such compensatory effects. Although the entropy-enthalpy compensation has been often noted, almost always in relation to weak molecular complexes in water solution, like the Michaelis complexes of enzymes, the explanations offered have favored as the cause some special properties of water [3] rather than the general thermodynamic relation of bond energy and bond stability, as I have done above. In fact so many vague explanations based on singular properties of water have been claimed for physicochemical and biological effects observed in water solutions that one is bound to recognize a certain "mystical" desire to credit water with properties that, to the disregard of more mundane explanations and logical consistency, often border on the miraculous [4].

INTERPRETATION OF THE EXPERIMENTAL RESULTS

Table 2 lists the strength of protein-protein and protein-solvent bonds that reproduce the entropy contributions to the free energy of association to better than 1%, when the bond enthalpy balance is exactly equal to the experimental enthalpy change. These computations have been made employing equations (1) and (2), together with the general relation between changes in bond enthalpy and the associated entropy according to equations (4) to (8) and depicted in figure 2. The computations for trimer and tetramers have been made assuming equal contributions to the experimental enthalpies and entropies from three and four interfaces respectively, and carrying out the computations as for a dimer. Figures in parentheses are computed by disregarding the enthalpy error and minimizing the entropy error, employing the one-variable procedure that results from the use of relation (9).
As the entropy associated to a given bond energy depends upon the number of bonds, we have adopted the simplest possible premise: The number M of such bonds is taken to be the same as the molecules of water involved in the reaction (1). In this way we interpret as an independent "bond" the

average energy linking two molecules of water together (3.5 x 2= 7 kcal/mol), and also the average interaction energy of a protein surface over an area equal to the cross section of a water molecule. Thus defined, the average energy of the protein-protein bonds in Table 2 is 1.82±0.78 kcal mol^{-1}, that of the protein-water bonds 4.39±0.45 kcal mol^{-1}.

TABLE 2. Calculated energies, in kcal, of the protein-protein (P-P) and protein-solvent (P-W) bonds, that satify the energetic data of Table 1 with an entropy error of δ(T.ΔS) kcal mol^{-1} and zero enthalpy error.

	P-P energy	P-W energy	δ(T.ΔS)
Yeast Hexo Kinase	0.76 (0.78)	3.84 (3.89)	0.21
Trp.Synthase	1.25 (1.25)	4.12 (4.12)	0.04
Phosphorylase A Dimer	2.39 (2.60)	4.68 (4.80)	0.18
Rubisco	1.59 (1.57)	4.29 (4.18)	0.15
Allophyco-cyanine	1.36 (1.35)	4.18 (4.17)	0.08
GAPDH	2.94 (2.92)	4.94 (4.96)	0.03
Phosphorylase A Tetramer	2.43 (2.42)	4.71 (4.71)	0.09

Figure 4 places these average values together with that of the bonds holding a pair of water molecules on a plot of energy and entropy similar to that of figure 2, and from the placements it follows that the whole of the entropy change in the association arises from the conversion of P-W bonds into P-P bonds. The conclusion is then inescapable that the *interaction of the putative "hydrophobic" intersubunit surfaces and the water molecules involves bonds that are of a strength comparable to those between the water molecules themselves*. From the relations between the bond strength and the change in the number of complexions (eqs. (6) and (8), and figure 2), it follows that the entropy changes associated with the replacement of the water-protein bonds of 4.5 kcal by water-water bonds of 7 kcal cannot contribute an appreciable entropy change, and that entropy changes large enough to explain the entropy-driven aggregation of protein subunits arise by replacement of protein-solvent bonds by the weaker protein-protein bonds.

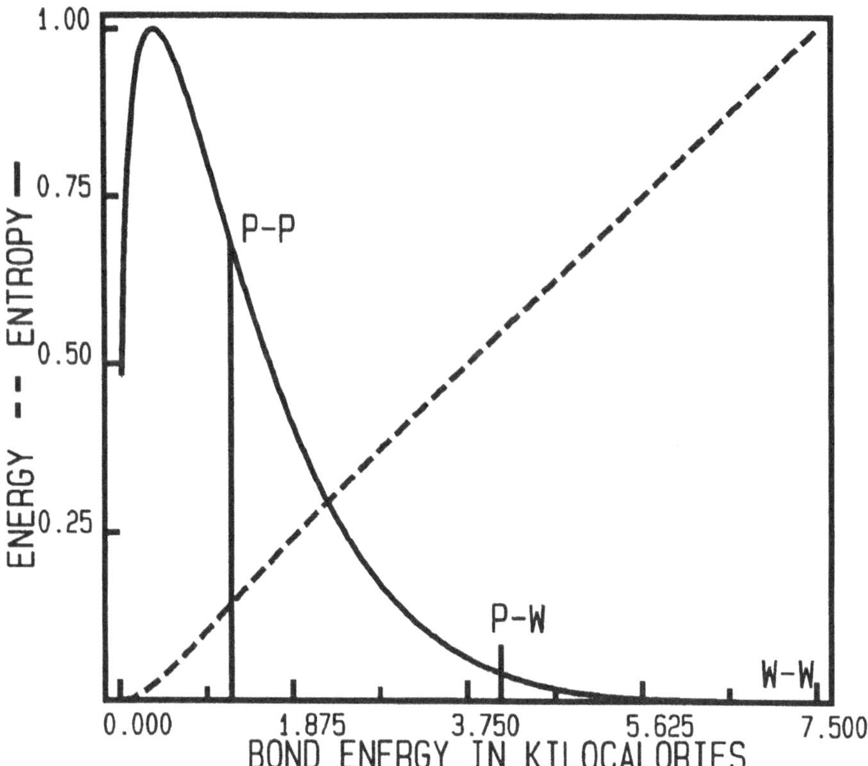

Figure 4. A plot like that of figure 2, with E in kcal/mol. Vertical lines indicate the entropy contributions from protein-protein, (P-P) protein solvent, (P-W) and solvent-solvent, (W-W) as derived from the data of Table 2.

THE LIMITING CONDITIONS FOR AN ENTROPY DRIVEN REACTION

From equations (6) and (8) as well as figure 4, it follows that for the entropy change to be of magnitude sufficient to determine the direction of a chemical equilibrium two conditions must prevail: First, the enthalpy change on reaction must be only a small fraction of the changes in enthalpy that affect reactants and products, that is the enthalpy change must be the small difference between two much larger quantities. Second, the number of bonds involved must be sufficiently large, and the energy of the weaker bonds must be small enough to provide a sufficient entropy change. These conditions are evident in the case of protein associations: In a dimer protein made up of monomers of 35 kDa each, the surface of contact of each of two cubic subunits would be some 1200 A^2. Each surface will contact on dissociation 120 molecules of water, and the formation of the new water-protein surface will involve breaking some 240 hydrogen bonds, amounting to a positive enthalpy change of *ca.* 840 kcal. The observed enthalpies of association show that this is balanced, within 2% of the total, by the difference between the enthalpy derived from inter-subunit bonds and that from the bonds of the exposed protein surfaces with the water. Also, the number of protein-protein bonds involved is large. If we take for this number a value larger than that of the water molecules involved we compute, proceeding as indicated above, a smaller E_{P-P}, and consequently a more positive enthalpy, but a larger entropy contribution. The resulting computed free energy would remain much the same, another example of the energy-entropy compensation. The data of table 2 show that the values of the P-P and P-W bonds obtained by a double variational procedure over the two quantities follow the relation

$$E_{P-W}=3.5+E_{P-P}/2 \qquad (9)$$

This is a simple consequence of the constancy attributed to the water-water bonds and the number of bonds per surface, so that any energy changes that reflect in the decrease or increase of one of the bond energies, E_{P-P} or E_{P-W}, must, on account of energy conservation, produce a compensatory change in the other. It is therefore possible to replace the double variational procedure used to obtain the bond energies shown in Table 2 by a procedure involving a single variable and the fixed relation (9), and if we choose to minimize the entropy error alone we obtain the set of P-W and P-P values shown in parentheses in Table 2. These values do not differ substantially from those obtained by the double variational procedure, and the similarity of results arises from the very large degree of compensation of the energies that must exist in all cases in which the direction of the equilibrium is determined by entropy changes. We note that the qualitative conclusions that may be derived from the calculated bond strengths are in agreement with the known physical properties of proteins:

1. The considerable strengths of the bonds between the subunit surfaces and water, which approaches that of the bonds between the water molecules themselves, is not surprising as both experiment [5] and computation [6] have demonstrated that the protein interior has considerable polarity. Moreover the changes in protein-water interactions that arise as a consequence of the conformational drift must

be in the direction of increasing interactions of the permanent dipoles of water and the peptide bonds, as discussed in my first lecture.

2. Kinetic observations of all kinds [7-9], confirmed by calculations of molecular dynamics [10], demonstrate the existence of many weak bonds within proteins. Their conversion into the stronger protein-solvent bonds is sufficient to account for the positive enthalpy change that opposes the association of the subunits.

PRESSURE AND TEMPERATURE CHANGES; GIBBS AND HELMHOLZ FREE ENERGIES

We have computed the strengths of the protein-protein and protein-solvent bonds by means of theory that takes E to be a constant. The free energy calculated from constant E values presupposes a constant volume and is therefore the Helmholz free energy, while the experimental free energy, obtained from data at constant pressure and variable temperature, is the Gibbs free energy. An estimation of the *relative* changes in E brought about by changes in volume resulting from those of pressure and temperature should permit us to derive the Gibbs free energy as a function of pressure, and thus point out the manner in which application of pressure leads to the destabilization of the association. Temperature and pressure produce changes in E through an alteration of the distances of interaction of the particles through thermal expansion or pressure compression of the interacting elements, the atoms or molecules. To determine the value of the bond energy E, that appears in equations (4) to (8), we require a relation that gives the dependence of the particle interactions upon their distance or volume. With this purpose we consider the interaction energy as the result of an attraction that decays with a power t^* of the distance and a repulsion that decays with a higher power of the distance s^*, in the manner originally proposed by Born. We replace the two customary arbitrary constants by two determinable parameters: the energy E_o and intermolecular distance r_o that correspond to a minimum of potential energy, found at temperature T_o and pressure p_o [18]. Moreover we can replace the distance r_o by a characteristic volume $V_o=r_o^3$ of the system, and obtain

$$E=E_o[(V_o/V)^s-s/t(V_o/V)^t]/(1-s/t) \tag{10}$$

where evidently $s=s^*/3$ and $t=t^*/3$. Equation (10) describes the relative volume changes of a homogeneous liquid at temperature T and pressure p with respect to V_o the volume at temperature T_o and pressure p_o. The attraction exponent t is known to be 2 ($t^*=6$) for interaction of mutually induced dipoles and t=1 (t^*=3) for permanent dipoles. Induction effects by permanent dipoles may be expected to have t comprised between 1 and 2. The repulsion exponent s may be determined by the best fit of the compression curve of liquids to equation (10) over the range of s values.

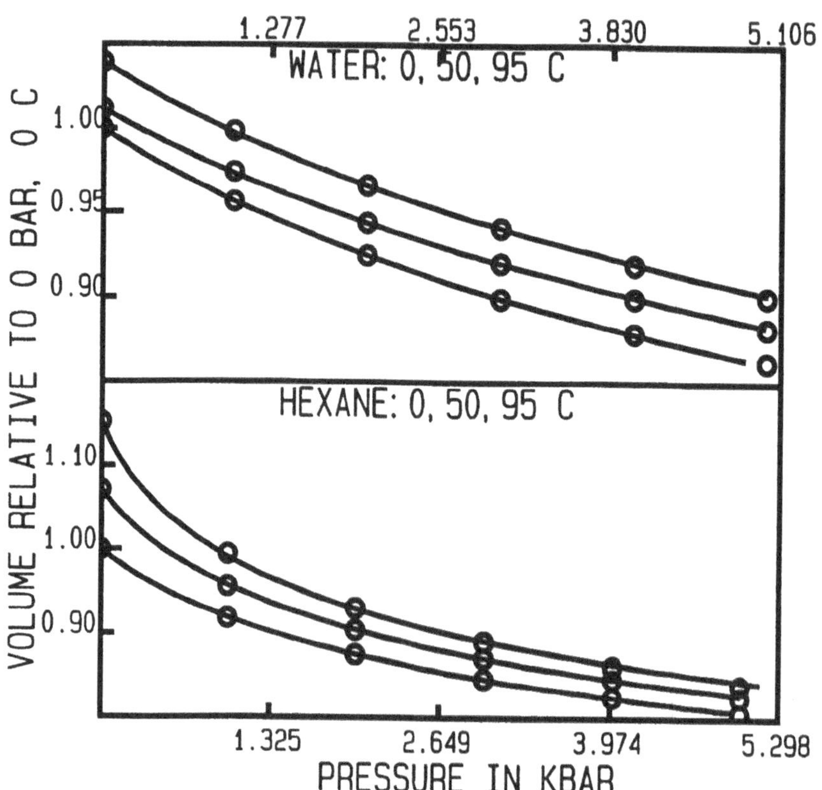

Figure 5. The circles are experimental values of the relative volumes of hexane and water according to Bridgman[11]. The lines are the volumes dependence upon the pressure calculated by eq.(10), as described in text with E_o, and the exponents s and t, as parameters.

Figure 5 show the agreement of the experimental values for the volumes of hexane and water determined by Bridgman [8] in the range of 1 to 5,000 atmospheres at the temperatures of 0, 50 and 95 °C, and those computed by equation (10) with T_o=0 °C and p_o=1 bar. s=10 for hexane and s=8 for water provide the best agreement between observed and computed pressures at which a volume V is observed. For these s values the coefficient of variation of the computed volume-pressure relation was generally closer than 1% to the experimental values.

Introducing the known coefficients s and t in equation (10) we can compute the changes in the value of E with temperature and pressure for bonds that have the apolar character of hexane or the dipolar character of water. We consider that both P-P and P-W bonds have complementary

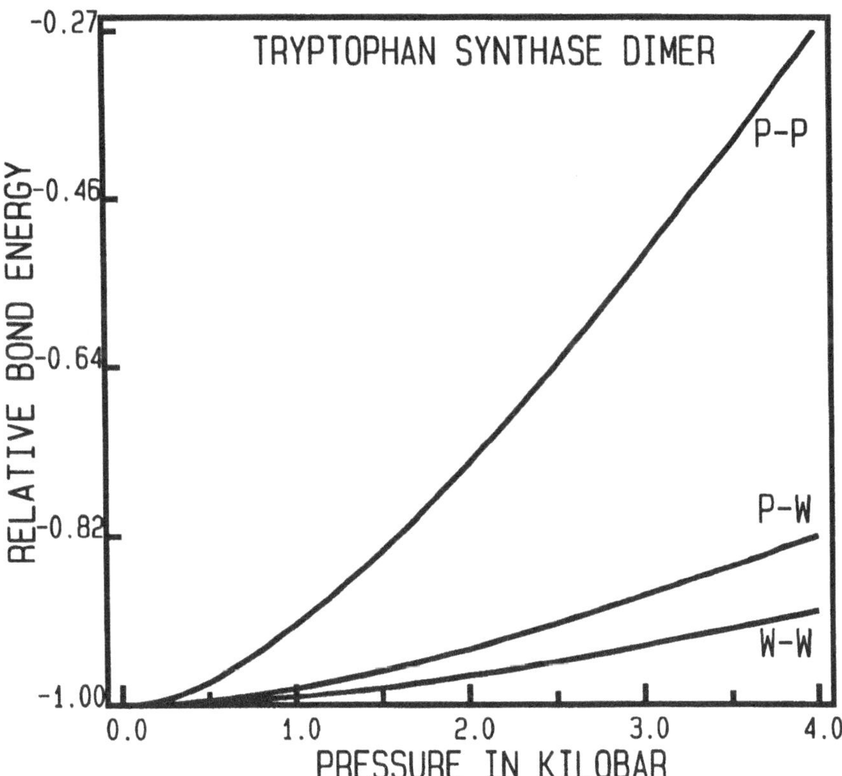

Figure 6. Relative decrease in energy of the bonds as a function of applied pressure at 0 °C, for the bond energies of tryptophan synthase β_2 dimer (Table 2).

---

fractions of apolar and dipolar character, and that the dependence of E on the expansivity and compressibility of these bonds are best represented by the equation

$$E= r.E_{(hexane)}+(1-r).E_{(water)} \tag{11}$$

that gives the fractions r and 1-r of truly apolar and dipolar character respectively. The relative changes in E_{hexane} and E_{water} are shown in figure 5. Evidently $r_{W-W}=0$, which leaves us with two arbitrary parameters r_{P-P} and r_{P-W}. These parameters are fixed by the requirement to produce, for a given protein concentration, profiles of dissociation against applied pressure that must closely agree with experiment on three counts:

1. The pressure of middissociation (α=0.5);
2. A slope in the plot of ΔG against p that reproduces $dG/dp=\Delta V$, the change in volume upon dissociation;

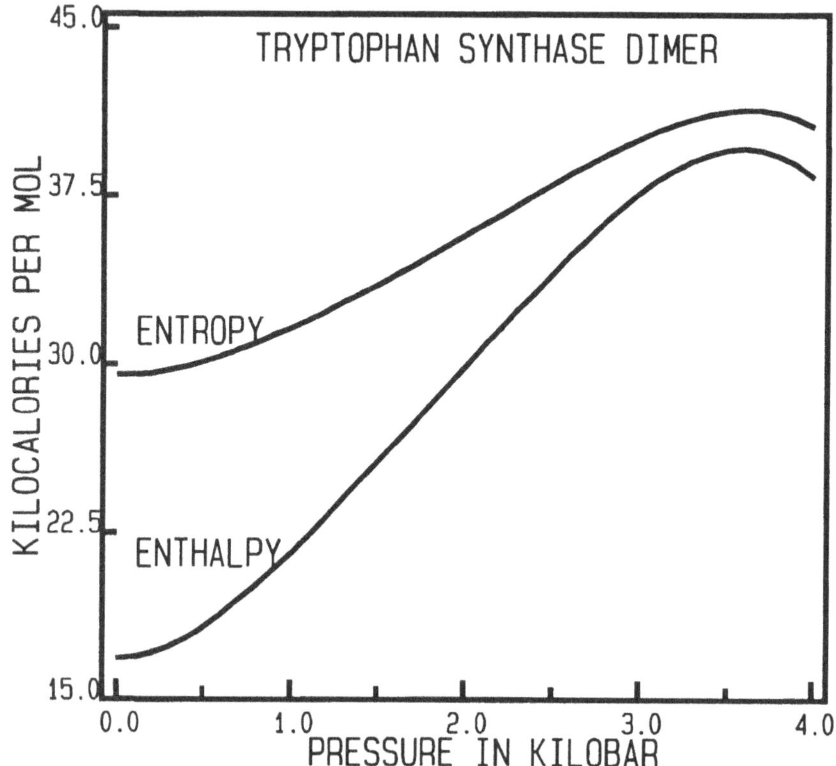

Figure 7. Changes in energy and entropy of association for the same bond energies as Figure 6.

---

3. A uniform increase in stability with temperature in the range of -20 to 20 ^{o}C.

Computations were made employing the fixed bond energy relations of equation (9) as valid at 0 ^{o}C and atmospheric pressure, and varying $E_O(P-P)$ between 0.75 and 2.5 kcal mol^{-1}. It became evident in the calculations that the three criteria just mentioned can be fulfilled when the fractional apolar character of the bonds varies within limits, but that there must necessarily be a difference in apolar character, always favoring the P-P bonds.
Additionally the apolar character of the P-W bonds never exceeds 0.27, even when the apolar character of the P-P

bonds is set to 1. A list of the correlated apolar characters of the bonds is shown in Table 3.

TABLE 3. Correlated character of the apolar fraction, r of the P-P and P-W interfaces. The figures were computed for T_o=0 °C, p_o= 1 bar and P-P=1.25 kcal mol^{-1}, and are those that accord with the three criteria discussed in text.

P-P	P-W
1	0.268
0.9	0.223
0.8	0.184
0.7	0.145
0.6	0.111
0.5	0.080
0.4	0.053
0.3	0.028
0.2	0.006

Even more important than the numerical agreement with the experimental values is the light that the computations throw on the reasons for the increased stability with temperature and the decreased stability with pressure, features that have been observed in all the aggregates so far studied. These depend on characteristic differences of the molecular interactions of dipolar and apolar character as to thermal expansion and compression. We can clearly discern the microscopic circumstances that result in the dissociation of the aggregates when pressure is applied: As the volume decreases on compression, the Born repulsion increasingly predominates over the attraction. The weaker the bonds the stronger the compression and the larger the loss of energy. As a result, the bond destabilization is largest for the P-P bonds, smaller for the P-W bonds and smallest for the W-W bonds (Fig.6). Figure 7 gives the changes in entropy and enthalpy that follow the compression at the three separate interfaces. While both increase, the enthalpy does so much more steeply, owning to the preferential destabilization of the weaker P-P bonds. We are able in this way to offer a simple explanation of the mechanism by which application of pressure leads to dissociation: Compression of the weaker and longer bonds results in preferential destabilization of the subunit interactions and shifts the equilibrium towards the formation of the shorter subunit-water bonds. We no longer need to justify the dissociation of the aggregate by pressure by appeal to the formal principle of Le Chatelier, but can explain the dissociation as the result of the differential destabilization of the bonds upon compression. At a pressure of about 3 kbar the free energy of association becomes nearly constant and independent of pressure. As the character of intersubunit bonds and protein-solvent bonds may not differ much from protein to protein we expect the dissociation of protein aggregates to occur at pressures under 3 kbar, if it is to occur at all.

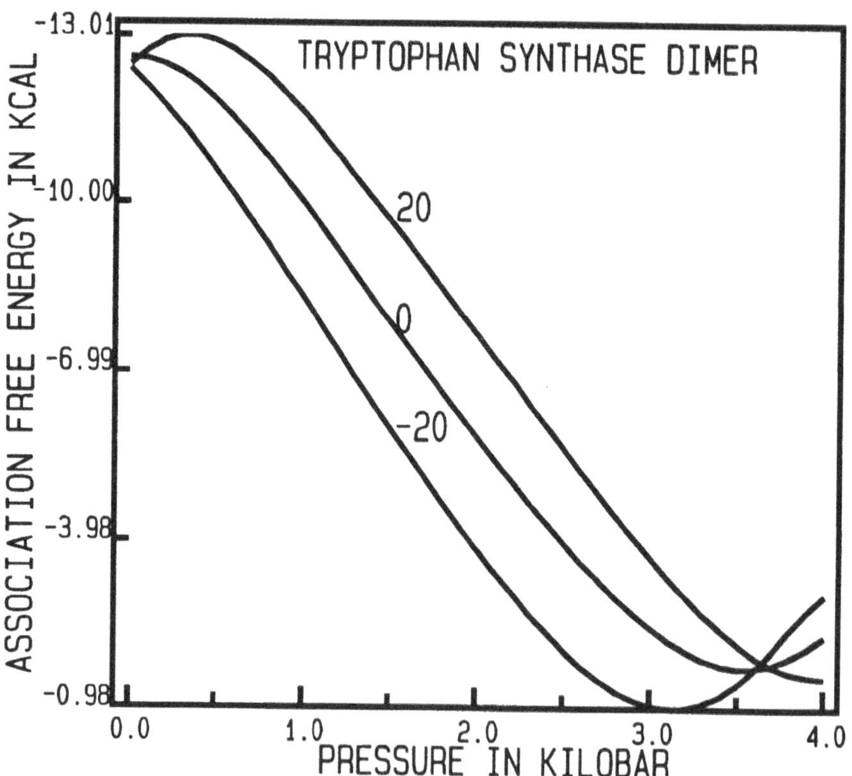

Figure 8. The stabilization to pressure owing to increase in temperature, for the same bond energies of figures 6 and 7.

The effects of the pressure at three different temperatures, -20, 0 and 20 ^{o}C are shown in figure 8 by plots of the free energy of association against the applied pressure. Application of pressure must reverse the expansion at the bonds, before reaching the range of distances at which repulsion leads to dissociation, and as the expansivity is greatest for the P-P bonds, the curves of free energy against pressure of figure 7 show a progressive displacement towards higher pressures as temperature is raised. In this way the increase in stability towards high pressure with increase in temperature is explained. The increase in stability with temperature at atmospheric pressure results both from the decrease in the energy of the bonds owing to expansion and the decrease in E/RT. All bonds are similarly affected but for the P-P bonds the enthalpy-entropy compensation (figure 3) is the most pronounced, and the altered energetic balance results in increased stability of the aggregates with temperature.

It is worthy of note that we could not have reached these conclusions as regards the effects of temperature and pressure by considering enthalpy and entropy as two separate components of the internal energy, but only by examining the microscopic circumstances of their mutual influence.

These computations, that take into account the relations of the energy of the bonds with their expansivity and compressibility show that the thermal stabilization and pressure destabilization of protein associations are not due to a peculiar characteristic of certain "hydrophobic" bonds, but the simple result of differences in the strengths of the bonds involved in the equilibrium, *and nothing else*. This is a sufficient reason to consider that the "hydrophobic bond" is not only a worthless but also a misleading concept.

CONCLUSIONS

A theory has been presented here of the relations of bond energy and entropy that can be applied to those cases where bonds exchanged in a chemical equilibrium are large in number and approximately homogeneous in energy, as it happens in the equilibrium of an oligomer and the constitutive monomers. The theory permits the computation of the energies of the protein-protein and protein-solvent bonds involved in these reactions starting from the experimental free energy of association and its thermal coefficient. It leads to the conclusion that the entropy-driven character of protein associations, and the increased stability with temperature results from the considerably number of weak protein-protein bonds rather than from solvent contributions, as often postulated in the past. This latter view could be held while it was widely believed that proteins were low entropy entities and the large entropy increases observed on association were to be found elsewhere. Observation and calculation [7-10] have dismissed the "low entropy protein", and this study of oligomer associations shows that protein entropy is in fact the determinant factor in these reactions.

Because the interactions within each subunit that determine the folding of the polypeptide chain have a very large contribution from the same types of bonds that are involved in the association of subunits it is possible to infer that the globularity of protein subunits is largely determined by the increase in entropy owning to the weaker protein-protein bonds as against the stronger water-protein bonds. These views are so diametrically opposed to the ones previously held on these matters as to beg reexamination of protein functions like folding, secretion, motility and even catalysis in its light.

REFERENCES

1. C.Truesdell (1966) *The Elements of Continuum Mechanics*, Springer-Verlag New York p234.

2. Equations (10) and (9) give the same result when W is a large number but for small numbers of complexions equation (10) is to be used, as discussed in Yaglom A.M. and Yaglom I.M. (1959), *Probabilite et Information* Dunod, Paris pp40-45.

3. Lumry R. & Gregory R.B. (1986) in *The Fluctuating Enzyme*, G.R.Welch ed. Wyley-Interscience, New York pp.58-77.
Lumry R. & Shyamala R. (1970) 'Enthalpy-entropy compensation in water solutions of proteins and small molecules: A ubiquitous property of water',Biopolymers **9**, 1125-1227.

4. Franks F. (1981) *Polywater,* MIT Press, cambridge USA.
Rousseau D.L. & Porto S.P.S. (1970) Polywater: Polymer or artifact?', Science **167**,1715-1718.
Hildebrand J.(1981) 'A history of solution theory' , Annu.Rev.Phys.Chem. **32**, 1-32, particularly p19: Solutions of non reacting gases.
Davenas E.,..., Benveniste J. (1988) 'Human basophile degranulation triggered by very dilute antiserum against IgE', Nature. **333**, 816-818. ("very dilute" refers to dilutions down to 1 in 10^{120}!).
Maddox J., Randi J. & Stewart W.W. (1988) 'High dilution experiments a delusion', Nature **334**, 287-291.

5. Macgregor R.B. & Weber G. (1986) 'Estimation of the polarity of the protein interior by optical spectroscopy' , Nature **319**, 70-73.

6 . Warshel A., Russell S.T. & Churg A.K. (1984), Macroscopic models for studies of electrostatic interactions in proteins: Limitations and applicability ,'Proc.Natl. Acad.Sci.USA **81**,4785-4789

7. Lakowicz J.R. & Weber G. (1973) 'Quenching of protein fluorescence by oxygen. Detection of structural fluctuations in proteins on the nanosecond time scale' Biochemistry **12**,4171-4179.

8. Wagner G., Demarco A. & Wuthrich K. (1976) 'Fast

motions of phenyl alanine residues in proteins determined by proton NMR', Biophys.Struct. Mechanism. **2**, 139-185.
Doster W., Cusack S. & Petry W.(1989) 'Dynamical transition of myoglobin revealed by inelastic neutron scattering', Nature **337**,754-756.

9. Lakowicz J.R. & Weber G. (1980) 'Nanoseond segmental mobilities of tryptophan residues in proteins observed by lifetime-resolved fluorescence anisotropies', Biophys.J. **10**, 591-601.
Gratton E., Alcala R.J. & Marriott G. (1986)' Rotations of tryptophan residues in proteins',Biochem.Soc. Transac. **14**, 6785-6788.

10. Karplus M. & McCammon J.A.(1981) 'The internal dynamics of globular proteins', CRC Critical Reviews of Biochemistry, **9**, 293-349

11. Bridgman P.W. (1931) *The Physics of High Pressure,* Dover pp128-130.

12. Silva J.L., Miles E.W & Weber G.(1986) 'Pressure dissociation and conformational drift of the β dimer of tryptophan synthase', Biochemistry **25**,5781-5786.

13. Ruan K. & Weber G. (1988) 'Dissociation of yeast hexokinase by hydrostatic pressure' Biochemistry **27**, 3295-3301.

14. Erijman L., Lorimer G.H. & Weber G. (1991), 'Reversible dissociation and conformational stability of dimeric ribulose diphosphate carboxylase/oxidase', submitted for publication.

15. Ruan K. & Weber G.(1989) 'Hysteresis and conformational drift of pressure-dissociated glyceraldehydephosphate dehydrogenase' Biochemistry **28**, 2144-2153

16. Ruan K. & Weber G.(1992) 'Pressure-induced dissociation of glycogen phosphorylase A of rabbit muscle' Submitted for publication.

17. Foguel D., & Weber G. (1991) 'Pressure dissociation of allophycocyanine trimer and cold denaturation of the monomers', in preparation.

18. Weber G. (1992) *Protein Interactions*, Chapman & Hall, New York & London. pp156-157.

HIGH PRESSURE VIBRATIONAL SPECTROSCOPIC STUDIES OF AQUEOUS BIOLOGICAL SYSTEMS : FROM MODEL SYSTEMS TO INTACT TISSUES.

Patrick T. T. Wong
Steacie Institute for Molecular Sciences
National Research Council of Canada
Ottawa, Ontario
Canada K1A 0R6

ABSTRACT. A review is given for our high-pressure vibrational spectroscopic studies on a wide range of biological systems from aqueous biomolecular assemblies to intact biological tissues and cells. Our work in this area has been focused on the study of structural and dynamic properties in model biomembranes, intermolecular interactions between various exogenous molecules and membrane lipids, pressure effects on the secondary and tertiary structure and hydrogen isotope exchange in proteins and polypeptides, hyperbaric manipulation of enzyme kinetics, protein aggregation and dissociation, and cell anomalies at the molecular level in various diseases including human cancers. These studies have been classified into five categories and examples from each category are given here for illustration.

INTRODUCTION:

High-pressure vibrational spectroscopy was first introduced to the study of biomolecular systems in our laboratory one decade ago (1,2). It is now well recognized that high-pressure vibrational spectroscopy is one of the most powerful physical methods for the study of biological phenomena at the molecular level with enhanced resolution (1-6). Some results of our high-pressure vibrational spectroscopic studies on model biomembranes have been reviewed in the last NATO conference on high-pressure chemistry and biochemistry (4). We have summarized our findings on the structural and dynamic properties of model

R. Winter and J. Jonas (eds.), High Pressure Chemistry, Biochemistry and Materials Science, 511–543.

biomembranes with simple saturated lipids. Since then, the high-pressure vibrational spectroscopic work in our laboratory progressed to studies of the structural and dynamic properties of model biomembranes with more complex lipids, the interactions between various exogenous molecules and membrane lipids, the secondary and the tertiary structure of proteins and polypeptides and the kinetics in proteins and polypeptides. Moreover, we have overcome the technical and methodological difficulties (7-9) and developed the high-pressure vibrational spectroscopic technique to the study of biological tissues and isolated cells (9-16). This development allowed us to investigate the molecular bases of a wide range of biological and biomedical problems. In the present lecture, examples of our studies in five categories are given to demonstrate the potential of the high-pressure vibrational spectroscopy for the study of the structural and dynamic properties at the molecular level in biological systems and in life science.

STRUCTURE ASPECTS IN LAMELLAR MEMBRANES:

Model biomembranes are prepared by dispersion of lipids into excess water to form closed multilamellar bilayers. In addition to the aqueous bilayers of the common saturated lipids (4), we have studied the structural and dynamic properties of model biomembranes prepared from a series of specific lipids such as ether-linked phospholipids (17), cis and trans unsaturated phospholipids (18-20), mixed-chain phospholipids (21), phosphatidylcholines containing ω-cyclohexyl fatty acyl chains (22), glycolipids (23,24), and sphingolipids (25). Results from the high-pressure vibrational spectroscopic study of a lamellar membrane consisting of unsaturated 1,2-dioleoyl phosphatidylcholine (DOPC) (20) are given here for illustration.

DOPC is a phospholipid with a rigid cis double bond in the middle of each of the two hydrocarbon chains (see figure 1). At ambient conditions, DOPC aqueous bilayers are in the liquid crystalline state (26) in which the conformational and orientational structures of the hydrocarbon chains are highly disordered. It has been shown (4) that external pressure can order a disordered liquid crystalline phase and induces a transition from the disordered liquid crystalline phase into an ordered gel phase. This phase transition is indicated by discontinuous changes in the infrared and Raman spectra (4).

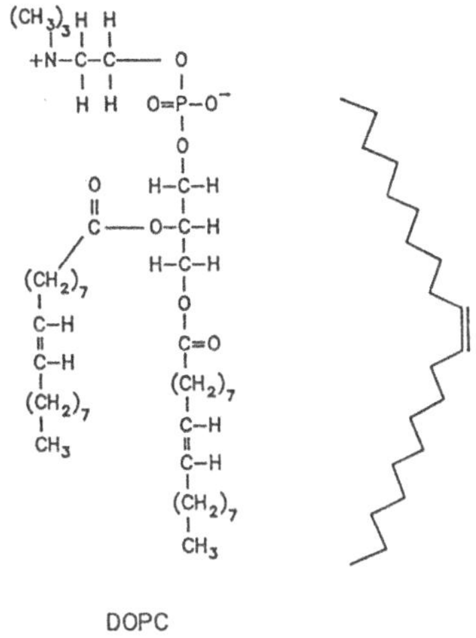

Fig. 1. Molecular structure of1,2-dioleoyl phosphatidyl-choline.

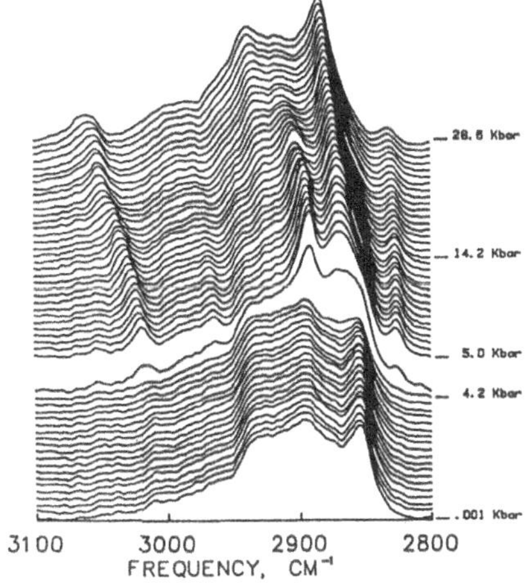

Fig. 2. Raman spectra of aqueous DOPC bilayers in the region of the CH stretching vibrations recorded under increasing pressure.

Dramatic changes in the infrared and Raman spectra of DOPC bilayers and discontinuities in the pressure dependencies of frequencies of all the infrared and Raman bands were observed at 5 kbar. This indicates that the liquid crystalline DOPC undergoes a phase transition into the gel phase at this pressure. For instance, the Raman spectra in the frequency region of the CH stretching vibrations shown in figure 2 exhibit distinct changes at 5 kbar. The spectra in this region below and above 5 kbar are typical those of the liquid crystalline phase and the gel phase of lipid bilayers, respectively (1-6). Above 5 kbar, all the infrared and Raman bands become narrower and the intensities of the all-trans C-C stretching bands increase dramatically while the gauche C-C stretching band at 1080 cm^{-1} disappears in the Raman spectra (data not shown, see ref. 20). These results suggest that the conformational structure of the hydrocarbon chains is highly ordered and the hydrocarbon chain segments on both sides of the cis double bond are highly extended. Consequently, the two zig-zag CH_2 chain segments on both sides of the cis double bond in each oleoyl chain form a bent structure as that shown in figure 1. Since reorientational fluctuations of a bent chain require more space than those of a straight and disordered chain, reorientational fluctuations are forbidden above 5 kbar in the gel phase.

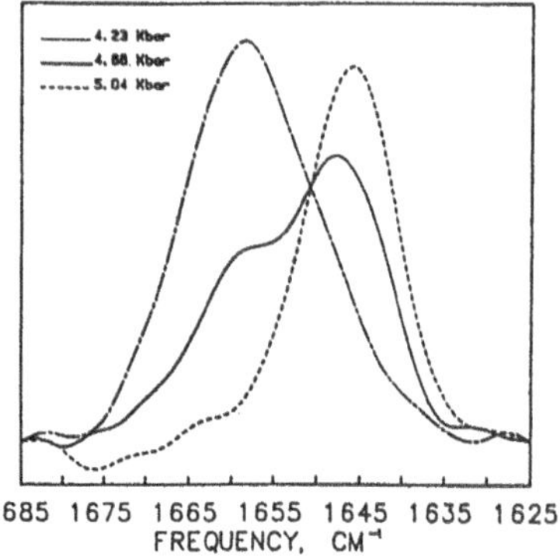

Fig. 3. Representative Raman spectra of the νC=C band near the critical pressure.

The bent configuration of the hydrocarbon chains in the gel phase of DOPC is also evident from the changes at 5 kbar in other regions of the infrared and Raman spectra. For examples, a decrease in the frequency of the C=C double bond stretching mode takes place above 5 kbar in the gel phase (see figure 3). The decrease in this frequency at 5 kbar is consistent with the interpretation in terms of an elongation of the C=C double bond. At a pressure as high as 5 kbar, compression along the direction normal to the bilayer interface causes a further bending of the two zig-zag chain segments at the level of the cis double bond. On the other hand, this pressure-induced deformation is opposed by the strong repulsion between the two hydrogen atoms of the CH_2 groups on both sides of the double bond. They are very close to each other in the bent configuration. The balance between the compression-enhanced chain bending and the van der Waals repulsion between the adjacent CH_2 groups on both sides of the double bond results in a slight elongation of the double bond.

If the two neighboring sn-1 and sn-2 hydrocarbon chains of each DOPC molecule in the gel phase are fully extended and form a bent structure at the double bonds, these two chains must be parallel to each other. Otherwise, the volume of the lipid bilayers would increase. This is not allowed at high pressure (see figure 4 E). For parallel methylene chains, there will be no correlation field splitting of any vibration mode of the methylene chains (27). Indeed, correlation field splitting is absent in any of the methylene chain modes at pressure as high as 36 kbar, such as the CH_2 bending mode shown in figure 5. In DOPC the sn-1 and sn-2 chains have nonequivalent conformations. The positions of the cis double bonds of the sn-1 and sn-2 chains with respect to the head group are not the same

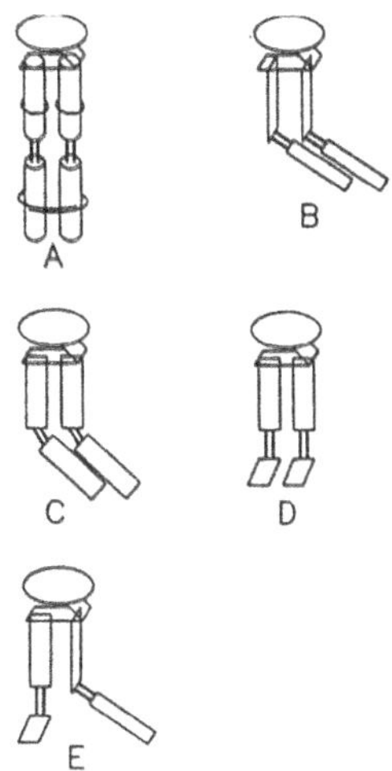

Fig. 4. Schematic illustration of possible interchain packing in DOPC.

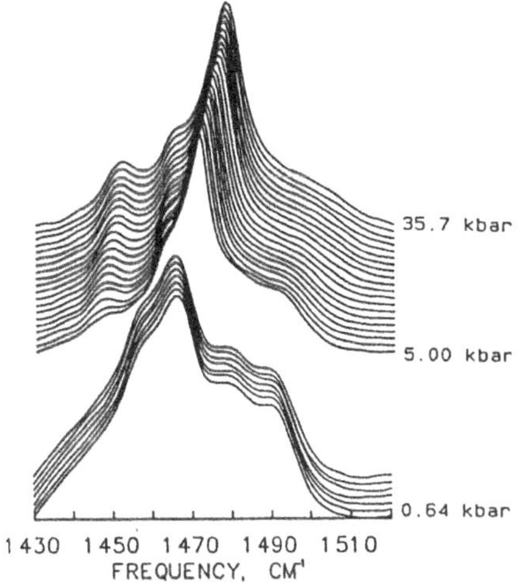

Fig. 5. Stacked contour plots of the infrared spectra of DOPC in the CH_2 bending mode region.

(see figure 1). If this molecular configuration is to be maintained in the gel phase of DOPC above 5 kbar, the bending in the sn-1 chain must occur at a position further away from the bilayer interface than that in the sn-2 chain. In this case, gaps would be generated in the gel phase between the sn-1 and sn-2 chains. Consequently, the overall volume of the bilayers would increase. This is also forbidden at high pressure. Therefore, in order to form a bent configuration in both chains, and at the same time maintain the closest packing, the C=C double bonds in the sn-1 and sn-2 chains must align with each other. The only way to achieve this is to move the sn-1 chain towards the interface and to align the C=C double bond of the sn-1 chain with that of the sn-2 chain by rotating the C-C bonds in the glycerol moiety in the head group. Such a structural change indeed takes place at the transition pressure. This is evident from the changes in the infrared and Raman spectra in the ester C=O stretching mode. The infrared C=O stretching band is broad in the liquid crystalline phase. It consists of two component bands at 1743 and 1727 cm^{-1} which are due to the stretching modes of the nonhydrogen bonded sn-1 C=O group and that of the hydrogen bonded sn-2

C=O group, respectively (28). Above 5 kbar , the overall C=O stretching band becomes very narrow and shifts to lower frequency equivalent to that of the hydrogen bonded sn-2 C=O stretching band in the liquid crystalline phase (see figure 6). These results indicate that above 5 kbar the environment of the sn-1 C=O group becomes similar to that of the sn-2 C=O group. This suggests that the sn-1 C=O group moves closer to the bilayer interface. It is now at the same level as the sn-2 C=O group.

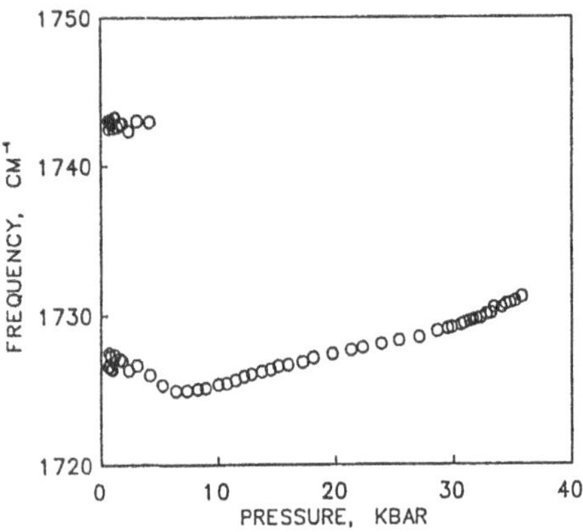

Fig. 6. Pressure dependence of the infrared frequency of the νC=O band of DOPC

Several possible packings of the bent chains in DOPC are shown schematically in figure 4. Figure 4A depicts the structure of the conformationally and orientationally disordered DOPC molecule in the liquid crystalline phase, whereas figures B-D represent three possible packings of the bent chains in the gel phase. A 120° rotation of the zig-zag chain segments on both sides of the cis double bond is required for the interchain packings shown in figure B and D. The conformation of these chains has been referred to as skew/skew (18,29). Thus, the double bonds in the structures B and D are no longer cis double bonds and the HC=CH plane in each chain is 120° out of the plane of the bent chain. The true cis double bond characteristics are found only in structure C. The spectral parameters observed and their pressure dependencies indeed favor structure C for the pressure-induced gel phase of DOPC. First, the frequency of the olefinic stretching band is extremely sensitive to pressure. The frequency shift of this band is > 2 cm^{-1} per kbar in the pressure range 5-10 kbar (20), which is unusually large and very seldom observed for an internal mode (1-6). This large pressure effect results from the bent geometry of the oleoyl chain at the cis double bond, which leads to a very close contact of the =C-H bond with the neighboring chain.

Consequently, at high pressure the interchain interactions experienced by the olefinic bond become much stronger than those of any other functional groups. If the cis characteristics of the olefinic bonds are removed and the =C-H bonds are tilted out of the way as in the skew/skew conformations of the structures B and D, the pressure-induced interchain interactions experienced by the olefinic bonds would be considerably reduced. The pressure-induced shift of the =C-H stretching frequency should be comparable with that of the CH_2 stretching modes.

The CH out-of-plane wagging mode of the =C-H group is softened by external pressure as indicated by the decrease in the frequency of this =C-H band with increasing pressure (20). The softening of this mode by pressure strongly suggests that pressure induces a structural deformation of the olefinic group in DOPC. In this pressure-induced deformation, the =C-H bonds rotate out of the plane of the bent chains, following the eigenvector of the out-of-plane wagging mode. As a consequence of this rotation, the strong interchain interactions at the olefinic groups are relaxed. In figure 4B and 4D, the olefinic groups are already rotated out of the plane of the bent chains by 120°. A further rotation of this functional group would result in a strengthening of the strong interchain interactions at the olefinic groups instead of a relaxation. Thus, this olefinic rotation would be disallowed at elevated pressures. Therefore, all these results suggest that the most plausible interchain packing in the pressure-induced gel phase of DOPC is that represented by figure 4C.

TWO-DIMENSIONAL SOLUTIONS IN LIPID BILAYERS:

When lipids are dispersed in water, two-dimensional ordered bilayers are formed. In lipid bilayers, the hydrophilic groups of lipid molecules are in contact with bulk water and the hydrophobic hydrocarbon chains are in the bilayer interior. The apolar interior of lipid bilayers is a suitable environment for solubilization of hydrophobic molecules. Conversely, the partition of polar molecules into the interior bilayer matrix is unfavorable. However, they can enter into the interfacial region of the bilayers or bind to the polar head groups. Therefore, the understanding of the solvent-solute interactions in two-dimensional solutions in lipid bilayers may provide solutions to the problems of biomembrane function and membrane biochemistry.

The changes in structural and dynamic properties resulting from the interactions between a series of two-dimensional lipid solvents and various solutes such as anesthetics (30-32), alkanes (33), cholesterol (34,35), toxins (36), environmentally sensitive fluorescent probes (37), polypeptides (38) and proteins (39) have been studied in our laboratory by high-pressure vibrational spectroscopy. These results have shown how high pressure vibrational spectroscopy leads to the precise location of the solute in lipid bilayers and reveals the effects of solutes on the structural and dynamic properties of solvents and vice versa. For example, the effects of a local anesthetic tetracaine (TTC) on the structural and dynamic properties of several lipids are given in detail here for illustration.

It is well known (30-32) that when TTC in either its uncharged form (high pH) or its charged form (low pH) is mixed with lipid bilayers, it is partitioned into the bilayer matrix. The uncharged form of TTC penetrates more deeply into the bilayers than the charged form. The most obvious effects of intercalated TTC on the structure of lipid bilayer solvents can be derived from the correlation field splitting of the CH_2 bending mode (δCH_2) of the methylene chains in lipids.

According to the first order perturbation theory (40-42), the exciton level of a lipid bilayer system with two nonequivalent methylene chains per unit cell (a and b chains) will split into two branches. Thus, each vibrational mode will split into two with the following frequencies:

$$\nu_I = \nu_o + \frac{1}{8\pi^2 \nu_o c}\left(\sum_{i=1}^{2N}\left(\frac{\partial^2 U_0}{\partial q_i^{\,2}}\right)_0 + \sum_{j=1}^{N}\left(\frac{\partial^2 U_1}{\partial q_a \partial q_{a_j}}\right)_0 + \sum_{j=1}^{N}\left(\frac{\partial^2 U_2}{\partial q_a \partial q_{b_j}}\right)_0\right) \qquad [1]$$

$$\nu_{II} = \nu_o + \frac{1}{8\pi^2 \nu_o c}\left(\sum_{i=1}^{2N}\left(\frac{\partial^2 U_0}{\partial q_i^{\,2}}\right)_0 + \sum_{j=1}^{N}\left(\frac{\partial^2 U_1}{\partial q_a \partial q_{a_j}}\right)_0 - \sum_{j=1}^{N}\left(\frac{\partial^2 U_2}{\partial q_a \partial q_{b_j}}\right)_0\right) \qquad [2]$$

where ν_0 is the harmonic frequency of each chain vibrational mode, U_0 is the pair perturbation potential function, U_1 and U_2 are the pair perturbation potential functions between equivalent and nonequivalent chains, respectively, q is normal coordinate and the subscripts a and b represent sites a and b of the methylene chains. The first term in these equations is the potential energy changes of the vibrational modes of individual chains arising from the perturbation of neighboring stationary chains. The second and the third terms are the potential energy changes due to the perturbation of intramolecular displacements of orientationally equivalent and nonequivalent neighboring chains, respectively. The splitting of each frequency of the isolated chain into ν_I and ν_{II} components in the bilayers is the so called correlation field splitting which is:

$$\Delta\nu = \frac{1}{4\pi^2\nu_0 c}\sum_{j=1}^{N}\left(\frac{\partial^2 U_2}{\partial q_a \partial q_{b_j}}\right)_0 \qquad [3]$$

If the orientation of all the methylene chains in a bilayer system is parallel to each other, the third term in equations [1] and [2] as well as $\Delta\nu$ in equation [3] equals zero. Consequently, there is no correlation field splitting in the methylene vibrational modes in this lipid system. In the following two circumstances, there will be no correlation field splitting in the spectra: (i) The conformation of the methylene chains is highly disordered due to the presence of a large number of gauche bonds, and thus the coupling of the vibrational modes between neighboring chains is random and weak. (ii) The methylene chains are conformationally highly ordered and fully extended, but the orientation of these fully extended chains is disordered due to reorientational fluctuations and the torsion/twisting motions of the chains. These fluctuations and motions are usually observed at low pressures or high temperatures. In these cases, only broadening rather than correlation field splitting in the vibrational bands of methylene chains will be observed.

It has been demonstrated (1,3-6) that the conformational disorder, the reorientational fluctuations and the torsion/twisting motions of the methylene chains in lipid bilayers can be ordered and dampened by an external pressure due to the lateral compression perpendicular to the chain axis. Consequently, at high enough pressure, these disordered structures can be removed. The correlation field splitting in the vibrational modes of the methylene chains would appear, provided that the equilibrium orientations of neighboring chains are nonequivalent. For more orientationally disordered chains, a higher pressure is required to stop these fluctuations and motions. Thus, the correlation field splitting pressure at which the splitting appears is higher. Consequently, the order/disorder dynamics of the methylene chains in lipid bilayers can be determined by the magnitude of the correlation field splitting pressure.

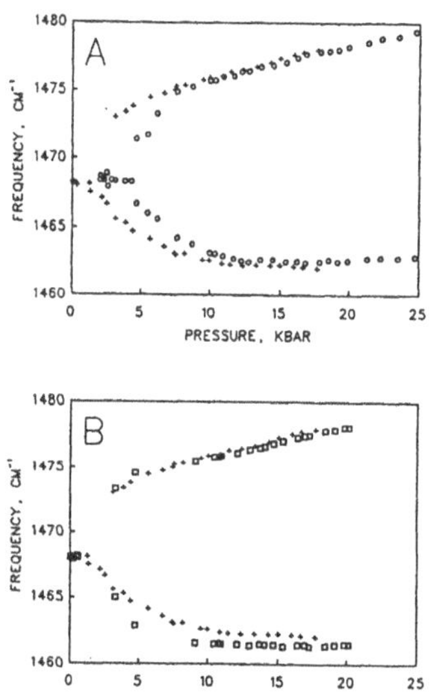

Fig. 7. Comparison of the pressure dependence of the frequencies of the CH_2 bending mode for (+) DMPC with that of (A)(o) DMPC + TTC, pH 9.5 and (B) (□) DMPC + TTC, pH 5.5.

Figure 7 compares the pressure dependence of the δCH_2 band of 1,2-dimyristoryl-sn-glycero-3-phosphocholine (DMPC) bilayers in the presence and absence of TTC. The anesthetic was incorporated in either its charged form at pH 5.5 or its uncharged form at pH 9.5. The critical pressure for the transition from the liquid-crystalline phase to the gel phase in DMPC is 0.15 kbar at 28 ^{o}C (2). Therefore, application of a very small amount of pressure to DMPC in the liquid-crystalline phase will induce a transition to the gel phase. As the pressure is increased, one observes a pressure-induced correlation field splitting of the δCH_2 mode, which at low-pressure gives rise to only a singlet at ~1470 cm^{-1}. It is evident from figure 8 that the pressure profile of the δCH_2 band is quite different among the three DMPC samples. A visual inspection of the spectra in figure 8C show that the "valley" between the two component bands of the pressure-induced

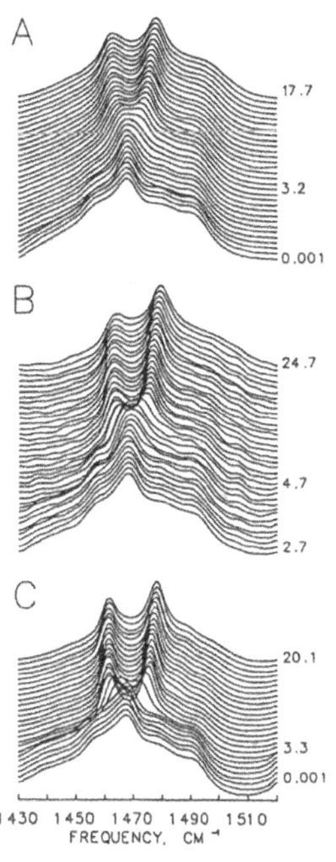

Fig. 8. Stack contour plots of the infrared spectra at increasing pressure of aqueous DMPC bilayers in the CH_2 bending region for (A) DMPC, (B) DMPC + TTC, pH 9.5, and (C) DMPC + TTC, pH 5.5.

correlation field splitting is particularly pronounced for DMPC in the presence of charged TTC but much less for pure DMPC. A pressure dependence of the δCH_2 mode similar to that observed for DMPC in the presence of charged TTC has been observed for various lipid bilayer systems (17, 43). They are known to form a lamellar gel phase with complete interdigitation of the hydrocarbon chains under the conditions of temperature and pressure where the correlation field splitting is observed. Therefore, figure 8C suggests that the incorporation of charged anesthetic in DMPC bilayers induces the formation of an interdigitated lamellar gel phase (43). Simon et al. (44) suggest that the one requirement for the induction of the interdigitated phase is that the guest molecule must reside at the membrane-water interface and must not extend too deeply into the acyl chain region. This explains why there is no interdigitation in the presence of uncharged TTC (see figure 8B). It penetrates too deeply into the hydrophobic region to allow a stable interdigitated phase.

The pressure dependencies of the frequencies of the δCH_2 band for DMPC in the absence and presence of uncharged TTC are shown in figure 7A. From this figure, the pressure at which the correlation field splitting becomes apparent can easily be determined. A well-defined correlation field component band becomes apparent at 3.2 kbar for pure DMPC and at 4.7 kbar for DMPC with TTC at pH 9.5. Any further increase in pressure will then result in a relatively rapid, nonlinear increase in the magnitude of this splitting. The correlation field splitting pressure is much higher for DMPC in the presence of uncharged TTC compared to that for pure DMPC, while the magnitude of the correlation field splitting is smaller for the former system. The smaller correlation field splitting for DMPC with uncharged TTC indicates smaller interchain interactions due to the decrease in both the orientational and conformational order. This disorder results from the intercalation of the anesthetic between the lipid acyl chains. The higher correlation field splitting pressure in the presence of uncharged TTC also indicates that the DMPC acyl chains are orientationally more disordered in the presence of uncharged TTC. Thus, a higher pressure is necessary to stop the acyl chain reorientational fluctuations and torsion/twisting motions and induce a correlation field splitting. The pressure dependencies of the frequencies of the δCH_2 mode components for pure DMPC and DMPC in the presence of charged TTC at 5.5 pH (figure 7B) are very similar. The magnitude of the splitting is, however, slightly larger in the presence of charged TTC. These results support the conclusion of the formation of an interdigitated lamellar gel phase in DMPC in the presence of charged TTC.

The pressure dependence of the CH_2 bending and rocking bands for 1,2-di-o-hexadecyl-sn-glycero-3-phosphocholine (DHPC) in the presence of uncharged TTC and those for pure aqueous DHPC bilayers are presented in figure 9. The pronounced "valley" between the correlation field components of the CH_2 bending and rocking bands for pure DHPC is due to the formation of an interdigitated lamellar gel phase (43). The pressure dependencies of the CH_2 bending and rocking mode bands for DHPC in the presence of uncharged TTC are very similar to those obtained for pure DHPC and are characteristic of lipids in an interdigitated gel phase. Moreover, the pressure dependencies of the frequencies of the δCH_2 band and its correlation field component for DHPC do not vary significantly upon the addition of uncharged TTC. These results, therefore, suggest that the incorporation of uncharged TTC into DHPC does not prevent the formation of an interdigitated lamellar gel phase.

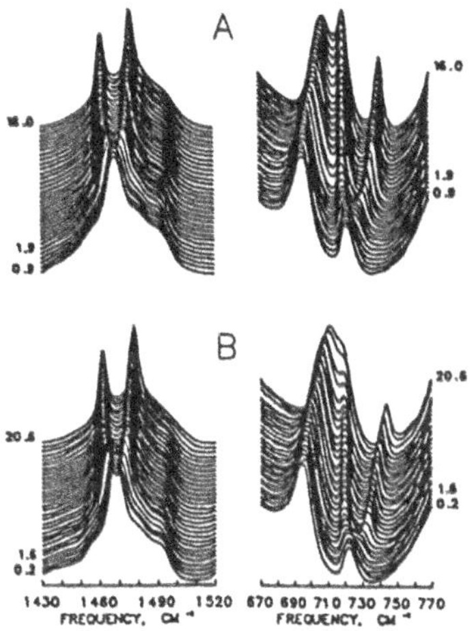

Fig. 9. Stacked contour plots of the infrared spectra of aqueous DHPC bilayers in the CH_2 bending (left) and in the CH_2 rocking region (right) for (A) DHPC and B) DHPC + TTC, pH 9.5. The band at 695 cm^{-1} is the phonon band of α-quartz used for the pressure calibration.

The pressure dependencies of the frequencies of the methylene bending mode components of 1,2-dimyristoyl-sn-glycero-3-phospho-L-serine (DMPS)in the absence and presence of TTC are shown in figure 10. A well-defined correlation field component band becomes apparent at 3.7 kbar for pure DMPS and at 2.8 kbar for both DMPS + TTC, pH 5.5, and DMPS + TTC, pH 9.5. The lower

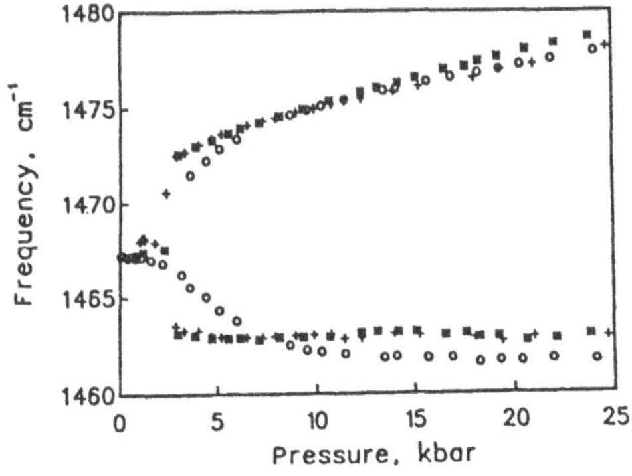

Fig. 10. Pressure dependencies of the frequencies of the CH_2 bending mode for DMPS (o), DMPS + TTC, pH 5.5 (■), and DMPS + TTC, pH 9.5 (+).

pressure required for the splitting in the presence of anesthetic indicates that the incorporation of TTC into DMPS bilayers increases the orientational ordering of

DMPS acyl chains in the gel phase. Thus, a lower pressure is necessary to stop the acyl chain reorientational fluctuations and induce a correlation field splitting. For the pure lipid, a further increase in pressure results in a relative rapid, nonlinear increase in the magnitude of this splitting while a sudden discontinuity with slight increase thereafter is observed in the presence of anesthetic. This suggests that in the pure lipid bilayers, the pressure-induced ordering of the acyl chains is gradual. The presence of TTC results in a large increase in ordering, then remains relatively constant with further increase in pressure. Apparently, the effects of TTC on the structure of DMPS are quite different from that of DMPC and DHPC. This is the result of strong interactions of TTC with the C=O and CO_2^- groups in DMPS (32).

STRUCTURE AND KINETICS IN POLYPEPTIDES AND PROTEINS:

The strong infrared band in the frequency region 1600-1700 cm^{-1} referred to as the amide I band is due to the in-plane C=O stretching vibration weakly coupled with C-N stretching and in-plane N-H bending of the amide groups in proteins and polypeptides (45-47). The peak maximum of the amide I band is sensitive to the conformational structure in polypeptides and proteins (45-47). Peak maximum is near 1650 cm^{-1} for α-helical structure. Two amide I bands at ~1635 and ~1685 cm^{-1} are due to β-sheet structure. The band at ~1685 cm^{-1} is mainly due to the amide I mode of the antiparallel β-sheet. The amide I bands of unordered random coils and turns are at ~1645 cm^{-1} and ~ 1665 cm^{-1}, respectively. Since each globular protein molecule contains segments of different substructures, the amide I band of a globular protein usually appears as a broad band with several maxima. The changes in the relative intensities of these maxima have been widely used for monitoring the changes in the conformational substructures in globular proteins (45-47).

Pressure-induced conformational changes in proteins (18, 48-51) and polypeptides (52) have been systematically studied in our laboratory by means of infrared and Raman spectroscopy. We have also studied the protective effect of model biomembranes against pressure-induced conformational changes in proteins and polypeptides (38). One of the most important results from these high pressure studies is that the correlation between the conformational state of proteins and their kinetic processes such as those in hydrogen isotope exchange and enzyme catalysis has been established (53, 54).

Infrared spectroscopy is a classical technique for the study of hydrogen/deuterium (H/D) exchange in proteins. The exchange rate in protein is measured by following the decrease in the intensity of the amide II band with time. This infrared band shifts to a frequency of 1458 cm^{-1} when the proton atoms of the amide group are replaced by deuterium atoms. One of the advantages of infrared spectroscopy over other physical methods for the H/D exchange study is that the conformational structure of the protein can be monitored simultaneously during the measurement of the H/D exchange rate constants. In the present lecture, results of the hydrogen isotope exchange kinetics in chymotrypsinogen are given.

Figure 11 shows the peak intensity ratio between amide II and amide I band as a function of time at several pressures. This intensity ratio before exchange occurs is 0.45 (47,55). With this value, the percentage of the exchanged labile protons in protein molecules can be calculated as a function of time. As seen in figure 11, the overall exchange rate increases with increasing pressure. However, the effect of pressure on the overall exchange rate is small below 3.5 kbar and becomes very large above 3.5 kbar. According to the standard graphic method, in

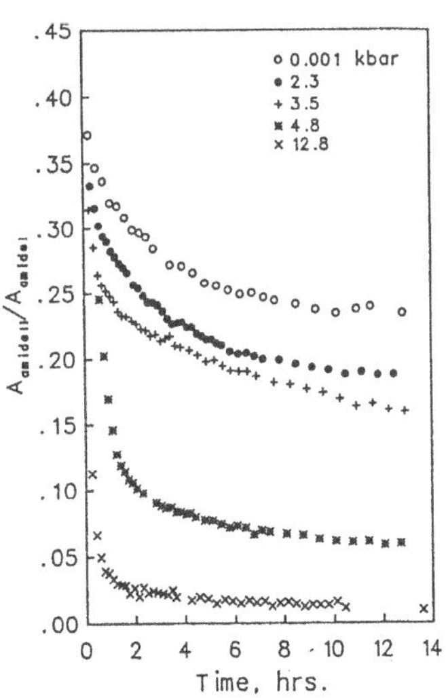

Fig. 11. Time dependencies of infrared intensity ratio between the amide II and amide I band of chymotrypsinogen at several pressures.

addition to the "extremely fast exchange" which occurs within 10 minutes, the first order H/D exchange rate can be distributed into two regions in many proteins

(56,57) i.e., the fast and slow exchange regions. In chymotrypsinogen, the fast exchange region is within the time period from 10 minutes to about 4 hours and the slow exchange is after 4 hours. The fast exchange rate constants at various pressures thus obtained are listed in Table 1 together with the percentage of labile protons which exchange within 10 minutes or do not exchange after 24 hours (hard-to-exchange protons).

TABLE 1 H/D exchange rate constants and percentage of unexchanged and extremely fast exchanged protons in chymotrypsinogen at several pressures.

Pressure	Hard to Exchange Proton	Proton Exchanged within 10 min	Exchange rate
kbar	%	%	k, min^{-1}
0.001	52	17	$3.7x10^{-2}$
2.3	42	26	$4.0x10^{-2}$
3.5	31	30	$4.8x10^{-2}$
4.8	11	36	$7.2x10^{-2}$
12.8	1	75	$1.9x10^{-1}$
28.9	32	47	$8.9x10^{-2}$

It is well known that pressure induces conformational structure changes in globular proteins(48-52). Figure 12 shows the deconvolved amide I band spectra of chymotrypsinogen at several pressures. Apparently, the amide I band consists of several overlapping bands with maxima at 1636, 1654, 1667 and 1686 cm^{-1} at pressures below 3.5 kbar. Therefore, the amide backbone in chymotrypsinogen consists of segments with the substructure of β-sheet, α-helical and unordered random coil and turns. The strongest component band is at 1636 cm^{-1}, which indicates that β-sheet conformation predominates over other substructures (45-47).

At pressures below 3.5 kbar, the intensity distribution among these component bands is hardly changed. Consequently, the conformational substructures in the protein are not affected by external pressure below 3.5 kbar. Starting from 3.5 kbar, the intensity in the frequency region between 1640 and 1650 cm^{-1} (random coil) and that near 1667 cm^{-1} (turns) increase with increasing pressure at the expense of the intensities of the component bands at 1636, 1686 (β-sheet) and 1654 cm^{-1} (α-helix). These results indicate that a pressure-induced denaturation associated with changes in conformational substructure in aqueous chymotrypsinogen starts at 3.5 kbar. In this denaturation process, the contributions of the unordered random coil and turns increase at the expense of those of the β-sheet and α-helical structures.

At 12.8 kbar the protein is completely denaturated and the amide backbones are in the random coil and turn structure and the remaining α-helix and β-sheet segments are largely distorted (49).

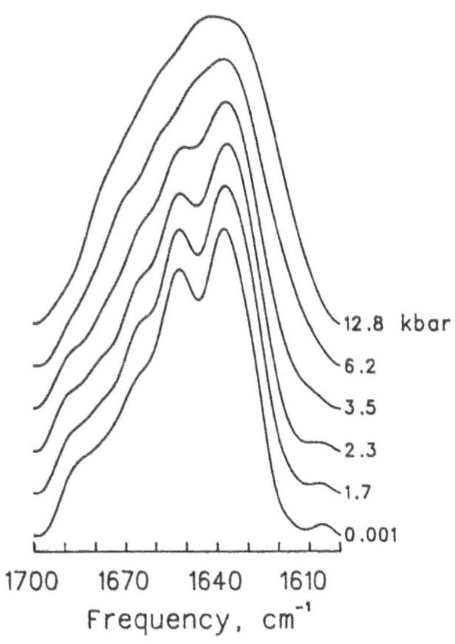

Fig. 12. Deconvoluted infrared spectra of chymotrypsinogen in the amide I band region at several pressures using a band width of 15 cm^{-1} and a resolution enhancement factor of 1.7.

As shown earlier, the exchange rate below 3.5 kbar changes only slightly with pressure and the exchange rate increases dramatically with pressure above 3.5 kbar at which the conformational structure of the protein starts to change. It is well established (46, 58-62) that the rate of H/D exchange of random coil segments and unordered turns is the fastest and that of β-sheet segment is the slowest. Therefore, the increase in the exchange rate constant with pressure at pressures above 3.5 kbar is the result of pressure-induced conversion of the amide backbone in the protein from α-helical and β-sheet conformation into the fast exchanged random coil and turns.

At 28.9 kbar, although the protein is completely denatured, the H/D exchange rate decreases instead of increases from that at 12.8 kbar (see figure 13 and Table 1). At 28.9 kbar bulk D_2O transforms into ice VII. The mobility of D_2O molecules in the ice VII state is much smaller than that in the liquid state. Therefore, the H/D exchange of labile protons in protein molecules with D_2O molecules in ice VII is expected to be prohibited. However, it is interesting that the H/D exchange still takes place continuously with time at 28.9 kbar. Therefore, there must be some D_2O molecules in this system which do not take part in the formation of ice VII at 28.9 kbar. Indeed, this is evident from the infrared spectra of the uncoupled OH stretching mode of HOD molecules which are present in the commercial D_2O samples.

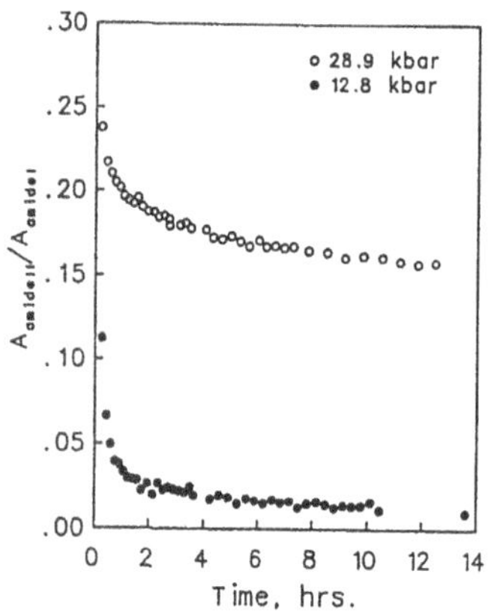

Fig. 13. Time dependencies of infrared intensity ratio between the amide II and amide I band of chymotrypsinogen at 12.8 and 28.9 kbar.

The uncoupled OH stretching mode of ice VII appears as a very sharp band at about 3390 cm^{-1} and that of the bound HOD molecules gives rise to a broad band extended from 3100 cm^{-1} to 3700 cm^{-1} (21, 63). Both the sharp ice VII band and the broad bound HOD band coexist in the infrared spectra of chymotrypsinogen at 28.9 kbar (see figure 14). Therefore, when solid chymotrypsinogen is dissolved in D_2O, some D_2O molecules bind them to the protein molecules immediately and do not take part in the formation of ice VII with bulk D_2O at 28.9 kbar.

Since the integrated intensity of the broad band of the bound HOD is comparable with that of the sharp ice VII band, a considerable amount of D_2O molecules are bound to the protein molecules when the protein is dissolved in D_2O. Consequently, in addition to the surface bound D_2O, some D_2O molecules penetrate into the interior cavities of the protein molecules. All these surface and interior bound D_2O molecules do not take part in the formation of ice VII at 28.9 kbar. The H/D exchange at 28.9 kbar takes place between these bound D_2O and the labile protons in the protein. Three distinct theories on the mechanism of H/D exchange in proteins have been proposed (58, 59, 64-66). The present results indicate that the H/D exchange mechanism in chymotrypsinogen is consistent with that of the "regional melting" model given by Kassiakoff (59) and is inconsistent with the "local unfolding" model and the "solvent penetration" mechanism (58, 64-66).

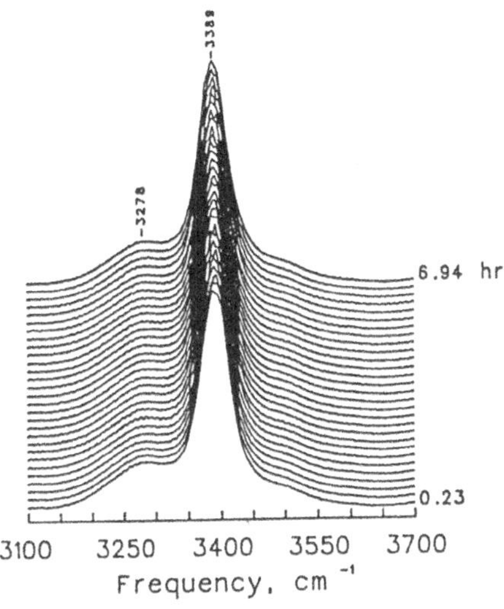

Fig. 14. A stack plot of the uncoupled OH stretching band of a D_2O solution of chymotrypsinogen at 28.9 kbar as a function of time.

Theoretical interests in volume changes for overall enzyme reaction and for intermediate steps in mechanism of enzyme reactions have attracted a number of studies in the area of pressure effects on enzyme activity and reaction rate (67-73). In our laboratory, high pressure infrared spectroscopy was applied for the first time in a series of such enzyme kinetic studies. This technique developed in our laboratory for enzyme kinetic study can reach much higher pressure than any of those reported (67-73). The results of one of our high-pressure enzyme kinetic studies, the conversion of p-nitrophenol phosphate disodium into p-nitrophenol by alkaline phosphatase, are given here. Contradictory results in the activation volume and pressure effects on the enzyme activity of this classical enzyme reaction have been reported. Penniston(67) and De Smedt et al. (68) have shown that the enzyme reaction of alkaline phosphatase is inhibited by pressure, whereas Greaney and Somero (69) and Kitano et al. (70) have reported a negative activation volume and thus pressure stimulates this enzyme reaction.

The rate constants of the enzyme conversion from p-nitrophenol phosphate to p-nitrophenol were determined at various pressures up to 14 kbar by monitoring the decrease with time in the intensity of the infrared band of the substrate at 887.6 cm^{-1} at these pressures. The rate constants thus obtained are plotted versus pressure in figure 15. The rate constant increases almost linearly with increasing pressure up to 7.2 kbar and then decreases abruptly at 8.3 kbar. Above 13.2 kbar the rate constant reaches zero. Therefore, the enzyme activity is stimulated by increasing pressure until 8.3 kbar at which the enzyme activity is dramatically inhibited. It is completely inhibited above 13.2 kbar.

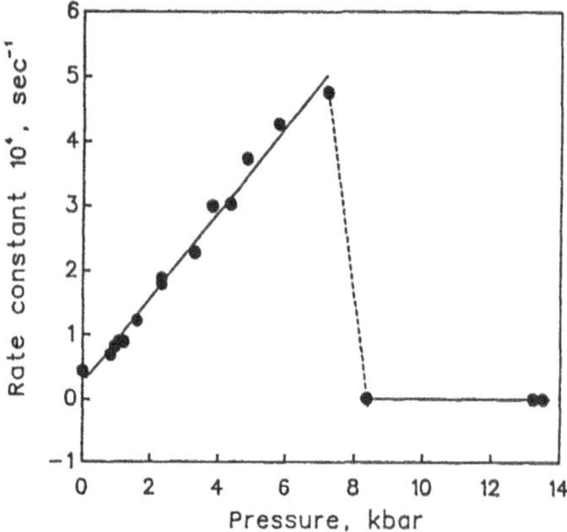

Fig. 15. Pressure dependence of the enzyme reaction rate constant.

The shape and intensity distribution of the amide I infrared bands of this enzyme are similar in the pressure range 0.001-7.2 kbar. The amide I band in this pressure range appears as a relatively narrow and symmetric band with the peak position at 1651 cm^{-1}, indicating the conformational structure of this enzyme in the pressure range 0.001-7.2 kbar is dominated by α-helix (45-47) Dramatic changes in both band shape and intensity distribution in the amide I band start at 8.3 kbar. At this pressure, the intensity of the component band near 1635 cm^{-1} increases at the expense of the intensity near 1651 cm^{-1}. This suggests that some α-helical segments in the enzyme molecules are converted into β-sheet structure by an external pressure of 8.3 kbar. This conversion is enhanced when pressure is further increased. It is clear from this conformational study that the lost of enzyme activity of alkaline phosphatase at pressures above 8.3 kbar is the result of denaturation of the enzyme arising from conformational changes in some regions of the enzyme molecules from α-helix to β-sheet structure.

According to transition state theory (74), the activation volume of the first order enzyme reaction can be calculated from the slope of the ln k / p plot, where k and p are rate constant and pressure, respectively. Figure 16 shows the ln k / p plot. A change in the slope is observed at about 3 kbar. Consequently, two activation volumes are obtained. They are -15.6 $cm^3 mol^{-1}$ (<3 kbar) and -4.4 $cm^3 mol^{-1}$ (> 3 kbar). Penniston (67) suggests that the activity of multimeric enzymes is inhibited (positive activation volume) whereas that of monomeric enzymes is stimulated (negative activation volume) by external pressure. The inhibition of activity of multimeric enzymes is interpreted in terms of pressure dissociation of multimeric enzyme in which their activities are highly dependent upon their protein-protein interactions among the corresponding subunits.

Alkaline phosphatase is a dimer in which the subunits are held together by noncovalent bonds (75). Our work showed this multimeric enzyme exhibits

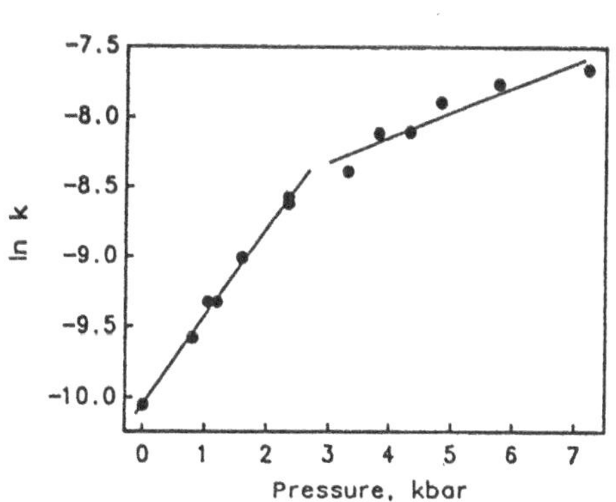

Fig. 16. Pressure dependence of ln k.

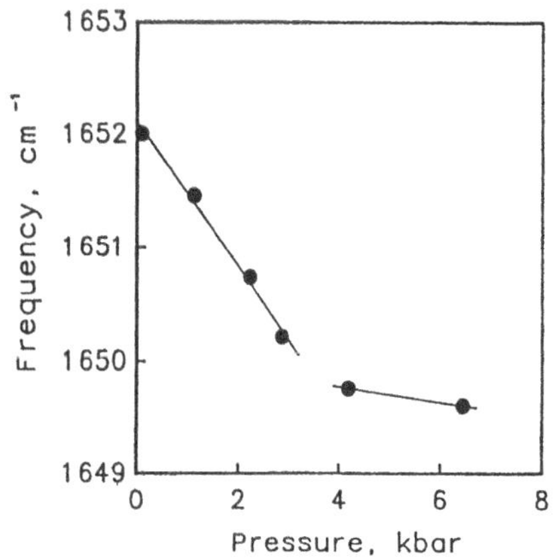

Fig. 17. Pressure dependence of the center of gravity of the amide I band of the enzyme

negative activation volume even in the pressure range above 3 kbar. At this pressure, most multimeric enzymes with noncovalent bonds have dissociated completely by hydration (76). Therefore, the present findings demonstrate that not all the multimeric enzymes exhibit positive activation volume and the activity of multimeric enzymes is not always inhibited by the application of hydrostatic pressure. Moreover, the protein-protein interactions in multimeric enzymes are not the major determinant for enzyme activity. These results also show that pressure-induced dissociation of the dimeric phosphatase enzyme only contributes to a change in activation volume in the pressure range above 3 kbar. Above this pressure, the dissociation of most protein oligomers is more or less complete (76).

Figure 17 shows the pressure dependence of the peak position of the amide I band measured according to its center of gravity. Apparently, a breakpoint between 3 and 4 kbar is observed. As mentioned earlier, the band shape and intensity distribution of the amide I band remain unchanged with increasing pressure in the pressure range 0.001-7.2 kbar. Therefore, the decrease in the frequency of this band from 0.001 to 7.2 kbar is primarily due to the pressure-induced strengthening of the hydrogen-bonds in the amide groups of the enzyme molecules (1, 45-47). Thus, the α-helical structure in the molecules becomes more rigid and the packing becomes tighter when the pressure increases. This pressure effect on the hydrogen bond strength is larger in the pressure range 0.001-3 kbar than in the pressure range 3-8 kbar and shows a discontinuity at 3 kbar. Note that this discontinuity in pressure enhancement of hydrogen bond strength coincides with the breakpoint in the activation volume (see figure 16). Consequently, one may deduce that the pressure-induced strengthening of the overall hydrogen bonding results in a negative activation volume and a pressure-induced increase in enzyme activity in the range 0.001-7.2 kbar.

DISSOCIATION AND AGGREGATION OF POLYPEPTIDES AND PROTEINS:

The effects of pressure on the dissociation of noncovalently bonded oligomeric proteins have been under active investigation (67). It has been reported that increasing hydrostatic pressure promotes protein dissociation through the hydration of salt bridges. On the other hand, we have found that at very high pressures, protein aggregation is promoted due to the formation of the intermolecular hydrogen bond networks among the amide groups of neighboring molecules (48, 49). For these protein aggregates, a low-frequency amide I band below 1620 cm^{-1} appears in the infrared spectra. With this infrared marker we have found that many human diseases have protein aggregation, as found in

Mallory Body of liver diseases and in Rothental fiber of brain diseases (77 and unpublished work).

To help ascertain the role of pressure in promoting protein dissociation due to salt bridge hydration, we have studied the effect of pressure on a model system consisting of an aggregated mixture of two homopolypeptides, poly-L-lysine (PLys) and poly-L-glutamate (PGlu), by FT-IR spectroscopy.

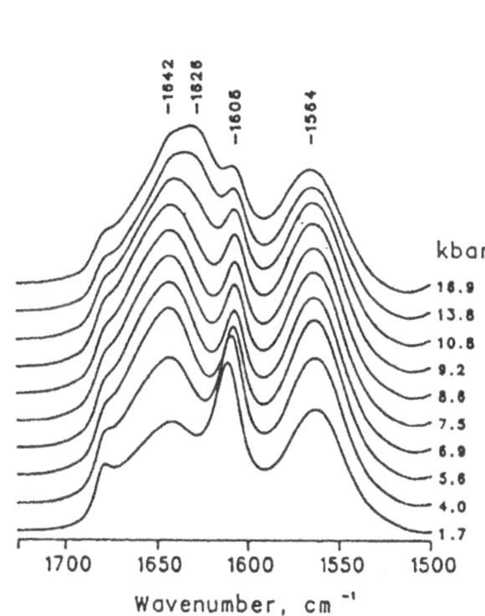

Fig. 18. A stack plot of the infrared spectra of the aqueous PLys/PGlu mixture in the amide I band region as a function of increasing pressure.

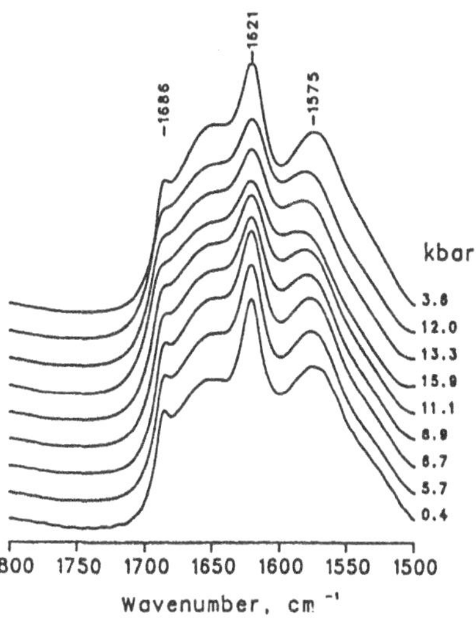

Fig. 19. A stack plot of the infrared spectra of the anhydrous solid of the PLys/PGlu mixture in the amide I band region as a function of increasing pressure.

When aqueous solutions of PLys and PGlu are mixed, salt bridges are readily formed between the negatively charged carboxylate groups of PGlu and the positively charged amino groups of PLys and a white gel-like aggregate is seen. The infrared marker of the intermolecular hydrogen-bond network involving the backbone amide groups appears in the infrared spectrum of this aggregate (see figure 18). When the pressure on this gel-like mixture is gradually increased, the opaque aggregate suspension becomes clearer and is eventually transformed into a transparent solution. In the infrared spectrum of this transparent solution the intermolecular hydrogen bond marker disappears. When pressure is reduced, the solution becomes cloudy and finally opaque, demonstrating the reversibility of this event. It has been found that the aggregate can also be solubulized in the presence of Ca^{2+} and Mg^{2+} ions (78). These results demonstrate that in this model system, salt bridges play a crucial role in promoting polypeptide aggregation. Therefore, with this model system we may then investigate the effects of pressure on salt bridge dissociation by monitoring the pressure-induced changes in the intermolecular hydrogen bond marker of the polypeptide aggregates by FT-IR spectroscopy, thereby gaining further insight about the mechanism by which pressure affects protein dissociation.

The infrared spectrum of a suspension of the aggregate reveals three bands in the amide I region at 1684 (w), 1645(w, br) and 1611 cm^{-1} (see figure 18), indicating an antiparallel β-sheet-like structure. The hydrogen bonds between the two chains are very strong, as indicated by the extremely low frequency of the amide I band (1611 cm^{-1}). The broad band at 1645 cm^{-1} indicates the presence of other substructures with random coil or α-helical conformations, as also observed for the individual homopolypeptides under ambient conditions (52 and unpublished results). Gradual increase in the hydrostatic pressure applied to a PLys/PGlu suspension in D_2O results in a decrease in the intensity of the 1611 cm^{-1} and 1684 cm^{-1} bands and an increase in the intensity of the 1645 cm^{-1} band (figure 18). Between 4 and 5.6 kbar the relative peak intensities of the 1611 and 1645 cm-1 bands are reversed, with the 1645 cm^{-1} band becoming the dominant band in the amide I region. The intensity of the 1611 cm^{-1} band continues to decrease with increasing pressure and becomes a weak shoulder at approximately 17 kbar. The amide I band observed at 1611 cm^{-1} near ambient pressure also shifts to 1606 cm^{-1} with increasing pressure. The 1645 cm^{-1} band continues to grow relative to the 1611 cm^{-1} band and gradually becomes more asymmetrical with increasing pressure. Above 10 kbar the 1645 cm^{-1} band splits into two bands at 1626 and 1638 cm^{-1}. Depressurization of the sample results in the decrease in the intensities of the 1626 and 1638 cm^{-1} bands and an increase in the intensities of the 1611 and 1684 cm^{-1} bands.

The above results suggest that the application of increasing hydrostatic pressure to a PLys/PGlu aggregate results in the breaking of salt bridges as evidenced by the dissociation of the hydrogen bonds between the peptides. The hydrogen bond dissociation is revealed by the decrease in the intensities of the bands at 1611 and 1684 cm^{-1}. The remaining hydrogen bonds become stronger with increasing pressure, which cause a shift of the amide I band from 1611 cm-1 to 1606 cm^{-1}. The decrease in the band intensity at 1611 cm^{-1} is also accompanied by an increase in the intensity of the 1645 cm^{-1} band. It indicates an increase in the random coil population of the mixture, attributed to the dissociation of the peptide aggregate with increasing pressure. With increasing pressure above 10 kbar, the random coil population decreases. The formation of α-helical structures is indicated by the appearance of the shoulder at approximately 1640 cm^{-1} (45-47). It is likely from the free PGlu in solution subjected to high pressure. The formation of some β-sheet structure is evidenced by the appearance of a band at 1626 cm^{-1}. It can be attributed to free PLys undergoing a transition from random coil to β-sheet under pressure (52). All of the above processes are reversed upon gradual depressurization of the sample, and the initial spectrum is obtained at ambient pressure.

To ascertain the role of water in hydration of the salt bridges and the subsequent separation of the two peptides with increasing pressure, a second series of pressure experiments were performed on a dry PGlu/PLys aggregate. At atmospheric pressure the drying of the suspension results in the shift of the component of the amide I band in the spectrum from 1684 and 1611 cm^{-1} to 1686 and 1621 cm^{-1}, respectively. The shift to higher frequency is a result of the loss of hydrogen bonding between the amide groups and the water molecules. The position and relative intensities of the 1686 and 1621 cm^{-1} components of the amide I band are not affected significantly as the increasing pressure is applied to the dry PGlu/PLys aggregate (figure 19). This indicates that pressure can not dissociate the salt bridges without the presence of water.

The above pressure experiments demonstrate that increasing pressure results in dissociation of the homopolypeptide aggregate and that water molecules play a crucial role in driving the polypeptide apart. These experiments also demonstrate that once separated, the homopolypeptide independently undergo conformational changes with increasing pressure.

BIOLOGICAL AND BIOMEDICAL STUDIES:

While molecular biology has made revolutionary contributions towards our understanding of the life and regulation of the cell, additional approaches will be needed to study biological phenomena with enhanced resolution. Pressure-tuning infrared spectroscopy has been shown to be a powerful method to study not only the structure of chemical compounds but also their relationship to surrounding molecules (1). Therefore, this method may be applied to investigate structural properties at the molecular level in biological tissues and cells. The application of pressure-tuning infrared spectroscopy to biological tissues and cells has been hampered by problems in sample preparation for optimal spectra acquisition. By resolving technical and methodological problems (7-9), we are now able to apply this technique to the study of biological tissues and isolated cells (9-16).

The main features in the infrared spectra of biological tissues and cells are similar to those in the infrared spectrum of a mixture with proper composition of proteins, nucleic acids, lipids and carbohydrates. These are the main constituents in biological tissues and cells (79). Bands in the infrared spectra of tissues and cells have been assigned by comparing the infrared spectra of tissues and cells with those of nuclei, DNA, RNA, lipids and proteins isolated from tissues and cells (13).

The advances in the pressure-tuning infrared spectroscopic technique for biological tissues and cells allow us to apply this technique to investigate the molecular bases of a wide range of biological and biomedical problems. We have studied alcohol-induced liver diseases (11), carcinogen-induced liver tumors in mice (14), interactions between anesthetics and nerves (30), Mallory body in liver diseases (77), cultured human cell lines (16), and structural modifications at the molecular level in human cancers (9, 10, 12). Unpublished studies in this area include: (i) Rothental fiber in brain diseases, (ii) cell anomalies in infected thymus glands, salivary glands and Hela cells, (iii) effects of hepatectomies and carcinogens on liver in mice, (iv) aging tissues, (v) interactions of anti cancer drugs with biomembranes and DNA, (vi) interactions of toxins with biomembranes, (vii) differentiation between resistant and sensitive ovarian and lung cells to cisplatin, and (viii) differentiation between resistant and sensitive malignant tumors to radiotherapy. In this lecture, some results on human cancers are presented.

The infrared spectra of a pair of normal and malignant colon tissues in the frequency region 1000-1350 cm^{-1} are shown in figure 20. The spectrum of the normal colon tissue and its changes in the malignant tissue are similar to those of other tissues such as tissues of esophagus, stomach, skin and breast.

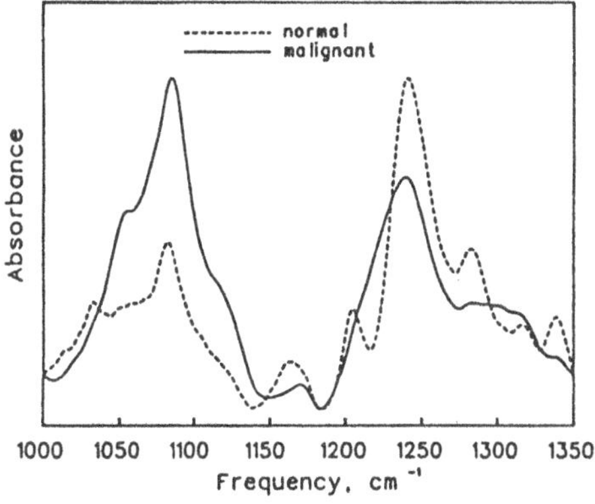

Fig. 20. Infrared spectra of a pair of normal and malignant colon tissue sections in the frequency region 1000-1350 cm^{-1}.

Figure 21 shows the infrared spectra of exfoliated cells from a pair of normal and malignant cervical tissues in the frequency region 950-1360 cm^{-1}. These spectra are similar to those of other glycogen-rich tissues such as the liver and vagina tissues.

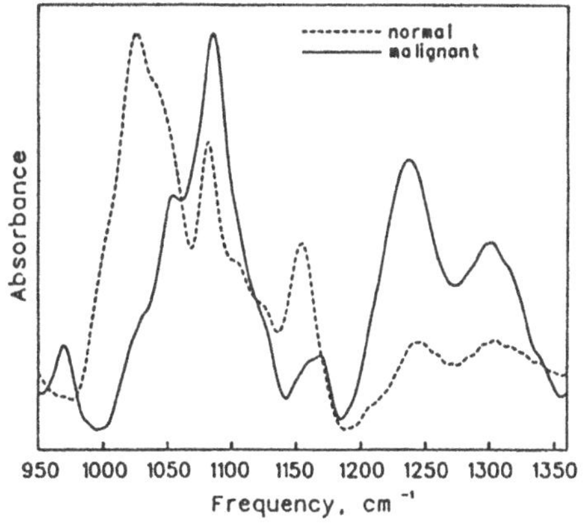

Fig. 21 Infrared spectra of normal and malignant cervical cells in the frequency region 950-1360 cm^{-1}.

By mean of pressure-tuning infrared spectroscopic studies, we found that many structural changes at the molecular level are common to all the cancers that we have studied so far. For instance, the number of hydrogen-bonded phosphodiester groups in DNA increases dramatically in cancer cells as indicated by the increase in the intensity of the low-frequency component band of the antisymmetric phosphate stretching mode of the phosphodiester groups (10, 12). The frequency of this low-frequency component band decreases with increasing

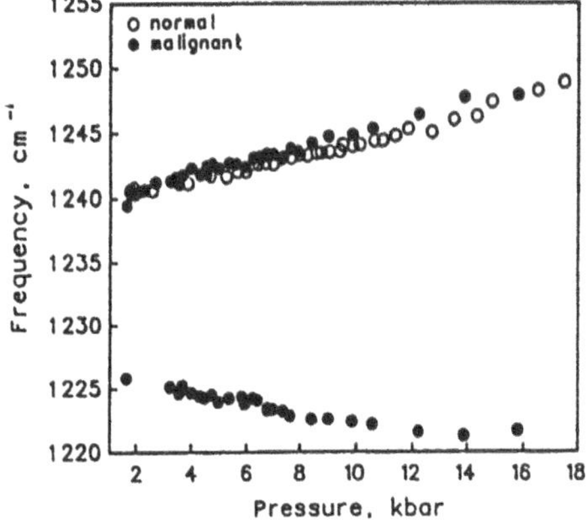

Fig. 22. Pressure dependencies of the frequency of the asymmetric phosphate stretching mode. The band from malignant tissue has two component bands.

pressure (figure 22), which provides further evidence for the hydrogen-bonded characteristics of this component band.

The interchain interactions between the neighboring nucleic acid molecules are enhanced due to a closer interchain packing in cancer cells. This is evident from the increase in the stretching frequency of the phosphodiester groups in cancerous tissues and cells (9, 12).

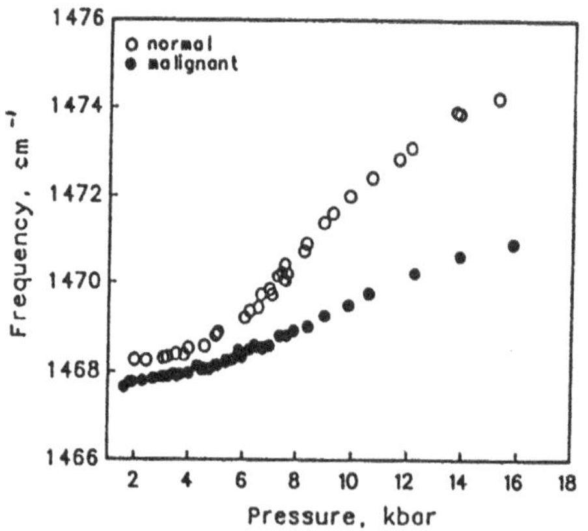

Fig. 23. Pressure dependence of the bending mode frequency of the methylene chain in membrane lipids.

The methylene chains of the membrane lipids become more disordered in cancer cells as shown by the smaller pressure-induced frequency shift of the vibrational modes of the methylene chains in cancer cells (see figure 23). There are many other structural changes at the molecular level in cancerous tissues and cells, which are given elsewhere (9-12, 80).

Results from our studies have demonstrated that there are a number of advantages for using pressure-tuning infrared spectroscopic technique to study biological tissues and cells. Firstly, only tiny amounts (less than 0.01 mg) of unprocessed (i.e., without staining and fixation) tissue sections and cells are needed. Secondly, molecules are examined in their natural state in the intact cell and therefore, their physical state and interactions with other molecules in the cells can be studied. For instance, changes in hydrogen bonding of nucleic acid in cancer could not be evaluated in isolated nucleic acids, because they would be stripped of histones and form hydrogen bonds with solvent in the process. Finally, several molecules can be monitored simultaneously. Aspects of the amount, structure and interactions of proteins, lipids nucleic acids and carbohydrates can be evaluated in a single tissue.

Our findings in cancers have prompted the consideration whether infrared spectroscopy has any value as a diagnostic tool in cancer and other diseases. There are obvious advantages in using infrared spectroscopy in diagnosis. It is rapid, inexpensive, automatable and requires minimal amounts of samples. It also compares most favorably with the existing methods in terms of cost and accuracy. A role as an adjunct for screening of large-volume histological examinations, such as cervical or bronchial smears, is also possible.

In conclusion, our studies have demonstrated that pressure-tuning vibrational spectroscopy is a powerful research tool in molecular biology and life science, and possesses the potential to be utilized in clinical diagnosis.

REFERENCES

1. P. T. T. Wong in "Vibrational Spectra and Structure," J. R. Durig , Ed., Elsevier, Amsterdam **16**, 357 (1987).
2. P. T. T. Wong, W. F. Murphy, and H. H. Mantsch, J. Chem. Phys. **76**, 5230 (1982).
3. P. T. T. Wong, Ann. Rev. Biophys. Bioeng. **13**,1 (1984).
4. P. T. T. Wong, in "High Pressure Chemistry and Biochemistry," NATO ASI Ser. Ser. C Math. Phys. Sci. **197**, 381 (1987).

5. P. T. T. Wong, in " Current Perspective in High Pressure Biology," H. W. Jannash, R. E. Marquis and A. M. Zimmerman, Eds., Academic Press, London, pp.287-314 (1987).
6. P. T. T. Wong, D. J. Siminovitch and H. H. Mantsch, Biochim. Biophys. Acta, **947**, 139 (1988).
7. P. T. T. Wong, US Patent No. 4970396 (1990).
8. P. T. T. Wong, US Patent No. 4980551 (1990).
9. P. T. T. Wong, and B. Rigas, Appl. Spectrosco. **44**, 1715 (1990).
10. B. Rigas, S. Morgello, I. S. Goldman and P. T. T. Wong, Proc. Natl. Acad. Sci. USA **87**, 8140 (1990).
11. H. Takahashi, S. W. French, and P. T. T. Wong, Alcohol Clin. Exp. Rev., **15**, 219 (1991).
12. P. T. T. Wong, R. K. Wong, T. A. Caputo, T. A. Godwin and B. Rigas, Proc. Natl. Acad. Sci. USA, **88**, 10988 (1991).
13. P. T. T. Wong, E. D. Papavassiliou and B. Rigas, Appl. Spectrosco. **45**, 1563 (1991).
14. P. T. T. Wong, M. Cadrin and S. W. French, Exp. Mol. Pathology, **55**, 269 (1991).
15. M. Cadrin, S. W. French, and P. T. T. Wong, Exp. Mol. Pathology, **55**, 170 (1991).
16. B. Rigas and P. T. T. Wong, Cancer Research, **52**, 84 (1992).
17. D. J. Siminovitch, P. T. T. Wong and H. H. Mantsch, Biophys. J. **51**, 465 (1987).
18. D. J. Siminovitch, P. T. T. Wong and H. H. Mantsch, Biochemistry, **26**, 3277 (1987).
19. D. J. Siminovitch, P. T. T. Wong, R. Berchtold and H. H. Mantsch, Chem. Phys. Lipids, **46**, 79 (1988).
20. P. T. T. Wong and H. H. Mantsch, Biophys. J., **54**, 781 (1988).
21. P. T. T. Wong and C. Huang, Biochemistry, **28**, 1259 (1989).
22. W. Hubner, P. T. T. Wong and H. H. Mantsch, Biochim. Biophys. Acta, **1027**, 229 (1990).
23. R. N. A. H. Lewis, D. A. Mannock, R. N. McElhaney, P. T. T. Wong and H. H. Mantsch, Biochemistry, **29**, 8933 (1990).
24. C. T. Choma and P. T. T. Wong, Chem. Phys. Lipids, **61**, 131 (1992).

25. S. Tupper, N. Tanphaichitr and P. T. T. Wong, Biochemistry, (1992) in press.
26. R. W. Williams and D. Chapman Prog. Chem. Fats other Lipids, **11**, 3 (1970).
27. R. G. Snyder, J. Mol. Spectrosc. **7**, 116 (1961).
28. P. T. T. Wong and H. H. Mantsch, Chem. Phys. Lipids, **46**, 213 (1988).
29. Y. Koyama and K. Ikeda, Chem. Phys. Lipids, **26**, 149 (1980).
30. M. Auger, H. C. Jarrell, I. C. P. Smith, P. T. T. Wong, D. J. Siminovitch and H. H. Mantsch, Biochemistry, **26**, 8513 (1987).
31 M. Auger, H. C. Jarrell, I. C. P. Smith, D. J. Siminovitch, H. H. Mantsch and P. T. T. Wong, Biochemistry, **27**, 6086 (1988).
32. M. Auger, I. C. P. Smith, H. H. Mantsch, and P. T. T. Wong, Biochemistry, **29**, 2008 (1990).
33. P. T. T. Wong and D. Zakim, J. Phys. Chem. **94**, 5052 (1990).
34. P. T. T. Wong, S. E. Capes and H. H. Mantsch, Biochim. Biophys. Acta, **980**, 37 (1989).
35. P. T. T. Wong, T. E. Chagwedera and H. H. Mantsch, Biophys. J., **56**, 845 (1989).
36. D. Zakim and P. T. T. Wong, Biochemistry, **29**, 2003 (1990).
37. P. L. G. Chong, S. E. Capes and P. T. T. Wong, Biochemistry, **28**, 8358 (1989).
38. D. Carrier, H. H. Mantsch and P. T. T. Wong, Biochemistry, **29**, 254 (1990).
39. R. B. Philp. D. J. McIver and P. T. T. Wong, Biochim. Biophys. Acta, **1021**, 91 (1990).
40. A. S. Davydov, Theory of Molecular Excitons, McGraw Hill, New York, (1962).
41. D. P. Graig and S. H. Walmsley, Excitons in Molecular Crystals, W. A. Benjamin, Inc., (1968).
42. R. M. Hexter, J. Chem. Phys. **33**, 1833 (1970).
43. D. J. Siminovitch, P. T. T. Wong and H. H. Mantsch, Biochim. Biophys. Acta, **900**, 163 (1987).
44. S. A. Simon, T. J. McIntosh and M. L. Hines, in "Molecular and Cellular Mechanisms of Anesthetics," S. H. Roth, and K. W. Miller, Eds., Plenum, New York, pp. 297-308 (1986).
45. H. Susi in "Structure and Stability of Biological Macromolecules," S. N. Timsshell and C. D. Fasman, Eds. Marcel Dekker, New York (1969).

46. F. S. Parker, "Applications of Infrared Spectroscopy in Biochemistry, Biology and Medicine," Plenum Press, New York (1971).
47. F. S. Parker "Applications of Infrared, Raman and Resonance Raman Spectroscopy in Biochemistry," Plenum Press, New York (1983).
48. K. Heremans and P. T. T. Wong, Chem. Phys. Letters, **118**, 101 (1985).
49. P. T. T. Wong and K. Heremans, Biochim. Biophys. Acta, **956**, 1 (1988).
50. P. T. T. Wong, I. S. Girons, Y. Guillou, G. N. Cohen, O. Barzu and H. H. Mantsch, Biochim. Biophys. Acta, **996**, 260 (1989).
51. A. Muga, W. K. Surewicz, P. T. T. Wong and H. H. Mantsch, Biochemistry, **29**, 2925 (1990).
52. D. Carrier, H. H. Mantsch and P. T. T. Wong, Biopolymers, **29**, 837 (1990).
53. P. T. T. Wong, Can. J. Chem., **69**, 1699 (1991).
54. P. T. T. Wong and D. W. Armstrong, Biochim. Biophys. Acta, (1992), in press.
55. L. Willumsen, Compt. Rend. Trav. Lab. Carlsberg, **36**, 327 (1968).
56. A. Hvidt, Compt. Rend. Trav. Lab. Carlsberg, **33**, 475 (1963).
57. K. R. Bhaskar and F. S. Parker, J. Biol. Chem. **245**, 3302 (1970).
58. C. K. Woodward and B. D. Hilton, Ann. Rev. Biophys. Bioeng. **8**, 99 (1979).
59. A. A. Kossiakoff, Nature, **296**, 713 (1982).
60. H. B. Bensusan and S. O. Nielsen, Biochemistry, **3**, 1367 (1964).
61. E. R. Blout, C. DeLoze and A. Asadourian, J. Am. Chem. Soc., **83**, 1895 (1961).
62. G. Wagner and K. Wthrich, L. Mol. Biol., **134**, 75 (1979).
63. P. T. T. Wong and D. J. Moffatt, Appl. Spectrosco. **41**, 1070 (1987).
64. A. D. Barksdatl and A. Rosenberg, Methods Biochem. Anal., **28**, 135 (1982).
65. C. K. Woodward, I. Simon and E. Tchsen. Mol. Cell. Biochem., **48**, 135 (1982).
66. S. W. Englander and J. J. Englander, in "Structure and Dynamics: Nucleic Acids and Proteins," I. Clementi and R. H. Sarma, Eds., Academic Press, New York (1983).
67. J. T. Penniston, Arch. Biochem. Biophys., **142**, 322 (1971).
68. H. De Smedt, R. Borghgraef, F. Ceuterick and K. Heremans, Biochim. Biophys. Acta, **556**, 479 (1979).
69. G. S. Greaney and G. N. Somero, Biochemistry, **18**, 5322 (1979).
70. H. Kitano, K. Nakamura and N. Ise, J. Appli. Biochem., **4**, 34 (1982).

71. P. L. G. Chong, P. A. G. Fortes and D. M. Jameson, J. Biol. Chem., **27**, 14484 (1985).

72. P. Butz, K. O. Greulich and H. Ludwig, Biochemistry, **27**, 1556 (1988).

73. K. Heremans and F. Wuytack, FEBS letters **117**, 161 (1980).

74. E. Morild, in "Advances in Protein Chemistry," C. B. Anfinsen, J. T. Edsall and F. M. Richards, Eds., Academic Press, New York, Vol. **34**, pp.93-167 (1981).

75. I. M. Klotz and P. W. Darnall, Science, **166,** 126 (1969).

76. G. Weber, in "High Pressure Chemistry and Biochemistry" NATO ASI Series C, Vol. **197**, 401 (1987).

77. K. Kachi, P. T. T. Wong, M. Cadrin and S. W. French, J. Cell Biol., **115**, 45a (1991).

78. A. A. Ismail and H. H. Mantsch, Biochim. Biophys. Acta, (1992), in press.

79. P. Sheeler and D. E. Bianchi, "Cell Biology: Structure, Biochemistry and Function," 2nd Ed., Wiley, New York, (1980).

80. P. T. T. Wong and B. Rigas, Can. Chem. News, Nov./Dec. 14 (1991).

VOLUMETRIC PROPERTIES OF MODEL BIOMEMBRANES

R. WINTER and M. BÖTTNER
Ruhr-University of Bochum
Institute of Physical Chemistry
Universitätsstraße 150
D-4630 Bochum 1, F.R.G.

ABSTRACT. In this contribution volumetric measurements on dilute aqueous dispersions of the model biomembrane dimyristoylphosphatidylcholine at temperatures ranging from 0 to 90 ^{0}C and at pressures from ambient up to 2500 bar are presented. The thermal expansion coefficient, the isothermal compressibility, and the lipid volume fluctuations of the model membrane in its various thermotropic and barotropic phases and an accurate temperature-pressure phase diagram for the liquid-crystalline to gel phase transformation of the lipid could be obtained from these experiments. In addition to the results of the pure phospholipid system, the influence of the local anaesthetic tetracaine on the thermodynamic properties and the phase behaviour of the membrane is discussed. As tetracaine can be viewed as a model system for amphiphilic molecules, these results provide insight into the general understanding of the physico-chemical action of many amphiphilic molecules on membranes. The biological relevance of these kind of studies is discussed.

1. INTRODUCTION

The polymorphism of model biomembrane systems has been the subject of many studies, both experimentally and theoretically in recent years [1]. In particular, the lamellar phases that are formed by phospholipids in excess water have attracted considerable interest because they can serve as model systems for biological membranes. Besides the variety of lyotropic phases, phospholipids also exhibit a variety of thermotropic phase transitions, such as the temperature

R. Winter and J. Jonas (eds.), High Pressure Chemistry, Biochemistry and Materials Science, 545–560.

dependent lamellar gel to gel ($L_{\beta'}$-$P_{\beta'}$) pretransition and the gel to liquid-crystalline ($P_{\beta'}$-L_{α}) main transition at higher temperatures (see Figure 1).

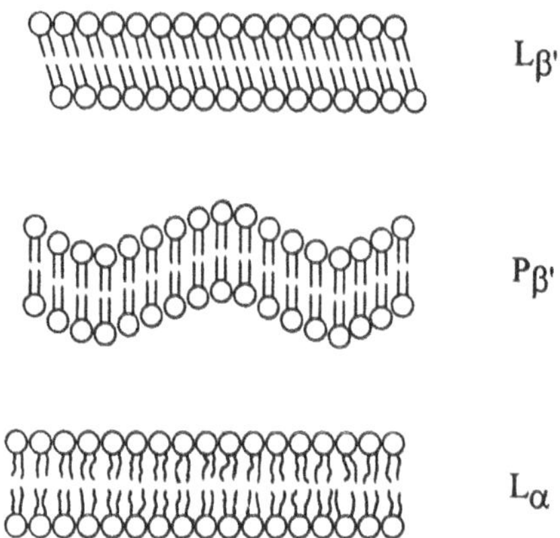

Fig. 1 Structures of phospholipid bilayers in their lamellar thermotropic phases (L_{α} liquid-crystalline phase, $P_{\beta'}$ "ripple" gel phase, $L_{\beta'}$ planar bilayer gel phase).

In the liquid-crystalline state, the hydrocarbon chains of the lipid bilayers are conformationally disordered, i.e. the acyl chains undergo extensive trans-gauche isomerizations, but the average chain orientation is perpendicular to the bilayer surface. This liquid-like state has been proposed as a requirement of optimal biological function of biomembranes. In the gel phases, the hydrocarbon chains are extended and relatively ordered. However, the lipid molecules can differ in bilayer surface structure and lipid chain packing. For dimyristoylphosphatidylcholine (DMPC) it has been shown that the $P_{\beta'}$-gel phase, which exists between 14 ^{o}C and 24 ^{o}C, has a two dimensional lattice structure in which the lipid bilayers are distorted by a periodic ripple in the plane of the lamellae, whereas the lower temperature $L_{\beta'}$ gel phase exhibits a planar bilayer surface.

In additon to these thermotropic phase transitions, further pressure-induced phases have been observed (for a review, see e.g. [3-5]). Besides the theoretical interest in high pressure phase behaviour of amphiphilic molecules, high pressure is also of considerable physiological significance. For example, pressure studies on lipid systems are of interest in understanding the physiology of deep-sea organisms, the sensitivity of excitable cell membranes to pressure, and the antagonistic action of pressure to anaesthetic action, which has been observed in vivo [6,7].

These phospholipid phase transitions have attracted a lot of experimental attention, not only because they exhibit interesting complex phenomena themselves, but also because they intimately reflect the molecular interactions of the membrane. Moreover, the study of the nature of the lipid bilayer phase transitions and, in particular, how they are affected by the incorporation of other species interacting with these membranes, such as cholesterol, proteins, ions and anaesthetics, may help in understanding membrane systems and function on a molecular level.

For a complete analysis and understanding of the phase behaviour of lipid dispersions, temperature dependent volumetric data are required at different pressures. However, only a few studies have been performed so far and these were mainly restricted to low pressures and to pure lipid systems, only [8-24]. Recent theoretical work has elucidated the strong need for experimental volumetric data [25-27]. Compressibility and expansibility data, that can be obtained from these measurements, are essential physical characteristics reflecting intermolecular interactions and dynamic processes and are of great importance in the understanding of the structure-activity relationship of biological systems. Moreover, dilatometric techniques can detect higher order phase transitions which generally cannot be determined by calorimetric techniques, and they are a very useful means to determine phase diagrams, because they neither perturb the lipid structure with a probe molecule nor do they depart significantly from thermal equilibrium measurements like many calorimetric techniques do.

2. EXPERIMENTAL TECHNIQUE

The PVT-measurements were carried out with a home-built apparatus, which is schematically shown in Figure 2. The high pressure cell is made from stainless steel and it has an internal boring of 20 mm. It has been designed for measurements in the temperature range from 0 - 90 ^{o}C and at pressures from ambient up to 2500 bar. The cell is closed at the top end by a high pressure flange which is sealed by a viton O-ring. A 5 wt.-% mixture of lipid in destilled water is prepared, degassed and filled into the cell holding about 7.5 cm^3. A bellow made from stainless steel separates the sample volume from the pressurizing medium. The volume change of the lipid dispersion is measured as a function of pressure and temperature by the elongation of the bellow which is determined by an inductive method outside the pressure vessel. The elongation is monitored with the aid of a micrometer screw via the displacement of a small ferromagnetic ball, which is mounted on a thin molybdenum rod sitting inside a high pressure 1/4" pipe, and which is connected to the bottom end of the bellow. The pressure is applied by means of a screw-type pressure generator and recorded by a Heise pressure-gauge. Temperature control is achieved by circulating water from a thermostat through the outside jacket of the pressure cell.

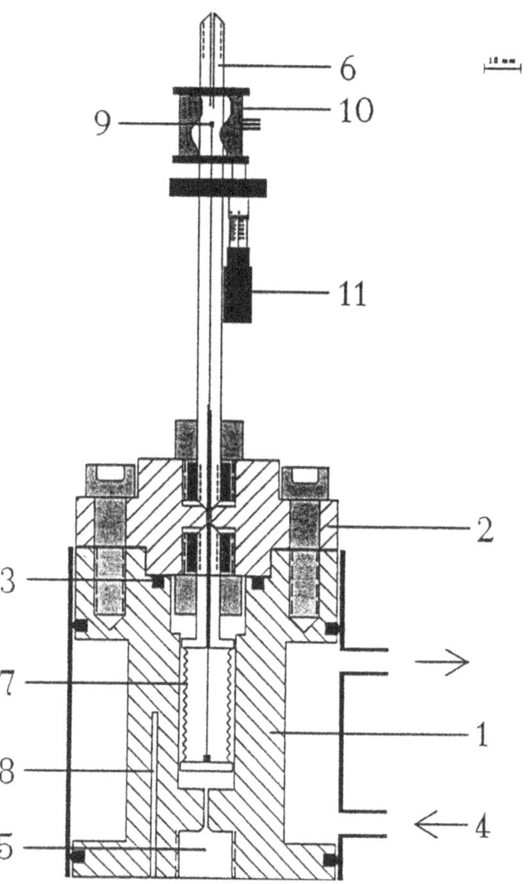

Fig. 2 Cross section of the volumetric cell (1 pressure cell, 2 top flange, 3 viton O-ring, 4 thermostating jacket, 5 high pressure nut, 6 high pressure pipe, 7 bellow, 8 bore for thermocouple, 9 magnetic ball, 10 inductive coil, 11 micrometer screw).

In order to obtain absolute volumetric data, calibration measurements with a sample of well-known PVT-data have to be performed. For that purpose we have used water because its PVT-data are known in a wide range of pressures and temperatures with high accuracy [28,29]. The apparent specific volume V_L^* of the dispersed lipid is obtained by substracting the corresponding water value, knowing the density of water and the mass fraction of the lipid in the dispersion. It normally is more appropriate to report partial specific volumes when dealing with solutions. However, for bilayer dispersions in excess water there is no essential difference between partial and apparent specific volumes [19].

3. RESULTS AND DISCUSSION

3.1 PURE PHOSPHOLIPID DISPERSIONS

Figure 3 shows data for the apparent specific lipid volume V_L^* of a 5 wt.-% DMPC dispersion as a function of temperature. With increasing temperature, the lipid volume increases. The small change of V_L^* near 14 °C corresponds to a small volume change in course of the $L_{\beta'}$ - $P_{\beta'}$ gel to gel transition [16,19]. Clearly, the change in volume at the gel to liquid-crystalline phase transition around 24 °C can be seen, which is accompanied by a 3 % volume change ($\Delta V_m = 0.028$ cm^3/g $\hat{=}$ 32 Å^3/molecule). The volume change in course of the main transition is mainly due to changes of the chain cross-sectional area, because chain disorder drastically increases at the transition point. Beyond the main transition, a continous increase in lipid volume in the disordered liquid-crystalline phase is observed. These data are typical for bilayer systems and have been observed in other phospholipid systems, too.

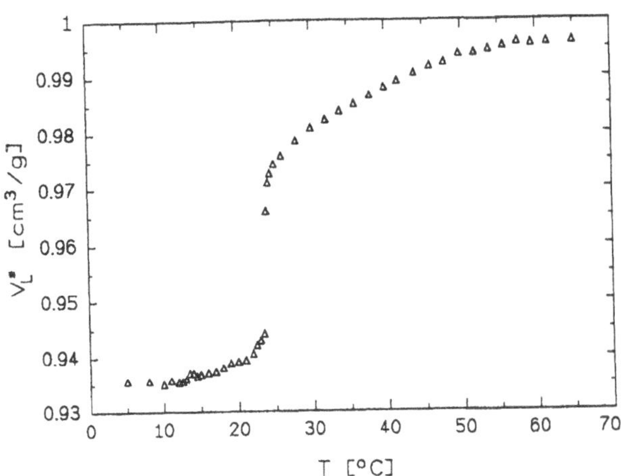

Fig. 3 The apparent specific volume V_L^* of DMPC in H_2O as a function of temperature.

Figure 4 exhibits the isobaric thermal expansion coefficient α_p of the lipid over the whole temperature range measured, which has been obtained from the data shown in Figure 3. With $\alpha_p = 6 \cdot 10^{-4}$ °C^{-1} near 15 °C, the expansion coefficient of the lipid in the gel phase is about half of that in the liquid-crystalline phase ($\alpha_p = 11 \cdot 10^{-4}$ °C^{-1} at T = 30 °C). At the main transition, α_p drastically increases.

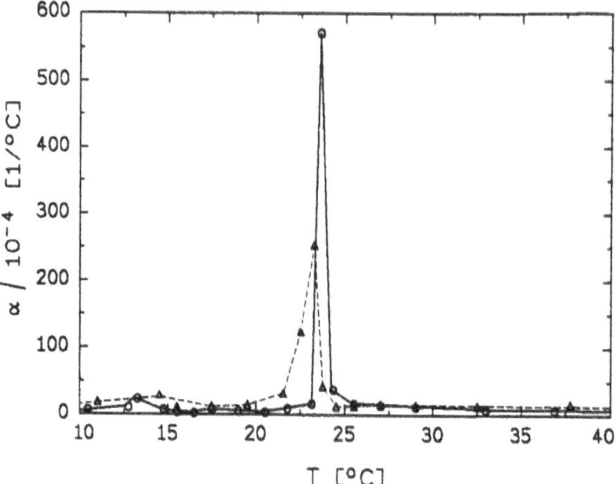

Fig. 4 Thermal expansion coefficient α_p of pure DMPC (o) and 3 mol-% TTC/DMPC (▲) in H_2O as a function of temperature (p = 1 bar).

As an example of a pressure dependent volumetric study, Figure 5 displays V_L*(p) of DMPC at T = 30 ^oC. By increasing the pressure, the phase transition into the ordered gel-phase can be induced, as can be seen from the abrupt decrease of the lipid volume within a rather narrow pressure range of about 10 bar around 265 bar.

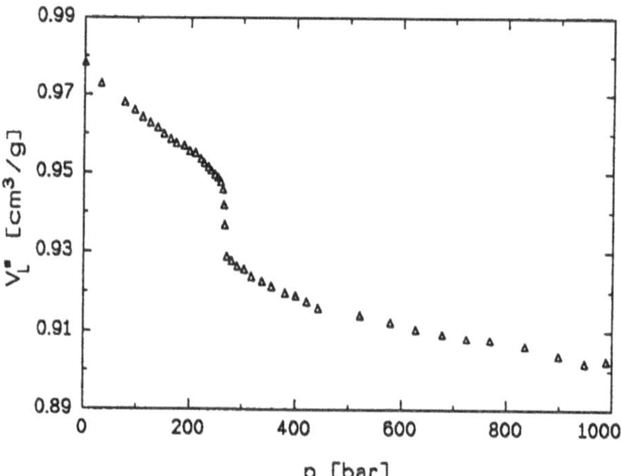

Fig. 5 The apparent specific volume V_L* of DMPC in H_2O as a function of pressure at T = 30 ^oC.

Figure 6 shows the measured specific lipid volume changes of DMPC dispersions at the main transition as a function of pressure. The volume change at the main transition seems to decrease only slightly with increasing temperature along the main transition line.

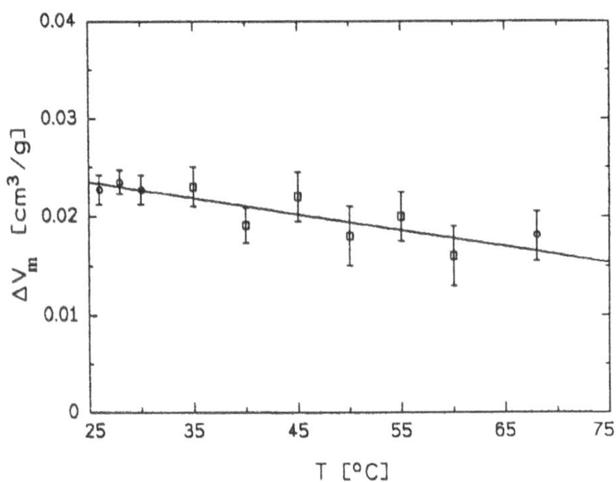

Fig. 6 The pressure-induced volume change ΔV_m at the main transition for pure DMPC dispersions taken at different temperatures.

In an elementary interpretation of the data, the lipid volume can be partitioned into two parts: $V_L = V_C + V_H$, where V_C denotes the volume of the hydrocarbon chains, and V_H the volume of the head group region. From X-ray measurements it has been concluded that $V_H \approx 340$ Å^3 per molecule for DMPC in the whole temperature region [13]. Using the volumetric data, one can calculate the average volume per CH_2-group, V_{CH_2}, in the hydrocarbon chain region as a function of temperature. The calculation shows that the mean volume of the methylen group increases ca. 1.5 Å^3 at the transition point. Outside the transition region, V_{CH_2} only slightly increases with increasing temperature.

From the transition midpoints of the isotherms like those in Figure 5, the PT-phase diagram of the main transition has been constructed as shown in Figure 7. Up to pressures around 1200 bar, a linear increase of the phase transition curve with 22 °C/kbar is observed, whereas for higher pressures, clearly a negative curvature is visible. A positive transition slope has also been observed for various other phospholipid systems (see e.g. [2-5,30,31]).

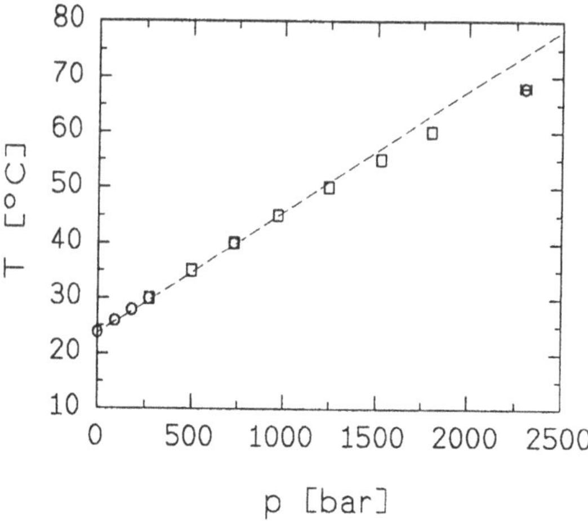

Fig. 7 Transition temperature of the main phase transition of DMPC multilamellar vesicles in H_2O as a function of pressure. The full line is only a guide to the eye demonstrating the decrease in slope for the higher pressures or temperatures, respectively.

Assuming a first-order phase transition for the liquid-crystalline to gel transformation, the entropy and enthalpy changes at the main transition can be calculated, employing the Clausius-Clapeyron equation: $dp/dT = \Delta S_m / \Delta V_m = \Delta H_m/(T_m V_m)$. With the measured values of T_m (1bar) = 24 °C and ΔV_m(1bar) = 19 cm^3/mol, an enthalpy change of ΔH_m = 26 kJ/mol and an entropy change of ΔS_m = 86 Jmol^{-1}K^{-1} is obtained. These values are in very good agreement with direct calorimetric measurements at ambient pressure. With increasing temperature or pressure, respectively, the enthalpy and entropy changes at the main transition vary only slightly.

From the volumetric data of Figure 5, also the isothermal compressibility coefficient χ_T of pure DMPC can be obtained. Figure 8 shows χ_T(p) of DMPC for two selected temperatures. It appears that the compressiblity of the $P_{\beta'}$ gel phase is substantially lower than that of the L_α-phase (for e.g. T = 30 °C: $\chi_T(P_{\beta'}) = 4.7 \cdot 10^{-5}$ bar^{-1} and $\chi_T(L_\alpha) = 11 \cdot 10^{-5}$ bar^{-1}). As can be also seen from Figure 8, χ_T drastically increases approaching the main transition point, does not show a sharp discontinuity, however. Actually, there has not been observed a sharp discontinuity in any thermodynamic or thermomechanic property at the phospholipid main phase transition. This has led to the proposal that the main transition might be pseudo-critical, i.e. in principle of first order but very close to a critical end point and consequently influenced by thermal density fluctuations.

This behaviour could be very well described theoretically, employing a multi-state Pink-Potts model [25-27].

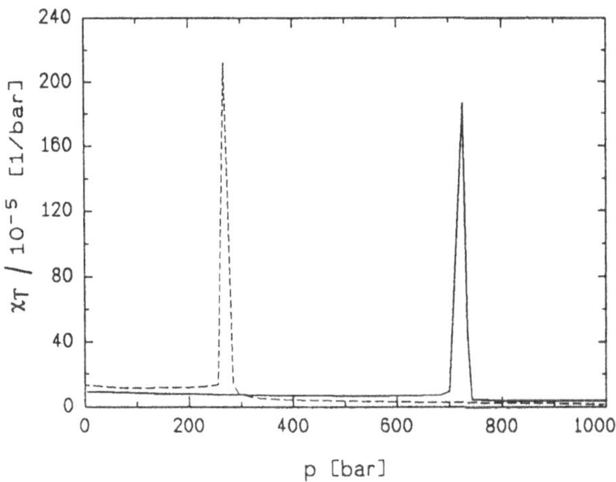

Fig. 8 Isothermal compressibility χ_T of DMPC in H_2O at T = 30 ^{o}C (---) and T = 40 ^{o}C (—) as a function of pressure.

An additional important quantity that can be obtained from the isothermal compressibility data is the mean fluctuation of the lipid volume in its different phases. The mean square volume fluctuations are given by $<V_L^2> = k_B T V_L \chi_T$, where T is the absolute temperature, k_B the Boltzmann constant and V_L the volume of the lipid [32]. The relative root mean square fluctuation of the lipid volume $\Delta V_{L,rel} = (<\Delta V_L^2>)^{1/2}/V_L$ in the liquid-crystalline phase is about 7 %, it increases up to about 29 % at the main transition and decreases to about 4 % in the gel phase. In comparison to volume fluctuations of other biochemical systems, like proteins [33,34], these volume fluctuations in lipid systems are large, they are smaller, however, in comparison to those of pure water. This fact may be of significant biological relevance, since in biological cells many proteins are immersed in lipid bilayer membranes.

3.2 THE INFLUENCE OF THE LOCAL ANAESTHETIC TETRACAINE

The molecular mechanisms of the anaesthetic action on nerve membranes are still poorly understood. It is still controversial whether the membrane proteins or the surrounding lipid matrix are the primary target sites of anaesthetic action [35,36]. An important key in understanding the molecular mechanisms of anaesthesia might be the antagonistic effect of hyperbaric pressures against anaesthetic action, which has been observed in vivo (see e.g. [37,38]). In order to elucidate the effect of the local anaesthetic on the properties of the pure lipid matrix, only, we performed experiments on the model membrane DMPC. Following the discovery that anaesthesia can be reversed by application of hydrostatic pressure, suggests to study also the pressure effect on the volumetric properties of the anaesthetic-lipid model membrane system. Most of our experiments on TTC-DMPC dispersions have been performed at pH = 9.5, where the anaesthetic is in its uncharged form. By ^{2}H-NMR studies it has been directly shown [39] that the anaesthetic intercalates partially into the bilayer at that pH-value (see Figure 9) and estimates of the molecular order parameter indicate that it is located with its long axis approximately parallel to the director of the fatty acyl chains.

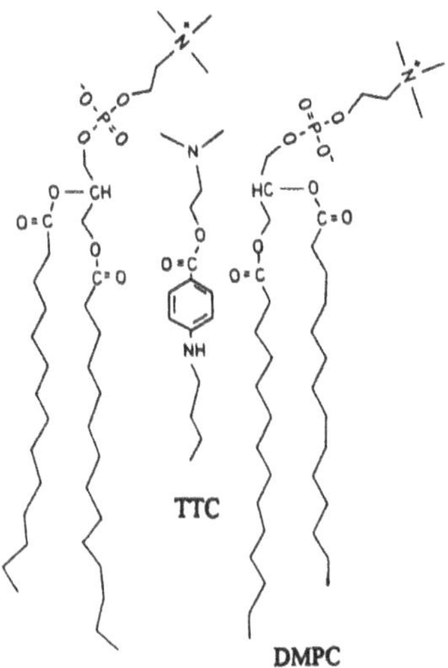

Fig. 9 Schematic drawing of the position of the local anaesthetic tetracaine (TTC) in DMPC at pH = 9.5 [39].

Figure 10 exhibits the temperature dependence of the apparent specific lipid volume V_L* of a DMPC dispersion containing 3 mol-% TTC in comparison to that of the pure lipid system. It is clearly visible, that the main transition is shifted towards a lower temperature by the addition of the local anaesthetic. The volume change is about 15 % smaller than that of the pure lipid system. At higher concentrations, the main transition is shifted towards still lower temperatures and the transition width increases. Obviously, chain disorder increases drastically upon addition of TTC by increasing the number of gauche conformers of the lipid hydrocarbon chains.

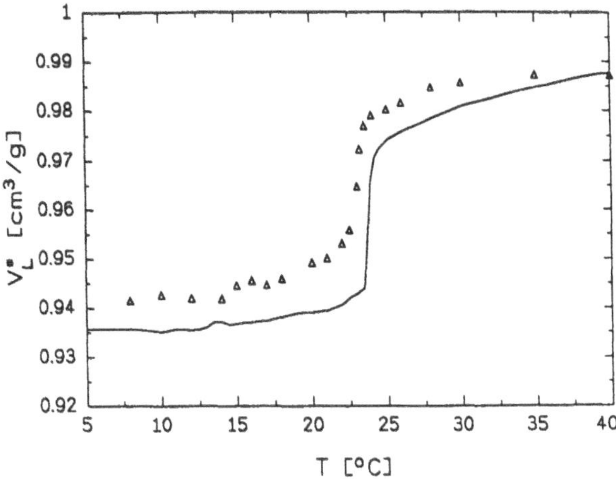

Fig.10 Temperature dependence of the apparent specific lipid volume V_L* of a pure DMPC (full line) and 3 mol-% TTC/DMPC (Δ) dispersion at 1 bar and pH= 9.5.

The expansion coefficient α_p of the TTC containing sample is shown in Figure 4. By the addition of TTC, α_p of the lipid system drastically increases, both in the liquid-crystalline and in the gel phase, e.g. about 50 % in the liquid-crystalline and 100 % in the gel phase by addition of 3 mol-% TTC. With increasing anaesthetic concentration, α_p further increases and dramatically enhanced values of α_p are found below and above the transition point.

The pressure dependent results for the 3 mol-% tetracaine containing sample are exhibited in Figure 11. The incorporation of the anaesthetic into the DMPC bilayer causes an about 20 % decrease of ΔV_m relative to that of the pure lipid system, it broadens the transition about 4-fold and shifts the pressure-induced liquid-crystalline to gel phase transition towards higher pressures. At concentrations of more than about 10 mol-% TTC in DMPC, the pressure-induced main transition is no more visible.

The isothermal compressibility of the lipid system is presented in Figure 12 as a function of pressure. Larger values for the compressibilities are obtained in both lipid phases by addition of 3 mol-% TTC, pointing again to the overall disordering of the membrane system. χ_T is drastically reduced at the main transition, however enhanced in the direct neighbourhood of the transition, similar to the behaviour found for the expansion coefficient in the transition region.

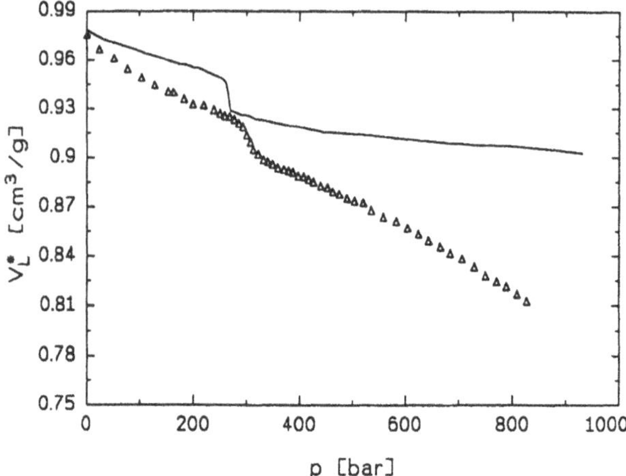

Fig.11 Comparison of the apparent specific volumes V_L* of a DMPC (—) and a 3 mol-% TTC/DMPC (▲) multilamellar dispersion as a function of pressure at T = 30 °C, pH= 9.5.

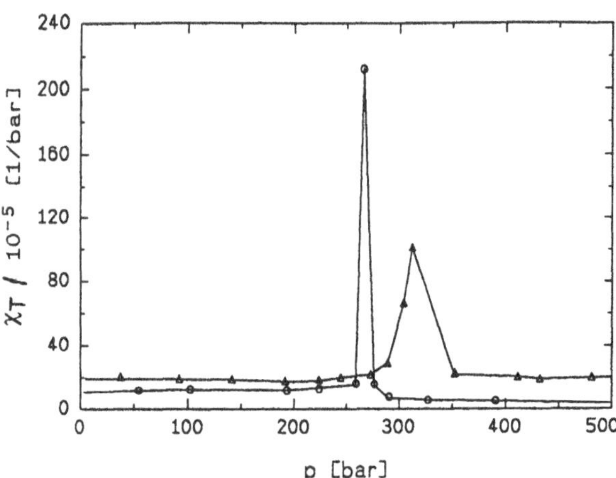

Fig.12 Isothermal compressibility χ_T of a pure DMPC (o) and 3 mol-% TTC/DMPC (Δ) dispersion at e.g. T = 30 °C and pH = 9.5.

The addition of 3 mol-% of the local anaesthetic leads to an increase of the relative lipid volume fluctuations $\Delta V_{L,rel}$ up to about 9 % in both lipid phases. At the transition point, $\Delta V_{L,rel}$ is diminished in comparison to the value of the pure lipid system, however. In a theoretical modelling of these kind of model systems [25-27], it could be shown, that the binding of foreign molecules in bilayer membranes might couple strongly to the thermal density and concentration fluctuations at the main phase transition, thus leading to the strong enhancement of χ_T and $\Delta V_{L,rel}$, which has been observed experimentally. These findings are also of important biochemical relevance, as in bilayer membranes, strong density or concentration fluctuations are related to the transmembrane permeability of ions and small molecules.

Though the biochemical action of local anaesthetics is still controversial as to whether or not the action is lipid mediated, one thing which is clear, however, is that local anaesthetics do strongly interact with the lipid membranes and change their physical properties, to an extent which often can be correlated with the potency of the agent [40].

ACKNOWLEDGEMENTS

We gratefully acknowledge financial support from the Deutsche Forschungsgemeinschaft.

4. REFERENCES

[1] G. Cevc and D. Marsh (1987) Phospholipid Bilayers, John Wiley & Sons, New York

[2] P.T.T. Wong, D.J. Siminovitch and H.H. Mantsch, Structure and Properties of Model Membranes: New Knowledge from High-Pressure Vibrational Spectroscopy, Biochimica et Biophysica Acta **947**, 139-171 (1988)

[3] R. Winter and W.-C. Pilgrim, A SANS Study of High Pressure Phase Transitions in Model Biomembranes, Ber. Bunsenges. Phys. Chem. **93**, 708-717 (1989)

[4] D.A. Driscoll, J. Jonas and A. Jonas, High Pressure ^{2}H Nuclear Magnetic Resonance Study of the Gel Phases of Dipalmitoylphosphatidylcholine, Chemistry and Physics of Lipids **58**, 97-104 (1991)

[5] L.F. Braganza and D.L. Worcester, Hydrostatic Pressure Induces Hydrocarbon Chain Interdigitation in Single-Component Phospholipid Bilayers, Biochemistry **25**, 2591-2596 (1986)

[6] J. Drouet, J.J. Risso and J.C. Rostain (eds.) (1990) Proceedings of the IInd International Meeting on High Pressure Biology, Toulon, August 19-22 (1990)

[7] H.W. Jannasch, R.E. Marquis and A.M. Zimmerman (1987) Current Perspectives in Pressure Biology, Academic Press, London

[8] J.F. Nagle, Lipid Bilayer Phase Transitions: Density Measurements and Theory, Proc. Nat. Acad. Sci. **70**, 3443-3444 (1973)

[9] K.R. Srinivasan, R.L. Kay and J.F. Nagle, The Pressure Dependence of the Lipid Bilayer Phase Transition, Biochemistry **13**, 3494-3496 (1974)

[10] N.I. Liu and R.L. Kay, Redetermination of the Pressure Dependence of the Lipid Bilayer Phase Transition, Biochemistry **16**, 3484-3486 (1977)

[11] A.G. MacDonald, A Dilatometric Investigation of the Effects of General Anaesthetics, Alcohols and Hydrostatic Pressure on the Phase Transition in Smectic Mesophases of Dipalmitoyl Phosphatidylcholine, Biochimica et Biophysica Acta **507**, 26-37 (1978)

[12] J.F. Nagle and H.L. Scott, Lateral Compressibility of Lipid Mono- and Bilayers. Theory of Membrane Permeability, Biochimica et Biophysica Acta **513**, 236-243 (1978)

[13] J. F. Nagle and D. A. Wilkinson, Density Measurements and Molecular Interactions, Biophys. J. **23**, 159-175 (1978)

[14] D.I. Melchior, F.J. Scavitto and J.M. Steim, Dilatometry of Dipalmitoyllecithin-Cholesterol Bilayers, Biochemistry **19**, 4828-4834 (1980)

[15] N.D. Russell and P.J. Collings, High Pressure Measurements in Phospholipid Bilayers Using Adiabatic Compression, J. Chem. Phys. **77**, 5766-5770 (1982)

[16] G. Schmidt and W. Knoll, Densitometric Characterization of Aqueous Lipid Dispersions, Ber. Bunsenges. Phys. Chem. **89**, 36-43 (1985)

[17] N. Vennemann, M.D. Lechner, T. Henkel and W. Knoll, Densitometric Characterization of the Main Phase Transition of Dimyristoyl-Phosphatidylcholine between 0.1 and 40 MPa, Ber. Bunsenges. Phys. Chem. **90**, 888-891 (1986)

[18] R.E. Tosh and P.J. Collings, High Pressure Volumetric Measurements in Dipalmitoylphosphatidylcholine Bilayers, Biochimica et Biophysica Acta **859**, 10-14 (1986)

[19] M.C. Wiener, S. Tristram-Nagle, D.A. Wilkinson, L.E. Campbell and J.F. Nagle, Specific Volumes of Lipids in Fully Hydrated Bilayer Dispersions, Biochimica et Biophysica Acta **938**, 135-142 (1988)

[20] J.F. Nagle and M.C. Wiener, Structure of Fully Hydrated Bilayer Dispersions, Biochimica et Biophysica Acta **942**, 1-10 (1988)

[21] A. Raudino, F. Zuccarello, C. La Rosa and G. Buemi, Thermal Expansion and Compressibility of Phospholipid Vesicles. Experimental Determination and Theoretical Modeling, J. Phys. Chem. **94**, 4217-4223 (1990)

[22] C. La Rosa and D. Grasso, Isothermal Compressibility of Phospholipid Vesicles: A New Fast Experimental Approach, Il Nuovo Cimento **12** D, 1213-1218 (1990)

[23] S.F. Scarlata, Compression of Lipid Membranes as Observed at Varying Membrane Positions, Biophys. J. **60**, 334-340 (1991)

[24] S. Utoh and T. Takemura, Phase Transition of Lipid Multilamellar Aqueous Suspension under Pressure. I. Investigation of Phase Diagram of Dipalmitoyl Phosphatidylcholine Biomembrane by High Pressure-DTA and -Dilatometry, Jap. J. Appl. Phys. **24**, 356-360 (1985)

[25] O.G. Mouritsen, Theoretical Models of Phospholipid Phase Transitions, Chemistry and Physics of Lipids **57**, 179-194 (1991)

[26] J.H. Ipsen, K. Jorgensen and O.G. Mouritsen, Density Fluctuations in Saturated Phospholipid Bilayers Increase as the Acyl-Chain Length Decreases, Biophys. J. **58**, 1099-1107 (1990)

[27] J.H. Ipsen, O.G. Mouritsen and M. Bloom, Relationsships between Lipid Membranes Area Hydrophobic Thickness and Acyl-Chain Orientational Order, Biophys. J. **57**, 405-412 (1990)

[28] B. Cho, K.-Y. Choi and S.-D. Choi, Thermodynamic Properties of Liquid Water up to 8000 bar and between 25 and 150 ^{o}C, Phys. Chem. Liq. **23**, 151-161 (1991)

[29] R. Hilbert, K. Tödheide and E.U. Franck, PVT Data for Water in the Ranges 20 to 600 ^{o}C and 100 to 4000 bar, Ber. Bunsenges. Phys. Chem. **85**, 636-643 (1981)

[30] W.D. Turley and H.W. Offen, Fluorescence Detection of Gel-Gel Transitions in DMPC Vesicles at High Pressures, J. Phys. Chem. **89**, 3962-3964 (1985)

[31] P.T.T. Wong and H.H. Mantsch, Effects of Hydrostatic Pressure on the Molecular Structure and Endothermic Phase Transitions of Phosphatidylcholine Bilayers: A Raman Scattering Study, Biochemistry, **24**, 4091-4096 (1985)

[32] L.D. Landau and E.M. Lifschitz (1987) Statistische Physik, Akademie -Verlag, Berlin

[33] K. Gekko and Y. Hasegawa, Biochemistry **25**, 6563 (1986)

[34] K. Gekko and Y. Hasegawa, Effect of Temperature on the Compressibility of Native Globular Proteins, J. Phys. Chem. **93**, 426-429 (1986)

[35] I. Ueda and H. Kamaya, Molecular Mechanisms of Anesthesia, Anesth. Analg. **63**, 929-945 (1984)

[36] N.P. Franks and W.R. Lieb, What is the Molecular Nature of General Anaesthetic Target Sites?, Trends in Pharmacological Sciences **8**, 169-174 (1987)

[37] F.H. Johnson and E.A. Flagler, Hydrostatic Pressure Reversal of Narcosis in Tadpoles, Science **112**, 91 (1950); ibid **92**, 112 (1951)

[38] M.J. Lever, K.W. Miller, W.D.M. Paton and E.B. Smith, Pressure Reversal of Anaesthesia, Nature **231**, 368-371 (1971)

[39] M. Auger, H.C. Jarell and I.C.P. Smith, Interaction of the Local Anaesthetic Tetracaine with Membranes Containing Phosphatidylcholine and Cholesterol. A ^{2}H NMR Study, Biochemistry **27**, 4660-4667 (1988)

[40] I. Ueda, C. Tashiro and K. Arakawa, Depression of Phase Transition Temperature in a Model Cell Membrane by Local Anesthetics, Anesthesiology **46**, 327-332 (1977)

EFFECTS OF PRESSURE ON LARGE MULTIMERIC PROTEINS AND VIRUSES

JERSON L. SILVA
Departamento de Bioquimica
Instituto de Ciencias Biomedicas.
Universidade Federal do Rio de Janeiro
Rio de Janeiro, RJ 21910
Brazil

ABSTRACT. We review the use of hydrostatic pressure to study protein-protein interactions of large aggregates and viruses. The pressure dissociation of many-subunit structures presents an apparent violation of the action mass law that is explained by heterogeneity of the energies of association. After decompression, the dissociated species get trapped in metastable conformations that may never recover the original form. The degree of reversibility of pressure effects varies in different viruses. In several viruses, non-infective particles are formed after a cycle of compression and decompression. These results demonstrate the potential of utilization of hydrostatic pressure to prepare non-infectious whole virus particles that are highly immunogenic.

1. Introduction

Most biological functions are carried out by oligomeric or multimeric proteins. It is commonly accepted that subunit assembly is a spontaneous step following the synthesis of the constituent polypeptides at the ribosomes. However we know little about the forces that govern the interaction between subunits. These forces share the same nature of the ones responsible for intramolecular protein folding. Intermolecular and intramolecular interactions in protein are only understood in general terms. Until the late 70's, we were completely restricted by the lack of a method to study subunit association without affecting grossly the conformation of the protein. During the last decade, we have apparently circumvented this problem by utilizing hydrostatic pressure as a perturbation of the equilibria. The dissociation of oligomeric proteins by hydrostatic pressure has permitted to study in detail the equilibria between the aggregates and their constitutive subunits. We review in this paper the use of high pressure to study protein-protein interactions of large aggregates and icosahedral viruses.

Hydrostatic pressure has been utilized to follow the denaturation and subunit dissociation of proteins (Heremans, 1982; Weber and Drickamer, 1983; Weber, 1987, Weber, 1992). The studies with small dimers and tetramers have shed light to the understanding of the forces governing the intersubunit interactions (Weber, a chapter in this volume). We briefly describe how the studies in a dimer can lead to important information on the thermodynamics and mechanism of subunit association.

Many new concepts have emerged from the last decade studies on the pressure stability of proteins. One of the most important discoveries is that in the dissociation equilibria of oligomers, distinct effects appear which have no parallel in the associations of small molecules. In many dimers and tetramers, a relatively stable defective reassociated form, or even a loss of aggregation capacity is observed after pressure dissociation and decompression (King and Weber, 1986; Silva *et al.*, 1986). In

R. Winter and J. Jonas (eds.), High Pressure Chemistry, Biochemistry and Materials Science, 561–578.

larger protein aggregates and viruses, the latter effects become amplified (Dreyfus *et al.*, 1988; Silva and Weber, 1988; Bonafe *et al.*, 1991). All these effects are explained as the result of a "conformational drift" that follows the substitution of intersubunit contacts by solvent-subunit contacts on dissociation, and its complete or partial reversion on association (Weber, 1992).

The pressure studies on large multisubunit proteins and viruses also furnish unique results. The changes in conformation of the dissociated polypeptides result in the formation of intermediate states of assembly that require specific conditions in the medium to reassociate. Another general feature in the pressure dissociation of large aggregates is the lack of dependence on protein concentration even at conditions whereas the reaction is completely reversible. The heterogeneity becomes more marked as the number of particles in the aggregate increases and in the viruses the equilibrium is completely deterministic, conferring virtual individuality to the particles (Silva and Weber, 1988; DaPoian *et al.*,1993). We review the cases where these properties have been observed and their biological implications.

Finally, we discuss some potential biotechnological application of the inactivation of virus particles by hydrostatic pressure.

2. Hydrostatic Pressure as a Thermodynamic Tool to Study Folding and Association Reactions in Proteins

The effects of pressure on the interaction between segments of two separated subunits or between intramolecular protein segments are expected to be of the same nature. The Gibbs Free Energy and the equilibrium constant for either reaction will depend on the volume change according to:

$$K_d(p) = K_{do} \exp(p\Delta V/RT) \qquad (1)$$

If we introduce the degree of extent of reaction at pressure p (α_p), we can deduce the following general equation for a dissociation or a denaturation process:

$$\ln(\alpha_p{}^n/(1-\alpha_p)) = p(\Delta V_p/RT) + \ln(K_{do}/n^nC^{(n-1)}) \qquad (2)$$

Where C is the total protein concentration and K_{do} is the dissociation or denaturation constant. In the case of a denaturation process, such as the case of Fig 1A, the order of reaction (n) is equal to 1, and, therefore, the equilibrium does not depend on protein concentration. In the case of a dissociation process n > 1, and equation 2 holds for the thermodynamic equilibrium of an aggregate and its n equal constitutive subunits is given by the relation

$$K = n^nC^{(n-1)}\alpha^n/(1-\alpha) \qquad (3)$$

where α is the degree of dissociation of the aggregate, and C the molar concentration of protein as aggregate. The dependence of α upon C reflects the existence of a dynamic microscopic equilibrium between aggregates and subunits. For the dimer-monomer reaction, equation 2 becomes:

$$\ln(\alpha_p{}^2/(1-\alpha_p)) = p(\Delta V_p/RT) + \ln(K_{do}/4C) \qquad (4)$$

Equations 2 and 4 permit the calculation of the standard volume change from measurements at a fixed protein concentration, at a series of different pressures, and because of the variable pressure we designate the volume thus obtained as ΔV_p. In a dissociation process, a change in protein concentration form C_1 to C_2 at a fixed pressure results in a parallel displacement Δp of the plot of ln K(p) versus p along the pressure axis. At any fixed value of degree of dissociation , this shift Δp in pressure upon change in concentration is given by

$$\Delta p = (n-1)\,(RT/\Delta V_C).\ln(C_2/C_1) \qquad (5)$$

The subscript C in ΔV_C is intended to draw attention to the procedure of its determination by the shift of the pressure dissociation profile with concentration.

Since dissociation has an order of reaction (n) higher than one, the dependence on protein concentration furnishes an additional variable that can be controlled and makes association equilibria studies more amenable to analysis. In contrast, denaturation reactions are independent of protein concentration.

3. Pressure Dissociation of Simple Dimeric Systems

3.1. ARC REPRESSOR - A MODEL FOR DIMER DISSOCIATION

Arc repressor is a small (13,000 M_r), dimeric DNA binding protein that proved to be ideal for address the problems of protein folding and protein association. Arc repressor dimer reversibly dissociates into subunits (Silva et al., 1992b) and there exists complete agreement between the thermodynamic parameters obtained from the dissociation induced by dilution and that induced by hydrostatic pressure. The dissociation of Arc was followed by different spectroscopic techniques: fluorescence polarization, energy of the fluorescence emission of the single tryptophan at position 14 (Silva et al., 1992b) and more recently by high resolution (^{1}H) NMR (Peng et al., 1993). The dissociation curves were translated to lower pressures by the decrease in protein concentration (Figure 1) according to equation 5. The dissociation reaction was completely reversible as evidenced by fluorescence spectroscopy (Fig. 1) and by gel filtration chromatography. The good agreement between ΔV_p and ΔV_C values demonstrates the stochastic character of the dissociation reaction, different from the dissociation of large protein aggregates (Silva et al., 1989).

Since the subunits of Arc repressor are intertwined in the small structure of the dimer (Breg et al., 1990), it was anticipated that the conformational stability of Arc would depend to a large extent on the interaction between subunits. The anomalously steep response to changes in concentration by dilution (Silva et al., 1992b) indicates that conformational drift occur as a result of dissociation of Arc repressor. Hydrodynamic data and studies on bis-ANS binding revealed that the pressure-dissociated Arc repressor has the properties of a molten globule. Bis-ANS is a fluorophore that binds non-covalently to non-polar segments of proteins, especially in proximity to positive charges. Denaturation promoted by urea or high temperature was accompanied by a dramatic decrease in bis-ANS binding (Silva et al., 1992b). On the other hand, there was a large increase in fluorescence of bis-ANS promoted by binding to the pressure-dissociated subunit. These results demonstrate that the pressure-dissociated form (A^*) has a different conformation from the denatured one (U). The conformational changes that follow dissociation induced by pressure are more limited than those caused by urea or high temperature. Binding of bis-ANS or ANS is a characteristic of proteins in the molten-globule state (Goto and Fink; 1989) in contrast to the absence of binding by unfolded states. The protein in the molten globule state is compact and has a high degree of secondary structure, but has several non-polar side chains exposed to water (Pytsin, 1987; Creighton, 1990). Arc repressor in the drifted dissociated state (A^*) binds bis-ANS and has partial secondary structure (Peng et al., 1993) suggesting that it has a molten-globule conformation. The pressure-dissociated subunit of Arc is compact as measured by its rotational diffusion. The fluorescence anisotropy of the monomer of Arc repressor (A^*), and therefore its hydrated molecular volume, is much smaller than that of the urea-denatured form (Silva et al., 1992b). The small anisotropy of the pressure-dissociated Arc suggests that it is in a very compact conformation, which is another feature of a molten-globule state. In contrast, the high anisotropy of urea-denatured Arc is more consistent with a random-coil conformation. More recently, it has been

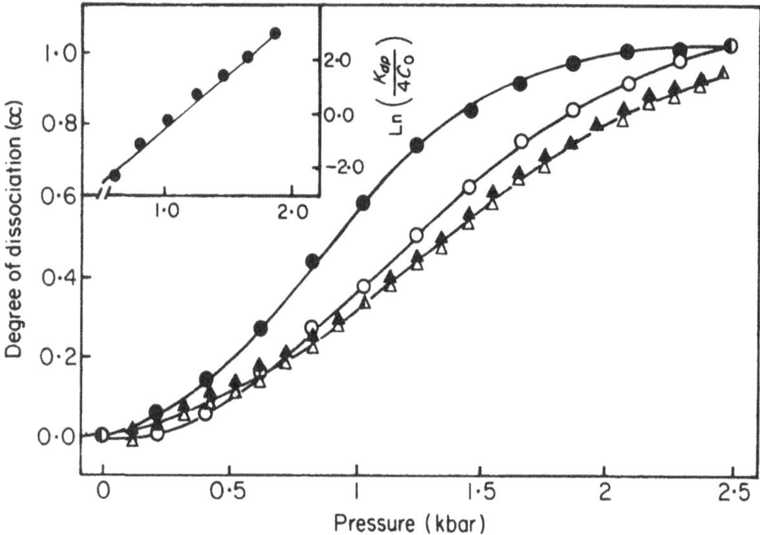

Figure 1: Effect of Arc protein concentration on pressure dissociation during compression. (●) 1 μM Arc repressor; (○) 5 μM Arc repressor; (△) 10 μM Arc repressor. Data were also obtained for 10 μM Arc during decompression (▲). Excitation wavelength was 280 nm, and emission was monitored from 300 to 420 nm. Inset: Plot of ln ($K_{dp}/4C_o$) *vs.* pressure for 1 μM Arc repressor. (Silva et al., 1992b).

TABLE 1: VOLUME CHANGES AND HYDRATION ON DISSOCIATION

DIMER*	MOLAR VOLUME CHANGE ($ml.mol^{-1}$)	SPECIFIC VOLUME CHANGE ($\mu l.g^{-1}$)	# H_2O (**)	# H_2O/aa residue
ENOLASE[a] (80,000 Mr)	55	0.688	18	0.027
HEXOKINASE[b] (96,000 Mr)	120	1.25	40	0.050
B_2 SUBUNIT[c] TRP SYNTHASE (90,000 Mr)	170	1.88	56	0.075
ARC REPRESSOR[d] (13,000 Mr)	100	7.69	33	0.311
R17 COAT PTN DIMER[e] (27,500 Mr)	130	4.73	43	0.188

(*) Volume changes from [a]Paladini and Weber (1981), [b]Ruan and Weber (1988), [c]Silva et al. (1986), [d]Silva et al. (1992) and [e]Da Poian et al. (1992). (**) To calculate the number of water molecules bound on dissociation, a cross section of 10 A^2 was considered for the molecule of water and a linear contraction on dissociation of 0.5 A was considered.

found that Arc monomer partially keeps the secondary structure as measured by two-dimensional (^{1}NMR) spectroscopy under pressure (Peng et al., 1993). Lifetime data of two different probes (tryptophan and bis-ANS) showed a more heterogeneous distribution of states for the monomer when compared to the dimer (Silva et al., 1992b). The existence of a heterogeneous population of conformations is also a property found in the partially-denatured compact states (Bycroft et al., 1990). (^{1}NMR) spectroscopy also demonstrates that Arc monomers (A^*) exist as collection of many conformations (Peng et al., 1993).

Similar properties to those described for Arc repressor have been found in other small dimers. R17 phage capsid dimer also had a large specific volume change on dissociation (Table 1) . The dissociated monomer was not stable and had the properties of a molten-globule or collapsed state (Da Poian et al., 1993a). The dimer of liver alcohol dehydrogenase was very stable but in the presence of sub-denaturing concentrations of guanidine, the pressure-dissociated subunits had properties of a collapsed state (Silva et al., in preparation).

We conclude that the dissociation of small dimers leads to an altered conformation that is probably the "physiological" state of the protein before association to the other subunit. Hydrostatic pressure only displaces the equilibrium based on the Le Chatelier principle. Pressure perturbation leads to a state that is similar to the one in equilibrium with the native state and, therefore, of great relevance for the understanding of protein interactions.

3.2. VOLUME CHANGES OF ASSOCIATION AND HYDRATION

The volume change of association is an important thermodynamic parameter and is related to the nature of the bonds in the interface between subunits. Table 1 lists the volume changes for several dimer-monomer dissociation systems. The volume changes obtained for different dimers are in a rather restricted range (50 to 150 ml/mol), and the free energies of association are also limited to the region of -9 to -11 kcal/mol. However the molecular weight of the different proteins is very different (10,000 to 100,000 Mr). Therefore, when the volume change is normalized to the molecular weight of the dimer, which furnishes the specific volume change, the value obtained for small dimers, like Arc repressor or R17 coat protein, are much higher than for other dimers (Table 1). This large change in volume per mass of protein found in the dissociation of small dimers can be explained by a high degree of interaction of buried aminoacid side chains with the solvent under dissociation. The hydration of charges that are involved in salt bridges (electrosctrition) or the hydration of polar and non-polar groups result in volume contraction. The more disordered state of the dissociated subunits when they lose the interface contact favors a higher degree of hydration. Table 1 also shows the number of water molecules bound on dissociation per number of aminoacid residues for each protein. In the case of Arc and R17 coat protein dimers, the hydration on dissociation was much higher. In all the dimers depicted in Table 1, the dissociation was accompanied by conformational drift of the subunits. However, it is clear that the effects are more drastic in the smaller proteins where the subunit interface is a substantial fraction of the whole structure. Probably, the molten state results from the penetration of water inside the protein. It can be generalized that the drifted state of dissociated monomers is characterized by the "melting" of the residues that participate in the subunit contacts.

Therefore a molten-state is apparently generated when the dispersion forces in the interface are destabilized and replaced by stronger interactions (permanent dipole-induced dipole). The inclusion of cosolvents or high concentrations of salts in the medium would restrain the binding of water to the protein, counteracting the effect of pressure. The inclusion of 30% glycerol shifted the half-pressure dissociation of Arc repressor by 980 bar, corresponding to a stabilization of the free energy of association of -2.39 kcal/mol (Silva, 1993). Similar effects of glycerol were observed in the pressure dissociation of F1-ATPase (Dreyfus et al., 1988) and of annelid hemoglobins (Bonafe et al., 1991). Similar to glycerol, 1M NaCl stabilized the dimeric state of Arc repressor (Silva, 1993). The hydration of Na^+ and Cl^- ions prevents the binding of water to the dissociated-subunits. Therefore, more compression work ($p\Delta V$) is required to produce dissociation. It is interesting that in all these cases, a slight decrease of the volume change occurs.

A large specific volume change and the stabilization of the associated state by glycerol and NaCl demonstrate that the drifted state of dissociated monomers is characterized by binding of water to the subunit contacts resulting in the "molten" of the residues in the interface (Silva, 1993). These data suggest that for small dimers, the dissociation lead to a molten-globule conformation that is probably the "physiological" state of the protein before association to the other subunit.

4. Deterministic dissociation of large multisubunit proteins and viruses

4.1. ANOMALOUS PRESSURE DISSOCIATION OF LARGE PROTEIN AGGREGATES

The pressure stability studies on many-subunit aggregates have led to important observations that bears upon the concepts of biological individuality and biological specificity. Studies on giant hemoglobins (erythrocruorins) (Silva et al., 1989; Bonafe et al., 1991), hemocyanins (Gomes et al., 1990; Bonafe et al., 1993) and several icosahedral viruses (Silva and Weber, 1988; Da Poian et al., 1993a) show that there are anomalies in the dissociation pattern when compared to the dissociation of dimers. The erythrocruorins are hemeproteins that belong to the class of multi-subunit hemoglobins whose most important characteristic is a two-tiered hexagonal structure of total molecular weight above 3 million. The erythrocruorin particles dissociate to small molecular weight subunits (Silva et al., 1989). An illustration of the dissociation reaction of the *Glossoscolex paulistus* hemoglobin is shown in Figure 2. After complete dissociation at 2.5 kbar followed by decompression, all the erythrocruorins studied continued mostly dissociated (Figure 3). However, complete reassembly occurred if the pressure-induced dissociation was smaller than 75%.

Calcium and glycerol stabilized the quaternary structure of the hemoglobin against pressure dissociation (Bonafe et al., 1991). The plot of the half-pressure dissociation ($p_{1/2}$) as a function of glycerol concentration results in a straight line, with a stabilizing free energy of -278 cal (mol of dissociating subunit)$^{-1}$ (mol of added glycerol)$^{-1}$ L^{-1} (Figure 4). This linear dependence suggests that the stabilization conferred by glycerol is due to the preferential hydration and therefore exclusion of glycerol from the assembly domains similar to the mechanisms of glycerol protection against denaturation of several proteins (Gekko and Timasheff, 1981). This effect is referred as a non-specific interaction. However, more specific interactions can not be ruled out to contribute to the reassembly process. Glycerol also shifts the $p_{1/2}$ of F1-ATPase to higher pressures and also prevents the irreversible inactivation (wrong reassociation) of the soluble enzyme (Dreyfus et al., 1988). Preferential exclusion may explain the $p_{1/2}$ shift but more specific interactions might be necessary to explain the modulation of reassembly found in F1-ATPase and *G. paulistus* hemoglobin.

Calcium and glycerol also promoted a large increase in the yield of recovery of fully assembled hemoglobin at the expense of the partially dissociated (one-twelfth subunit) and fully dissociated forms (Fig. 3). Addition of calcium after return to atmospheric pressure only increased recovery of the fully associated form in a long time-scale (many days). The existence of time-dependent changes in the conformation of the dissociated subunits is suggested to explain the partial association to one-twelfth sub aggregates (drifted forms) that lack the ability to reassemble to native hemoglobin (Silva et al., 1989; Bonafe et al., 1991). The promotion of reassembly by non-proteic factors (calcium and glycerol) is suggested to occur by preventing the formation of wrong intermediate forms (drifted one-twelfth subunits). These "assembly effectors" act during the assembly process but, once the complex is totally formed, they do not need to be present to maintain the protein associated.

Didecamer hemocyanin from Gastropoda presented a similar behavior to the dissociation of giant hemoglobins (Bonafe, 1993). A huge hysteresis between the compression and decompression curves was found indicating that the subunit-subunit affinity of dissociated hemocyanin is much lower than that of associated subunit. A large fraction of the dissociated products drifts to a "non-associable" conformation if the dissociation is proceeded in the absence of Ca and at high pHs, generating stable

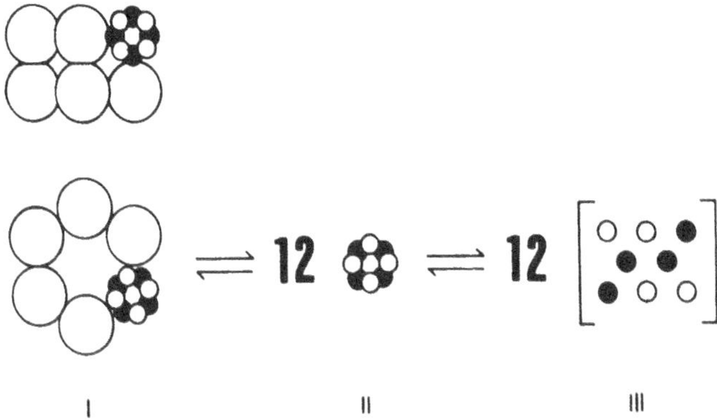

Figure 2. Schematic representation of dissociation process of hemoglobin of *Glossoscolex paulistus* hemoglobin. Native form in side and top-views (I), constituted by 12 principal subunits (II). Each principal subunit consists of several smaller subunits (III). (Bonafe et al., 1991).

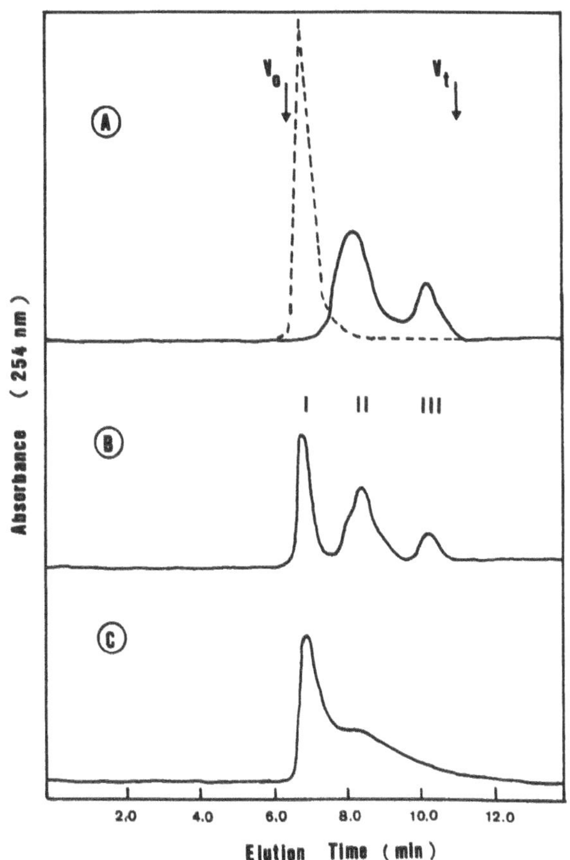

Figure 3. Effect of calcium and glycerol on the reassembly of hemoglobin. The elution of the protein from the filtration column GPC300 was monitored by the absorption at 254 nm. (A) The dashed profile is for 2.0 mg/ml hemoglobin that was incubated at atmospheric pressure. The continuous line is for 2.0 mg/ml hemoglobin that was incubated at 2.5 kbars for one hour in the absence of calcium and glycerol. (B) 2.0 mg/ml hemoglobin was incubated at 2.5 kbars for one hour in the presence of 30 mM calcium chloride. (C) As in (B), except that the protein was incubated under pressure in the presence of 30% glycerol plus 30 mM $CaCl_2$. (Bonafe et al., 1991).

intermediate states of assembly, lacking the ability to form decamers and didecamers. These intermediates are constituted predominantly of dimers and they bind oxygen reversibly exhibiting higher affinity for oxygen. The dissociation process shows high reversibility at low pH (6.8 - 5.3) or in the presence of calcium at mmolar concentrations. The pressure-dissociation curves shifted to higher pressures as the pH of the hemocyanin solution was decreased. The derived stabilizing free energy ($\Delta\Delta G$) was 28 kcal . (mol of added $H^+)^{-1}$. Calcium or proton switch the conformation back to the associable state that finally generates the assembled structure, reinforcing the idea that some special factors are required in the sites for assembly of large multimeric proteins.

4.2. THE DETERMINISTIC DISSOCIATION OF LARGE AGGREGATES AND VIRUSES

The absence of concentration effect on the pressure dissociation of several multimeric proteins and viruses shows a departure of the stochastic behavior for the dissociation equilibrium of these systems. Figure 5 illustrates the anomalous dissociation of *G. paulistus* hemoglobin. Lack of concentration dependence has been found in several systems (Table 2). For tetrameric proteins, such as lactate dehydrogenase (King and Weber, 1986) and glyceraldehydephosphate dehydrogenase (Ruan and Weber, 1989), the observed $\Delta p_{1/2}$ was about one fourth of the expected value. Only dimers have shown $\Delta p_{1/2}$ values similar to those expected: β_2-tryptophan synthase (Silva *et al.*, 1986), yeast hexokinase (Ruan and Weber, 1988) and the Arc repressor (Silva et al., 1992b) exhibit the typical behavior of stochastic chemical equilibria with respect to the concentration dependence of the dissociation curves.

The simplest explanation for this anómalous behavior is that a solution of a large aggregate constitutes a heterogeneous population (Ruan and Weber, 1989; Silva et al., 1989). The pressure-dissociation equilibria result from a deterministic equilibrium. The designation of "deterministic equilibrium" for these cases of independence upon the concentration of the particles seems justified by the similarity with the behavior of macroscopic bodies. In either case the new equilibrium after an external perturbation arises from individual characteristics of each object that are retained over times much longer than those necessary for equilibration. Each particle responds to pressure independently of the others and, at any given pressure, is in one of two states, whole or dissociated, which persists for times long compared with the duration of the experimental procedure. In all these large aggregates and viruses the thermodynamic dependence of the degree of dissociation upon the applied pressure (eq. 2) is followed just as well as in the dimers.

From the lack of concentration dependence, all these large aggregates are present in solution in a broad distribution of free energies of association rather than an homogeneous state. A "local" equilibria can be conceived at each characteristic pressure with very low exchange between the fractions. Erijman and Weber (1991) have demonstrated using transfer of electronic excitation energy (sensitized fluorescence) that pressure increases the rate of dissociation predominantly in dimers and decrease the rate of association of tetramers. Inversely of observing in dimers, tetramers show the presence of individual properties that can be retained over periods of time much longer than the equilibration time of the dissociation. Tetramers have a dependence upon the concentration that is much smaller than expected for a dynamic equilibrium and the energy transfer show that these phenomena depend upon the interconversion rates of the members of a heterogeneous population with respect to the time in which the experiments are performed. These observations indicate that the heterogeneity is already observed in tetramers (Erijman and Weber, 1991) and becomes more drastic in large aggregates such as erythrocruorin (Silva et al., 1989), hemocyanin (Bonafe et al., 1993) and virus particles (Silva and Weber, 1988; Da Poian et al., 1993a).

Dissociation studies of hemocyanin using salts and alkaline pHs have evidenced an apparent violation of the mass action law (van Holde and Miller, 1982). It has been proposed (Engelborghs and Lontie, 1973, Siezen and van Driel 1973) that those molluscan hemocyanins consist of a large population of forms, each having a unique and very sharp pH zone of dissociation. At any pH, there would be a population of molecules which is completely dissociated, a population which is still intact,

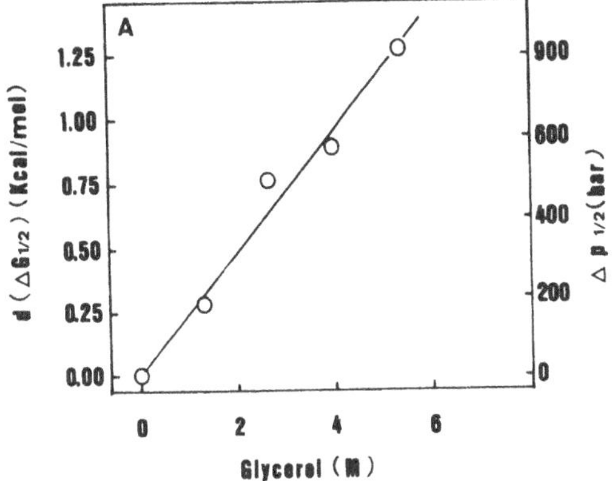

Figure 4. Effect of glycerol concentrations on the stabilizing free energy of association $(d(\Delta G_{1/2}))$. (Bonafe et al., 1991)

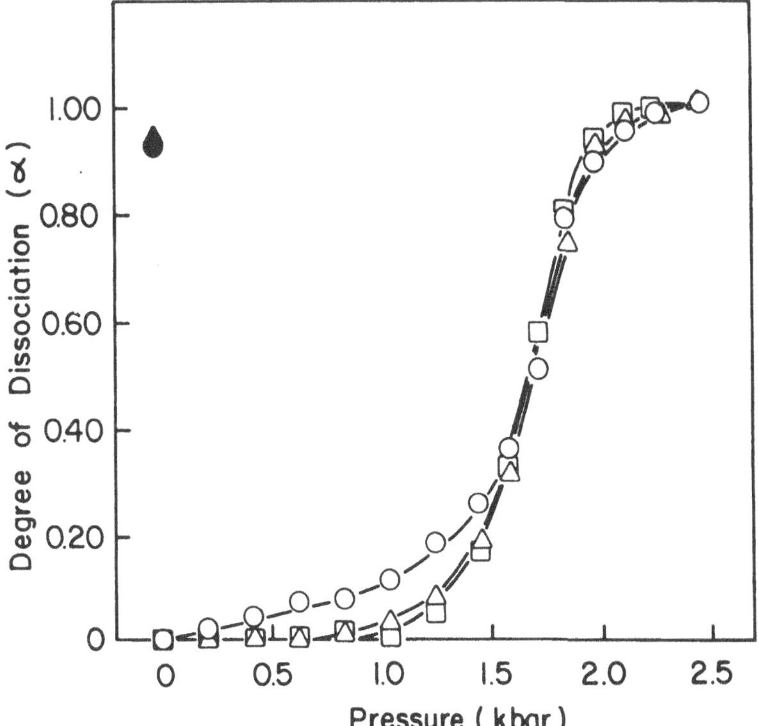

Figure 5. Effect of protein concentration on the pressure dissociation of *G. paulistus* hemoglobin. Plot of degree of dissociation *vs.* pressure at three protein concentrations, 0.2 mg/ml (○), 0.5 mg/ml (□) and 2 mg/ml (△). The full symbols correspond to the values of degree of dissociation after return to atmospheric pressure for the respective hemoglobin concentrations. (Silva et al., 1989).

and only a small population which is in equilibrium. Similarly, the pressure-induced dissociation of Gastropod hemocyanin results from a deterministic equilibrium in which each particle responds to pressure independently of the others and, at any given pressure, is in one of two states, whole or dissociated, which persists for long times when compared with the duration of the experiment. The deterministic behavior was partially attenuated in presence of Ca^{2+}. We provide a "practical thermodynamic" picture for the heterogeneity of the dissociation of hemocyanin. Each species dissociate with a specific characteristic pressure pi, which is the ratio $p_i = \Delta G_i / \Delta V_i$. For each external pressure applied to the system, there is a split in two populations: the ones with $p_i \leq p_{ext}$ undergo dissociation and contributes to compose the dissociate fraction (α); the ones with $p_i > p_{ext}$ does not undergo dissociation and represents the associated fraction ($1 - \alpha$).

TABLE 2. Comparison between the observed and expected values of $\Delta p_{1/2}$ (bar) for different oligomeric proteins:

	Expected $\Delta p_{1/2}$	Observed $\Delta p_{1/2}$
Dimers:		
. tryptophan synthase B_2 subunit[a]	310	330
. hexokinase[b]	450	330
. Arc repressor[c]	515	547
Tetramers:		
. lactate dehydrogenase[d]	400	130
. GAPDH[e]	784	160
Large Protein Assemblies:		
. Brome Mosaic virus[f]	2000	200
. *G. paulistus* hemoglobin[g]	1200	< 50
. hemocyanin[h]	1250	< 50
. R17 Phage(2.5 M urea)[i]	1200	< 100
. R17 Phage(4.5 M urea)[i]	1200	580

$\Delta p_{1/2}$ is the difference in half-dissociation pressure ($p_{1/2}$) when the protein concentration was changed 10-fold ($C_2/C_1 = 10$). The data were obtained from the literature: [a] Silva et al., 1986; [b] Ruan and Weber, 1988; [c] Silva et al., 1992b; [d] King and Weber, 1986; [e] Ruan and Weber, 1989; [f] Silva and Weber, 1988; [g] Silva et al., 1989; [h] Bonafe et al., 1993; [i] Da Poian et al., 1993a.

4.3. DISSOCIATION OF VIRUSES BY HYDROSTATIC PRESSURE

Viruses are remarkably stable structures. Their survival depends on the ability to protect for long periods of time the genetical material against the surroundings. Therefore, stability against different perturbations is the only defense until they reach their target hosts. On the other hand, the particles need to carry information to release or expose the nucleic acids when bound or inside the host cells. This rather opposing ability must reside in specific switches that have not been identified. Another important question concerning the border between inanimate and living matter is the degree of singularity of a virus particle. In other words, do viruses behave as stochastic molecules or more similar to macroscopic bodies? As previously discussed, the pressure studies on viruses and other multimeric proteins do show a departure of the stochastic behavior (Silva and Weber, 1988; Ruan and Weber, 1989; Silva *et al*, 1989; Bonafe et al., 1993).

The assembly of virus particles is not completely understood. The structure of several

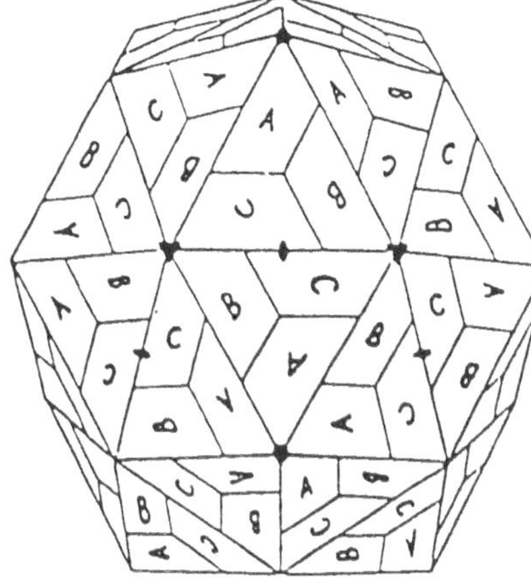

Figure 6. A T= 3 Icosahedral Virus Particle. The subunits labeled A, B, and C have the same amino acid sequence but are in different environments.

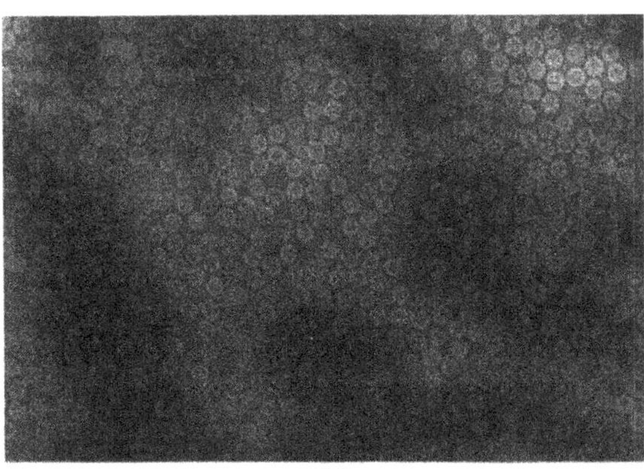

(a)

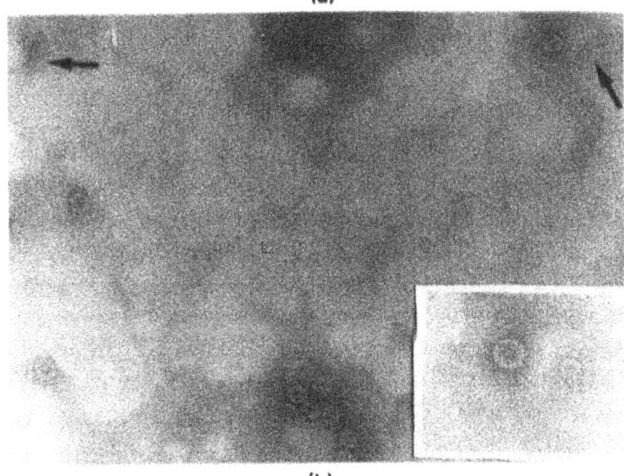

(b)

Figure 7. Electron microscopy of Brome Mosaic Virus fixed at atmospheric pressure (a) or under 1.4 kbar (b). Inset: comparison of T=3 and T=1 particles. (Silva and Weber, 1988).

icosahedral viruses shows that capsid proteins often occur in non-equivalent environments (Rossmann and Johnson, 1989). This explains the accommodation of more than three asymmetric units per face of the icosahedron as illustrated in Figure 6. However, the main paradox is to explain how equal subunits assume different conformations in the icosahedral lattice. Hydrostatic pressure is a powerful tool to address these questions on virus assembly. Several viruses are perturbed by pressure in the range 1-3 kbar. The first pressure study on an icosahedral virus was on Brome Mosaic Virus (BMV) (Silva and Weber, 1988). The pressure dissociation of BMV was measured by spectroscopic techniques and electron microscopy of samples fixed under pressure (Fig. 7). BMV particles undergo reversible dissociation until a pressure of 1.4 kbar, that promotes 75% dissociation, but reassembly steeply decreases at higher pressures (Silva and Weber, 1988). At higher pressures, the conformation of the dissociated capsid protein is sufficiently affected and generates unspecific aggregates under decompression. This results from the decrease in affinity to form capsids and it has been proposed that the reverse of the conformational drift (loss of subunit affinity) can account for the assembly of complex particles with different conformations.

All the virus particles studied by pressure exhibited a lack of dependence on protein concentration for the dissociation curves similar to those found with large aggregates. The explanation is also given by the heterogeneity of the free energies of association, which virtually confers individuality to the virus particles. This deterministic behavior is probably very important for the survival of a fraction of the particles when they are exposed to critical perturbations in the environment.

The dissociation of RNA bacteriophages (Da Poian et al., 1993a) and some plant viruses (commoviruses) (Da Poian et al., 1993b) present special features. Pressures up to 2.5 kbar promote only 5-10% dissociation of the whole particles and, only in the presence of urea (1.0-5.0 M), pressure promotes complete, reversible dissociation of the virus particles. At the lower urea concentrations reversible dissociation of R17 virus particles shows no dependence on protein concentration indicating a high degree of heterogeneity of the particles, but higher urea concentrations, 2.5-5.0 M, result in progressive restoration of the protein concentration dependence of the pressure dissociation (Table 2). At still higher urea concentrations, 5.0-8.0 M, irreversible dissociation of virus takes place at atmospheric pressure. Different to the dissociation of whole particles, the isolated dimers of the capsid protein of R17 fully dissociates in the absence of urea and the dissociation is dependent on protein concentration to the extent predicted for a stochastic equilibrium (Da Poian et al., 1993a). Dimers are also much less stable than the whole virus both to dissociation by pressure or urea.

Da Poian et al. (1993a) propose a theoretical model that accounts for the progressive effects of urea, and for the restoration of the concentration dependence of the dissociation of whole particles. The experiments demonstrate that the "thermodynamic individuality" of the virus particles arises in conformational differences in the assembled viruses, and that there is a direct relation between the stability of the particles and their heterogeneity.

There are several other cases of aggregates that are not sensitive to pressures up to 3.0 kbar at room temperature. Because of the entropic character of the association reactions, partial or complete dissociation can be achieved by lowering the temperature at high pressure. Subzero temperatures can be achieved by the decrease in freezing point of water under pressure. Phycobilisomes (PBS) from cyanobacteria dissociate from the photosynthetic membranes at high pressures and room temperature but their integrity is maintained (Foguel et al., 1992). However, high pressure and cold temperatures elicit the disassembly of the PBS into subunits. One of the components of PBS, the APC trimer, only dissociates at high pressure and at subzero temperatures (Foguel and Weber, 1993). The particles of cowpea mosaic virus (CPMV) are also resistant to pressures up to 3.0 kbar at room temperature (Da Poian et al., 1993b). Disassembly of the particles can be achieved by a combination of high pressure and sub-denaturing concentrations. Dissociation only occurs by lowering the temperature to subzero values. The same phenomenon is observed with the coat protein shells of bacteriophage P22 (Prevelige, King, Weber, and Silva, unpublished results). In this latter case, reassembly does not occur by warming and decompressing, unless the scaffolding protein is added to chaperone the association process. Also in this study, it is observed that the isolated monomers of the capsid protein are unexpectedly denatured by lower pressures at room temperature. On contrary, when the monomers are in the icosahedral lattice, they become much more stable to pressure, and from the low temperature behavior, its stability is provided by a large increase in entropy. The decrease in entropy due to the decrease in the degrees of rotational freedom by associating hundreds of monomers in the capsid is overwhelmed by the increase in the entropy of replacing the strong bonds of protein-water interactions by the weak bonds of protein-protein interactions (see Weber, in this Volume).

5. Pressure inactivation of viruses as the basis for killed vaccines

5.1. PRESSURE INACTIVATION OF VESICULAR STOMATITIS VIRUS AND ROTAVIRUS

The studies described in the previous chapters have demonstrated the reversible pressure dissociation of oligomeric proteins, large aggregates and viruses. The formation of non-infective particles after a cycle of compression and decompression has been demonstrated in more complex viruses such as Vesicular Stomatitis Virus (Silva *et al.*, 1992a) and rotaviruses (Pontes et al, 1993). The formation of altered aggregates after a cycle of compression and decompression have generated the idea of using hydrostatic pressure as a means to suppress virus infectivity, while preserving, or perhaps improving the immunogenic properties. The formation of imperfect virus particles after a cycle of compression and decompression has been demonstrated in simian virus 40 (Silva and Weber, unpublished), rotavirus (Pontes et al, 1993) and VSV (Silva et al., 1992a) by electron microscopy, gel filtration and spectroscopic observations.

We have recently reported the inactivation by pressure of vesicular stomatitis virus, a membrane enveloped virus (Silva et al., 1992a). The pressure-inactivated particles were as immunogenic as the native ones. A pressure of 2.6 kbar applied for 12 hours reduced infectivity of VSV by a factor of 10^4 (Table 3). The antibodies against pressurized material were as effective as those against the intact virus when measured by their neutralization titer (Table 4).

Fluorescence measurements demonstrated that application of pressure to VSV results in perturbations of the particle interactions that permit binding of specific molecular probes. Electron microscopy showed that the membrane of the pressurized virus was partially preserved, presenting the spike pattern of the membrane G-protein. Unlike the icosahedral viruses dissociation into smaller particles was not observed, but a constant change in the morphology was the presence of a bulge in the surface of the pressurized virus indicating a displacement of the capsid subunits, retained under the lipid and protein membrane (Figure 8).

Pressure-induced inactivation was also observed in rotavirus (Pontes et al, 1993). Rotavirus is responsible for a significant fraction of the estimated 5-10 million fatal causes of diarrheal diseases that occur in developing countries each year. A rotavirus vaccine would have a significant impact on reducing the morbidity and mortality from rotavirus in infant and young children. A pressure of 2.5 kbar applied for 30 min. reduced infectivity by a factor of 10^5 and the antibodies against pressurized material were as effective as those against the intact virus when measured by their neutralization titer. Gel filtration and electron microscopy studies showed that after pressure treatment the particles were very similar to the control. The electron microscopy revealed a subtle change after pressure incubation that consisted in a discontinuity in the second layer of protein in the capsid.

The pressure effects observed in vesicular stomatitis virus (Silva et al., 1992a) and rotavirus (Pontes et al, 1993) are related to those observed in other oligomeric proteins and icosahedral viruses. In the oligomeric proteins, and some of the viruses, dissociation of protein subunits takes place under pressure and is followed by complete or partial reassociation on decompression. The differences observed in the reversibility of the dissociation and in the regain of the native properties are determined by the number of subunits and the complexity of their interactions in the original aggregates. When out of contact with each other the subunits undergo conformational drift. Reversal of the conformational drift takes place progressively more slowly the higher the structural requirements and the number of

TABLE 3: PRESSURE-INACTIVATION OF VESICULAR STOMATITIS VIRUS

Condition	Titer (PFU/ml) A	B	C	D
Native Virus	1.82×10^7	4.0×10^6	3.9×10^6	2.0×10^7
(atmospheric pressure)				
2.6 kbar (3 hrs)	1.41×10^5			
2.6 kbar (5 hrs)	1.80×10^4			
2.6 kbar (12 hrs)	1.80×10^3	$< 2.5 \times 10^1$	$< 2.5 \times 10^1$	75

A-Non-purified virus preparation.

B-Non-purified virus preparation diluted 10-fold in 50 mM Tris-HCl.

C-The same preparation of B, except that was kept at room temperature for 2 weeks after the incubation 1 bar or 2.6 kbar.

D-Purified preparation of VSV. The stock concentration was 3.4 mg/ml and it was diluted in 50 mM Tris-HCl buffer (pH 7.5) to a final concentration of 100 μg/ml.All the pressure treatments were performed at 20°C. (Silva et al., 1992a).

TABLE 4: NEUTRALIZATION ASSAY OF ANTIBODY ELICITED BY PRESSURE-INACTIVATED VESICULAR STOMATITIS VIRUS

Dilution	Control	Pre-immune	Anti-Native VSV	Anti-Pressurized VSV
1:2	5.0×10^8	5.3×10^8	$< 2.5 \times 10^1$	$< 2.5 \times 10^1$
1:20	5.8×10^8		$< 2.5 \times 10^1$	$< 2.5 \times 10^1$
1:200	5.7×10^8		1.6×10^6	1.0×10^6
1:2000	5.8×10^8		1.6×10^8	1.4×10^8

A high titer of VSV (5.0×10^8 pfu/ml) was incubated for 30 minutes with a gamma-globulin fraction (Ab) diluted as indicated. The stock concentrations of gamma-globulin were 4.8 mg/ml (pre-immune), 5.1 mg/ml (anti-native VSV) and 4.9 mg/ml (anti-pressurized VSV). (Silva et al., 1992a).

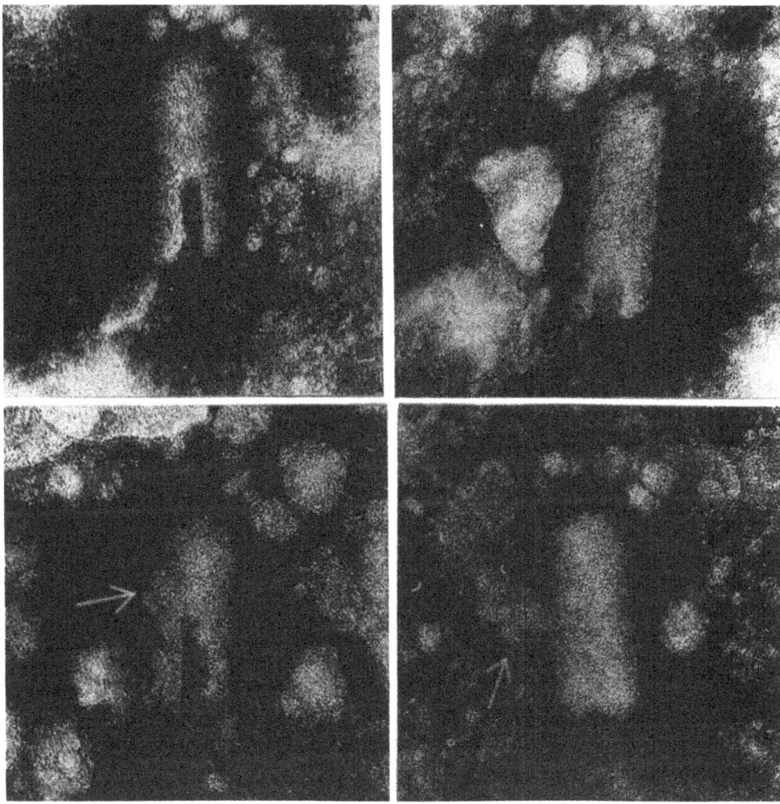

Figure 8. Electron microscopy of pressurized VSV negative stained with phosphotungstic acid. VSV (1 x 10^7 pfu/ml) was incubated 12 hours at atmospheric pressure (A and B) or at 2.6 kbar (C and D). The samples were negative stained with phosphotungstic acid. The arrows indicate the protrusions in the pressurized virus. (Silva et al., 1992a)

contacts of each subunit with its neighbors in the original article.

5.2. A "PASCALIZED" VACCINE

The use of hydrostatic pressure as a virus inactivation method may fulfill the two basic requirements to produce an efficient vaccine: 1. The need for whole particles to effectively elicit immunization; 2. The need for complete elimination of infectivity. Immunization is the most efficient way of preventing infectious diseases in humans and animals. There are basically two types of anti-viral vaccines in use: 1. based on live (attenuated) particles and 2. based on killed (inactivated) whole-virion particles. In addition, there is the vaccine against hepatitis B that is based on a genetically engineered sub-virion component. The present live and killed vaccines have advantages and disadvantages. As discussed recently by Budowsky (1991), the live vaccines has many risks to lead to complications. Since it requires the reproduction of virus in the organism, it can generate many postvaccinal reactions, such as damage to the host cell chromosomes, suppression of immune response, autoimmune diseases, neuropathologic disorders and teratogenic malformations. Budowsky (1991) suggests that properly prepared killed vaccines must be used in place of living ones.

The killed vaccines have their own drawbacks. In some cases, chemical inactivation by formalin has provided highly efficient immunization (Salk's polio vaccine). However, in other cases it has resulted, not only in insufficient immunity protection, but in more severe forms of the disease upon exposure to wild virus (e.g. Measles, Respiratory Syncitial Virus). These problems apparently occurred because formalin treatment destroyed one of the antigens. Similar reactions are unlikely to arise when pressure is used. As the inactivation of the viruses results exclusively from the altered physical disposition of the parts, without changes in covalent chemistry one would not anticipate a loss of the immunogenic properties.

In the particular case of a membrane-enveloped virus, an *enhancement* of immunogenicity of the whole particle after pressure treatment may exist due to the exposure of antigenic sites buried in the membrane or hidden by strong subunit interactions. It can be expected that natural selection favors the virus varieties in which antibody-eliciting epitopes are masked. The irregular conformational drift of the subunits will lead to their unmasking, resulting then in an increase in desirable immunogenic properties. Both rotavirus and VSV are already strongly immunogenic as intact particles, so that the question of the effect of the conformational drift on the immunogenicity remains open. Its answer will require a study of the immune response to poorly immunogenic viruses before and after pressurization.

In conclusion, the observations on several viruses indicate that hydrostatic pressure provides an efficient means of virus inactivation without loss of the immunogenicity. Studies are in progress to test the applicability of this conclusion to viruses that are poorly immunogenic. If generally valid, the high pressure method could provide a simple and inexpensive means to produce effective anti-viral vaccines. Similar to the large interest of Biotechnology industry in food products processed by pressure ("Pascalized Foods") (see Heremans, in this Volume), we hope that "Pascalized Vaccines" prove to be useful in combating human and animal virus diseases.

6. Acknowledgements

This work was supported by grants from Conselho Nacional de Desenvolvimento Cientifico e Tecnologico (CNPq), Financiadora de Estudos e Projetos (FINEP) of Brazil and by a Grant from the European Economic Community (CtI*-CT90-0857) to JLS. My special acknowledge to Professor Gregorio Weber who initiated me in the high pressure field and for his constant inspiration and support during all these years.

7. References

Bonafe, C. F. S., Villas-Boas, M., Suarez, M. C., and Silva, J. L. (1991). Reassembly of a large protein promoted by nonprotein factors. Effects of calcium and glycerol on the association of extracellular hemoglobin. J. Biol. Chem. 266, 13210-13216.

Bonafe, C. F. S., Araujo, J. R., and Silva, J. L. (1993). Intermediate states of assembly in the deterministic dissociation of gastropod hemocyanin by hydrostatic pressure. Submitted.

Breg, J. N., Opheusdsen, J. H. J., Burgering, M. J. M. Boelens, R., and Kaptein, R. (1990). Structure of Arc repressor in solution: evidence for a family of b-sheet DNA-binding proteins. Nature 346, 586-589.

Budowsky, E. I. (1991). Problems and prospects for preparation of killed antiviral vaccines. Adv. Virus Res. 39, 255-290.

Bycroft, M., Matouscheck, A., Kellis, J. T. Jr., Serrano, L., and Fersht, A. R. (1990). Detection and characterization of a folding intermediate in barnase by NMR. Nature 346, 488-490.

Creighton, T. E. (1990). Protein folding. Biochem. J. 270, 1-16.

Da Poian, A., Oliveira, A., Gaspar, L. P., Silva, J. L. and Weber, G. (1993a). Reversible pressure dissociation of R17 bactriophage: The physical individuality of virus particles. Submitted.

Da Poian, A., Weber, G., Johnson, J. E. and Silva, J. L. (1993b). Reversible pressure dissociation of commoviruses: Implication for Virus Assembly and Disassembly. In preparation.

Dreyfus, G., Guimaraes-Motta, H., and Silva, J. L. (1988). Effects of hydrostatic pressure on the mitochondrial ATP synthase. Biochemistry 27, 6704-6710.

Engelborghs, Y., and Lontie, R. (1973). Dissociation of *Helix pomatia* haemocyanin under the inlfuence of alkali salts. J. Mol. Biol. 77, 577-587.

Erijman, L., and Weber, G. (1991). Oligomeric Protein Associations: Transition from stochastic to deterministic equilibrium. Biochemistry 30, 1595-1599.

Foguel, D., Chaloub, R. M., Silva, J. L., Crofts, A., and Weber, G. (1992). Pressure and low temperature effects on the fluorescence emission spectra and lifetimes of the photosynthetic components of cyanobacteria. Biophys. J., in press.

Foguel, D. and Weber, G. (1993). Pressure and low temperature dissociation of allophycocyanin. Submitted.

Gekko, K., and Timasheff, S. N. (1981). Biochemistry 20, 4667-4676.

Gomes, F. C., Pereira, E. R., Bonafe, C. F. S. and Silva, J.L. (1990). Dissociation of the hemocyanin of Megalobulimulus ovatus (Gastropoda) induced by hydrostatic pressure. In "Invertebrate Dioxygen Carriers" (Preaux, G., and Lontie, R.) Leuven University Press, Louvain, pp 315-318.

Goto, Y., and Fink, A. L. (1989). Conformational states of b-Lactamase: Molten-globule states at acidic and alkaline pH with high salt. Biochemistry 28, 945-952.

Heremans, K. A. H.(1982). High pressure effects upon proteins and other biomolecules. Annu. Rev. Biophys. Bioeng. 11, 1-21.

Paladini, A. A. and Weber, G. (1981). Pressure-induced reversible dissociation of enolase. Biochemistry 20, 2587-2593.

Peng, X., Jonas, J., and Silva, J. (1993). Molten globule conformation of Arc repressor monomers determined by high pressure. (^{1}H) NMR spectroscopy. Submitted

Pontes, L., Villas-Boas, M., Barreto, A. L. G. and Silva, J. L. (1993). Pressure inactivation of rotaviruses as the basis for a killed vaccine. Submitted.

Pytsyn, O. B. (1987). Protein folding: Hypotheses and experiments. J. Protein Chem. 6, 277-293.

Rossmann, M. G. and Jonhson, J. E. (1989). Icosahedral RNA virus structure. Ann. Rev. Biochem. 58, 533-573.

Ruan, K., and Weber, G. (1988). Dissociation of yeast hexokinase by hydrostatic pressure. Biochemistry 27, 3295-3301.

Ruan, K., and Weber, G. (1989). Hysteresis and conformational drift of pressure-dissociated glyceraldehyde phosphate dehydrogenase. Biochemistry 28, 2144-2153.

Siezen, R. J., and van Driel, R. (1973). Structure and properties of hemocyanins. VIII.

Microheterogeneity of α-hemocyanin of *Helix pomatia*. Biochim. Biophys. Acta 295, 131-139.

Silva (1993). The drifted and molten-globule state of proteins under pressure. Braz. J. Med. Biol. Res., in press.

Silva, J. L., Luan, P, Glaser, M., Voss, E. and Weber, G. (1992a). Effects of hydrostatic pressure on a membrane-enveloped virus: high immunogenicity of the pressure-inactivated virus. J. Virol. 66, 2111-2117.

Silva, J. L., Miles, E. W. and Weber, G. (1986). Pressure dissociation and conformational drift of the beta dimer of tryptophan synthase. Biochemistry 25, 5780-5786.

Silva, J. L., Silveira, C. F., Correia, A. and Pontes, L. (1992). Dissociation of a native dimer to a molten globule monomer. Effects of pressure and dilution on the association equilibrium of Arc repressor. J. Mol. Biol. 223, 545-555.

Silva, J. L., Villas-Boas, M., Bonafe, C. F. S., Meirelles, N. C. (1989). Anomalous pressure dissociation of large protein aggregates. Lack of concentration and irreversibility at extreme degrees of dissociation of extracellular hemoglobin. J. Biol. Chem. 264, 15863-15868.

Silva, J. L. and Weber, G. (1988). Pressure-induced dissociation of brome mosaic virus. J. Mol. Biol. 199, 149-161.

van Holde, K. E., and Miller, K. I. (1982). Haemocyanins. Q. Rev.Biophys. 15, 1-129.

Weber, G. (1986). Phenomenological description of the association of protein subunits subjected to conformational drift. Effects of dilution and of hydrostatic pressure. Biochemistry 25, 3626-3631.

Weber G. (1987). Dissociation of oligomeric proteins by hydrostatic pressure. In "High Pressure Chemistry and Biochemistry", NATO ASI series. Math. Phys. Sci. vol. 97. (R.van Eldyk and J.Jonas, Eds). Reidel, Dordrecht.pp 401-420.

Weber, G. (1992). Protein Interactions, Chapman and Hall, New York & London.

Weber, G. and Drickamer, H. G. (1983). The effects of high pressure upon proteins and other biomolecules. Q. Rev. Biophys. I16, 89-112.

PRESSURE STUDIES ON PROTEIN-DNA INTERACTIONS

M. VILLAS-BOAS*, J. L. SILVA‡ and R. M. CLEGG*

‡Departamento de Bioquimica,
Instituto de Ciencias Biomedicas,
Universidade Federal do Rio de Janeiro,
21910 Rio de Janeiro, RJ, Brazil.

*Max Planck Institut f. biophys. Chemie
Abteilung Mol. Biol.
Postfach 2841
D-W Göttingen, Germany

ABSTRACT.

The use of hydrostatic pressure as a tool to study protein-DNA complexes is reported here for two systems: Arc repressor-DNA and nucleosomes. In the former case, it is shown that the magnitude of stabilization of the subunit interaction was correlated with the specificity of the protein-DNA interaction. Pressure-dissociation studies of Arc repressor in the presence of several synthetic DNAs of the same size (24 base pairs) were performed and the largest free energy stabilization of the subunit association was found for the operator DNA sequence. These results demonstrate the importance of free-energy linkage for the protein-DNA recognition process. In a second case, we examine the high pressure perturbation of the intact reconstituted nucleosome structure. The single SH group of H3 histones was labeled specifically with the polar sensitive fluorophore acrylodan and the spectral shift of the emission fluorescence was observed at different pressures and ionic strengths. The emission spectrum shifts to the red when high pressure is applied, suggesting a disassembly of the complex in response to the application of pressure, with a concomitant exposure of the acrylodan fluorophore to the solvent. This is consistent with a major structural change of the closely juxtaposed H3 histones in the intact nucleosomal complex, exposing their contacting faces to the solvent.

1. Introduction

DNA-binding proteins play a critical role in many cellular functions in both procaryotic and eucaryotic organisms, eg during replication, transcription and genetic recombination. Some DNA-binding proteins interact largely non-specifically with DNA and participate mostly as structural elements: for instance the histones provide the basic structural elements of the chromosome on which all higher order chromosome structures are based, and one role of the nuclear matrix proteins seems to be for providing the correct

R. Winter and J. Jonas (eds.), High Pressure Chemistry, Biochemistry and Materials Science, 579–602.

environment to coordinate the complex series of events that take place in the nucleus during the cell cycle. The functions of many DNA-binding proteins are more sequence-specific; this is particularly evident for the transcription regulators. The action spectrum of other DNA-interacting proteins involves both specific and non-specific interactions, as well as catalytic activity, such as RNA polymerase. In all these cases, the protein-DNA interactions are dynamically exchanging during the cell cycle, and one can anticipate that these processes involve many interchanges of protein-protein and DNA-protein interactions during the cell cycle. A common feature of many of the DNA-interacting proteins is that they are oligomeric structures. For instance, histones are octameric assemblies and most transcription repressors are dimeric or tetrameric proteins. The oligomeric structure of these proteins is apparently very important for their function.

Hydrostatic pressure is a unique tool for studying protein subunit interactions and protein-DNA interactions. The great advantage of hydrostatic pressure over other thermodynamic perturbants is that it can efficiently promote dissociation of many oligomeric proteins (and higher order structures such as viruses) (Silva & Weber, 1988; Weber, 1987) without major effects on the secondary structures of proteins. Pressures of up to 3.0 kbar appear to have minimal effects on the properties of single-chain proteins (Heremans, 1982; Weber & Drickamer, 1983), and hydrostatic pressure up to this limit can be employed generally to provide specific and reliable information concerning protein-protein interactions.

We present pressure studies with two DNA-protein systems, one involving the relatively non-specific DNA-binding nucleosomes, and the other involving a protein belonging to the group of sequence-specific DNA binding proteins, the Arc repressor and its operator DNA.

2. Experimental Procedures

2.1. ARC REPRESSOR

Arc repressor was expressed in E. coli and purified as previously described (Vershon et al., 1986; Silva et al., 1992; Silva & Silveira, 1992). Unless noted, experiments were performed at 20ºC in a standard buffer: 0.05 M Tris HCl (pH 7.5), 100 mM NaCl.
Fluorescence spectra were recorded on an ISS 200 spectrofluorometer (Champaign, IL) and anisotropy measurements were made in a polarization fluorescence instrument built in Rio de Janeiro (ICL100). The high pressure apparatus has been previously described (Silva et al., 1992).
Fluorescence spectra were quantified by the center of spectral mass:

$$< v_p >= \Sigma\, v_i\, F_i / \Sigma\, F_i \qquad (1)$$

where F_i is the fluorescence emitted at wavenumber v_i. The degree of dissociation (α_p) is related to $<v_p>$ by the expression:

$$\alpha_p = (1 + Q\,(<v_p> - <v_f>) / (<v_i> - <v_p>)) \qquad (2)$$

where Q is the ratio of the quantum yield of dissociated and associated forms, $< v_p >$ is the center of spectral mass at pressure p, and $<v_f>$ and $<v_i>$ are the corresponding quantities for dissociated and associated forms (Silva et al., 1989).

2.2. NUCLEOSOMES

Chromatin (from chicken erythrocytes) was partially digested with micrococcal nuclease producing a series of oligomers of nucleosomes. The four core histones were then purified. The DNA was separated according to size; this DNA and the purified labeled (see below) histones were then reconstituted in the presence of high salt concentration and urea. The sample was dialysed against decreasing salt concentration buffers until all salt and urea were removed. The samples were stored in ice.

The only cysteine residue in the core protein complex is at position 110 of the H3 histones from chicken erythrocytes, and it is possible to specifically label this amino acid with fluorescence probes (Ashikawa et al., 1982; Dieterich et al., 1977; Dieterich et al., 1979; Eshaghpour et al., 1980). Fluorescence techniques facilitate the study of the static and dynamic responses of nucleosomes to varying conditions, also in intact chromosomes. We have chosen the fluorescence probe acrylodan that is a sulfhydryl specific fluorescent label exceptionally sensitive to the polar environment of the dye. The fluorescence emission spectrum is shifted to the blue as the surrounding environment of the dye becomes less polar. This spectral characteristic provides us with a useful experimental handle to study the structural changes of nucleosomes as we show below. The basic nucleosomal structure still forms in the presence of stoichiometric labeling of this amino acid with acrylodan (data not shown). These reconstituted labeled histones are then used to carry out the high pressure fluorescence experiments described below.

All measurements on nucleosomes were made in Göttingen. The fluorescence measurements in the pressure cell were made on an instrument constructed in this laboratory, and the spectra for the ion titrations were collected with an SLM (8000S, Urbana) instrument. The data were collected and analyzed with a Macintosh IIci computer (Apple) equipped with software (Lab VIEW) and an ADC data acquisition system (Lab-NB) from National Instruments (Austin). The pressure equipment is from Nova Swiss (Effretikon); the quartz sample cell (200 μl volume) is constructed here, and the optical cell (sapphire windows) has been modified. All measurements were taken at 21 °C.

3. Results and Discussion

3.1. ARC REPRESSOR: FREE ENERGY COUPLING BETWEEN SUBUNIT-SUBUNIT AND SUBUNIT-OPERATOR

Arc repressor is a small DNA-binding protein dimer (M_r= 13,000), and belongs to a group of proteins called transcription repressors; they are generally dimeric or tetrameric proteins and their oligomeric structure is important for their function (Brennan & Matthews, 1989; Schleif, 1988). The Arc repressor represses transcription from the P_{ant} promoter of *Salmonella* bacteriophage P22 (Vershon et al., 1985). The Arc repressor belongs to a family of proteins that have an antiparallel β-sheet as the interfacial DNA-binding motif (Breg et al., 1990; Knight et al., 1989; Phillips, 1991). A tertiary structure model for Arc has been proposed based upon homology between Arc and the *E. coli* Met repressor and on two-dimensional NMR data (Breg et al., 1990). The proposed structure consists of an intertwined dimer, in which residues 8-14 of different monomers form an antiparallel β-sheet. It has been shown that the 21-bp operator can accommodate two dimers (Brown et al., 1990). Arc repressor dimer reversibly dissociates into subunits (Silva et al., 1992); the dissociation can be facilitated either by dilution at constant pressure or by an increase in pressure at a fixed protein concentration. Dissociated Arc is compact with its nonpolar core partially exposed and has the conformational properties of a molten globule (Silva et al., 1992).

We are interested in the relation between the specificity of operator DNA recognition and the strengthening of subunit-subunit interactions in Arc repressor under equilibrium conditions (Silva & Silveira, 1992). To evaluate the role of subunit interaction in the protein-DNA recognition process, the effects of several DNA molecules on the free energy of association of Arc repressor have been measured. Here we describe pressure-dissociation studies of Arc repressor in the presence of several synthetic DNAs of the same size (24 base pairs).

3.1.1. *Pressure dissociation of Arc repressor.* Arc repressor has only one tryptophan (amino acid #14) at the interface between subunits (Breg et al., 1990; Silva et al., 1992). The low polarity of the Trp14 environment when Arc repressor is in the dimeric state results in a blue-shifted fluorescence emission compared to a solvated tryptophan. The average energy of emission (center of spectral mass) decreased by 1400 cm^{-1} after complete dissociation (Figure 1A). Figure1B shows the dissociation of Arc repressor as measured by the decrease of the center of spectral mass and by fluorescence anisotropy. The steady-state anisotropy for the dimer at atmospheric pressure was 0.059; this low value is expected from the small size of the protein and from some contribution of segmental motion (Silva et al., 1992). The anisotropy decreased to 0.026 at 2.5 kbar, suggesting a decrease in the rotational hydrodynamic radius. The pressure-induced anisotropy changes for native Arc correspond well to the changes in the center of spectral mass of Trp14 emission, and confirms the correlation between the emission at each pressure and the degree of dissociation (Silva et al., 1992). The small anisotropy of the pressure-dissociated Arc demonstrates that it has a very compact conformation, suggesting, together with other evidence, that the Arc monomer is a molten globule. Figure 1B also shows that the fluorescence anisotropy increases when Arc repressor is denatured by urea. This result suggests that the urea-denatured form of Arc repressor is an expanded molecule in contrast to the compact conformation of pressure-dissociated form.

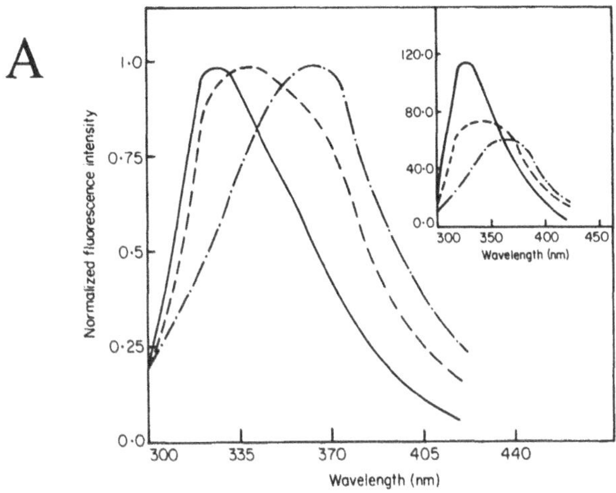

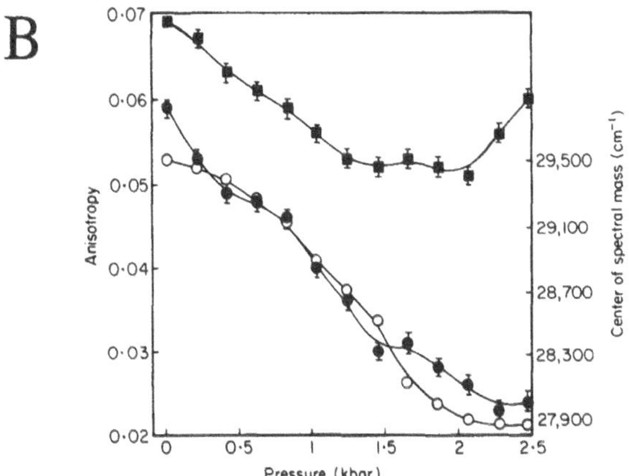

Figure 1- Effect of pressure on the intrinsic fluorescence emission and anisotropy of Arc repressor.

(1A) Normalized intrinsic fluorescence spectra of Arc repressor (1μM) at atmospheric pressure (—), at 1.0 kbar (— —) and at 2.5 kbar (—·), excited at 280 nm, in 0.05 M Tris

HCl (pH 7.5), 0.1 M NaCl. Inset: non-normalized emission spectra of Arc repressor at atmospheric pressure (—), at 1.0 kbar (— —) and at 2.5 kbar (—·).
(1B) Effect of pressure on the center of spectral mass of Trp emission (o) and on the Trp fluorescence anisotropy (●) of Arc repressor at the conditions of Fig 1A (0.05 M Tris HCl (pH 7.5), 0.1 M NaCl, absence of urea). Protein concentration was 2 μM. For the anisotropy measurements, excitation was at 290 nm and emission was viewed through a WG 20 filter. The data (■) shows the anisotropy values when 5 M urea was added. Modified from (Silva et al., 1992).

3.1.2. *Effects of DNA on the pressure dissociation of Arc repressor.* The effects of different 24 bp DNA sequences on the dissociation of Arc repressor into monomers were evaluated. Sufficiently high concentrations of DNA and of the Arc protein were used to guarantee complete binding of protein to DNA at atmospheric pressure. The increase in polarization on binding DNA (not shown) confirmed that Arc protein was bound to DNA at atmospheric pressure.

Figure 2 shows the dissociation curves in the absence and in the presence of different DNAs.

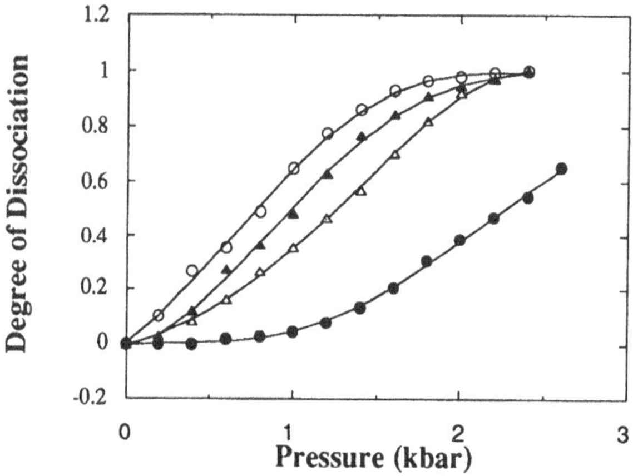

Figure 2- Effects of Different DNAs on the Dissociation of Arc Repressor. Degrees of dissociation were calculated from measurements of center of spectral mass of intrinsic fluorescence (see Experimental Procedures). The pressure-induced dissociation of 1 μM Arc repressor was measured in the absence of DNA (o); or in the presence of 1 μM operator DNA (●); 1 μM $(dG\text{-}dC)_{24}$ (▲); or 1 μM $(dA\text{-}dT)_{24}$ (△).

The sequence containing the 21-base operator stabilized the Arc subunit interaction (Figure 1B), practically preventing its dissociation by pressure, and the Arc repressor binds with very high affinity to this sequence (Vershon et al., 1986) (Figure 1B). In the absence of DNA, 90% dissociation was achieved at 1600 bar whereas at this same pressure no significant dissociation occurred in the presence of the operator (Figure 2). Nonspecific DNAs were much less effective than the operator in stabilizing the Arc subunit interaction (Figure 2). We have recently found that a plasmid sequence (pBR322) shifted the $p_{1/2}$ by only 530 bar, which corresponds to a stabilization of -0.63 kcal/mol for the subunit interaction, an effect much smaller than that provided by the operator sequence (Silva & Silveira, 1992). Arc repressor binds to plasmid pBR322 with an affinity 100-fold lower than its affinity for operator DNA (Vershon et al., 1986). The sequence $(dA\text{-}dT)_{24}$ promoted a considerable stabilization (Figure 2), similar to that previously found with the poly (dA-dT) (Silva & Silveira, 1992). The sequence $(dG\text{-}dC)_{24}$ conferred much less protection against the pressure-dissociation (Figure 2).

The magnitude of stabilization of the subunit interaction was determined from the expression:

$$\delta G_s = - [(p_{1/2})_D \cdot \Delta V_D/n_D - (p_{1/2})_o \cdot \Delta V_o/n_o] \qquad (3)$$

where δG_s is the stabilizing free energy per subunit, $(p_{1/2})_D$ and $(p_{1/2})_o$ are respectively the midpoints of pressure dissociation of the DNA-liganded and unliganded proteins, and $\Delta V_D/n_D$ and $\Delta V_o/n_o$ are the volume changes of association per subunit for DNA-liganded and unliganded Arc, respectively. Both $\Delta V_D/n_D$ and $\Delta V_o/n_o$ are divided by the respective number of subunits, n_D and n_o. In the absence of DNA, Arc is a dimer but it forms a tetramer when bound to operator (Brown et al., 1990; Silva & Silveira, 1992). ΔV was obtained from the slope of the plot of ln $(\alpha_p^2/(1 - \alpha_p))$ *vs* pressure for the unliganded protein (Silva et al., 1992) and from ln $(\alpha_p^4/(1 - \alpha_p))$ *vs* pressure for DNA-liganded protein. The fourth power is justified for the DNA-liganded protein in view of the previous evidence that two dimers bind to the 21-base operator (Brown et al., 1990; Silva & Silveira, 1992). The values of $\Delta V_D/n_D$ for different DNAs were similar to $\Delta V_o/n_o$ (absence of DNA) (between 45 and 50 ml/mol). Therefore Equation 1 can be simplified to the form:

$$\delta G_s = - [(p_{1/2})_D - (p_{1/2})_o] \cdot \Delta V = - \Delta p_{1/2} \cdot \Delta V \qquad (4)$$

where $\Delta V = \Delta V_D/n_D = \Delta V_o/n_o$

Operator DNA shifted the $p_{1/2}$ by about 1700 bar (Figure 2) which according to Eq. 4 corresponds to an increase of 2.1 kcal/mol in the free energy of association per subunit. If the operator accommodates two dimers of Arc repressor (Brown et al., 1990), the total free energy of stabilization is 8.4 kcal/mol.

The values of free energy of stabilization (δG_s) derived from $\Delta p_{1/2}$ for the nonspecific DNAs are much smaller than that obtained for operator. Among the nonspecific DNAs, $(dA\text{-}dT)_{24}$ promoted the largest stabilization (-0.88 kcal/mol). This result is consistent with the predominance of the bases A and T in the 21-base operator sequence (Vershon et al., 1989). Poly (dG-m^5·dC) did not produce stabilization of subunit interaction in Arc (Silva & Silveira, 1992).

3.1.3. *Free-energy linkage between subunit association and DNA binding*. The difference in stability of the various Arc-DNA complexes shows that the magnitude of the stabilization of the subunit interaction is determined by the sequence of the DNA bound by the repressor. The stabilization results from the linkage between the free energies of subunit association and DNA binding. The greatest free-energy coupling occurs for operator DNA. We also found that this free-energy coupling does not occur with the mutant protein PL8 (Silva & Silveira, 1992). The PL8 has a single amino acid replacement Pro8 —> Leu that promotes a loss of DNA specificity (Vershon et al., 1986). In the absence of DNA, PL8 is also a dimer. The binding of either the operator sequence, the synthetic DNAs or plasmid pBR322 did not affect the pressure dissociation of PL8 dimer indicating the absence of free-energy coupling (Silva & Silveira, 1992). The coupling between subunit association and ligand binding, as the one described here for Arc-DNA interaction, is a feature of many oligomeric systems (Weber, 1992).

Figure 3 shows a free-energy diagram for the coupled reactions of Arc subunit association and binding to DNA. This diagram is a simplified version of the reaction scheme published elsewhere (Silva & Silveira, 1992). A DNA with the potential binding of two Arc dimers (D_4) was considered, given that for both operator and poly (dA-dT) a stoichiometry of two dimers per 21-bp DNA was obtained. In this diagram, only the species that are more likely to be populated were included.

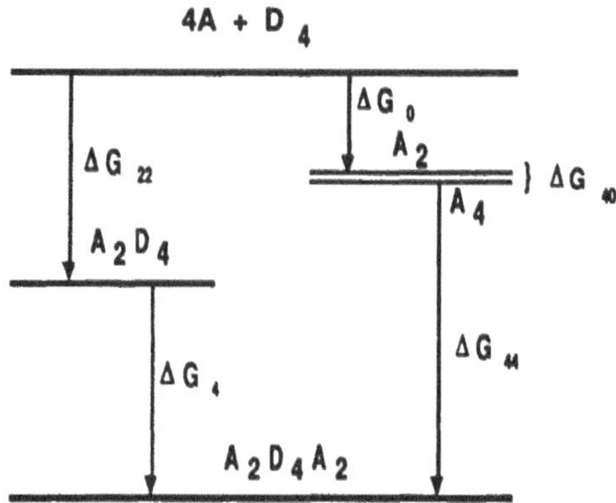

Figure 3. Free-energy diagram of the reaction pathway of Arc repressor and DNA D4.

Utilizing free-energy conservation assumptions, we can deduce the following relationships from the diagram of Figure 3:

$$\Delta G_{44} = (\Delta G_{22} + \Delta G_4) - (\Delta G_o + \Delta G_{40}) \qquad (5)$$

The free energy of association of dimer to tetramer in the absence of DNA (ΔG_{40}) is much smaller than ΔG_o and can be disregarded. The sum ($\Delta G_{22} + \Delta G_4$) is the total free energy of subunit association in the presence of DNA (ΔG_D) . Therefore, it follows from eq. 5;

$$\Delta G_{44} = (\Delta G_D - \Delta G_o) = - \Delta p_{1/2} \cdot \Delta V \qquad (6)$$

and from eq. 4 , we infer that the stabilizing free energy ($\delta G_s = - \Delta p_{1/2} \cdot \Delta V$) is equal to the free energy of stability (ΔG_{44}) of the protein-DNA complex ($A_2D_4A_2$).

We conclude that the difference in stability of the protein-protein interactions could drive the final stability of the DNA-protein interaction. This free-energy coupling is probably entropic and may reflect the correct positioning of the second subunit that is promoted by binding to the first subunit. Recently, it has been shown that coupling entropy plays an important role in allosteric enzymes (Reinhart et al., 1989). If the specific DNA-binding sites exist in tandem arrays, oligomerization with order higher than four may indeed occur. This seems to be the case for Met repressor of *E. coli* a protein with structural homology to Arc repressor (Breg et al., 1990; Knight et al., 1989; Phillips, 1991).

Half-maximal binding of Arc to operator sequence DNA occurs at a protein concentration of 10^{-10} M (Vershon et al., 1986) a concentration considerably lower than the dimer-monomer dissociation constant (10^{-8} M) (Silva et al., 1992). Thus, we expected that most of the stabilization conferred by operator DNA should be coupled to the strengthening of the interaction between subunits in the dimer (A-A). Mutational studies indicate that the aminoacid residues in the intersubunit β-sheet (A-A interface) are important for the specificity of DNA binding (Breg et al., 1990; Knight & Sauer, 1992). A component of the stabilization could also be related to the interaction between dimers (A_2-A_2) in the complex $A_2D_4A_2$. This latter interaction would be similar to that found between lysozyme molecules on the surface of thyroglobulin (Rawitch & Weber, 1972).

The model of the structure of Arc repressor (Breg et al., 1990) and the structure of other related repressors (Phillips, 1991) suggest the use of intersubunit β-strands for sequence recognition. The large free-energy linkage between DNA binding and subunit association elicit the attractive possibility of a mutual coupling between the intersubunit contacts and the two strands of DNA. In this way, not only the preferential binding to the operator, but the mechanism of repression can be rationalized. The Arc repressor promotes repression not by preventing the binding of RNA polymerase to the promoter but rather by blocking its isomerization from a closed to open complex (Vershon et al., 1987). This step requires melting the DNA strands. A mutual coupling implies that the increase in interaction between the two subunits of Arc repressor bound to DNA promotes an increase in the interaction between the two DNA strands. Therefore, the mutual coupling could prevent isomerization of bound RNA polymerase from the closed to open complex.

The magnitude of the Arc subunit stabilization promoted by the operator is large enough to account for the much higher affinity of the operator relative to nonspecific DNA sequences. Our results further imply that a thermodynamic constraint ($\delta G_s = \Delta G_{(n)} - \Delta G_{(0)}$) may furnish the basis for the DNA recognition process. When a mutation compromises the operator DNA specificity, it affects δG_s, as in the case of the PL8 mutant. Further studies on other sequence-specific DNA-binding proteins will be necessary to determine the generallity of our findings.

3.2. NUCLEOSOMES: STRUCTURAL CHANGES INDUCED BY HIGH PRESSURE AND IONS

The basic structure of chromatin consists of successive levels of DNA organization mediated by a number of proteins. Our understanding concerning the conformational flexibility of the chromosome (for an extensive review of all structural aspects of chromatin, see (van Holde, 1989)), and its significance for the biological role of chromatin, is only rudimentary, partially due to the difficulty of applying the usual physical techniques to study such a complex system. The extremely complex organization of chromatin is not static; the chromatin undergoes structural transitions during the cell cycle and this requires a flexible structure that can respond dynamically to different situations(Gasser & Laemmli, 1987). For instance, particular parts of the genome must become active as a template at well defined times(Reeves, 1984). The fundamental mechanisms by which this is achieved is still largely a mystery. Information concerning the molecular factors responsible for particular activity and the coordination between different parts of the chromosomal apparatus is becoming rapidly available on a qualitative and descriptive level. The complex interplay between the parts of the cellular machinery show a fascinating kinetic and structural coordination. However, the physical mechanisms by which even the simpler reactions are carried out are largely unknown, and models involve considerable speculation. Methods are needed to investigate structural interconversions and changes that are applicable to the complex multicomponent biological systems as well as to the pure isolated components. Then the knowledge of the isolated systems can be extended to the more biologically relevant milieu. Pressure is a comparatively mild and selective variable, and it is especially useful for perturbing multicomponent molecular systems without affecting secondary structures of macromolecules, such as proteins (Weber & Drickamer, 1983). The combination of pressure and fluorescence provides an experimental approach sensitive and selective enough to investigate specific interactions of biological molecules at several levels of complexity.

The nucleosomal complex: The nucleosomal complex is the underlying structural feature of eucaryotic chromosomes(Kornberg & Klug, 1981). The functional nucleosome is comprised of an octamer protein core consisting of two identical tetramers; each tetramer is composed of four separate polypeptide chains: the H2A, H2B, H3 and H4 chains. Duplex DNA is wrapped around the octamer in a superhelical fashion encircling the protein core. Another protein, the H1 histone, is bound to the outside surface of this structure and appears to hold together the DNA-core assembly. The H1 histone is not considered part of the core nucleosome structure and is absent in the studies reported here. Nucleosome structures are often arranged along the DNA molecule with a repeat distance of ~200 base pairs; 146 basepairs of DNA are tightly bound to the octamer histone core forming a relatively nuclease resistant complex. This structure has been exceptionally stable throughout evolution.

The nucleosomes and the separated histones can be purified from a variety of different nuclei preparations. It is possible to disassemble the nucleosome complex, separate the histones, and reassemble the histones with DNA from another source to form reconstituted nucleosomes; these reconstituted nucleosomes retain many of the major structural properties of the original nucleosomes. These are excellent samples to study the equilibrium structures and kinetic characteristics of their conformational changes. It is possible to obtain reproducibly a large amount of pure, homogeneous reconstituted system suitable for sensitive spectroscopic physical studies.

Nucleosome ion dependent structures: DNA is a highly negatively charged polyelectrolyte, and histones are strongly basic proteins. The nucleosome complex is expected to respond sensitively to the ionic strength of the solvent. The different structural variants of this DNA-protein complex, and the dynamic interconversion between the different ion dependent states, constitutes one of the major defining characteristics of intact nucleosomes. In the last 20 years, a major effort has been expended to study the ion dependent conformational transitions of nucleosomes, with the hope that knowledge of these structural transitions will lead to a better understanding of the dynamic nature and biological function of nucleosomal complexes within the chromosome.

Pressure sensitivity of the nucleosome structure: Pressure, coupled with fluorescence spectroscopy, has proved to be especially valuable for obtaining crucial information concerning the stability and interconversions of complex aggregated systems (Bonafe et al., 1991; Silva et al., 1989; Silva & Weber, 1988; Weber, 1987). The highly electrostatically charged nature of the nucleosomal components, and the aggregated state of the core histones, leads to the expectation that hydrostatic pressure would perturb significantly the structure of intact nucleosomes. This is found to be the case, and here we report several characteristics of pressure induced transitions of reconstituted nucleosomes observed by changes in the fluorescence spectrum of the covalently attached acrylodan(Prendergast et al., 1983) probe. The role of the ionic strength on the type and extent of these structural transitions is investigated, and the pressure induced conformational transitions are correlated with the purely salt induced conformational changes.

3.2.1. *Sensitivity of the structure to the ionic strength.* The general properties of our nucleosomes labeled with acrylodan were investigated to compare our results to previously reported properties of nucleosomes, eg the reaction of nucleosomes with ionic strength (Cary et al., ; Dieterich et al., 1979; Stacks & Schumaker, 1979). Representative results are shown in Figure 4, where we show the response of the fluorescence spectra to low (Figure 4 top) and high (Figure 4 bottom) ionic strength variations.

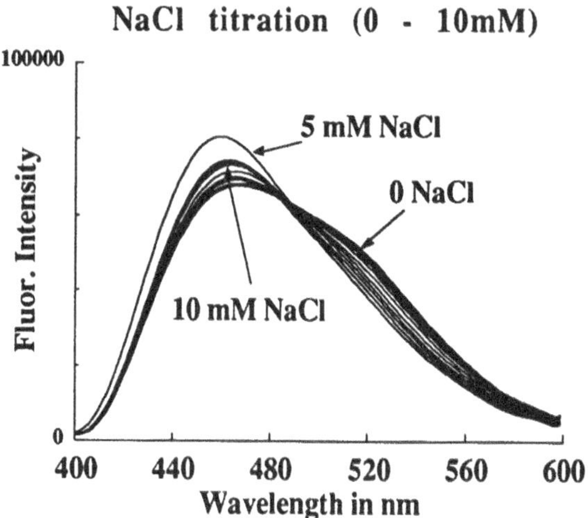

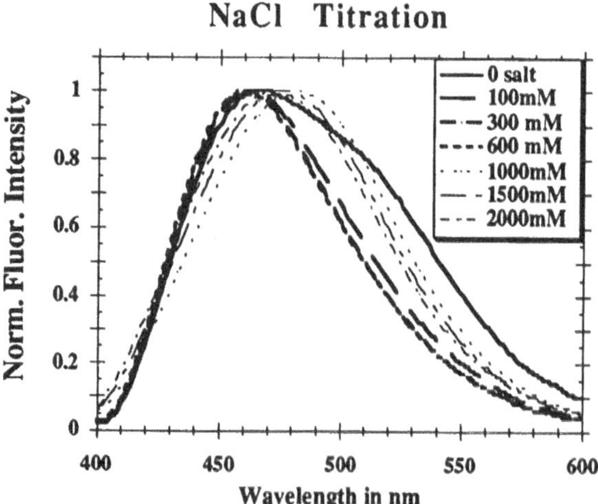

Figure 4. top - *very low salt:* Changes in the acrylodan nucleosome structure as reported by the acrylodan fluorescence spectrum induced by salt concentrations in the 0 to 10 mM concentration range. Below 0.5 mM [NaCl] there are at least two spectral components (arrow for 0 NaCl), and the shape and intensity of the fluorescence emission spectrum does not change in this very low salt concentration range from 0 to 0.5 mM [NaCl]. Two underlying peaks are centered at about 520 and 460 nm. With increasing [NaCl] from 1 mM to 10 mM, there is a gradual change in the shape of the fluorescence spectrum: the red component (~520 nm) decreases essentially to zero; the peak of the

blue component spectrum remains constant. Thus acrylodan reports on the low salt transition of reconstituted nucleosomes that is known to occur between 1-3 mM salt (see text). At these very low salt conditions the repulsions between the charges on the negatively charged DNA polyelectrolyte, and between the positively charged basic histones, apparently leads to a partially unfolded structure of the nucleosome. This is indicated by the partial red spectral shift from 460 nm (characteristic of the tightly folded, closed structure at 10 to 100 mM [NaCl], see Figure 5) to ~520 nm below 0.5 mM salt; a red shift is to be expected if the dye leaves the nonpolar environment of the folded nucleosome, and comes increasingly into contact with the polar solvent. However, at [NaCl] < 1 mM we expect the interaction between the positive histones and the negative DNA to be especially strong, so that the proteins will remain attached to the DNA; possibly this electrostatic attraction is a major driving force in the unfolding reaction, in addition to the repulsions between the like charges.
Figure 4 bottom - *high salt:*Varying [NaCl] from 0.1 to 2.0 molar, we see another major structural transition at about 600 mM NaCl. Between 0.6 and 1 M [NaCl] there is a red spectral shift from 460 nm to 480 nm. In addition the fluorescence is quenched at higher salt concentrations. The red spectral shift of the fluorescence emission promoted by high salt conditions alone is less the shift observed either for the low salt conditions (see Figure 4 top) or at high pressure (Figure 5).

A conformational change at very low ionic strength (< 3 mM) and another structural change at about 500-600 mM NaCl can be seen by following shifts in the energy of emission of the acrylodan (Figure 4). At very low salt, the nucleosomes are in a partially extended state caused by electrostatic repulsion. Two conformations are seen clearly from 0 to 3 mM added NaCl, and from 3 to 10 mM the spectrum shifts gradually to the blue (Figure 4, top). According to the fluorescence spectra, there is a stable tight conformation existing from 10 mM to ~600 mM NaCl . The emission spectrum then shifts considerably toward the red as the salt is increased to 2000 mM (Figure 4, bottom). The maximum fluorescence emission, with its maximum of 460 nm at low salt concentrations, shifts to 480 nm as the salt concentration is increased. These two major structural changes have been reported previously to occur with intact nucleosomes (Dieterich et al., 1979; van Holde, 1989). Thus the acrylodan molecule covalently attached to the SH group of the H3 histone appears to be a good probe to discern important structural changes in the nucleosomal complex.

3.2.2. *Structural changes of intact labeled nucleosomes induced by pressure.* We have taken advantage of the spectral shifts of the acrylodan emission spectrum observed at different ion concentrations to observe the pressure induced conformational changes in the nucleosome structures. As we show, the extent and nature of the induced pressure changes can be correlated with the major known ion-dependent unfolding processes thought to occur in nucleosome structures. The sensitivity of the system to pressure, and the magnitude and form of the spectroscopic shifts, were dependent upon the ionic strength (Figure 5) and the presence of cosolvents (such as glycerol, data not shown). We have determined at least three major structure components with identifiable spectral components, and it appears that pressure (together with specific ionic conditions) is able to mediate the interconversion between these nucleosome states. The spectroscopic changes observed at low ionic strength are not reversible; however, at intermediate ionic strengths the interconversions are completely reversible, Figure 5. Details of these results are now presented.

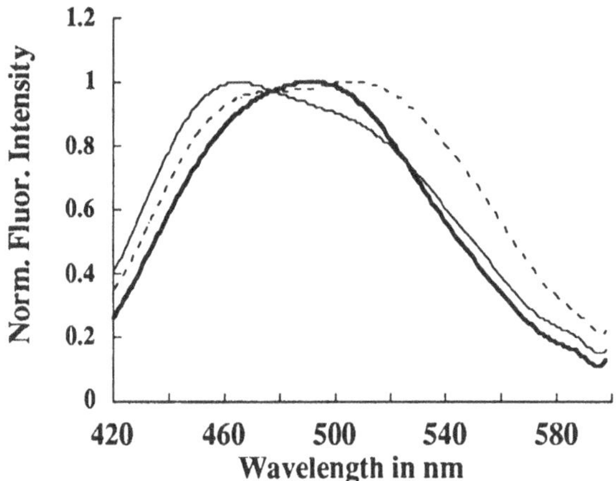

Figure 5a. *pressure with no added salt* - (——) = atmospheric pressure, (____) = 2000 bar, (- - - -) = return to atmospheric pressure. The presence of at least two populations is clearly observable. The shape of the emission spectrum is changed dramatically when applying 2000 bar pressure. The spectrum at 2000 bar indicates a conformational change of the nucleosome with increasing contact of the acrylodan dye with the solvent (~500 nm peak). However, the process appears to involve more than two states because there is not simply an increase of the 520 nm component with a concomitant decrease of the 460 nm component. At 2000 bar there may be another structural state of the DNA-histone complex, not seen at the other salt conditions. A shoulder of the 460 nm component remains at the high pressure. Returning to atmospheric pressure, the proportion of 520 and 460 nm components has been altered (see text). The conformational change instigated by high pressure at very low salt is not completely reversible (at least not on the time scale of this experiment).

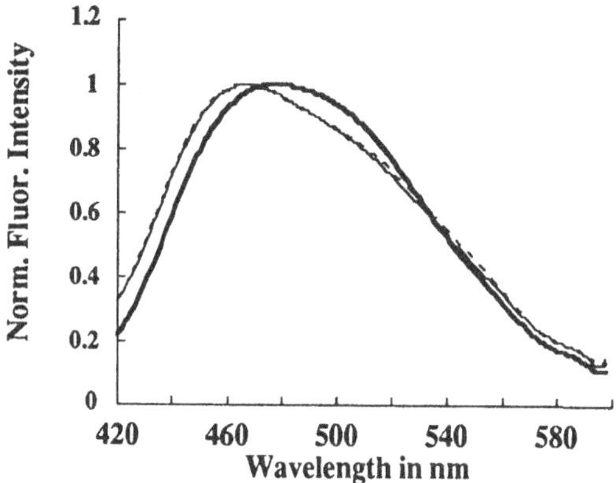

Figure 5b. *- 3 mM [NACl]:* (——) = atmospheric pressure, (——) = 2000 bar, (- - - -) = return to atmospheric pressure. At 3 mM [NaCl] the 520 nm component (seen with no added salt, Figure 4 top) is considerably reduced. The high pressure perturbation shows only a slight red shift and the effect is completely reversible.

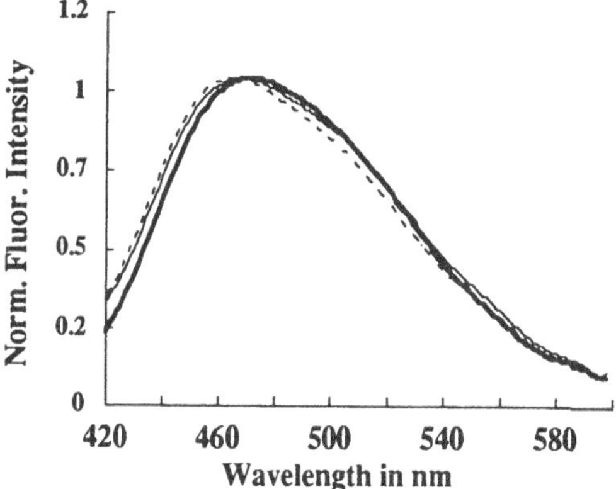

***Figure 5c** - 0.1 M [NACl]:*(——) = atmospheric pressure, (____) = 2000 bar, (- - - -) = return to atmospheric pressure. The spectrum at one bar represents a typical spectrum of fully intact nucleosomes. The peak is at ~460 nm, and at this ionic strength the structure is quite resistant to pressure perturbations. The small pressure sensitivity that is observable is completely reversible. Under these ionic conditions the reconstituted nucleosomes form a stable tight complex, and the H3 placement within the nucleosome structure is refractory to pressure perturbation.

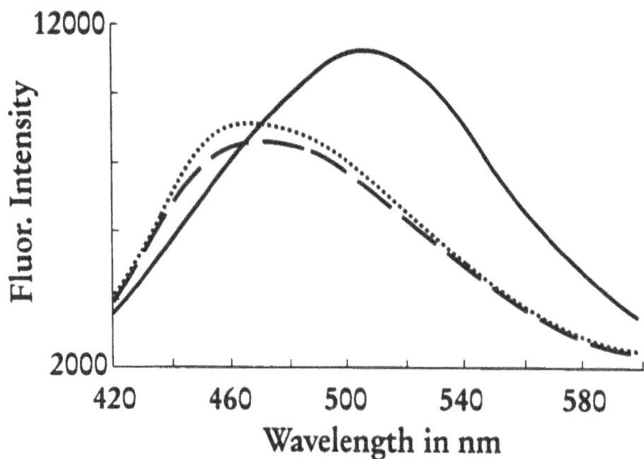

Figure 5d - *0.5 M [NaCl]:* The spectrum is shifted from ~470 nm at 1 bar (dotted line) to almost 520 nm (thick solid line) at 2000 bar pressure, and back to 470 nm when returning to 1 bar (dashed curve). This indicates a considerable rearrangement of the H3 histones from an apparent intact compact structure to a more open structure, allowing much more contact of the dye with the solvent at high pressure. We see that the nucleosome structure responds very sensitively to the application of high pressure. A spectrum taken immediately after the application of the high pressure shows a kinetic intermediate spectrum demonstrating that the "unfolding" process is not rapid on the time scale of several minutes at this salt concentration (data not shown). The slow kinetics at this salt concentration highlights the possibility that the inability to significantly perturb the structure with high pressure at lower ionic strengths (eg 0.1 M) may be due to a kinetic activation barrier, rather than an equilibrium effect. This must be checked in future experiments. The pressure induced spectral change is reversible after completion of the kinetics.

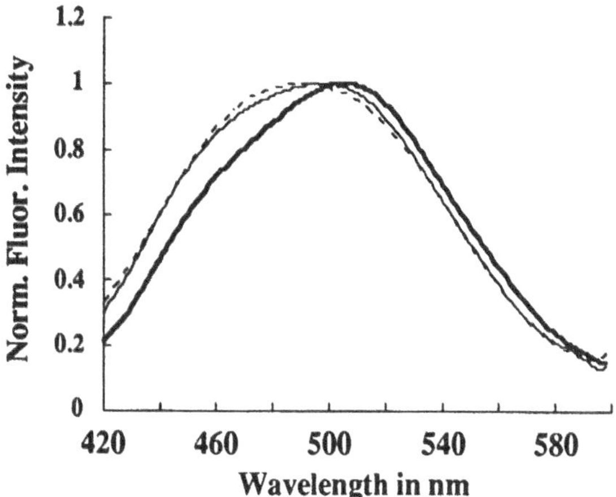

Figure 5e. *-1.0 M [NaCl]:* (——) = atmospheric pressure, (___) = 2000 bar, (- - - -) = return to atmospheric pressure. The response of the nucleosome structure to high pressure at 1 M salt is similar in many respects to that at 0.5 M salt (Figure 5d). However, the transition begins from a state corresponding to the completed high salt transition conformation at 1 bar (see Figure 4 bottom), and progresses at high pressure to a state even more accessible to the solvent. The spectral changes are reversible on the time scale of the pressure experiment.

The pressure sensitivity varies with the ionic strength of the solution: In Figure 5 a-e it is seen that upon increasing the hydrostatic pressure a red shift of the fluorescence emission spectra results; this red shift is more extensive than observed when varying the high salt alone at atmospheric pressure. In the case of increasing pressure at constant [NaCl] the 460 nm peak shifts to 520 nm, indicating a more extensive exposure of the acrylodan fluorophore to the polar environment at higher hydrostatic pressure, and consequently a major conformational change in the molecular complex (Figure 5 b and c). At a [NaCl] between 3 mM and 1000 mM this structural alteration returns to the initial point immediately after pressure release as one can see in Figure 5 (b through e), where the spectra are shown, before and after pressure treatment. At 500 mM [NaCl] the pressure sensitivity is maximum and the greatest spectral shifts are observed (Figure 5d); in addition, in the salt concentration range of ~500 mM we observe slow kinetic processes following a pressure change, as the nucleosomes relax to new pressure conditions (kinetics not shown). Certain kinetic components of these pressure induced conformational changes are on the minute time scale; these kinetic studies are being pursued presently and more rapid dynamic processes can be investigated using a rapid pressure-jump method (Clegg & Maxfield, 1976).

The present data is consistent with the simple model of two molecular conformations represented by two distinct maximum fluorescence emission peaks. One of the species has a maximum fluorescence emission at ~460 nm and corresponds to fully assembled nucleosome particles. The other conformational state exhibits its maximum emission about 520 nm and very probably corresponds to a dissociated nucleosome complex. The proteins, at least the H3 molecule, are probably still bound to the nucleic acid molecule, since the pressure induced spectral shifts are more extensive in the absence of DNA (Figure 6). This corroborates earlier conclusions (Dieterich et al., 1979). Thus we can interpret the pressure induced spectral changes of the reconstituted nucleosomes by assuming an equilibrium between two interchangeable states of the DNA -histone complex. The equilibrium extent of this reaction is poised and modified by the solution salt concentration. In higher salt conditions and atmospheric pressure (see Figure 4, bottom), the spectral peak shifts only to 480 nm, and this indicates that the structural changes induced by even the highest ion concentrations in the absence of high pressure are less extensive than in the presence of high pressure at lower ion concentrations (Figure 5d).

Similar experiments have been carried out at very low salt concentrations with labeled reconstituted nucleosomes (Figure 5a) and we observed distinct high pressure induced transitions. The spectral characteristics indicate different molecular conformational changes than found by higher salt concentration. The results for this salt concentration range are summarized in the conclusions, below.

The pressure-induced red shift of the acrylodan spectrum from solutions containing histones alone labeled with acrylodan is reminiscent of the effect in the presence of DNA; however, the red shift is more extensive. This is shown in Figure 6. At 3mM salt concentration the histones fluoresce with a maximum at 500 nm; this spectrum may represent fluorescence emission of $(H3\text{-}H4)_2$ and higher complexes (Benedict et al., 1984; D'Anna & Isenberg, 1974; Godfrey et al., 1980). This is considerably higher than the 460 nm component observed under the same conditions in the presence of DNA, and indicates that the acrylodan face of the H3 molecule is more available to the solvent than in the intact nucleosome complex. Upon the application of high pressure the spectrum shifts to 530 nm and this change is reversible. This is consistent with a dissociation of the histone aggregates, and some of the steps involved in the dissociation of the histones alone may partially contribute to the response of the nucleosomes to high pressure. An extensive study using pressure to perturb the equilibrium of core histone complexes — in the absence of DNA — labeled at the H3-SH position with the dansyl fluorophore has been reported (Royer et al., 1989).

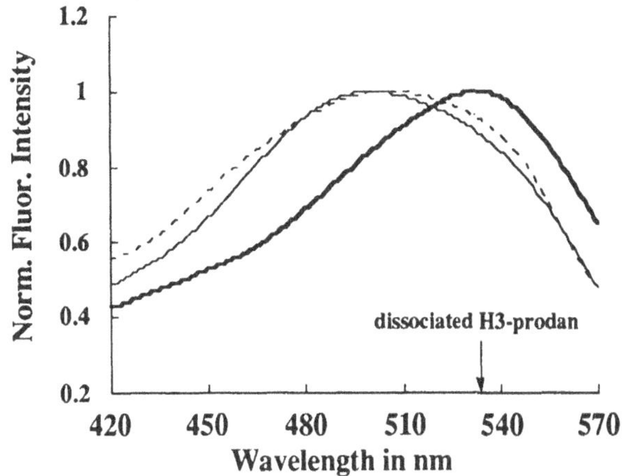

Figure 6. *3 mM [NaCl] (core histones without DNA):* (——) = atmospheric pressure, (____) = 2000 bar, (- - - -) = return to atmospheric pressure. The red shift of the acrylodan fluorescence in response to pressure perturbation is most dramatic when the DNA is absent. The ~540 nm peak under high pressure is observed only in the absence of DNA. This also demonstrates that the histones are probably not dissociated completely in the presence of DNA, even under high pressure (see text).

3.2.3. *Comparison of the pressure-induced structural changes at different ionic strengths*. This is the first report concerning high pressure perturbations of intact reconstituted nucleosomes. We have labeled specifically the single SH group of the H3 histones with the polar sensitive fluorophore acrylodan and observed the spectral shift of the emission fluorescence spectrum as a function of ionic strength and pressure. We can correlate the structural changes induced by pressure with the ionic strength profile observed by us and by others (Dieterich et al., 1979). The emission spectrum shifts to the red when high pressure is applied, indicating a disassembly of the complex in response to the application of pressure, with a concomitant exposure of the acrylodan fluorophore to the solvent. This is consistent with a major structural change of the closely juxtaposed H3 histones in the intact nucleosomal complex, opening up their contacting faces to the solvent. This conformational change appears to be more extensive when induced by pressure that in response to only the presence of elevated ionic strength, judging from the extent of the spectral shifts. The extent of the structural changes in response to pressure perturbation depends on the ionic strength of the solution. At physiological ion concentrations (~100 mM) the nucleosome complex is most stable, showing only minor responses to elevated pressure. At very dilute salt concentrations (< 3mM) the complex appears to open up even in the absence of high pressure, probably due to large electrostatic repulsions between the closely spaced positive charges of the proteins, and the electrostatic repulsions of the negative charges on the DNA polyelectrolyte (Dieterich et al., 1979; Libertini & Small, 1982; Libertini & Small, 1987). The application of high pressure in these low salt conditions induces structural changes that are not reversible; the fluorescence spectrum is shifted somewhat to the blue after the pressure is returned to 1 bar from 2000 bar. This could be interpreted as an annealing of the nucleosome complex at high pressure in these low salt conditions, resulting in an even more compact structure than is formed by the procedure used to prepare the reconstituted nucleosomes involving slow dialysis from high salt and urea concentrations (see Methods, above). We are presently determining the equilibrium constants and the change in molar volumes corresponding to the separate conformational changes. The slow structural rearrangements following pressure perturbation at intermediate salt concentrations are being investigated in an thorough kinetic study.

4. General Discussion and Conclusions

In the last two decades molecular biology has undergone extensive development, especially concerning the techniques of recombinant DNA; protein engineering has evolved into a commonplace tool in basic and applied research. However, although one can engineer genes, there are many aspects of their mechanisms and control that we do not understand. The genetic "participants" consist of a great number of specialized proteins that perform their duties (gene activation, repression, replication, transcription, etc.) dynamically, and the interactions between the different components play an important role in procaryotic and eucaryotic organisms. Therefore, any method that can perturb a biological system in a controlled way will contribute valuable biophysical information about these complex molecular systems. We believe that high pressure can fulfill many of the requirements for such systems and we provide two encouraging studies that represent two extremes of protein-DNA interactions: one, a sequence-specific procaryotic repressor, the Arc repressor, and the other a less specific basic structural element of the eucaryotic chromosome, the nucleosome. Both complexes form very stable protein-DNA interactions, but the protein-protein interactions display very different characteristics.

In the case of Arc repressor, the interaction with the DNA is dramatically affected as the protein subunits are dissociated. The disruption with pressure of the dimeric unit bound to the DNA leads to the dissociation of the protein from the DNA. We show that this constitutes the basis for the specific sequence recognition of DNA-sequences and for the higher affinity of the operator for the repressor. Therefore, the free-energy coupling between the protein-protein interaction and the DNA binding accounts for the high specificity of Arc-operator DNA interaction.

Histones also interact with very high affinity with DNA. Our results show that when pressure perturbs an intact nucleosome complex, the interactions between the histones change significantly. The structural changes of the nucleosomes are different from the linked interactions of the Arc repressor; the loosened structure of the histone components remains bound to the DNA. The changes in chromatin that occur during the cell cycle are probably of a similar nature to the transition induced by pressure.

The effects of salt on the subunit stability is also different for Arc repressor and nucleosomes. Whereas the interaction between subunits of the Arc repressor monomers is stronger at high salt concentrations (Silva et al., in press), the histone-histone interactions are loosened at higher ionic strengths.

The future of pressure studies on protein-DNA complexes is very promising. We are extending these studies to multisubunit proteins that have catalytic activity on binding to DNA. We have found that the RNA polymerase complex is disassembled by pressure (Clegg & Silva, manuscript in preparation) and the character of the pressure inactivation of the RNA polymerase activity supports the view that a multitude of protein conformers with differing pressure sensitivities exist under normal conditions. We expect in the near future to extend these studies to more complex assemblies such as isolated chromosomes and even whole cells.

5. Acknowledgments

This work was supported by a Grant from the European Economic Community (CtI*-CT90-0857) to RMC and JLS. We also thank Annelies Zechel for technical assistance with the histone preparations.

REFERENCES

Ashikawa, I., Nishimura, Y., Tsubol, M. & Zama, M. (1982) *J. Biochem. 92*, 1425-1430.

Benedict, R. C., Moudrianakis, E. N. & Ackers, G. K. (1984) *Biochemistry 23*, 1214-1218.

Bonafe, C. F. S., Villas-Boas, M. & Silva, J. L. (1991) *J. Biol. Chem. 266*, 13210-13216.

Breg, J. N., Opheusdsen, J. H. J., Burgering, M. J. M., Boelens, R. & Kaptein, R. (1990) *Nature 346*, 586-589.

Brennan, R. G. & Matthews, B. W. (1989) *TIBS 14*, 286-289.

Brown, B. M., Bowie, J. U. & Sauer, R. T. (1990) *Biochemistry 29*, 11189-11195.

Cary, P. D., Moss, T. & Bradbury, E. M. *Eur. J. Biochem. 89*, 475-482.

Clegg, R. M. & Maxfield, B. W. (1976) *Rev. Sci. Instrum. 47*, 1383-1393.

D'Anna, J. A. & Isenberg, I. (1974) *Biochemistry 13*, 4992-4997.

Dieterich, A. E., Axel, R. & Cantor, C. R. (1977) *Cold Spring Harbor Symp. Quant Biol. 42*, 199-206.

Dieterich, A. E., Axel, R. & Cantor, C. R. (1979) *J. Mol. Biol. 129*, 587-602.

Eshaghpour, H., Dieterich, A. E., Cantor, C. R. & Crothers, D. M. (1980) *Biochemistry 19*, 1797-1805.

Gasser, S. M. & Laemmli, U. K. (1987) *Trends in Genetics 3*, 16-22.

Godfrey, J. E., Eickbush, T. H. & Moudrianakis, E. N. (1980) *Biochemistry 19*, 1339-1346.

Heremans, K. A. H. (1982) *Annu. Rev. Biophys. Bioeng. 11*, 1-21.

Knight, K. L., Bowie, J. U., Vershon, A., K., Kelley, R. D. & Sauer, R. T. (1989) *J. Biol Chem. 264*, 3639-3642.

Knight, K. L. & Sauer, R. T. (1992) *EMBO J. 11*, 215-223.

Kornberg, R. D. & Klug, A. (1981) *Scientific American 1981*, 48-60.

Libertini, L. J. & Small, E. W. (1982) *Biochemistry 21*, 3327-3334.

Libertini, L. J. & Small, E. W. (1987) *Nucl. Acid Res. 15*, 6655-6664.

Phillips, S. E. V. (1991) *Current Opinion in Struct. Biol. 1*, 89-98.

Prendergast, F. G., Meyer, M., Carlson, G. L., Iida, S. . & Potter, J. D. (1983) *J. Biol. Chem. 258*, 7541-7544.

Rawitch, A. B. & Weber, G. (1972) *J. Biol. Chem. 247*, 680-685.

Reeves, R. (1984) *Biochem. Biophys. Acta 782*, 343-393.

Reinhart, G. D., Hartleip, S. B. & Symcox, M. M. (1989) *Proc. Natl. Acad. Sci. U.S.A. 86*, 4032-4036.

Royer, C. A., Rusch, R. M. & Scarlata, S. F. (1989) *Biochemistry 28*, 6631-6637.

Schleif, R. (1988) *Science 241*, 1182-1187.

Silva, J. L. & Silveira, C. F. (1992) *Proc. Natl. Acad. Sci., U.S.A. in press*,

Silva, J. L., Silveira, C. F., Correia, A. & Pontes, L. (1992) *J. Mol. Biol. 223*, 545-555.

Silva, J. L., Villas-Boas, M., Bonafe, C. F. S. & Meirelles, N. (1989) *J. Biol. Chem. 264*, 15863-15868.

Silva, J. L. & Weber, G. (1988) *J. Mol. Biol. 199*, 149-159.

Stacks, P. C. & Schumaker, V. N. (1979) *Nucleic Acids Res. 7*, 2457-2467.

van Holde, K. E., *Chromatin* (Springer-Verlag, New York, 1989).

Vershon, A. K., Bowie, J. U., Karplus, T. & Sauer, R. T. (1986) *Prot.: Struct. Funct. & Genet. 1*, 302-311.

Vershon, A. K., Kelley, R. D. & Sauer, R. T. (1989) *J. Biol. Chem. 25*, 3267.

Vershon, A. K., Liao, S.-M., McClure, W. R. & Sauer, R. T. (1987) *J. Mol. Biol. 195*, 323-331.

Vershon, A. K., Youderian, P., Susskind, M. M. & Sauer, R. T. (1985) *J. Biol. Chem.*

260, 12124-12129.

Weber, G. (1987) in *High Pressure Chemistry and Biochemistry* (van Eldyk, R. & J., J., Ed.) pp. 401-420, Reidel, Dordrecht.

Weber, G., *Protein Interactions* (Chapman & Hall, New York & London, 1992).

Weber, G. & Drickamer, H. G. (1983) *Q. Rev. Biophys. 16*, 89-112.

THE EFFECT OF THE LIPID MATRIX ON THE RESPONSE OF A MEMBRANE ENZYME TO HIGH PRESSURE

DAVID ZAKIM and JURAJ KAVECANSKY
Division of Digestive Diseases
Cornell University Medical College
1300 York Avenue
New York, New York 10021
U. S. A.

ABSTRACT. A partially purified but delipidated preparation of the microsomal enzyme UDP-glucuronosyltransferase was reinserted into unilamellar bilayers of DMPC or DOPC and activity determined as a function of pressure in the range of 0.001 to 2.2 kbar. Application of pressure had a single, simple effect on delipidated enzyme; activity declined in accord with the effect expected for an enzyme with a positive volume of activation. By contrast, pressure had complex effects on enzyme embedded in DMPC or DOPC. Several new functional states of the enzyme were generated as pressure was increased. These effects of pressure were not an inherent property of UDP-glucuronosyltransferase but depended on the lipid environment into which the enzyme was inserted. The current data demonstrate the utility of high pressure experiments as a tool for probing the mechanism of lipid-dependent regulation of membrane enzymes.

Introduction

The functional states of enzymes solvated by water are determined by the properties of the water phase (e.g. pH, ionic strength) and by interactions between water-soluble molecules and specific sites on the enzymes. These regulatory factors also affect the functional states of enzymes that are integral components of membranes. But in the latter case, the physical and chemical features of the lipid regions of membranes also determine how a given enzyme will function [for reviews see refs. 1-3]. Only a relatively small number of membrane enzymes have been examined from the perspective of regulation by the lipid bilayer. It is apparent, nevertheless, that the activities of some membrane enzymes depend on specific interactions with the polar regions of membrane phospholipids while for others function is modulated by the length and unsaturation of the polymethylene chains [1-15]. Little is known in detail, however, as to what physical and chemical features of the polar region (or apolar region) account for regulating membrane enzymes and how this regulation is effected.

Are Membrane Enzymes Regulated by the Bulk Phase Properties of Lipid Bilayers ?

Most interest in lipid-dependent regulation of membrane enzymes has focused on regulation by the polymethylene chains [10-15]. The most important data, in the context of this issue, are those showing that Arrhenius plots for many membrane enzymes display discontinuities coinciding with temperatures for the main phase transitions (gel to liquid crystalline transition) of bilayers (or biological membranes) containing the enzyme being studied [1, 2, 9-14]. In view of the large (approximately 10-fold [16]) and discontinuous change in the viscosity of bilayers at the liquid

R. Winter and J. Jonas (eds.), High Pressure Chemistry, Biochemistry and Materials Science, 603–617.

crystal to gel phase transition, the discontinuities in relevant Arrhenius plots have been interpreted to mean that membrane enzymes are regulated by the bulk phase viscosity of lipid bilayers, which is the prevailing view of how membrane lipids regulate the functional states of membrane enzymes. There are serious difficulties, however, with this interpretation of the physical significance of discontinuities of Arrhenius plots of membrane enzymes. Thus, in association with the liquid crystal to gel transition, the span of bilayers increases; and the microscopic details of packing between chains is altered [17]. In addition, membrane enzymes can phase separate in gel phases. Any, all, or none of these discontinuous changes in bilayers could account for the observed discontinuities in Arrhenius plots of membrane enzymes. In the absence of techniques for separating potential contributions to regulation of enzymes from the different types of changes that occur at the main phase transitions of bilayers, it seems difficult if not impossible to ascribe a specific mechanism as the basis for discontinuities in Arrhenius plots of membrane enzymes. There is ample evidence too that these uncertainties are not theoretical issues.

Some membrane enzymes display discontinuities in Arrhenius plots at the phase transition temperatures of bilayers only under some but not all conditions [18]. Other membrane enzymes that display discontinuities of Arrhenius plots corresponding to temperatures for the main phase transitions of the lipid matrix, display the same discontinuity at the same temperature in the absence of lipids [19, 20]. Some enzymes behave like Ca-ATPase, which has a discontinuous Arrhenius plot at the temperature of the main phase transition of the lipid environment [10], but appears not to be regulated by the viscosity of membrane lipids under other conditions [21]. Also, when examined carefully many Arrhenius plots for membrane enzymes are more compatible with the interpretation that the plots bend continuously, suggesting a thermally driven equilibrium between two different conformational states, rather than an abrupt change in function - due to any mechanism - at the main phase transition [22]. Finally, we think it is important to note that the interpretation of the significance of discontinuities of Arrhenius plots at temperatures for the main phase transition of bilayers, i.e., that an abrupt change in the viscosity of the bilayer damps catalysis according to the Kramers equation, has been extended generally to the question of lipid-dependent regulation of membrane enzymes in liquid crystalline phospholipids - the condition under which these enzymes function in living cells. There is no evidence to justify this hypothesis.

Alternate Views of the Mechanism of Lipid-Dependent regulation of Membrane Enzymes

The idea that the viscosity of bilayers is the key property of lipids for regulating membrane enzymes implies that the chemistry of catalysis in gel and liquid crystal phases is identical but that rates are different because of the variation in the mechanical properties of the medium solvating the enzyme. This follows from a large number of studies of aqueous-soluble enzymes that are sensitive to the viscosity of water [23-25]. But a recent theoretical analysis of the problem shows that the differences in energies of activation of membrane enzymes in gel and liquid crystalline phases are too large to be accounted for by the different mechanical properties of gel and liquid crystalline phases [26]. It seems then that the data showing discontinuities in Arrhenius plots of membrane enzymes are not truly compatible with the interpretation that discontinuous changes in bulk phase viscosity account for the changes observed in enzyme function in association with the liquid crystal to gel transition. The analysis of published data in the context of a rigorous application of the Kramers equation implies instead that lipid-enzyme interactions can modulate the stabilities of different conformational isomers of membrane enzymes [26, 27] and that this is the fundamental mechanism for the lipid-dependent regulation of the membrane enzymes that have been studied to date. The relative stabilities of different states could reflect the effects of variations in the microscopic details of packing between transmembrane domains of membrane enzymes and different types of polymethylene chains. Obviously, there is no reason to believe that a single mechanism will apply to all membrane enzymes. On the other hand, lipid-dependent regulation of the stabilities of different conformational isomers that have variable kinetic and regulatory properties is attractive. It provides a more variable and thereby more powerful mechanism for regulating enzymes than the non-specific mechanical hypothesis; and, possibly, it could account for the great diversity of phospholipid species in membranes of higher organisms as compared with bacteria, which remains unexplained.

Theory of the Effect of Pressure on the Function of Membrane Enzymes

A major problem in studying the effects of variations in the structure of polymethylene chains on the functional states of membrane enzymes is that one cannot change the chemical nature of the polymethylene chains of a bilayer without altering the temperature dependence of the viscosity, the temperature for the main phase transition, etc. For, as the chains are varied, both the chemical and physical properties of the membrane are altered. Hence, in most published work that has examined the function of membrane enzymes in different liquid crystalline bilayers, it is impossible to separate possible effects secondary to the change in chemical composition of the bilayers from effects, of the change in composition, on bulk phase viscosity.

An interesting and potentially useful feature of lipid bilayers, in the context of this difficulty, is that viscosity is pressure-dependent over a range of pressures at which membrane enzymes are likely to retain function. High pressure can be used to vary the viscosity of lipid bilayers in the absence of changes in composition. Moreover, the pressure-dependence of viscosity is related to the compressibility of a bilayer, which in turn depends on composition , i.e., the pressure-dependence of η for bilayers depends on compressibility. Independent of differences in the activity of a given enzyme, in a series of lipids with varying lipid compositions at ambient pressure, a given increase in pressure will change the viscosity of a bilayer of defined composition in a highly predictable way. And there is a precise relationship between k_{cat} and viscosity of the bilayer in the case that the activity of an enzyme is viscosity-dependent: $k_{cat} = A/\eta \; e^{-\Delta Ea/RT}$, where Ea is the energy of activation. In theory, then, one can use the effects of high pressure on bilayers to test the applicability of the Kramers equation [36] to the function of enzymes in liquid crystalline phases. Another potentially useful feature of pressure is that it forces systems to adopt conformations of least volume. Given the absence of a basis for energetic interactions between hydrophobic surfaces of transmembrane domains of membrane proteins and polymethylene chains, any specific effect of polymethylene chain composition on the conformations of transmembrane domains of enzymes are likely to be related to geometric constraints on allowable modes for packing between chains and between chains and protein domains. Pressure can be expected to have dramatic effects on this packing and to amplify subtle differences in the manner of packing of transmembrane domains of enzymes with polymethylene chains. Therefore, high pressure would appear to be an excellent technique for examining the basis for the effects of variation in the polymethylene chain composition of bilayers on the functional states of membrane enzymes.

High pressure, however, has potentially complex effects on the activity of membrane enzymes because of the numerous mechanisms by which high pressure can alter reaction rates. Thus, high pressure can diminish the activity of an enzyme catalyzed reaction by dissociating oligomeric enzymes [28]. Activity also will decrease as pressure is increased for an enzyme with a positive volume of activation. And in the case of a membrane enzyme there is the possibility that activity will decrease as pressure is increased because of pressure-induced increases in the viscosity of the bilayer region of the membrane.

The best evidence indicates that many membrane enzymes are oligomers [c.f. 29]. These are unlikely to be dissociated by high pressure, however, because they are present at extremely high concentrations as compared with enzymes solvated by water. The concentrations of enzymes in reconstituted lipid-enzyme preparations, for example, are on the order of 1 mole enzyme per several hundred moles phospholipid. Pressure will not dissociate oligomeric, aqueous-soluble enzyme at concentrations this high [28]. Hence, we do not think that pressure-dependent dissociation of membrane enzymes will confound studies of membrane enzymes.

Obviously, pressure will diminish k_{cat} in the case that the catalyzed reaction has a positive volume of activation. And as mentioned already, it is anticipated that pressure will inhibit membrane enzymes via pressure-dependent increases in the viscosity of bilayers in which the enzyme is embedded. But, a tacit assumption of the hypothesis that viscosity of the bilayer regulates membrane enzymes is that the reaction coordinate of an enzyme-catalyzed reaction is independent of the nature of the lipid environment. This means that the volume of activation of an enzyme-catalyzed reaction also will be independent of the lipid environment. Therefore, it should be possible to use the effect of pressure on membrane viscosity in the liquid crystal state to test the mechanical hypothesis for regulation of membrane enzymes.

The Properties of UDP-glucuronosyltransferase

The membrane enzymes of interest to this laboratory have been the family of UDP-glucuronosyltransferases. These enzymes are most abundant in the microsomal fraction of the cell. Extensive studies of one pure isoform, for which acidic phenols like p-nitrophenol are the best substrates, have shown that the enzyme is lipid-dependent. Pure, delipidated enzyme has a residual activity [6], but activity increases when the enzyme is inserted into a suitable lipid matrix [6]. Although the activity of pure, reconstituted enzyme depends on the distribution of charge within the polar region of phospholipids, the enzyme does not appear to require polar regions of specific chemical composition [30, 31]. Activity does depend, however, on the nature of the polymethylene chains of phospholipids [6-8]. The data in TABLE 1 show, for example, an approximate 700-fold range of activity for pure, delipidated enzyme inserted into micelles of lysophosphatidylcholine with varying polymethylene chains. The activation of delipidated enzyme by lysophosphatidylcholines is not a non-specific effect of detergents, but depends on specific features present within the polar region of phospholipids [30, 31]. The activity of UDP-glucuronosyltransferase is lower in bilayers as compared with micelles (compare TABLES 1 and 2), but activity in bilayers also depends on the polymethylene chain composition of phosphatidylcholines. The data in these TABLES are not strictly comparable because they were obtained with different pure preparations of the enzyme. But measurements of activity with a single preparation of pure enzyme show that k_{cat} is 8-fold greater in micelles of oleoyl-lysophosphatidylcholine as compared with bilayers of dioleoylphosphatidylcholine (DOPC) [8].

In addition to modulation of k_{cat}, lipids regulate several other properties of UDP-glucuronosyltransferase, such as response to allosteric effectors [32], apparently negatively cooperative

TABLE 1. Activity of the p-nitrophenol-metabolizing isoform of UDP-glucuronosyltransferase in micelles of lysophosphatidylcholine with varying polymethylene chain composition. Activities (μmol min^{-1} mg^{-1}) were measured at 30°C and extrapolated to infinite concentrations of substrates. Data are from [7, 8].

Polymethylene chain	V_{max}
none	0.07
C14	0.91
C16	3.7
C18	3.8
C18:1cis	67.2

interactions between binding sites for UDP-glucuronic acid [32], thermal stability to denaturation [33], and the temperature for onset of a rapidly reversible interconversion between an active and inactive state of the enzyme [34]. Interestingly, all the lipid-dependent effects demonstrable in a system comprising pure enzyme and a single type of phospholipid have their counterparts for enzyme in situ in microsomes [35, 36], which lends credence to the idea that the function of UDP-glucuronosyltransferase could be regulated in vivo by interactions with the lipid environment.

TABLE 2. Activity of the p-nitrophenol-metabolizing isoform of UDP-glucuronosyltransferase in bilayers of phosphatidylcholine with varying polymethylene chain composition. Activities (μmol min^{-1} mg^{-1}) were measured as in TABLE 1. The ratio of lipid to enzyme (mol/mol) was about 600/1.

Polymethylene chains	V_{max}
diC18	0.22
diC18:1trans	1. 0
diC18:1cis	1.6
diC18:3	4.0

The data in TABLES 1 and 2 appear to fit the conventional wisdom for lipid-dependent regulation of membrane enzymes: activity of UDP-glucuronosyltransferase is greatest in the lipid environment for which the packing of the polymethylene chains is least dense. Thus, activity is far greater in micelles than in bilayers with the same polymethylene and polar group composition; activity is greater in unsaturated versus saturated chains. On the other hand, several of the effects of lipids on the regulatory characteristics of the enzyme do not fit with this idea. The enzyme for instance is more stable to thermal denaturation in saturated as compared with unsaturated phospholipids [33]; and it is difficult to rationalize the effects of different lipids on the allosteric properties of the enzyme in the context of mechanical regulation by the viscosity of the membrane environment. More extensive studies of the values of V_{max} (or k_{cat}) as a function of the polymethylene chain composition of phosphatidylcholines also have shown that the apparent relation in TABLES 1 and 2, between bulk phase viscosity and k_{cat}, begins to break down as more lipids are examined (Rotenberg & Zakim, in preparation). We have begun to believe, therefore, that the critical feature of bilayers for regulating the properties of UDP-glucuronosyltransferase is not the viscosity of the bilayer.

Effects of High Pressure on the Function of UDP-glucuronosyltransferase

The pressure-dependence of η is much greater for DOPC as compared with DMPC [37]. If the activity of UDP-glucuronosyltransferase were modulated only by the viscosity of bilayers, we would expect activity to decline more with pressure for enzyme in bilayers of DOPC than in DMPC, independently of differences in activities at ambient pressure. Differences in the pressure titration of enzyme in these two environments hence should be due differences in the compressibility of the bilayers.

STUDIES OF ENZYME INSERTED INTO BILAYERS OF DMPC

Methods. For technical reasons we could not assay UDP-glucuronosyltransferase with p-nitrophenol as aglycone at high pressure, which depends on changes in absorbance at 400 nm. We used a fluorescent assay instead, employing 1-naphthol as aglycone. The isoform of UDP-glucuronosyltransferase that glucuronidates p-nitrophenol glucuronidates 1-naphthol as well, but at about 1% of the rate with p-nitrophenol [7]. An isoform separate from the p-nitrophenol-metabolizing enzyme accounts for nearly all the glucuronidation of 1-naphthol in microsomes [unpublished data]. The latter enzyme has not been studied as carefully as the p-nitrophenol isoform, but the properties regulated by lipids appear to be the same for both isoforms.

A partially purified mixture of isoforms of UDP-glucuronosyltransferase was delipidated as in Hochman et al. [6]. The enzyme was then inserted into unilammelar bilayers of DMPC as in Scotto and Zakim [38] and into unilamellar bilayers of DOPC as in Kasahara and Hinkle [39]. Density gradient separation of lipids, enzyme, and lipid-enzyme complexes in prior studies, with both methods for inserting enzyme into preformed bilayers, have shown essentially complete recovery of enzyme as lipid-enzyme complexes [33]. Activities of UDP-glucuronosyltransferase were measured with 1-naphthol as aglycone as in Dannenberg et al. [40]. The pressure cell and method of applying and measuring pressure are described in Dannenberg et al. [40]. All measurements were at 37°C using 1 mM UDP-glucuronic acid and 0.04 mM 1-naphthol. These concentrations of substrates do not give the value of k_{cat} for the enzyme [41]. Therefore, the pressure titrations do not yield volumes of activation for the catalytic step. This limitation does not alter the conclusions that can be drawn form the data to be presented.

Studies in the range of 0.001 to 0.6 kbar. The data in Figure 1 (closed circles) show a pressure titration of enzyme in DMPC for a system solvated by H_2O. Also shown are the effects of pressure on the activity of delipidated enzyme (open triangles). The latter data indicate that enzyme activity fell as pressure was increased. At pressure > 0.2 kbar, the activity of the delipidated enzyme was too small to measure accurately; and these data are not shown. By contrast with delipidated enzyme, enzyme in DMPC was more active. It also responded to pressure in a manner completely different from that of the delipidated enzyme. Enzyme in DMPC was activated by pressure up to 0.6 kbar, but the pressure-dependence of the activation did not fit the expression $[d\ln k/dP]_T = -\Delta V^{\#}/RT$, where the volume of activation is negative, because the data show a non-linear response of log activity to increasing pressure. A simple explanation for the pressure-induced activation of UDP-glucuronosyltransferase is that the enzyme has a low pressure state (E_1) and a high pressure state (E_2) and that E_2 has greater activity than E_1. The equilibrium between these states would shift to the right, e.g., E_1 --> E_2, as pressure is increased because the total

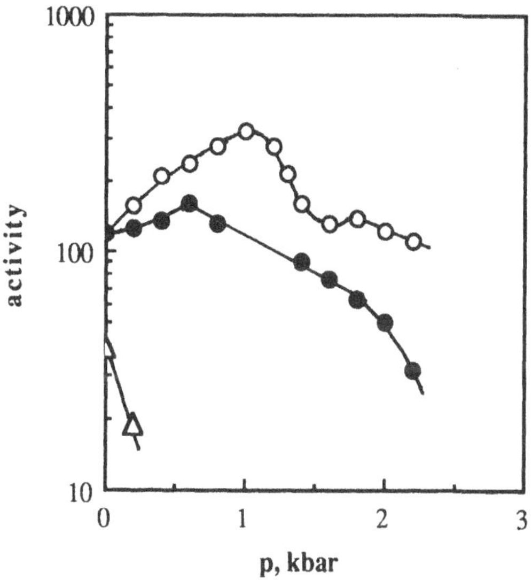

Figure 1. Pressure titration of the activity of a delipidated preparation (Δ) of the 1-naphthol metabolizing isoform of UDP-glucuronosyltransferase and after insertion into bilayers of DMPC solvated by H_2O (•) or D_2O (o). The ratio of lipid/enzyme (mol/mol) was approximately 600/1. Activities were measured at 37°C and are given as nmol min^{-1} mg^{-1}.

volume of the system enzyme + DMPC would be minimized by this transition. We expect, however, that the equilibrium constant for this reaction will also be linear when plotted as lnK versus pressure. So the data appear not to fit this explanation either. The non-linearity of the pressure-dependent activation can be rationalized, however, by proposing that the volumes of activation of the putative states E_1 and E_2 are non-identical. We believe, therefore, that the most reasonable explanation for the activation of UDP-glucuronosyltransferase between 0.001 and 0.6 kbar, for enzyme in DMPC, is that application of pressure caused a change in the conformational state and thereby the functional state of the enzyme. This conclusion is supported too by observations that pressures higher than 0.6 kbar clearly led to transitions between states of the enzyme (see below).

If we look no further than the data for delipidated enzyme and those in Figure 1 for enzyme-DMPC complexes, in the range of 0.001 to 0.6 kbar, several important conclusions can be drawn about the effect of lipids on the functional state of UDP-glucuronosyltransferase. First, the fundamental effect of the lipid environment on UDP-glucuronosyltransferase must be subtle and not represent a global reorganization of the enzyme in that the delipidated enzyme retains activity. Second, the response of the enzyme to pressure depends on the presence of a lipid environment. Third, the general conclusion that DMPC alters the reaction pathway for glucuronidation is independent of the mechanistic basis for the pressure-induced activation up to 0.6 kbar. Thus, the environment of DMPC either changed the reaction pathway for glucuronidation from one with a positive volume of activation to one with a negative volume of activation; or the interactions between UDP-glucuronosyltransferase and DMPC determine how the enzyme will respond to increases in pressure. In either case, the data are not compatible with the idea that lipids influence the function of the enzyme simply because of the viscosity of the polymethylene chains.

The significance of the inflection at 0.6 kbar. The pressure-induced activation of UDP-glucuronosyltransferase reached a maximum at 0.6 kbar. Activity then declined at higher pressures and the plot of log activity versus pressure was linear until 1.8 kbar. There are two possible mechanisms for the inflection at 0.6 kbar. The first is that all the UDP-glucuronosyltransferase is converted to the high pressure state (E_2) at 0.6 kbar; and above this pressure the decline in activity is due to the effect of pressure on activity of an enzyme for which the volume of activation is positive. The second possibility is that the inflection at 0.6 kbar reflects the response of UDP-glucuronosyltransferase to a change in the state of the DMPC bilayer from the liquid crystal to gel phase. This possibility occurs because the data in Figure 1 were obtained at 37°C; and DMPC undergoes an isothermal transition from liquid crystal to gel, at 37°C, between 0.6 and 0.7 kbar [Winter, personal communication]. We can exclude the latter possibility as the cause for the inflection at 0.6 kbar. The experiments on which this conclusion is based are shown in open circles in Figure 1 and represent the pressure titration of enzyme in DMPC complexes that are solvated by D_2O. Thus, the temperature for the liquid crystal to gel phase transition is barely affected by substituting D_2O for H_2O, i.e., this transition will be at about 0.7 kbar for DMPC solvated by H_2O or D_2O. Yet the pressure titration for the enzyme-lipid complexes in D_2O shows (open circles, Figure 1) that the inflection is at about 1.0 kbar, which is considerably above the pressure need to induce the gel phase at 37°C. We can conclude then that UDP-glucuronosyltransferase is insensitive to the isothermal phase transition within bilayers of DMPC that are solvated by D_2O. There is no reason to believe that this would not be so for enzyme in DMPC solvated by H_2O. Indeed, the continuity in activity at the inflection point (0.6 kbar) for the data in closed circles supports this idea. For in the case that the enzyme is sensitive to the viscosity of the bilayer, the Kramers' equation predicts an abrupt 10-fold decrease in activity for a small change in pressure. Therefore, we believe that the inflection at 0.6 kbar for the system solvated by H_2O reflects that all the UDP-glucuronosyltransferase is in the E_2 state at 0.6 kbar and that the pressure-dependent decline in activity at $P > 0.6$ kbar indicates a positive volume of activation for E_2.

It is important to note that there are significant differences between the pressure titrations of enzyme-lipid complexes in H_2O and D_2O. The pressure-induced activation is greater for the system hydrated by D_2O versus H_2O, and the pressure for maximum activation is shifted to 1.0 kbar in the former medium. We do not understand the basis for these effects of D_2O, but this uncertainty does not mitigate the conclusions drawn about the physical significance of the inflection point at 0.6 kbar. Of interest was that the activities and response of delipidated enzyme to pressure were the same in H_2O and D_2O.

Studies at $P > 1.6$ kbar. At $P > 1.8$ kbar there was a second inflection point for the enzyme-DMPC complexes solvated by H_2O. There was a second inflection at 1.6 kbar for enzyme-DMPC complexes solvated by D_2O. In the former case, the inflection reflects a change in the volume of activation of the enzyme. Thus, the pressure-dependence of activity was steeper at $P > 1.8$ kbar than it was between 0.6 and 1.6 kbar. The onset of the transition between high pressure states of the enzyme with different volumes of activation was abrupt at 1.6 kbar. The transition was not complete, however, at the highest pressures tested. For enzyme-DMPC complexes solvated by D_2O, there were apparently two separate changes in the state of the enzyme between 1.6 and about 2.0 kbar; and at P > 1.8 kbar, the apparent volume of activation was smaller than at lower pressures. In addition, the transition between different high pressure states was completed over a small range of pressures for enzyme in D_2O.

STUDIES OF ENZYME INSERTED INTO BILAYERS OF DOPC

The data for the effects of pressure on the function of UDP-glucuronosyltransferase in DMPC indicate that the response of UDP-glucuronosyltransferase to pressure depended on the nature of the lipid environment but not in a way that can be understood in the context of the effects of pressure on the viscosity of the lipid environment. The key question from the perspective of understanding the basis for lipid-dependent regulation of membrane enzymes, however, is the extent to which the effects of pressure, as for example in Figure 1, are determined by the detailed nature of the lipid environment of UDP-glucuronosyltransferase. That is, the effects of pressure for enzyme in DMPC could be generic and not related to the demonstrable effects of different phospholipids on the function of the enzyme at ambient pressure. We addressed this issue by examining the effect of pressure on enzyme activity after inserting it into DOPC. These data are shown in Figure 2.

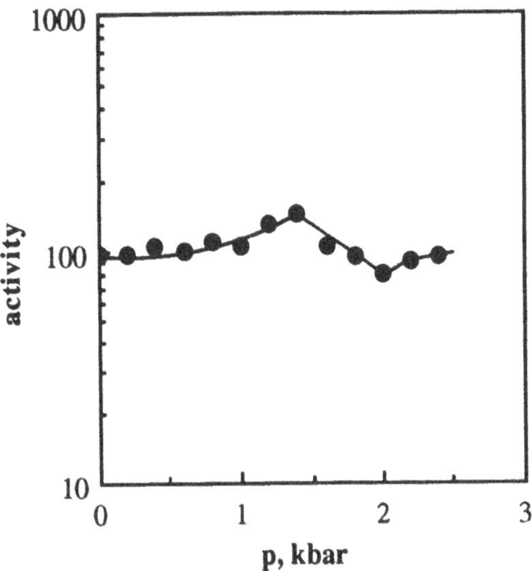

Figure 2. Pressure titration of the activity of the 1-naphthol metabolizing isoform of UDP-glucuronosyltransferase inserted into bilayers of DOPC. Details of the experiments were the same as in Figure 1.

We note first that activity of UDP-glucuronosyltransferase is somewhat smaller in DOPC as compared with DMPC. This result in itself is important because it is different from the effects of these lipids on the activity of the p-nitrophenol isoform of UDP-glucuronosyltransferase [42]. Activity for this isoform is higher in DOPC versus DMPC, indicating that a given lipid has differential effects on the state of selective members of the family of UDP-glucuronosyltransferases. But, if lipid-dependence reflected only the impact of the bulk viscosity of the bilayer on enzyme function, the relative activities of all isoforms of UDP-glucuronosyltransferase, as a function of polymethylene chain composition, should be the same. In fact, because the bulk phase viscosity of the lipid environment is independent of the nature of the membrane enzyme when the latter is present in the membrane as a dilute solution, we would expect to find the same relative activities, as a function of polymethylene chain environment, for all lipid-dependent membrane enzymes. Exceptions to this rule would occur for enzymes for which activity is proportional to $1/\eta^m$, where m is less than 1. But even in this instance the stratification of activities of different lipid-dependent enzymes, as a function of lipid composition of a given series of lipids, should not be altered qualitatively. This obviously is not the case for the two isoforms of UDP-glucuronosyltransferase for which data are available.

As compared with the enzyme-DMPC complexes, the pressure for maximum activation of UDP-glucuronosyltransferase in DOPC (solvated by H_2O) was shifted to 1.4 kbar. Since bilayers of DOPC are more compressible than DMPC, the upward shift in the pressure for the first inflection in the pressure titration in DOPC, as compared with DMPC, could be explained simply if we assume that the low and high pressure states of UDP-glucuronosyltransferase were the same in DMPC and DOPC. Higher pressure would then be needed for enzyme in DOPC versus DMPC to stabilize the high pressure state relative to the low pressure state. We note too that the negative slope of log activity versus P was nearly the same for enzyme in DOPC at $P > 1.4$ kbar and enzyme in DMPC at pressures between 0.6 and 1.8 kbar. This result is contrary to expectation for regulation of activity by the viscosity of the bilayers because DOPC is more compressible than DMPC. The viscosity of DOPC changes to a greater extent than DMPC for application of the same pressure [37]. Most important, however, is the crucial difference in the behavior of UDP-glucuronosyltransferase at high pressure in DOPC versus DMPC. Enzyme in DOPC is activated at 1.8 kbar. Although there was a transition between functional states of UDP-glucuronosyltransferase in DMPC at close to this pressure, the effect of high pressure (in the range of 1.6 kbar) was fundamentally different for enzyme in DOPC as compared with DMPC. These results establish that the stabilities of different conformational states of UDP-glucuronosyltransferase, that have varying kinetic parameters, are a function of the composition of the polymethylene chain environment of the enzyme.

The temperature for the main phase transition of unilamellar bilayers of DOPC is approximately -20°C; the pressure-dependence of the transition is about the same for DOPC and DMPC. DOPC thus was in the liquid crystal state for all pressures used in Figure 2. None of the inflections for enzyme in DOPC can be attributed to a phase change within the lipids.

Physical Significance of the Pressure Data

The data presented above show that the activity of UDP-glucuronosyltransferase is not modulated by the bulk phase viscosity of the bilayer environment of the enzyme. The data show, for example, that the activities of two different isoforms are differentially affected by insertion into bilayers of DMPC or DOPC. Thus, whereas the p-nitrophenol isoform is more active in DOPC than in DMPC, the reverse is true for the 1-naphthol isoform. Additional evidence for the lack of regulation of UDP-glucuronosyltransferase by bilayer viscosity is that the 1-naphthol isoform is insensitive to the liquid crystal to gel phase transition of DMPC. Finally, the difference in the compressibilities of bilayers of DMPC and DOPC does not correlate with the effect of pressure on activities in different bilayers. Therefore, the pressure data provide several lines of evidence against the hypothesis that the lipid-dependence of UDP-glucuronosyltransferase reflects the non-specific regulation of the enzyme by the bulk phase properties of the lipid matrix. More importantly, the current data show that application of pressure effected changes in the functional state of UDP-glucuronosyltransferase and that the nature of these changes was not an inherent property of the enzyme. That is, the relative stabilities of different high pressure states of the enzyme are determined in part by the polymethylene chain composition of the phosphatidylcholine bilayer in which the enzyme is located. This specific result is not surprising in view of the known effects of pressure on the properties of aqueous-soluble enzymes [43], but it has not been considered as relevant to the problem of the lipid-dependent regulation of membrane enzymes.

Application of pressure causes systems to adjust thermodynamically by reducing the total volume of all components; and the response to pressure of an aqueous-soluble protein will depend on the properties of water. Exposure of charged groups, in the high pressure state, is favored, for example, because of the electrostrictive effect. As mentioned already, there is no basis for proposing that the details of folding of the hydrophobic surface of a transmembrane domain of a membrane protein will be directed by specific, energetic interactions with the surrounding polymethylene chains of a bilayer. Interactions in this type of system are likely to depend on geometric constraints on the modes for packing between large molecules. But constraints imposed on a protein by allowable modes of packing between adjacent polymethylene chains and between these chains and the surface of a protein could provide for specific modulation of the conformation of proteins by different polymethylene environments. Studies of pure bilayers show for example that the packing of DOPC at high pressure is quite different from that of DMPC [44]. So it is not unexpected that the packing of a protein could also differ in these two types of environments. This appears to be so for UDP-glucuronosyltransferase in that pressure-driven transitions between states depend on the specific features of the polymethylene chains interacting with the enzyme. We believe, therefore, that the current data provide a mechanism for lipid-dependent regulation of membrane enzymes that is specific and that depends on the microscopic details of packing between polymethylene chains and enzymes. Perhaps some of the poorly understood features of the polymethylene composition of mammalian membranes have consequences for the microscopic regulation of membrane enzymes via differences in the microscopic interactions between these chains and membrane enzymes. The complexity of species of phospholipids in mammalian membranes could reflect the need to regulate many different membrane proteins via specific lipid-protein interactions.

Acknowledgements

Work from the authors laboratory was supported by a grant from the NSF. The current address of J. K. is Department of Pharmacology, University of Cincinnati, Cincinnati, Ohio.

References

1. Kimelberg, H. K. (1977) The influence of membrane fluidity on the activity of membrane-bound enzymes. Cell Surface Revs. 3, 205-293.

2. Houslay, M. D. and Gordon, L. M. (1983) The activity of adenylate cyclase is regulated by the nature of its lipid environment. Curr. Top. Membr. Transp. 18, 179-231.

3. Carruthers, A and Melchior, D. L. (1986) How bilayer lipids affect membrane protein activity. Trends Biochem. Sci. 11, 331-335.

4. Isaacson, Y. A., Deroo, P. W., Rosenthal, A. F., Bittman, R., McIntyre, J. O., Bock, H-G., Gazzotti, P., and Fleischer, S. (1979) The structural specificity of lecithin for activation of purified D-β-hydroxybutyrate apodehydrogenase. J. Biol. Chem. 254, 117-126.

5. Kovatchev, S., Vaz, W. L. C., and Eibl, H. (1981) Lipid dependence of the membrane bound D-lactate dehydrogenase of Eschericia coli. J. Biol. Chem. 256, 10369-10374.

6. Hochman, Y., Vessey, D. A., and Zakim, D. (1981) A kinetic mechanism for modulation of the activity of microsomal UDP-glucuronosyltransferase by phospholipids. Effects of lysophosphatidylcholines. J. Biol. Chem. 256, 4783-4788.

7. Magdalou, J., Hochman, Y., and Zakim, D. (1982) Factors modulating the catalytic specificity of a pure form of UDP-glucuronosyltransferase. J. Biol. Chem. 257, 13624-13629.

8. Hochman Y. and Zakim, D. (1983) A comparison of the kinetic properties of two different forms of microsomal UDP-glucuronosyltransferase. J. Biol. Chem. 257, 4143-4146.

9. Kimelberg, H. K. and Papahadjopoulos, D. (1974) Effects of phospholipid chain fluidity, phase transitions, and cholesterol on (Na + K)-stimulated adenosine triphosphatase. J. Biol. Chem. 249, 1071-1080.

10. Squire, T. C., Bigelow, D. J., and Thomas, D. D. (1988) Lipid fluidity directly modulates the overall protein rotational mobility of the Ca-ATPase in the sarcoplasmic reticulum. J. Biol. Chem. 263, 9178-9186.

11. Silvius, J. R. and McElhaney, R. N. (1980) Membrane lipid physical state and the modulation of the Na^+,Mg^{2+}-ATPase activity in Acholeplasma laidlawii B. Proc. Natl. Acad. Sci. U.S.A. 77, 1255-1259.

12. Whitesell, R. R., Regen, D. M., Beth, A. H., Pelletier, D. K., and Abumrad, N. A. (1989) Activation energy of the slowest step in the glucose carrier cycle: break at 23°C and the correlation with membrane fluidity. Biochemistry 28, 5618-5625.

13. Dornmair, K. and Jahnig, F. (1989) Internal dynamics of lactose permease. Proc. Natl. Acad. Sci. U. S. A. 86, 9827-9831.

14. Foot, M., Cruz, T. F., and Clandinin, M. T. (1983) Effect of dietary lipid on synaptosomal acetylcholinesterase activity. Biochem. J. 211, 507-509.

15. Heron, D. S., Shinitzky, M., Hershkowitz, M., and Samuel, D. (1980) Lipid fluidity markedly modulates the binding of serotonin to mouse brain membranes. Proc. Natl. Acad. Sci. U. S. A. 77, 7463-7467.

16. Kwok, R. and Evans, E. (1981) Thermoelasticity of large lecithin bilayer vesicles. Biophys. J. 21, 637-652.

17. Cevc, G. and Marsh, D. M. (1987) Phospholipid bilayers: physical principals and models. Wiley-Interscience, N. Y.

18. Carruthers, A. and Melchior, D. L. (1984) Human erythrocyte transporter activity is govern-ed by bilayer lipid composition in reconstituted vesicles. Biochemistry 23, 6901-6911.

19. Dean, W. L. and Tanford, C. (1978) Properties of a delipidated detergent-activated Ca-ATPase Biochemistry 17, 1683-1690.

20. DeSmedt, H., Borghgraef, F., Ceuterick, F., and Heremans, K. (1979) Pressure effects on lipid-protein interactions in Na-K-ATPase. Biochim. Biophys. Acta 556, 479-489.

21. East, J. M., Jones, O. T., Simmonds, A. C., and Lee, A. G. (1984) Membrane fluidity is not an important physiological regulator of the (Ca^{2+}-Mg^{2+})-dependent ATPase of sarcoplasmic reticulum. J. Biol. Chem. 259, 8070-8071.

22. McElhaney, R. N. (1982) Effects of membrane lipids on transport and enzymic activity. Curr. Top. Membr. Transp. 17, 317-380.

23. Gavish, B. and Werber, M. M. (1979) Viscosity-dependent structural fluctuations in enzyme catalysis. Biochemistry 18, 1269-1275.

24. Beece, D., Eisenstein, L., Frauenfelder, H., Good, D., Marden, M. C., Reinisch, L., Reynolds, A. H., Sorensen, L. B., and Yue, K. T. (1980) Solvent viscosity and protein dynamics. Biochemistry 19, 5147-5157.

25. Doster, W. (1983) Viscosity scaling and protein-dynamics. Biophys. Chem. 17, 97-103.

26. Zakim, D. and Scarlata, S. F. (1992) Are membrane enzymes regulated by the viscosity of the membrane environment? Biochemistry, in press.

27. Vessey, D. A. and Zakim, D. (1971) Regulation of microsomal enzymes by phospholipids II. Activation of hepatic uridine diphosphate-glucuronyltransferase. J. Biol. Chem. 246,4649-4656.

28. Weber, G. and Drickamer, H. G. (1983) The effect of high pressure upon proteins and other biomolecules. Q. Revs. Biophys. 16, 89-112.

29. Vessey, D. A. and Kempner, E. S. (1989) In situ structural analysis of microsomal UDP-glucuronosyltransferase by radiation inactivation. J. Biol. Chem. 264, 6334-6338.

30. Zakim, D., Cantor, M., and Eibl, H. (1988) Phospholipids and UDP-glucuronosyltransferase. Structure/function relationships. J. Biol. Chem. 263, 5164-5169.

31. Zakim, D. and Eibl, H. (1992) The influence of charge and the distribution of charge in the polar region of phospholipids on the activity of UDP-glucuronosyltransfer-ase. J. Biol. Chem. 267, 13166-13170.

32. Hochman, Y., Kelley, M., and Zakim, D. (1983) Modulation of the number of ligand binding sites of UDP-glucuronosyltransferase by the gel to liquid-crystal phase transition of phosphatidylcholines. J. Biol. Chem. 258, 6509-6516.

33. Rotenberg, M. and Zakim, D. (1989) Effects of phospholipids on the thermal stability of microsomal UDP-glucuronosyltransferase. Biochemistry 28, 8577-8582.

34. Rotenberg, M. and Zakim, D. (1991) Effects of cholesterol on the function and thermotropic properties of pure UDP-glucuronosyltransferase. J. Biol. Chem. 266, 4159-4161.

35. Dannenberg, A. J., Rotenberg, M., and Zakim, D. (1989) Regulation of UDP-glucuronosyltransferase by lipid-protein interactions. Comparison of the thermotropic properties of pure reconstituted enzyme with microsomal enzyme. J. Biol. Chem. 264, 238-242.

36. Zakim, D. and Dannenberg, A. J. (1992) How does the microsomal membrane regulate UD P-glucuronosyltransferases? Biochem. Pharmacol. 43, 1385-1393.

37. Scarlata, S. F. (1989) Evaluation of the thermal coefficient of the resistance to fluorophore rotation in model membranes. Biophys. J. 55, 1215-1223.

38. Scotto, A. W. and Zakim, D. (1985) Reconstitution of membrane proteins. Spontaneous association of integral membrane proteins with preformed unilamellar lipid bilayers. Biochemistry 24, 4066-4075.

39. Kasahara, M. and hinkle, P. C. (1977) Reconstitution and purification of the D-glucose trans porter from human erythrocytes. J. Biol. Chem. 252, 7384-7390.

40. Dannenberg, A. J., Kavecansky, J., Scarlata, S. F., and Zakim, D. (1990) Organization of microsomal UDP-glucuronosyltransferase . Activation by treatment at high pressure. Biochemistry 29, 5961-5967.

41. Vessey, D. A. and Zakim, D. (1972) Regulation of microsomal enzymes by phospholipids V. Kinetic studies of hepatic uridine diphosphate-glucuronyltransferase. J. Biol. Chem. 247, 3023-3028.

42. Scotto, A. W. and Zakim, D. (1988) Reconstitution of membrane proteins. Spontaneous incorporation of integral membrane proteins into preformed bilayers of pure phospholipid. J. Biol. Chem. 263, 18500-18506.

43. Morild, E. (1981) The theory of pressure effects of enzymes. Adv. Prot. Chem. 34, 93-166.

44. Siminovitch, D. J., Wong, P. T. T., Berchtold, R., and Mantsch, H. H. (1988) A comparison of the effect of one and two mono-unsaturated acyl chains on the structure of phospholipid bilayers: a high pressure infrared spectroscopic study. Chem. Phys. Lipids 46, 79-87.

DISSOCIATION OF LARGE OLIGOMERIC PROTEINS BY HIGH HYDROSTATIC PRESSURE: DYNAMIC LIGHT SCATTERING STUDIES

GREGORY REINHART[1],
ENRICO GRATTON[2] and
WILLIAM W. MANTULIN[2]
[1]*University of Oklahoma, Department of Chemistry and Biochemistry, Norman, OK 73019; USA and*
[2]*Laboratory for Fluorescence Dynamics, Department of Physics, University of Illinois at Urbana–Champaign, Urbana, IL 61801; USA*

ABSTRACT. In the study of oligomeric protein association, the combined approach of high hydrostatic pressure perturbation with optical spectroscopic detection has provided great insight into structure and dynamics of these complex systems. Various methods offer advantages in specific cases. For example, a decrease in the light scattering intensity tracks the pressure induced dissociation of oligomeric proteins. However, optical artifacts complicate the interpretation of light scattering experiments under pressure. Dynamic light scattering offers the possibility of directly detecting changes in the translational diffusion coefficient, rather than the total scattered intensity associated with oligomer dissociation. The translational diffusion coefficient of the protein is related to its molecular weight and hydrodynamic volume, both of which change with oligomer dissociation. Dynamic light scattering offers greater sensitivity for the study of very large oligomers, that are not readily accessible using other spectroscopic methods. In addition, dynamic light scattering is conveniently used in conjunction with high hydrostatic pressure perturbation. We have performed dynamic light scattering experiments on the pressure dissociation of hemocyanin (gastropod), a very large molecular assembly (approximately 8×10^6 D). Under ambient pressure conditions, dynamic light scattering measurements show a heterogeneous oligomer population. The application of about 2 kbar of pressure strongly changes the dynamic light scattering spectrum by depleting most of the low frequency spectral components (large molecular weight oligomers). The formation of large oligomers is stabilized by calcium ions and consequently in 10 mM calcium 2 kbar of pressure provides only a small perturbation to self-association. Our results provide an initial indication that dynamic light scattering in conjunction with high hydrostatic pressure perturbation techniques offers a unique and sensitive approach to the study large oligomeric proteins.

1. Introduction

The application of high hydrostatic pressure perturbation, coupled with spectroscopic detection, especially of the highly sensitive and specific fluorescence signal, has provided a fertile area of research into association reactions of oligomeric proteins. The laboratory of Professor Gregorio Weber has been especially active in developing experimental approaches, providing a theoretical (thermodynamic and kinetic) framework for data analysis and assembling a large repertoire of dimeric and tetrameric proteins to measure pressure induced changes in volume, free energies of association, conformational drift and

R. Winter and J. Jonas (eds.), High Pressure Chemistry, Biochemistry and Materials Science, 619–626.

deterministic equilibrium (see for example, Ruan and Weber, 1988, 1989; Erijman and Weber 1991; and Weber 1987). While we continue to use these powerful tools and approaches in our research, the study of large oligomeric proteins, where fluorescence approaches are less applicable, prompted us to consider the adaptation of our fluorescence instrumentation to photon correlation spectroscopy (or dynamic light scattering, DLS). Optical design of the hydrostatic high pressure vessel and the spectrofluorimeter constrain the detection optics to a 90° geometry. A laser light source and a digital spectrum analyzer provide the modifications necessary for DLS measurements on the spectrofluorimeter. DLS measurements, which came into prominence with the advent of powerful laser light sources, offer the opportunity to determine the diffusional behavior of macromolecules (Dubin et al. 1967; and Pecora, 1972). Importantly, DLS measurements are compatible with high pressure perturbation studies as shown by the polymer studies of Nyström et al. (1991) or Freeman et al. (1990) and the diamond anvil work of Herbst et al. (1992). We have examined the dissociation of a large oligomeric protein: hemocyanin from gastropod. Hemocyanin (HC) forms large aggregates of approximately 8 x 10^6 D that are amenable to pressure dissociation in the 2 kbar range (Bonafe et al. 1993). The large oligomers are stabilized by low pH and calcium. In our initial attempts at DLS measurements on this system, we identified relative changes in scattering behavior upon the application of high hydrostatic pressure.

2. Materials and Methods

2.1. INSTRUMENTATION

Figure 1 shows a block diagram of the modified spectrofluorimeter used in the DLS measurements. The laser light source was a continuous wave HeCd (Liconix model 4240NB) at 424 nm. The laser was focused onto a round quartz cuvette inside a high hydrostatic pressure optical vessel with excitation and exit optical ports at 90° (Paladini and Weber, 1981). The pressure transducing fluid surrounding the cuvette was ethyl alcohol. The exit optical port was restricted to a pinhole, so that only a small defined volume in the cuvette was observed. The automated pumping system was provided by Nova Swiss and pressure was recorded on a Heise gauge. The scattered light was collected by a photomultiplier tube (Hamamatsu R928) with associated ISS, Inc. (Champaign, IL) electronics and amplifiers. The dynamic signal or spectrum analyzer with averaging capabilities (Hewlett Packard 3561 A) stored and transferred the trace to a computer. Typical acquisition times ranged from 5-10 minutes. The computer (486 Gateway, Model 2000) and associated plotter (Hewlett Packard Paint Jet XL) provided output and analysis capabilities.

2.1.1. *Hemocyanin.* Hemocyanin from the Brazilian gastropod *Megalobulimulus ovatus* was a generous gift from Dr. J. Silva. Its isolation, purification and associative properties are described by Bonafe et al. (1993). The sample was prepared in Tris buffer (pH 7.6) at 0.5 mg/ml with low absorbance (A(280)=0.06). Under these conditions the application of 2 kbar hydrostatic pressure is sufficient to dissociate the large HC self-associated structure (approximate MW 8×10^6) with a midpoint near 1.2 kbar.

3. Results

Figure 2 shows the power spectrum of the photon correlation intensity as a function of frequency in the HC compression cycle at pressures ranging from 1 atm to 2 kbar. The highest intensity is achieved at 1 atm and each successive application of pressure, in increments of 500 bar, reduces the intensity of the spectrum. The buffer blank does not provide a significant signal. The low frequency signal corresponds to larger oligomers. To highlight the pressure induced changes, Figure 3 shows the intensity difference plot (1 atm data less the intensity profile at pressure). There are several oligomer size distributions present in the frequency range examined. The application of hydrostatic pressure reduces the relative population of the larger oligomers (low frequency signals). It is known that changes in pH and calcium ion concentration stabilize and favor the formation of large oligomers (Bonafe et al. 1993). Figure 4 presents the intensity difference spectrum of HC in the presence of 10 mM calcium ion in the pressure range of 1 atm to 2 kbar. At the lower pressures the distribution of oligomeric species is not modified, i.e. the associative properties are indeed stabilized. At 2 kbar, the curve indicates some perturbation of the self-association of HC in line with the spectroscopic and chromatographic results of Bonafe et al. (1993).

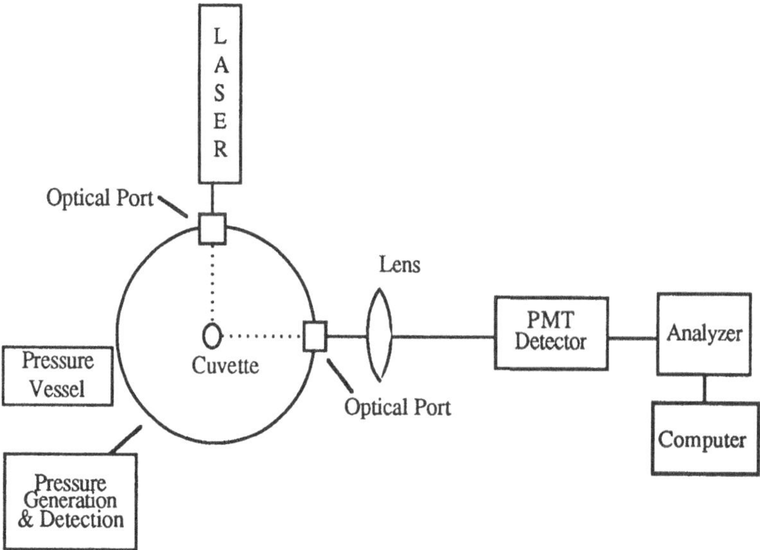

Figure 1: Laser Light Source: HeCd laser (424 nm); High Pressure Vessel: optical vessel design with a 90° orientation between exciting and scattered light coming from a quartz cuvette; Pressure System: aautomated pumping system connected to the pressure vessel and a pressure gauge by steel tubing filled with ethyl alcohol for pressure transduction; Detector: photomultiplier and associated electronics and amplifiers; Signal Analyzer: dynamic signal (spectrum) analyzer with averaging capabilities; Computer and Plotter.

4. Discussion

The amplitude of scattered light yields information about macromolecular size, whereas the spectral width of the scattered light represents translational motion (neglecting the smaller contribution of rotational motion) associated with macromolecular diffusion. The macromolecular concentration fluctuations within the small volume observed obey the diffusion equation and they decay exponentially at a rate Γ given by:

$$\Gamma = D\left(\frac{2\pi}{\lambda}\right)^2 = DK^2$$

where D (cm^2/s) is the diffusion coefficient and K the scattering fluctuation wave vector at wavelength λ. Exponential decay implies a Lorentzian curve for the power spectrum of the scattered light:

$$I(\nu) \simeq \frac{DK^2/2\pi}{(\nu - \nu_0)^2 + (DK^2/2\pi)^2}$$

where ($\nu - \nu_0$) corresponds to the frequency difference of line broadening with a half-width of 2 $DK^2/2\pi$ (Dubin et al., 1967). One calculates the macromolecule's D directly from the half-width of the Lorentzian curve. The self-association of HC results in large oligomers (up to approximately 8×10^6 D) formed from smaller units of 4.5×10^5 D (Bonafe et al., 1993). The data in Figures 2 and 3 show a heterogeneous size distribution of HC oligomers resulting from the varying diffusion constants of different oligomeric species. Because of this size heterogeneity it is difficult to fit the data with a single Lorentzian curve. Nonetheless, it is qualitatively possible to say that the large oligomers exhibit a half width of less than 1 KHz and that the successive application of hydrostatic pressure incrementally broadens the spectral width to less than 10 KHz. The current experimental arrangement resulted in an insufficient data density at low frequencies to quantitatively establish D. Nonetheless, the decrease in scattering intensity with applied hydrostatic pressure demonstrates the depopulation of large oligomers and the evolution of smaller aggregates. The data in Figure 4, which show only a slight pressure effect on the scattering intensity and spectrum, are consistent with the stabilization of HC by calcium ions as previously demonstrated by Bonafe et al. (1993). From the preliminary data in Figures 2-4, we conclude that pressure induced changes in self-association of large oligomeric proteins can be detected in a modified fluorimeter.

5. Summary

Our initial attempts at performing DLS in a modified fluorometer with a laser light source, show that we can detect relative changes in diffusion coefficient as induced by application of hydrostatic pressure. In our opinion, most research grade fluorimeters could be upgraded to include this additional capability. Because the HC oligomeric system investigated proved to be heterogeneous in particle size we were unable to extract quantitative diffusion constants for the distribution of sizes. However, our results clearly show that high hydrostatic pressure induces changes in self association of large oligomeric proteins and results in different diffusional behavior. As we gain more experience with the experimental arrangement and data analysis approaches, we anticipate that a quantitative determination of D for large oligomers will emerge. The instrumentation described in this article should be able to resolve D ranging from 10^{-6} (Durbin et al., 1967) to 10^{-10} cm^2/s. These D values correspond to typical molecular weights of large and small proteins.

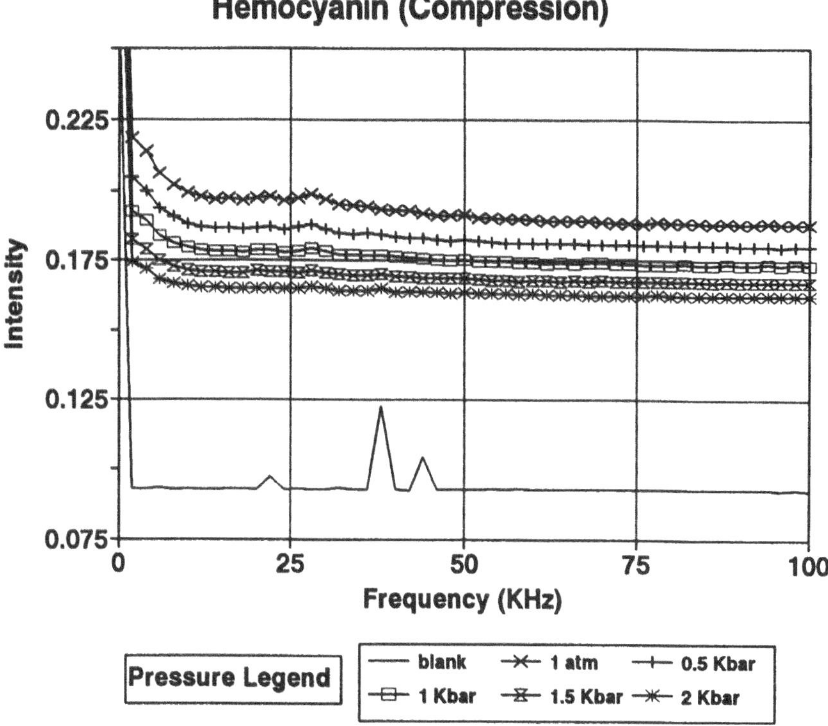

Figure 2: The power spectrum of the HC compression cycle shows the change in the photon correlation intensity as a function of frequency at various pressures. The most intense signal is at 1 atm and each succeeding spectrum (lower intensity) represents an incremental increase in pressure of 500 bar to a maximum of 2 kbar (5 curves). The buffer blank signal is low. The instrument does not respond at zero frequency and the apparent spike in the intensity profile at that point is artifactual.

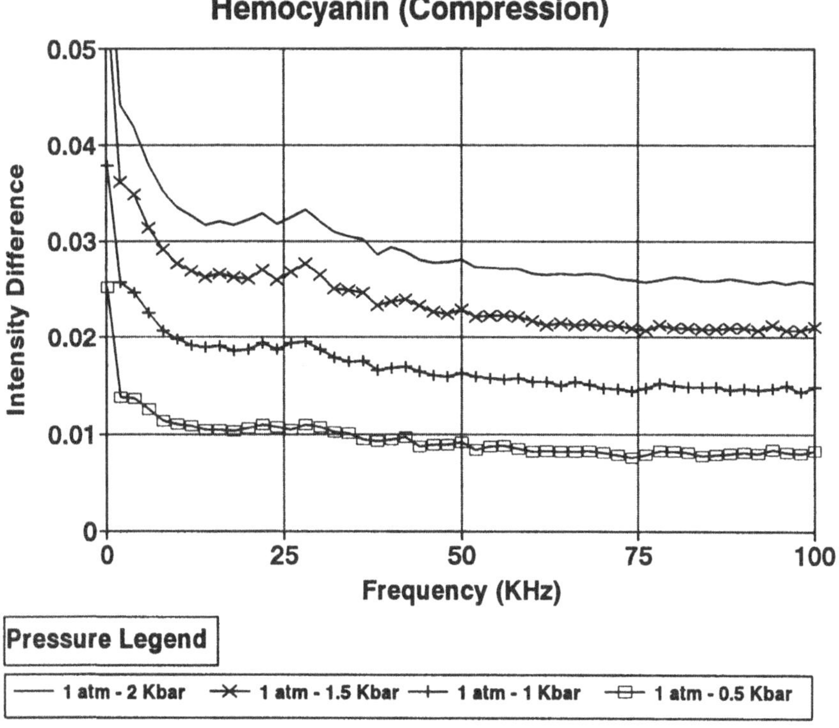

Figure 3: The intensity difference plot (from the data in Figure 2) shows that HC is heterogeneous in its size distribution with peaks at low and intermediate frequencies. As the compression is increased both the large and intermediate size particles are dissociated into smaller entities with faster diffusional rates.

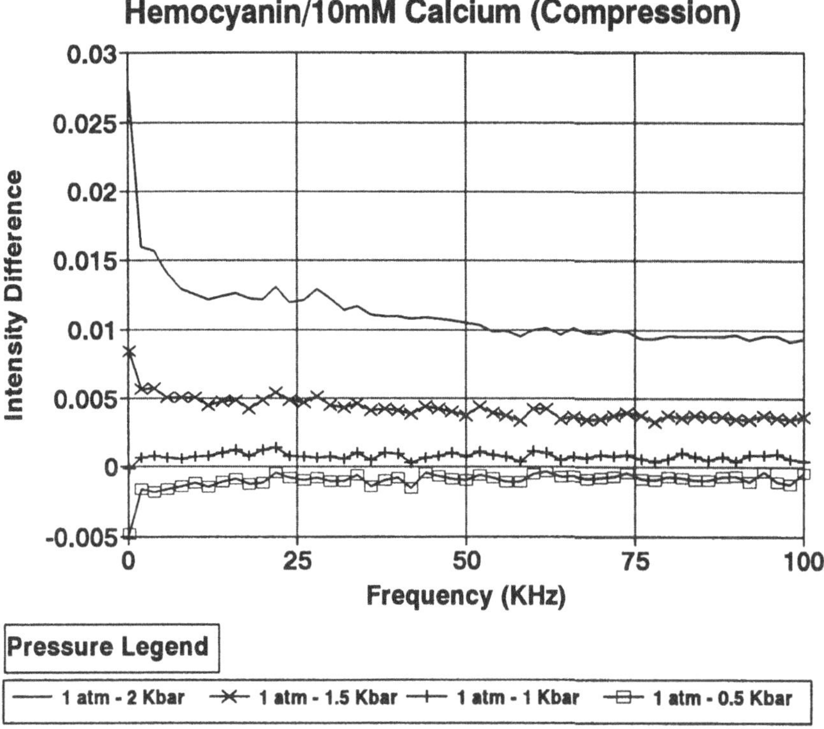

Figure 4: The compression of HC in the presence of 10 mM calcium ion shows the stabilizing effect of the salt. Only at the higher pressure does the DLS indicate a change in the distribution of diffusional rates.

6. Acknowledgments

The authors thank Dr. J. Silva for the generous gift of the hemocyanin. The authors acknowledge the manuscript preparation by C. Johnson and J. Butzow. The Laboratory for Fluorescence Dynamics is supported by NIH RR03155 and UIUC.

7. References

Bonafe, C. F. S., Araujo, J. R., and Silva, J. L. (submitted 1993). Deterministic dissociation of gastropod hemocyanin by pressure: intermediate states of assembly and reassemble process mediated by Ca^{2+} and H^{+}. *Biochemistry*.

Erijman, L., and Weber, G. (1991). Oligomeric protein associations: transition from stochastic to deterministic equilibrium. *Biochemistry 30*, 1595-1599.

Freeman, B. D., Soane, D. S. and Denn, M. M. (1990). Effect of hydrostatic pressure on polystyrene diffusivity in toluene. *Marcomolecules, 23*, 245-251.

Herbst, C. A., King, Jr., H. E., Gao, Z. and Ou-Yang, H. D. (1992). Dynamic light scattering measurements of high pressure viscosity utilizing a diamond anvil cell. *J. Appl. Phys., 72*, 838-844.

Dubin, S. B., Lunacek, J. H. and Benedek, G. B. (1967). Observation of the spectrum of light scattered by solutions of biological macromolecules. *Proc. Natl. Acad. Sci., Physics, 57*, 1164-1171.

Pecora, R. (1972). Quasi-elastic light scattering from macromolecules. *Annual Review of Biophys. and Bioengr., 1*, 257-276.

Ford, Jr., N. C. (1985). Light scattering apparatus. *Dynamic Light Scattering*. ed. R. Percora, Plenum Press, New York, 7-58.

Nyström, B., Olafsen, K., and Roots, J. (1991). Effect of pressure on hydrodynamic properties of poly (ethylene oxide) in deuterated and in non-deuterated solvents. *Polymer, 32*, 904-908.

Paladini, A. A., and Weber, G. (1981). Absolute measurements of fluorescence polarization at high pressures. *Rev. Sci. Instrum., 52*, 419-427.

Ruan, K., and Weber, G. (1988). Dissociation of yeast hexokinase by hydrostatic pressure. *Biochemistry, 27*, 3295-3301.

Ruan, K., and Weber, G. (1989). Hysteresis and conformational drift of pressure-dissociated glyceraldehydephosphate dehydrogenase. *Biochemistry, 28*, 2144-2153.

Weber, G. (1987). Dissociation of oligomeric proteins by hydrostatic pressure. R. van Eldik, and J. Jonas eds., D. Reidel Publishing Company. *High Pressure Chem. and Biochem.*, 401-420.

COMMENTS ON TRENDS IN HIGH PRESSURE RESEARCH

J. JONAS: -

One can identify two main goals of this session dealing with current trends in high pressure research. First, one would like to illustrate on specific examples the wide range of problems that can be studied by high pressure techniques. Secondly, these comments should inform the researcher about promising developments in various subfields, and possible future trends. Indeed, these comments cover an impressive array of research activities from studies of pressure effects on LED, superconductors, expanded metals, to mechanistic and synthetic studies in organic and inorganic chemistry, and high pressure experiments on biochemical systems including membranes and proteins. Even the recent efforts of industrial use of pressure in food science are mentioned. One may ask what are the main reasons for the expansion and diversity of the high pressure field. The answer lies in the tremendous progress in scientific instrumentation and its adaptation for work at high pressures. As a matter of fact, nearly all important experimental techniques used by chemists, biochemists and materials scientists can be used under experimental conditions of high pressure. Another point to make is related to the continued extension of the accessible pressure and temperature ranges. Both progress in experimental techniques and theoretical methods allows one to attack problems dealing with complex systems including biological systems. In addition, the high pressure experiments help many applied technological fields in their transition from purely empirical trial-and-error approaches to more logical schemes of research based on the atomic and molecular level understanding of the complex system or process studied. This ASI highlights well the richness and strength of the high pressure field and points towards continued new developments which guarantee an exciting future of the high pressure research.

H. G. DRICKAMER: -

Since materials science is one of the areas covered at this Institute, it seems desirable to make a few comments about future prospects in this general area of applications of high pressure techniques. These comments reflect my own research interests and, therefore, I do not discuss pressure effects on the properties of metals *per se*, or in general, pressure effects on mechanical properties, although processes like hydrostatic extrusion are important. The interaction of hydrogen with metals and the formation of metal hydrides represents an area where there is continuing work of high quality. It is possible that there are opportunities for work on the formation of nitrides or similar compounds at high pressure. The formation of new structural phases under pressure is an ubiquitous phenomenon and one with some theoretical as well as practical interest, e.g. the growth of synthetic diamond. Unless these new phases have a theoretical significance or a specific practical application, this does not seem to be a growth field. One aspect of interest from the standpoint of basic understanding is the conversion of crystalline materials to an amorphous phase with pressure. A variety of materials from ice to quartz including a solid solution of $Ca(NO_3)_2$ and $Na(NO_3)$ has been shown to undergo this transition reversibly. The growth of crystals from supercritical water is well established. It seems possible that one could grow, e.g. organic and metallorganic crystals from other supercritical fluids.

R. Winter and J. Jonas (eds.), High Pressure Chemistry, Biochemistry and Materials Science, 627–634.

High pressure is becoming an increasingly important tool in polymer science especially to probe internal structure and motion. Optical, especially luminescent, probes have been used to establish the "free volume" and will ultimately be of importance for characterizing local environments in polymers in a more sophisticated way. A comparison of the behavior of probes dissolved in the polymer with the same probe attached to the chain gives information about local chain motion. Polymers are increasingly considered for electronic devices to take advantage of the electronic properties of organic molecules in a medium more stable than an organic crystal. In all these efforts, the effect of pressure on the optical, electrical and magnetic properties is very valuable, frequently essential, for an understanding of the electronic phenomena. There has been only a limited number of pressure studies involving these probes, and this seems to be a very fruitful area for expansion.

Pressure also has an important role in the somewhat related field of inorganic glasses. In particular high pressure Raman and IR spectra can establish a relationship between volume and the distribution of bond energies and force constants. For many applications glasses are doped with transition or rare earth metals. The effect of pressure on the absorption peak energies and line shapes and of the luminescent intensities, energies, and lifetimes compared with comparable data for the crystalline state could be an important factor in characterizing local sites, and the distribution of such sites in glasses of both scientific and technological interest.

An area where pressure has had a continuing and increasing importance is applied semiconductor physics. Over thirty years ago the measurement of the effect of pressure on the gap between the top of the valence band and the bottom of the conduction band (the absorption edge) established the value of the deformation potential, a property necessary for many calculations, and also was fundamental for description of the band structure of Si, Ge and the III-V and II-VI compounds. The development of layered materials with layer thickness of 5-10 nm has permitted pressure investigations of individual states in these bands. The two most significant developments in more recent semiconductor science are the invention of the light emitting diode (LED) and the quantum well laser. With the latter device it is possible to generate watts of power (cw) in a laser only a mm or so in linear dimensions. So far these emitting devices are limited to light in the near IR or red end of the spectrum because of the necessity of using direct gap semiconductors. With pressure it has been possible to move the emission into the yellow-orange region of the spectrum, to show the limitations on any particular composition of semiconductor as well as to establish the criteria necessary for an ambient pressure device to move into the yellow and ultimately into the green part of the spectrum. Such a device would be worth hundreds of millions of dollars. It should also be possible to develop a pressure tunable semiconductor laser where the wavelength of the output could be varied by changing the applied pressure. In an earlier presentation this week it was demonstrated how pressure could be used to characterize the properties of semiconductor phosphors. This technique could be expected to cover a large fraction of the phosphors which are currently of technological importance.

Finally I should like to mention high pressure electrochemistry, and particularly the characterization of electrode surfaces covered with adsorbed molecules or polymers as a field where there has only been exploratory work to indicate a wide range of research opportunities.

D. J. DUNSTAN: -

The diamond-anvil cell (DAC) is now a well-established tool for generating high pressure that can be used by any laboratory as an adjunct to many different experiments. There is no longer any mystique in its use in optical experiments. So far, however, this is only true of a very simple gasket and sample configuration: a plain hole, with unsupported samples floating in the hole. The need for experiments requiring more elaborate configurations will inspire - in fact, is already inspiring - more elaborate microengineering in the sample space. I believe that the application of microphotolithography and other nanoengineering techniques will allow the construction of part or entire experimental apparatus on a scale small enough for inclusion in the sample space of a DAC.

An example, about which there has been much discussion at this meeting, is the possibility of putting the RF coil of an NMR experiment inside the sample space. One nanoengineering technique that could be used is the construction of the coil by deposition directly onto one of the diamond culets of alternate partial rings of metal and of insulator films. A free-standing coil could also be constructed by photolithography on a suitable former, electrodeposition and etching.

Any advance along these lines requires electrical feedthroughs into the sample space. In my presentations at this meeting I did not discuss electrical feedthroughs since they are still very much in the development stage. Many laboratories have reported some work done with them; no laboratories have yet reported a large body of work done routinely with electrical feedthroughs. Four requirements must be met before this will be possible. Firstly, feedthroughs must be compatible with a reasonable pressure range, say up to 500 kbar. Secondly, they must be made using simple technology, otherwise they will be restricted to specialist laboratories. Thirdly, they must be reliable. This does not necessarily mean that the probability of success in an individual experiment must be close to 100%, but it does mean that all systematic difficulties will have been solved, so that one can approach the experiment with complete confidence of eventual success. Fourthly, the sample should be in a hydrostatic medium as in optical experiments.

No laboratory has met all four conditions. What is very encouraging is that every condition has been met somewhere, and many laboratories have met two or three of the conditions. Thus, the groups of Besson (Paris) and of Spain (Colorado) have met the first and fourth conditions; Yu at Berkeley has largely met the first three conditions. My own group at Surrey has made good progress towards the last three conditions. The third condition has been fulfilled by King and Tozer at Exxon and by the Amsterdam group (Hemmes, van Eenige), together with considerable progress towards the other three conditions. It is tempting to speculate that by combining the best practice from each method and each group a simple and reliable way of making electrical feedthroughs to a sample under high hydrostatic pressure in the DAC will soon be achieved, enabling a much wider range of high-pressure experiments.

RINKE J. WIJNGAARDEN: -

Speaking about the future trends in the research on the "old" superconductors under pressure, one of the key compounds is hydrogen. The theoretically predicted conditions necessary for the formation of superconducting atomic metallic hydrogen are (nearly) within experimental reach. Experiments to study the superconducting properties will be extremely difficult. Given the participation of a few groups which have very advanced experimental capabilities, progress may nevertheless be expected. At much lower pressures investigations of the metal hydrides may give interesting results. In our group a keen interest exists in $Pd_{1-x}Ag_xH$ and VH_3. Speaking on the "old" superconductors in general, it is interesting to note, that relatively few experiments on the pressure dependence of the (upper) critical have been done.

Speaking about future trends in the research on cuprate perovskite high temperature superconductors, pressure will be both important for preparation and for the investigation of the physical properties. As an example, after the discovery of $YBa_2Cu_3O_7$, it proved possible to synthesize $YBa_2Cu_4O_9$ under high oxygen pressure. Since $YBa_2Cu_5O_9$ was found as interstitial phase in $YBa_2Cu_3O_7$, one may hope that this compound might also be synthesized under high pressure.

High pressure experiments may yield important information on the correlation between the frequencies of elemental excitations (phonons, excitons, plasmons, etc.) and T_c, and hence may point to the mechanism responsible for electron pairing. In our group we have proved that in high pressure experiments it is important to reach pressures of ~20 GPa. The superconducting properties are a roughly parabolic function of charge carrier density and a pressure of ~20 GPa is needed to map out a significant portion of this parabola. The comparison of such experimental results with theories may give important information. Regrettably not all theories give explicit expressions from which a predicted pressure dependence of T_c and H_{c2} can be deduced. This, incidentally, was the main reason to concentrate on a phenomenological approach in my talk. Nevertheless I hope to have demonstrated that such an approach already gives very valuable information.

MARVIN ROSS: -

Computational materials physics, which includes high pressure applications, in the past dealt mainly with relatively simple systems. The primary interest was in calculating such properties as equations of state, phase transitions and shock phenomena. Only a relatively small number of particles (less than several hundred) could be treated by molecular dynamics and ab-initio calculations were limited to solids of a few atoms per unit cell. The major players were those with access to CRAY computers.

At the present time a virtual revolution is taking place in large scale computing. This is due to the general availability of powerful desktop computers and advanced software programs for carrying out calculations for systems with large numbers of atoms. Distributed computing, consisting of networks of desktop computers, have enormously increased the potential for even larger scale simulations. In the next few years, we can expect to see more rapid advances in parallel computers in which many processors are coupled to work interactively. These machines will allow for molecular dynamics calculations for millions of atoms and ab-initio electron band calculations for hundreds of atoms. The appropriate new software and numerical methods are currently being developed to support industrial applications that are increasingly concerned with the high tech mesoscale systems characteristic of advanced electronics systems. These require modeling on an atomistic scale and involve molecular dynamics and electronic structure calculations. Thus high pressure science, which does not have the resources to develop such methods, will have the opportunity to benefit from these developments and use them to understand many of the complex chemical and biochemical problems that are now being attacked experimentally. Some of these involving the application of pressure, were discussed at this ASI, and include protein structure and phase transitions, lubrication, chemical kinetics, as well as many other topics previously considered far too complicated to be seriously considered as possible computational research topics.

RUDI VAN ELDIK: -

My comments deal first in mechanistic studies and then briefly mention synthetic applications of high pressure approach.

Contributions at the ASI clearly demonstrated the importance to construct volume profiles for the mechanistic interpretation of $\Delta V^{\#}$ data. The combination of $\Delta V^{\#}$ and $\bar{V}$ or $\Delta \bar{V}$ data allow an analysis of the chemical process in terms of volume changes along the reaction coordinate. In this way it is possible to come to a better mechanistic understanding of the observed effect of pressure on the kinetics of the process under study. Such a volume profile analysis should also account for the elementary steps that represent the essential reactions that occur in the overall process. This can usually only be accomplished in a kinetic way by analyzing the process on different timescales by a systematic pressure dependence study of each elementary step in order to construct the overall volume profile. Such examples have now been reported in the literature.

Contributions at this ASI have demonstrated that it is appropriate to distinguish between instrinsic, solvational and steric contributions towards $\Delta V^{\#}$. The solvational contribution can usually be obtained from a systematic solvent dependence study. The steric contribution can only be obtained from a series of measurements in which the steric hindrance is changed systematically. It is unknown at this point whether steric hindrance will also affect the position of the transition state in terms of "early" or "late". Series of studies have shown that there is good correlation between the activation enthalpy and the location of the transition state in terms of volume changes. For a particular system, fast reactions will be characterized by a low activation enthalpy and an "early" transition state in terms of volume changes, whereas a slow reaction will have a high activation enthalpy and a "late" transition state. This seems to be especially true for systems in which the reactivity is controlled by electronic factors. A future objective will be to correlate energy and volume changes along the reaction coordinate in a 3D way, i.e. to combine a conventional energy profile with a volume profile along a common reaction coordinate. Numerous examples in inorganic, organic and organometallic chemistry were presented where synthetic processes can be tuned with the aid of pressure. This creates the possibility to force certain reactions that do not occur at ambient pressure, or to affect the product distribution (isomer ratio) through the application of pressure. Pressure can be used to come to more favorable reaction conditions by which for instance side reactions can be prevented, especially in the case of heat sensitive reactants or products. A number of very surprising effects caused by the application of high pressure techniques in synthetic chemistry were reported at the ASI. These require further investigation in order to understand the reasons for these effects. One may also point out that a number of interesting industrial applications were reported, viz. in the food industry (Heremans), medical applications (Wong), study of lubricants (Jonas), treatment of waste with supercritical fluids (Franck).

GREGORIO WEBER: -

High pressure will be much used by biochemists and biophysicists in the future, as commercial instruments for the purpose are developed and become better known. Of the three modes of perturbation of natural systems that we can presently use high pressure is unique in that it does not change the energy of the system, like temperature, or alters the chemical composition. The study of all phenomena that depend on low-energy interactions, that is virtually all biological phenomena, will benefit from the pressure perturbation method.

Certain problems of structure and dynamics of proteins can greatly benefit from an examination at temperatures below 0°C. At a pressure of 1-2 kbar one can experiment at temperatures from -10 to -20°C without changing appreciably the properties of water (viscosity, dielectric constant, etc.). The cold denaturation of proteins can be studied under equilibrium conditions by this means. The same property of water will permit improved analytical separations by chromatography and electrophoresis at low temperature.

The observation that pressure interferes with complex biological functions (relief of anesthesia, nerve conduction) at values much lower than those necessary to produce observable changes in isolated systems is still largely unexplained and will certainly receive much attention.

J. JONAS: -

As illustrated by the lectures and contributed talks during this ASI, there have recently been a major expansion of high pressure research providing unique information about systems of interest to a wide range of scientific disciplines. Since nuclear magnetic resonance has been applied to a wide spectrum of problems in chemistry, biochemistry and materials science it is not surprising to find that high pressure NMR techniques have also had many applications. Clearly, the high information content of NMR experiments combined with high pressure capability creates a powerful tool in the hands of chemists, biochemists and material scientists.

In particular, one can predict an increased use of advanced high resolution, high pressure NMR techniques to biochemical systems. The high information content of the many advanced NMR techniques including 2D-NMR techniques such as NOESY, COSY and ROESY have yet to be fully explored in the high pressure NMR experiments. The possibility of determining the three-dimensional structure of proteins during the pressure induced dissociation and/or unfolding of proteins opened a new direction in high pressure NMR spectroscopy dealing with pressure effects on biochemical systems.

K. HEREMANS: -

Applications of high pressure technology for the processing of food has a long history. In the United States, Hite and coworkers have made extensive investigations on the possibility to preserve fruit and vegetables by killing microorganisms with high pressure treatment of about 500 Mpa for 30 minutes. The experiments by Bridgman on egg white showed that pressure treatment resulted in coagulation of the white with an appearance like that of a boiled egg. This suggested that microorganisms are killed by the action of pressure on the proteins.

On the other hand there remains the distinct possibility that pressure may act upon the colloidal constituents of biological material. This possibility has been suggested a long time ago but has not been further explored as extensively as the protein denaturation hypothesis. The colloidal hypothesis was put forward after the observation that the pressure required to kill bacteria is inversely proportional to their complexity. Such studies have been performed by many groups, but in particular in France by Macheboeuf and Basset. Studies on proteins, enzymes, immunoglobulins and the survival of bacteria, spores and viruses were reported.

In Japan the tradition in high pressure research on biological systems started with the paper of Kiyama and Yanagimoto that appeared in 1951 followed by the fundamental work of Suzuki. It took about another 30 years before the trend-setting paper of Hayashi appeared. In 1989 a Japanese R&D association was founded to stimulate cooperation between industry, universities and government. In 1990 several products were put on the market: strawberry jam, grapefruit juice, etc. The society organized meetings to discuss progress in the field. The papers were presented and published in Japanese. The recent developments, including the developments in Japan, have been discussed at the First European meeting on High Pressure and Biotechnology which was held at La Grande Motte in France. The first research project has now been set up on a European scale with the participation of twelve research teams from industry as well as from universities to study basic aspects of high pressure food processing. Future developments will decide whether *Bridgmanization*, as the process may be called in honor of the man who made the field of high pressure since the beginning of this century, will be a contribution as vital to the quality of food as the process of *Pasteurization* is now.

However, the use of pressure as an alternative to temperature treatment, has brought about the need for fundamental studies on the pressure-temperature behavior of macromolecular food constituents such as proteins, lipids and polysaccharides. Pressure treatment of food materials, like temperature treatment, involves the inactivation of microorganisms and enzymes. In most cases this is not only a thermodynamic but a kinetic problem as well. In addition, the macroscopic appearance of food is of primary importance. In most cases information is lacking as to the correlation between macroscopic appearance of processed food and the molecular conformation of the constituents as a function of temperature and pressure. The mechanisms of protein gelation and the sol-gel behavior of polysaccharides is far from being understood. Of considerable attraction is the fact the pressure induced gels of food preserve their natural color and flavor compared to temperature induced gels.

R. WINTER: -

In recent years, a variety of pressure studies on model biomembrane systems have been performed. Besides the biological relevance of these studies (high pressure nervous syndrome, life in the deep sea, pressure reversal of anesthesia, etc.), pressure has been used as a physico-chemical tool. This is because pressure provides a ready means of separating the effects of temperature and density, and pressure changes the intermolecular distances. The investigation of the temperature and pressure dependence of thermodynamic and dynamic properties thus leads to a more complete understanding of biochemical systems.

Recently, a great deal of effort has been devoted to the development of theoretical models of biopolymers in solution, not only from a fundamental scientific point of view, but also because of its possible industrial and clinical applications. For the reasons given above, pressure dependent studies provide a unique opportunity to test these theoretical models of biochemical systems. Only a few theoretical attempts have been undertaken so far to describe membrane properties, such as membrane structure and the phase transformations occurring in these systems. Now that more experimental data on the pressure and temperature dependence of thermodynamic and dynamic properties of membrane and membrane protein systems are available, new theoretical concepts could be developed and tested, which might lead to a deeper understanding and a possible prediction of biopolymer behavior in solution. In the future, also kinetic studies of biopolymer phase transitions using pressure jump techniques should be performed, which might help to detect intermediate structures in biochemical membrane processes and to describe biopolymer phase transformations theoretically.

SUBJECT INDEX

AUTHOR INDEX

LIST OF PARTICIPANTS

M. Adams
SERC
Rutherford Lab. ISIS
Chilton, Didcot
Oxon OX11 OQX
U.K.

Prof. Dr. A. Avranas
Laboratory of Phys. Chemistry
Department of Chemistry
Aristotle University
GR-54006 Thessaloniki
Greece

Prof. Dr. F. Barocchi
Universita di Firenze
Dipartimento di Fisica
Largo E. Fermi 2
50125 Firenze
Italy

C. Bernsdorff
Ruhr-Universität Bochum
Physikalische Chemie II
Postfach 102148
D-W-4630 Bochum
F.R.G.

Dr. T. Bleha
Polymer Institute
Slovak Academy of Sciences
Dubravka cesta
84236 Bratislava
CZECH

M. Böttner
Universität Marburg
Fachbereich Physik. Chemie
Hans-Meerwein-Straße
D-W-3550 Marburg/Lahn
F.R.G.

Prof. Dr. M. Buback
Universität Göttingen
Institut für Physikalische Chemie
Tammannstraße 6
D-W-3400 Göttingen
F.R.G.

Prof. Dr. A. Calimli
Ankara University
Deptm. of Chemical Engineering
06100 Tandogan
Ankara
Turkey

M. Celli
Universita di Firenze
Dipartimento di Fisica
Largo E. Fermi 2
50125 Firenze
Italy

Prof. Dr. P.L.-G. Chong
Department of Biochemistry
Meharry Medical College
1005 David B. Todd Blvd.
Nashville
Tennessee 37208
U.S.A

Prof. Dr. M.L.A. da Cunha
Depart. de Ciencia dos Materials
Faculdade de Ciencias e Tecnologia
da Universidade Nova de Lisboa
Quinta da Torre
2825 Monte de Caparica
Portugal

Prof. Dr. H. Drickamer
Dep. of Chemical Engineering
University of Illinois
1209 W. California
105 Roger Adams Lab
Urbana, IL 61801
U.S.A.

B. Dudzinski
Polish Academy of Sciences
Centre of Molecular and
Macromolecular Sciences
90-363 Lodz
Sienkiewicza 112
Poland

Dr. D.J. Dunstan
Department of Physics
University of Surrey
Guildford Surrey GU2 5XH
U.K.

Prof. Dr. R. van Eldik
Institut für Anorganische Chemie
Universität Witten/Herdecke
Stockumer Str. 10
D-W-5810 Witten
F.R.G.

Dr. D. Foguel
University of Illinois
Deptm. of Biochemistry
399 Roger Adams Lab
Urbana, IL 61801
U.S.A.

Prof. Dr. E.U. Franck
Inst. für Physikalische Chemie
Universität (TH) Karlsruhe
Kaiserstraße 12
7500 Karlsruhe
F.R.G.

Dr. U. Frey
Chemistry Department
Gordon House
29, Gordon Square
London WC1H OPP
U.K.

I. Gatzke
Ruhr-Universität
Universitätsklinik
D-W-6430 Bochum
F.R.G.

Dr. D.G. Gillies
Department of Chemistry
University of Surrey
Guildford
Surrey GU2 5XH
U.K.

K. Goossens
Department of Chemistry
Katholieke Universiteit Leuven
Lab. of Chem. & Biol. Dynamics
Celestijnenlaan 200 D
B-3001 Leuven
Belgium

K. Halvorson
Dep. of Chemistry and Biochemistry
Arizona State University
Tempe
AZ 85287-1604
U.S.A.

Prof. Dr. F. Hensel
Universität Marburg
FB Physikalische Chemie
Hans-Meerwein-Straße
D-W-3550 Marburg/Lahn
F.R.G.

Prof. Dr. K. Heremans
Department of Chemistry
Katholieke Universiteit Leuven
Lab. of Chem. & Biol. Dynamics
Celestijnenlaan 200 D
B-3001 Leuven
Belgium

E. Janssen
Department of Chemistry
Katholieke Universiteit Leuven
Lab. of Chem. & Biol. Dynamics
Celestijnenlaan 200 D
B-3001 Leuven
Belgium

Dr. G. Jenner
EHICS DCOA
Lab. de Piezochimie Organique
B.P. 296
1, Rue Blaise Pascal
F-67008 Strasbourg Cedex
France

Prof. Dr. A. Jonas
University of Illinois
Deptm. of Biochemistry
South Mathews Ave.
Urbana, Il 61801
U.S.A.

Prof. Dr. J. Jonas
University of Illinois
School of Chemical Sciences
South Mathews Ave.
Urbana, Il 61801
U.S.A.

J. Justino
Centro de Quimica Estrutural
Complexo I, Inst. Superior Tecnico
1096 Lisboa Cedex
Portugal

G. Karakas
Middle East Tech. University
Chemical Eng. Department
Inönü Bulvari Ankara 065331
Turkey

Dr. H.E. King
Exxon Research and
Engineering Company
Annandale, NJ 08801
U.S.A.

A. Landwehr
Ruhr-Universität Bochum
Physikalische Chemie II
Postfach 102148
D-W-4630 Bochum
F.R.G.

Dr. M. Mackowiak
Institute of Molecular Physics
Polish Academy of Sciences
Smoluchowskiego 17/19
PL-60179 Poznan
Poland

Prof. Dr. W. Mantulin
University of Illinois
Loomis Lab
Urbana, IL 61801
U.S.A.

S.E. Mustonen
Univ. of California
11933 Darlington Av.
Los Angeles, CA 90049
U.S.A.

Prof. Dr. M.F. Nicol
Dep. of Chemistry and Biochemistry
405 Hilgard Avenue
Los Angeles
California 90024-1569
U.S.A.

Prof. Dr. E. Nies
University of Technology
Lab. of Polymer Techn. & Chemistry
Den Dolech 2
P.O. Box 513
5600 MB Eindhoven
The Netherlands

Dr. R.B. Novac
Institute of Atomic Physics
IFTAR Lab-22
P.O. Box MG-7
Bucharest-Macurele
Romania

Chr. Ott
Universität Göttingen
Inst. für Organische Chemie
Tammannstraße 2
D-W-3400 Göttingen
F.R.G.

Prof. Dr. N. Papadopoulos
Lab. of Physical Chemistry
University of Thessaloniki
54006 Thessaloniki
Greece

Prof. Dr. L.V. Pinheiro
Faculdade de Farmacia da
Universidade de Lisboa
Av. das Forcas Armadas
1600 Lisboa
Portugal

Dr. G. Pirianov
Nation. Centre of Oncology
6 Plovdivsco pole Street
Sofia - 1156
Bulgaria

G. Pratesi
Universita di Firenze
Dipartimento de Fisica
Largo E. Fermi 2
50125 Firenze
Italy

K.E. Prehoda
Department of Chemistry
Univ. of Wisconsin-Madison
Madison, WI 53706-1569
U.S.A.

Dr. F.E. Prieto
Instituto de Fisica UNAM
Apartado Postal 20 364
Delegacion Alvaro Obregon
01000 Mexico
D.F. Mexico

J. Rathenow
Universität Marburg
Fachbereich Physikalische Chemie
Hans-Meerwein-Straße
D-W-3550 Marburg/Lahn
F.R.G.

Dr. M. Ross
Lawrence Livermore Natl. Lab.
P.O. Box 808/L-299
2000 East Avenue
Livermore, CA 94550
U.S.A.

Prof. Dr. M. Sampoli
Dip. Energetics
University of Florence
Via Santa Marta 3
50139 Firenze
Italy

B. Sampoli
Deptm. of Chemistry
University of Florence
50139 Firenze
Italy

Dr. H.W. Scheeren
Lab. of Organic Chemistry
Cath. University Toernooivela
6525 E.D. Nijmegen
The Netherlands

M. Schick
Lab. für Physikalische Chemie
ETH Zürich
Universitätstr. 22
CH-8092 Zürich
Schweiz

P. Schneider
Max Planck Institut
für biophys. Chemie
P.O. Box 2841
D-W-3400 Göttingen
F.R.G.

Dr. D. Schwarzer
Max Planck Institut
für biophys. Chemie
Am Faßberg
D-W-3400 Göttingen
F.R.G.

Prof. Dr. J. Silva
University of Illinois
Deptm. of Biochemistry
399 Roger Adams Lab
Urbana, IL 61801
U.S.A.

J. Sly
Department of Physics
University of Surrey
Guildford Surrey GU2 5XH
England
U.K.

M. Stolz
Ruhr-Universität Bochum
Physikalische Chemie II
Postfach 102148
D-W-4630 Bochum
F.R.G.

Prof. Dr. C. Viana
Deptm. of Chemistry
Faculdade de Ciencias
R. da Escola Politecnica
1200 Lisboa
Portugal

Dr. M. Villas-Boas
c/o MPI für biophys. Chemie
P.O. Box 2841
D-W-3400 Göttingen
F.R.G.

S. Wallen
Department of Chemistry
University of Illinois
Urbana, IL 61801
U.S.A.

Prof. Dr. G. Weber
Department of Biochemistry
University of Illinois
1209 W. California
397 Roger Adams Lab
Urbana, IL 61801
U.S.A.

Prof. Dr. R. Winter
Ruhr-Universität Bochum
Physikalische Chemie II
Postfach 102148
D-W 4630 Bochum
F.R.G.

Dr. R.J. Wijngaarden
Vrije Universiteit
Faculteit Natuurkunde
en Sterrenkunde
De Boelelaan 1081
1081 HV Amsterdam
The Netherlands

Prof. Dr. P.T.T. Wong
National Research Council
100 Sussex Dr.
Ottawa
Ontario K1A OR6
Canada

A. Zahl
Inst. für Anorgan. Chemie
Universität Witten/Herdecke
Stockumer Str. 10
D-W-5810 Witten
F.R.G.

Prof. Dr. D. Zakim
The New York Hospital
Cornell Medical Center
525 East 68th Street
New York, NY 10021
U.S.A.

Prof. Dr. T.W. Zerda
Texas Christian University
Physics Department
P.O. Box 32915
Fort Worth, TX 76129
U.S.A.

X. Zhang
Inst. für Anorgan. Chemie
Universität Witten/Herdecke
Stockumer Str. 10
D-W-5810 Witten
F.R.G.

The manufacturer's authorised representative in the EU is Springer Nature Customer Service Centre GmbH, Europaplatz 3, 69115 Heidelberg, Germany. If you have any concerns regarding our products, please contact ProductSafety@springernature.com

Printed and bound by CPI Group (UK) Ltd, Croydon, CR0 4YY
24/04/2026
02096348-0012